Biology, Ecology and Culture of Grey Mullet (Mugilidae)

Biology, Ecology and Culture of Grey Mullet (Mugilidae)

Editors

Donatella Crosetti

ISPRA (Italian Institute for Environmental Protection & Research)
Aquaculture Department
Rome
Italy

Stephen Blaber

CSIRO Oceans & Atmosphere
Queensland
Australia

CRC Press is an imprint of the
Taylor & Francis Group, an **informa** business
A SCIENCE PUBLISHERS BOOK

Cover credits

- Mullet roe drying in the sun in Taiwan. Photo by I. Chiu Liao.
- Thicklip mullets in a *Posidonia oceanica* meadow in Menorca (Baleares Islands, Spain). Photo by Manuel Elices.
- Grey mullets from aquaculture. Photo by I. Chiu Liao.

CRC Press
Taylor & Francis Group
6000 Broken Sound Parkway NW, Suite 300
Boca Raton, FL 33487-2742

Printed on acid-free paper
Version Date: 20151014

International Standard Book Number-13: 978-1-4822-5212-5 (Hardback)

Library of Congress Cataloging-in-Publication Data

Biology, ecology and culture of grey mullet (Mugilidae) / Donatella Crosetti, Stephen Blaber, editors.
pages cm
"A CRC title."
Includes bibliographical references and index.
ISBN 978-1-4822-5212-5 (hardcover : alk. paper) 1. Grey mullets. 2. Marine fishes. I. Crosetti, Donatella, editor. II. Blaber, Stephen J. M., editor.

QL638.M8B56 2015
597'.7--dc23 2015028953

Visit the Taylor & Francis Web site at
http://www.taylorandfrancis.com

and the CRC Press Web site at
http://www.crcpress.com

Preface

The Mugilidae, commonly known as grey mullets, are one of the most ubiquitous teleost families in coastal waters of the world. They occur in most temperate, sub-tropical and tropical waters in both hemispheres. As a family they have an extraordinary adaptability, which has resulted in species that are found mainly in the clear and pristine waters of coral reefs to those that prefer highly turbid estuarine and fresh waters. Some species can even survive in some of the most polluted waters in the world, e.g., in the harbour at Viskhapatnam in India (Blaber 2000). Wherever grey mullet occur they often dominate the fish fauna and due to their primarily detritivorous feeding occupy a unique position in the food web. In some areas their species diversity may be very high, e.g., in the St. Lucia coastal lake system in South-East Africa at least 10 species are sympatric (Blaber 1976). Not surprisingly these fish are economically important in most regions, particularly the worldwide species *Mugil cephalus*, which forms the basis of significant commercial fisheries in developed parts of the world such as Australia and the U.S.A., as well as vital artisanal and subsistence fisheries in developing countries. Mullets are also cultured in many regions of the world, both in extensive systems, such as the more or less confined coastal lagoon areas in the Mediterranean region, and in semi-intensive and intensive systems, often in polyculture with other species, though culture is still based on the collection of wild fry, as no induced spawning is practiced at a commercial level. Egypt is by far the greatest producer of cultured grey mullets, with 84% of the world mullet aquaculture production (138,143 tonnes in 2013, FAO 2015).

Mugilidae taxonomy and nomenclature have been revised several times, and a critical revision is ongoing at present, with the new information provided by molecular tools which certainly represent a great challenge to the traditional morphologically based taxonomy (see Chapters 1 and 2). Indeed Mugilidae are very conservative in morphological traits, a characteristic of the family which may have lead in the past to misidentifications and erroneous synonymies among taxa, especially from specimens from regions of the world that are far apart. Many taxonomic issues are still being debated and will probably lead in the near future to a total upset of the family taxonomy and nomenclature. For practical reasons, the Mugilidae nomenclature used in the Eschmeyer 'Catalog of fishes' (2015) was adopted for all chapters of this book, and possible eventual synonyms or new species names cited by other authors are reported in brackets.

The last comprehensive review of the Mugilidae was published more than 30 years ago in 1981 (Oren 1981). Although this book concentrated mainly on aquaculture, it also provided syntheses of much of the biological and ecological knowledge available at the time. A subsequent book by Hussenot and Gauthier (2005) published in French provided valuable information on the European grey mullets. An excellent review of the biology, genetics, ecology and fisheries of *M. cephalus* was recently published (Whitfield et al. 2012). This review was one of the outputs of the EU funded project MUGIL (see Chapter 21) and deals with all the most important issues of *M. cephalus* biology and ecology. It is restricted however, to this one species. Recent advances in knowledge, including great leaps forward in ecological and biological information from many tropical developing countries, more intensive taxonomic investigations and biogeographical studies coupled with advances in genetic techniques, and major advances in applied aquaculture, indicate that a new review of what is known about the Mugilidae is overdue.

The present volume hopes to go some way towards filling this gap. It is divided into two sections, the first dealing with biology, ecology and systematics, and the second with culture and fisheries.

The first two chapters are concerned with taxonomy and systematics. Chapter 1 by González-Castro and Ghasemzadeh reviews and discusses the present status of mugilid taxonomy around the world, and shows how both the traditional and new techniques and tools can be used to identify the many species in this family. Unfortunately, the close similarities in the morphology and anatomy of most grey mullet species have made difficult the tasks of inferring phylogenetic relationships and evolution, as well as the identification of species and genera. Hence there has been little consensus on the systematics of the family. In Chapter 2, Durand describes how genetic polymorphisms, which constitute a valid and powerful alternative to morphology, can be used to test the prevailing phylogenetic assumptions based upon morphological traits. He demonstrates the implications of recent molecular phylogeny for the taxonomy of the Mugilidae, concluding that there is more and more molecular evidence that the species diversity of the Mugilidae is greatly underestimated.

The next five chapters provide much new information about the biogeography and distribution of Mugilidae in different regions of the world. In Chapter 3, Barletta and Dantas document the situation in the Americas; in Chapter 4, the biogeography of Mugilidae in India, South-East and East Asia is described by Shen and Durand and the same is done for Australia and Oceania by Ghasemzadeh in Chapter 5; in Chapter 6, Durand and Whitfield describe the biogeography and distribution of African Mugilidae; the biogeography of Mugilidae in the Mediterranean, Europe and the North-East Atlantic is explained by Turan in Chapter 7.

Biological and ecological information is provided in the next five chapters. The musculoskeletal anatomy of the flathead grey mullet *Mugil cephalus* is described in great detail in Chapter 8 by Ghasemzadeh and this chapter contains several line drawings of the most important skeletal bones. Chapter 8 thus provides a sound basis for future comparisons with the osteology of other mullet species. The variously described mud-eating, iliophagy, detritus feeding, deposit feeding and interface feeding habits of grey mullet are detailed by Cardona in Chapter 9. Age and growth are described by Ibáñez in Chapter 10, reproduction by González-Castro and Minos in Chapter 11, the biology of fry and juveniles by Koutrakis in Chapter 12, and their remarkable adaptations to salinity and their osmoregulation are discussed by Nordlie in Chapter 13. The very significant ecological role of grey mullet in coastal waters and estuaries around the world is described by Whitfield in Chapter 14. Rossi, Crosetti and Livi have provided a very eloquent overview in Chapter 15 of research on the genetics of Mugilidae, with particular reference to *Mugil cephalus*.

The second part of the book begins with Chapter 16 by Crosetti on the current status of mullet capture fisheries and their aquaculture. This is followed by Chapter 17 by Prosser on capture methods and commercial fisheries, and then Chapter 18 by Leber et al. on culture-based stock enhancement, with particular reference to Hawaii. The next two Chapters 19 and 20 provide detailed case studies of the culture industries in Taiwan and Egypt by Liao, Chao and Tseng and Sadek respectively. The book concludes with a chapter about the MUGIL project, which involved scientists from eight countries collaborating to document what was known in 2009 about all aspects of *Mugil cephalus*.

References

Blaber, S.J.M. 1976. The food and feeding ecology of Mugilidae in the St. Lucia Lake system. Biological Journal of the Linnean Society, London. 8: 267–277.

Blaber, S.J.M. 2000. Tropical Estuarine Fishes: Ecology, Exploitation and Conservation. Blackwell, Oxford 372pp.

Eschmeyer, W.N. 2015. Catalog of Fishes: Genera, Species, References. (http://research.calacademy.org/research/ichthyology/catalog/fishcatmain.asp). Electronic version accessed on 15/01/2015.

FAO. 2015. FAO FishStat. Aquaculture production 1957–2013 (www.fao.org). Electronic version accessed on 20/03/2015.

Hussenot, J. and D. Gauthier. 2005. Les Mulets des Mers d'Europe (Synthèse des connaissances sur les bases biologiques et les techniques d'aquaculture). QUAE, IFREMER, France.

Oren, O.H. 1981. Aquaculture of Grey Mullets. Serial: International Biological Programme (United Kingdom), no. 26.

Whitfield, A.K., J. Panfili and J.D. Durand. 2012. A global review of the cosmopolitan flathead mullet *Mugil cephalus* Linnaeus 1758 (Teleostei: Mugilidae), with emphasis on the biology, genetics, ecology and fisheries aspects of this apparent species complex. Rev. Fish. Biol. Fish 22: 641–681.

Acknowledgements

The editors would like to thank most sincerely all the authors for their contributions. The hard work that has gone into the research and reviewing of the literature is evident in each of the chapters. It has been a truly collaborative effort with scientists coming together from around the world to analyse what is known about this important worldwide family of fishes. Donatella Crosetti is thankful to Giovanna Marino, the head of the Aquaculture Department in ISPRA (Italian Institute for Environmental Protection and Research), who believed in this project and included it in the activities of the Aquaculture Department, and would like to acknowledge ISPRA and FAO (Food and Agriculture Organization of the United Nations) Libraries for their support in collecting bibliographic references. Stephen Blaber is grateful to CSIRO (Commonwealth Scientific and Industrial Research Organisation) Oceans and Atmosphere in Brisbane, Australia, for research facilities. We are also very grateful to all at CRC Press for their help and forbearance in seeing this book to fruition.

Contents

CHAPTER 1

Morphology and Morphometry Based Taxonomy of Mugilidae

Mariano González-Castro[1,*] and *Javad Ghasemzadeh*[2]

Introduction

The Critical State of the Taxonomy of Mugilidae

Members of the family Mugilidae, generally known as mullets, are coastal marine fishes with a worldwide distribution including all temperate, subtropical and tropical seas. They not only inhabit offshore and coastal waters, but also depending on the species, spend part or even their whole life cycle in coastal lagoons, lakes and/or rivers. A considerable period of time has passed since the last book was published on Mugilidae (Oren 1981). Many important and critical changes related to the taxonomy and systematics of this family have taken place since this last publication. We may perhaps be in the 'middle of a revolution', with regard to the phylogeny and taxonomy of mullets. New and more efficient methodologies have developed in the past few decades, which improve the accurate discrimination of taxa; for example the sequencing of mitochondrial and/or nuclear genes (molecular taxonomy) and the geometric morphometrics (a relatively novel discipline which is based on the use of anatomical landmarks in order to evaluate differences in the shape of organisms). As could be expected, as a consequence of the application of these new techniques, new results have been obtained on Mugilidae, such as the appearance of cryptic species, but also conflicts have arisen at the generic and even subfamily levels. Table 1.1 summarizes the nominal genera of Mugilidae, in chronological order of appearance, with their status according to different authors.

Much more work has to be done in order to clarify and consolidate the taxonomy and systematics of Mugilidae. Hence, the aim of this chapter is to review and discuss the present status of mugilid taxonomy around the world, and also to show both the traditional and new tools that can be employed to identify/discriminate these fishes.

What are Mullets?

Fishes of the family Mugilidae belong to Actinopterygii, which is the class that groups the highest number of species, has the most recent expansion and manifests more notable evolutionary lines toward both slender

[1] Laboratorio de Biotaxonomía Morfológica y molecular de peces, Instituto de Investigaciones Marinas y Costeras IIMyC-CONICET, UNMdP, Mar del Plata, Argentina.
Email: gocastro@mdp.edu.ar; gocastro@gmail.com
[2] Faculty of Marine Sciences, Chabahar Maritime University, Iran.
Email: jghasemz@yahoo.com.au; ghasemzadeh@cmu.ac.ir
* Corresponding author

Table 1.1. Nominal genera of the Mugilidae in chronological order of appearance, with their status according to Schultz, Senou, Thomson, Ghasemzadeh, Durand et al. and Eschmeyer and Fong.

Genus	Author and date	Type species	Genus assigned by author (date)					
			Schultz (1946)	**Senou (1988)**	**Thomson (1997)**	**Ghasemzadeh (1998)**	**Durand et al. (2012b)**	**Eschmeyer and Fong (2015)**
Mugil	Linnaeus 1758	*Mugil cephalus* Linnaeus 1758	*Mugil*	*Mugil*	*Mugil*	*Mugil*	*Mugil*	*Mugil*
Chelon	Artedi 1793	*Mugil chelo* Cuvier 1829	*Chelon*	*Chelon*	*Chelon*	*Chelon*	*Chelon*	*Chelon*
Cephalus	Lacepède 1799	*Mugil cephalus* Linnaeus 1758	*Mugil*	*Mugil*	*Mugil*	*Mugil*	*Mugil*	*Mugil*
Agonostomus	Bennett 1832	*Agonostomus telfairii* Bennett 1832	*Agonostomus*	*Agonostomus*	*Agonostomus*	*Agonostomus*	*Agonostomus*	*Agonostomus*
Cestraeus	Valenciennes 1836	*Cestraeus plicatilis* Valenciennes 1836	*Cestraeus*	*Cestraeus*	*Cestraeus*	*Cestraeus*	*Cestraeus*	*Cestraeus*
Dajaus	Valenciennes 1836	*Mugil monticola* Bancroft 1834	*Agonostomus*	*Agonostomus*	*Agonostomus*	*Agonostomus*	*Dajaus*	*Agonostomus*
Nestis	Valenciennes 1836	*Nestis cyprinoides* Valenciennes 1836	*Agonostomus*	*Agonostomus*	*Agonostomus*	*Agonostomus*	*Agonostomus*	*Agonostomus*
Arnion	Gistel 1848	*Mugil cephalus* Linnaeus 1758	*Mugil*	*Mugil*	*Mugil*	*Mugil*	*Mugil*	*Mugil*
Ello	Gistel 1848	*Mugil cephalus* Linnaeus 1758	*Mugil*	*Mugil*	*Mugil*	*Mugil*	*Mugil*	*Mugil*
Joturus	Poey 1860	*Joturus pichardi* Poey 1860	*Joturus*	*Joturus*	*Joturus*	*Joturus*	*Joturus*	*Joturus*
Myxus	Günther 1861	*Myxus elongatus* Günther 1861	*Myxus*	*Chelon*	*Myxus*	*Myxus*	*Myxus*	*Myxus*
Chaenomugil	Gill 1863	*Mugil proboscideus* Günther 1861	*Chaenomugil*	*Chaenomugil*	*Chaenomugil*	*Chaenomugil*	*Chaenomugil*	*Chaenomugil*
Rhinomugil	Gill 1863	*Mugil corsula* Hamilton 1822	*Rhinomugil*	*Rhinomugil*	*Rhinomugil*	*Rhinomugil*	*Rhinomugil*	*Rhinomugil*
Gonostomyxus	Macdonald 1869	*Gonostomyxus loaloa* Macdonald 1869	*Cestraeus*	*Cestraeus*	*Cestraeus*	*Cestraeus*	*Cestraeus*	*Cestraeus*
Neomyxus	Steindachner 1878	*Myxus* (Neomyxus) *sclateri* Steindachner 1878	*Neomyxus*	*Neomyxus*	*Chaenomugil*	*Neomyxus*	*Neomyxus*	*Neomyxus*
Querimana	Jordan and Gilbert 1883	*Myxus harengus* Günther 1861	*Mugil*	*Mugil*	*Mugil*	*Mugil*	*Mugil*	*Mugil*
Aeschrichthys	Macleay 1883	*Aeschrichthys goldiei* Macleay 1883	*Cestraeus*	*Cestraeus*	*Cestraeus*	*Cestraeus*	*Cestraeus*	*Cestraeus*

Liza	Jordan and Swain 1884	*Mugil capito* Cuvier 1829	*Chelon*	*Chelon*	*Liza*	*Liza*	*Chelon*	*Liza*
Trachystoma	Ogilby 1888	*Trachystoma multidens* Ogilby 1888	*Trachystoma*	*Chelon*	*Myxus*	*Trachystoma*	*Trachystoma*	*Trachystoma*
Neomugil	Vaillant 1894	*Neomugil digueti* Vaillant 1894	*Agonostomus*	*Agonostomus*	*Agonostomus*	*Agonostomus*	*Dajaus*	*Agonostomus*
Oedalechilus	Fowler 1903	*Mugil labeo* Cuvier 1829	*Chelon*	*Oedalechilus*	*Oedalechilus*	*Oedalechilus*	*Oedalechilus*	*Oedalechilus*
Squalomugil	Ogilby 1908	*Mugil nasutus* de Vis 1883	*Rhinomugil*	*Rhinomugil*	*Rhinomugil*	*Rhinomugil*	*Rhinomugil*	*Squalomugil*
Xenorhynchichthys	Regan 1908	*Joturus stipes* Jordan and Gilbert 1882	*Joturus*	*Joturus*	*Joturus*	*Joturus*	*Joturus*	*Joturus*
Ellochelon	Whitley 1930	*Mugil vaigiensis* Quoy and Gaimard 1825	*Chelon*	*Ellochelon*	*Liza*	*Ellochelon*	*Ellochelon*	*Ellochelon*
Protomugil	Popov 1930	*Mugil saliens* Risso 1810	-	*Chelon*	*Liza*	*Liza*	*Chelon*	*Liza*
Sicamugil	Fowler 1939	*Mugil hamiltoni* Day 1869	*Trachystoma*	*Sicamugil*	*Sicamugil*	*Sicamugil*	*Sicamugil*	*Sicamugil*
Gracilimugil	Whitley 1941	*Mugil ramsayi* Macleay 1883	*Trachystoma*	*Chelon*	*Liza*	*Gracilimugil*	*Gracilimugil*	*Liza*
Moolgarda	Whitley 1945	*Moolgarda pura* Whitley 1945	-	*Moolgarda*	*Valamugil*	*Valamugil*	-	*Moolgarda*
Planiliza	Whitley 1945	*Moolgarda* (Planiliza) *ordensis* Whitley 1945	-	*Chelon*	*Liza*	*Liza*	*Planiliza*	*Liza*
Aldrichetta	Whitley 1945	*Mugil forsteri* Valenciennes 1836	-	*Aldrichetta*	*Aldrichetta*	*Aldrichetta*	*Aldrichetta*	*Aldrichetta*
Xenomugil	Schultz 1946	*Mugil thoburni* Jordan and Starks 1896	*Xenomugil*	*Mugil*	*Mugil*	*Mugil*	*Mugil*	*Xenomugil*
Crenimugil	Schultz 1946	*Mugil crenilabis* Forskal 1775	*Crenimugil*	*Crenimugil*	*Crenimugil*	*Crenimugil*	*Crenimugil*	*Crenimugil*
Oxymugil	Whitley 1948	*Mugil acutus* Valenciennes 1836	-	*Chelon*	*Liza*	*Liza*	*Planiliza*	*Liza*
Pteromugil	Smith 1948	*Mugil diadema* Gilchrist and Thompson 1911	-	*Chelon*	*Liza*	*Liza*	*Planiliza*	*Liza*
Strializa	Smith 1948	*Mugil canaliculatus* Smith 1935	*Chelon*	*Chelon*	*Liza*	*Liza*	*Chelon*	*Liza*
Valamugil	Smith 1948	*Mugil seheli* Forsska˚ l 1775	-	*Moolgarda*	*Valamugil*	*Valamugil*	*Crenimugil*	*Valamugil*

Table 1.1. contd....

Table 1.1. contd.

Genus	Author and date	Type species	Genus assigned by author (date)					
			Schultz (1946)	Senou (1988)	Thomson (1997)	Ghasemzadeh (1998)	Durand et al. (2012b)	Eschmeyer and Fong (2015)
Plicomugil	Schultz 1953	*Mugil labiosus* Valenciennes 1836	-	*Oedalechilus*	*Oedalechilus*	*Oedalechilus*	*Plicomugil*	*Oedalechilus*
Osteomugil	Luther 1977	*Mugil cunnesius* Valenciennes 1836	-	*Moolgarda*	*Valamugil*	*Valamugil*	*Valamugil*	*Osteomugil*
Minimugil	Senou 1988	*Mugil cascasia* Hamilton 1822	-	*Minimugil*	*Sicamugil*	*Sicamugil*	*Minimugil*	*Sicamugil*
Paracrenimugil	Senou 1988	*Mugil heterocheilos* Bleeker 1855	-	*Paracreni-mugil*	*Crenimugil*	*Crenimugil*	*ND*	*Crenimugil*
Pseudoliza	Senou 1988	*Mugil parmatus* Cantator 1849	-	*Pseudoliza*	*Liza*	*Paramugil*	*Planiliza*	*Paramugil*
Paramugil	Ghasemzadeh 1998	*Mugil parmatus* Cantator 1849	-	*Pseudoliza*	*Valamugil*	*Paramugil*	*Planiliza*	*Paramugil*
Neochelon	Durand et al. 2012b	*Mugil falcipinnis* Valenciennes 1836	-	*Chelon*	*Liza*	-	*Neochelon*	*Neochelon*
Parachelon	Durand et al. 2012b	*Mugil grandisquamis* Valenciennes 1836	-	*Chelon*	*Liza*	*Liza*	*Parachelon*	*Parachelon*
Pseudomyxus	Durand et al. 2012b	*Mugil capensis* Valenciennes 1836	-	*Chelon*	*Myxus*	*Myxus*	*Pseudomyxus*	*Pseudomyxus*

and faster forms (Nelson 2006, Cousseau 2010). There has been, and there is still, much disagreement concerning the evolutionary relationships of the order Mugiliformes, represented by a single family. While the monophyly of this family has never been challenged, phylogenetic placement of this enigmatic assemblage has been a long standing problem of systematic ichthyology (Stiassny 1993). Berg (1940) placed the Atherinidae, Mugilidae and Sphyraenidae in the order Mugiliformes, but at the Subperciformes level. Subsequently, Greenwood et al. (1966) and Nelson (1984) reviewed the subordinal status of these three families, and placed them in the order Perciformes. Later Nelson (1994, 2006) placed them in the order Mugiliformes.

The species of Mugilidae are characterized not only by both a remarkably uniform external morphology, but also a scarcely less so internal anatomy. This can be demonstrated by a comparison of the attributes commonly employed to identify mullets, as the number of scales, fin spines and fin rays, and measurements of body proportions (González-Castro 2007). They are medium to large-sized fishes, reaching a maximum size of 120 cm standard length, but commonly to about 30 cm standard length; subcylindrical body; head often broad and flattened dorsally (rounded in *Agonostomus* and *Joturus*) (Harrison and Senou 1999).

Mullets have two widely separated dorsal fins, the first of four spines and the second one usually with an unbranched ray (often called a spine) and six–10 branched rays. The pelvic fins are sub-abdominal, with a spine and five branched rays. The anal fin has two–three spines and eight–12 branched rays. The lateral line is absent, and adult mullets usually have ctenoid scales. The mouth is of moderate size, with small (labial) or missing teeth. The gill arches are usually long. They have a muscular stomach and an extremely long intestine. They have 24–26 vertebrae (Nelson 2006).

Traditionally, the features of diagnostic value for Mugilidae included: the structure of scales, the relative position of the nostrils, the number and shape of the gill rakers, the form of the preorbital, the relative lengths of the paired fins and of their axillary scales and the position of origin of the various fins, the presence or absence of an adipose eyelid and the degree of intrusion over the eye, as well as the number of pyloric caeca and the relative length of the intestine (Thomson 1997). More recently the body shape, and also the scales have been analyzed by means of geometric morphometrics. They proved to be useful as a discriminating tool at the specific and population levels (Corti and Crosetti 1996, Heras et al. 2006, Ibáñez et al. 2007, González-Castro et al. 2012). Mullet also possess a characteristic oral and branchial filter-feeding-mechanism involving gill rakers and a specialized pharyngobranchial organ comprising a large, denticulate pharyngeal pad and pharyngeal sulcus on each side of the pharyngobranchial chamber (Harrison 2002).

A Historical Overview of the Diagnostic Osteo-Morphological Features used in the Main Reviews of the Genera of Mullets

Schultz (1946) made a comprehensive revision of the genera of Mugilidae. He paid attention to the taxonomic importance of mouth parts and other qualitative characters such as the position (inferior or terminal) of the mouth, the relative thickness of the lips, the degree of lips' coverage by papillae and crenulations, the nature of the upper attachment of the maxilla, the curvature and degree of exposure of the posterior angle of the maxilla, the morphology and distribution of teeth, and the presence or absence of the symphysial knob. On the basis of these characters he recognized a total of 13 genera (including three new genera which were created/described by him) namely: *Cestraeus* Valenciennes 1836; *Joturus* Poey 1860; *Rhinomugil* Gill 1863; *Agonostomus* Bennett 1831; *Chaenomugil* Gill 1863; *Neomyxus* Steindachner 1878; *Xenomugil* Schultz 1946; *Crenimugil* Schultz 1946; *Mugil* Linnaeus 1758; *Myxus* Günther 1861; *Chelon* Artedi 1793; *Trachystoma* Ogilby 1888 and *Heteromugil* Schultz 1946. Figure 1.1 shows the possible relationships of the genera of Mugilidae according to Schultz (1946).

Smith (1948) conducted a generic revision of the South African mullets, and applied the characters used by Schultz. He confirmed the taxonomic value of the mouthparts, but noted that Schultz did not examine world-wide representatives. Smith (1948) added five more genera to those described by Schultz (1946). Again Schultz (1953) reviewed his own, and Smith's work, and after making corrections and additions, accepted 14 genera as valid.

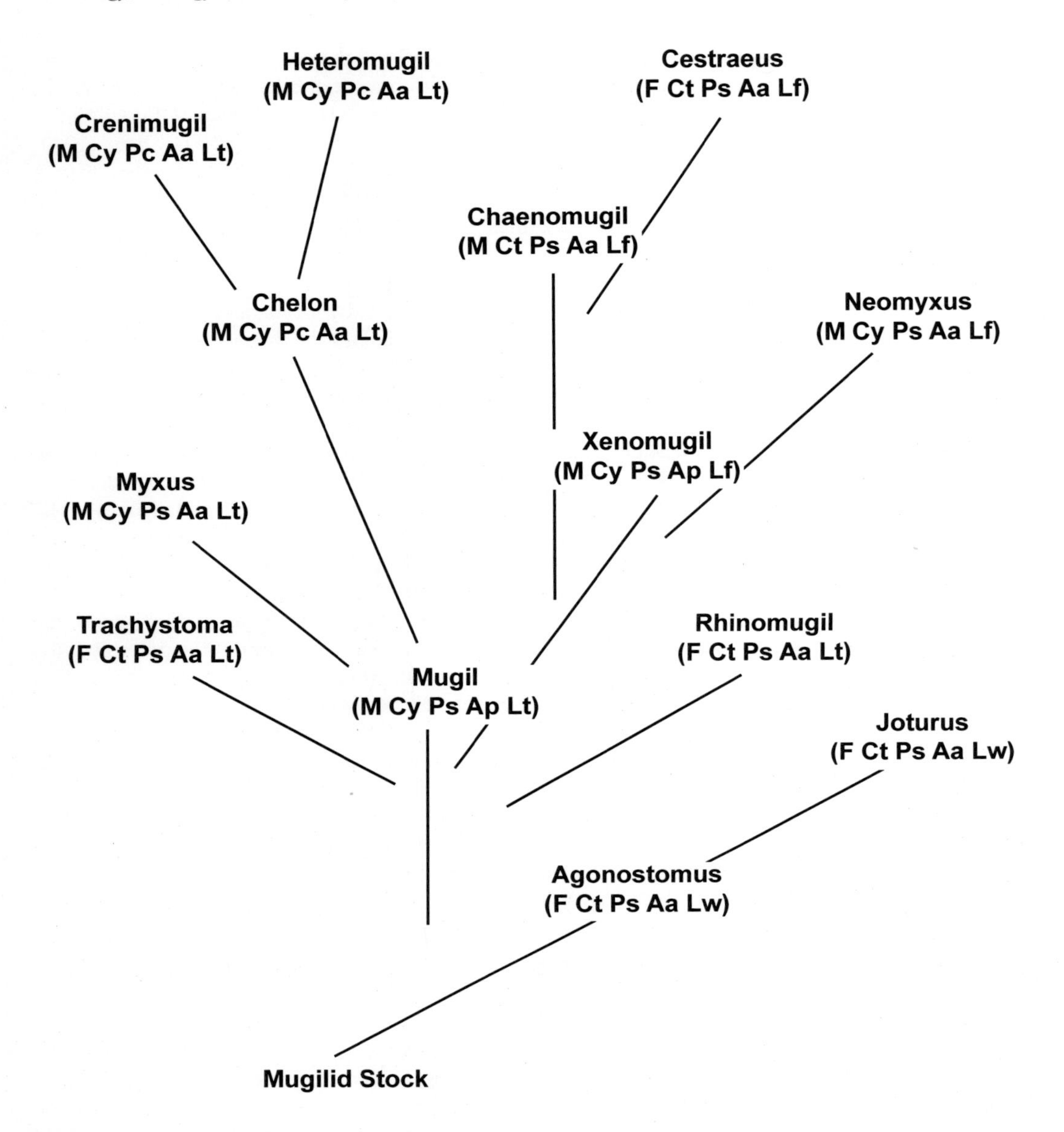

Figure 1.1. Diagram of the possible relationships of genera of the Mugilidae. Letters in the parentheses under each genus indicate some of the characters used in forming an opinion on the general evolutionary trends (Aa.-Adipose eyelid absent; Ap.-Adipose eyelid present; Ct.-Ctenoid scales; Cy.-Cycloid scales; F.-Freshwater habitat; Lf.-Lower lip folded downward; Lt.-Lower lip with thin edge projecting forward; Lw.-Lower lip thickish; M.-Marine habitat and entering brackish waters; Pc.-Front edge of preorbital straight and the maxillary and premaxillary extending in the same general line as front edge of preorbital; Ps.-Front edge of preorbital concave or deeply notched, the maxillary and premaxillary bent at an abrupt angle posteriorly, and exposed below preorbital) (from Schultz 1946).

Ingham's (1952, unpublished thesis) comprehensive revision of the biology and taxonomy of mullets (based principally on examination of the material in the British Museum of Natural History) recognized 67 species in 13 genera, and indicated that another 11 species were possibly valid.

Thomson (1954a) conducted a revision of the mullets of Australian waters and adjacent seas, based on characters of mouth parts, dentition, digestive system, morphometric and meristics. He listed 38 nominal genera (excluding fossils) world-wide, of which 13 genera were recognized as valid, and described

17 species belonging to nine genera in Australia and the South Pacific region. Thomson (1981) considered 64 species in 14 genera (of 282 nominal species) as valid, and presented a detailed description of distinctive characters and diagnostic features useful for recognition of mullet species.

Senou (1988, unpubl. thesis) reviewed the phylogenetic relationships of mullets, using osteological and morphological characters. He recognised 44 species (17 genera) in two subfamilies, Agonostominae (with three genera) and Mugilinae (14 genera).

Thomson (1997) undertook a world-wide revision of the family Mugilidae. He included some new features of diagnostic value such as the structure of the scales, the relative position of the nostrils, the number and form of the gill rakers, the shape of the preorbital, the relative length of the paired fins and of their axillary scales, the position of the origin of the various fins, the presence or absence of the adipose eyelid and the degree of its intrusion over the eye, as well as the number of pyloric caeca and the relative length of the intestine. This author recognized as valid 14 of 40 described genera, and 62 of 280 nominal species. He also introduced a new species (*Liza mandapamensis*), and classified 18 species as *inquerenda*.

Thomson (1997) accepted the division of the family Mugilidae by Jordan and Evermann (1896) into two subfamilies of Agonostominae and Mugilinae, mainly based on the character of presence or absence of sessile teeth on the jaws, the shape of the lower jaw and the degree of complexity of the stomach. The genera *Joturus* Poey 1860, *Agonostomus* Bennett 1831, *Cestraeus* Valenciennes 1836 and *Aldrichetta* Whitley 1945 belonged to the subfamily Agonostominae while the subfamily Mugilinae included the genera *Rhinomugil* Gill 1863, *Sicamugil* Fowler 1939, *Chaenomugil* Gill 1863, Oedalechilus Fowler 1903, *Crenimugil* Schultz 1946, *Chelon* Artedi 1793, *Mugil* Linnaeus 1758, *Myxus* Günther 1861b, *Valamugil* Smith 1948 and *Liza* Jordan and Swain 1884.

Ghasemzadeh (1998) reviewed the systematics, phylogeny and distribution of Indo-Pacific and Australian mullets, using morphological and osteological data. He recognized 18 of the 40 world-wide described genera as valid and described a new genus *Paramugil* (Ghasemzadeh et al. 2004). He also described 27 species belonging to 14 genera in Indo-Pacific and Australian waters.

Eschmeyer and Fong (2015) and Eschmeyer (2015) performed the last revision of the systematics of Mugilidae. They recognized 20 genera and 71 species. The species names of the present book will refer to this catalogue. The following genera are recognized as valid: *Cestraeus* Valenciennes 1836, *Agonostomus* Bennett 1832, *Aldrichetta* Whitley 1945, *Mugil* Linnaeus 1758, *Chaenomugil* Gill 1863, *Chelon* Artedi 1793, *Crenimugil* Schultz 1946, *Ellochelon* Whitley 1930, *Liza* Jordan and Swain 1884, *Joturus* Poey 1860, *Moolgarda* Whitley 1945, *Myxus* Günther 1861, *Neochelon* Durand et al. 2012b, *Neomyxus* Steindachner 1878, *Oedalechilus* Fowler 1903, *Parachelon* Durand et al. 2012b, *Pseudomyxus* Durand et al. 2012b, *Rhinomugil* Gill 1863, *Sicamugil* Fowler, 1939, *Trachystoma* Ogilby 1888.

Morphological and Meristic Diagnostic Characters Traditionally Employed for Taxonomic Determination of Mugilidae

The identification and taxonomy of mullets has relied on external morphology, meristics, morphometrics and the structure of some internal organs. The remarkably uniform external morphology of mullets has resulted in continuous confusion in their identification and classification. Further taxonomic confusion has been due to the wide range of variability in characters examined, and slight diagnostic differences between species (Ghasemzadeh 1998).

Characters which have been used by different authors include dentition (Ebeling 1957, 1961, Thomson 1975, Farrugio 1977), scales (Cockerell 1913, Jacot 1920, Pillay 1951, Thomson 1981, Chervinski 1984, Liu and Shen 1991, Ibáñez et al. 2011), number of pyloric caeca (Perlmutter et al. 1957, Hotta and Tung 1966, Luther 1977), the alimentary tract (Thomson 1966), intestinal convolution (Hotta 1955), osteology (Ishiyama 1951, Hotta and Tung 1966, Sunny 1971, Kobelkowsky and Resendez 1972, Luther 1977, Senou 1988, Ghasemzadeh 1998), otoliths (Morovic 1953), morphology of the cephalic lateral line canals (Song 1981), pharyngobranchial organ (Harrison and Howes 1991), and dentition, pigmentation and melanophore patterns in identification of fry and juveniles (van der Elst and Wallace 1976, Cambrony 1984, Reay and Cornell 1988, Serventi et al. 1996, Minos et al. 2002). Following a brief description, characters and features of diagnostic value which are commonly used in identification and taxonomy of mullets are given.

Adipose Eyelid

The adipose membrane is not a third eyelid, but a fatty deposition on the head around the eyes which can be present or absent, depending on the genera. This tissue is transparent in life, and becomes opaque on death. As used in most keys and descriptions, the extent of the development of the adipose eyelid refers only to mature specimens (Thomson 1954a). This tissue is not developed in newly hatched fish, and does not become apparent until a length of 4 or 5 cm, after which the area of the eye covered may continually increase during life (e.g., *Mugil cephalus*) or remain relatively insignificant as in some species of the genus *Liza*. The terminology for recording the extent of the development of the adipose eyelid is not very exact, and most authors use the terminology which was suggested by Thomson (1954a). The term 'absent' indicates that no trace of an adipose eyelid could be found. The terms 'obsolescent' or 'rudimentary' refer to any stage between a rim around the eye to a lid covering up to a third of the iris; and 'present' indicates the development beyond a mere rim, so as to cover a measurable portion of the eye. Thus, this character is commonly employed to differentiate between genera. Thomson (1981) points out that the occurrence of varying width within different genera suggests that the genera diverged after the development of the adipose tissue in the Mugilinae subfamily, with subsequent independent trends to obsolescence of this characteristic.

Pyloric Caeca

The number of pyloric caeca varies among mullet species, and can be of some taxonomic importance, especially among different genera. The primitive number of two pyloric caeca is found throughout the subfamily Agonostominae and in *Trachystoma*, *Gracilimugil*, *Neomyxus*, *Myxus* (except *Myxus capensis*), *Mugil*, *Sicamugil* and *Chaenomugil* of Mugilinae (Thomson 1997, Ghasemzadeh 1998). In other genera the number of pyloric caeca varies between three and 48 (usually, between five and 10). However, its counting assumes the dissection of the specimen, so this meristic characteristic is not useful in eviscerated (i.e., museum) specimens. Also, it requires extra time in order to perform the dissection, which makes this feature unsuitable for quick taxonomic identification on the field.

Normally, the number of pyloric caeca varies within a certain range in specimens of the same species, but it is usual to find well differentiated species of the same genus sharing the same number of pyloric caeca.

Teeth

The teeth are important anatomical elements which can be employed as diagnostic features in taxonomy of mullets. In the plesiomorphic Agonostominae the jaw teeth are of proximal or sessile type (Jordan and Evermann 1896, Fink 1981, Thomson 1997, Ghasemzadeh 1998), and borne directly on the premaxilla and dentary bones. In Mugilinae, the jaw teeth are minute and labial, and are borne on the distal end of flexible and closely packed fibrous strands, which are proximally joined to the premaxilla and dentary, and supported by labial tissue (Thomson 1997, Ghasemzadeh 1998). In many species of mullet, only a single row of teeth is developed which is referred to as primary teeth (Ebeling 1957), but in others there may be several inner rows (termed secondary teeth). In some species the form of the primary and secondary teeth are different, and since distal-type teeth are loosely attached to the underlying bone, they are presumably often lost and replaced (Ebeling 1957). However, in mullets there is a tendency for teeth to be lost with age, and aged toothless specimens are known (Thomson 1981, Ghasemzadeh 1998). There is a great variation in the shape of mullet teeth (ciliform, setiform, caniniform, bicuspid, tricuspid and multicuspid). The shape of the teeth and the pattern of dentition have been widely employed in taxonomic and systematic studies of mullets (Schultz 1946, Ebeling 1957, 1961, Thomson 1954b, 1975, 1997, Farrugio 1977, Menezes 1983, Ghasemzadeh 1998, Harrison and Senou 1999, Harrison et al. 2007).

Stomach Shape

Stomach and caeca can be seen by cutting the fish along the abdomen and removing the liver (Harrison and Senou 1999). As a general rule, the morphology of the stomach shows several differences between the Agonostominae and Mugilinae. It is a simple U-shaped sac of thin wall in the former, with the exception of the genus *Aldrichetta*, which exhibit thicker walls than in other Agonostominae (Thomson 1997). In Mugilinae, the stomach is usually divisible into a thin-walled cardiac crop and a very thick-walled biconical pyloric gizzard. This thick-walled, muscular stomach is a site of mechanical action used to break down algal cell walls. Bacteria, blue-green algae, diatoms and macroalgae that have been ingested with sand or other sedimentary material are triturated in this gizzard-like organ.

Head

The head as a whole is an informative organ from the taxonomic point of view, normally employed in any identification key of mullets. Although in mullets the head is often broad and flattened or gently convex dorsally, a wide variation in shape and relative size can be observed amongst the species of Mugilidae. The positional relationships among the different anatomical elements (jaws, nostrils, lips, eyes, opercular and preorbital bones, jugular space, etc.) and also their form (shape plus size), generates a variety of head shapes which can be used to aid taxonomic identification at the species level. At the geometric-morphometric level, many studies have discriminated species of mullets based on shape variables related to the head. For example, Heras et al. (2006) found that the one–two and one–four variables, which are inter-landmark distances belonging to the first box truss (that represents the head), were important measurements for the specific discrimination between *Mugil curema* and *M. cephalus*. These results agree with previous work (Ibáñez-Aguirre and Lleonart 1996) where lineal morphometry, based on cephalic length and length to anal fin, differentiated both species. Moreover, González-Castro et al. (2012) found that the inter-landmark distances of the first box-truss, that represent the head shape, contributed significantly not only to the discrimination amongst seven species of the genus *Liza* and *Mugil* but also to plausible cryptic species of both *M. curema* and *M. cephalus* species complexes.

Mouth and Lips

The mouth in Mugilidae is normally of small/moderate size. It is terminal, although sometimes subterminal. The mouth gape is a ratio that has been employed in the past for taxonomic purposes. It is defined by Thomson (1997) as MW/ML, where: MW is the Mouth Width from mouth corner to mouth corner and, ML is the Mouth Length from the anterior tip of the lip to the posterior corner of the mouth opening.

Lips may be narrow or thick, smooth, lamellate or papillate. The upper lip may be terminal, or it may be overhung by a projection of the snout. The mouth features such as being terminal or subterminal, mouth gape, lips shape, dentition and ornamentation, and angle of dentary symphysis are all characters which have been used in taxonomy of mullets.

Jaws

In Mugilidae jaw structure is basically of the percoid type, distinguished by the premaxillary having short pedicels and a shaft which, in Agonostominae is widest at its mid-length and pointed at its distal end, whereas in the Mugilinae the shaft is broadest at the blade-like distal end (Thomson 1997, Ghasemzadeh 1998). In some genera of Mugilinae the edge of the premaxilla remains more or less parallel with the line of the mouth gape, but in others it curves down behind the corner of the mouth. The maxilla lies behind the premaxilla and at its upper end attaches to the ethmoid by a ligament. The degree of protrusibility of the mouth is largely governed by the degree of mobility of the maxilla, because they are locked to the premaxillary pedicels via the maxillary processes which also fuse in the midline. When the mouth is closed, the premaxillary pedicels retreat under the nasal bones (Thomson 1997). The maxillary and premaxillary bones may be almost straight as in most genera of Agonostominae, or may bend downwards posteriorly as in members of the subfamily Mugilinae. When the latter occurs, the posterior end of the maxilla is usually visible when the mouth is closed (Thomson 1954a). This feature has been used by most taxonomists as a diagnostic character to identify some genera of mullets.

The anterior edge of the lower jaw consists of a pair of dentary bones which are joined together at the dentary symphysis. Each dentary bone has a horizontal arm which is edentulous or variably toothed, and another ventral arm with a fossa for the insertion and articulation of the angular bone (Ghasemzadeh 1998). The osteology of jaws has also been used by many authors in taxonomy and classification of mullets (Ishiyama 1951, Senou 1988, Ghasemzadeh 1998, Harrison 2002, Ghasemzadeh et al. 2004).

Preorbitals

The preorbitals are a pair of triangular bones, situated obliquely in front of the eyes. The anterior edge of these bones are elongate and denticulate, and depending on the genus or species of mullets may be notched, curved or straight (Ghasemzadeh 1998).

Nostrils

The nostrils may be variously placed in different species of mullets. In some species, the nostrils are nearer to each other than the posterior is to the eye or the anterior to the lip; in other species their position may be different (Thomson 1997). The posterior nostril usually reaches just above the level of the upper rim of the eye, but in a few species is higher. On the other hand, in *Rhinomugil squamipinnis* (Swainson) the posterior nostril is displaced to the level of the lower half of the eye (Thomson 1997).

Pharyngobranchial Organ

The structure of the pharyngobranchial organ (PBO) of Mediterranean mullet was studied by Capanna et al. (1974). They presented an account of its anatomy, histology, dentition and possible complex filtering function (for feeding on small benthic particles) with some photographic images of the skeletal components of the PBO. Harrison and Howes (1991) reviewed the PBO of mullet, and gave a detailed description of its structure, associated musculature and dentition, ontogeny, possible function, and its taxonomic utility among the genera of mullets.

Scales

Three types of scales can be observed in adult mullets: cycloid scales, as in *Myxus elongatus*, ctenoid scales, as in *Ellochelon vaigiensis,* and ctenoid scales with a digitated membranous hind border, as in *Valamugil* spp. The morphology and morphometry of scales has been employed for identifying genera, species and populations within Mugilidae (Ibáñez et al. 2007, Ibáñez et al. 2011).

Axillary and Obbasal Scales

The presence or absence of an axillary scale is another feature which is used in taxonomy of mullets. Thomson (1954a) defined the term axillary scale only for the elongated scale occurring at the base of the pectoral fins. He termed the elongated scales occurring at the base of the first dorsal and ventral fins as dorsal and ventral obbasals.

Meristic Characters

Number of Scales in the Lateral and Transverse Series

Traditionally, the number of scales in the lateral series (Ll) can be counted over the left side of specimens, from the scale located just behind the head (i.e., immediately above the insertion of the pectoral fin) to the caudal flexure (hypural plate limit). Its number varies approximately from 24 (*Ellochelon vaigiensis* or *Liza luciae*) to almost 63 in *Aldrichetta forsteri*. This meristic character is usually employed as a prominent diagnostic feature. However it is common to find overlapping scale counts in mullets at the intrageneric level (Thomson 1997, Ghasemzadeh 1998, Harrison 2002, González-Castro et al. 2008, González-Castro et al. 2012).

The transverse scale count (tr) can be interpreted as the number of scales between the origin of the first dorsal fin and the origin of the pelvic fin. However, some authors have used this count starting from the second dorsal fin to the origin of the anal fin base, which is less common. Transverse scale counts vary from eight to 10 (i.e., *Ellochelon vaigiensis*, *Liza grandisquamis* and *Liza luciae*), with a mode of 11 (some *Mugil*, *Liza* and *Valamugil* spp.), to a maximum value of 19 (*Aldrichetta forsteri*).

Number of Spines and Rays of Paired and Unpaired Fins

The first dorsal fin consists of four spines which is one of the most diagnostic characters of Mugilidae. Each spine is supported by a single basal pterygiophore. The first three spines are placed very close to each other, while the fourth spine is well-separated from them. The second dorsal fin consists of seven–10 rays in different genera of mullets, ranging from seven to eight rays in *Rhinomugil*; eight rays in *Mugil* and *Crenimugil*; nine rays in *Myxus, Trachystoma, Gracilimugil, Ellochelon, Liza, Paramugil* and *Valamugil*; nine to 10 rays in *Aldrichetta* and *Gracilimugil* and 10 rays in *Neomyxus* (Thomson 1997, Ghasemzadeh 1998). The anterior most ray of this fin is frequently mistaken for a spine. In fact it is a short and slender ray which is often unbranched and only segmented near its tip in adults (Ghasemzadeh 1998). The anal fin has three spines in most genera of mullets, except *Neomyxus* which has two spines and *Gracilimugil* which has three–four spines. The number of anal fin rays is eight in *Ellochelon, Rhinomugil* and some *Mugil* species; nine–11 rays in *Gracilimugil* and *Neomyxus* and some *Mugil*; 11–13 rays in *Aldrichetta* and nine rays in the rest of genera of Mugilidae (Ghasemzadeh 1998). Pectoral fins have one spine and 14–20 rays in different genera of mullets. Pelvic fins have the typical number of one spine and five rays.

Morphometric Differentiation of Mullets

The Disadvantage of the 'Size Effect' in Morphometric Analysis: Influence of Size due to Allometric Growth

Most of the variability in a set of multivariate morphometric data from natural populations is due to individual size. In morphometrics, size must be considered as a contingent source of variability since it is associated with individual growth and the aim of such studies is usually focused on shape that must be size-free. In the general case of allometric growth (one type of ontogenetic variation), there is a variation in shape related to the variation in size (Lleonart et al. 2000). Hence body size is usually a confounding factor in any morphometric analysis. When specimens under study belong to different populations, and especially to different age classes, then is to be expected that size generates an important bias in taxa-discrimination (González-Castro et al. 2007).

The influence of size due to allometric growth may be removed by appropriate statistical procedures (Gould 1966, cited in Lleonart et al. 2000). There are numerous normalization methods whose aim is to eliminate the size effect in the context of allometric growth. However, some of the most popular methods have critical shortcomings that lead to misinterpretation of the results (Lleonart et al. 2000). Among these, the ratio of every measurement to the one chosen as the independent variable effectively reduces all the individuals to the same size, but does not remove the undesired size effect because they maintain their size-dependent shape due to allometry. In other words, it is only valid if growth is isometric (i.e., shape does not change with size).

There are other methods to eliminate this size effect, such as 'shearing' (Humphries et al. 1981) or 'Burnaby's method for size correction' (Burnaby 1966), which combine various techniques (i.e., regression) with a Principal Component Analysis (PCA) subsequently extracting the first principal component under the widespread view that this component represents the size in the PCA. However, it was observed that both size and shape are embedded in the first component (Mosimann 1970, Sprent 1972, Humphries et al. 1981). Lleonart et al. (2000) present a novel normalization-technique to scale data that exhibit an allometric growth. The method is theoretically derived from the equation of allometric growth. This normalization procedure is, consequently, compatible with allometry. It completely removes all the information related to size, not only scaling all individuals to the same size, but also adjusting their shape to that which they would have in their new size.

The normalization procedure of Lleonart et al. (2000) has been employed successfully in several works related to Mugilidae taxonomy published in the last two decades (Ibáñez-Aguirre and Lleonart 1996, Cousseau et al. 2005, Heras et al. 2006, González-Castro et al. 2008, González-Castro et al. 2012).

Different Kinds of Morphometric Variables and Different Morphometric Approaches

Linear Morphometric Measurements

Linear morphometric measurements (LMMs) are the 'traditional' measures employed on fishes. Among the most commonly in use in Mugilidae are: standard length (SL), head length (HL), head width (HW), snout length (Sn), pectoral fin length (PL), predorsal 1 distance (pD1d), predorsal 2 distance (pD2d), preventral distance (pVd), preanal distance (pAd), body height (BH) (Fig. 1.2).

Traditionally, LMMs were usually used to calculate ratios (i.e., percentages of some corporal variable over the total length), which were then employed to perform uni/multivariate analysis (without any consideration of the shape variation related to the size change). However, there are several biases and weaknesses inherent in traditional character sets: (1) Most characters tend to be aligned with the longitudinal axis, thus a large amount of the data are repetitious while other information (i.e., variation in oblique directions) is lacking; (2) Coverage of form is highly uneven by region as well as by orientation (dense in some areas of the body and sparse in others); (3) Some morphological points, such as the tip of the snout and the posterior end of the vertebral column, are used repeatedly. Any uncertainty in the positions of these morphological features will be propagated through a series of measurements; (4) Some LMMs are 'extremal' rather than 'anatomical' (i.e., greatest body depth), and therefore their placement may not be homologous from form to form; (5) Many measurements extend over much of the body. Long distances are usually employed in the traditional data sets, but are less informative than short ones (Strauss and Bookstein 1982, González-Castro 2007).

LMMs have proved to be useful however, if the size effect due to allometric growth is removed prior to the multivariate analysis. In this respect, the normalization procedure of Lleonart et al. (2000), followed by multivariate analysis (PCA; Discriminant Analysis, DA), has been employed in some works related to taxonomy or comparative morphometrics of Mugilidae (Ibáñez-Aguirre and Lleonart 1996, Cousseau

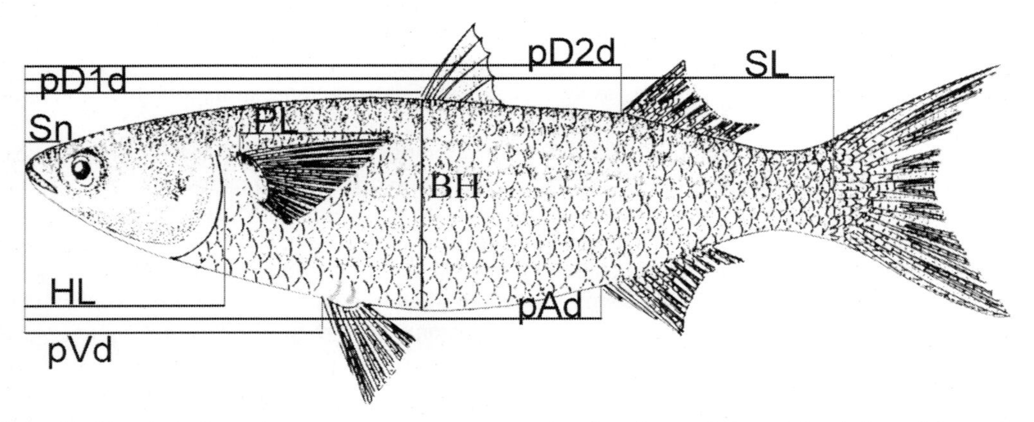

Figure 1.2. Linear morphometric measurements commonly employed on the morphological descriptions of the species of Mugilidae.

et al. 2005, Ibáñez-Aguirre et al. 2006, González-Castro et al. 2012) and also in studies related to growth analysis by means of morphometry (Minos et al. 1995, Ibáñez-Aguirre et al. 1999).

The use of morphometric characters to distinguish young stages of grey mullet species is a method with low accuracy due to major changes in body proportions (allometry), which occur in these stages (Thomson 1981). In this case, the use of the same body size of individuals of the compared groups (species) overcomes the problem, but the findings on this size are limited. According to Katselis et al. (2006) (who analyzed the variation in eight morphometric characteristics of the fry of four grey mullet species: *Liza aurata, Liza saliens, Chelon labrosus* and *Mugil cephalus*) "...this problem has been overcome with the use of the total length class of 20–35 mm for all species". According to DA classification, 92.7% of the specimens examined in this study were correctly classified into the four species.

Interlandmarks Distances Based on Box-Truss

Anatomical landmarks are true homologous points identified by some consistent feature of the local morphology. This implies that, when we establish a set of landmarks in two different forms to be compared (species, populations, morphs, etc.), by definition these landmarks must be located without any doubt in both morphs and have correspondence (biological homology) among forms (Bookstein 1991).

Strauss and Bookstein (1982) proposed a protocol for character selection, the truss network, which enforces systematic coverage of the form and largely overcomes the disadvantages of traditional data sets. This protocol systematically detects shape differences in oblique as well horizontal and vertical directions and archives the configuration of landmarks so that the form may be reconstructed (mapped) from the set of distances among landmarks (i.e., to obtain Cartesian coordinates for landmarks). Analyses of landmark data are usually based either on distance between selected pairs of landmarks or on the coordinates of the landmarks.

González-Castro (2007) (Fig. 1.3) and González-Castro et al. (2008, 2012) combined the Box-truss concepts of Strauss and Bookstein (1982), with the technique (for removing the influence of size due to allometric growth) of Lleonart et al. (2000) followed by multivariate analysis (PCA, DA), suggesting a new approach on the taxonomic analysis of mullets, but also of other fishes. This combination of methodologies represents a concise and low-cost way to successfully distinguish/discriminate species (or even populations) of Mugilidae.

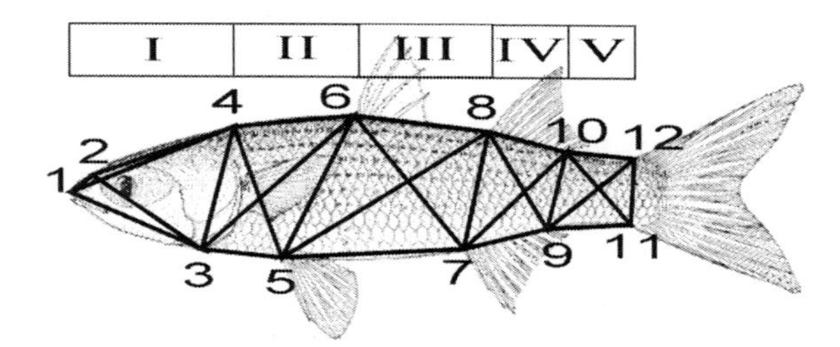

Figure 1.3. Box truss (Roman numerals) showing the interlandmark distances, based on 12 anatomical landmarks, proposed for the morphometric analysis of Mugilidae (from González-Castro 2007).

Coordinate Data

Morphometric techniques have 'evolved' in the last few decades in parallel with the introduction of promissory methods for archiving forms of organisms (Rohlf 1990). This discipline experienced a major revolution through the invention of coordinate-based methods, the discovery of the statistical theory of shape, and the computational realization of deformation grids (Mitteroecker and Gunz 2009). This new morphometric approach has been termed geometric morphometrics as it preserves the geometry of the landmark configurations throughout the analysis. Cartesian coordinates obtained from anatomical landmarks are the keystones on which geometric morphometrics are based. These coordinate data can come from several sources, such as a digital camera, digitizing tablet or indirectly by reconstruction of landmarks from Box-trusses (Bookstein et al. 1985, Rohlf and Marcus 1993, González-Castro et al. 2012).

Corti and Crosetti (1996) performed the first geometric morphometric analysis on the grey mullet *Mugil cephalus*. Based on Partial Warps scores the authors described the shape differences and characterize 10 populations of this species, Galapagos being the most morphometrically distinct.

Ibáñez et al. (2007) performed a geometric morphometric analysis of fish scales for identifying genera, species and local populations within the Mugilidae. Fish scale form was least effective in discriminating populations from nearby areas, better when populations are more geographically dispersed, and best between species and genera. Scale form variation reflected previous genetic studies that differentiated congeneric *M. cephalus* and *M. curema*.

Recently, González-Castro et al. (2012) based on the Cartesian coordinates of 12 anatomical landmarks reconstructed from distance measurements among the landmarks (based on a Box-truss scheme) performed a geometric morphometric analysis of the body shape of six representative species of Mugilidae: *M. cephalus* Linnaeus 1758; *M. liza* Valenciennes 1836; *M. curema* Valenciennes 1836; *M. hospes* Jordan and Culver 1895; *Liza aurata* (Risso 1810); *L. ramada* (Risso 1826). Morphometry allowed discrimination not only among the six species, but also the American and European 'populations' of *Mugil cephalus* and the North and South American *Mugil curema*. Although some overlap among samples was detected, the DA (Cross-validated Discriminant Analysis) correctly classified 83.8% of the fishes according to their body shape.

The Barcode of Life Initiative as a Complementary-Genetic Tool for Non-Geneticist Mullet Taxonomists

Resolving species boundaries between closely related fish species, or families characterized by external uniformity (as in the Mugilidae is) is notoriously difficult. Such species can generally only be diagnosed based on few characters that often have a host of problems. For example, the morphological or meristic differences between these species may be very slight, difficult to describe and applicable only to a punctual ontogenetic stage. In the reverse situation, we can look at populations that are polymorphic with regard to characters that are normally diagnostic for species. The current trend in the field of taxonomy is defined as 'integrative taxonomy': to combine morphological and meristic analysis with newer disciplines, such as molecular genetics or geometric morphometrics (Dayrat 2005, González-Castro et al. 2008, Padial et al. 2010). Thus, it is possible to obtain comparable results, creating a synergistic effect and more robust conclusions. As was commented earlier in this chapter, mullet taxonomy is in a crisis. Overcoming this crisis is likely to be related to the integration of morphological/metrical and molecular disciplines.

The use of a universally accepted short DNA sequence for identification of species (DNA barcoding or Barcode) has been proposed for application across all forms of life, within the Barcode of life Initiative. DNA barcoding may be an efficient aid to traditional taxonomy, designed to facilitate fast and accurate species identification (Hebert et al. 2003a,b, Hebert and Gregory 2005, Miller 2007). The fragment of 648 base pairs (bp) of the mitochondrial gene cytochrome c oxidase subunit 1 (COI) is the primary sequence of DNA barcoding for species of the animal kingdom (Hebert et al. 2003a). It is based on the premise that every species will probably have a unique DNA barcode and that genetic variation between species exceeds that within species (Hebert et al. 2003a,b). The primary goal of barcoding focuses on the assembly of reference sequence libraries derived from expert-identified voucher specimens in order to develop reliable molecular tools for species identification in nature.

Let us presume that on the date (September 2014), if we enter into the BOLD (Barcode of Life Data Systems) Public Data Portal (http://www.boldsystems.org/index.php/), and write 'Mugiliformes' in the link of 'Taxonomy', we will find the following information: 1149 published records, forming 112 BINs (clusters) related to 102 species with barcodes, with specimens from 43 countries, deposited in 37 institutions. Of these records, 893 have species names, and represent 83 public species. The Barcode Index Number

(BIN) system is a persistent registry for animal OTUs (Operational Taxonomic Units) recognized through sequence variation in the COI DNA barcode region. Since OTUs show high concordance with species, this system can be used to verify species identifications (Ratnasingham and Hebert 2013). Therefore, these data can give us an overview of the current state of the Barcode related to the taxonomy of Mugilidae. One hundred and twelve BINs suggest the existence of at least 112 species of Mugilidae already Barcoded, a record which is quite distant from the 72 species recognized by Nelson (2006) or the 71 species reported in Eschmeyer (2015) catalogue. We can also enter the public record list, download public sequences (in order to compare it with our sequences) and look at the record details for each specimen/sequence, or align the sequences and perform a Neighbour Joining Analysis.

Current Taxonomic Status and Conflicts in Fishes Belonging to the Mugilidae

Taxonomic Conflicts at the Generic Level

The Genera Chelon *Artedi, 1793 and* Liza *Jordan and Swain, 1884*

The generic name of *Chelon* is proposed on page 118 of an appendix to volume IV of the 1793 edition of Artedi's '*Synonymia nominum piscium*'. There is no description or type specimens mentioned and also no evidence whether the proposed names are binomial. Subsequently Jordan and Evermann (1917) designated *Mugil chelo* Cuvier 1829 as the possible type, and this was accepted by Schultz (1946), which according to Trewavas and Ingham (1972) is definitive, confirming the earlier tentative one. Trewavas and Ingham (1972) argued that since Röse (in: Walbaum 1793) cited pre-Linnaean names, the description and interpretation of subsequent authors like Rondelet and Gesner (in: Gudger 1934) apply at least mainly to *Mugil labeo* Cuvier. They also pointed out that these authors may also have confused *M. chelo* Cuvier with *M. labeo* in their description. Trewavas and Ingham (1972) also argued that if we accept that Röse's citations covering two species, even confusing the tautonomous name '*Chelo*' or '*Chelon*', the designation of type-species is left to a subsequent author, so according to the International Code of Zoological Nomenclature of 1961, Schultz's designation may be accepted.

Schultz (1946) however, included all the species belonging to the genera *Liza, Valamugil*, *Ellochelon* and *Oedalechilus* in *Chelon*. Earlier Oshima (1922) recognized the genus *Chelon* Röse, and assigned *Mugil crenilabis* (Forsskål 1775) to this genus. Schultz (1953) recognized *Oedalechilus* (Fowler 1903) as a valid genus, but still considered *Liza, Valamugil* and *Ellochelon* as synonymies of *Chelon*. Trewavas and Ingham (1972) considered *Mugil chelo* Cuvier 1829, as a synonym of *M. labrosus* Risso 1826, and concluded that the species of *Crenimugil* display closer affinity to *Chelon labrosus* than to species of *Liza*. Some authors like Taylor (1964), Senou (1988) and Randall (1995), have used the generic name of *Chelon* instead of *Liza*, without any explanation to elucidate their decision. Ghasemzadeh (1998), stated that the nomenclatural issue of date and authorship of *Chelon* and the confusion behind the history of the name is complex and should be addressed first, and the subject of whether *Chelon* is a synonym of *Liza* remains unresolved and requires more detailed taxonomic discussion.

Valamugil *Smith, 1948* vs Moolgarda *Whitley, 1945.* Osteomugil?

Whitley (1945) established the genus *Moolgarda*. His generic description was based on an orthotype of *Moolgarda pura* which was probably a specimen of *Valamugil buchananai* or (unlikely) *V. cunnesius* which are abundant and frequently reported in coastal shallow waters of his type locality region (Point Cloates, western Australia). Whitley's description corresponds with *Valamugil* Smith 1948, especially his reference to the large pectoral axillary scales of the paired fins; obsolescent adipose eyefold, barely covering one third of the eye posteriorly, upper jaw terminal; upper lip moderately thick with microscopic cilia or entire, not papillose, jaws toothless; and also origins of second dorsal and anal fins about opposite each other, or anal fin slightly anterior. These characters are diagnostic for *Valamugil*. In the next paragraph Whitley mentioned that the closest group to the new genus was *Liza*. Whitley (1945) also compared Günther's (1861) description of *Mugil capito* Cuvier, as the genotype of *Liza*, with his Australian fish, and referred to the hidden maxillary in his fish, which is another diagnostic character of *Valamugil*. Furthermore Whitley also observed some differences in the angle of the mandible, proportions of head and body, and scale counts, to propose a new genus for the Australian fish.

Whitley (1945) also suggested that two mainly eastern Australian species, '*Mugil' argenteus* Quoy and Gaimard, and *M. compressus* Günther, may tentatively be included in *Moolgarda*. He did not have any specimens of those two fishes at hand, and consulted Günther's (1861) description, which did not agree with the new genus. Therefore, his description was based on a species of *Valamugil*. Unfortunately, no type specimen of *Moolgarda pura* was retained in either the Australian Museum, the western Australian Museum, or another institution, and most authors use *Valamugil* Smith 1948, instead of *Moolgarda* Whitley 1945, which is reasonable according to the zoological code.

Luther (1977) studied some genera and species of Indian mullets using osteology of the vertebral column, degree of adipose eyelid development on the orbit, visibility of the end of maxilla and number of pyloric caeca. He erected the new genus *Osteomugil* based on a single specimen of *Mugil cunnesius* Valenciennes 1836. In his remarks Luther (1977) stated that his new genus has some affinity to *Valamugil*. Subsequent studies by Thomson (1997), Senou (1988) and Ghasemzadeh (1998) proved the synonymy of *Osteomugil* with *Valamugil*.

Taxonomic Conflicts at the Species Level

The Mugil curema *Species Complex and* Mugil rubrioculus *nova* sp.

The white mullet *Mugil curema* Valenciennes 1836 is a widely distributed mullet. This species inhabits the Pacific coast of America from the Gulf of California to North Chile; and the Atlantic coast of America from Cape Cod to Argentina and the west coast of Africa from Gambia to the Congo (Menezes 1983, Thomson 1997, Harrison 2002, González-Castro et al. 2006).

Mugil curema was 'traditionally' considered a conspicuous species, well differentiated from its congeners by its meristic counts and morphological characters. However, some taxonomic confusion has occurred during the last four decades in both North and South America, as evidenced by the long discussions which arose around them, the validity or synonymy of *Mugil gaimardianus* Desmarest 1831, *Mugil brasiliensis* Günther 1861 (currently both invalid species), *Mugil rubrioculus* sp. nov. (Harrison et al. 2007) and its taxonomic/morphological relationship with the white mullet *Mugil curema* (Alvarez-Lajonchere 1975, Menezes 1983, Godinho et al. 1988, Cervigón 1993, Nirchio et al. 2003, Harrison et al. 2007). Moreover, some inconsistencies between the identification keys and the field characters have been observed for specimens of the white mullet around the American continent (Menezes 1983, González-Castro et al. 2006, M. González-Castro, pers. comm.).

It was in the last decade that some work shed light on this apparent taxonomic uncertainty: *Mugil curema* is undoubtedly a species complex (Nirchio and Cipriano 2005, Heras et al. 2006, Fraga et al. 2007, Heras et al. 2009, Durand et al. 2012a). This haplogroup is apparently monophyletic (Durand et al. 2012b), but it includes *Mugil incilis*, which is easily meristically and morphologically identifiable from *Mugil curema* (Thomson 1997, Harrison 2002). The *Mugil curema* species complex would be constituted by at least four lineages, which include three different karyotypes (more information is given in Chapter 15—Rossi et al. 2015). Interestingly, in an assessment of lineal versus landmark-based morphometry for discriminating species of Mugilidae, González-Castro et al. (2012) showed that the three morphometrics approaches employed separated *M. curema* specimens in two groups (Argentinean and Mexican samples), suggesting they may constitute different species. The variables (interlandmarks distances) responsible for these differences were mostly located in the head (Box-truss I) and in the segment of the body delimited by the ventral, first/second and anal fins (Box-trusses III and IV) (Fig. 1.3). On the other hand, meristics counts do not show significant differences between both groups (González-Castro et al. 2012) and instead, the colouration pattern seems to be useful in order to contribute to the specific determination of this species complex (González-Castro et al. 2006, Harrison et al. 2007, M. González-Castro, pers. comm.).

Harrison et al. (2007) presented karyological and morphological evidence for a mullet in Venezuelan coastal waters that does not conform to the description of any other species from the western central Atlantic and has the feature of a red eye that was often used by earlier authors to define nominal *M. gaimardianus*. These authors establish a valid name for the species, *Mugil rubrioculus* n. sp. Surprisingly, Durand et al. (2012a), found a phylogroup of the *Mugil curema* species complex which has 2N = 48 (the same number of chromosomes as *Mugil rubrioculus*). *Mugil rubrioculus*, *Mugil hospes* and the *Mugil curema* species complex (including *M. incilis*) would constitute one of the two Sub-Clades of the monophyletic genus *Mugil*, according to Heras et al. (2009) and Durand et al. (2012a). In the future, much work should to be done in order to morphologically differentiate, and assign a specific name to, each of the remaining three lineages of the *Mugil curema* species complex.

Mugil cephalus: *The Biggest Species Complex of Mugilidae, or Just a Cosmopolitan Species?*

The flathead mullet *Mugil cephalus* Linnaeus 1758 is the type species of the genus *Mugil* and undoubtedly the most studied mullet. A recent global review of this species concentrated on the biology, genetics, ecology and fisheries aspects (Whitfield et al. 2012). *Mugil cephalus* is the most widespread species of the Mugilidae: the species has been recorded in coastal and estuarine waters of temperate, subtropical and tropical regions, mainly between latitudes 42°N and 42°S (Thomson 1997, Harrison 2002, Nelson 2006, González-Castro et al. 2008, Durand et al. 2012a, Whitfield et al. 2012).

Despite its global spread in both hemispheres, *M. cephalus* has a discontinuous distribution. Questions regarding its taxonomic status have been raised in many genetic studies, most of which suggest *Mugil cephalus* is a species complex (Crosetti et al. 1994, Rossi et al. 1998a, Rocha-Olivares et al. 2000, Fraga et al. 2007, González-Castro 2007, González-Castro et al. 2008, Heras et al. 2009, Jamandre et al. 2009). Recently, Durand et al. (2012a) postulated that a *Mugil cephalus* species complex would be constituted by 14 parallel lineages that included the *M. liza* lineage and 13 other lineages, all currently designated as *M. cephalus*. Generally, each lineage has a regional distribution, whereas in some instances, different lineages co-exist at a single locality. Shen et al. (2011) recorded three lineages for Taiwan; another example is New Caledonia where two lineages were sampled, one of which was also sampled in New Zealand, the other one (L3) also occurred in Fiji and Taiwan (Durand et al. 2012a).

The huge number of results obtained strongly suggest that the '*Mugil cephalus* species-complex' is comprised of at least 14 biological species, including the mitochondrial lineage of *M. cephalus* (Linnaeus 1758) sampled in the Mediterranean (the type-locality), and *M. liza* (for further details and discussion see Chapter 2—Durand 2015 and Chapter 15—Rossi et al. 2015). To delimit, describe and give scientific names to these cryptic species could be the biggest challenge facing the taxonomy of mullets.

References

Alvarez-Lajonchere, L.S. 1975. Estudio systematico de *Mugil brasiliensis*, *Mugil gaimardianus* y *Mugil curema*. Ciencias 14: 1–18.

Artedi, P. 1793. Generapiscium in quibus sistema totumichthyologiaeproponitur cum classibus, ordinibus, generumcharacteribus, specierumdifferentiis, observationibusplurimis. Redactisspeciebus 242, ad genera 52. Ichthyologiaepars 3. (2nd ed.) pp. 293–432, Röse: Gypsewald.

Bennett, E.T. 1832. Observations on a collection of fishes from the Mauritius presented by Mr. Telfair, with characters of new genera and species. Proc. Zool. Soc. Lond. Part I: 165–169.

Berg, L.S. 1940. Classification of fishes, both recent and fossil. Trav. Inst. Zool. Acad. Sci. URSS 5: 87–517.

Bookstein, F.L. 1991. Morphometric Tools for Landmark Data, Cambridge University Press, Cambridge, UK.

Bookstein, F.L., R.L. Chernoff, J.M. Elder, Jr. Humpries, G.R. Smith and R.E. Strauss. 1985. Morphometrics in Evolutionary Biology: The Geometry of Size and Shape Change, with Examples from Fishes. Academy of Natural Sciences of Philadelphia Special Publication 15.

Burnaby, T.P. 1966. Growth-invariant discriminant functions and generalized distances. Biometrics 22: 96–107.

Cambrony, M. 1984. Identification et périodicité du recrutement des juvéniles de Mugilidae dans les étangs littoraux du Languedoc-Roussillon. Vie Milieu 34: 221–227.

Capanna, E., S. Cataudella and G. Monaco. 1974. The pharyngeal structure of Mediterranean Mugilidae. Monitore Zoologico Italiano-Ital. J. Zool. 8: 29–46.

Cervigón, F. 1993. Los peces marinos de Venezuela. Volumen II. Fundación Científica Los Roques. Caracas, Venezuela.

Chervinski, J. 1984. Using scales for identification of four Mugilidae species. Aquaculture 38: 79–81.

Cockerell, T.D.A. 1913. The scales of some Queensland fishes. Mem. Qld. Mus. 2: 51–9.

Corti, M. and D. Crosetti. 1996. Geographic variation in the grey mullet: a geometric morphometric analysis using partial warp scores. J. Fish Biol. 48: 255–269.

Cousseau, M.B. (ed.). 2010. Ictiología. Aspectos Fundamentales. La vida de los peces sudamericanos. Editorial Universitaria de Mar del Plata (EUDEM), Mar del Plata.

Cousseau, M.B., M. González-Castro, D.E. Figueroa and A.E. Gosztony. 2005. Does *Mugil liza* Valenciennes 1836 (Teleostei: Mugiliformes) occur in Argentinean waters? Rev. Biol. Mar. Oceanogr. 40: 133–140.

Crosetti, D., W.S. Nelson and J.C. Avise. 1994. Pronounced genetic structure of mitochondrial DNA among populations of the circumglobally distributed grey mullet (*Mugil cephalus*). J. Fish Biol. 44: 47–58.

Dayrat, B. 2005. Towards integrative taxonomy. Biol. J. Linn. Soc. 85: 407–415.

Durand, J.D. 2015. Implications of molecular phylogeny for the taxonomy of Mugilidae. *In*: D. Crosetti and S. Blaber (eds). Biology, Ecology and Culture of Grey Mullet (Mugilidae). CRC Press, Boca Raton, USA (this book).

Durand, J.D., K.N. Shen, W.J. Chen, B.W. Jamandre, H. Blel, K. Diop, M. Nirchio, F.J. García de León, A.K. Whitfield, C.W. Chang and P. Borsa. 2012a. Systematics of the grey mullet (Teleostei: Mugiliformes: Mugilidae): molecular phylogenetic evidence challenges two centuries of morphology-based taxonomic studies. Mol. Phylogen. Evol. 64: 73–92.

Durand, J.D., W.J. Chen, K.N. Shen, C. Fu and P. Borsa. 2012b. Genus-level taxonomic changes implied by the mitochondrial phylogeny of grey mullets (Teleostei: Mugilidae). Comp. Rend. Biol. 335: 687–697.

Ebeling, A.W. 1957. The dentition of eastern Pacific mullets, with special reference to adaptation and taxonomy. Copeia 3: 173–185.

Ebeling, A.W. 1961. *Mugil galapagensis*, a new mullet from the Galapagos islands, with notes on related species and a key to the Mugilidae of the eastern Pacific. Copeia 3: 295–304.

Eschmeyer, W.N. (ed.). 2015. Catalog of Fishes: Genera, Species, References. (http://research.calacademy.org/research/ichthyology/catalog/fishcatmain.asp). Electronic version accessed 10 February 2015.

Eschmeyer, W.N. and J.D. Fong (eds.). 2015. Catalog of Fishes: species by family/subfamily. (http://researcharchive.calacademy.org/research/ichthyology/catalog/SpeciesByFamily.asp). Electronic version accessed 10-02-2015.

Farrugio, H. 1977. Clés commentées pour la détermination des adultes et des alevins de Mugilidae de Tunisie. Cybium 2: 57–73.

Fink, W.L. 1981. Ontogeny and phylogeny of tooth attachment modes in actinopterygian fishes. J. Morph. 167: 167–184.

Fowler, H.W. 1903. New and little known Mugilidae and Sphyraenidae, Proc. Acad. Nat. Sci. Philadelphia 55: 743–752.

Fowler, H.W. 1939. A small collection of fishes from Saigon, French Indo-China. Notul. Nat. Acad. Philad. 8: 1–6.

Fraga, E., H. Schneider, M. Nirchio, E. Santa-Brigida, L.F. Rodrigues-Filho and I. Sampaio. 2007. Molecular phylogenetic analyses of mullets (Mugilidae, Mugiliformes) based on two mitochondrial genes. J. Appl. Ichthyol. 23: 598–604.

Ghasemzadeh, J. 1998. Phylogeny and systematics of Indo-Pacific mullets (Teleostei: Mugilidae) with special reference to the mullets of Australia. Ph.D. dissertation, Macquarie University, Sydney.

Ghasemzadeh, J., W. Ivantsoff and Aarn. 2004. Historical overview of mugilid systematics with description of *Paramugil* (Teleostei: Mugiliformes: Mugilidae), new genus. Aqua. J. Ichthyol. Aquat. Biol. 8: 9–22.

Gill, T.N. 1863. Descriptive enumeration of a collection of fishes from the western coast of Central America, presented to the Smithsonian Institution by Captain John M. Dow. Proc. Acad. Nat. Sci. Philad. 15: 162–174.

Gistel, J. 1848. Naturgeschichte des Thierreichs für höhere Schulen, Hoffman'sche Verlags-Buchhandlung, Stuttgart.

Godinho, H.M., P.C.S. Serralheiro and J.D. Scorvo Filho. 1988. Review and discussion of papers about the species of the genus *Mugil* (Teleostei, Perciformes, Mugilidae) of the Brazilian coast (Lat. 3°S–33°S). B. Inst. Pesca, Sao Paulo 15: 67–80.

González-Castro, M. 2007. Los peces representantes de la Familia Mugilidae en Argentina. Ph.D. thesis. Universidad Nacional de Mar del Plata, Argentina.

González-Castro, M., J.M. Díaz de Astarloa and M.B. Cousseau. 2006. First record of a tropical affinity mullet, *Mugil curema* (Actinopterygii: Mugilidae), in a temperate Southwestern Atlantic coastal lagoon. Cybium 30: 90–91.

González-Castro, M., S. Heras, M.B. Cousseau and M.I. Roldán. 2008. Assessing species validity of *Mugil platanus* Günther, 1880 in relation to *Mugil cephalus* Linnaeus, 1758 (Actinopterygii). Ital. J. Zool. 75: 319–325.

González-Castro, M., A.L. Ibáñez, S. Heras, M.I. Roldán and M.B. Cousseau. 2012. Assesment of lineal versus landmarks-based morphometry for discriminating species of Mugilidae (Actinopterygii). Zool. Stud. 51: 1515–1528.

Greenwood, P.H., D.E. Rosen, S.H. Weitzman and G.S. Myers. 1966. Phyletic studies of teleostean fishes, with a provisional classification of living forms. Bull. Am. Mus. Nat. Hist. 131: 339–456.

Gudger, E.W. 1934. The five great naturalists of the sixteenth century; Belon, Rondelet, Salviani, Gesner and Aldrovandi: a chapter in the history of Ichthyology. Isis New York 22: 21–40.

Günther, A. 1861. Catalogue of the Acanthopterygian Fishes in the British Museum. Vol. 3. Gobiidae, Discoboli, Pediculati, Blenniidae, Labyrinthici, Mugilidae, Notacanthi. Trustees, British Museum, London.

Harrison, I.J. 2002. Mugilidae. pp. 1071–1085. *In*: K. Carpenter (ed.). FAO Species Identification Guide for Fisheries Purposes. The Living Marine Resources of the Western Central Atlantic. Vol. 2. Bony Fishes part 1 (Acipenseridae to Grammatidae). FAO, Rome.

Harrison, I.J. and G.J. Howes. 1991. The pharyngobranchial organ of mugilid fishes; its structure, variability, ontogeny, possible function and taxonomic utility. Bull. Br. Mus. Nat. Hist. (Zool.) 57: 111–132.

Harrison, I.J. and H. Senou. 1999. Mugilidae. pp. 2069–2108. *In*: K.E. Carpenter and V.H. Niem (eds.). FAO Species Identification Guide for Fishery Purposes. The Living Marine Resources of the Western Central Pacific, Vol. 4. Bony Fishes Part 2 (Mugilidae to Carangidae). FAO, Rome.

Harrison, I.J., M. Nirchio, C. Olivera, E. Ron and J. Gaviria. 2007. A new species of mullet (Teleostei: Mugilidae) from Venezuela, with a discussion on the taxonomy of *Mugil gaimardianus*. J. Fish Biol. 71: 76–97.

Hebert, P.D.N. and T.R. Gregory. 2005. The promise of DNA barcoding for taxonomy. Syst. Biol. 54: 852–859.

Hebert, P.D.N., A. Cywinska, S.L. Ball and J.R. deWaard. 2003a. Biological identifications through DNA barcodes. P. Roy. Soc. Lond. B Bio. 270: 313–321.

Hebert, P.D.N., S. Ratnasingham and J.R. deWaard. 2003b. Barcoding animal life: cytochrome c oxidase subunit 1 divergences among closely related species. P. Roy. Soc. Lond. B Bio. 270: S96–S99.

Heras, S., M. González-Castro and M.I. Roldan. 2006. *Mugil curema* in Argentinean waters: Combined morphological and molecular approach. Aquaculture 261: 473–478.

Heras, S., M.I. Roldán and M. González-Castro. 2009. Molecular phylogeny of Mugilidae fishes revised. Rev. Fish Biol. Fish 19: 217–231.

Hotta, H. 1955. On the mature mugilid fish from Kabashima, Nagasaki Pref., Japan, with additional notes on the intestinal convolution of Mugilidae. Jpn. J. Ichthyol. 4: 162–169.

Hotta, H. and I.S. Tung. 1966. Identification of fishes of the family Mugilidae based on the pyloric caeca and the position on inserted first interneural spine. Jpn. J. Ichthyol. 14: 62–66 (In Japanese, English summary and key).

Humphries, J.M., F.L. Bookstein, B. Chernoff, G.R. Smith, R.L. Elder and S.G. Poss. 1981. Multivariate discrimination by shape in relation to size. Sust. Zool. 30: 291–308.

Ibáñez-Aguirre, A.L. and J. Lleonart. 1996. Relative growth and comparative morphometrics of *Mugil cephalus* L. and *M. curema* V. in the Gulf of Mexico. Sci. Mar. 60: 361–368.

Ibáñez-Aguirre, A.L. and M. Gallardo-Cabello. 2004. Reproduction of *Mugil cephalus* and *M. curema* (Pisces: Mugilidae) from a coastal lagoon in the Gulf of Mexico. Bull. Mar. Sci. 75: 37–49.

Ibáñez-Aguirre, A.L., M. Gallardo-Cabello and J. Chiappa Carrara. 1999. Growth análisis of striped mullet *Mugil cephalus* and white mullet, *M. curema* (Pisces: Mugilidae), in the Gulf of Mexico. Fish Bull. 97: 861–872.

Ibáñez-Aguirre, A.L., E. Cabral-Solis, M. Gallardo-Cabello and E. Espino-Barr. 2006. Comparative morphometrics of two populations of *Mugil curema* (Pisces: Mugilidae) on the Atlantic and Mexican Pacific coasts. Sci. Mar. 70: 139–145.

Ibáñez, A.L., I.G. Cowx and P. O'Higgins. 2007. Geometric morphometric analysis of fish scales for identifying genera, species and local populations within the Mugilidae. Can. J. Fish Aquat. Sci. 64: 1091–1100.

Ibáñez, A.L., M. González-Castro and E. Pachecho. 2011. First record of *Mugil hospes* in the Gulf of Mexico and its identification from *M. curema* using ctenii. J. Fish Biol. 78: 386–390.

Ingham, S.E. 1952. The biology and taxonomy of the Mugilidae. Unpublished M.S. thesis in the Library of the British Museum of Natural History.

Ishiyama, R. 1951. Revision of the Japanese mugilid fishes, especially based upon the osteological characters of the cranium. Jap. J. Ichthyol. 1: 238–250.

Jacot, A.P. 1920. Age, growth and scale characters of the mullet *Mugil cephalus* and *Mugil curema*. Trans. Amer. Micr. Soc. 39: 199–229.

Jamandre, B.W., J.D. Durand and W.N. Tzeng. 2009. Phylogeography of the flathead mullet *Mugil cephalus* in the north-west Pacific as inferred from the mtDNA control region. J. fish Biol. 75: 393–407.

Jordan, D.S. and B.W. Evermann. 1896. The Fishes of North and Middle America: a descriptive catalogue of the species of fish-like vertebrates found in the waters of America, north of the isthmus of Panama. Part 1. Bull. U.S. Nat. Mus. 47: 1–1240.

Jordan, D.S. and B.W. Evermann. 1917. The Genera of Fishes. Stanford University Press, California.

Jordan, D.S. and C.H. Gilbert. 1882. List of fishes collected at Panama by Captain John M. Dow, now in the United States National Museum. Proceedings of the United States National Museum 5: 373–378.

Jordan, D.S. and J. Swain. 1884. A review of the American species of marine Mugilidae. Proc. U.S. Nat. Mus. 7: 261–275.

Katselis, G., G. Hotos, G. Minos and K. Vidalis. 2006. Phenotypic affinities on fry of four Mediterranean grey mullet species. Turk. J. Fish Aquat. Sc. 6: 49–55.

Kobelkowsky, A.D. and A. Resendez. 1972. Estudio comparative del endosqueleto de *Mugil cephalus* y *Mugil curema* (Pisces, Perciformes) An Inst. Biol. Univ. Nac. Auton. Mex. Ser. Cienc. Mar. Limnol. 43: 33–81.

Lacepède, B.G.E. 1799. Histoire naturelle des poissons. Tome second, Plassan, Paris.

Linnaeus, C. 1758. Systemanaturae per regna tria naturae, secundum classes, ordines, genera, species, cum characteribus, differentiis, synonymis, locis, Tomus I, editio decima, reformata, Holmiae, Stockholm.

Lleonart, J., J. Salat and G.J. Torres. 2000. Removing allometrics effects of body size in morphological analysis. J. Theor. Biol. 205: 85–93.

Liu, C.H. and S.C. Shen. 1991. Lepidology of the Mugilid fishes. J. Taiwan Mus. 44: 321–357.

Luther, G. 1977. New characters for consideration in the taxonomic appraisal of grey mullets. J. Mar. Biol. Assoc. India 19: 1–9.

Macdonald, J.D. 1869. On the characters of a type of a proposed new genus of Mugilidae inhabiting the freshwaters of Viti Levu, Feejee group; with a brief account of the native mode of capturing it. Proc. Gen. Meetings Sci. Business Zool. Soc. Lond. 1: 38–40.

Macleay, W. 1883. On a new and remarkable fish of the family Mugilidae from the interior of New Guinea, Proc. Linn. Soc. New South Wales 8: 2–6.

Menezes, N.A. 1983. Guia prático para conhecimento e identificaçao das tainhas e paratis (Pisces: Mugilidae) do litoral Brasileiro. Revta. Bras. Zool., S. Paulo 2: 1–12.

Miller, S.E. 2007. DNA barcoding and the renaissance of taxonomy. P. Natl. Acad. Sci. USA 104: 4775–4776.

Minos, G., G. Katselis, P. Kaspiris and I. Ondrias. 1995. Comparison of the change in morphological pattern during the growth in length of the grey mullets *Liza ramada* and *Liza saliens* from western Greece. Fish Res. 23: 143–155.

Minos, G., G. Katselis, I. Ondrias and I.J. Harrison. 2002. Use of melanophore patterns on the ventral side of the head to identify fry of grey mullets (Teleostei: Mugilidae). Isr. J. Aquacul.-Bamid. 54: 12–26.

Mitteroecker, P. and P. Gunz. 2009. Advances in Geometric Morphometrics. Evol. Biol. 36: 235–247.

Morovic, D. 1953. Sur la determination des mugesadriatiquesd'apres la forme de l'otolith sagitta. Notes Inst. Oceanogr. Split 9: 1–7.

Mosimann, J.E. 1970. Size allometry: size and shape variables with characterizations of the lognormal and generalized gamma distributions. J. Am. Stat. Assoc. 65: 930–945.

Nelson, J.S. 1984. Fishes of the World, 2nd Edition. John Wiley and Sons, New York.

Nelson, J.S. 1994. Fishes of the World, 3rd Edition. John Wiley and Sons, Inc., New York.

Nelson, J.S. 2006. Fishes of the World, 4th Edition. John Wiley and Sons, New York. 601pp.

Nirchio, M. and R. Cipriano. 2005. Cytogenetical and morphological features reveal significant differences among Venezuelan and Brazilian samples of *Mugil curema* (Teleostei: Mugilidae). Neotrop. Ichthyol. 3: 107–110.

Nirchio, M., F. Cervigón, J.I.R. Porto, J.E. Pérez, J.A. Gómez and J. Villalaz. 2003. Karyotype supporting *Mugil curema* Valenciennes, 1836 and *Mugil gaimardianus* Desmarest, 1831 (Mugilidae: Teleostei) as two valid nominal species. Sci. Mar. 67: 113–115.

Ogilby, J.D. 1888. On a new genus and species of Australian Mugilidae. Proc. Gen. Meetings Sci. Business Zool. Soc. Lond. 4: 614–616.

Ogilby, J.D. 1908. New or little known fishes in the Queensland Museum. Ann. Queensl. Mus. 9: 1–41.

Oren, O.H. (ed.). 1981. Aquaculture of Grey Mullets. International Biol. Prog. 26, Cambridge University Press, Cambridge.

Oshima, M. 1922. A review of the fishes of the family Mugilidae found in the waters of Formosa. Ann. Carn. Mus. 13: 240–259.

Padial, J.M., A. Miralles, I. De la Riva and M. Vences. 2010. The integrative future of taxonomy. Frontiers in Zoology 7: 16. www.frontiersinzoology.com/content/7/1/16.

Perlmutter, A., L. Bogard and I. Pruginin. 1957. Use of the estuarine and sea fish of the family Mugilidae (grey mullets) for pond culture in Israel. Proc. Fish. Counc. Mediterr., FAO 4: 289–304.

Pillay, T.V.R. 1951. Structure and development of the scales of five species of grey mullets of Bengal. Proc. Nat. Inst. Sci. India 17: 413–24.

Poey, F. 1860. Memorias sobre la historia natural de la Isla de Cuba, acompañadas de sumarios latinos y extractos en francés, Tomo 2, Barcina, La Habana.

Popov, A.M. 1930. Kefali (Mugilidae) ebropi s opisaniem novogo vida iz Tichookeanoskich vod SSSR [Mullets of Europe (Mugilidae) with descriptions of a new species from the Pacific Ocean], Trudy Sevastopol'skoi Biologicheskoi Stantii Akademii nauk SSSR 2: 47–125.

Randall, J.E. 1995. Coastal Fishes of Oman. University of Hawaii Press, Honolulu, Hawaii.

Ratnasingham, S. and P.D.N. Hebert. 2013. A DNA-Based Registry for All Animal Species: The Barcode Index Number (BIN) System. PLoS One 8: e66213.

Reay, P.J. and V. Cornell. 1988. Identification of grey mullet (Teleostei: Mugilidae) Juveniles from British waters. J. Fish Biol. 32: 95–99.

Regan, C.T. 1908. A collection of freshwater fishes made by Mr. C.F. Underwood in Costa Rica. Ann. Magazine Nat. Hist. 2: 455–464.

Rocha-Olivares, A., N.M. Garber and K.C. Stuck. 2000. High genetic diversity, large inter-oceanic divergence and historical demography of the striped mullet. J. Fish Biol. 57: 1134–1149.

Rohlf, F.J. 1990. Morphometrics. Annu. Rev. Ecol. Syst. 21: 299–316.

Rohlf, F.J. and L.F. Marcus. 1993. A revolution in morphometrics. Tree 8: 129–132.

Rossi, A.R., M. Capula, D. Crosetti, L. Sola and D.E. Campton. 1998a. Allozyme variation in global populations of striped mullet, *Mugil cephalus* (Pisces: Mugilidae). Mar. Biol. 131: 203–212.

Rossi, A.R., D. Crosetti and S. Livi. 2015. Genetics of Mugilidae. *In*: D. Crosetti and S. Blaber (eds). Biology, Ecology and Culture of Grey Mullet (Mugilidae). CRC Press, Boca Raton, USA (this book).

Schultz, L.P. 1946. A revision of the genera of mullets, fishes of the family Mugilidae, with descriptions of three new genera. Proc. U.S. Nat. Mus. 96: 377–395.

Schultz, L.P. 1953. Family Mugilidae. *In*: L.P. Schultz, E.S. Herald, E.A. Lachner, A.D. Welander and L.P. Woods (eds.). Fishes of the Marshal and Marianas Islands, Vol. 1. Bull. U.S. Nat. Mus. 202: 310–22.

Senou, H. 1988. Phylogenetic Interrelationships of the Mullets (Pisces: Mugilidae) Tokyo University. Ph.D. Thesis (In Japanese).

Serventi, M., I.J. Harrison, P. Torriecelli and G. Gandolfi. 1996. The use of pigmentation and morphological characters to identify Italian mullet fry. J. Fish Biol. 49: 1163–1173.

Shen, K.N., B.W. Jamandre, C.C. Hsu, W.N. Tzeng and J.D. Durand. 2011. Plio-Pleistocene sea level and temperature fluctuations in the northwestern Pacific promoted speciation in the globally-distributed flathead mullet *Mugil cephalus*. BMC Evol. Biol. 11: Article No 83.

Smith, J.L.B. 1948. A generic revision of the Mugilid fishes of South Africa. Ann. Mag. Nat. Hist. 14: 833–843.

Song, J.K. 1981. Chinese mugilid fishes and morphology of their cephalic lateral line canals. Sinozoologia 1: 9–21.

Sprent, P. 1972. The mathematics of size and shape. Biometrics 28: 23–27.

Steindachner, F. 1878. Ichthyologische Beiträge (VII), Sitzungsberichte derkaiserlichen Akademie der Wissenschaften, Mathematisch-Naturwissenschaftliche Classe 78: 377–400.
Stiassny, M. 1993. What are grey mullets? Bull. Mar. Sci. 52: 197–219.
Strauss, R.E. and F.L. Bookstein. 1982. The truss: Body form reconstruction in morphometrics. Syst. Zool. 131: 113–135.
Sunny, K.G. 1971. Morphology of the vertebral column of *Mugil macrolepis* (Smith). Bull. Dep. Mar. Oceanogr. Univ. Cochin. 5: 101–108.
Taylor, W.R. 1964. Fishes of Arnhem Land. Records of the American Australian Scientific Expedition to Arnhem land 4: 45–308.
Thomson, J.M. 1954a. The Mugilidae of Australia and adjacent seas. Aust. J. Mar. Freshwater Res. 5: 70–131.
Thomson, J.M. 1954b. The organs of feeding and the food of some Australian Mullet. Aust. J. Mar. Freshwater Res. 5: 469–85.
Thomson, J.M. 1966. The grey Mullets. Annu. Rev. Oceangr. Mar. Biol. 4: 301–335.
Thomson, J.M. 1975. The dentition of grey mullets. Aquaculture 5: 108.
Thomson, J.M. 1981. The taxonomy of grey mullets. pp. 1–12. *In*: O.H. Oren (ed.). Aquaculture of Grey Mullets. IBP, 26, Cambridge University Press, Cambridge.
Thomson, J.M. 1997. The Mugilidae of the world. Mem. Queens. Mus. 41: 457–562.
Trewavas, E. and S.E. Ingham. 1972. A key to the species of Mugilidae (Pisces) in the north-eastern Atlantic and Mediterranean, with explanatory notes. Jour. Zool. Lond. 167: 15–29.
Vaillant, L.L. 1894. Sur une collection de poissons recueillie en Basse-Californie et dans le Golfe par M. Le´on Diguet, Bull. Soc. Philomatique Paris 6: 69–75.
Valenciennes, A. 1836. Mugiloides. pp. 1–127. *In*: G. Cuvier and A. Valenciennes (eds.). Histoire Naturelle des Poissons. Vol. 11. F.G. Levrault, Paris & Strasbourg.
van der Elst, R.P. and J.H. Wallace. 1976. Identification of the juvenile mullet of the east coast of South Africa. J. Fish Biol. 9: 371–374.
Walbaum, J.J. 1793. *Petri Artedi renovati,* pars 4. Petri Artedi sueci genera piscium in quibus systematotum ichthyologia eproponitur cum classibus, ordinibus, generumcharacteribus, specierumdifferentiis, observationibus plurimis. Redactis speciebus 242 ad genera 52. Ichthyologiae Pars III. Grypeswaldiae. 723 pp., 3 pls.
Whitley, G.P. 1930. Five new generic names for Australian fishes. Austr. Zool. 6: 250–251.
Whitley, G.P. 1941. Ichthyological notes and illustrations. Austr. Zool. 10: 1–52.
Whitley, G.P. 1945. New sharks and fishes from Western Australia Part 2. Austr. Zool. 11: 1–43.
Whitley, G.P. 1948. New sharks and fishes from Western Australia Part 4. Austr. Zool. 11: 259–276.
Whitfield, A.K., J. Panfili and J.D. Durand. 2012. A global review of the cosmopolitan flathead mullet *Mugil cephalus* Linnaeus 1758 (Teleostei: Mugilidae), with emphasis on the biology, genetics, ecology and fisheries aspects of this apparent species complex. Rev. Fish Biol. Fish 22: 641–681.

CHAPTER 2

Implications of Molecular Phylogeny for the Taxonomy of Mugilidae

Jean-Dominique Durand

"La première famille dont nous traçons l'histoire dans ce volume est une de celles qui nous ont donné le plus de peine, à M. Cuvier et à moi."

"*... la similitude, on peut dire désepérante, de tous ces poissons, attache à leur synonymie et à l'expression de leurs caractères des difficultés tout aussi insurmontables....*"

Cuvier and Valenciennes 1836

Introduction

Since the first attempt at Mugilidae systematics by Cuvier and Valenciennes (1836), the number of species and genera, and their phylogenetic relationships, have been constantly debated. In successive revisions, an increasing number of morpho-anatomical traits have been considered. The profound morpho-anatomical similarity of Mugilidae species, and the difficulty of interpreting anatomical differences from an evolutionary perspective, render the situation very complex in terms of making phylogenetic inferences and clearly identifying species and genera. As a result, there is no consensus on Mugilidae systematics, and taxonomic inconsistencies persist (see Chapter 1—González-Castro and Ghasemzadeh 2015).

Genetic polymorphisms constitute a valid and powerful alternative to morphology that can be used to test the prevailing phylogenetic assumptions based upon morphological traits. Early applications of genetics to Mugilidae had this objective, of resolving taxonomic problems with species identification and phylogenetic relationships. The first study used karyotype similarities, and variables such as the number of chromosomes and the position of their centromere (Cataudella et al. 1974), to investigate the validity of Mugilidae genera in the Mediterranean Sea. The Mediterranean mullet species had the same number of chromosomes (2n = 48), but their morphology differed, leading Cataudella et al. (1974) to group species under three cytotaxonomic categories that were largely consistent with the taxonomy based on anatomical traits. Subsequent cytogenetic studies also identified 2n = 48 chromosomes in Mugilidae (for a review Rossi et al. 1996) except in *Mugil curema* whose populations in the USA and Brazil have 2n = 28 chromosomes (LeGrande and Fitzsimons 1976, Cipriano et al. 2002, Nirchio et al. 2005) and 2n = 24 in Venezuela (Nirchio and Cequea 1998, Nirchio et al. 2001, 2003). Even though such cytogenetic

IRD, Centre for Marine Biodiversity, Exploitation and Conservation, Université Montpellier 2, CC 093, Place Eugène Bataillon, 34095 Montpellier Cedex 5, France.
Email: jean-dominique.durand@ird.fr

features question the taxonomic status of these populations, no definitive conclusions could be made (Rossi et al. 2005).

Later cytogenetic techniques were improved by staining and fluorescent *in situ* hybridization (FISH) with several types of DNA probes, which permitted observation of finer features such as the Nucleolus Organizer Regions (NORs) (reviewed in Sola et al. 2007). This technique was used to investigate geographic variation of genetic differentiation of *Mugil cephalus* (Rossi et al. 1996) and phylogenetic relationships among Mugilidae species (for a review see Sola et al. 2008). No cytogenetic polymorphisms were observed among worldwide samples of *M. cephalus* (Rossi et al. 1996), whereas variations in the location of genes for 18S rRNA and 5S rRNA, and the composition of the constitutive heterochromatin, were observed among the previous cytotypes of Mugilidae species (Sola et al. 2008). These cytogenetic features were used to reconsider cytotaxonomic relationships, but evolutionary interpretation of the variation in these traits was still dependent upon prevailing morpho-anatomical hypotheses.

In the same period, allozyme electrophoresis was used to identify species (Herzberg and Pasteur 1975, Callegarini and Basaglia 1978) and, by increasing loci and tissues, the phylogenetic relationships among species (Autem and Bonhomme 1980, Menezes et al. 1992, Rossi et al. 1998a, Papasotiropoulos et al. 2001, Rossi et al. 2004, Turan et al. 2005, Blel et al. 2008). This finally led to analysis of the genetic diversity/ population structure of *Mugil cephalus* (Campton and Mahmoudi 1991, Rossi et al. 1998b, Huang et al. 2001). With few exceptions (Menezes et al. 1992, Lee et al. 1995, Liu et al. 2010, Rossi et al. 1998a), all phylogenetic investigations were exclusively focused on Mediterranean species, which greatly limited their impact for understanding the systematics of the family. The low levels of genetic polymorphism that are typically recovered with allozyme loci however prevented the application of these techniques to wider species sampling.

The extensive development of Polymerase Chain Reaction (PCR) methods in the 1990s allowed the most significant advance in Mugilidae systematics. The PCR methods greatly facilitated access to DNA sequence polymorphisms, providing direct insights into the evolutionary history of families and species. Initial molecular phylogenetic studies were limited to a single sequence portion of a gene, with the objective of clarifying phylogenetic relationships among Mediterranean mullet species (Caldara et al. 1996, Rossi et al. 2004, Papasotiropoulos et al. 2007, Imsiridou et al. 2007, Erguden et al. 2010). Later studies were extended to more gene portions and other species of Mugilidae (Fraga et al. 2007, Semina et al. 2007, Aurelle et al. 2008, Heras et al. 2009, Liu et al. 2010, Durand et al. 2012a, Siccha-Ramirez et al. 2014, Xia et al. submitted). Molecular techniques were also used to investigate species boundaries and population genetic structure (Crosetti et al. 1993, 1994, Rocha-Olivares et al. 2000, 2005, Heras et al. 2006, Liu et al. 2007, Liu et al. 2009, Jamandre et al. 2009, Ke et al. 2009, Jamandre et al. 2010, Shen et al. 2011, Sun et al. 2012, Durand et al. 2013, Krüeck et al. 2013, McMahan et al. 2013, Mai et al. 2014). All these studies shed new light on the phylogeny and diversity of the Mugilidae, leading some authors to propose large revisions of their taxonomy (Durand et al. 2012b, Xia et al. submitted). This revision of Mugilidae taxonomy, based on the results of molecular phylogenetic and population genetic studies, is presented in the following sections. The Mugilidae diversity consists of more than 91 mitochondrial lineages, corresponding to 53 morphological species and 38 putative species, which form 25 genera, five tribes and four subfamilies.

Impact of Recent Molecular Phylogeny on Taxonomy

Subfamilies

The inference of systematic relationships of the family based upon morpho-anatomical characters has lead to conflicting hypotheses (see Fig. 1 in Durand et al. 2012a). The only area of agreement concerned the phylogenetic position of the genera *Agonostomus* and *Joturus*, usually presented as basal in all phylogenies (Schultz 1946, Senou 1988, Harrison and Howes 1991, Thomson 1997, Ghasemzadeh 1998). Some authors included *Cestraeus* and *Aldrichetta* among these plesiomorphic genera (Harrison and Howes 1991, Thomson 1997, Ghasemzadeh 1998) forming the subfamily Agonostominae *sensu* Thomson (1997); all other species belong to the subfamily Mugilinae *sensu* Thomson (1997).

The first comprehensive molecular systematics of the Mugilidae, using phylogenetic analyses of nucleotide sequence variation at three mitochondrial loci, highlighted seven major lineages that radiated early, to all current forms, from a common ancestor (Durand et al. 2012a). While *Joturus*, *Agonostomus* and *Cestraeus* belonged to a unique lineage, they are also closely related to genera of more recent origin in phylogenies based upon morpho-anatomical characters, such as *Mugil*, *Chaenomugil*, *Myxus* and *Neomyxus*. The low resolution of deep nodes of the tree, however limits phylogenetic interpretations and tests of morpho-anatomical hypotheses. Recently, Xia et al. (submitted) used additional loci (mitochondrial and nuclear) to gain better resolution of the phylogenetic tree (Fig. 2.1). They demonstrated that Mugilidae genera belong to four clades, considered to represent subfamilies (Xia et al. submitted). These subfamilies were identified by the combination of six morpho-anatomical traits: 1. the scale type for all or a majority of body scales, 2. the position of the pelvic fin tip relative to the first dorsal fin, 3. the position of the jaw end relative to the mouth gape (JM), 4. the shape of the preorbital frontal edge, 5. the number of pyloric caeca and 6. the maxilla below the mouth corner when the mouth is closed.

Myxinae Xia, Durand and Fu submitted

Type Genus. *Myxus* Günther, 1861

The Myxinae subfamily sensu Xia et al. (submitted) is composed of two monotypic genera, *Myxus* and *Neomyxus*. This subfamily is basal in the phylogenetic tree suggesting that the two genera are the most plesiomorphic. This contrasts with the general conclusion of reviews based on morpho-anatomical traits, such as Harrison and Howes (1991), which considered the morphology of the Pharyngo-Branchial Organ (PBO) of *Myxus* and *Neomyxus* as one of the most derived. Beyond the molecular evidence, this subfamily is characterized by the following combination of diagnostic morpho-anatomical traits: cycloid scales, a pelvic fin tip that barely reaches the vertical from first dorsal fin origin, upper jaw position above the line of mouth gape, presence of only two pyloric caeca.

This subfamily is endemic to the Pacific, occurring along the eastern coast of Australia and in the central Pacific.

Mugilinae Xia, Durand and Fu Submitted

Type Genus. *Mugil* Linnaeus, 1758

The new-recombined Mugilinae subfamily *sensu* Xia et al. (submitted) is composed of six genera: *Mugil*, *Chaenomugil*, *Agonostomus*, *Dajaus*, *Joturus* and *Cestraeus*. The genus *Cestraeus* is the most divergent while other genera belong to two evolutionary lineages: one consisting of *Joturus*, *Dajaus* and *Agonostomus* and a second consisting of *Mugil* and *Chaenomugil* (Fig. 2.1). With the exception of *Aldrichetta*, all genera considered as plesiomorphic by Thomson (1997) and assigned to the Agonostominae subfamily fall within this molecular subfamily. Despite this congruent result, the presence in this clade of the genera *Mugil* and *Chaenomugil* justified the new recombination of the subfamily Mugilinae. Beyond the molecular evidence, this subfamily is characterized by a combination of ctenoid scales (with the exception *Mugil capurii*), a pelvic fin tip that reaches the vertical from spines I–IV of the first dorsal fin, a shape of the preorbital front edge that is not notched, presence of only two pyloric caeca and the maxilla below the mouth is not visible when the mouth is closed (Xia et al. submitted). The Mugilinae subfamily occurs in all tropical and subtropical waters of the world, but with a phylogenetic diversity that is higher along the shores of the Americas.

Rhinomugilinae Xia, Durand and Fu submitted

Type Genus. *Rhinomugil* Gill, 1863

The Rhinomugilinae subfamily *sensu* Xia et al. (submitted) comprises 11 genera that belong to four evolutionary lineages, considered as tribes by Xia et al. (submitted) (Fig. 2.1). This subfamily is

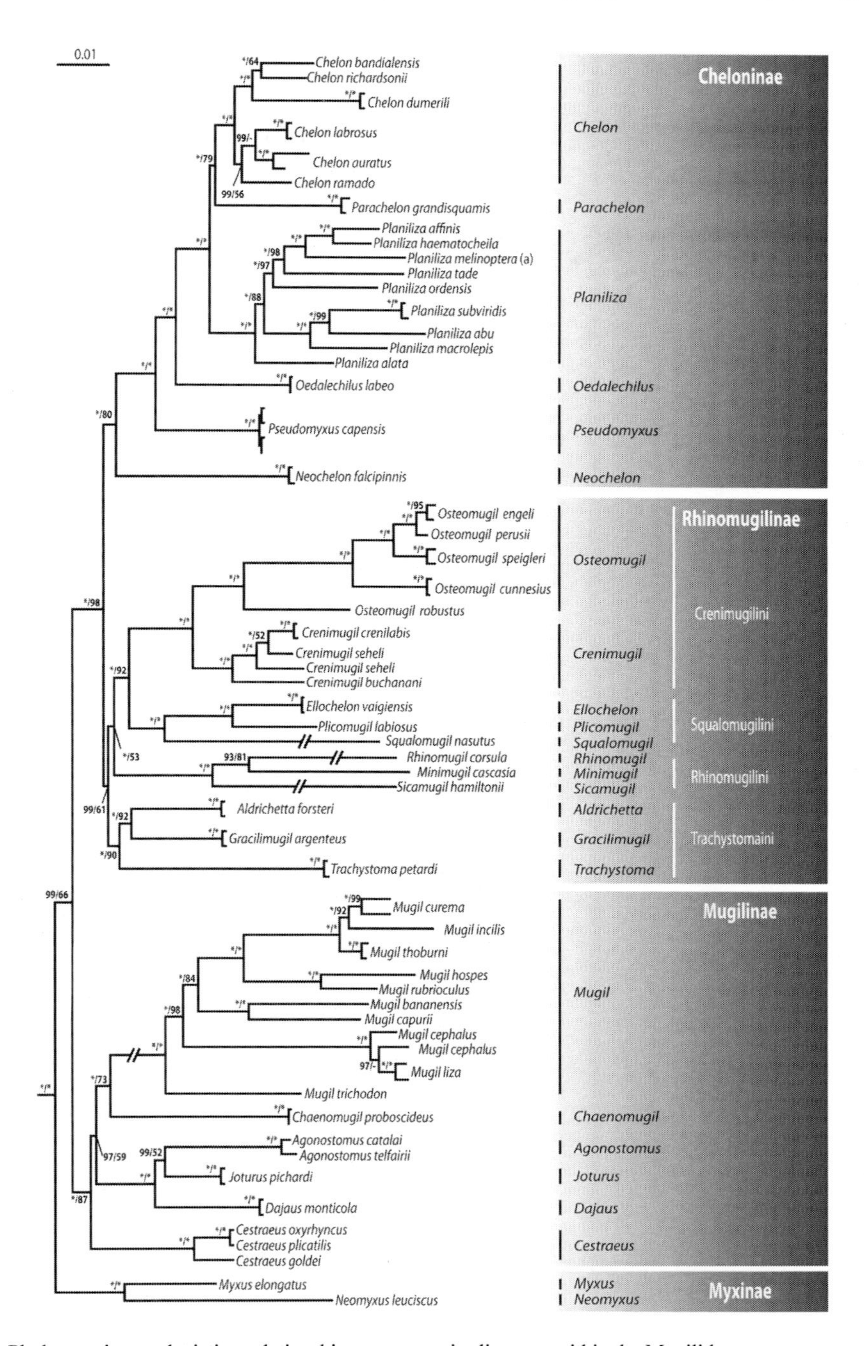

Figure 2.1. Phylogenetic tree depicting relationships among major lineages within the Mugilidae.

Relationships were inferred using partitioned Bayesian and maximum likelihood (ML) analyses of 12,945 bp un-ambiguous sequences from twelve nuclear loci (9,843 bp) and three mitochondrial genes (3,102 bp) (Xia et al. submitted). With exception of *Mugil*, *Sicamugil hamiltoni*, *Rhinomugil corsula* and *Squalomugil nasutus*, branch length is proportional to the number of substitutions under an optimal substitution model for each partition of concatenated mitochondrial and nuclear gene data (provided in Xia et al. submitted). Numbers on the branches are Bayesian posterior probabilities for the Bayesian analyses and ML bootstrap values (in %, from 1000 replicates). Asterisks indicate nodes with a posteriori probability from partitioned Bayesian analysis of 1 and a ML bootstrap of 100%, '-' indicate that the ML bootstrap value is less than 50%. Out-group taxa are not shown. (a) correspond to the specimen MNHN-IC-2011-0212 named *Paramugil parmatus* in Durand et al. (2012a) but re-identified by Ghasemzadeh as *Planiliza melinoptera*. In white, subfamilies and tribes names proposed by Xia et al. (submitted).

phylogeneticaly closer to the Cheloninae *sensu* Xia et al. (submitted) and presents wide morpho-anatomical diversity that precludes the identification of diagnostic morpho-anatomical traits. This subfamily occurs in the Indo-Pacific and is divided into four tribes.

Trachystomaini Xia, Durand and Fu Submitted

The Trachystomaini tribe consists of three monotypic genera: *Trachystoma*, *Gracilimugil* and *Aldrichetta*. The short length of mucus canals on scales is the main characteristic of this tribe within the Rhinomugilinae subfamily. This tribe occurs exclusively in the South West Pacific.

Rhinomugilini Xia, Durand and Fu Submitted

The Rhinomugilini tribe consists of three monotypic genera: *Rhinomugil*, *Sicamugil* and *Minimugil*. It is differentiated from all other tribes by the following combination of morpho anatomical traits: two pyloric caeca, no scale free area on the top of the head, and absence of endopterygoid teeth. This tribe occurs in freshwaters of India, Bangladesh and Myanmar.

Squalomugilini Xia, Durand and Fu Submitted

The Squalomugilini tribe consists of three genera: *Squalomugil*, *Plicomugil* and *Ellochelon*. This tribe is differentiated from all others by possessing four or more pyloric caeca (more than 14 in *Ellochelon* and *Squalomugil*), an emarginated or truncated caudal fin, a second dorsal fin origin at vertical $\geq 2/3$ along the anal fin base. This tribe occurs in the Indian Ocean and the West Pacific.

Crenimugilini Xia, Durand and Fu Submitted

The Crenimugilini tribe consists of two genera: *Crenimugil* and *Osteomugil*. All members of this tribe have distinctive long pectoral axillary scales, and scales with a membranous, digitated hind margin (Fig. 2.2). It is widely distributed in the Indo-Pacific.

Cheloninae Xia, Durand and Fu Submitted

Type Genus. *Chelon* Artedi, 1793

The subfamily Cheloninae comprises six genera: *Neochelon*, *Oedalechilus*, *Pseudomyxus*, *Planiliza*, *Parachelon* and *Chelon*. All genera are considered of recent origin in phylogenies based on morpho-anatomical traits (Schultz 1946, Harrison and Howes 1991, Thomson 1997). This is in agreement with the molecular phylogeny; this subfamily and the Rhinomugilinae are the most recently diverged (Xia et al. submitted). Within the Cheloninae, the genus *Neochelon* is the most divergent, followed by *Pseudomyxus* and then *Oedalechilus*. *Planiliza*, *Parachelon* and *Chelon* diverged more recently from a common ancestor (Fig. 2.1). The Cheloninae subfamily is characterized by ctenoid scales (with the exception of *Neochelon* and *Pseudomyxus*), a pelvic fin tip that reaches the vertical from spines I–IV, a jaw end below the line of mouth gape, a preorbital front edge that is notched, a number of pyloric caeca > 14, and a maxilla below the mouth corner that is visible when the mouth is closed. This subfamily occurs in the East Atlantic, in the Mediterranean Sea and the Indo-Pacific. It is absent from American continental waters (East Pacific, West Atlantic).

Genera

Cuvier and Valenciennes (1836) produced the first major taxonomic study of the Mugilidae, based on a major worldwide sampling. They assigned Mugilidae to four genera: *Mugil*, *Dajaus*, *Cestraeus* and *Nestis*.

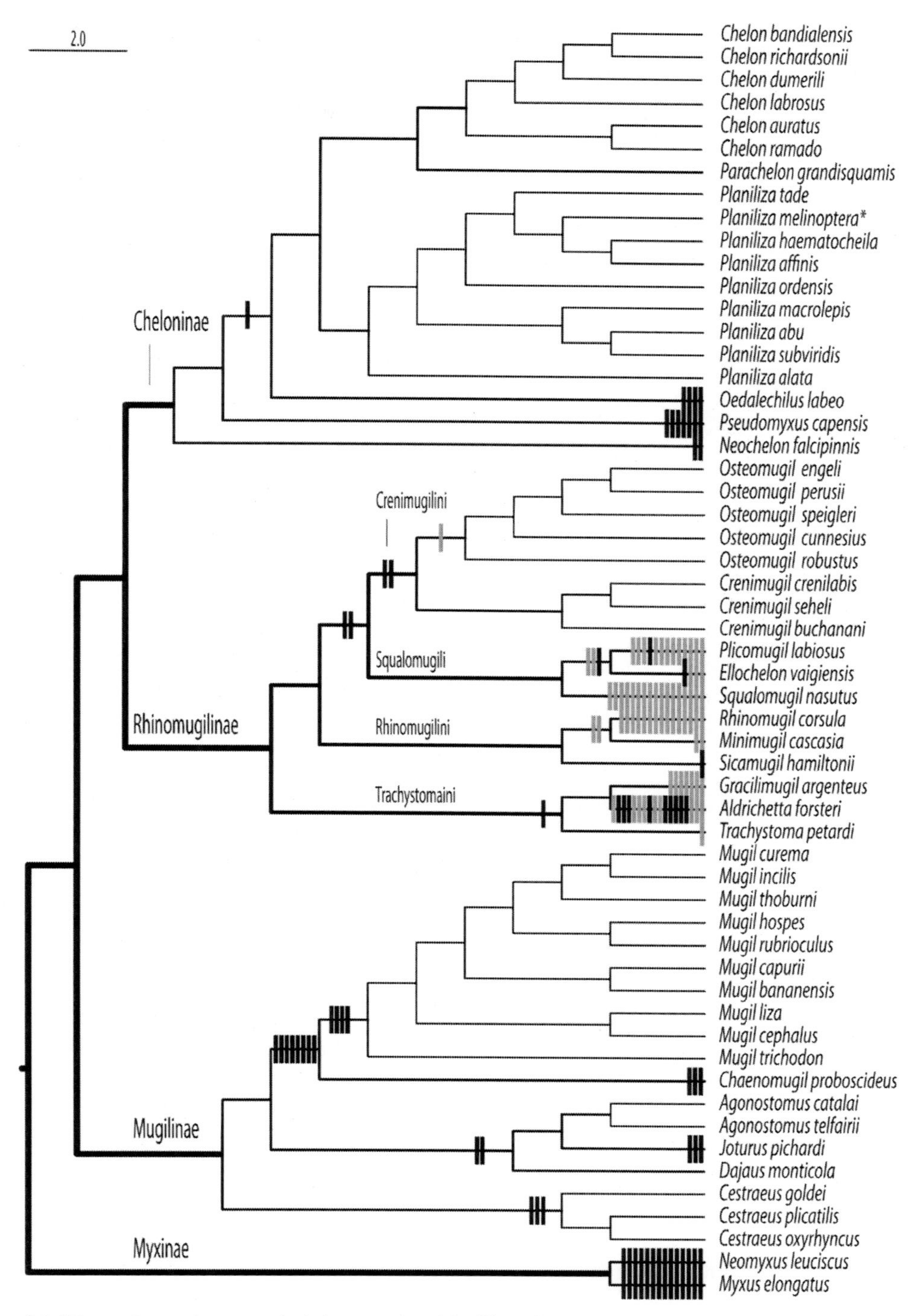

Figure 2.2. Diagnostic morpho-anatomical characters in subfamilies, tribes and genera of the Mugilidae family.

There were highlighted among 68 morpho-anatomical traits inferred on the basis of the likelihood reconstruction of ancestral character state using Mesquite 2.75 (Maddison and Maddison 2011, Xia et al. submitted). Bars correspond to diagnostic traits of genera belonging to the same subfamily (black bars) or tribe (grey bars). Branch thickness corresponds to the taxonomical ranks: subfamily, tribe, genus, species. * corresponds to the specimen MNHN-IC-2011-0212 named *Paramugil parmatus* in Durand et al. (2012a) but re-identified by Ghasemzadeh as *Planiliza melinoptera.*

Later, Günther (1861) only accepted three genera as valid based upon the presence and disposition of the teeth: *Mugil*; *Agonostoma* and *Myxus*; *Nestis* and *Dajaus* being junior synonyms of *Agonostoma*. After these initial classifications, new genera were continuously described, up to 30, before the taxonomic revision proposed by Schultz (1946). Based upon mouth anatomy, Schultz (1946) reduced the number of genera to 13, while Smith (1948) recognized 16. Later, Schultz (1953) reaffirmed the validity of 13 genera and described one more. In 1988, Senou listed 15 genera that were only partially consistent with those previously accepted, while describing three new ones. None of these new genera were considered in subsequent classifications, including those provided by Harrison and Howes (1991), Thomson (1997) or Ghasemzadeh (1998). Among these, discrepancies remained concerning nomenclature and phylogenetic relationships inferred from morpho-anatomical traits. It was in this context that Durand et al. (2012a) and, more recently, Xia et al. (submitted), investigated phylogenetic relationships using DNA sequence polymorphisms. These molecular phylogenies, using a large sample of species representative of global Mugilidae diversity, permitted tests of morpho-anatomical assumptions and, more importantly, proposed a revised classification, as described below.

Agonostomus Bennett, 1832

Type species. *Agonostomus telfairii* Bennett 1832 (holotype BMNH 1861.8.14.9). Mauritius, Mascarenes, South-western Indian Ocean.

All molecular phylogenetic reconstructions, using either mitochondrial or a combination of mitochondrial and nuclear gene polymorphisms, have highlighted the paraphyly of *Agonostomus* with respect to *Joturus* (Durand et al. 2012a,b, Xia et al. submitted, Fig. 2.1). Because the type species of the genus *Agonostomus* is the South-West Indian species *A. telfairii*, Durand et al. (2012b) suggested placing American *Agonostomus* species under a different genus name, namely *Dajaus*, to conserve the monotypic genus *Joturus*. This suggestion is in agreement with Cuvier and Valenciennes (1836). *Dajaus* is the brother genus of *Joturus* and *Agonostomus* inside the Mugilinae subfamily (Xia et al. submitted, Fig. 2.1). There are no morpho-anatomical synapomorphies that characterize *Agonostomus* and *Dajaus* genera, but their allopatric distributions and phylogenetic positions argue for the validity of the two genera. *Agonostomus* comprises two species *A. telfairii* and *A. catalai* and occurs exclusively in the South West Indian Ocean (Comores, Mayottes, Madagascar, Réunion and Mauritius).

Aldrichetta Whitley, 1945

Type species. *Mugil forsteri* Valenciennes 1836. No types known.

All molecular phylogenetic reconstructions (Durand et al. 2012a,b, Xia et al. submitted) are in agreement with a taxonomy based on morphological and anatomical traits (Eschmeyer and Fong 2014). The *Aldrichetta* genus is monotypic and the brother genus of *Gracilimugil*, part of the Trachystomaini tribe inside the Rhinomugilinae subfamily (Xia et al. submitted, Fig. 2.1). *Aldrichetta* is supported by 19 diagnostic morpho-anatomical traits, 10 of which are at the subfamily rank: jaw end on line of gape, preorbital front edge not notched, mid-gape at level of or below lower rim of the eye, nine soft rays in the second dorsal fin, 58 or more scales in the longitudinal series, 18 or more scales in the transverse series, 18 or more scales in the longitudinal series reached by the tip of the pectoral fin when laid back, 19 or more scales between the operculum and the vertical from the origin of the first dorsal fin, 37 or more scales between the operculum and the vertical from the origin of the second dorsal fin (Xia et al. submitted, Fig. 2.2). It is restricted to the temperate coastal waters of Australia and New Zealand (Thomson 1997).

Cestraeus Valenciennes, in Cuvier and Valenciennes, 1836

Type species. *Cestraeus plicatilis* Valenciennes, in Cuvier and Valenciennes 1836 (holotype MNHN A-2894). Sulawesi, Indonesia.

All molecular phylogenetic reconstructions (Durand et al. 2012a,b, Xia et al. submitted) are in agreement with the taxonomy based on morphology and anatomy (Eschmeyer and Fong 2014). The *Cestraeus* genus is represented by three species: *Cestraeus goldiei* Macleay 1883, *Cestraeus oxyrhynchus* Valenciennes, in Cuvier and Valenciennes 1836, and *Cestraeus plicatilis* Valenciennes, in Cuvier and Valenciennes 1836. It belongs to the Mugilinae subfamily *sensu* Xia et al. (submitted, Fig. 2.1) and it is supported by three diagnostic morpho-anatomical traits: fleshy lobes over end of upper jaws, mid-gape at level of or below lower rim of eye, 20 to 21 rays in the pectoral fin (Thomson 1997, Xia et al. submitted, Fig. 2.2). The genus *Cestraeus* is present in the Indo-Malay-Papua archipelago, in New Caledonia and in Fiji (Thomson 1997).

Chaenomugil Gill, 1863

Type species. *Mugil proboscideus* Günther 1861 (syntype BMNH 1860.7.21.22). Pacific coast of Central America.

All molecular phylogenetic reconstructions (Durand et al. 2012a,b, Xia et al. submitted) are in agreement with the taxonomy based on morphological and anatomical traits (Eschmeyer and Fong 2014). The *Chaenomugil* genus is monotypic and the brother genus of *Mugil*, inside the Mugilinae subfamily *sensu* Xia et al. (submitted, Fig. 2.1). In its subfamily, it is characterized by three diagnostic morpho-anatomical traits: edge of the lower lip permanently turned down, mouth corner reaching vertical from anterior nostrils or a little behind, first dorsal fin origin nearer caudal base than to snout tip (Thomson 1997, Xia et al. submitted, Fig. 2.2). *Chaenomugil* occurs in the eastern Pacific, from Baja California to Peru (Thomson 1997).

Chelon Artedi, 1793

Type species. *Mugil chelo* Cuvier 1829 (lectotype MNHN 0000-6400, paralectotypes: MNHN A-3588 to 3589; A-3596; A-3599; A.3602 to 3603; A-3775; A-4651; A-4693; A-4697). Brest, France.

Phylogenetic reconstructions based on genetic data have provided contrasting views of phylogenetic relationships between the genera *Liza* and *Chelon.* While *Chelon* and *Liza* species belong to two different clades disputing the validity of these genera (Autem and Bonhomme 1980, Papasotiropoulos et al. 2001, Murgia et al. 2002, Blel et al. 2008), others studies have pointed out the paraphyly of the genus *Liza* with *Chelon* (Caldara et al. 1996, Papasotiropoulos et al. 2002, Rossi et al. 2004, Turan et al. 2005, Gornung et al. 2007, Imsiridou et al. 2007, Papasotiropoulos et al. 2007, Aurelle et al. 2008, Heras et al. 2009). However, these studies were usually limited to few species (Mediterranean ones) or based on allozyme markers with limited variation. Recent molecular phylogenies based on large species samples (Durand et al. 2012a,b) or numerous mitochondrial and nuclear sequence polymorphisms (Xia et al. submitted) have clearly demonstrated the paraphyly of both *Liza* and *Chelon* genera (Durand et al. 2012a,b, Xia et al. submitted). Considering the phylogenetic tree and the position of *Chelon* and *Liza* type species *Mugil chelo* and *Mugil capito* (currently *C. ramado*) respectively, Durand et al. (2012b) synonymized *Liza* with *Chelon*, resurrected *Gracilimugil* and *Planiliza* (see hereafter) and created three new genera. *Chelon* is the brother genus of genera *Parachelon* and *Planiliza* (Fig. 2.1) and consists of at least nine species (see hereafter and Chapter 6): *C. auratus, C. bandialensis, C. bispinosus, C. dumerili, C. labrosus, C. ramado, C. richardsonii, C. saliens,* and *C. tricuspidens*. No diagnostic morpho-anatomical traits has been highlighted in *Chelon*, among other genera belonging to the subfamily Cheloninae (Xia et al. submitted). This genus occurs exclusively in temperate and tropical waters of the East Atlantic Ocean, Mediterranean Sea and the temperate waters of South Africa.

Crenimugil Schultz, 1946

Type species. *Mugil crenilabis* Forsskål 1775. No types known.

Molecular phylogenetic reconstructions based on mitochondrial sequences (Durand et al. 2012a) or a combination of mitochondrial and nuclear gene sequences (Xia et al. submitted), both highlighted the paraphyly of *Moolgarda* and *Valamugil* with *Crenimugil*. Based upon the fact that *Moolgarda* is both a *nomen nudum* and a *nomen dubium* (Thomson 1997; for more details see Ghasemzadeh 1998), and the principle of priority, Durand et al. (2012b) synonymized *Moolgarda* and *Valamugil* with *Crenimugil*. *Crenimugil* is the brother genus of *Osteomugil*, part of the Crenimugilini tribe within the Rhinomugilinae subfamily (Xia et al. submitted, Fig. 2.1). *Crenimugil* consists of three nominal species: *C. crenilabis*, *C. buchanani* and *C. seheli* but the species diversity of this genus is probably strongly underestimated (see hereafter, Table 2.1). No diagnostic morpho-anatomical traits were revealed when 68 morpho-anatomical characters were plotted onto the molecular phylogenetic tree (Xia et al. submitted, Fig. 2.2). The genus *Crenimugil* has a wide Indo-West Pacific distribution.

Table 2.1. Polyphyletic Mugilidae species in the molecular phylogenetic trees of Durand et al. (2012a,b).

Species	NL	Sp.	%DinterL COI	%DintraL COI	NSyn/L COI (a)	Gene Isol	Range
Chelon dumerili	2	M	6.6	[0-0.1]	5/11	na	Pa
Crenimugil seheli	3	P	[5.5-9.3]	[0.1-0.14]	8/4/11	na	S
Dajaus monticola	3	M	[7.4-14.6]	[0-0.2]	25/20/na	na	A/S
Ellochelon vaigiensis	2	M	5.8	0	32/na	na	?
Mugil cephalus	14	P	[1-5.9]	[0-0.6]	3/0/3/5/2/2/2/0/1/0/1/1/12/3	na/Yes (b, c)	A/S
Mugil curema	4	P	[3.3-5.6]	[0.1-0.5]	7/4/5/5	Yes (d)	A/S
Mugil rubrioculus	2	M	5.5	na	na	na	A
Osteomugil cunnesius	3	P	[10.4-12.9]	[0-0.7]	10/na/na	na	A
Planiliza alata	2	P	13.1	[0-0.2]	1/8	na	A
Planiliza macrolepis	2	M	3.9	[0-0.11]	2/1	na	A
Planiliza melinoptera	2	P	13.7	0	3/na	na	S
Planiliza tade	2	P	14.3	na	na	na	?
	41						

All lineages (41) are putative cryptic species considering the monophyly (M)/paraphyly (P) of the nominal species (Sp.), the ratio between the level of interlineage divergence (%DinterL) and the intralineage nucleotide diversity (%DintraL) estimated using the nucleotide polymorphism of the cytochrome oxydase I (COI) fragment, the number of synapomorphies per lineage (NSyn/L), the genetic isolation and the distribution range (Pa: parapatric, S: sympatric, A: allopatric). NL: number of lineage, na: not available. (a) Durand and Borsa (2015), (b) Shen et al. (2011), (c) Krüeck et al. (2013), (d) Durand et al. (2012a).

Dajaus Valenciennes, 1836, in Cuvier and Valenciennes, 1836

Type species. *Mugil monticola* Bancroft 1834. No types preserved.

Valenciennes in Cuvier and Valenciennes (1836) described the genus *Dajaus* for *Mugil monticola* present in the West Indies, which was later synonymized with *Agonostomus* by Günther (1861). However, paraphyly of the genus *Agonostomus* justifies the resurrection of the genus *Dajaus* (Durand et al. 2012b). *Dajaus* is the brother genus of *Agonostomus* and *Joturus* inside the Mugilinae subfamily (Xia et al. submitted, Fig. 2.1). No diagnostic morpho-anatomical traits were revealed (Xia et al. submitted, Fig. 2.2). Only one nominal species has been described, *Dajaus monticola*, but recent genetic investigations indicated the existence of three to four cryptic species (Durand et al. 2012a, McMahan et al. 2013, Table 2.1). The genus *Dajaus* occurs in rivers of the West Indies and Americas, from Florida to Venezuela and California to the Galapagos Islands (Thomson 1997).

Ellochelon Whitley, 1930

Type species. *Mugil vaigiensis* Quoy and Gaimard 1825 (holotype MNHN A-3641). Pulau Waigeo, Papua Barat province, Indonesia, western Pacific.

All molecular phylogenetic reconstructions (Durand et al. 2012a,b, Xia et al. submitted) are in agreement with taxonomy based on morphological and anatomical traits (Eschmeyer and Fong 2014). The genus *Ellochelon* is the brother genus of *Plicomugil*, part of the Squalomugilini tribe within the Rhinomugilinae subfamily (Xia et al. submitted, Fig. 2.1). *Ellochelon* is supported by three diagnostic morpho-anatomical traits among genera of the Squalomugilini tribe, and one among genera of the Rhinomugilinae subfamily: interorbital shape flat (Xia et al. submitted). *Ellochelon* would be monotypic but there is some evidence of cryptic diversity that question species composition in this genus (see hereafter, Table 2.1). This genus has a wide Indo-West Pacific distribution, from Natal in South Africa to Tahiti (Thomson 1997).

Gracilimugil Whitley, 1941

Type species. *Mugil ramsayi* Macleay 1883 (syntypes: AMS IA.5944-46). Burdekin River, Queensland, Australia.

Most recent Mugilidae taxonomic revisions based on morpho-anatomical traits contest the findings of Whitley (1941) or Ghasemzadeh (1998), and assign *Mugil argenteus* (the senior synonym of *Mugil ramsayi*) to the genus *Chelon* (Senou 1988) or *Liza* (Thomson 1997, Harrison and Senou 1997, Kottelat 2013, Eschmeyer and Fong 2014). However, molecular phylogenetic reconstructions (Durand et al. 2012a,b, Xia et al. submitted) clearly demonstrate that *Mugil argenteus* is an independent evolutionary lineage justifying validity of the genus *Gracilimugil* (Durand et al. 2012b). It is a monotypic genus, brother of *Aldrichetta,* within the Trachystomaini tribe in the Rhinomugilinae subfamily (Xia et al. submitted, Fig. 2.1). Eight diagnostic morpho-anatomical traits characterized *Gracilimugil* among genera of its tribe: adipose eyelid reaching the rim of eye, tendon flange 1/2–2/3 down maxilla shaft, pads over maxilla and the tendon to the mouth, tongue keeled, mouth gape horizontal or slightly oblique, two valves in pharyngobranchial organ, gills rakers very long (Xia et al. submitted, Fig. 2.2). This genus occurs in western Australia from Cardwell in Queensland to the Moor River (Thomson 1997).

Joturus Poey, 1860

Type species. *Joturus pichardi* Poey 1860 (holotype: MCZ 23886, possible type: USNM 132429). Río Almendares, near Havana, Cuba.

All molecular phylogenetic reconstructions (Durand et al. 2012a,b, Xia et al. submitted) are in agreement with taxonomy based on morphological and anatomical traits (Eschmeyer and Fong 2014). The genus *Joturus* is the brother genus of *Dajaus* and *Agonostomus* within the Mugilinae subfamily (Fig. 2.1). It is supported by three diagnostic morpho-anatomical traits: upper lip recessed under snout, nine soft rays in the second dorsal fin, 11 soft rays in the anal fin (Xia et al. submitted, Fig. 2.2). This genus would be monotypic, but there is evidence of cryptic species in nominal species on both sides of the American continent, such as *Dajaus monticola* (Durand et al. 2012a, McMahan et al. 2012, Table 2.1). This genus occurs on both the Pacific and the Atlantic Coasts of the American continent, from Mexico to Panama, and in the Caribbean Sea (Thomson 1997).

Minimugil Senou, 1988

Type species. *Mugil cascasia* Hamilton 1822 (no types known). Rivers of northern Bengal.

Fowler (1939) created the genus *Sicamugil* for two small freshwater species *Mugil hamiltoni* (type species) and *Mugil cascasia.* While most recent revisions based on morpho-anatomical traits have subscribed to this view (Thomson 1997, Eschmeyer and Fong 2014), Senou (1998) considered that the morpho-anatomical differences between these two species are greater than a congeneric level and created the genus *Minimugil*, with *Mugil cascasia* as type species. Molecular phylogenetic reconstructions (Durand et al. 2012a,b, Xia et al. submitted) are in agreement with Senou's view because *Mugil cascasia* is phylogeneticaly closer to *Rhinomugil corsula* than to *Sicamugil hamiltoni*. It is the brother genus of *Rhinomugil*, part of the Rhinomugilini tribe within the Rhinomugilinae subfamily (Xia et al. submitted, Fig. 2.1). Among genera of the tribe, this genus is supported by two diagnostic morpho-anatomical traits: first dorsal fin origin nearer snout tip than to caudal base, height of the 2nd dorsal fin equal as the 1st one (Xia et al. submitted, Fig. 2.2). The only known species of this genus, *M. cascasia,* is distributed in the Ganges River and its tributaries (Thomson 1997).

Mugil Linnaeus, 1758

Type species. *Mugil cephalus* Linnaeus 1758 (possible syntypes: NRM 43, 44, 143). European sea, Europe.

All phylogenetic reconstructions (Durand et al. 2012a,b, Xia et al. submitted) are in agreement with taxonomy based on morphological and anatomical traits compiled by Eschmeyer and Fong (2014). The genus *Mugil* is the brother genus of *Chaenomugil* inside the Mugilinae subfamily (Xia et al. submitted, Fig. 2.1). It is supported by four diagnostic morpho-anatomical traits: nostrils nearer lip and eye than to each other, adipose tissue on face intruding over eye to pupil, distinct and long pectoral axillary scale, gill rakers long (Xia et al. submitted, Fig. 2.2). This genus comprises 13 nominal species: *M. bananensis, M. broussonetii, M. capurii, M. cephalus, M. curema, M. curvidens, M. hospes, M. incilis, M. liza, M. rubrioculus, M. setosus, M. thoburni, M. trichodon.* The species diversity of this genus is probably largely underestimated however, because cryptic species have been assumed for some species presenting large distribution ranges that encompass well known biogeographic barriers (see hereafter, Table 2.1). The genus *Mugil* has a worldwide distribution with the exception of Arctic and Antarctic seas (Thomson 1997).

Myxus Günther, 1861

Type species. *Myxus elongatus* Günther 1861 (syntypes: BMNH 1847.6.17.33 and 1847.10.22.16). Coast of Australia.

All phylogenetic reconstructions (Durand et al. 2012a,b, Xia et al. submitted) demonstrate that *Myxus elongatus*, the type species of the genus *Myxus*, has no close phylogenetic relationships with any other Mugilidae species, especially not to *Mugil capensis* and *Mugil petardi* that previously were assigned to the same genus (Thomson 1997, Harrison and Senou 1997). Consequently, *Myxus* would be a monotypic genus, brother of the genus *Neomyxus* inside the *Myxinae* subfamily (Xia et al. submitted, Fig. 2.1). This genus is supported by 14 diagnostic morpho-anatomical traits: maxima below mouth corner when mouth closed visible, preorbital front edge notched, preorbital filling space between lip and eye, posterior nostril not reaching above level of upper rim of eye, lower lip or its edge folded down absent, lip groove present, mouth corner reaching horizontal above lower rim of eye, vomer teeth present, palatine teeth present, seven-eight soft rays in the 2nd dorsal fin, caudal fin forked, height of the 1st dorsal fin equal to the 2nd dorsal fin, multicanaliculate scale absent, wide sulcus in the pharyngobranchial organ (Xia et al. submitted, Fig. 2.2). The genus *Myxus* is restricted to temperate waters of Australia.

Neochelon Durand, Chen, Shen, Fu and Borsa, 2012

Type species. *Mugil falcipinnis* Valenciennes in Cuvier and Valenciennes 1836 (syntypes: MNHN A-3728, A-3729). Senegal.

While *Mugil falcipinnis* has been considered as part of the genus *Liza* or *Chelon* in all taxonomic revisions based on morpho-anatomical traits (Harrison and Howes 1991, Albaret 1992, Thomson 1997), its unique placement in the Mugilidae phylogenetic tree (Durand et al. 2012a) lead Durand et al. (2012b) to create the genus *Neochelon.* In the phylogenetic tree that combines mitochondrial and nuclear gene sequences (Fig. 2.1), *Neochelon* is the most divergent genus among all genera of the Cheloninae subfamily (Xia et al. submitted). It is supported by two diagnostic morpho-anatomical traits: 11 soft rays in the anal fin, scales with a membranous, digitated hind margin (Xia et al. submitted, Fig. 2.2). *Neochelon falcipinnis* the unique species of the genus *Neochelon*, occurs in West Africa from Saint-Louis in northern Senegal to Congo (Thomson 1997).

Neomyxus Steindachner, 1878

Type species. *Myxus (Neomyxus) sclateri* Steindachner 1878 (syntypes: NMW 67168, 77884, 82505). Gilbert Islands and Hawaiian Islands.

All phylogenetic reconstructions (Durand et al. 2012a,b, Xia et al. submitted) are in agreement with the taxonomy based on morphological and anatomical traits (Eschmeyer and Fong 2014). *Neomyxus* is the brother genus of *Myxus*, within the *Myxinae* subfamily (Xia et al. submitted, Fig. 2.1). This genus is supported by 14 diagnostic morpho-anatomical traits: maxima below the mouth corner when mouth closed not visible, preorbital front edge not notched, preorbital not filling space between lip and eye, posterior nostril reaching above level of upper rim of eye, lower lip or its edge folded down present, lip groove absent, mouth corner reaching horizontal at or below lower rim of eye, vomerine teeth absent, palatine teeth absent, nine soft rays in the 2nd dorsal fin, caudal fin emarginated, height of the 1st dorsal fin lower than the 2nd dorsal fin, multicanaliculate scale present, narrow sulcus in the pharyngobranchial organ (Xia et al. submitted, Fig. 2.2). The only representative of this genus, *N. leuciscus*, occurs around islands of the central Pacific, from the southern Japanese and Hawaiian islands to Samoa (Thomson 1997).

Oedalechilus Fowler, 1903

Type species. *Mugil labeo* Cuvier 1829 (lectotype: MNHN A-3606, Paralectotypes: MNHN A-3607, A-4654). Mediterranean Sea.

All phylogenetic reconstructions (Durand et al. 2012a,b, Xia et al. submitted) stress the paraphyly of *Oedalechilus*, when *Mugil labeo* and *Mugil labiosus* are assigned to the same genus, as proposed by several authors on the basis of morphological and anatomical similarities (Senou 1988, Thomson 1997, Senou 2002). In fact, the type species *O. labeo* has no close phylogenetic relationships with any species within a clade comprising various genera, including *Chelon* (Fig. 2.1). Consequently, *Oedalechilus* is a monotypic genus belonging to the Cheloninae subfamily (Xia et al. submitted). This genus is supported by four diagnostic morpho-anatomical traits: posterior nostril reaching above level of upper rim of the eye, lower lip or its edge folded down, a single pair of shelf-like fold inside the mouth corner, labial teeth absent (Xia et al. submitted, Fig. 2.2). The only representative of this genus, *O. labeo*, occurs in the western Mediterranean Sea and the Azores archipelago (Thomson 1997).

Osteomugil Lüther, 1977

Type species. *Mugil cunnesius* Valenciennes 1836 (syntypes: MNHN A-4636 Moluccas, A-3701-02 Mumbai, A-3726-27 Malabar, B-2678 [ex A-3702], B-2629 [ex A-3701], 1992-0561 [ex A-3727]). Coromandel coast, India; Molucca Islands, Indonesia; Mumbai, India.

The genus *Osteomugil* was created by Lüther (1977) on the basis of some osteological characters that differentiate *Mugil cunnesius* (the type species) and possibly *Mugil perusii* and *Mugil engeli* from other species belonging to *Liza* (synonymized here with *Chelon*), *Valamugil* (synonymized here with *Crenimugil*) and *Ellochelon*. Phylogenetic reconstructions (Durand et al. 2012a, Xia et al. submitted) based on various gene sequences demonstrated that all *Osteomugil* species belong to the same clade, leading Durand et al. (2012b) to resurrect this genus, which had been synonymized with *Valamugil* by Thomson (1997). *Osteomugil* is the brother genus of *Crenimugil* sensus Durand et al. (2012b), part of the Crenimugilini tribe within the Rhinomugilinae subfamily (Xia et al. submitted, Fig. 2.1). Within its subfamily, *Osteomugil* is characterized by a combination of morpho-anatomical characters: the two synapomorphies of the Crenimugilini tribe and the adipose eyelid reaching iris of the eye (Xia et al. submitted, Fig. 2.2). The *Osteomugil* genus consists of five nominal species: *O. cunnesius*, *O. engeli*, *O. robustus*, *O. speigleiri*, and *O. perusii*. However, species diversity in the genus is probably underestimated (Table 2.1). It is widespread across the Indo-Pacific, from Africa to the Marquesas and Tuamotu Islands, north to southern Japan.

Parachelon Durand, Chen, Shen, Fu and Borsa, 2012

Type species. *Mugil grandisquamis* Valenciennes in Cuvier and Valenciennes 1836 (lectotype: MNHN A-3743, paralectotypes: MNHN A-3744, A-3745). Gorée, Senegal.

In all recent taxonomic reviews based on morpho-anatomical traits, *Mugil grandisquamis* is assigned to the genus *Liza* or *Chelon* (Albaret 1992, Senou 1988, Thomson 1997). However, its unique placement in the mitochondrial phylogenetic tree (Durand et al. 2012a) justifies the creation of the genus *Parachelon* (Durand et al. 2012b). Xia et al. (submitted, Fig. 2.1) confirmed this placement. Plotting 68 morpho-anatomical traits on the molecular tree did not however, reveal any diagnostic traits (Xia et al. submitted). Nevertheless, Harrison and Howes (1991) noticed that the pharyngobranchial organ morphology of *L. grandisquamis* is more similar to species from other genera *Crenimugil* (*C. seheli*), *Ellochelon* (*E. vaigiensis*), and *Paramugil* (*P. parmatus*) than to any other *Liza* species (*Chelon* and *Planiliza*). *Parachelon* is the brother genus of *Chelon* and *Planiliza* within the Cheloninae subfamily (Xia et al. submitted, Fig. 2.1). The only known species of the genus, *Parachelon grandisquamis* occurs in West Africa, from Senegal to Nigeria (Thomson 1997).

Planiliza Whitley, 1945

Type species. *Moolgarda (Planiliza) ordensis* Whitley 1945. Type by original designation.

All Indo-Pacific species previously assigned to *Chelon*, *Liza* and its synonym *Planiliza* genera constituted a strongly supported clade in molecular phylogenies (Durand et al. 2012a, Xia et al. submitted, Fig. 2.1). Durand et al. (2012b) resurrected the genus *Planiliza* for species in this clade because type species of genera *Liza* and *Chelon* belong to another clade. *Planiliza* is the brother genus of genera *Chelon* and *Parachelon* within the Cheloninae subfamily (Xia et al. submitted). Xia et al. (submitted) did not identify diagnostic traits among 68 morpho-anatomical traits (Fig. 2.2), but a distribution range limited to the Indo-Pacific Ocean clearly distinguishes them from their closest relatives (*Chelon* and *Parachelon*) that are present in the Atlantic, Mediterranean and temperate waters off South Africa. This genus is one of the most diversified of the Mugilidae and further phylogenetic investigation is needed to determine species diversity. The phylogenetic tree provided by Durand et al. (2012a) included several undetermined or paraphyletic species (see hereafter, Table 2.1). Based on molecular evidence, the following nominal species belong to this genus: *P. abu*, *P. affinis* (synonymized with *P. lauvergnii* in Eschmeyer and Fong 2014's revision), *P. alata*, *P. carinata*, *P. haematocheila*, *P. klunzingeri*, *P. melinoptera*, *P. subviridis*, *P. macrolepis*, *P. ordensis*, *P. tade*. The genus is widely distributed in the Indo-Pacific from the Red Sea to Oceania. It has been observed recently in the Mediterranean Sea (*P. carinata*) and in the Black Sea (*P. haematocheila*) due, respectively, to migration through the Suez Canal or to introduction for commercial purposes.

Plicomugil Schultz, 1953

Type species. *Mugil labiosus* Valenciennes 1836 (syntypes: MNHN A-3616, A-3617). Red Sea; Mumbai, India.

The genus *Plicomugil* was created by Schultz (1953) who considered that *Mugil labiosus* shows outstanding development of the mouthparts that does not overlap those of other species, especially with *Mugil labeo* that had been considered congeneric. This genus was however, constantly synonymized with *Oedalechilus* in all later taxonomic revisions (Thomson 1997, Senou 1998, Ghasemzadeh 1998, Senou 2002, Eschmeyer and Fong 2014) despite some contrasting anatomical evidence (Harrison and Howes 1991). In molecular phylogenetic reconstructions (Durand et al. 2012a, Xia et al. submitted) *M. labiosus* shows no close phylogenetic relationships with *O. labeo*, which justified the resurrection of *Plicomugil* (Durand et al. 2012b). It is the brother genus of *Ellochelon* part of the Squalomugilini tribe within the Rhinomugilinae subfamily (Xia et al. submitted, Fig. 2.1). Among genera of the tribe, it is supported by 13 diagnostic morpho-anatomical traits; one still diagnostic when considering all genera of the subfamily: four pairs of shelf-like folds inside the mouth corner (Xia et al. submitted, Fig. 2.2). This monotypic genus occurs in the Red Sea and Indo-Pacific from East Africa to the Marshall Islands and from southern Japan to Queensland, Australia (Eschmeyer and Fong 2014).

Pseudomyxus Durand, Chen, Shen, Fu and Borsa, 2012

Type species. *Mugil capensis* Valenciennes in Cuvier and Valenciennes 1836 (syntypes: MNHN A-4643, A-4700). Cape of Good Hope, South Africa.

In all taxonomic revisions based on morpho-anatomical traits, *Mugil capensis* has been considered part of the genus *Myxus* (Smith and Smith 1986, Heemstra and Heemstra 2004, Thomson 1997). Molecular phylogenetic reconstructions (Durand et al. 2012a,b, Xia et al. submitted) demonstrated however, that this species has no close phylogenetic relationships with *Myxus elongatus* and, among the subfamily Cheloninae *sensu* Xia et al. (submitted), is one of the most divergent species (Fig. 2.1). On the basis of the molecular phylogeny, Durand et al. (2012b) created the genus *Pseudomyxus*. This genus is also supported by seven diagnostic morpho-anatomical traits: adipose eyelid absent, tendon flange $< 1/2$ way down the maxilla shaft, pads over the lower end of the maxilla absent, or only over the tendon to the mouth, lip groove present, mouth gape moderately oblique, nine soft rays in the second dorsal fin, short length of mucus canals on scales (Thomson 1997, Xia et al. submitted, Fig. 2.2). The genus *Pseudomyxus* is monotypic and occurs in South Africa (Thomson 1997).

Rhinomugil Gill, 1863

Type species. *Mugil corsula* Hamilton 1822. No types known.

In all taxonomic revisions based on morpho-anatomical traits, the genus *Rhinomugil* comprises two species: *Mugil corsula* (the type species) and *Mugil nasutus* (Schultz 1946, Thomson 1997, Senou 1998, Eschmeyer 2014). Recent molecular phylogenetic reconstructions (Durand et al. 2012a,b, Xia et al. submitted) demonstrated however, that these two species do not belong to the same genus or even the same tribe (Xia et al. submitted, Fig. 2.1). Because the type species of *Rhinomugil* is *Mugil corsula*, only this species is maintained in this genus which is brother of *Minimugil*, part of the Rhinomugilini tribe within the Rhinomugilinae subfamily (Fig. 2.1). Among genera of the tribe, the *Rhinomugil* genus is supported by 15 diagnostic morpho-anatomical traits: adipose eyelid extending over iris, preorbital not filling space between lip and eye, interorbital concave, eyes raised above dorsal contour of head, opercular spin absent, upper lip recessed under snout, mouth gape very slightly oblique, mouth corner at level below lower rim of eye, mouth corner reaching vertical at or behind posterior nostril, tongue teeth absent, caudal fin slightly forked, second dorsal fin origin at vertical $\geq 2/3$ along anal fin base, pectoral fin past tip of pelvic spine when laid back, axillary scale short, large or moderate size of denticulate area in the pharyngobranchial organ (Thomson 1997, Xia et al. submitted, Fig. 2.2). This genus occurs in freshwaters of the Indian subcontinent (Thomson 1997).

Squalomugil Ogilby, 1908

Type species. *Mugil nasutus* De Vis 1883 (holotype: QM I.120, non-types: AMS I.12693). Cardwell, Rockingham Bay, Queensland, Australia.

Ogilby (1908) created the *Squalomugil* genus, apparently in ignorance of Gill's work (Thomson 1997). No subsequent taxonomic revision considered this genus as valid (Schultz 1946, Thomson 1997, Senou 1998, Eschmeyer and Fong 2014), except Taylor (1964) on the basis of the different position of the nostrils between *Mugil nasutus* and *Mugil corsula*. As mentioned in the *Rhinomugil* section however, all molecular phylogenetic reconstructions (Durand et al. 2012a,b, Xia et al. submitted) have justified the validity of *Squalomugil*, being closer to the type species of genera such as *Ellochelon* and *Plicomugil* than to the type species of *Rhinomugil* (Fig. 2.1). It forms with its brothers genera the Squalomugilini tribe inside the Rhinomugilinae subfamily (Xia et al. submitted). Among genera of its tribe, it is supported by 18 diagnostic morpho-anatomical traits: maxilla below the mouth corner when mouth closed was not visible, nostrils farther to each other than to eye or lip, adipose eyelid extending over iris, no scale free area on top of the head, pads over the lower end of the maxilla and over the tendon to the mouth absent or only over the tendon to the mouth, preorbital not filling space between lip and eye, interorbital concave, eyes raised above dorsal contour of head, lip groove present, upper lip recessed under snout, mouth corner at level below lower rim of eye, no vomer, endopterygoid, palatine and tongue teeth, upper insertion of pectoral fin at or below mid-eye level, axillary scale reaching < ½ along pelvic spine, no multicanaliculate scale (Thomson 1997, Xia et al. submitted, Fig. 2.2). The only known species of this genus, *S. nasutus*, occurs in tropical Australia and the southern shores of New Guinea (Thomson 1997).

Sicamugil Fowler, 1939

Type species. *Mugil hamiltoni* Day 1869 (syntypes or Day specimens: AMS B.7993; BMNH 1889.2.1.3724-3725; MCZ 17525; NMW 67653; ZSI F11401, A.355, B.150). Irrawaddy River, Pegu, and other rivers of Myanmar.

Molecular phylogenetic reconstructions (Durand et al. 2012a,b, Xia et al. submitted) have demonstrated that the genus *Sicamugil* is paraphyletic when *M. cascasia* is considered congeneric. For this reason, Durand et al. (2012b) maintained *Sicamugil* for *S. hamiltoni*, the type species of the genus. The *Sicamugil* genus is part of the Rhinomugilini tribe within the Rhinomugilinae subfamily (Xia et al. submitted, Fig. 2.1). It is supported by one diagnostic morpho-anatomical traits: the axillary scale does not reach the base of the spine IV (Xia et al. submitted, Fig. 2.2). The only known species of this genus, *S. hamiltoni*, occurs in rivers of Myanmar (Thomson 1997).

Trachystoma Ogilby, 1888

Type species. *Trachystoma multidens* Ogilby 1888 (no types known). Brackish water at Keruah River mouth, Port Stephens, Australia.

Due to the close external morphological similarity of *Trachystoma multidens* (junior synonym of *Mugil petardi*) to *Myxus elongates*, the type species of the genus *Myxus*, some taxonomical revisions have considered *Trachystoma* as a junior synonym of *Myxus* (Thomson 1997, Harrison and Senou 1997). In molecular phylogenetic reconstructions (Durand et al. 2012a,b, Xia et al. submitted), the species *Mugil petardi* formed a distinct and unique clade, which confirmed the peculiar systematic status of the monotypic genus *Trachystoma*. It is the brother genus of two genera, *Aldrichetta* and *Gracilimugil*, which occur in the same biogeographic area, within the Trachystomaini tribe of the Rhinomugilinae subfamily

(Xia et al. submitted, Fig. 2.1). Among genera of the tribe, the genus *Trachystoma* is supported by one diagnostic morpho-anatomical traits: in the first dorsal fin, spine I equal or longer than spine II (Xia et al. submitted, Fig. 2.2). *T. petardi*, the only species in the genus, inhabits the rivers of eastern Australia, from Queensland to New South Wales (Thomson 1997).

Other Genera

In this chapter, the validity of *Paracrenimugil* Senou 1988 and *Paramugil* Ghasemzadeh 1998 has not been evaluated because there are no molecular phylogenetic reconstructions available that included specimens of these genera. Senou (1988) created the monotypic genus *Paracrenimugil* with regard to a phylogenetic reconstruction based on 46 morphological characters. These placed *Mugil heterocheilos* Bleeker 1855 basal in a clade that included two subclades corresponding to the genera *Crenimugil* and *Osteomugil* (Durand et al. 2012b). Concerning the *Paramugil* genus, it was created by Ghasemzadeh (1998) on the basis of 18 diagnostic morphological and osteological differences that distinguished *Mugil parmatus* Cantator 1849 (type species) and *Mugil georgii* Ogilby 1897 from other species belonging to *Liza*, *Valamugil* and/or *Mugil* genera. Before Ghasemzadeh (1998), Senou (1988) had already suggested that *Mugil parmatus* Cantator 1849 belonged to a specific genus and had created the monotypic genus *Pseudoliza*. Despite this morpho-anatomical evidence, Durand et al. (2012b) synonymized *Paramugil* with *Planiliza* on the basis of the molecular phylogenetic tree of Durand et al. (2012a). However, Ghasemzadeh recently studied the morphology of the specimen MNHN-IC-2011-0212 identified as *Liza parmata* in the study of Durand et al. (2012a) and he concluded that it was misidentified, as its morpho-anatomical and meristic traits corresponded to *Planiliza melinoptera*. Consequently, new specimens of *Mugil parmatus* are needed to determine the phylogenetic relationships of this species within the Mugilidae.

Species

Numerous species have been described over the last two centuries. According to Thomson (1954) however, mugilid species diversity has probably been much overestimated because most of the earlier taxonomic work relied on the examination of specimens collected locally, without comparing these to morphologically similar species described elsewhere. In his last systematic revision, Thomson (1997) accepted only 62 species as valid among the existing 280 nominal species. Delimiting species boundaries in the Mugilidae is tricky because morpho-anatomical traits present ranges of variation that frequently overlap between taxa. In this context, DNA-based approaches can be used to identify species within taxa that have been overlooked or that present low levels of morpho-anatomical variation (Petit and Excoffier 2009, Zou et al. 2011, Kekkonen and Hebert 2014). Integration of criteria inferred from phylogeny, phylogeography and population genetics studies could produce primary species hypotheses that can be further tested, using morphological, ecological, behavioural, and geographic criteria. Among DNA-based criteria, monophyly, the presence of fixed mutations (character-based DNA barcoding), the geographic distribution of genetic diversity, and reproductive isolation, are species properties in various species concepts (de Queiroz 2007). Despite their varied perspectives and occasional incompatibilities, according to de Queiroz (2007), "a unified species concept can be achieved by treating existence as a separately evolving metapopulation lineage as the only necessary property of species" and other criteria such as different lines of evidence (operational criteria) relevant to assessing lineage separation. Considering these operational criteria, the species diversity of the Mugilidae is probably much greater than assumed by Thomson (1997). Using a sample of 257 individuals from 53 recognized species, Durand et al. (2012a,b) highlighted 91 lineages or Operational Taxonomic Units (OTUs) in the Mugilidae phylogenetic tree. Seven of these OTUs were in samples of undescribed species (because the morpho-anatomical description was either not available or did not match any species description). Among recognized species, 12 were polyphyletic and harboured from two to 14 lineages with levels of divergence that greatly exceed the average intraspecific differentiation or distance (D, Kimura's two-parameter model; Kimura 1980), which is estimated with the COI marker to be 0.35% in fishes (Ward et al. 2009, Table 2.1). Although some authors contest the level of divergence as a species criterion (Ferguson 2002), up to seven of the polyphyletic species were paraphyletic, which is the first line of evidence for the presence of cryptic species in Mugilidae. When paraphyly is not proven, there is some evidence of reproductive isolation, such as demonstrated for some sympatric lineages within the nominal species *Mugil cephalus* (Shen et al. 2011, Krüeck et al. 2013). Similarly, among four lineages observed in *M. curema*, two are in allopatry, separated by well known biogeographic barriers (the American continent and the Atlantic Ocean) while those potentially in sympatry present two different cytotypes (2n = 24 and 2n = 28) which probably prevent interbreeding (Durand et al. 2012a).

To conclude, there is increasing molecular evidence that the species diversity of the Mugilidae is greatly underestimated (Durand and Borsa 2015). Morphometry and anatomy are sometimes useful to describe mugilid diversity, as for *Mugil rubrioculus* in 2007 (Harrison et al. 2007), but there is no doubt that molecular approaches are very valuable, while not the only possible approaches, in providing rapid advances in knowledge of this family. In this context, DNA barcoding programmes such as FISHBOL (http://www.fishbol.org) represent an excellent opportunity to reveal putative cryptic species. However, it would first be necessary to evaluate the sequence variability of the *COI* fragment used in barcoding programmes to know if it is able to identify putative cryptic species highlighted in this present study.

Acknowledgements

I thank Donatella Crosetti and Stephen J.M. Blaber, editors of this book, for inviting me to write this chapter on the molecular taxonomy of the Mugilidae. This chapter was only possible thanks to the close collaboration of the following researchers: Kang-Ning Shen, Philippe Borsa, Rong Xia and Cuizhang Fu. I am very grateful to them. Last I am indebted to David McKenzie for the editing of my text.

References

Albaret, J.J. 1992. Mugilidae. pp. 780–788. *In*: C. Lévêque, D. Paugy and G.G. Teugels (eds.). The Fresh and Brackish Water Fishes of West Africa, Volume 2. Musée Royal de l'Afrique Centrale, Tervuren, Belgique and O.R.S.T.O.M., Paris, France, Collection Faune Tropicale n° XXVIII.

Aurelle, D., R.M. Barthelemy, J.P. Quignard, M. Trabelsi and E. Faure. 2008. Molecular Phylogeny of Mugilidae (Teleostei: Perciformes). Open Mar. Biol. J. 2: 29–37.

Autem, M. and F. Bonhomme. 1980. Eléments de systématique biochimique chez les Mugilidés de Méditerranée. Biochem. Syst. Ecol. 8: 305–308.

Blel, H., N. Chatti, R. Besbes, S. Farjallah, A. Elouaer, H. Guerbej and K. Said. 2008. Phylogenetic relationships in grey mullets (Mugilidae) in a Tunisian lagoon. Aquac. Res. 39: 268–275.

Caldara, F., L. Bargelloni, L. Ostellari, E. Penzo, L. Colombo and T. Patarnello. 1996. Molecular phylogeny of grey mullets based on mitochondrial DNA sequence analysis: evidence of a differential rate of evolution at the intrafamily level. Mol. Phylogenet. Evol. 6: 416–424.

Callegarini, C. and F. Basaglia. 1978. Biochemical characteristics of mugilids in the lagoons of the Po delta. Boll. Zool. 45: 35–40.

Campton, D.E. and B. Mahmoudi. 1991. Allozyme variation and population structure of striped mullet (*Mugil cephalus*) in Florida. Copeia 485–492.

Cataudella, S., M.V. Civitelli and E. Capanna. 1974. Chromosome complements of the Mediterranean mullets (Pisces: Perciformes). Caryologia 27: 93–105.

Cipriano, R.R., M.M. Cestari and A.S. Fenocchio. 2002. Levantamento citogenético de peixes marinhos do litoral do Paraná. IX Simposio de Citogenética e Genética de Peixes, Maringa, Brasil.

Crosetti, D., J.C. Avise, F. Placidi, A.R. Rossi and L. Sola. 1993. Geographic variability in the grey mullet *Mugil cephalus*: preliminary results of mtDNA and chromosome analyses. Aquaculture 111: 95–101.

Crosetti, D., W.S. Nelson and J. Avise. 1994. Pronounced genetic structure of mitochondrial DNA among populations of the circumglobally distributed grey mullet (*Mugil cephalus* Linnaeus). J. Fish Biol. 44: 47–58.

Cuvier, G. and A. Valenciennes. 1836. Histoire naturelle des poissons. Tome onzième. Levrault, Paris, xx+506 pp., pls. 307–343.

de Queiroz, K. 2007. Species concepts and species delimitation. Syst. Biol. 56: 879–886.

Durand, J.D. and P. Borsa. 2015. Mitochondrial phylogeny of grey mullets (Acanthopterygii: Mugilidae) suggests high proportion of cryptic species. C. R. Biol. 338: 266–277.

Durand, J.D., W.J. Chen, K.N. Shen, C. Fu and P. Borsa. 2012a. Genus-level taxonomic changes implied by the mitochondrial phylogeny of grey mullets (Teleostei: Mugilidae). C. R. Biol. 335: 687–697.

Durand, J.D., K.N. Shen, W.J. Chen, B.W. Jamandre, H. Blel, K. Diop, M. Nirchio, F.J. Garcia de Leon, A.K. Whitfield, C.W. Chang and P. Borsa. 2012b. Systematics of the grey mullets (Teleostei: Mugiliformes: Mugilidae): Molecular phylogenetic evidence challenges two centuries of morphology-based taxonomy. Mol. Phylogenet. Evol. 64: 73–92.

Durand, J.D., H. Blel, K.N. Shen, E.T. Koutrakis and B. Guinand. 2013. Population genetic structure of *Mugil cephalus* in the Mediterranean and Black Seas: a single mitochondrial clade and many nuclear barriers. Mar. Ecol. Prog. Ser. 474: 243–261.

Erguden, D., M. Gurlek, D. Yaglioglu and C. Turan. 2010. Genetic identification and taxonomic relationship of Mediterranean Mugilid species based on mitochondrial 16S rDNA sequence data. J. Anim. Vet. Adv. 9: 336–341.

Eschmeyer, W.N. (ed.). 2014. Catalog of Fishes: Genera, Species, References. (http://research.calacademy.org/research/ichthyology/catalog/fishcatmain.asp). Electronic version accessed 01-09-2014.

Eschmeyer, W.N. and J.D. Fong. 2014. Species by Family/Subfamily. (http://researcharchive.calacademy.org/research/ichthyology/catalog/SpeciesByFamily.asp). Electronic version accessed 01-09-2014.

Ferguson, J.W.H. 2002. On the use of genetic divergence for identifying species. Biol. J. Linn. Soc. 75: 509–516.

Fraga, E., H. Schneider, M. Nirchio, E. Santa-Brigida, L.F. Rodrigues-Filho and I. Sampaio. 2007. Molecular phylogenetic analyses of mullets (Mugilidae, Mugiliformes) based on two mitochondrial genes. J. Appl. Ichthyol. 23: 598–604.

Ghasemzadeh, J. 1998. Phylogeny and systematics of Indo-Pacific mullets (Teleostei: Mugilidae) with special reference to the mullets of Australia. Macquarie University. PhD Dissertation, 397 pages.

Gornung, E., P. Colangelo and F. Annesi. 2007. 5S ribosomal RNA genes in six species of Mediterranean grey mullets: genomic organization and phylogenetic inference. Genome 50: 787–795.

Günther, A. 1861. Catalogue of the acanthopterygian fishes in the collection of the British Museum. Vol. 3. British Museum, London, xxv+586 pp.

Harrison, I.J. and G.J. Howes. 1991. The pharyngobranchial organ of mugilid fishes; its structure, variability, ontogeny, possible function and taxonomic utility. Bull. Br. Mus. Nat. Hist. Zool. 57: 111–132.

Harrison, I.J. and H. Senou. 1997. Order Mugiliformes. Mugilidae. Mullets. pp. 2069–2108. *In*: K.E. Carpenter and V.H. Niem (eds.). FAO Species Identification Guide for Fishery Purposes. The Living Marine Resources of the Western Central Pacific. Volume 4. Bony Fishes part 2 (Mugilidae to Carangidae). FAO, Rome.

Harrison, I.J., M. Nirchio, C. Oliveira, E. Ron and J. Gaviria. 2007. A new species of mullet (Teleostei: Mugilidae) from Venezuela, with a discussion on the taxonomy of *Mugil gaimardianus*. J. Fish Biol. 71: 76–97.

Heemstra, P.C. and E. Heemstra. 2004. Coastal fishes of southern Africa. NISC, Grahamstown & SAIAB, Grahamstown, xxiv+488 pp.

Heras, S., M. González-Castro and M.I. Roldan. 2006. *Mugil curema* in Argentinean waters: Combined morphological and molecular approach. Aquaculture 261: 473–478.

Heras, S., M.I. Roldan and M. González-Castro. 2009. Molecular phylogeny of Mugilidae fishes revised. Rev. Fish Biol. Fish 19: 217–231.

Herzberg, A. and R. Pasteur. 1975. The identification of grey mullet species by disc electrophoresis. Aquaculture 5: 99–106.

Huang, C., C. Weng and S. Lee. 2001. Distinguishing two types of gray mullet, *Mugil cephalus* L. (Mugiliformes: Mugilidae), by using glucose-6-phosphate isomerase (GPI) allozymes with special reference to enzyme activities. J. Comp. Physiol. B 171: 387–394.

Imsiridou, A., G. Minos, V. Katsares, N. Karaiskou and A. Tsiora. 2007. Genetic identification and phylogenetic inferences in different Mugilidae species using 5S rDNA markers. Aquac. Res. 38: 1370–1379.

Jamandre, B.W., J.D. Durand and W.N. Tzeng. 2009. Phylogeography of the flathead mullet *Mugil cephalus* in the north-west Pacific as inferred from the mtDNA control region. J. Fish Biol. 75: 393–407.

Jamandre, B.W., J.D. Durand, K.N. Shen and W.N. Tzeng. 2010. Differences in evolutionary patterns and variability between mtDNA cytochrome b and control region in two types of *Mugil cephalus* species complex in northwest Pacific. J. Fish Soc. Taiwan 37: 163–172.

Ke, H.M., W.W. Lin and H.W. Kao. 2009. Genetic diversity and differentiation of grey mullet (*Mugil cephalus*) in the coastal waters of Taiwan. Zool. Sci. 26: 421–428.

Kekkonen, M. and P.D.N. Hebert. 2014. DNA barcode-based delineation of putative species: efficient start for taxonomic workflows. Mol. Ecol. Res. 14: 706–715.

Kimura, M. 1980. A simple method for estimating evolutionary rates of base substitutions through comparative studies of nucleotides sequences. J. Mol. Evol. 16: 111–120.

Kottelat, M. 2013. The Fishes of the inland waters of southeast Asia: A catalogue and core bibliography of the Fishes known to occur in freshwaters, mangroves and estueries. Raff. Bull. Zool. Supp. 27: 1–663.

Krüeck, N.C., D.I. Innes and J.R. Ovenden. 2013. New SNPs for population genetic analysis reveal possible cryptic speciation of eastern Australian sea mullet (*Mugil cephalus*). Mol. Ecol. Res. 13: 715–725.

Lee, S.C., J.T. Chang and Y.Y. Tsu. 1995. Genetic relationships of four Taiwan mullets (Pisces: Perciformes: Mugilidae). J. Fish Biol. 46: 159–162.

LeGrande, W.H. and J.M. Fitzsimons. 1976. Karyology of the mullets *Mugil curema* and *M. cephalus* (Perciformes: Mugilidae) from Louisiana. Copeia 1976: 388–391.

Liu, J.X., T.X. Gao, S.F. Wu and Y.P. Zhang. 2007. Pleistocene isolation in the Northwestern Pacific marginal seas and limited dispersal in a marine fish, *Chelon haematocheilus* (Temminck and Schlegel, 1845). Mol. Ecol. 16: 275–288.

Liu, J.Y., C.L. Brown and T.B. Yang. 2009. Population genetic structure and historical demography of grey mullet, *Mugil cephalus*, along the coast of China, inferred by analysis of the mitochondrial control region. Biochem. Syst. Ecol. 37: 556–566.

Liu, J.Y., C.L. Brown and T.B. Yang. 2010. Phylogenetic relationships of mullets (Mugilidae) in China Seas based on partial sequences of two mitochondrial genes. Biochem. Syst. Ecol. 38: 647–655.

Lüther, G. 1977. New characters for consideration in the taxonomy appraisal of grey mullets. J. Mar. Biol. Ass. India 169: 1–9.

Maddison, W.P. and D.R. Maddison. 2011. Mesquite: a modular system for evolutionary analysis. Version 2.75 <http://mesquiteproject.org>.

Mai, A.C.G., C.I. Miño, L.F.F. Marins, C. Monteiro-Neto, L. Miranda, P.R. Schwingel, V.M. Lemos, M. González-Castro, J.P. Castello and J.P. Vieira. 2014. Microsatellite variation and genetic structuring in *Mugil liza* (Teleostei: Mugilidae) populations from Argentina and Brazil. Est. Coast. Shelf Sci. 149: 80–86.

McMahan, C.D., M.P. Davis, O. Dominguez-Dominguez, F.J. Garcia de Leon, I. Doadrio and K.R. Piller. 2013. From the mountains to the sea: phylogeography and cryptic diversity within the mountain mullet, *Agonostomus monticola* (Teleostei: Mugilidae). J. Biogeogr. 40: 894–904.

Menezes, M.R., M. Martins and S. Naik. 1992. Interspecific genetic divergence in grey mullets from Goa region. Aquaculture 105: 117–129.

Murgia, R., G. Tola, S.N. Archer, S. Vallerga and J. Hirano. 2002. Genetic identification of grey mullet species (Mugilidae) by analysis of mitochondrial DNA sequence: application to identify the origin of processed ovary products (*bottarga*). Mar. Biotechnol. 4: 119–126.

Nirchio, M. and H. Cequea. 1998. Karyology of *Mugil liza* and *M. curema* from Venezuela. Bol. Invest. Mar. Cost. 27: 45–50.

Nirchio, M., D. González and J.E. Pérez. 2001. Estudio citogenético de *Mugil curema* y *M. liza* (Pisces: Mugilidae): Regiones organizadoras del nucleolo. Bol. Inst. Oceanogr. Venez. 40: 3–7.

Nirchio, M., F. Cervigon, J.I. Revelo Porto, J.E. Perez, J.A. Gomez and J. Villalaz. 2003. Karyotype supporting *Mugil curema* Valenciennes, 1836 and *Mugil gaimardianus* Desmarest, 1831 (Mugilidae: Teleostei) as two valid nominal species. Sci. Mar. 67: 113–115.

Nirchio, M., R. Cipriano, M. Cestari and A. Fenocchio. 2005. Cytogenetical and morphological features reveal significant differences among Venezuelan and Brazilian samples of *Mugil curema*. Neotrop. Ichthyol. 3: 107–110.

Ogilby, J.D. 1908. New or little known fishes in the Queensland Museum. Ann. Queensland Mus. 9: 3–41.

Papasotiropoulos, V., E. Klossa-Kilia, G. Kilias and S. Alahiotis. 2001. Genetic divergence and phylogenetic relationships in grey mullets (Teleostei: Mugilidae) using allozyme data. Biochem. Genet. 39: 155–68.

Papasotiropoulos, V., E. Klossa-Kilia, S. Alahiotis and G. Kilias. 2007. Molecular phylogeny of grey mullets (Teleostei: Mugilidae) in Greece: evidence from sequence analysis of mtDNA segments. Biochem. Genet. 45: 623–636.

Papasotiropoulos, V., E. Klossa-Kilia, G. Kilias and S. Alahiotis. 2002. Genetic divergence and phylogenetic relationships in grey mullets (Teleostei: Mugilidae) based on PCR RFLP analysis of mtDNA segments. Biochem. Genet. 40: 71–86.

Petit, R.J. and L. Excoffier. 2009. Gene flow and species delimitation. TREE 24: 386–393.

Rocha-Olivares, A., N.M. Garber and K.C. Stuck. 2000. High genetic diversity, large interoceanic divergence and historical demography of the striped mullet. J. Fish Biol. 57: 1134–1149.

Rocha-Olivares, A., N.M. Garber, A.F. Garber and K.C. Stuck. 2005. Structure of the mitochondrial control region and flanking tRNA genes of *Mugil cephalus*. Hidrobiológica 15: 139–149.

Rossi, A.R., D. Crosetti and E. Gornung. 1996. Cytogenetic analysis of global populations of *Mugil cephalus* (striped mullet) by different staining techniques and fluorescent *in situ* hybridization. Heredity 76: 77–82.

Rossi, A.R., M. Capula, D. Crosetti, L. Sola and D.E. Campton. 1998a. Allozyme variation in global populations of striped mullet, *Mugil cephalus* (Pisces: Mugilidae). Mar. Biol. 131: 203–212.

Rossi, A.R., M. Capula, D. Crosetti, D.E. Campton and L. Sola. 1998b. Genetic divergence and phylogenetic inferences in five species of Mugilidae (Pisces: Perciformes). Mar. Biol. 131: 213–218.

Rossi, A.R., A. Ungaro, S. De Innocentiis, D. Crosetti and L. Sola. 2004. Phylogenetic analysis of Mediterranean Mugilids by allozymes and 16S mt-rRNA genes investigation: are the Mediterranean species of *Liza* monophyletic? Biochem. Genet. 42: 301–315.

Rossi, A.R., E. Gornung, L. Sola and M. Nirchio. 2005. Comparative molecular cytogenetic analysis of two congeneric species, *Mugil curema* and *M. liza* (Pisces: Mugiliformes), characterized by significant karyotype diversity. Genetica 125: 27–32.

Schultz, L.P. 1946. A revision of the genera of mullets, fishes of the family Mugilidae, with descriptions of three new genera. Proc. U.S. Natl. Mus. 96: 377–395.

Schultz, L.P. 1953. Family Mugilidae. *In*: L.P. Schultz, E.S. Herald, E.A. Lachner, A.D. Welander and L.P. Woods (eds.). Fishes of the Marshall and Marianas Islands, Vol. I. Families from Asymmetrontidae through Siganidae. Bull. U.S. Natl. Mus. 202: 310–322.

Semina, A.V., N.E. Polyakova and V.A. Brykov. 2007. Analysis of mitochondrial DNA: taxonomic and phylogenetic relationships in two fish taxa (Pisces: Mugilidae and Cyprinidae). Biochemistry 72: 1349–1355.

Senou, H. 1988. Phylogenetic Interrelationships of the Mullets (Pisces: Mugilidae). Tokyo University, Tokyo. 172 pages.

Senou, H. 2002. Mugilidae. pp. 537–541. *In*: T. Nakabo (ed.). Fishes of Japan with Pictorial Keys to the Species, English Edition. Tokai University Press, Tokyo.

Shen, K.N., B.W. Jamandre, C.C. Hsu, W.N. Tzeng and J.D. Durand. 2011. Plio-Pleistocene sea level and temperature fluctuations in the northwestern Pacific promoted speciation in the globally-distributed flathead mullet *Mugil cephalus*. BMC Evol. Biol. 11: 83.

Siccha-Ramirez, R., N.A. Menezes, M. Nirchio, F. Foresti and C. Oliveira. 2014. Molecular identification of mullet species of the Atlantic South Caribbean and South America and the phylogeographic analysis of *Mugil liza*. Rev. Fish Biol. Fish 22: 86–96.

Smith, J.L.B. 1948. A generic revision of the mugilid fishes of South Africa. Ann. Mag. Nat. Hist. 14: 833–843.

Smith, M.M. and J.L.B. Smith. 1986. Mugilidae. pp. 714–720. *In*: M.M. Smith and P.C. Heemstra (eds.). Smiths' Sea Fishes. Springer-Verlag, Berlin.

Sola, L., E. Gornung, M.E. Mannarelli and A.R. Rossi. 2007. Chromosomal evolution in Mugilidae, Mugilomorpha: an overview. pp. 165–194. *In*: E. Pisano, C. Ozouf-Costaz, F. Foresti and B.G. Kapoor (eds.). Fish Cytogenetics. Science Publishers, Enfield, NH, USA.

Sola, L., M. Nirchio and A.R. Rossi. 2008. The past and the future of cytogenetics of Mugilidae: an updated overview. Bol. Inst. Oceanogr. Venezuela 47: 27–33.

Sun, P., Z.H. Shi, F. Yin and S.M. Peng. 2012. Genetic variation analysis of *Mugil cephalus* in China Sea based on mitochondrial COI Gene Sequences. Biochem. Genet. 50: 180–191.

Taylor, W.R. 1964. Fishes of Arnhem Land. Records of the American-Australian Scientific Expedition to Arnhem Land 4: 44–307, pls. 1–68.

Thomson, J.M. 1954. The Mugilidae of Australia and adjacent seas. Austr. J. Mar. Freshw. Res. 5: 70–131.

Thomson, J.M. 1997. The Mugilidae of the world. Mem. Queensl. Mus. 43: 457–562.

Turan, C., M. Caliskan and H. Kucuktas. 2005. Phylogenetic relationships of nine mullet species (Mugilidae) in the Mediterranean Sea. Hydrobiologia 532: 45–51.

Ward, R.D., R. Hanner and P.D.N. Hebert. 2009. The campaign to DNA barcode all fishes, FISH-BOL. J. Fish Biol. 74: 329–356.

Whitley, G.P. 1941. Ichthyological notes and illustrations. Aust. Zool. 10: 1–52.

Xia, R., J.D. Durand and C. Fu. Submitted. Multilocus resolution of the Mugilidae phylogeny (Teleostei: Mugiliformes): Implication on the family's taxonomy.

Zou, S., Q. Li and L. Kong. 2011. Monophyly, distance and character-based multigene barcoding reveal extraordinary cryptic diversity in *Nassarius*: a complex and dangerous community. PLOS ONE 7: e47276.

CHAPTER 3

Biogeography and Distribution of Mugilidae in the Americas

Mário Barletta[1,*] and *David Valença Dantas*[1,2]

American Biogeographic Shelf Regions and Provinces

The American continent stretches for about 14000 km from north to south, spanning over 127° of latitude (72° N to 55° S) and covering approximately 42.5 million km^2 (Kohn and Cohen 1998). The continent is bordered by the Atlantic Ocean on the east and the Pacific on the west. The latitudinal range confers important biogeographical and ecological characteristics to this landmass. The most obvious is the freshwater outflow from the Amazon River to the Atlantic that apparently acts as a biogeographical barrier separating the fish fauna of Brazil and the Caribbean (Briggs 1995). According to Briggs (1995), the distribution patterns of aquatic species in local situations may be affected by factors such as food, shelter, salinity and dissolved oxygen, but on a global or oceanic scale, temperature primarily controls their distribution. The widespread patterns demonstrated by many species indicate that the surface of the ocean is subdivided into four zones: the Tropical Zone (TR), the Warm-Temperate Zone (W-T), the Cold-Temperate Zone (C-T), and the Cold or Polar Zone (C), each with their respective regions and zoogeographical provinces (Briggs 1995). However, the upsurge of more information on phylogeography, palaeontological research, earth movements and sea level changes in the last 20 years required a new arrangement of regions and provinces (Briggs and Bowen 2012). Moreover, the provinces proposed by Briggs (1995) have recently been subdivided into ecoregions to address the appropriate scale of conservation efforts (Spalding et al. 2007). Biogeographic classifications are essential for developing ecologically representative systems of protected areas, as required by international agreements for conservation and regional fisheries management. The biogeographic classification proposed by Spalding et al. (2007) focuses on coastal and shelf waters, combining benthic and shelf-dependent pelagic biotas, representing waters where most of the marine biodiversity is confined, human attention is higher, and where there is often a complex synergy of threats—far greater than in offshore waters. According to the biogeographic division of Spalding et al. (2007), the American continent contains five large realms (Temperate northern Atlantic, Temperate northern Pacific, Tropical Atlantic, Tropical eastern Pacific and Temperate South America), each with its respective provinces and ecoregions (Fig. 3.1, Table 3.1).

[1] Laboratório de Ecologia e Gerenciamento Costeiro e Estuarino (LEGECE), Departamento de Oceanografia, Universidade Federal de Pernambuco (UFPE), Recife, Pernambuco, Brazil.
[2] Departamento de Engenharia de Pesca, Centro de Educação Superior da Região Sul (CERES), Universidade do Estado de Santa Catarina (UDESC), Laguna, Santa Catarina, Brazil.
* Corresponding author: barletta@ufpe.br

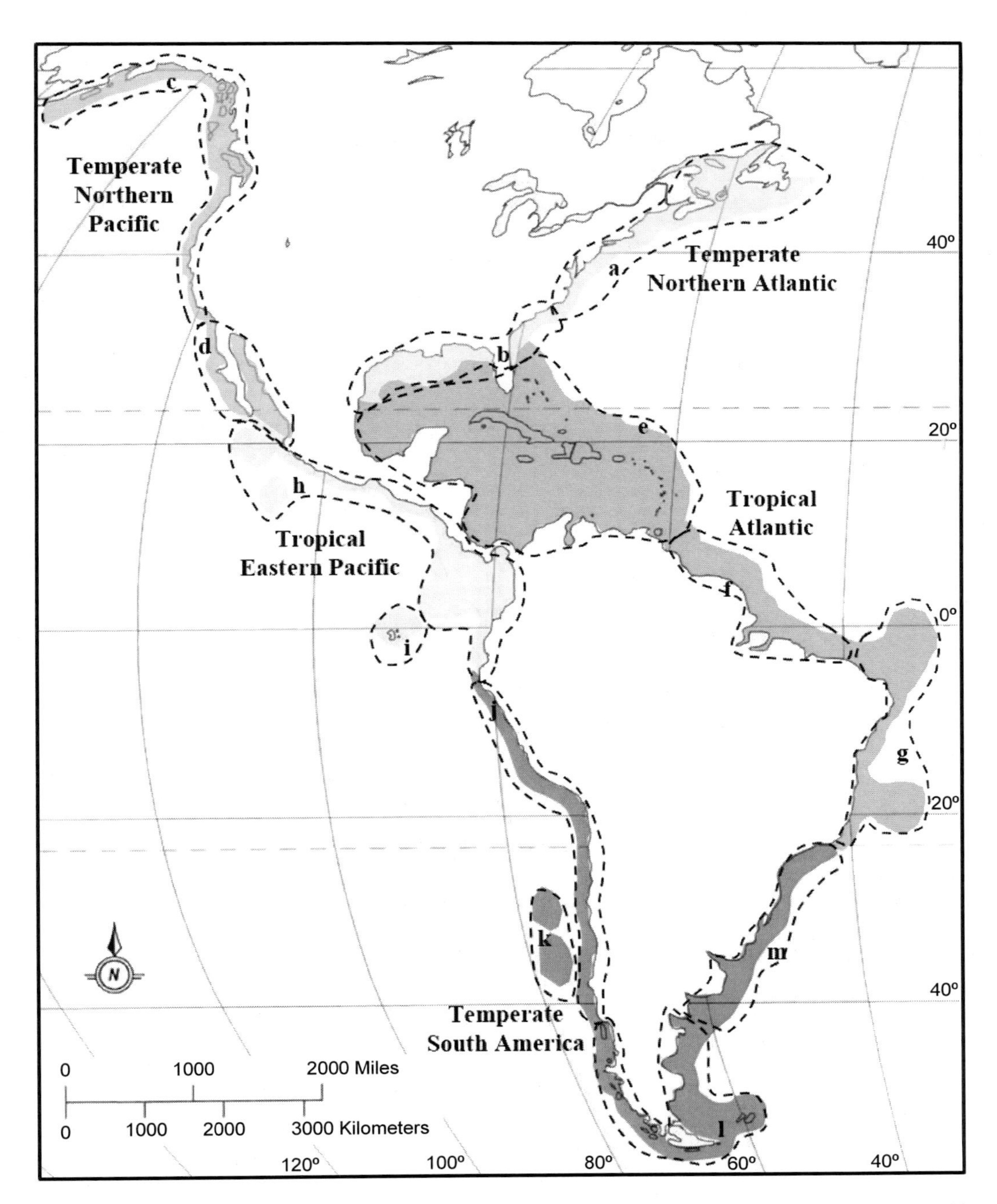

Figure 3.1. Marine realms and provinces in the America continent (North, Central and South). The realms are indicated in different shades and provinces in letters delimitated by draw lines: (a) Cold Temperate Northwest Atlantic; (b) Warm Temperate Northwest Atlantic; (c) Cold Temperate Northeast Pacific; (d) Warm Temperate Northeast Pacific; (e) Tropical Northwestern Atlantic; (f) North Brazil Shelf; (g) Tropical Southwestern Atlantic; (h) Tropical East Pacific; (i) Galapagos; (j) Warm Temperate Southeastern Pacific; (k) Juan Fernández and Desventuradas; (l) Megallanic; (m) Warm Temperate Southwestern Atlantic.

Table 3.1. Marine Realms, Provinces and Ecoregions in the American Continent (North, Central and South), modified from Spalding et al. (2007). The realms and provinces areas are indicated in Fig. 3.1.

Realm	Province	Ecoregion
Temperate Northern Atlantic	CTNAP	Gulf of St. Lawrence-Eastern Scotian Shelf (48° 32′ 38″ N; 64° 31′ 54.1″ W)
		Southern Grand Banks-South Newfoundland (46° 19′ 59.9 N; 50° 43′ 22.8″ W)
		Scotian Shelf (42° 51′ 38.5″ N; 62° 11′ 14.6″ W)
		Gulf of Maine/Bay of Fundy (43° 23′ 55.7″ N; 68° 13′ 26″ W)
		Virginian (39° 20′ 24″ N; 74° 52′ 33.6″ W)
	WTNAP	Carolinian (32° 44′ 44.2″ N; 78°51′ 9.7″ W)
		Northern Gulf of Mexico (29° 5′ 31.2″ N; 92° 11′ 57.8 W)
Temperate Northern Pacific	CTNPP	Aleutian Islands (52° 4′ 28.6″ N; 26° 28′ 34.3″ W)
		Gulf of Alaska (58° 26′ 0.2″ N; 149° 56′ 2.4″ W)
		North American Pacific Fijordland (55° 9′ 39.6″ N; 131° 15′ 32.4″ W)
		Puget Trough/Georgia Basin (48° 53′ 41.3″ N; 121° 39′ 43.2″ W)
		Oregon, Washington, Vancouver Coast and Shelf (45° 11′ 41.6″ N; 125° 12′ 25.2" W)
		Northern California (37° 5′ 48.5″ N; 122° 17′ 56.4″ W)
	WTNPP	Southern California Bight (30° 5′ 26.2″ N; 117° 47′ 2.4″ W)
		Cortezian (27° 30′ 7.6″ N; 109° 21′ 10.8″ W)
		Magdalena Transition (24° 10′ 18.8″ N; 112° 59′ 24″ W)
Tropical Atlantic	TNAP	Bermuda (32° 17′ 24.4″ N; 64° 45′47.5″ W)
		Bahamian (24° 8′ 37.3″ N; 73° 49′ 10.2″ W)
		Eastern Caribbean (16° 16′ 47.6″ N; 61° 15′ 42.5″ W)
		Greater Antilles (19° 1′ 57.7″ N; 76° 1′ 44″ W)
		Southern Caribbean (10° 49′ 16.3″N; 68° 17′ 47.4″ W)
		Western Caribbean (17° 50′ 15.7″ N; 86° 55′ 6.2″ W)
		Southern Gulf of Mexico (21° 40′ 15.6″ N; 92° 56′ 22.9″ W)
		Floridian (26° 28′ 1.2″ N; 82° 32′ 54.6″ W)
	NBSP	Guianan (6° 31′ 32.9″ N; 57° 17′ 1.3″ W)
		Amazonia (1° 3′ 52.3″ S; 48°31′ 19.9″ W)
	TSAP	São Pedro and Sao Paulo Islands (0° 21′ 12.1″ N; 29° 52′ 15.6″ W)
		Fernando de Noronha and Atoll das Rocas (3° 52′ 50.1″ S; 32° 10′ 54.5″ W)
		Northeastern Brazil (7° 2′ 50″ S; 37° 3′ 4.3″ W)
		The Eastern Brazil (18° 22′ 21.4″ S; 38° 55′ 40.8′ W)
		Trindade and Martin Vaz Islands (20° 31′ 48″S; 30° 5′ 41.3″ W)

Table 3.1. contd....

Table 3.1. contd....

Realm	Province	Ecoregion
Tropical Eastern Pacific	TEPP	Revillagigedos (18° 20′ 2.4″ N; 113° 10′ 12″ W)
		Clipperton (10° 22′ 19.6″ N; 109° 10′ 8.4″ W)
		Mexican Tropical Pacific (17° 24′ 47.2″ N; 102° 25′ 44.4″ W)
		Chiapas-Nicaragua (13° 21′ 1.1″ N; 90° 54′ 2.9″ W)
Tropical Eastern Pacific	TEPP	Nicoya (7° 57′ 4.3″ N; 84° 34′ 11.6″ W)
		Cocos Islands (5° 13′ 23.9″ N; 87° 22′ 6.2″ W)
		Panama Bight (3° 26′ 49.7″ N; 78° 54′ 49″ W)
		Guayaquil (3° 9′ 29.1″ S; 80° 44′ 11.8″ W)
	GP	Northern Galapagos Islands (2° 54′2.1″ N; 91° 58′ 45.1″ W)
		Eastern Galapagos Islands (1° 4′ 37.4″ S; 88° 47′ 47.4″ W)
		Western Galapagos Islands (0° 59′ 5.9″ S; 93° 10′ 8″ W)
Temperate South America	WTSPP	Central Peru (8° 54′ 21.5″ S; 78° 52′ 31.1″ W)
		Humboldtian (18° 9′ 58.3″ S; 72° 34′ 56.3″ W)
		Central Chile (29° 6′ 58″ S; 71° 12′ 27.7″ W)
		Araucanian (37° 2′ 52.4″ S; 73° 4′ 3.4″ W)
	JFDP	Juan Fernández and Desventuradas (30° 32′ 12.8″ S; 79° 53′ 31.6″ W)
	WTSAP	Southeastern Brazil (24° 39′ 4″ S; 46° 25′ 32.9″ W)
		Rio Grande (30° 59′ 49.2″ S; 51° 0′ 15.8″ W)
		Rio de la Plata (32° 58′ 50.2″ S; 58° 23′ 12.5″ W)
		Uruguay-Buenos Aires Shelf (37° 39’ 27” S; 58° 1’ 17” W)
	MP	North Patagonian Gulfs (42° 53′ 19.7″ S; 65° 33′ 50″ W)
		Patagonian Shelf (49° 7′ 11.6″ S; 65° 2′ 51.4″ W)
		Malvinas/Falklands (51° 37′ 1.2″ S; 59° 3′ 38.9″ W)
		Channels and Fjords of Southern Chile (53° 23′ 35.9″ S; 71° 32′ 41.6″ W)
		Chiloense (43° 45′ 35.6″ S; 73° 43′ 28.9″ W)

* CTNAP (Cold Temperate North-West Atlantic Province); WTNAP (Warm Temperate North-West Atlantic Province); CTNPP (Cold Temperate North-East Pacific Province); WTNPP (Warm Temperate North-east Pacific Province); TNAP (Tropical North-western Atlantic Province); NBSP (North Brazil Shelf Province); TSAP (Tropical South-western Atlantic Province); TEPP (Tropical East Pacific Province); GP (Galapagos Province); WTSPP (Warm Temperate South-eastern Pacific); JFDP (Juan Fernández and Desventuradas Province); WTSAP (Warm Temperate South-western Atlantic Province); MP (Magellanic Province).

Temperate Northern Atlantic

In the American continent, the Temperate northern Atlantic Realm has two provinces, the Cold Temperate North-West Atlantic Province (CTNAP) and the Warm Temperate North-West Atlantic Province (WTNAP) (Fig. 3.1) (Spalding et al. 2007). The CTNAP (Fig. 3.1) extends from the Gulf of St. Lawrence, in the east of Canada, to the south of Virginia State in the United States, and presents five ecoregions (Spalding et al. 2007) (Table 3.1). The WTNAP (Fig. 3.1) extends from the continental shelf on the extreme south of Virginia State to the northern Gulf of Mexico, excluding the southern part of Florida, and has two ecoregions (Spalding et al. 2007) (Table 3.1). The projection of the South Equatorial Current north-west of South America increases its speed and receives the North Equatorial Current, then strongly running into the southern Caribbean and through to the Yucatán Channel. After passing through the Yucatán Channel, the main stream makes an abrupt turn to the east and forms the Florida Current, called the Gulf Stream as it leaves the coast of the eastern United States and heads across the North Atlantic. According to Briggs (1995), the Gulf Stream affects the distribution of shore animals in the western north Atlantic and because of its tendency to transport larvae and adults, many tropical specimens are left stranded along the shores of North-eastern North America.

Temperate Northern Pacific

In the American continent, the Temperate northern Pacific Realm has two provinces, the Cold Temperate North-East Pacific Province (CTNPP) and the Warm Temperate North-East Pacific Province (WTNPP) (Spalding et al. 2007). The CTNPP (Fig. 3.1) extends from the Aleutian Islands to the northern portion of California, and comprises six ecoregions (Table 3.1). The WTNPP (Fig. 3.1) extends from the southern coast of California to Baja California and the Gulf of California, with three ecoregions (Table 3.1). The California Current has a strong influence on the Temperate northern Pacific Realm and swings to the west off the southern part of Baja California, preventing the upsurge of tropical waters from Mexico and Central America.

Tropical Atlantic

In the American continent, the Tropical Atlantic Realm comprises three provinces, the Tropical North-western Atlantic Province (TNAP), the North Brazil Shelf Province (NBSP) and the Tropical South-western Atlantic Province (TSAP) (Spalding et al. 2007). The TNAP (Fig. 3.1) extends from Bermuda and Caribbean waters, including the southern portion of Florida, Florida Keys, and southern Gulf of Mexico, to the northern coast of South America, and has nine ecoregions (Spalding et al. 2007) (Table 3.1). The NBSP (Fig. 3.1) includes coastal and shelf waters of the North Brazil Shelf from north-east of Venezuela to Piauí State, Brazil, and contains two ecoregions (Table 3.1). The TSAP (Fig. 3.1) extends from the north-east coast of Brazil to the north-east coast of Rio de Janeiro State, including the Trindade e Martin Archipelago, and has five ecoregions (Table 3.1). The South Atlantic Ocean has some particular biogeographical features that influence the faunal distribution, and the most obvious are the freshwater outflows from the Amazon and Orinoco Rivers that apparently act as major biogeographical barriers separating the coastal fish fauna from Brazil and the Caribbean (Briggs 1995, Floeter and Gasparini 2000).

Tropical Eastern Pacific

The Tropical eastern Pacific Realm extends from the south-western portion of Mexico, including the Revillagigedos Archipelago, to north-western Peru, including the Cocos Islands, and has two provinces: the Tropical East Pacific Province (TEPP), with eight ecoregions, and the Galapagos Province (GP), with three ecoregions (Spalding et al. 2007) (Fig. 3.1, Table 3.1). To the south, the cold Peru (Humboldt) Current flows upto the coast of Peru and prevents the tropical waters from ordinarily extending more than ~ 3° below the Equator. However, to the north, the corresponding California Current makes its main swing to the west off the southern part of Baja California. Apparently, the discrepancy in the behaviour of the two main, cold, equatorial-bound currents of the eastern Pacific is due to the differences in coastline topography (Briggs 1995). Dispersal of species from the mainland on the east is limited by a deep-water gap of about 140 km, and tropical animals from the north cannot follow the edges of the peninsula of Baja California because of the lower water temperatures in those areas (Briggs 1995). The fish fauna at the end of the peninsula reflects the effects of this relative isolation. The number of species is reduced, there are several endemisms and a permanent risk of invasion by transpacific species (Briggs 1995).

Temperate South America

The Temperate South American Realm extends from Peru to South Chile along the Pacific coast, and from South-eastern Brazil to southern Chile on the Atlantic side, comprising four provinces: the Warm Temperate South-eastern Pacific Province (WTSPP), the Juan Fernández and Desventuradas Province (JFDP), the Warm Temperate South-western Atlantic Province (WTSAP) and the Magellanic Province (MP) (Spalding et al. 2007). The WTSPP (Fig. 3.1) extends from Peru to the Reloncaví Sound, south of Chile, and has four ecoregions (Table 3.1). The JFDP comprises two Archipelagos in the South Pacific on the coast of Chile, with an ecoregion of the same name (Table 3.1). The WTSAP (Fig. 3.1) extends from South-eastern Brazil to the Río Negro Province, in South-central Argentina, and has four ecoregions (Spalding et al. 2007) (Table 3.1). The MP (Fig. 3.1) comprises the southernmost portion of South America on both Atlantic and Pacific sides, from south of Río Negro Province, Argentina, to south of Puerto Montt, Chile, and contains five ecoregions (Spalding et al. 2007) (Table 3.1). The Peru Current that originates from the West Wind Drift, and flows northwards along the Peruvian coast influences the Pacific coast of the Temperate South American Realm. It is affected by south-westerly winds that cause an upwelling (Briggs 1995). This upwelling maintains cold surface water temperatures of about 3°C. On the other hand, the Atlantic coast of the Temperate South American Realm is influenced by the warm southward-flowing Brazil Current, and by the Falklands (Malvinas) Current from the south to north in the opposite direction (Briggs 1995).

Biogeography and Distribution of Mugilidae in America

The family Mugilidae (Order Mugiliformes) comprises species that are distributed in coastal waters in all tropical, subtropical and temperate seas. They are generally considered ecologically and economically important (Thomson 1997, Harrison 1995, 2002, Durand et al. 2012). Most of the Mugilidae species occur in the Indo-West Pacific region (Harrison 2002). Eschmeyer and Fong (2014) recognize 20 valid genera of Mugilidae, but only two of them (*Liza* and *Mugil*) represent 40% of the species richness within the family. According to the 'Catalog of Fishes' (Eschmeyer and Fong 2014, Eschmeyer 2014) for the American continent, on both Atlantic and Pacific coasts, there are 14 recognized species, distributed in five genera: *Agonostomus* (1 sp.), *Chaenomugil* (1 sp.), *Joturus* (1 sp.), *Mugil* (9 spp.) and *Xenomugil* (1 sp.). The species *M. cephalus* and *M. curema* have a widespread distribution and an unusual amount of variation between geographically distant populations, which certainly complicates the taxonomy of these species (Harrison et al. 2007). Nevertheless, the family Mugilidae has been the focus of several systematic revisions (Schultz 1946, Thomson 1954, Senou 1988, Thomson 1997, Harrison et al. 2007, Menezes et al. 2010, Rodrigues-Filho et al. 2011, Durand et al. 2012, 2013), cytogenetic (Nirchio et al. 2005, 2007, 2009, Rossi et al. 1996, 2005, Sola et al. 2008) and biogeographic studies (Nirchio et al. 2007, 2009, Menezes et al. 2010, Livi et al. 2011, Durand et al. 2012, 2013, MacMahan et al. 2013, Siccha-Ramirez et al. 2014). The distribution pattern of Mugilidae species in the American continent fits a general model of transisthmian germinate pairs. After being separated by the Isthmus of Panama (Tringali et al. 1999, Grant and Leslie 2001, Lessios 2008), Mugilidae developed divergences between Caribbean and Pacific lineages about three–four million years ago (Lessios 2008). Beyond the Isthmus of Panama, other geographical features and geological processes in the ocean basins of the Americas are described as having a large influence on fish fauna distribution and biogeographical patterns. Bellwood and Wainwright (2002) proposed major hard biogeographic barriers (e.g., Isthmus of Panama) and a number of soft barriers (e.g., eastern Pacific Barrier, Amazon-Orinoco, California, Salvador) and Briggs (1995) described the current patterns of the ocean basins that influenced the fauna distribution along the American coasts. According to Briggs (1995), the southern portion of the eastern Pacific Region is influenced by the cold Peru (Humboldt) Current, which runs all the way up the coast of Peru and prevents tropical waters from flowing beyond ~ 3°C. However, the northern portion of the eastern Pacific region corresponding to the California Current moves west of the southern part of Baja California (Briggs 1995). For the western Atlantic Region the eastward projection of Brazil splits the warm South Equatorial Current into two branches, one runs southwards (Brazilian Current) and remains within the South Atlantic system, and the other runs northwest parallel to the shore, accelerating when it encounters the flow of the North Equatorial Current (Briggs 1995). A detailed description of the distribution and some aspects of the biogeography of each Mugilidae species occurring in the American continent, in addition to the main ecological features are described below. The nomenclature for genera and species follows Eschmeyer and Fong (2014) and Eschmeyer (2014).

Agonostomus monticola (Bancroft 1834)—Mountain mullet

The mountain mullet *Agonostomus monticola* is distributed in freshwater streams of the Bahamas, Greater Antilles, the Atlantic and Pacific slopes of Central America and Colombia. Atlantic-bound rivers of Venezuela were also reported to have this species, as well as rivers and streams of the southern USA (Florida and Louisiana), South of Baja California and Galapagos Islands (Harrison 1995, 2002, Eschmeyer 2014) (Fig. 3.2). Adults inhabit freshwater streams and probably spawn in the lower reaches of rivers, or even in the sea, during the maximum rainfall periods. Larvae and juveniles are found at river mouths and in offshore waters. The mountain mullet are found in three zoogeographic realms. In the Tropical eastern Pacific realm, the species are found in both its provinces (Tropical East Pacific and Galapagos). In the Tropical Atlantic realm, mountain mullets are found in the Tropical North-western Atlantic province and, finally, in the Temperate North Atlantic realm of the Warm Temperate North-West Atlantic province. Recently, a study using mitochondrial and nuclear sequenced data indicates that four distinct lineages exist, corresponding to two groups in the Caribbean and Gulf of Mexico basins, and two others on the Pacific coast of Central America (McMahan et al. 2013). The authors suggest that the time of divergence for all four clades might have been early to mid-Miocene.

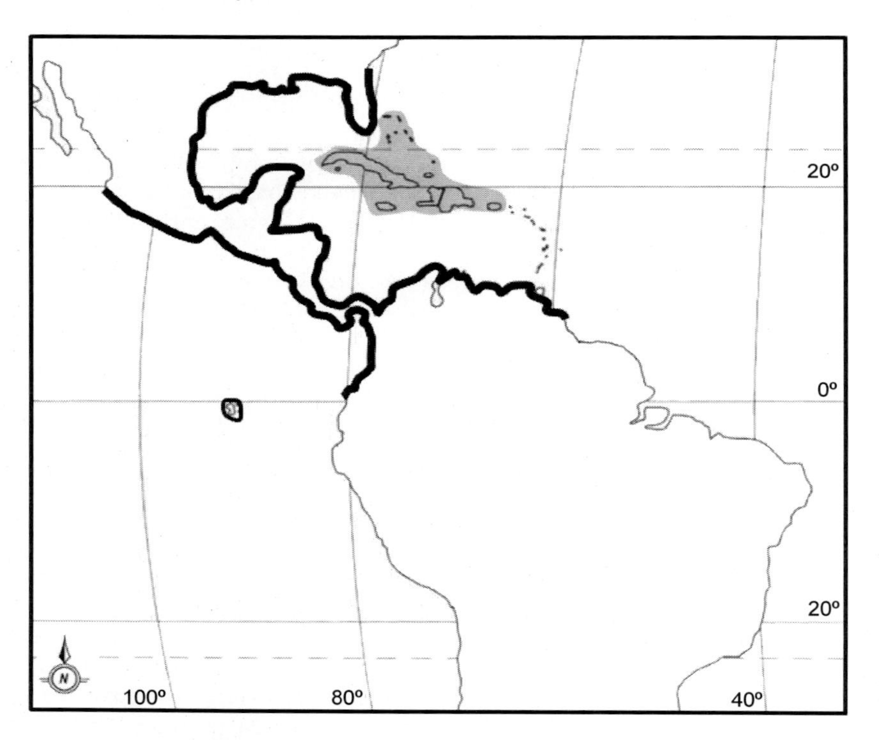

Figure 3.2. Distribution of the mountain mullet *Agonostomus monticola* (Brancroft 1834) in the America continent (North, Central and South).

Chaenomugil proboscideus (Günther 1861)—Snouted mullet

The snouted mullet *Chaenomugil proboscideus* is distributed along the eastern central Pacific from the Gulf of California to Panama (Fig. 3.3), including the southern area of the Warm Temperate North-East Pacific province, in the Temperate North Pacific realm, and the Tropical East Pacific and Galapagos provinces in the Tropical eastern Pacific realm. It also occurs in the Revillagigedo and Galapagos Islands (Harrison 1995, 2002). This species inhabits rocky littorals, feeding on algae that grow on rocks by scraping it off with specialized teeth (Harrison 1995). Despite some information about the distribution of the snouted mullet being available (Harrison 1995, 2002, Durand et al. 2012, Eschmeyer 2014), little is known of its biogeographic origins.

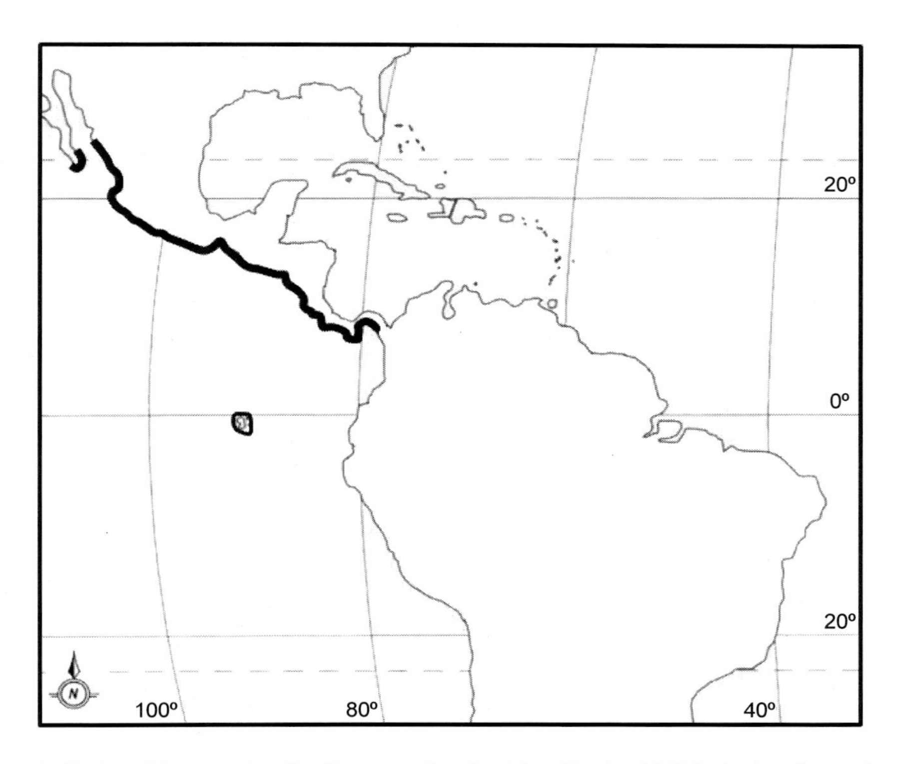

Figure 3.3. Distribution of the snouted mullet *Chaenomugil proboscideus* (Günther 1861) in the America continent (North, Central and South).

Joturus pichardi Poey 1860—Bobo mullet

The bobo mullet (*Joturus pichardi*) inhabits freshwater streams of the eastern central Pacific slopes, from the North of Mexico to Panama (Harrison 1995), on the Atlantic slopes of Central America and Colombia (Harrison 2002). It was also reported for the Bahamas, Greater Antilles and Lesser Antilles (Harrison 2002) (Fig. 3.4). Adults inhabit the upper reaches, in freshwater streams of rivers, but enter brackish waters (Harrison 1995, 2002). Spawning probably occurs in coastal lagoons or at sea (Harrison 1995). In the Tropical Atlantic realm, bobo mullets are found in the Tropical North-western Atlantic province, along the coastal waters of Bahamian, Greater Antilles, South-western Caribbean, western Caribbean and southern Gulf of Mexico ecoregions.

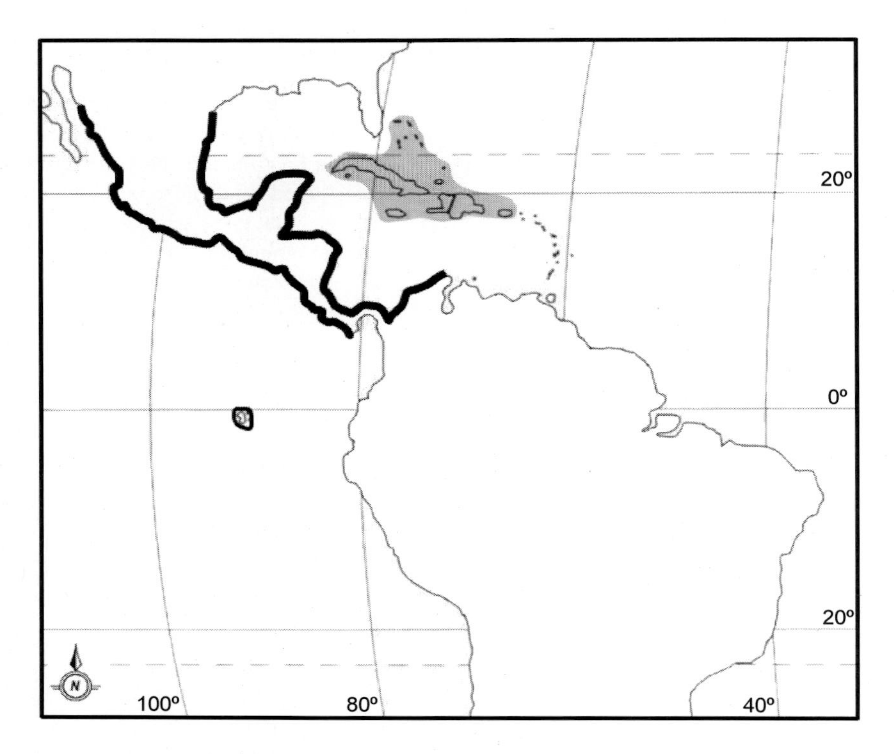

Figure 3.4. Distribution of the bobo mullet *Joturus pichardi* Poey 1860 in the America continent (North, Central and South).

Mugil cephalus L. 1758—Flathead grey mullet

The flathead grey mullet *Mugil cephalus* is distributed worldwide, from 42° N to 42° S, ranging across all biogeographic realms and provinces of the American continent. The only exceptions are the Cold Temperate Northeast Pacific province in the Pacific, and the Magellanic province at the tip of South America. In the western Atlantic, it was supposed to be distributed from Nova Scotia to Argentina, including the Gulf of Mexico (Harrison 2002) though according to Menezes et al. (2010) the name *M. cephalus* cannot be used for species occurring in the western south Atlantic. In the eastern Pacific, it is distributed from California to Chile, including the Gulf of California and Galapagos Islands (Harrison 1995) (Fig. 3.5). Adults are found in inshore marine waters, estuaries, lagoons and rivers (Harrison 2002). Given its wide distribution, and similar morphological and morphometric characteristics to closely related mullets, *M. cephalus* could been confused with species such as *M. liza* (Menezes et al. 2010) which is now regarded as a synonym of *M. platanus* (Fraga et al. 2007). Moreover, there is recent evidence that *M. cephalus* is part of a species complex together with 14 other *Mugil* species (Shen et al. 2011, Durand et al. 2012). A recently taxonomic review of the species of the genera *Mugil* from Atlantic South Caribbean and South America using morphological, cytogenetic and molecular data, excluded *Mugil cephalus* from the area (Menezes et al. 2015).

Mugil brevirostris (Ribeiro 1915)

The *Mugil brevirostris* is distributed from the northern (Amapá) to the southern Brazilian coast (Rio Grande do Sul) (Fig. 3.6) (Menezes et al. 2015). According to the authors, the species occurs in

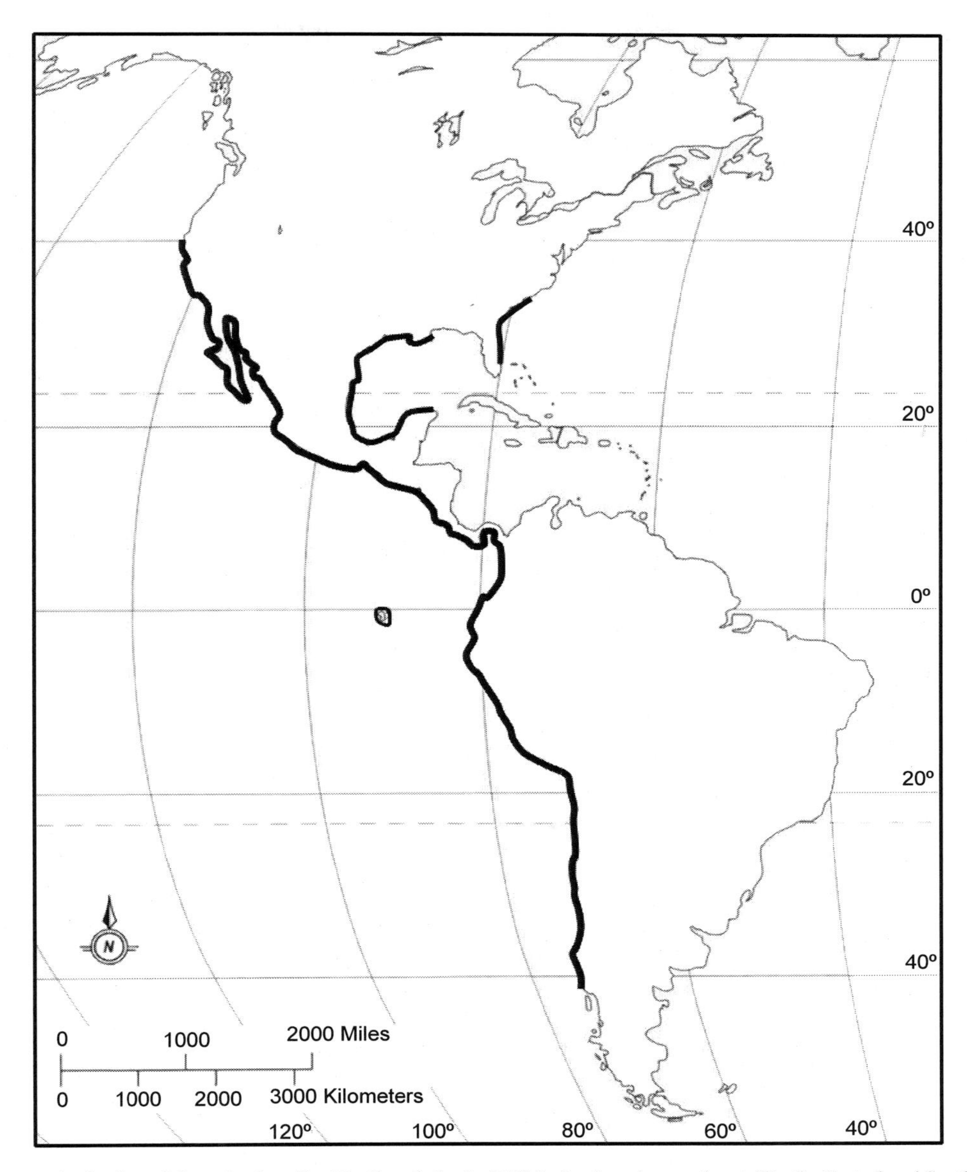

Figure 3.5. Distribution of the striped mullet *Mugil cephalus* L. 1758 in the America continent (North, Central and South).

the North Brazilian Shelf, Tropical Southwestern Atlantic and Warm Temperate Southwestern Atlantic provinces, found in freshwater, brackish and marine coastal waters. According to Eschmeyer (2014) the species is questionably a synonym of *Mugil trichodon*.

Mugil curema Valenciennes 1836—White mullet

The white mullet *Mugil curema* is distributed on both coasts of the Americas (Eschmeyer 2014). In the eastern Pacific, it occurs from Baja to Chile and Galapagos Islands (Cervigón et al. 1993, Harrison 2002, Eschmeyer 2014). The species is also reported for the western Atlantic, from Nova Scotia to northern Argentina, including Bermuda, the Caribbean and Gulf of Mexico (Harrison 1995) (Fig. 3.7). Adults inhabit inshore marine waters and estuaries, and they are not usually found in freshwater environments. Due to

Figure 3.6. Distribution of the *Mugil brevirostris* (Ribeiro 1915) in the America continent (North, Central and South).

the wide distribution of the white mullet, this species occurs in all ecoregions along the coastal waters of the American continent, except the ecoregions of the Cold Temperate North-East Pacific province and Magellanic provinces.

Mugil curvidens Valenciennes 1836—Dwarf mullet

The dwarf mullet *Mugil curvidens* inhabits shallow coastal waters of the western Atlantic from Bermuda, the Bahamas, the Antilles, and south of Rio de Janeiro, Brazil, including Ascension Island in the mid-Atlantic (Cervigón et al. 1993, Harrison 2002, Eschmeyer 2014) (Fig. 3.8). The distribution of the dwarf mullet is restricted to the Tropical Atlantic realm, occurring in the ecoregions of Bahamian, eastern Caribbean and Greater Antilles in the Tropical North-western Atlantic province, in the ecoregions of Guiana and

Figure 3.7. Distribution of the white mullet *Mugil curema* Valenciennes 1836 in the America continent (North, Central and South).

Amazonia in the North Brazil Shelf province, and in the ecoregions of North-eastern Brazil and eastern Brazil in the Tropical South-western Atlantic province. Information about the ecology and habitats of the dwarf mullet is scarce, but the species probably occurs in shallow coastal areas and is caught with other mullets with gillnets and cast nets (Cervigón et al. 1993, Harrison 2002).

Mugil hospes Jordan and Culver 1895—Hospe mullet

The hospe mullet *Mugil hospes* is distributed in the western Atlantic from Belize to Brazil and, in the eastern Pacific, from Mexico to Ecuador, with no records from the Gulf of California (Cervigón et al. 1993, Harrison 2002, Eschmeyer 2014) (Fig. 3.9). The presence of *M. hospes* in the West Indies is not confirmed

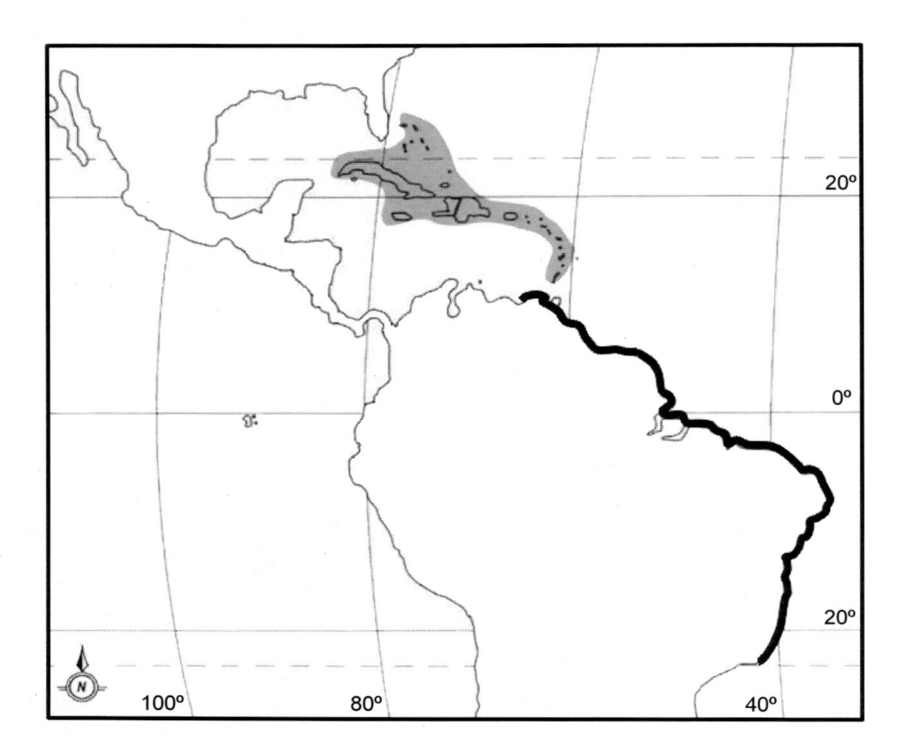

Figure 3.8. Distribution of the dwarf mullet *Mugil curvidens* Valenciennes 1836 in the America continent (North, Central and South).

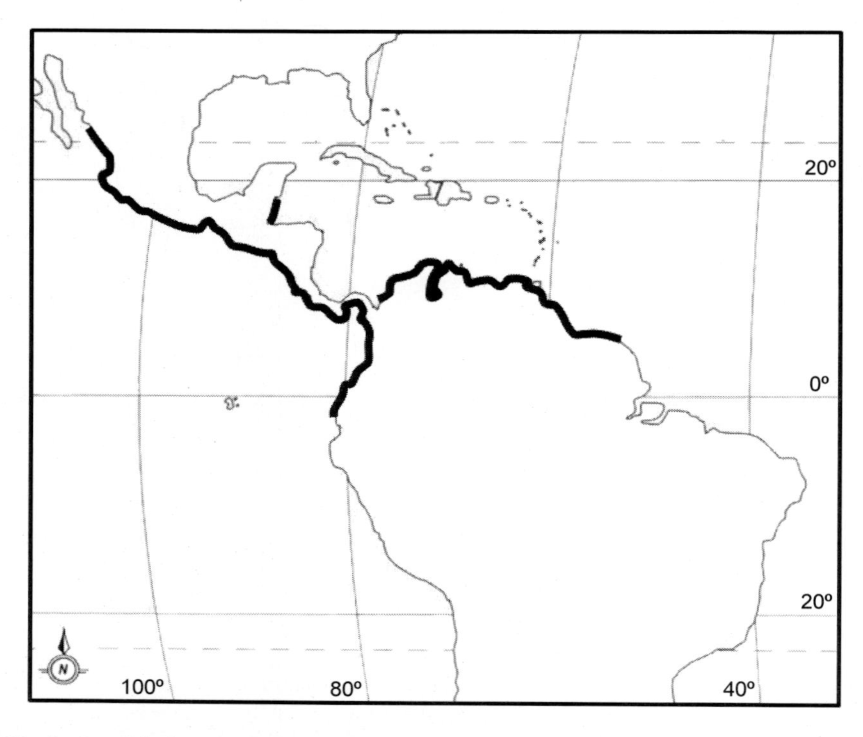

Figure 3.9. Distribution of the hospe mullet *Mugil hospes* Jordan and Culver 1895 in the America continent (North, Central and South).

(Harrison 2002). In addition, Menezes et al. (2015) contest the presence of the hospe mullet in the western South Atlantic based on the comparison of type material of this species with specimens from this area that indicated morphological differences. The distribution of the hospe mullet includes the ecoregions of the Tropical East Pacific province in the Tropical eastern Pacific realm, and in the Tropical Atlantic realm the ecoregions southern Caribbean, South-western Caribbean and western Caribbean in Tropical North-western Atlantic province, and the ecoregion Guianan on the North Brazil Shelf. Adults inhabit inshore marine waters and may enter river mouths, being commonly reported over sand and mud bottoms (Harrison 2002). It has relatively low commercial importance because of its small average size. Hospe mullets are caught occasionally with gillnets and/or beach seines (Cervigón et al. 1993, Harrison 2002).

Mugil incilis Hancock 1830—Parassi mullet

The parassi mullet *Mugil incilis* is distributed in the West Indies and Atlantic Coasts of Central and South America down to South-eastern Brazil, being also reported in the Caribbean for Haiti and Panama (Harrison 2002) (Fig. 3.10). The species is mainly found in brackish estuaries but also in marine and hyper-saline waters. During reproduction, adults gather in small schools at the mouth of coastal rivers and creeks (Cervigón et al. 1993, Harrison 2002). The distribution of the parassi mullet is restricted to the Tropical Atlantic realm. The species occurs in the ecoregions of Greater Antilles, southern Caribbean and South-western Caribbean, in the Tropical North-western Atlantic province; in all ecoregions of the North Brazilian Shelf province and in the Tropical South-western Atlantic province and, finally, in the South-eastern Brazil ecoregion in the Warm Temperate South-western Atlantic province.

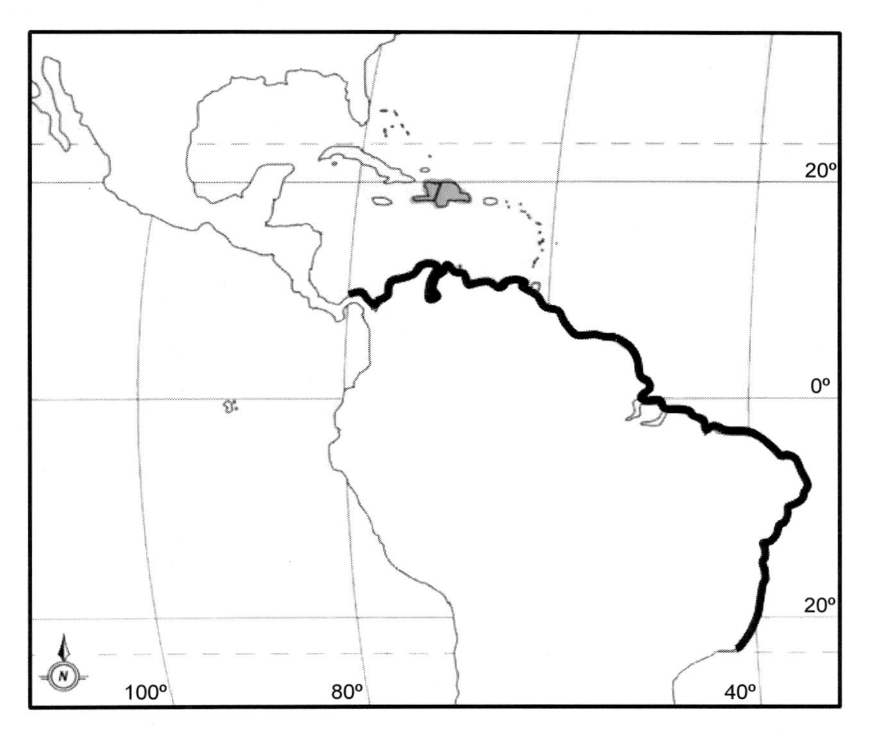

Figure 3.10. Distribution of the parassi mullet *Mugil incilis* Hancock 1830 in the America continent (North, Central and South).

Mugil liza Valenciennes 1836—Lebranche mullet

The lebranche mullet *Mugil liza* is widely distributed in the western Atlantic, in inshore waters around Bermuda and southern Florida, the West Indies, Bahamas, and throughout the Caribbean Sea all the way south to Argentina (Cervigón et al. 1993, Harrison 2002, Eschmeyer 2014) (Fig. 3.11). Adults inhabit inshore marine waters and brackish water lagoons; they may occasionally enter freshwater, but never ascend far, to the upper reaches of rivers (Cervigón et al. 1993, Harrison 2002, Eschmeyer 2014). In the Tropical Atlantic realm, the lebranche mullet occurs in the ecoregions of Bahamian, eastern Caribbean,

Figure 3.11. Distribution of the lebranche mullet *Mugil liza* Valenciennes 1836 in the America continent (North, Central and South).

Greater Antilles, southern Caribbean and South-western Caribbean of the Tropical North-western Atlantic province, and in all ecoregions of the North Brazilian Shelf and Tropical South-western Atlantic provinces. In the Temperate South America realm the species occurs in all ecoregions of the Warm Temperate South-western Atlantic province.

Mugil margaritae Menezes, Nirchio, Oliveira and Siccha-Ramirez 2015

The species *Mugil margaritae* was recently described by Menezes et al. (2015) based on morphological, cytogenetic and molecular data. The species is known from two localities along the Venezuelan coast (Fig. 3.12), in the Tropical Northwestern Atlantic province.

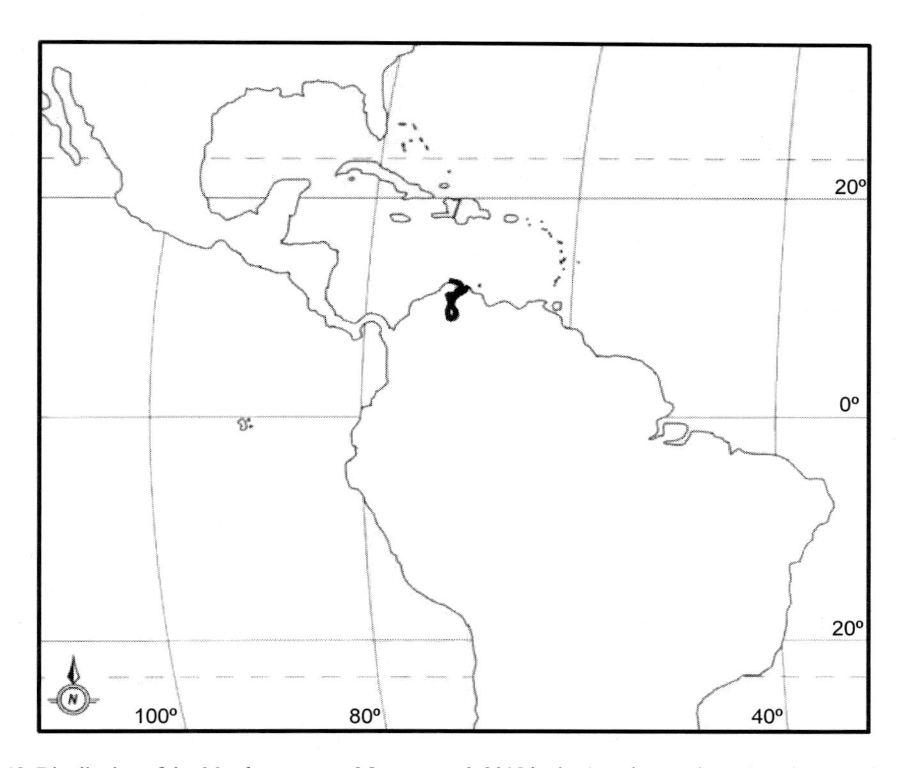

Figure 3.12. Distribution of the *Mugil margaritae* Menezes et al. 2015 in the America continent (North, Central and South).

Mugil rubrioculus Nirchio, Oliveira, Ron and Gaviria 2007—Redeye mullet

The redeye mullet *Mugil rubrioculus* is distributed along the western Atlantic and Caribbean Sea, between 26°57' N and 10°57' N (Harrison et al. 2007) (Fig. 3.13). It inhabits brackish waters over sandy and muddy substrata (Eschmeyer 2014). *Mugil rubrioculus* is restricted to the Tropical Atlantic realm, occurring in the ecoregions of Bahamian, Greater Antilles, southern Caribbean and South-western Caribbean of the Tropical North-western Atlantic province, and in all ecoregions of the North Brazilian Shelf. Some authors have used the name *Mugil gaimardianus* for a species from the western central Atlantic similar to *M. curema*, but distinct from it, with the iris having a reddish colour, longer pectoral fins (extending beyond the origin of the first dorsal fin) and a wider skull at the level of the sphenotics (Cervigón 1993, Menezes 1983). Recently, morphological studies of Harrison et al. (2007) reported that *M. gaimardianus* is not available for the western central Atlantic due to the difficult diagnosis based on the original description by

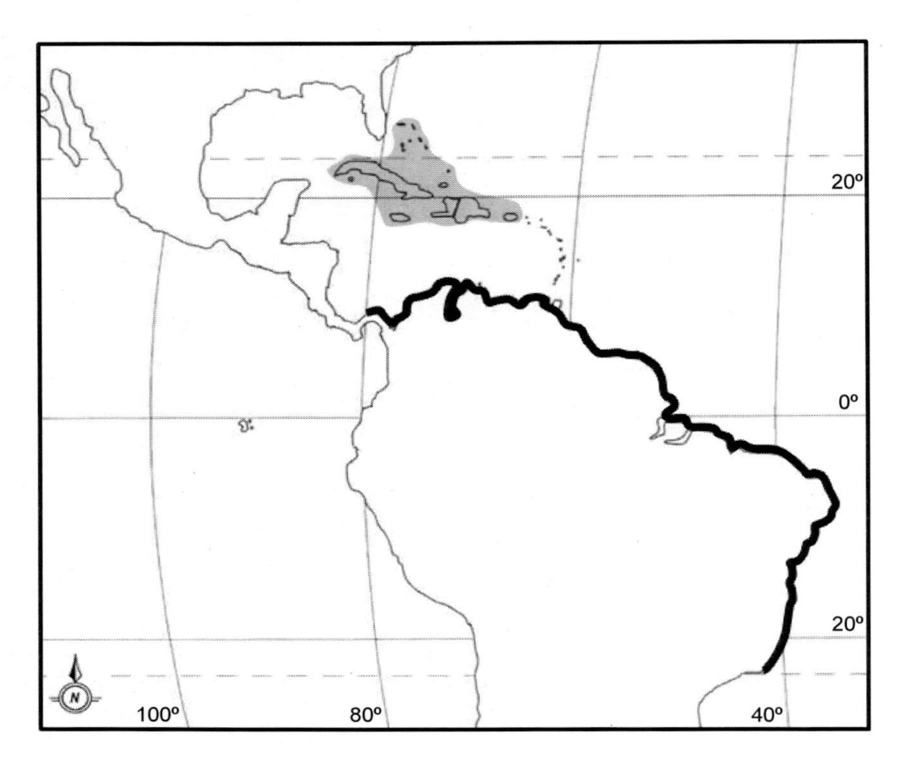

Figure 3.13. Distribution of the redeye mullet *Mugil rubrioculus* Nirchio et al. 2007 in the America continent (North, Central and South).

Desmarest, appearing in Bory de Saint-Vincent (1831). In addition, no type specimens of *M. gaimardianus* are known to exist (Alvarez-Lajonchere et al. 1992). Indeed many descriptions of *Mugil* species in the western central Atlantic appear to be misidentifications of others mullet species (e.g., *Mugil curema*, *M. rubrioculus*) (Harrison et al. 2007). According the authors, the species *M. gaimardianus*, from Panama and Margarita Island, Venezuela (Nirchio et al. 2003), were not conspecific with any other nominal species and the name had been suppressed. A new species description and name has been made by Harrison et al. (2007), with provision of the new name *Mugil rubrioculus*.

Mugil setosus Gilbert 1892—Liseta mullet

The liseta mullet (*Mugil setosus*) is distributed in marine coastal waters of the eastern Pacific in Mexico and the Revillagigedo Islands (Harrison 1995) (Fig. 3.14). The liseta mullet is restricted to the Tropical eastern Pacific realm, occurring in the ecoregions of Revillagigedos and Mexican Tropical Pacific of the Tropical East Pacific province.

Mugil trichodon Poey 1875—Fantail mullet

The fantail mullet *Mugil trichodon* is distributed From Bermuda and southern Florida, through the Gulf of Mexico, the Bahamas and Antilles, and as far south as North-eastern Brazil (Cervigón et al. 1993, Harrison 2002, Eschmeyer 2014) (Fig. 3.15). Adults inhabit inshore marine waters, brackish lagoons and river mouths, perhaps entering freshwaters (Cervigón et al. 1993, Harrison 2002, Eschmeyer and Fong 2014). *Mugil trichodon* occur in all ecoregions of all provinces of the Tropical Atlantic realm. In the Temperate North Atlantic, the species occurs in the northern Gulf of Mexico ecoregion of the Warm Temperate North-

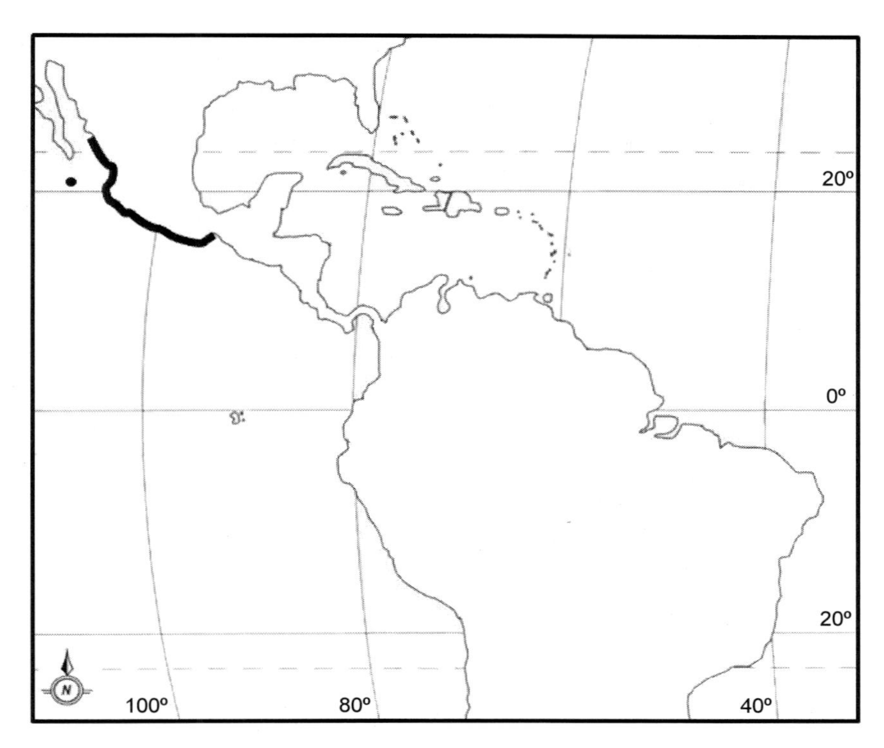

Figure 3.14. Distribution of the liseta mullet *Mugil setosus* Gilbert 1892 in the America continent (North, Central and South).

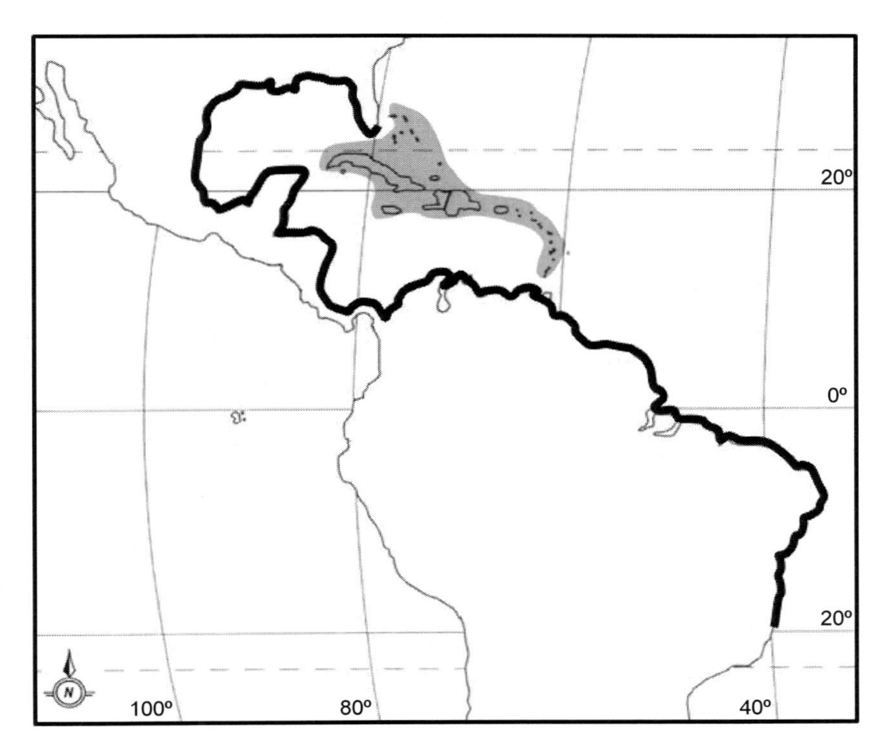

Figure 3.15. Distribution of the fantail mullet *Mugil trichodon* Poey 1875 in the America continent (North, Central and South).

west Atlantic province. According to Menezes et al. (2015) *Mugil trichodon* occurs only in the southern Caribbean area and along the coast of Venezuela and the previous records from Brazilian coasts are based on specimens actually belonging to *Mugil curvidens*.

Xenomugil thoburni (Jordan and Starks 1896)—Thoburn's mullet

The thoburn's mullet (*Xenomugil thoburni*) is distributed in the eastern Pacific, being recorded principally for the Galapagos Islands, but also existing from Guatemala to Panama, and Peru (Harrison 1995, Eschmeyer 2014) (Fig. 3.16). The species is restricted to the Tropical eastern Pacific realm, occurring in all ecoregions of the Galapagos province, and in the ecoregions of Chiapas-Nicaragua and Nicoya, in the Tropical East Pacific province.

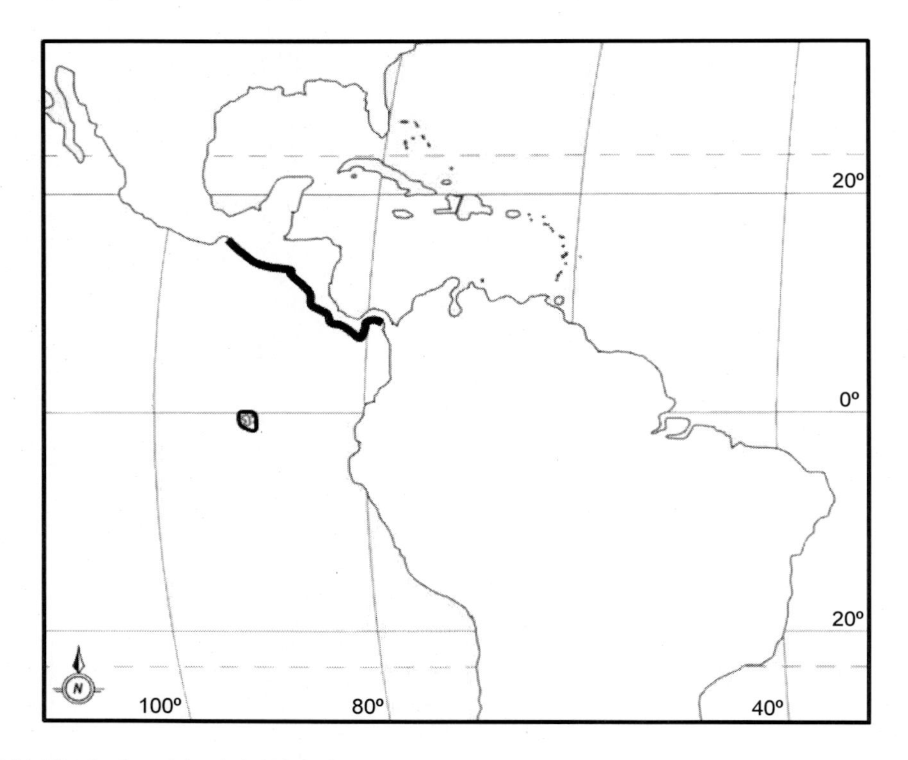

Figure 3.16. Distribution of the thoburn's mullet *Xenomugil thoburni* (Jordan and Starks 1896) in the America continent (North, Central and South).

Conclusions and Perspectives

The species of the family Mugilidae are found in marine inshore, estuaries and freshwater systems, and their abundance and distribution in tropical, subtropical and temperate latitudes of all continents makes them commercially important for fisheries and aquaculture in many regions of the world (Cervigón et al. 1993, Nash and Shehadeh 1980, Nelson 2006). The family Mugilidae includes over 70 species, though its taxonomy is quite complex and still controversial, despite several systematic revisions (see Chapter 1—González-Castro and Ghasemzadeh 2015, Chapter 2—Durand 2015, Chapter 15—Rossi et al. 2015). The distribution and biogeography of these species remain unclear, especially because of the difficulty in separating species based on their morphological characteristics. However, biogeographic characterization is essential for developing ecological and conservation planning for fish. Recent advances in molecular phylogenetic and phylogeographic analysis provide new tools for a better understanding of the distribution and biogeographic patterns of mullets of the world. International partnerships among scientists could provide more information to help understand the evolutionary relationships of Mugilidae species, and to serve as a framework for sustainable management of their fisheries and culture.

References

Alvarez-Lajonchere, L., E. Trewavas and G.J. Howes. 1992. *Mugil curema* and *M. liza* Valenciennes in Cuvier and Valenciennes 1836 (Osteichthyes, Perciformes): proposed conservation of the specific names. Bull. Zool. Nomencl. 49: 271–275.

Bellwood, D.R. and P.C. Wainwright. 2002. The history and biogeography of fishes on coral reefs. pp. 5–32. *In*: P.F. Sale (ed.). Coral Reef Fishes: Dynamics and Diversity in a Complex Ecosystem. Academic Press, Boston, MA, USA.

Briggs, J.C. 1995. Global Biogeography. Developments in Paleontology and Stratigraphy, Vol. 14. Elsevier, Amsterdam. 473p.

Briggs, J.C. and B.W. Bowen. 2012. A realignment of marine biogeographic provinces with particular reference to fish distribution. J. Biogeogr. 39: 12–30.

Bory de Saint-Vincent, J.B.G.M. 1831. Dictionnaire Classique d'Histoire Naturelle, Vol. 17. Paris: Rey & Graviers.

Cervigón, F. 1993. Los peces marinos de Venezuela, 2nd ed. Vol. II. Caracas, Venezuela: Fundación Científica Los Roques 1–499.

Cervigón, F., R. Cipriani, W. Fischer, L. Garibaldi, M. Hendrickx, A.J. Lemus, R. Márquez, J.M. Poutiers, G. Rabaina and B. Rodríguez. 1993. FAO Species Identification Sheets for Fishery Purposes. Field Guide to the Commercial Marine and Brackish-water Resources of the Northern Coast of South America. Rome, FAO. 513p.

Durand, J.D. 2015. Implications of molecular phylogeny for the taxonomy of Mugilidae. *In*: D. Crosetti and S.J.M. Blaber (eds.). Biology, Ecology and Culture of Grey Mullet (Mugilidae). CRC Press, Boca Raton, USA (this book).

Durand, J.-D., K.-N. Shen, W.-J. Chen, B.W. Jamandre, H. Blel, K. Diop, M. Nirchio, F.J. Garcia de León, A.K. Whitfield, C.-W. Chang, and P. Borsa. 2012. Systematic of the grey mullets (Teleostei: Mugiliformes: Mugilidae): Molecular phylogenetic evidence challenges two centuries of morphology-based taxonomy. Mol. Phylogenet. Evol. 64: 73–92.

Durand, J.D., H. Blel, K.N. Shen, E.T. Koutrakis and B. Guinand. 2013. Population genetic structure of *Mugil cephalus* in the Mediterranean and Black Seas: a single mitochondrial clade and many nuclear barriers. Mar. Ecol. Prog. Ser. 474: 243–261.

Eschmeyer, W.N. (ed.). 2014. Catalog of Fishes: Genera, Species, References. (http://researcharchive.calacademy.org/research/ichthyology/catalog/fishcatmain.asp). Electronic version accessed 20.04.2014.

Eschmeyer, W.N. and J.D. Fong (eds.). 2014. Catalog of Fishes: Species by Family/Subfamily. (http://researcharchive.calacademy.org/research/ichthyology/catalog/SpeciesByFamily.asp). Electronic version accessed 20.04.2014.

Floeter, S.R. and A. Gasparini. 2000. The south-western Atlantic reef fish fauna: composition and zoogeographic patterns. J. Fish Biol. 56: 1099–1114.

Fraga, E., H. Schneider, M. Nirchio, E. Santa-Brigida, L.F. Rodrigues-Filho and I. Sampaio. 2007. Molecular phylogenetic analyses of mullets (Mugilidae: Mugiliformes) based on two mitochondrial genes. J. Appl. Ichthyol. 23: 598–604.

González-Castro, M. and J. Ghasemzadeh. 2015. Morphology and morphometry based taxonomy of Mugilidae. *In*: D. Crosetti and S.J.M. Blaber (eds.). Biology, Ecology and Culture of Grey Mullet (Mugilidae). CRC Press, Boca Raton, USA (this book).

Grant, W.S. and R.W. Leslie. 2001. Inter-ocean dispersal is an important mechanism in the zoogeography of hakes (Pisces: *Merluccius* spp.). J. Biogeogr. 28: 699–721.

Harrison, I.J. 1995. Mugilidae. pp. 1293–1298. *In*: W. Fischer, F. Krupp, W. Schneider, C. Sommer, K.E. Carpenter and V.H. Niem (eds.). Guia FAO para la Identificación de Especies para los Fines de la Pesca. Pacifico Centro-Oriental, Vol. III. FAO, Roma.

Harrison, I.J. 2002. Mugilidae. pp. 1071–1085. *In*: K. Carpenter (ed.). FAO Species Identification Guide for Fishery Purposes. The Living Marine Resources of the Western Central Atlantic, Vol. 2. Bony Fishes Part 1 (Acipenseridae to Grammatidae). (Rome, FAO).

Harrison, I.J., M. Nirchio, C. Oliveira, E. Ron and J. Gaviria. 2007. A new species of mullet (Teleostei: Mugilidae) from Venezuela, with a discussion on taxonomy of *Mugil gaimardianus*. J. Fish Biol. 71: 76–97.

Kohn, A. and S.C. Cohen. 1998. South American Monogenea—list of species, hosts and geographical distribution. International J. Parasitol. 28: 1517–1554.

Lessios, H.A. 2008. The Great American Schism: divergence of marine organisms after the rise of the Central American Isthmus. Annu. Rev. Ecol. Evol. Syst. 39: 63–91.

Livi, S., L. Sola and D. Crosetti. 2011. Phylogeographic relationships among worldwide populations of the cosmopolitan marine species, the striped grey mullet (*Mugil cephalus*), investigated by partial cytochrome b gene sequences. Biochem. Syst. Ecol. 39: 121–131.

McMahan, C.D., P.D. Matthew, O. Domínguez-Domínguez, F.J. García-de-León, I. Doadrio and K.R. Piller. 2013. From the mountains to the sea: phylogeography and cryptic diversity within the mountain mullet, *Agonostomus monticola* (Teleostei: Mugilidae). J. Biogeogr. 40: 894–904.

Menezes, N.A. 1983. Guia prático para conhecimento e identificação das tainhas e paratis (Piscies: Mugilidae) do litoral brasileiro. Rev. Bras. Zool. 2: 1–12.

Menezes, N.A., C. Oliveira and M. Nirchio. 2010. An old taxonomic dilemma: the identity of the western South Atlantic lebranche mullet (Teleostei: Perciformes: Mugilidae). Zootaxa 2519: 59–68.

Menezes, N.A., M. Nirchio, C. Oliveira and R. Siccha-Ramirez. 2015. Taxonomic review of the species of *Mugil* (Teleostei: Perciformes: Mugilidae) from the Atlantic South Caribbean and South America, with integration of morphological, cytogenetic and molecular data. Zootaxa 3918: 001–038.

Nash, C.E. and Z.H. Shehadeh. 1980. Review of breeding and propagation techniques of grey mullet, *Mugil cephalus*. ICLARM Stud. Rev. 3: 1–87.

Nelson, J.S. 2006. Fishes of the World (4th edition). John Wiley and Sons, Inc., New York, NY. 601p.

Nirchio, M., F. Cervigón, J.I.R. Porto, J.E. Pérez, J.A. Gómez and J. Villalaz. 2003. Karyotype supporting *Mugil curema* Valenciennes, 1836 and *Mugil gaimardianus* Desmarest, 1831 (Mugilidae: Teleostei) as two valid nominal species. Sci. Mar. 67: 113–115.

Nirchio, M., E. Ron and A.R. Rossi. 2005. Karyological characterization of *Mugil trichodon* Poey 1876 (Pisces: Mugilidae). Sci. Mar. 69: 525–530.

Nirchio, M., C. Oliveira, I.A. Ferreira, J.E. Pérez, J.I. Gaviria, I. Harrison, A.R. Rossi and L. Sola. 2007. Comparative cytogenetic and allozyme analysis of *Mugil rubrioculus* and *M. curema* (Teleostei: Mugilidae) from Venezuela. Interciencia 32: 757–762.

Nirchio, M., C. Oliveira, I.A. Ferreira, C. Martins, A.R. Rossi and L. Sola. 2009. Classical and molecular cytogenetic characterization of *Agonostomus monticola*, a primitive species of Mugilidae (Mugiliformes). Genetica 135: 1–5.

Rodrigues-Filho, L.F.S., D.B. Cunha, M. Vallinoto, H. Schneider, I. Sampaio and E. Fraga. 2011. Polymerase chain reaction banding patterns of the 5S rDNA gene as a diagnostic tool for the discrimination of South American mullets of the genus *Mugil*. Aquac. Res. 42: 1117–1122.

Rossi, A.R., D. Crosetti, E. Gornung and L. Sola. 1996. Cytogenetic analysis of global populations of *Mugil cephalus* (striped mullet) by different staining techniques and fluorescent *in situ* hybridization. Heredity 76: 77–82.

Rossi, A.R., E. Gornung, L. Sola and M. Nirchio. 2005. Comparative molecular cytogenetic analysis of two congeneric species, *Mugil curema* and *M. liza* (Pisces: Mugiliformes), characterized by significant karyotype diversity. Genetica 125: 27–32.

Rossi, A.R., S. Livi and D. Crosetti. 2015. Genetics of Mugilidae. pp. 349–397. *In*: D. Crosetti and S.J.M. Blaber (eds). Biology, Ecology and Culture of Grey Mullet (Mugilidae). CRC Press, Boca Raton, USA (this book).

Schultz, L.P. 1946. A revision of the genera of mullets, fishes of the family Mugilidae, with description of three new genera. Proc. U.S. Natl. Mus. 96: 377–395.

Senou, H. 1988. Phylogenetic interrelationships of the Mullets (Pisces: Mugilidae). Ph.D. Dissertation. Tokyo University, Tokyo.

Shen, K.N., B.W. Jamandre, C.C. Hsu, W.N. Tzeng and J.D. Durand. 2011. Plio-Pleistocene sea level and temperature fluctuations in the northwestern Pacific promoted speciation in the globally-distributed flathead mullet *Mugil cephalus*. BMC Evol. Biol. 11: 83.

Siccha-Ramirez, R., N.A. Menezes, M. Nirchio, F. Foresti and C. Oliveira. 2014. Molecular identification of mullet species of the Atlantic South Caribbean and South America and the phylogeographic analysis of *Mugil liza*. Rev. Fish. Sci. Aquac. 22: 86–96.

Sola, L., M. Nirchio and A.R. Rossi. 2008. The past and the future of cytogenetics of Mugilidae: updated overview. Bol. Inst. Oceanogr. Venez. 47: 27–33.

Spalding, M.D., H.E. Fox, G.R. Allen, N. Davidson, Z.A. Ferdana, M. Finlayson, B.S. Halpern, M.A. Jorge, A. Lombana, S.A. Lourie, K.D. Martin, E. McManus, J. Molnar, C.A. Recchia and J. Robertson. 2007. Marine ecoregions of the world: a bioregionalization of coastal and shelf areas. BioScience 57: 573–583.

Thomson, J.M. 1954. The Mugilidae of Australia and adjacent seas. Aust. J. Mar. Fresh. Res. 5: 70–131.

Thomson, J.M. 1997. The Mugilidae of the world. Mem. Queensl. Mus. 41: 457–462.

Tringali, M.D., T.M. Bert, S. Seyoum, E. Bermingham and D. Bartolacci. 1999. Molecular phylogenetics and ecological diversification of the transisthmian fish genus *Centropomus* (Perciformes: Centropomidae). Mol. Phylogenet. Evol. 13: 1993–207.

CHAPTER 4

The Biogeography of Mugilidae in India, South-East and East Asia

Kang-Ning Shen[1,*] and *Jean-Dominique Durand*[2]

Introduction

India, South-East and East Asia harbour a high marine biodiversity, with a hotspot located between the Indo-Malayan and Philippines Archipelagos (IMPA). The origin of this biodiversity pattern is still debated, several hypotheses being advanced to explain this observation (Bellwood and Meyer 2009, Carpenter et al. 2011 and references therein). Among these hypotheses there are elevated local speciation rates (the centre of the origin hypothesis), a greater accumulation of species formed elsewhere (the centre of accumulation hypothesis), presence of refugia (the centre of endemism hypothesis) and the overlap of a distinct biogeographic ichthyofauna (the centre of overlap hypothesis) (Bellwood and Wainwright 2002, Mora et al. 2003, Barber et al. 2006, Bellwood and Meyer 2009, Carpenter et al. 2011, Hubert et al. 2012). In this context it is of prime importance to precisely depict the species diversity and its geographic distribution, since the quality of any biogeographic inferences are largely dependent on this knowledge. Unfortunately, recent phylogeographic molecular taxonomic investigations have shown that biodiversity is frequently underestimated and consequently species geographic distribution poorly evaluated (Zemlak et al. 2009, Carpenter et al. 2011, Hubert et al. 2012). This is especially true for species of the Mugilidae family since their conservative morphology and the paucity of useful taxonomic characters make taxonomy, biogeography and biological research challenging (Durand et al. 2012b). Recent genetic investigations (for a review see Chapter 2—Durand 2015 and Chapter 15—Rossi et al. 2015) may help solve questions of taxonomy and phylogenetics, and hence the biogeography of Mugilidae. This chapter reviews the diversity of Mugilidae species and their biogeography in India, South-East and East Asia.

Biogeographic Characteristics of India, South-East and East Asia

There is a biogeographic classification for the world's coastal and shelf areas (Marine Ecoregions of the World), which is of critical importance in supporting analyses of patterns in marine biodiversity (Spalding et al. 2007). Within these 12 marine realms, the Central Indo-Pacific realm is thought to have the highest

[1] Center of Excellence for the Oceans, National Taiwan Ocean University, Taiwan, Republic of China.
Email: knshen@ntou.edu.tw

[2] IRD, Center for Marine Biodiversity, Exploitation and Conservation, Université Montpellier 2, CC 093, Place Eugène Bataillon, 34095 Montpellier Cedex 5, France.
Email: jean-dominique.durand@ird.fr

* Corresponding author

marine biodiversity, with a hotspot located between the Indo-Malayan and Philippines Archipelagos (IMPA). Allen (2002) considered eastern Indonesia–southern Philippines as the most likely diversity centre. Kulbicki et al. (2004) also refer to a centre of fish diversity, which they call 'the Philippines–South China Sea–Indonesia triangle'. In contrast, Carpenter and Springer (2005) indicate two peaks for marine shore fishes that are more to the north (Central Philippines) and west (Malacca Strait). The northernmost boundary of the diversity centre of shore fish indicated by Allen (2002) is concordant with the boundary between the Indo-Malayan tropical sub-region and the Sino-Japanese subtropical sub-region, based on the 20°C winter surface isotherm (Zhang and Zhang 1986). Although these terminologies for hypothetical centres of marine diversity are different among authors, all evidence suggests that they are within the IMPA. In addition, the so-called plate endemics have received special attention in fish biogeography (Springer 1982, Myers 1989, Springer and Williams 1990) and species ranges may be confined to particular tectonic plates. When these plates move to or from each other, so do their faunas, which may cause them to join with each other or to split from each other.

India, South-East and East Asia lie on intersection of four tectonic plates (the Indo-Australian, Sunda, Eurasian, and the Philippines Sea plates) that experienced important volcanic activities from Cenozoic (Rangin et al. 1990, Hall 1996). The Indo-Australian Plate is the cause of many earthquakes and volcanism because it plunges beneath the edge of the Eurasian Plate. The Philippine Plate has also caused a lot of earthquakes and volcanic activity in the Philippines, because it is also pushing into the Eurasian Plate. Therefore, many of the islands in East and South-East Asia were raised from the tectonic activities and volcanism which provided habitat diversity for marine life. This historical geographic process associated with the contemporary features (the present day sea surface temperature and currents) is one of the most important factors in determining the current distribution of species.

During the Pleistocene, the alternation of glacials and interglacials caused large scale eustatic sea-level movements, which had a remarkable effect on the shore lines and coral reefs of the Indo-Malayan region, especially on and along the continental shelves (Chappell and Thom 1977, Voris 2000). Due to the effects of changing ice volume and distributions, the sea-level fluctuated worldwide through a range of 100–120 m, with vertical changes of up to 10–20 m per 1000 years (Chappell 1983, Masse and Montaggioni 2001, Siddall et al. 2003), and horizontal shoreline migrations across the broad shelves far exceeding 100 m per 10 years (Chappell and Thom 1977). Large shelf areas of the IMPA were exposed and therefore experienced a series of local marine extinctions during Pleistocene sea-level regressions (Springer and Williams 1990, Voris 2000). South-East Asian peninsulars and islands were above sea level during the recurring ice ages of the Pleistocene epoch, during which much of the world's water was frozen in glaciers. The exposed shelf, known as Sundaland, allowed ancient people and Asian animals to travel south to what are now the islands of South-East Asia. For marine fishes, the rise of Sundaland became a barrier for dispersal. During ice ages, five main refuges formed in this area (South China Sea, Celebes Sea, Sulu Sea, Banda Sea and Andaman Sea), which probably influenced the present day distribution and diversity of fishes (Fig. 4.1). Northward, sea level regression also deeply impacted the marine connectivity, as several marginal seas were isolated (Liu et al. 2007, Wang et al. 2008, Shen et al. 2011).

In addition to past geological events able to explain the present day marine species richness and distribution in India, South-East and East Asia, the high habitat diversity and distribution is an important factor to consider (Randall 1998). Mangroves and coral reef areas are two important coastal habitats in South-East Asia. Biogeographically, they show similar distributions that are latitudinally determined by the 20°C winter isotherm (McCoy and Heck 1976, Hogarth 2001). Mangroves are trees and shrubs that grow in the intertidal zones of estuaries, coastlines and islands and are found along the coasts of Africa, Asia, the Americas and Australia, mostly in tropical zones, but with the highest diversity in South-East Asia (Giri et al. 2011). Mangroves provide a habitat for many estuarine dependent species and Mugilidae is a typical example. Mangroves provide nursery habitat for many species, including commercial fish and crustaceans, and thus contribute to sustaining the local abundance of fish and shellfish populations. In Selangor, Malaysia and the Philippines, 119 and 128 species, respectively, were recorded as associated with mangrove ecosystems (Chong et al. 1990). Mangrove ecosystems in Papua New Guinea and the Solomon Islands have been found to provide important nurseries for sandy and muddy-bottom demersal

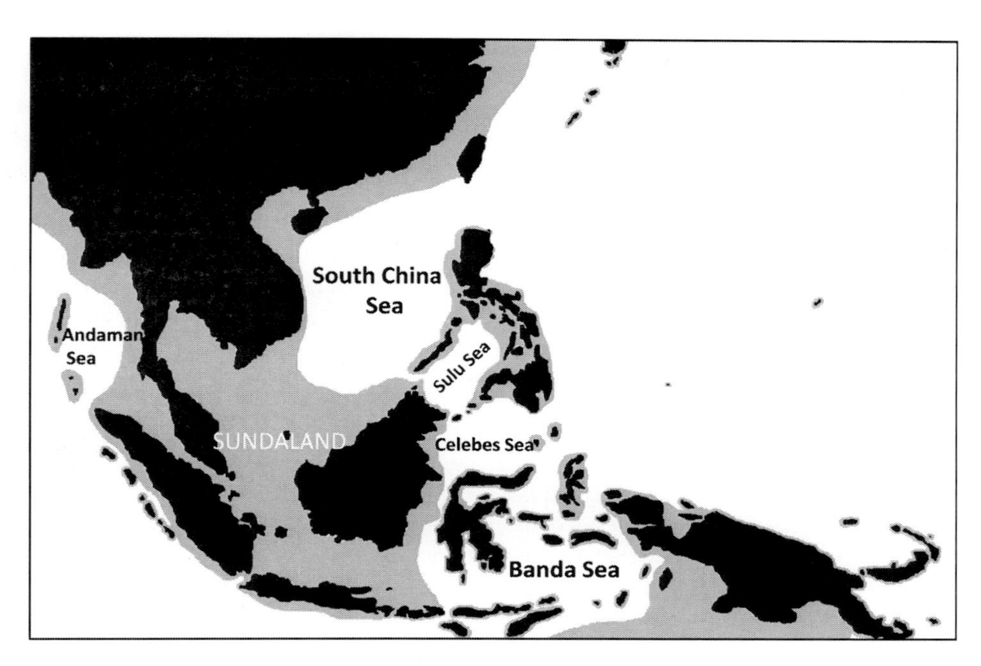

Figure 4.1. The Indo-Malay-Philippine Archipelago with the area shallower than the 120 m depth contour delineated in light grey. This area estimates the sea bottom that would have been exposed during Pleistocene sea-level lows. The shelf exposed during ice ages formed Sundaland in Southeast Asia.

and surface feeding species (Blaber and Milton 1990). The subtropical and tropical mangrove creeks of Taiwan also support high abundances of small fishes, especially Gobiidae, Mugilidae, Leiognathidae, and Cichlidaes (Kuo et al. 1999).

The Indonesian/Philippines archipelago also has the world's greatest concentration of reefs and the greatest coral diversity. Coral reefs are estimated to cover 284,300 km^2, just under 0.1% of the oceans' surface area. The Indo-Pacific region (including the Red Sea, Indian Ocean, South-East Asia and the Pacific) accounts for 91.9% of this amount (Spalding et al. 2001). The reef is topographically complex. Much like a rain forest, it has many strata and areas of strong shade, cast by the tall coral colonies. Because of the complexity, thousands of species of fish and invertebrates live in the reefs, which are by far the richest of marine habitats. There are several species of Mugilidae that are largely connected with coral reef habitats, for example, bluespot mullet *Moolgarda (Crenimugil) seheli*, bluetail mullet *Moolgarda (Crenimugil) buchanani*, fringelip mullet *Crenimugil crenilabis* and squaretail mullet *Ellochelon vaigiensis* ssl. All these historical and present environment characteristics have contributed to the high diversity of Mugilidae in South-East Asia.

The Diversity and Taxonomy of Mugilidae in India, South-East and East Asia

The Mugilidae, commonly referred to grey mullets, is a family of Teleostean fishes, class Actinopterygii, order Mugiliformes (Eschmeyer and Fong 2014). It includes many species which inhabit coastal, brackish and fresh waters of all tropical and temperate regions of the world. The family was previously included in the order Perciformes and is now considered as the only representative of Mugiliformes (see Nelson 1984). Within the family, several taxonomic revisions have been made, raised by the conservative morphology and the paucity of useful taxonomic characters (Nelson 1984, Corti and Crosetti 1996). Until recently the taxonomy of Mugilidae was based on morphological and morphometric characters, including fin rays and scale numbers, and various osteological characters especially in the mouth (Schultz 1946, Harrison and Howes 1991, Thomson 1997). Unfortunately, interspecific variability of meristic characters usually overlapped, while some anatomical characters considered to be of taxonomic value underwent marked changes with growth. As a consequence species determination keys are difficult to use and can easily lead to the misidentification of many species of grey mullet. This is particularly true in the Indo-Pacific area and more precisely in South-East Asia where a large part of the Mugilidae diversity occurs. In this context, molecular studies provide a powerful alternative to morpho-anatomic characters, able to highlight both phylogenetic relationships among taxa and species diversity. Recently, Durand et al. (2012b) proposed a broad taxonomic revision of the Mugilidae considering the molecular phylogeny of the family estimated from the sequence polymorphisms of three mitochondrial genes (Durand et al. 2012a). Even if no new genus was created in the East Indo-West Pacific area (Durand et al. 2012b), numerous putative species corresponding to deeply divergent mitochondrial lineages were flagged (Durand and Borsa in press). Considering both, species checklists available and some molecular data, a general review of the Mugilidae diversity of the East Indo-West Pacific is proposed hereafter.

Molecular data considered consists of Cytochrome C Oxidase subunit I (COI) sequences published by Durand et al. (2012a,b) and those available in Genbank, usually produced owing to DNA barcoding projects, part of the Fish Barcode of Life Initiative (FISH-BOL) (Ward et al. 2009). If the variability of the partial COI sequence is not enough to investigate the phylogenetics relationships among Mugilidae genus and species (Durand et al. 2012a) it is adequate to highlight mitochondrial lineages corresponding to either valid or putative species. The first result of the phylogeny built confirms the difficulty of identifying Mugilidae species on the basis of their morpho-anatomy since several sequences are probably mislabelled considering their placement in the phylogenetic trees (Figs. 4.2–4.4). For example, several sequences of Indian specimens, present label names that do not match the genus name of sequences belonging to the same clade or the species name of sequences sharing the same lineage (Figs. 4.2–4.4). This result clearly stresses the limit of the DNA barcoding approach, when the initial and fundamental step of species identification is somewhat neglected. To circumvent this difficulty we considered Durand et al. (2012a) sequences as reference as DNA sequences were produced in a phylogenetic context from a number of voucher specimens conserved in museums.

According to Eschmeyer (2014), a total of 31 species and 10 genera are present in the area. However recent genetic investigation demonstrated that some of the genera should be abandoned or resurrected as the species diversity would be largely underestimated (Durand and Borsa in press).

The nomenclature proposed hereafter followed Eschmeyer (2014)'s catalogue of fish, with in parentheses genera nomenclature proposed by Durand et al. (2012b) when different (Table 4.1 and Figs. 4.2–4.4). All molecular arguments as well as morpho-anatomically traits justifying the Durand et al. (2012b)'s nomenclature is described in Chapter 2 of this book (Durand 2015).

Genus Cestraeus *Valenciennes, 1836*

Goldie river mullet ***Cestraeus goldiei*** Macleay, 1883

Only one COI barcode is available for this species, but it represents an independent lineage close to *Cestraeus oxyrhynchus*, which is in agreement with the current taxonomy (Fig. 4.2).

Distribution: Indo-West Pacific from the Philippines (Cagayan) to New Caledonia, New Guinea (Fig. 4.5a, Table 4.1).

Habitat: Fresh and brackish waters, adults stay in rivers, mature adults probably spawn in brackish or marine waters.

Sharp-nosed river mullet ***Cestraeus oxyrhynchus*** Valenciennes, 1836

Only one COI barcode is available for this species, but it represents an independent lineage close to *Cestraeus goldei*, which is in agreement with the current taxonomy (Fig. 4.2).

Distribution: Indo-West Pacific from Indonesia to Fiji, north to the Philippines (Cagayan), south to Caledonia (Fig. 4.5b, Table 4.1).

Habitat: Found in both fresh and brackish waters, possibly ascending some way up rivers.

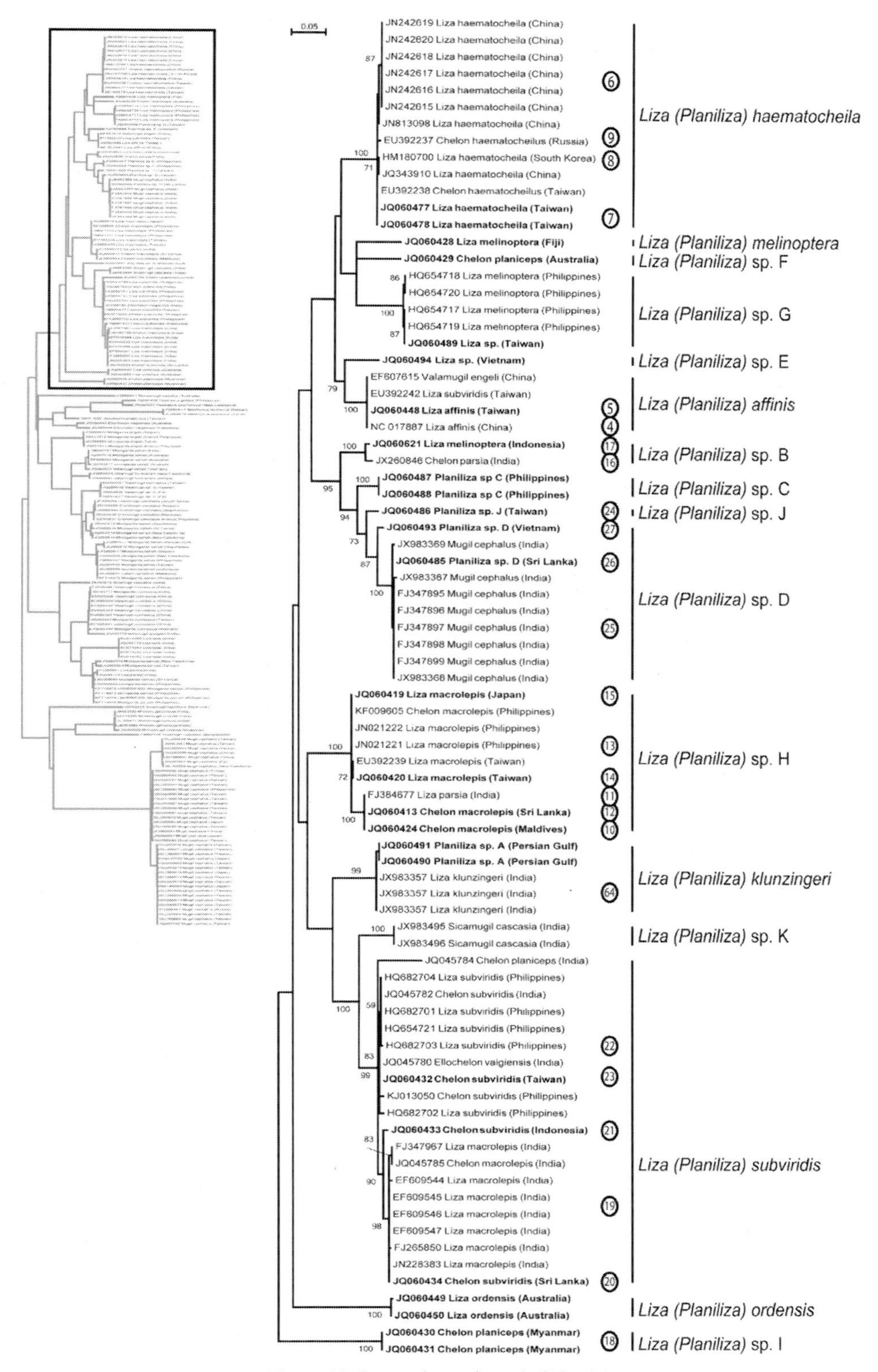

Figure 4.2. See caption at the end of Fig. 4.4.

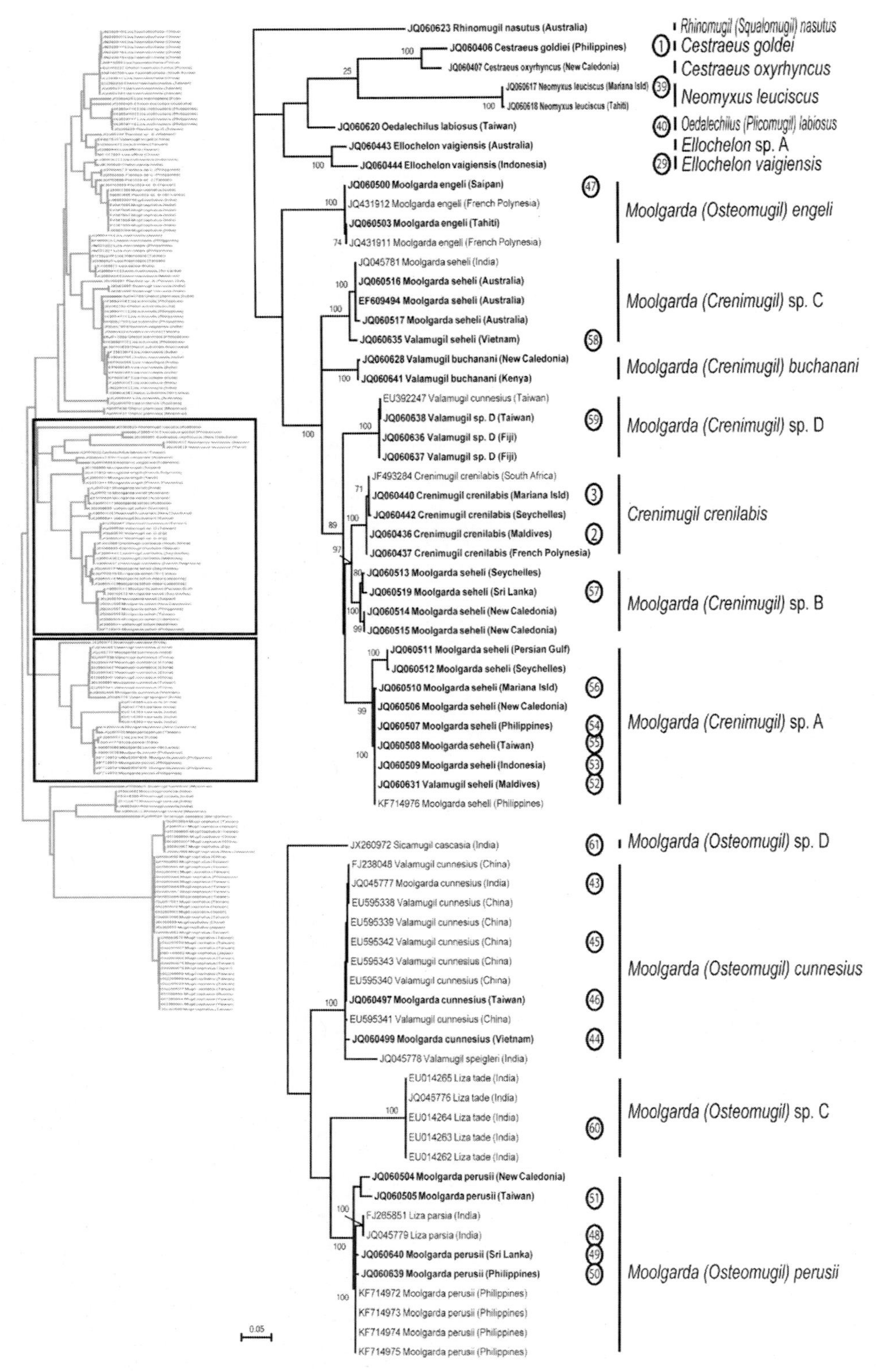

Figure 4.3. See caption at the end of Fig. 4.4.

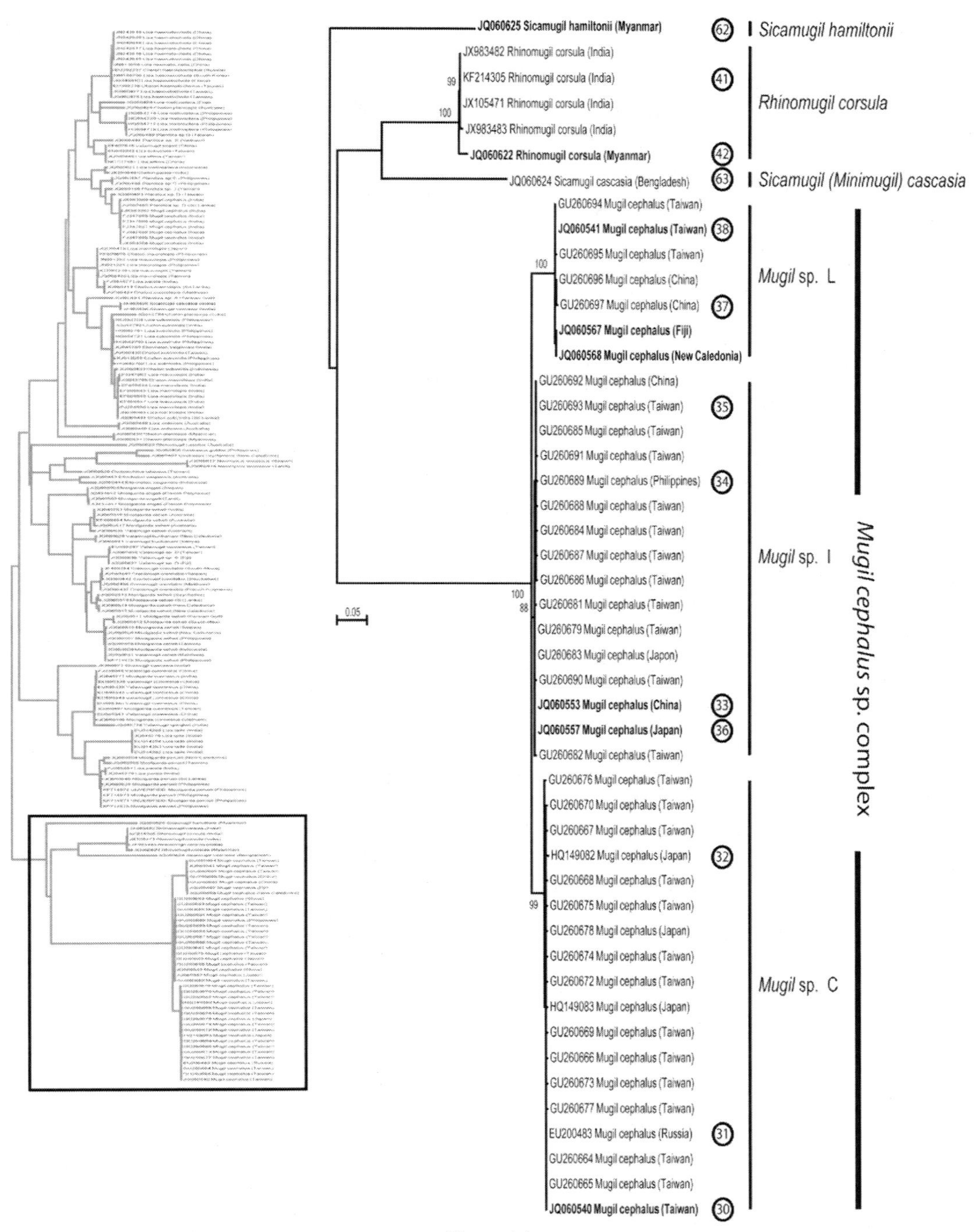

Figure 4.4

Figure 4.2–4.4. Maximum Likelihood phylogenetic tree of Mugilidae in the Indo-Pacific area. The NCBI accession no. and collected location are shown in the phylogenetic tree, in bold those published in Durand et al. (2012a). The nomenclature adopted followed Eschmeyer (2014)'s catalogue of fishes, and in parentheses, when different, the genus. The circled numbers at the tips of the tree correspond to sequence mentioned in the Table 4.1 and used to determine valid or putative species distribution range.

Table 4.1. Mugilidae species present in India, South-East and East Asia. Grey cells indicate the presence of the species while the number refers to the COI sequence used to build the phylogenetic trees (Figs. 4.2–4.4). "?" indicates that there are no DNA sequences available and thus that the taxonomic status was not evaluated in Durand et al. (2012b) *: putative species highlighted in phylogenetic trees presented in Figs. 4.2 to 4.4. A: Jones et al. (1981), B: Fischer and Whitehead (1974), C: Fischer and Bianchi (1984), D: Carpenter and Niem (1999), E: Kwun et al. (2013), F: Froese and Pauly (2014), G: Masuda et al. (1984), H: Myers and Donaldson (2003).

Eschmeyer 2014	Durand et al. 2012b Durand and Borsa 2015	Maldives	India	Sri Lanka	Bengladesh	Myanmar	Thailand	Cambodia	Vietnam	Malaysia	Singapore	Indonesia	Brunei	China	Philippines	Taiwan	Papua New Guinea	South Korea	North Korea	Russia	Japan	Mariana Islands
Cestraeus goldiei Macleay 1883	*Cestraeus goldiei*														1							
Cestraeus oxyrhynchus Valenciennes 1836	*Cestraeus oxyrhynchus*																					
Cestraeus plicatilis Valenciennes 1836	*Cestraeus plicatilis*																					
Crenimugil crenilabis Forsskål 1775	*Crenimugil crenilabis*	2																				3
Crenimugil heterocheilos Bleeker 1855	?																					
Liza affinis Günther 1861	*Planiliza affinis*													4		5						
Liza alata Steindachner 1892	*Planiliza ordensis*																					
Liza haematocheila Temminck and Schlegel 1845	*Planiliza haematocheila*													6		7		8		9		
Liza klunzingeri Day 1888	*Planiliza klunzingeri*		64																			
Liza macrolepis Smith 1846	*Planiliza* sp. H	10	11	12											13	14					15	
Liza mandapamensis Thomson 1997	?																					
Liza melinoptera Valenciennes 1836	*Planiliza melinoptera*																					
	Planiliza sp. B		16									17										
Liza parmatus Cantor 1849	?																					
Liza parsia Hamilton and Buchanan 1822	?																					
Liza tade Forsskål 1775	*Planiliza* sp. I					18																
	Planiliza sp. F																					
Liza subviridis Valenciennes 1836	*Planiliza subviridis*		19	20								21			22	23						
Liza sp.	*Planiliza* sp. J															24						
	Planiliza sp. D		25	26					27													
	Planiliza sp. K*		28																			
Ellochelon vaigiensis Quoy and Gaimard 1825	*Ellochelon vaigiensis*											29										
	Ellochelon sp. A																					
Mugil broussonnetii Valenciennes 1836	?																					
Mugil cephalus Linnaeus 1758	*Mugil* sp. C															30				31	32	
	Mugil sp. I													33	34	35					36	
	Mugil sp. L													37		38						
Neomyxus leuciscus Günther 1871	*Neomyxus leuciscus*																					39

Oedalechilus labiosus Valenciennes, 1836	*Plicomugil labiosus*															40						
Rhinomugil corsula Hamilton 1822	*Rhinomugil corsula*		41			42																
Rhinomugil nasutus DeVis, 1883	*Squalomugil nasutus*																					
Moolgarda buchanani Bleeker, 1853	*Crenimugil buchanani*																					
Moolgarda cunnesius Valenciennes, 1836	*Osteomugil cunnesius*		43						44					45		46						
Moolgarda engeli Bleeker, 1859	*Osteomugil engeli*																					47
Moolgarda perusii Valenciennes, 1836	*Osteomugil perusii*		48	49											50	51						
Moolgarda seheli Forsskål 1775	*Crenimugil* sp. A	52										53			54	55						56
	Crenimugil sp. B			57																		
	Crenimugil sp. C								58													
Moolgarda speigleri Bleeker 1859	?																					
Moolgarda sp.	*Crenimugil* sp. D															59						
	Osteomugil sp. C*		60																			
	Osteomugil sp. D*		61																			
Sicamugil hamiltonii Day 1869	*Sicamugil hamiltonii*					62																
Sicamugil cascasia Hamilton 1822	*Minimugil cascasia*				63																	
Ref.		A	B, C	B, C	B, C	D	D	D	D	D	D	D	D	D	D	D	D	E	F	F	G	H

*: putative species highlighted in phylogenetic trees presented in Fig. 4.1 & 4.2.

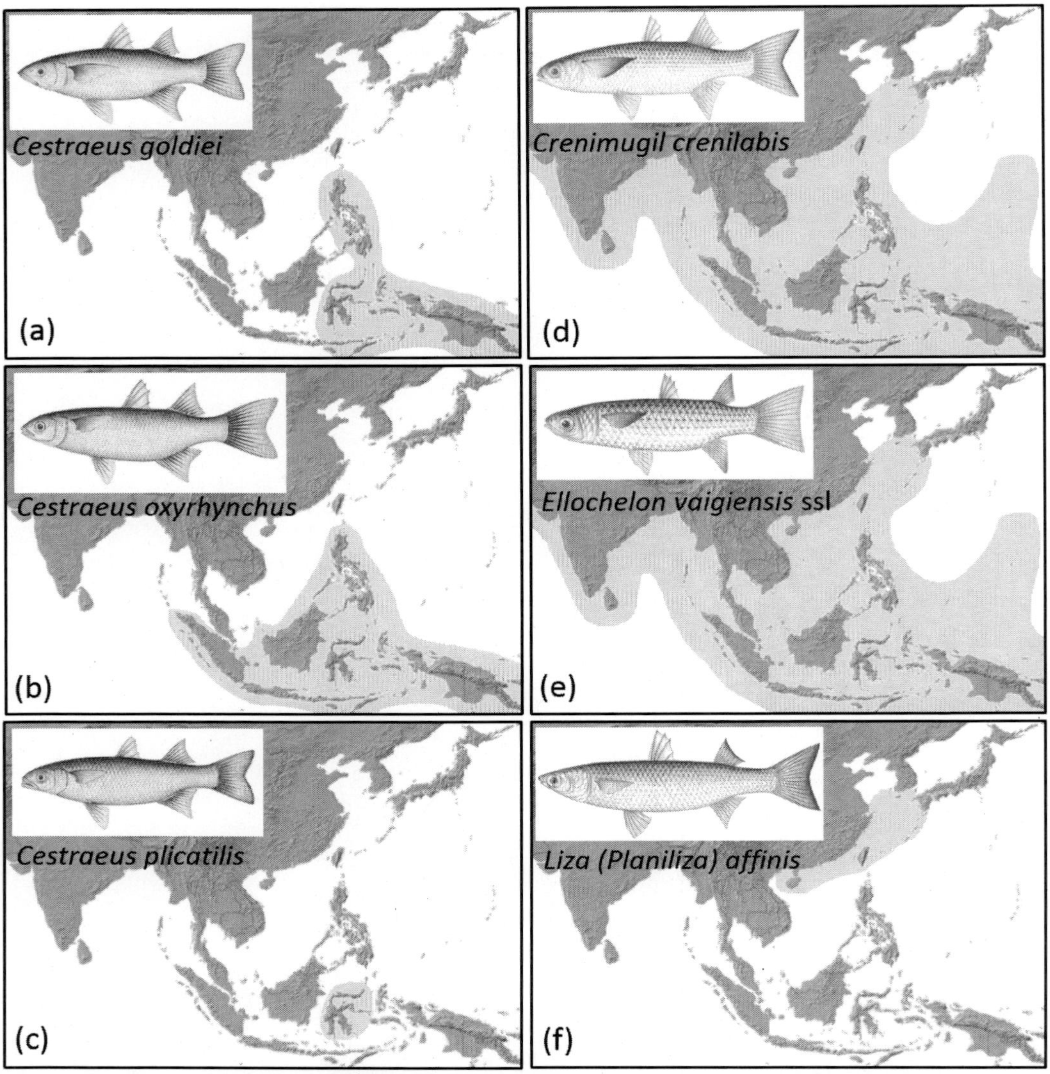

Cestraeus goldiei
(a)

Crenimugil crenilabis
(d)

Cestraeus oxyrhynchus
(b)

Ellochelon vaigiensis ssl
(e)

Cestraeus plicatilis
(c)

Liza (Planiliza) affinis
(f)

Liza (Planiliza) haematocheila
(g)

Liza (Planiliza) melinoptera ssl
(j)

Figure 4.5. contd....

Figure 4.5. contd.

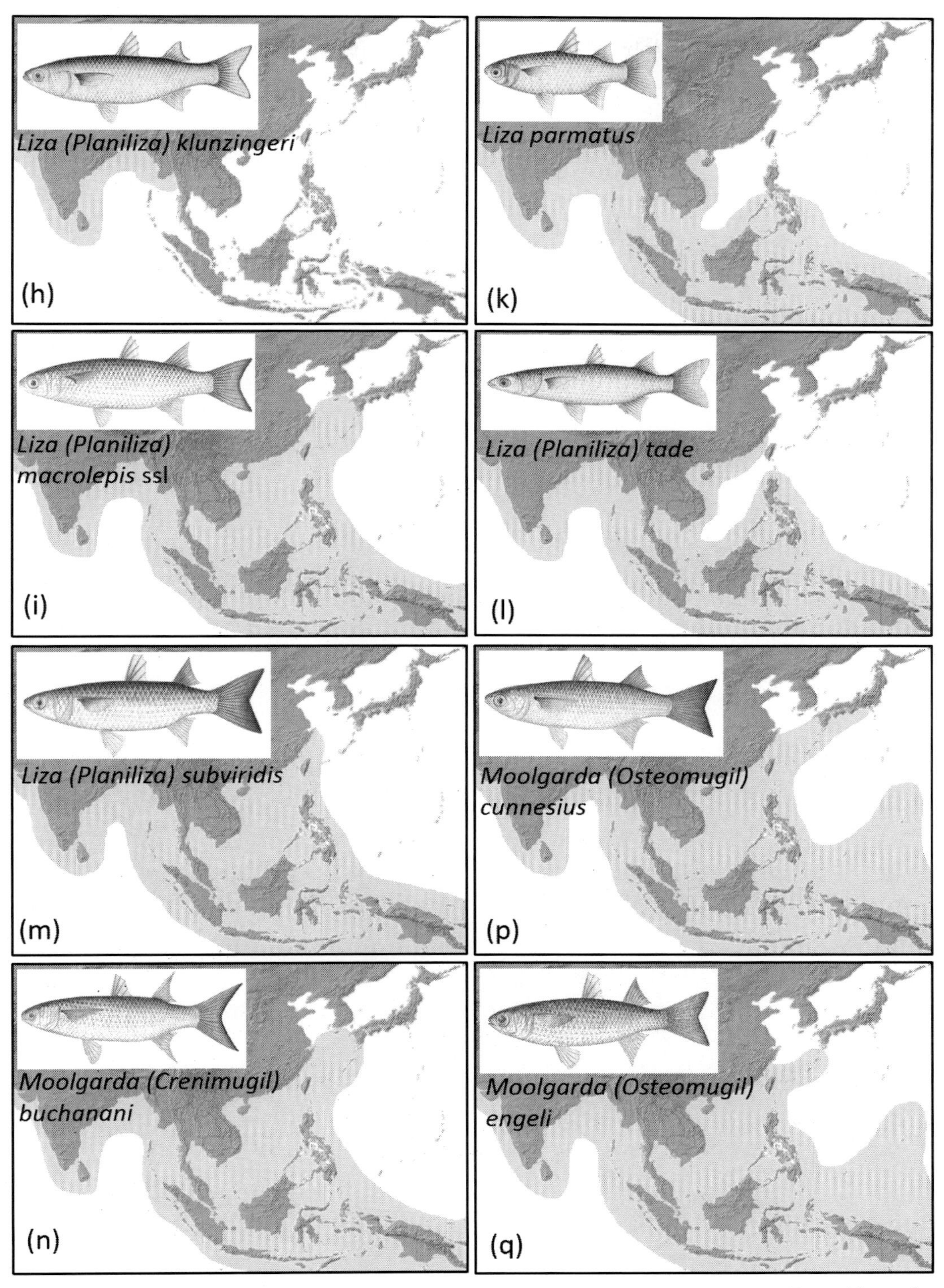

Figure 4.5. contd....

Figure 4.5. contd.

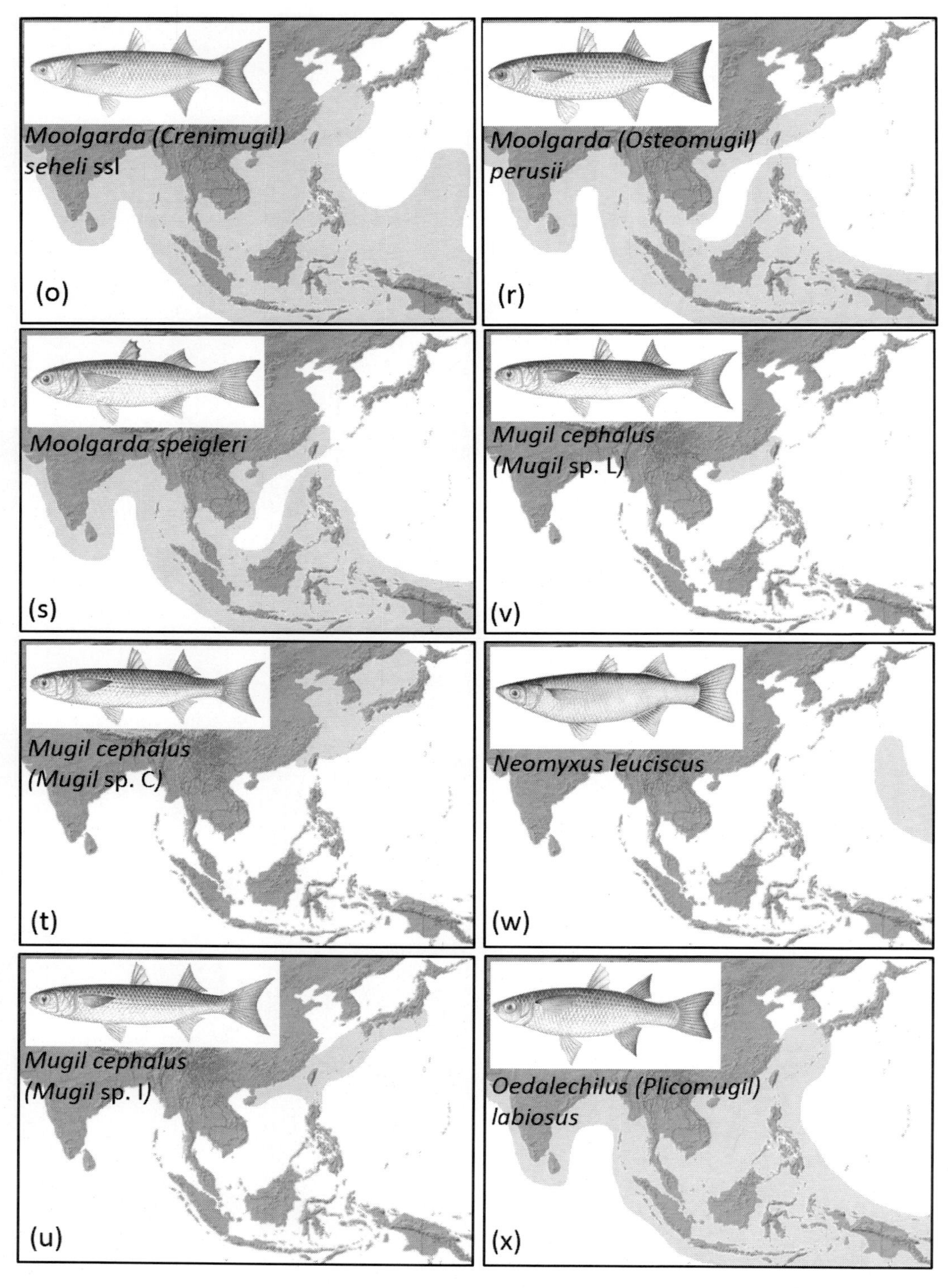

Figure 4.5. contd....

Figure 4.5. contd.

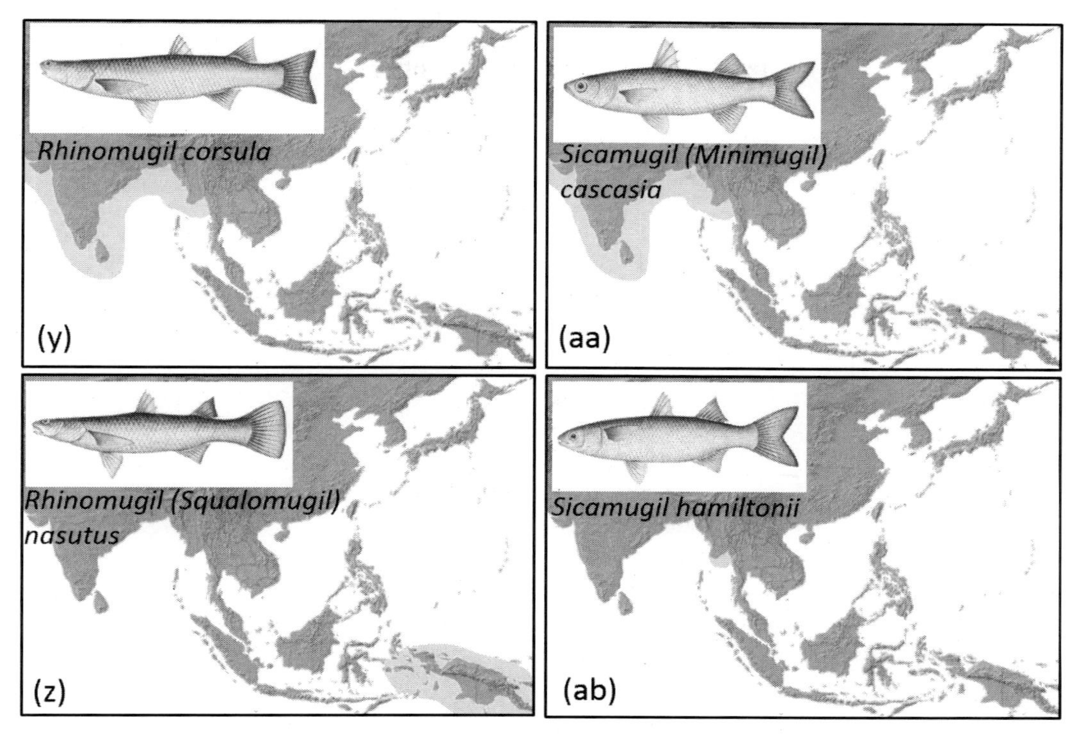

Figure 4.5. The distribution of different Mugilidae in the Indo-Pacific. The light grey area is based on the distribution area for each species modified from FAO and recently published data (see reference in Table 4.1 legend). The fish pictures were drawn by Taiwanese artist Mr. I-Lang Cheng. All the copyrights belong to National Museum of Maine Biology and Aquarium, Pingtung, Taiwan.

Lobed river mullet ***Cestraeus plicatilis*** Valenciennes, 1836

No COI barcodes are available for this species.

Distribution: Celebes, also New Caledonia, Vanuatu and Fiji (Fig. 4.5c). No any COI sequences of this species have been deposited in NCBI.

Habitat: Found in both fresh and brackish waters, possibly ascending some way up rivers.

Genus Crenimugil *Schultz, 1946*

According phylogenetic reconstruction based on mitochondrial and nuclear sequences, the genus *Crenimugil* also includes some species previously considered as members of the genus *Moolgarda*, such as *M. buchanani* and *M. seheli* (Durand et al. 2012b, Xia et al. submitted, Chapter 2—Durand 2015).

Fringelip mullet ***Crenimugil crenilabis*** Forsskål, 1775

All *Crenimugil crenilabis* COI sequences collected in various locations over its distribution range (South Africa, Seychelles, Maldives, Mariana Islands, and French Polynesia) form a unique lineage (Fig. 4.2) agreeing with the current taxonomy.

Distribution: Widespread throughout the tropical Indo-Pacific, from the Red Sea and Madagascar to Tuamotu Islands, south to Lord Howe Island and north to southern Japan (Fig. 4.5d, Table 4.1).

Habitat: Found in coastal waters. Common in tidepools and reef flats. Forms large schools in coral reef areas. Feeds on diatoms and algae on dead corals.

Half fringelip mullet ***Crenimugil heterocheilus*** Bleeker, 1855

No COI barcodes or any molecular data are available for this species while some taxonomic concern exists for this taxa (Durand et al. 2012b).

Distribution: Tropical Indo-Pacific from Indonesia and the Philippines to Vanuatu. North to southern Japanese Islands.

Habitat: Found in coastal waters, ascending rivers into fresh water, penetrating far upstream.

Genus Ellochelon *Whitley, 1930*

Squaretail mullet ***Ellochelon vaigiensis*** Quoy and Gaimard, 1824

Two deep mitochondrial lineages have been highlighted in *Ellochelon vaigiensis* (Durand et al. 2012a). Considering this result and the level of sequence divergence, Durand and Borsa (in press) proposed to retain the epithet *vaigiensis* for the lineage that includes specimens from Waigeo, the type-locality and from French Polynesia. The other lineage, represented by a specimen from an unknown location in Australia, has been provisionally assigned to a putative species named *Ellochelon* sp. A (Fig. 4.2).

Distribution: Throughout the Indo-Pacific from East Africa to Tuamotu Islands, north to southern Japan and Korea to South Great Barrier Reef and New Caledonia (Fig. 4.5e).

Habitat: Common along clear shallow coastal areas and protected sandy shores in lagoons, reef-flats, estuaries, and coastal creeks. Juveniles can be found in rice fields and mangroves. Usually forms schools. Spawning probably occurs at sea. Feeds on small algae, diatoms, benthic polychaetes, molluscs, crustaceans, as well as living and detrital organic matter, fry take copepods, rotifers and microalgae.

Genus Liza *Jordan and Swain, 1884*

Considering the mitochondrial and nuclear phylogeny of the Mugilidae family the genus *Liza* may be synonymized with the genera *Chelon* and *Planiliza* Whitley 1945 (Durand et al. 2012b, Xia et al. submitted, Chapter 2—Durand 2015). All Indo-Pacific ***Liza*** spp. belong to the genus *Planiliza.*

Eastern keelback mullet ***Liza (Planiliza) affinis*** Günther, 1861

Only one mitochondrial lineage has been recovered in this species (Fig. 4.3) which is in agreement with the taxonomy.

Distribution: From Japan (except northern Hokkaido) through to the Ryukyu Islands and Taiwan, and coasts from Shanghai to Peihar and Hainan Island (Fig. 4.5f).

Habitat: Inhabits inlet waters and muddy estuaries.

Diamond mullet ***Liza (Planiliza) ordensis*** Steindachner, 1892

Based on samples collected in northern Australia, only one lineage has been highlighted which is in agreement with the present taxonomy.

Distribution: Northern Australia, New Guinea and Tonga.

Habitat: Inhabits coastal waters and estuaries, sometimes ascending rivers into freshwater. Prefers slow moving waters and turbid muddy lagoons. Feeds on microalgae, detritus, terrestrial plant material, aquatic insects. Perhaps catadromous.

Redlip mullet ***Liza (Planiliza) haematocheila*** Temminck and Schlegel, 1845

All COI sequences labelled as *Liza (Planiliza) haematocheila* in Genbank form a unique lineage which is in agreement with the current taxonomy (Fig. 4.3).

Distribution: North-West Pacific, Japan, and from the Amur River southward to Xiamen through the Korean Peninsula (Fig. 4.5g).

Habitat: Adults inhabit shallow coastal waters as well as freshwater regions of rivers. Feed on small algae, diatoms, benthic polychaetes, crustaceans, molluscs, organic matter and detritus in sand and mud.

Klunzinger's mullet ***Liza (Planiliza) klunzingeri*** Day, 1888

COI sequences of *Liza (Planiliza) klunzingeri* available in Genbank from specimens collected in India belong to the same lineage of COI sequences obtained from *Liza* (*Planiliza*) sp. collected in the Persian Gulf (Table 4.1, Fig. 4.3). As the species identification of their specimen collected in the Persian Gulf was lacking, Durand and Borsa provisionally named this lineage *Liza* (*Planiliza*) sp. A. If Indian samples are not misidentified, this mitochondrial lineage may be considered as *Liza (Planiliza) klunzingeri.*

Distribution: Western Indian Ocean, Persian Gulf to India (Fig. 4.5h).

Habitat: Marine species and is an important resource in the Persian Gulf and the Oman Sea.

Largescale mullet ***Liza (Planiliza) macrolepis*** Smith, 1849

This nominal species has an important genetic diversity associated with a strong phylogeographic structure that suggests the presence of sibling species (Durand et al. 2012a). Pending further taxonomic investigation, Durand and Borsa (in press) suggest assigning the name *Liza (Planiliza) macrolepis* to the mitochondrial lineage present along the East African coast, while the mitochondrial lineage occurring from India to the Pacific is provisionally named *Liza (Planiliza)* sp. H.

Distribution: *Liza (Planiliza) macrolepis* ssl is common throughout most of the Indo-Pacific from East Africa and the Red Sea to Marquesas and Tuamoto Islands, north to Japan and Marianas Islands (Fig. 4.5i).

Habitat: Schools in shallow coastal waters, estuaries, harbours. Brackish water species. Forms large schools during spawning at sea. Feeds on small algae, diatoms, foraminifera, benthic polychaetes, crustaceans, molluscs, organic matter and detritus, fry take copepods and floating algae.

Indian mullet ***Liza mandapamensis*** Thomson, 1997

No COI DNA barcode is available for this species described recently by Thomson (1997) from a unique specimen (BPBM 20618).

Distribution: One locality: Kilakarei, south of Mandapam in South India.

Habitat: Freshwater (Eschmeyer 2014).

Otomebora mullet ***Liza (Planiliza) melinoptera*** Valenciennes, 1836

Molecular investigation revealed that this nominal species is polyphyletic suggesting the existence of cryptic species (Durand and Borsa in press). Pending further taxonomic investigation, *Liza (Planiliza) melinoptera* was conserved for the mitochondrial lineage observed in Fiji close to Vanikoro, the type locality of the species *melinoptera.* The other lineages from individuals collected in South Java, India and the Philippines (Fig. 4.3) are provisionally named *Liza (Planiliza)* sp. B and *Liza (Planiliza)* sp. G (Durand and Borsa in press, Fig. 4.3).

Distribution: Indo-Pacific from East Africa to the Marquesas Islands, north to South China Sea and south to tropical Australia and Tonga (Fig. 4.5j).

Habitat: Schools in shallow coastal waters, enters lagoons, estuaries and rivers. Feeds on plant detritus, microalgae, benthic organisms and organic matter in sand and mud. Juveniles feed on planktonic algae.

Broad-mouthed mullet ***Liza parmatus*** Cantor, 1849

No COI barcode is available for this species.

Distribution: Western Pacific: South China Sea southward to the Philippines, Indonesia New Guinea and North Australia (Fig. 4.5k).

Habitat: Found in seas, estuaries and rivers.

Goldspot mullet ***Liza parsia*** Hamilton, 1822

All *Liza parsia* COI sequences (No. 11 and 48 in Table 4.1) are misidentified and correspond to either *L. (P.) macrolepis* ssl or *Moolgarda (Osteomugil) perusii* (Figs. 4.2 and 4.3). No COI barcodes are available for this species.

Distribution: Indian Ocean, found along the coasts of Pakistan, India, Sri Lanka and Andaman Islands.

Habitat: Found in shallow coastal waters, estuaries, lagoons, and sometimes entering tidal rivers. Feed on diatoms, and other organic matter.

Tade grey mullet ***Liza (Planiliza) tade*** Valenciennes, 1836

In his last revision, Eschmeyer (2014) has synonymized *L. (P.) planiceps* with *L. (P.) tade*. However, two paraphyletic mitochondrial lineages were recovered for this taxa (Durand et al. 2012a). One of the two *L. (P.) tade* lineages concerned specimens sampled in Myanmar; the other lineage was sampled in northern Australia (Fig. 4.3). Considering this result and pending further taxonomic investigation, Durand and Borsa (2015) proposed naming provisionally these two lineages as *Planiliza* sp. I and *L. (P.)* sp. F (Fig. 4.3).

Distribution: This species is distributed throughout the Indo-Pacific: from the Red Sea in the west to China and Marianas Island in the north and Vanuatu in the south. Reports from tropical Australia might be misidentifications of other species (Fig. 4.5l).

Habitat: Usually marine, found in schools in shallow coastal waters and lagoons. Also enters estuaries and rivers to feed. Juveniles may be found in rice fields and mangrove swamps. Spawning occurs at sea. Feeds on floating algae, benthic organisms and organic material in sand and mud.

Greenback mullet ***Liza (Planiliza) subviridis*** Valenciennes, 1836

Durand et al. (2012a) highlighted only one mitochondrial lineage in their *Liza (Planiliza) subviridis* specimens collected in Sri Lanka, Indonesia and Taiwan. Including sequences from Genbank, this lineage is also observed in India and the Philippines (Fig. 4.3 and Table 4.1). If some sequences labelled as *L. (P.) subviridis* match a different species, there is no taxonomic question for this species, even if species misidentification clearly occurred in some DNA barcode studies.

Distribution: Indo-Pacific from Red Sea to Samoa, North to East China Sea (Fig. 4.5m).

Habitat: School in shallow coastal waters and enter lagoons, estuaries, and freshwater. Juveniles may be found in rice fields and mangroves. Adults feed on microalgae, filamentous algae, diatoms, and benthic detrital material taken in with sand and mud; fry eat zooplankton, diatoms, detrital material and inorganic sediments.

Genus **Moolgarda** *Whitley, 1945*

For reasons expressed in recent taxonomic reviews (Durand et al. 2012b, Xia et al. submitted, Chapter 2—Durand 2015) it has been proposed to invalidate the genus *Moolgarda.* Species belonging to this genus according Eschmeyer (2014) may be placed in the genera *Crenimugil* Schultz 1946 or *Osteomugil* Lüther 1977.

Bluetail mullet ***Moolgarda (Crenimugil) buchanani*** Bleeker, 1853

No DNA barcodes are available for this species in the area investigated here, but Durand et al. (2012a) published mitochondrial sequences of specimens collected in Kenya and New Caledonia (Fig. 4.2). Only one lineage has been identified for this species, which is in agreement with its taxonomic status.

Distribution: Indo-Pacific from South Africa through parts of Indonesia to parts of Melanesia and Micronesia, north to the Marianas Islands and South Japan (Fig. 4.5n).

Habitat: Inhabits shallow coastal waters. Young enter estuaries and lagoons. Feeds on algae, diatoms, detritus, and crustaceans. Usually schooling in coral reef areas.

Bluespot mullet ***Moolgarda (Crenimugil) seheli*** Forsskål, 1775

Molecular investigation stressed the important polyphyly of this species consisting of three paraphyletic lineages (Durand et al. 2012a, Fig. 4.2). Pending further taxonomic investigation, Durand and Borsa (2015) proposed naming these lineages *Moolgarda* (*Crenimugil*) sp. A, B and C.

Distribution: Widespread throughout the Indo-Pacific, from East Africa and the Red Sea to the Marquesas Islands, north to Japan and Hawaii, and south to southern Queensland and New Caledonia (Fig. 4.5o).

Habitat: Inhabits shallow coastal waters. Young enter estuaries and lagoons. Forms schools in coral reef areas. Feeds on algae, diatoms, detritus, and crustaceans.

Longarm mullet ***Moolgarda (Osteomugil) cunnesius*** Valenciennes, 1836

This species is polyphyletic, three mitochondrial lineages being highlighted in the molecular phylogeny published by Durand et al. (2012a). However, the type locality of the species being Moluccas, the lineage observed in Taiwan, Vietnam and now China and India may correspond to *Moolgarda (Osteomugil) cunnesius* (Fig. 4.2). Pending further taxonomic investigations, the two other lineages were observed, one in Australia and the other in South Africa, named *Moolgarda* (*Osteomugil*) sp. A and B respectively (Durand and Borsa 2015).

Distribution: Reliable reports are very rare due to earlier taxonomic confusion. Perhaps widespread from Red Sea to West Pacific (Fig. 4.5p).

Habitat: Found in shallow coastal waters and brackish estuaries. Feeds on algae, diatoms, detritus, and crustaceans in sand and mud.

Kanda ***Moolgarda (Osteomugil) engeli*** Bleeker, 1859

COI barcodes available for this species for specimens collected in Maraian Islands and Central Pacific form a unique lineage in agreement with the current taxonomic status (Fig. 4.2). Presence of this species in the '*Crenimugil*' clade/genus is not statistically supported and is clearly an artefact due to the short sequence fragment considered here.

Distribution: Widespread across Indo-Pacific, from Africa to Marquesas and Tuamotu Islands, north to southern Japan (Fig. 4.5q).

Habitat: Inhabits coastal waters, shallow lagoons, protected inlets, and over sandy to muddy areas of reef flats. Juveniles may enter rivers. Feeds on small algae, diatoms, and other organic matters in sand and mud.

Longfinned mullet ***Moolgarda (Osteomugil) perusii*** Valenciennes, 1836

All DNA sequences analyzed for this species whatever the sampling location (India, Sri Lanka, the Philippines, Taiwan and New Caledonia) belong to the same mitochondrial lineage (Fig. 4.2), which is in agreement with the current taxonomic status.

Distribution: Indo-West Pacific: Africa to the Mariana Islands (Fig. 4.5r).

Habitat: Inhabits coastal waters and estuaries. Schooling over mudflats. Feeds on algae, diatoms, detritus, and crustaceans in sand and mud.

Speigler's mullet ***Moolgarda speigleri*** Bleeker, 1858

In Genbank only one COI sequence is labelled as *Moolgarda speigleri,* but it is placed in the *Moolgarda (Osteomugil) perusii* clade (Fig. 4.2). This could either mirror a species misidentification or the synonymy of *Moolgarda speigleri* with *Moolgarda (Osteomugil) perusii.* Further investigations are required to clarify the taxonomic status of this species.

Distribution: Indo-West Pacific, from Baluchistan to Borneo and New Guinea, northwards up the Chinese coast (Fig. 4.5s).

Habitat: Schools in shallow coastal waters, estuaries, and backwaters. Feeds on small algae, diatoms, and other organic matters in with sand and mud, fry feed on copepods and floating algae.

Genus Mugil *Linnaeus, 1758*

Broussonnet's mullet ***Mugil broussonnetii*** Valenciennes, 1836.

No COI DNA barcode is available for this species.

Distribution: West Pacific. No any COI sequences of this species have been deposited in NCBI.

Habitat: Inhabits inshore marine waters.

Flathead mullet ***Mugil cephalus*** Linnaeus, 1758

All molecular investigations that targeted this species demonstrated the large polyphyly of this species at the worldwide scale (Crosetti et al. 1994, Livi et al. 2011, Durand et al. 2012a). In the North-West (NW) Pacific, three lineages have been identified (Ke et al. 2009, Shen et al. 2011) and thanks to independent nuclear loci it has been demonstrated that individuals belonging to these mitochondrial lineages are genetically isolated and correspond to distinct species (Shen et al. 2011). Based on these results, and pending further taxonomic investigations, Durand and Borsa (in press) proposed naming these NW Pacific species *Mugil* sp. L, C and I (Fig. 4.4).

Distribution: Worldwide in tropical, subtropical, and warm temperate waters, but less abundant in tropics and apparently rare in Indonesia. The three cryptic species found in the North-West Pacific present a partial sympatric distribution (Shen et al. 2011). *Mugil* sp. C: south of Taiwan Strait north to Japan Sea (Fig. 4.5t). *Mugil* sp. I: North to Japan and south to northern Philippines and west to Hainan (Fig. 4.5u). *Mugil* sp. L: from North Taiwan to South China. However, this species is also found in New Caledonia and Fiji suggesting a more tropical affinity than two other species (Fig. 4.5v).

Habitat: Inhabits inshore marine waters, estuaries, lagoons, and rivers, tolerant of water temperature from 12 to 30°C and salinities from hypersaline to freshwater. Adults roam, forming schools and sometimes jumping, but fish forage singly. First spawning in the third year, mature fish group in estuaries, form schools, and move out to sea to spawn in surface waters. Spent fish return to brackish estuaries and rivers. Two month old fry migrate to estuaries and swim upstream. Juveniles found in sheltered lagoons and bays. Adults feed on fine particulate material, detritus, and diatoms.

Genus Neomyxus *Steindachner, 1878*

Acute-jawed mullet ***Neomyxus leuciscus*** Günther, 1871

Molecular phylogenetic investigation did not reveal cryptic species in this taxa (Durand et al. 2012a, Fig. 4.2).

Distribution: Central Pacific, from southern Japanese and Hawaiian Islands in the north, south to Tubai and Ducie Islands, rare in Mariana Islands (Fig. 4.5w).

Habitat: In shallow coastal waters, around reef flats, tidepools, lagoons, drainage ditches, and docks. At night tends to move inshore to surface waters and shallows close to beaches.

Genus Oedalechilus *Fowler, 1903 / Genus* Plicomugil *Schultz, 1953 should be Resurrected for the Type Species* Mugil labiosus *(Durand et al. 2012b)*

Hornlip mullet ***Oedalechilus (Plicomugil) labiosus*** Valenciennes, 1836

Distribution: Widespread throughout the tropical Indo-Pacific, from the Red Sea and Madagascar to Samoa, north to southern Japan and Korea and south to the Great Barrier Reef (Fig. 4.5x).

Genus Rhinomugil *Gill, 1864*

Corsula ***Rhinomugil corsula*** Hamilton, 1822

All COI sequences availables in Genbank, from specimens collected in India and Myanmar form a unique mitochondrial lineage, which is in agreement with the current taxonomy.

Distribution: India, Bangladesh, Nepal and Myanmar (Rahman 1989) (Fig. 4.5y).

Habitat: Inhabits rivers and estuaries.

Shark mullet ***Rhinomugil (Squalomugil) nasutus*** De Vis, 1883

Molecular investigation demonstrated that this species does not form a monophyletic clade with *Rhinomugil corsula,* the type species of *Rhinomugil*, which lead Durand et al. (2012b) to resurrect the genus *Squalomugil* De Vis 1883 for the type species *Mugil nasutus*.

Distribution: New Guinea and tropical Australia (Fig. 4.5z).

Habitat: Usually found in small schools in muddy freshwaters and coastal waters (e.g., mangroves). Swimming at the surface with eyes and snout exposed. Perhaps feeding on surface algae and insects which come onto the surface, and also on muddy banks and bottoms. Capable of breathing air and wriggling over mudbanks for short distances.

Genus Sicamugil *Fowler, 1939*

Molecular investigation demonstrated the paraphyly of *Sicamugil cascasia* and *Sicamugil hamiltonii*, the type species of the genus *Sicamugil* Fowler, 1939; Consequently, Durand et al. (2012b) proposed resurrecting the genus *Minimugil* Senou 1988 for *Mugil cascasia*.

Yellowtail mullet ***Sicamugil (Minimugil) cascasia*** Hamilton, 1822

Distribution: Pakistan, India and Bangladesh (Talwar and Jhingran 1991) and Myanmar (Fig. 4.5aa).

Habitat: Inhabits upper reaches of rivers.

Burmese mullet ***Sicamugil hamiltonii*** Day, 1870

Distribution: Sittang and Irrawaddy River systems in Myanmar (Durand et al. 2012a) (Fig. 4.5ab).

Habitat: Inhabits rivers.

Biogeography and Habitat Exploitation

The boundaries of separated species ranges give much insight into past and present dispersal barriers. Many fishes confirm a past disjunction along the eastern side of the Sunda Shelf, which may be related to Holocene sea-level fluctuations, for example: butterflyfishes (Chaetodontidae) (McMillan and Palumbi 1995), Indian scad mackerel *Decapterus russelli* (Perrin and Borsa 2001), three-spot seahorse *Hippocampus trimaculatus* (Lourie and Vincent 2004) and goldband snapper *Pristipomoides multidens* (Ovenden et al. 2004). These results are in agreement with the importance of the latest sea-level low as an important factor for determining present species ranges and consequently for boundaries of species diversity patterns. However, the high mobility of most coastal or coral reef dependent Mugilidae species has resulted in new wide distributions throughout the Indo-Pacific since the latest sea-level low, for example *Liza (Planiliza) macrolepis* ssl, *Liza (Planiliza) subviridis*, *Crenimugil crenilabis*, *Ellochelon vaigiensis* ssl, *Moolgarda (Crenimugil) buchanani*, *Moolgarda (Crenimugil) seheli*, etc. Some species restricted to freshwater or estuary dependent have a narrower, localized distribution range, such as temperate NW Pacific Liza: *Liza (Planiliza) affinis*, *Liza (Planiliza) haematocheila*, Indian tropical *Liza* species *Liza (Planiliza) klunzungeri* and *Mugil cephalus* species complex (Whitfield et al. 2012).

The high diversity of habitats in this region provides a variety of environments and also variety of foods for Mugilidae. Most species are euryhaline, inhabiting coastal marine waters, brackish water lagoons, estuaries, and may enter freshwaters (particularly at young stages); some species more typically inhabit fresh water, but can also be found in brackish waters. Coastal species usually spawn offshore, and freshwater species spawn in brackish waters.

Mugilidae feed on algae, diatoms and detritus of bottom sediment or on the surface of mangroves roots, rocks and corals (Chapter 9—Cardona 2015). They swim in groups and have a nomadic lifestyle.

These groups range from a few to often more than a hundred fishes, swimming over large sandy areas. Mullets scoop up sand in search of algae, fish eggs and detritus.

The distribution and habitat usage are different among species. Most Mugilidae live in muddy and sandy estuaries, lagoons, mangroves and coastal waters, but some inhabit coral reef and rocky shore areas. Few live in the freshwater environment. *Ellochelon vaigiensis* ssl is common in clear mangrove areas and the boundary between shallow sand beach and coral reef. The juveniles of *Ellochelon vaigiensis* are sometimes found grazing on the surfaces of the roots of mangroves. *Moolgarda (Crenimugil) seheli* are also commonly found in schools around shallow coastal coral reefs. They graze the debris and diatoms on the sand or dead corals. *Liza (Planiliza) haematocheila, Liza (Planiliza) subviridis*, *Moolgarda (Osteomugil) cunnesius*, *Moolgarda (Osteomugil) perusii* and *Mugil cephalus* ssl, are more common in muddy mangroves, lagoons and estuaries where they graze on the mud.

Acknowledgements

Food and Agriculture Organization of the United Nations; Department of Ichthyology, California Academy of Sciences; Fishbase; Chih-Wei Chang, National Museum of Marine Biology and Aquarium, Taiwan for providing pictures of the fish. To Steve Blaber for editing in English. To the editors S. Blaber and D. Crosetti for their invitation to contribute towards this book.

References

Allen, G.R. 2002. Indo-Pacific coral-reef fishes as indicators of conservation hotspots. Proceedings Ninth International Coral Reef Symposium, Bali 2000, 2: 921–926.

Barber, P.H., M.V. Erdmann and S.R. Palumbi. 2006. Comparative phylogeography of three codistributed stomatopods: origins and timing of regional lineage diversification in the coral triangle. Evolution 60: 1825–1839.

Bellwood, D.R. and P.C. Wainwright. 2002. The history and biogeography of fishes on coral reefs. pp. 5–32. *In*: P.F. Sale (ed.). Coral Reef Fishes. Academic Press, New York.

Bellwood, D.R. and C.P. Meyer. 2009. Searching for heat in a marine biodiversity hotspot. J. Biogeogr. 36: 569–576.

Blaber, S.J. and D.A. Milton. 1990. Species composition, community structure and zoogeography of fishes of mangrove estuaries in the Solomon Islands. Mar. Biol. 105: 259–267.

Cardona, L. 2015. Food and feeding of Mugilidae. *In*: D. Crosetti and S.J.M. Blaber (eds.). Biology, Ecology and Culture of Grey Mullet (Mugilidae). CRC Press, Boca Raton, USA (this book).

Carpenter, K.E. and V.H. Niem (eds.). 1999. FAO Species Identification Sheets for Fishery Purposes. The Living Marine Resources of the Western Central Pacific, Volume 4. FAO, Rome.

Carpenter, K.E. and V.G. Springer. 2005. The center of the center of marine shorefish biodiversity: the Philippine Islands. Environ. Biol. Fish. 72: 467–480.

Carpenter, K.E., P.H. Barber, E.D. Crandall, M.C.A. Ablan-Lagman, Ambariyanto, G.N. Mahardika, B.M. Manjaji-Matsumoto, M.A. Juinio-Meñez, M.D. Santos, C.J. Starger and A.H.A. Toha. 2011. Comparative phylogeography of the coral triangle and implications for marine management. J. Mar. Biol. 2011: 1–14.

Chappell, J. 1983. Sea level changes and coral reef growth. pp. 46–55. *In*: D.J. Barnes (ed.). Perspectives on Coral Reefs. Brian Clouston, Manuha, Australia.

Chappell, J. and B.G. Thom. 1977. Sea levels and coasts. pp. 275–291. *In*: J. Allen, J. Golson and R. Jones (eds.). Sunda and Sahul: Prehistoric Studies in Southeast Asia, Melanesia and Australia. Academic Press, London.

Chong, V.C., A. Sasekumar, M.U.C. Leh and R. D'Cruz. 1990. The fish and prawn communities of a Malaysian coastal mangrove system, with comparison to adjacent mud flats and inshore waters. Estuar. Coast. Shelf S. 31: 703–722.

Corti, M. and D. Crosetti. 1996. Geographic variation in the grey mullet: geometric morphometric analysis using partial warp scores. J. Fish Biol. 48: 255–269.

Crosetti, D., W.S. Nelson and J.C. Avise. 1994. Pronounced genetic structure of mitochondrial DNA among populations of the circumglobally distributed grey mullet (*Mugil cephalus*). J. Fish Biol. 44: 47–58.

Durand, J.D. 2015. Implications of molecular phylogeny for the taxonomy of Mugilidae. *In*: D. Crosetti and S.J.M. Blaber (eds.). Biology, Ecology and Culture of Grey Mullet (Mugilidae). CRC Press, Boca Raton, USA (this book).

Durand, J.D. and P. Borsa. 2015. Mitochondrial phylogeny of grey mullets (Acanthopterygii: Mugilidae) suggests high proportion of cryptic species. C. R. Biol. 338: 266–277.

Durand, J.D., K.N. Shen, W.J. Chen, B.W. Jamandre, H. Blel, K. Diop, M. Nirchio, F.J. Garcia de León, A.K. Whitfield, C.W. Chang and P. Borsa. 2012a. Systematics of the grey mullets (Teleostei: Mugiliformes: Mugilidae): molecular phylogenetic evidence challenges two centuries of morphology-based taxonomic studies. Mol. Phylogenet. Evol. 64: 73–92.

Durand, J.D., W.J. Chen, K.N. Shen, C.Z. Fu and P. Borsa. 2012b. Genus-level taxonomic changes implied by the mitochondrial phylogeny of grey mullets (Teleostei: Mugilidae). C. R. Biol. 335: 687–697.

Eschmeyer, W.N. (ed.). 2014. Catalog of Fishes: Genera, Species, References. (http://research.calacademy.org/research/ichthyology/catalog/fishcatmain.asp). Electronic version accessed on 01/04/2014.

Eschmeyer, W.N. and J.D. Fong. 2014. Catalog of Fishes: Species by Family/Subfamily. (http://research.calacademy.org/research/ichthyology/catalog/SpeciesByFamily.asp). Electronic version accessed 01/04/2014.

Fischer, W. and P.J.P. Whitehead (eds.). 1974. FAO Species Identification Sheets for Fishery Purposes. Eastern Indian Ocean (fishing area 57) and Western Central Pacific (fishing area 71), Volume 3. FAO, Rome.

Fischer, W. and G. Bianchi (eds.). 1984. FAO Species Identification Sheets for Fishery Purposes. Western Indian Ocean (fishing area 51), Volume 3. FAO, Rome.

Froese, R. and D. Pauly (eds.). 2014. FishBase. World Wide Web electronic publication. www.fishbase.org, electronic version accessed 11/2014.

Giri, C., E. Ochieng, L.L. Tieszen, Z. Zhu, A. Singh, T. Loveland, J. Masek and N. Duke. 2011. Status and distribution of mangrove forests of the world using earth observation satellite data. Global Ecology and Biogeography 20: 154–159.

Hall, R. 1996. Reconstructing Cenozoic SE Asia. pp. 203–224. *In*: R. Hall and D. Blundell (eds.). Tectonic Evolution of SE Asia. Geological Society, London, Special Publications 106.

Harrison, I.J. and G.J. Howes. 1991. The pharyngobranchial organ of mugilid fishes. Its structure, variability, ontogeny, possible function and taxonomic utility. Bulletin of the British Museum of Natural History 57: 111–132.

Hogarth, P.J. 2001. Mangrove ecosystems. pp. 853–870. *In*: S.A. Levin (ed.). Encyclopedia of Biodiversity, Vol. 3. Academic Press, San Diego, CA.

Hubert, N., C.P. Meyer, H.J. Bruggemann, F. Guerin, R.J.L. Komeno, B. Espiau, R. Causse, J.T. Williams and S. Planes. 2012. Cryptic diversity in Indo-Pacific coral-reef fishes revealed by DNA-barcoding provides new support to the centre-of-overlap hypothesis. PLoS ONE 7(3): doi:10.1371/journal.pone.0028987.

Jones, S., M. Kumaran and M.A. Manikfan. 1981. On some fishes from the Maldives Part I: species known from the Laccadive Archipelago in the collections. Journal of the Marine Biological Association of India 23: 81–197.

Ke, H.M., W.W. Lin and H.W. Kao. 2009. Genetic diversity and differentiation of grey mullet (*Mugil cephalus*) in the coastal waters of Taiwan. Zoological Science 26: 421–428.

Kulbicki, M., P. Labrosse and J. Ferraris. 2004. Basic principles underlying research projects on the links between the ecology and the uses of coral reef fishes in the Pacific. pp. 119–158. *In*: L.E. Visser (ed.). Challenging Coasts. Transdisciplinary Excursions into Integrated Coastal Zone Development. Amsterdam University Press, Amsterdam.

Kuo, S.R., H.J. Lin and K.T. Shao. 1999. Fish assemblages in the mangrove creeks of northern and southern Taiwan. Estuaries 22: 1004–1015.

Kwun, H.J., Y.S. Song, S.H. Myoung and J.K. Kim. 2013. Two new records of juvenile *Oedalechilus labiosus* and *Ellochelon vaigiensis* (Mugiliformes: Mugilidae) from Jeju Island, Korea, as revealed by molecular analysis. Fisheries and Aquatic Sciences 16: 109–116.

Liu, J.X., T.X. Gao, S.F. Wu and Y.P. Zhang. 2007. Pleistocene isolation in the northwestern Pacific marginal seas and limited dispersal in a marine fish, *Chelon haematocheilus* (Temminck and Schlegel 1845). Molecular Ecology 16: 275–288.

Livi, S., L. Sola and D. Crosetti. 2011. Phylogeographic relationships among worldwide populations of the cosmopolitan marine species, the striped gray mullet (*Mugil cephalus*), investigated by partial cytochrome b gene sequences. Biochemical Systematics and Ecology 39: 121–131.

Lourie, S.A. and A.C.J. Vincent. 2004. A marine fish follows Wallace's Line: the phylogeography of the three-spot seahorse (*Hippocampus trimaculatus*, Syngnathidae, Teleostei) in Southeast Asia. Journal of Biogeography 31: 1975–1985.

Masse, J.P. and L.F. Montaggioni. 2001. Growth history of shallow-water carbonates: control of accommodation on ecological and depositional processes. International Journal of Earth Sciences 90: 452–469.

Masuda, H., K. Amaoka, C. Araga, T. Uyeno and T. Yoshino (eds.). 1984. The Fishes of the Japanese Archipelago. Tokyo University Press, Tokyo. 437p.

McCoy, E.D. and K.L. Heck. 1976. Biogeography of corals, seagrasses and mangroves: an alternative to the center of origin concept. Systematic Zoology 25: 201–210.

McMillan, W.O. and S.R. Palumbi. 1995. Concordant evolutionary patterns among Indo-West Pacific Butterflyfishes. Proceedings of the Royal Society B 260: 229–236.

Mora, C., P.M. Chittaro, P.F. Sale, J.P. Kritzer and S.A. Ludsin. 2003. Patterns and processes in reef fish diversity. Nature 421: 933–936.

Myers, R.F. 1989. Micronesian Reef Fishes. A Comprehensive Guide to the Coral Reef Fishes of Micronesia. Coral Graphics, Guam.

Myers, R.F. and T.J. Donaldson. 2003. The fishes of the Mariana Islands. Micronesica 35-36: 594–648.

Nelson, J.S. 1984. Fishes of the World, 2nd edition. John Wiley and Sons, New York, NY.

Ovenden, J.R., J. Salini, S. O'Connor and R. Street. 2004. Pronounced genetic population structure in a potentially vagile fish species (*Pristipomoides multidens*, Teleostei: Perciformes: Lutjanidae) from the East Indies triangle. Molecular Ecology 13: 1991–1999.

Perrin, C. and P. Borsa. 2001. Mitochondrial DNA analysis of the geographic structure of Indian scad mackerel in the Indo-Malay archipelago. Journal of Fisheries Biology 59: 1421–1426.

Rahman, A.K.A. 1989. Freshwater Fishes of Bangladesh. Zoological Society of Bangladesh. Department of Zoology, University of Dhaka. 364p.

Randall, J.E. 1998. Zoogeography of shore fishes of the Indo-Pacific region. Zoological Studies 37: 227–268.

Rangin, C., L. Jolivet and M. Pubellier. 1990. A simple model for the tectonic evolution of southeast Asia and Indonesian region for the past 43 My. Bulletin de la Société géologique de France 8: 889–905.

Rossi, A.R., S. Livi and D. Crosetti. 2015. Genetics of Mugilidae. *In*: D. Crosetti and S.J.M. Blaber (eds.). Biology, Ecology and Culture of Grey Mullet (Mugilidae). CRC Press, Boca Raton, USA (this book).

Schultz, L.P. 1946. A revision of the genera of mullets, fishes of the family Mugilidae, with descriptions of three new genera. Proceedings of the United States National Museum 96: 377–395.

Shen, K.N., B.W. Jamandre, C.C. Hsu, W.N. Tzeng and J.D. Durand. 2011. Plio-Pleistocene sea level and temperature fluctuations in the northwestern Pacific promoted speciation in the globally-distributed flathead mullet *Mugil cephalus*. BMC Evolutionary Biology 11: 83.

Siddall, M., E.J. Rohling, A. Almogi-Labin, C. Hemleben, D. Meischner, I. Schmelzer and D.A. Smeed. 2003. Sea-level fluctuations during the last glacial cycle. Nature 423: 853–858.

Spalding, M.D., C. Ravilious and E.P. Green. 2001. World Atlas of Coral Reefs. Berkeley, CA: University of California Press and UNEP/WCMC ISBN 0520232550.

Spalding, M.D., H.E. Fox, G.R. Allen, N. Davidson, Z.A. Ferdaña, M. Finlayson, B.S. Halpern, M.A. Jorge, A. Lombana, S.A. Lourie, K.D. Martin, E. McManus, J. Molnar, C.A. Recchia and J. Robertson. 2007. Marine ecoregions of the world: a bioregionalization of coast and shelf areas. BioScience 57: 573–583.

Springer, V.G. 1982. Pacific plate biogeography, with special reference to shore fishes. Smithsonian Contributions to Zoology 367: 1–182.

Springer, V.G. and J.T. Williams. 1990. Widely distributed Pacific plate endemics and lowered sea-level. Bulletin of Marine Science 47: 631–640.

Talwar, P.K. and A.G. Jhingran. 1991. Inland Fishes of India and Adjacent Countries, Volume 2. A.A. Balkema, Rotterdam.

Thomson, J.M. 1997. The Mugilidae of the world. Memoirs of the Queensland Museum 41: 457–562.

Voris, H.K. 2000. Maps of Pleistocene sea levels in Southeast Asia: shorelines, river systems and time durations. Journal of Biogeography 27: 1153–1167.

Wang, M., X. Zhang, T. Yang, Z. Han, T. Yanagimoto and T. Gao. 2008. Genetic diversity in the mtDNA control region and population structure in the *Sardinella zunasi* Bleeker. African Journal of Biotechnology 7: 4384–4392.

Ward, R.D., R. Hanner and P.D.N. Hebert. 2009. The campaign to DNA barcode all fishes, FISH-BOL. Journal of Fish Biology 74: 329–356.

Whitfield, A.K., J. Panfili and J.D. Durand. 2012. A global review of the cosmopolitan flathead mullet *Mugil cephalus* Linnaeus 1758 (Teleostei: Mugilidae), with emphasis on the biology, genetics, ecology and fisheries aspects of this apparent species complex. Reviews in Fish Biology and Fisheries 22: 641–681.

Zemlak, T.S., R.D. Ward, A.D. Connell, B.H. Holmes and P.D.N. Hebert. 2009. DNA barcoding reveals overlooked marine fishes. Molecular Ecology Resources 9: 237–242.

Zhang, Q.Y. and Y.Z. Zhang. 1986. The fish of the Taiwan Strait. pp. 465–470. *In*: T. Uyeno, R. Arai, T. Tainuchi and K. Matsuura (eds.). Indo-Pacific Biology: Proceedings of the Second International Conference on Indo-Pacific Fishes. Ichthyological Society of Japan, Tokyo.

CHAPTER 5

Biogeography and Distribution of Mugilidae in Australia and Oceania

Javad Ghasemzadeh

Introduction

Oceania is the name of the region consisting of island groups within the Central and South Pacific Ocean (Fig. 5.1). The term Oceania is often used more specifically to denote a continent comprising Australia and proximate islands or biogeographically as a synonym for either the Australasian ecozone (Wallacea and Australasia) or the Pacific ecozone (Melanesia, Polynesia, and Micronesia). It spans over 8.5 million square kilometres. Some of the countries included in Oceania are Australia, New Zealand, Tuvalu, Samoa, Tonga, Papua New Guinea, the Solomon Islands, Vanuatu, Fiji, Palau, Micronesia, the Marshall Islands, Kiribati and Nauru. Oceania also includes several dependencies and territories such as American Samoa, Johnston Atoll and French Polynesia.

Physical Geography of Oceania

In terms of its physical geography, the islands of Oceania are often divided into four different sub-regions based on the geological processes playing a role in their physical development. The first of these is Australia. It is separated because of its location in the middle of the Indo-Australian Plate and the fact that due to its location there was no mountain building during its development. Instead, Australia's current physical landscape features were formed mainly by erosion.

The second landscape category in Oceania is the islands found on the collision boundaries between the Earth's crustal plates. These are found specifically in the South Pacific. For example at the collision boundary between the Indo-Australian and Pacific plates are places like New Zealand, Papua New Guinea and the Solomon Islands. The North Pacific portion of Oceania also features these types of landscapes along the Eurasian and Pacific plates. These plate collisions are responsible for the formation of mountains like those in New Zealand which climb to over 3,000 metres.

Volcanic islands such as Fiji are the third category of landscape types found in Oceania. These islands typically rise from the seafloor through hotspots in the Pacific Ocean basin. Most of these areas consist of very small islands with high mountain ranges.

Faculty of Marine Sciences, Chabahar Maritime University, Iran.
Email: jghasemz@yahoo.com.au; ghasemzadeh@cmu.ac.ir

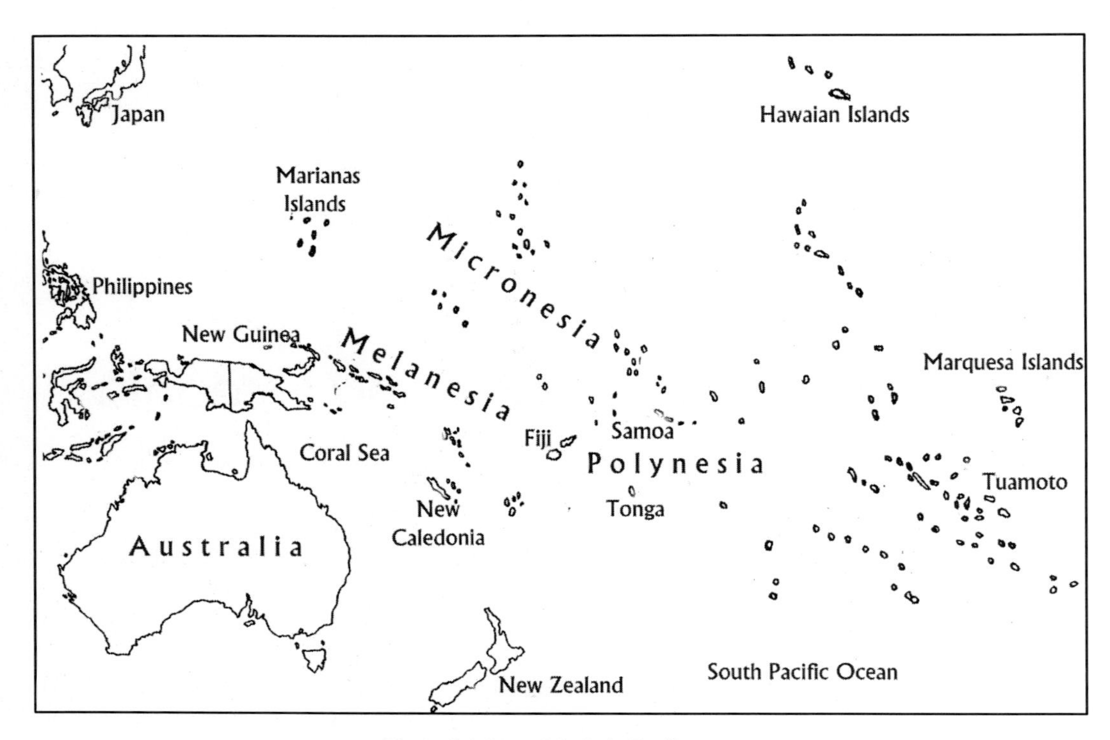

Figure 5.1. Map of the Indo-Pacific area.

Finally, coral reef islands and atolls such as Tuvalu are the last type of landscape found in Oceania. Atolls specifically are responsible for the formation of low-lying land regions, some with enclosed lagoons.

Climate of Oceania

Most of Oceania is divided into two climate zones. The first of these is temperate and the second is tropical. About half of Australia and all of New Zealand are within the temperate zone and most of the island areas in the Pacific are considered tropical. Oceania's temperate regions feature high levels of precipitation, cold winters and warm to hot summers. The tropical regions in Oceania are hot and wet year round.

In addition to these climatic zones, most of Oceania is impacted by continuous trade winds and sometimes hurricanes (called tropical cyclones in Oceania) which have historically caused catastrophic damage to countries and islands in the region.

Flora and Fauna of Oceania

Because most of Oceania is tropical or temperate there is an abundant amount of rainfall which produces tropical and temperate rainforests throughout the region. Tropical rainforests are common in some of the island countries located near the tropics, while temperate rainforests are common in New Zealand. In both of these types of forests there is a plethora of plant and animal species, making Oceania one of the world's most biodiverse regions.

It is important to note however, that not all of Oceania receives abundant rainfall and portions of the region are arid or semiarid. Australia, for example, features large areas of arid land which have little vegetation. In addition, El Niño has caused frequent droughts in recent decades in northern Australia and Papua New Guinea.

Oceania's fauna, like its flora, is also extremely biodiverse. Because much of the area consists of islands, unique species of birds, animals and insects evolved out of isolation from others. The presence of coral reefs such as the Great Barrier Reef and Kingman Reef also represent large areas of biodiversity and some are considered biodiversity hotspots.

Records of Mugilidae in Australia and Oceania

Mullets (Mugilidae) are an ecologically diverse group of spiny-rayed teleosts, with a worldwide distribution (except the polar regions). They inhabit the surface zone of oceans, estuaries and some inland freshwaters.

Berra (1981) listed the distribution of mullets as between latitudes 65° N to 50° S in Europe, Africa, Asia, Australia and New Zealand, and 40° N to 37° S in the east coast, and 30° N to 5° S in the west coast of the American continent. They occur therefore, in tropical, subtropical and temperate zones of the world. The flathead grey mullet *Mugil cephalus* Linnaeus is the only cosmopolitan species, occurring between latitudes 42° N and 42° S in all coastal waters of the world (Thomson 1963). Because of its ubiquity, size and desirability as a food fish, *M. cephalus* is one of the most important fishery species, particularly for developing countries, and inhabitants of thousands of islands in Oceania.

In Australia Macleay (1880) presented an account of the Australian mullets, and subsequently described new genera and species of Indo-Pacific mullets (Macleay 1884a, 1884b). Ogilby (1887) and Whitley (1930, 1941, 1945, 1948) described further genera and species. A comprehensive guide to the mullets of the Indo-Australian Archipelago was published by Weber and de Beaufort (1922). Thomson (1954) conducted a revision of the mullets of Australian waters and adjacent seas, and described 17 species belonging to nine genera in Australia and the South Pacific region. Munro (1956) recorded 20 species (eight genera) in Australian waters, and 16 species (six genera) from New Guinea (Munro 1967). Merrick and Schmida (1984) also reported nine genera and 17 species around Australia. Since Thomson's (1954) revision, much of the research on mullets has been regional, concerning aspects of the biology of one or more species. These studies include those of Harris (1968), Grant and Spain (1975a, 1975b, 1975c, 1977), Grant et al. (1977), Spain et al. (1980), Chubb et al. (1981), Lenanton et al. (1984) and Langi et al. (1990). Harrison and Senou (1999) produced a comprehensive FAO document on the mullets of Indo-West Indian Pacific region. They recognized 11 genera and 31 species in this region and represented detailed species identification keys, accounts on the diagnostic characters, habitat, biology, fisheries and distribution of 30 species occurring in this area.

Ghasemzadeh (1998, unpubl. thesis) reviewed the systematics, phylogeny and distribution of Indo-Pacific and Australian mullets, using morphological and osteological data. He recognised 18 of the 40 world-wide described genera as valid and described a new genus *Paramugil*[1] (Ghasemzadeh et al. 2004). He reported 27 species belonging to 14 genera of *Cestraeus, Aldrichetta, Myxus, Trachystoma, Gracilimugil,*[2] *Oedalechilus, Neomyxus, Ellochelon, Rhinomugil, Mugil, Paramugil,*[1] *Liza, Crenimugil*, and *Valamugil*[3] from Australia, Oceania and Indo-Pacific area (Ghasemzadeh 1998, unpubl. thesis). His phylogenetic analysis of Indo-Pacific mullets based on characters derived from external morphology, morphometric and meristic, osteology and splanchnology suggested that *Cestraeus* and *Aldrichetta* are the most plesiomorphic taxa, comprising with *Agonostomus* and *Joturus* the subfamily Agonostominae. The remaining genera are in the subfamily Mugilinae.

[1] Corresponding to *Liza* Jordan and Swain 1884. In Eschmeyer (2015).
[2] Corresponding to *Liza* Jordan and Swain 1884. In Eschmeyer (2015).
[3] Corresponding to *Moolgarda* Whitley 1945. In Eschmeyer (2015).

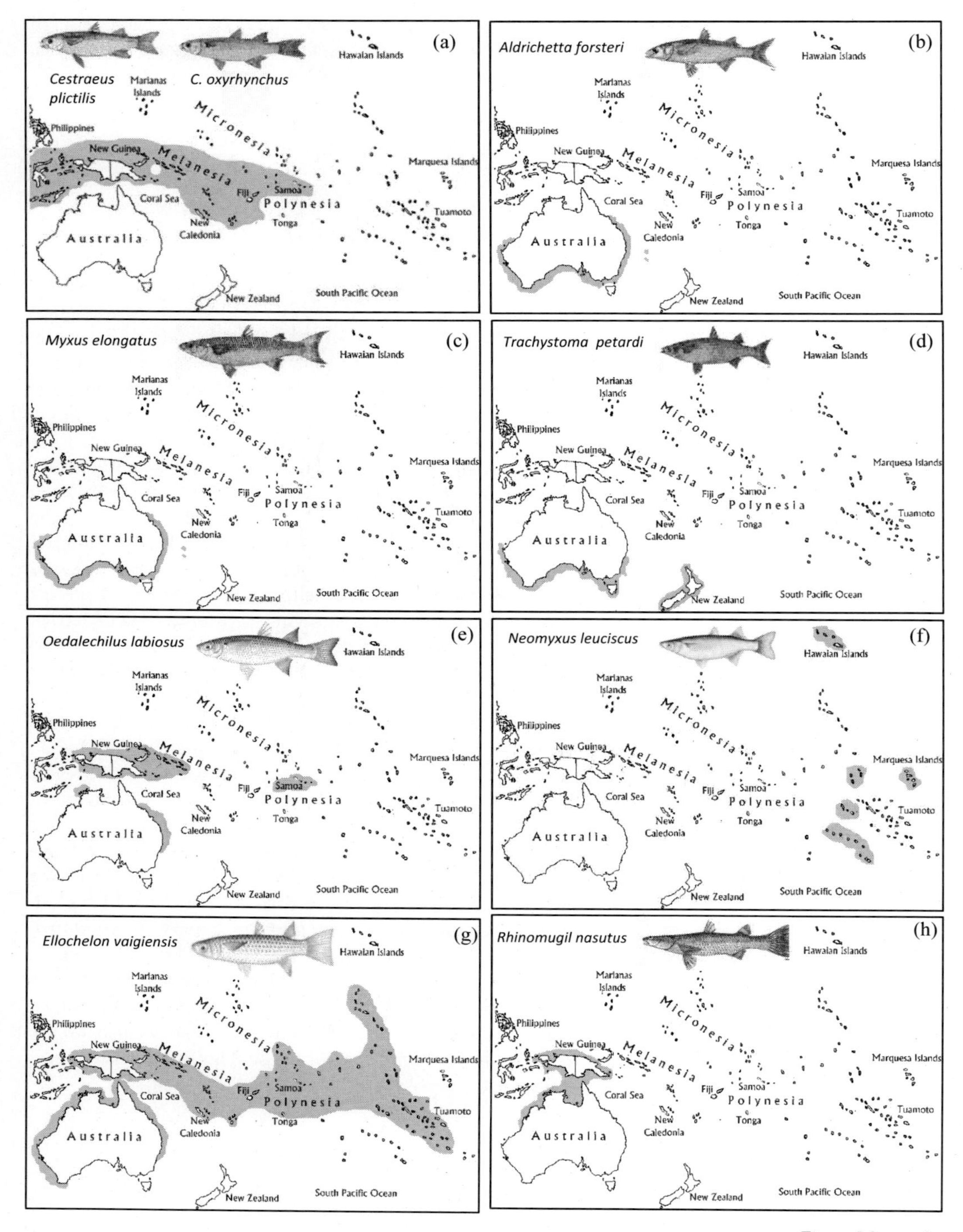

Figure 5.2. contd....

Figure 5.2. contd.

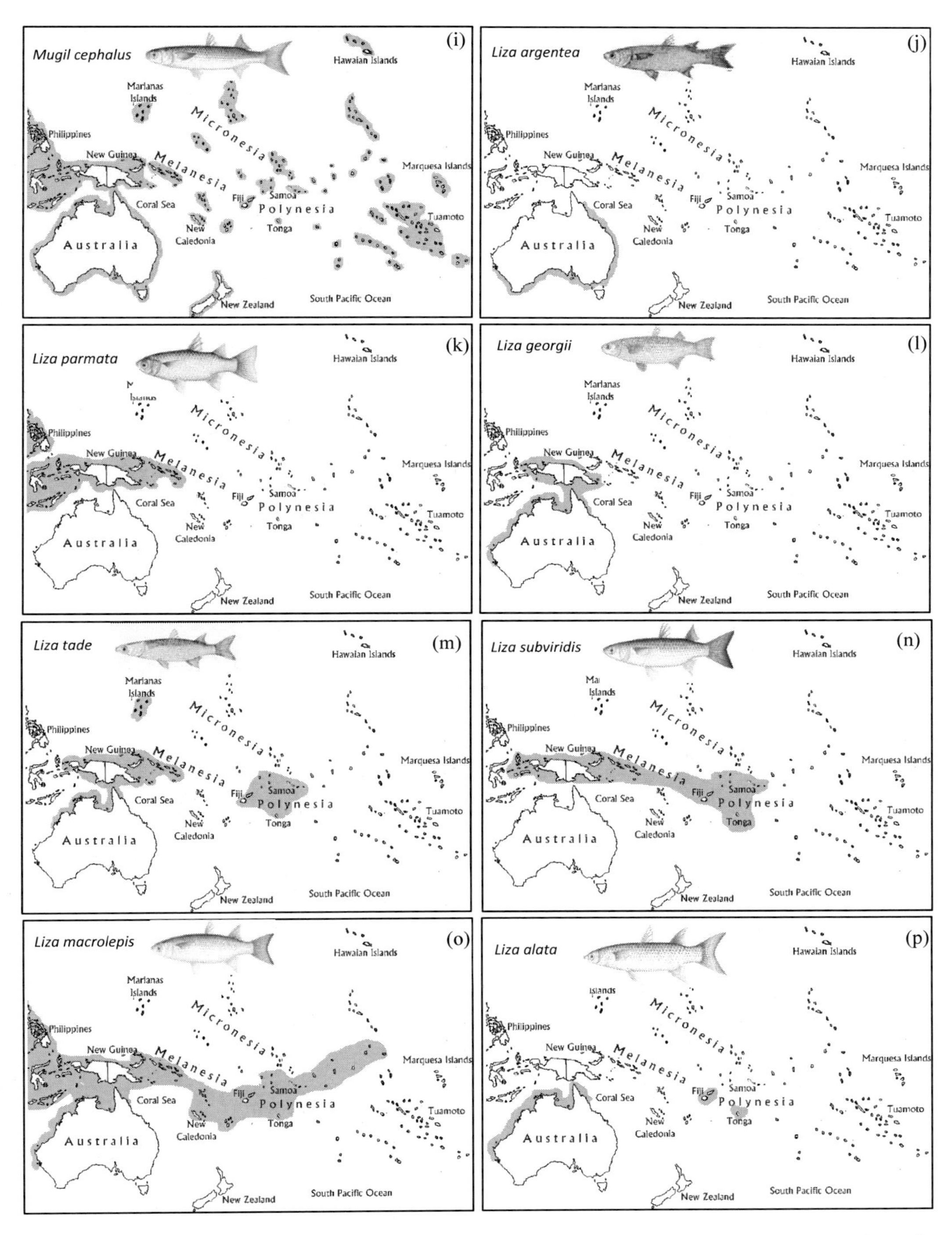

Figure 5.2. contd....

Figure 5.2. contd.

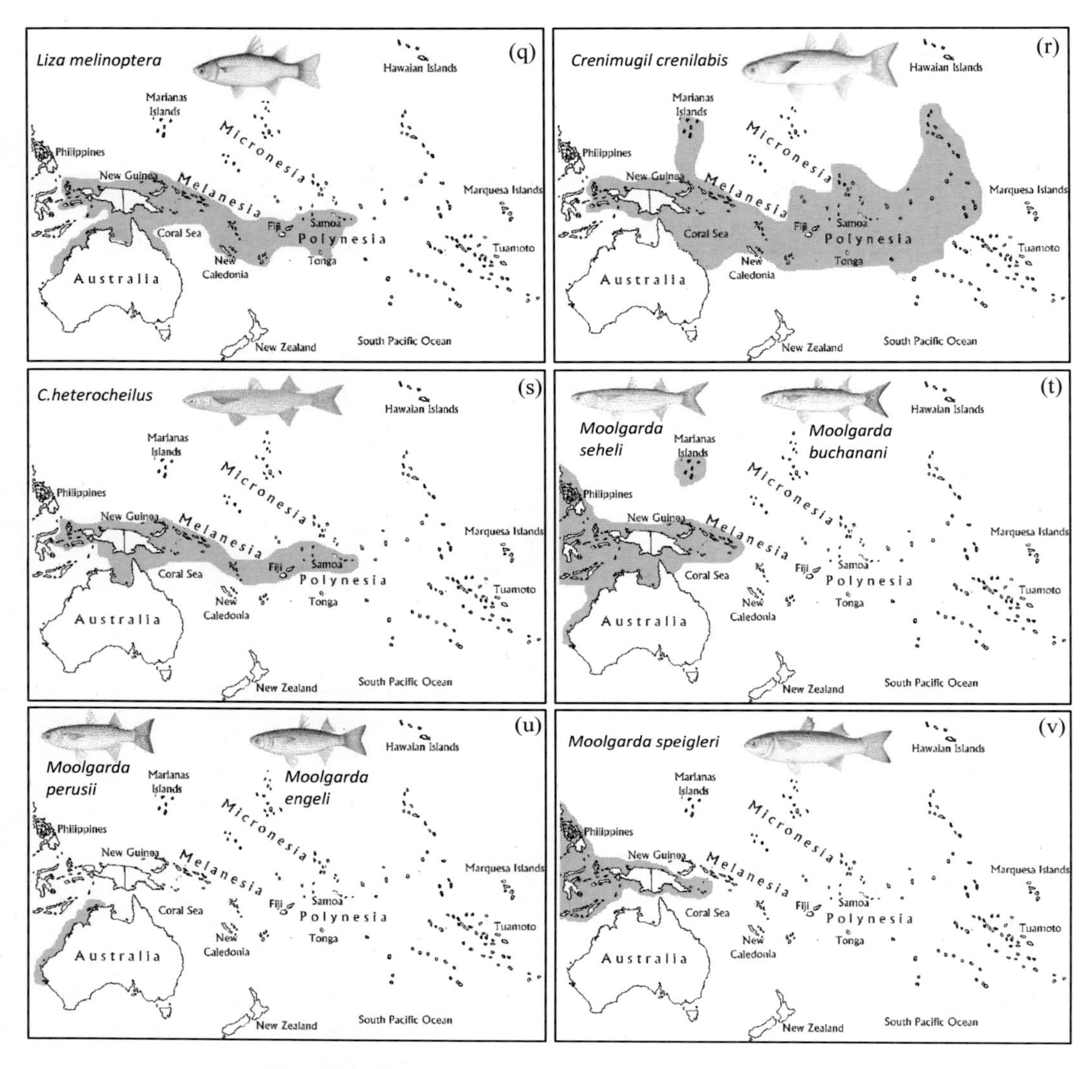

Figure 5.2. Distribution maps of mullets in Australia and Oceania.

According to Ghasemzadeh (1998, unpubl. thesis), 20 species of mullets belonging to 12 genera are found in Australian waters. The species *Trachystoma petardi, Myxus elongatus, Gracilimugil*[4] *argenteus*, and *Paramugil georgii,*[5] are restricted to Australian waters.[6]

[4] Corresponding to *Liza argentea* in Eschmeyer (2015).

[5] Corresponding to *Liza georgii* in Eschmeyer (2015).

[6] Recently molecular techniques and fish barcoding has proved a useful tool for identification of species of fishes such as mullets which are remarkably uniform, with continuous confusion in their classification (see Chapter 1—González-Castro and Ghasemzadeh 2015, Chapter 2—Durand 2015 and Chapter 15—Rossi et al. 2015). Molecular phylogenetic analysis can also help to discriminate the species that appear polyphyletic, cryptic or sibling species. Durand et al. (2012a,b) conducted a comprehensive study on the phylogenetic relationships of Mugilidae, using the mitochondrial-DNA sequences of a total of 257 individual mullet specimens belonging to 55 currently recognized species, and 20 mugilid genera currently

Footnote 6 contd....

A key to the identification of the species of mullets found in Australia and Oceania is reported in Table 5.1; a brief description and distribution range of each species follows hereafter.

Taxonomy, Distribution and Habitats of Mugilidae Species in Australia and Oceania

Genus ***Cestraeus*** Valenciennes, 1836

Lobed river mullet ***Cestraeus plicatilis*** Valenciennes, 1836

D1 IV; D2 9; A III, 9; pectoral fins 1 spur 18–20; LSS 38–44; TRS 12–13; GR 30–34/52–56; pyloric caeca 2; body depth 24.2–31% and head length 20.5–27.6% of SL.

Distribution: New Guinea down to New Caledonia and New Hebrides, Fiji and Samoa, Vanuatu, the Indonesian archipelago, also in the western Pacific from the Philippines in the north to the Celebes. It is not present in Australian waters.

Habitat: Fresh and brackish waters, possibly ascending some way up rivers.

Sharp-nosed river mullet ***Cestraeus oxyrhynchus*** Valenciennes, 1836

D1 IV; D2 9; A III, 9; pectoral fins 1 spur 18–20; LSS 43–45; TRS 12–13; GR 30–34/52–56; pyloric caeca 2; body depth 22.2–27.8% and head length 24–25% of SL.

Distribution: New Guinea down to New Caledonia and Fiji, the Indonesian archipelago, also in the western Pacific from the Philippines (Cagayan) in the north to Celebes. It is not present in Australian waters.

Habitat: Fresh and brackish waters, adults stay in rivers, mature adults probably spawn in brackish or marine waters.

Genus ***Aldrichetta*** Whitley, 1945

Yellow-eye mullet ***Aldrichetta forsteri*** Valenciennes, 1836

D1 IV; D2 9–10; A III, 11-13; pectoral fins 1 spur 14–15; LSS 54–63; TRS 17–19; GR 19–24/24–29; pyloric caeca 2; body depth 22.8–28.1% and head length 25.2–27.2% of SL.

Distribution: Restricted to temperate waters of Australia and New Zealand. In Australia it ranges from southern New South Wales to Victoria, Tasmania, South Australia and western Australia up to Geraldton.

Habitat: Shallow estuaries and close inshore waters.

Genus ***Myxus*** Günther, 1861

Sand mullet ***Myxus elongatus*** Günther, 1861

D1 IV; D2 9; A III, 9; pectoral fins 1 spur 15–16; LSS 42–45; TRS 13–14; GR 35–42/60–68; pyloric caeca 2; body depth 22.0–24.8% and head length 26.7–29% of SL.

Distribution: Restricted to temperate waters of Australia. It is recorded from South Queensland, New South Wales, Victoria, South Australia, western Australia, Lord Howe Island and Norfolk Island.

Habitat: Shallow estuaries and close inshore waters; schools enter freshwater and estuaries, but spawning in the sea. Feeds on microscopic organisms and organic detritus in the bottom of mud.

Footnote 6 contd....

recognized as valid. According to their results the Mugilidae family comprises seven major lineages that radiated early on from the ancestor to all current forms, and all genera that were represented by two species or more, except *Cestraeus*, were paraphyletic or polyphyletic (Durand et al. 2012a,b). They concluded that anatomical characters used for the systematics of mullets are poorly informative phylogenetically, and further molecular phylogenetic studies at the infra-generic level, are needed to elucidate the status of polyphyletic and cryptic species, etc. Hence in this chapter the current taxonomic names by Eschmeyer (2015) are used and the newly resurrected scientific names at generic and species level are left to further scientific research and discussion.

Genus ***Trachystoma*** Ogilby, 1887

Freshwater mullet, pink eye mullet ***Trachystoma petardi*** Castelnau, 1875

D1 IV; D2 9; A III, 9; pectoral fins 1 spur 14–15; LSS 47–50; TRS 14–16; GR 28–34/48–65; pyloric caeca 2; body depth 22.2–30% and head length 21.7–27.7% of SL.

Distribution: Restricted to the eastern coast of Australia from Georges River in New South Wales to Burnett River in southern Queensland. It usually inhabits freshwater, and is abundant in the Nepean River, Richmond River, Hawkesbury River, Georges River in New South Wales, and Nerang River in Queensland.

Habitat: Usually in freshwater rivers and their estuaries. Feeds on microscopic organisms and organic detritus at the bottom of mud.

Genus ***Oedalechilus*** Fowler, 1903

Hornlip mullet ***Oedalechilus labiosus*** Valenciennes, 1836

D1 IV; D2 9; A III, 9; pectoral fins 1 spur 16–18; LSS 33–36; TRS 11–12; GR 31–36/37–42; pyloric caeca 3; body depth 27–34.1% and head length 25.6–29.6% of SL.

Distribution: New Guinea, Solomon Islands, Samoa. In Australia it is recorded from the Great Barrier Reef, and Lizard Island around Queensland; Melville Bay and Brewer Island in the northern Territory; and Fenelon Island in western Australia.

Habitat: Found in shallow coastal waters, around coral reefs, and in harbours.

Genus ***Neomyxus*** Steindachner, 1878a

Acute-jawed mullet ***Neomyxus leuciscus*** Günther, 1871

D1 IV; D2 9; A II, 10–11; pectoral fins 1 spur 14–15; LSS 45–48; TRS 13–14; GR 31–35/37–65; pyloric caeca 2; body depth 23.6–26% and head length 24.6–29.6% of SL.

Distribution: In the Central Pacific from southern Japanese and Hawaiian Islands in the north, down to Christmas, Phoenix, Ellice, Fiji, Samoa, and Tonga Islands in the south, and Society, Cook, Tuamotu, Tubuai, Gambier, Duci and Marquesas Islands in the South western Pacific. Rare in the Marianas Islands, and not recorded in Australian waters.

Habitat: Inhabit shallow coastal waters, around reefs, tide pools, lagoons, ditches and docks. At night tends to move inshore to surface waters and shallows close to the beach.

Genus ***Ellochelon*** Whitley, 1930

Square-tail mullet-diamond scale mullet ***Ellochelon vaigiensis*** (Quay and Gaimard, 1824)

D1 IV; D2 9; A III, 8; pectoral fins 1 spur 15–17; LSS 26–28; TRS 8–10; GR 31–35/43–47; pyloric caeca 26–48; body depth 27.4–30.3% and head length 26.2–31.4% of SL.

Distribution: New Caledonia, New Guinea, New Hebrides, Fiji, Tonga, Samoa, Cook Islands, and Society Islands to Tuamotu. Marshall Islands, Phoenix Islands; south and south-east in Indonesia, Timor Sea, Coral Sea. In Australia It is recorded from North New South Wales, northwards to tropical waters of Queensland, Great Barrier Reef, northern Territory and western Australia.

Habitat: Common along shallow coastal areas and protected sandy shores in lagoons, reef flats, estuaries, coastal creeks, around reefs, tide pools, lagoons, ditches and docks. This species usually moves within the area of tidal, but may enter freshwater. Juveniles are common in mangroves and rice fields. Feeds on small algae, diatoms, benthic polychaets, molluscs, crustacean, living and detrital organic matter.

Genus ***Rhinomugil*** Gill, 1864

Shark mullet, mud mullet ***Rhinomugil nasutus*** De Vis, 1883

D1 IV; D2 8–9; A III, 8; pectoral fins 1 spur 13–14; LSS 28–31; TRS 9–10; GR 31–35/40–45; pyloric caeca 17–23; body depth 18.5–25.9% and head length 24.7–29.6% of SL.

Distribution: In freshwater rivers of tropical Australia, Indonesia and New Guinea. It is recorded mainly from Cardwell in Queensland, North Queensland, northern Territory and western Australia (Fitzroy River and Derby).

Habitat: Usually found in small schools in muddy freshwaters and coastal waters (mangroves). Swimming at the surface with eyes and snout exposed. Feeding on muddy banks and bottoms.

Genus ***Mugil*** Linnaeus, 1758

Grey mullet, Flathead mullet, Sea mullet ***Mugil cephalus*** Linnaeus, 1758

D1 IV; D2 9; A III, 8; pectoral fins 1 spur 15–17; LSS 38–43; TRS 13–14; GR 25–30/31–47; pyloric caeca 2; body depth 32–41.5% and head length 26.1–29.9% of SL.

Distribution: Distributed world-wide in tropical, subtropical, and warm temperate waters. Reported from all parts of the Indo-Pacific and Australia.

Habitat: Usually found in small schools in muddy freshwaters and coastal waters (mangroves). Swimming at the surface with eyes and snout exposed. Feeding on muddy banks and bottoms.

Genus ***Liza*** Jordan and Swain 1884 (***Gracilimugil*** Whitley, 1941; ***Paramugil*** Ghasemzadeh, Ivantsoff and Aarn, 2004)

Flat-tail mullet (Tiger mullet) ***Liza argentea*** Quoy and Gaimard, 1825 (***Gracilimugil argenteus*** Quoy and Gaimard, 1825)

D1 IV; D2 9–10; A III, 9–11; pectoral fins 1 spur 14–16; LSS 34–37; TRS 13–15; GR 42–45/100–130; pyloric caeca 2; body depth 26.2–31.6% and head length 27.7–33.7% of SL.

Distribution: Limited to the southern coasts of Australia. It is distributed from Cardwell in Queensland, south to New South Wales, Victoria, South Australia to the Moore River and perhaps up to Geraldton in western Australia. It is more abundant in New South Wales coastal and brackish waters, and is not recorded from Tasmania, North Queensland and northern Territory.

Habitat: Adults inhabit brackish waters. Abundant in estuaries; also found in shallow bogs and saline lagoons. Fish younger than one year can be found in freshwaters, but not when older. Feeds on benthic microscopic organisms and organic detritus in the bottom of muds.

Broad mouthed mullet ***Liza (Paramugil) parmata*** Cantor, 1850

D1 IV; D2 9; A III, 9; pectoral fins 1 spur 13–15; LSS 24–26; TRS 9–11; GR 27–35/40–95; pyloric caeca 7–15; body depth 26.4–29.9% and head length 24.2–32.0% of SL.

Distribution: Malayan and Indonesian Archipelago, Fiji, New Guinea and tropical Australia in Queensland and northern Territory. Also in the Philippines and China Sea.

Habitat: Usually found in small schools in seas, estuaries and rivers. Feeds on small algae and zooplankton and detritus.

Fantail mullet ***Liza (Paramugil) georgii*** Ogilby, 1897

D1 IV; D2 9; A III, 9; pectoral fins 1 spur 14–16; LSS 28–32; TRS 9–12; GR 30–35/38–58; pyloric caeca 18–26; body depth 28.4–32.8% and head length 25.4–32.5% of SL.

Distribution: This species is also confined to temperate and tropical coastal waters of Australia. It is recorded from the Georges River and Port Hacking in New South Wales, northwards in Queensland to the Gulf of Carpentaria, northern Queensland and northern Territory and adjacent islands. Its distribution continues to temperate waters of western Australia, where it is recorded down to the Fitzroy River delta.

Habitat: Inhabits coastal waters, estuaries and bays; young may enter freshwater rivers. Forming schools, shoaling over shallow-water sand flats, particularly at rising tide at dusk. Feeds on small algae, zooplankton, filtering benthic organisms in bottom sand, and decaying plant materials.

Tade mullet, Grey mullet ***Liza tade*** Forskål, 1775

D1 IV; D2 9; A III, 9; pectoral fins 1 spur 15–17; LSS 30–35; TRS 9–13; GR 36–45/47–65; pyloric caeca 5–6; body depth 22.4–30.5% and head length 21.3–29.7% of SL.

Distribution: In Oceania it is reported from Fiji, Samoa, Tonga, New Guinea and Indonesia to Borneo and Guam. It is also abundant in tropical Australia. It is widely distributed from Red Sea, South Africa, North Indian Ocean, Bay of Bengal, and all Indo-Pacific waters including, the Philippines, Thailand, Myanmar, China, Taiwan, Hong Kong and Korea.

Habitat: Usually marine, found in schools in shallow coastal waters, estuaries and lagoons. Juveniles may be found in rice fields and mangrove swamps. Spawning occurs offshore at sea. Feeds on floating algae, benthic organisms and organic material in sand and mud.

Greenback mullet ***Liza subviridis*** Valenciennes, 1836

D1 IV; D2 9; A III, 9; pectoral fins 1 spur 14–17; LSS 28–33; TRS 10–12; GR 30–36/42–62; pyloric caeca 5; body depth 26.3–31.9% and head length 25.3–34.6% of SL.

Distribution: Widely distributed from the Red Sea, Persian Gulf, South Africa, North Indian Ocean, Bay of Bengal, China Sea, and Japan to the western Pacific. It is reported from all Oceania and Indo-Pacific waters including Fiji, Samoa, Tonga, New Caledonia, Indonesia, New Guinea, the Philippines, Thailand, Myanmar, China, Taiwan, Hong Kong and Korea. In Australia it is abundant in the tropics (North Queensland, northern Territory, and North-western Australia), New South Wales and Queensland but not reported from Tasmania, Victoria, and South Australia.

Habitat: Inhabits shallow coastal waters, estuaries, bays and mangroves; Juveniles may be found in rice fields and freshwater embayments. Adults feed on microalgae, filamentous algae, diatoms, and benthic detrital material taken in with sand and mud.

Largescale mullet ***Liza macrolepis*** Smith, 1849

D1 IV; D2 9; A III, 9; pectoral fins 1 spur 14–16; LSS 30–35; TRS 10–12; GR 30–35/37–79; pyloric caeca 5–6; body depth 26.6–32.8% and head length 26.3–32.5% of SL.

Distribution: *Liza macrolepis* is widely distributed in the Indian Ocean and adjacent seas and most of Oceania and the Indo-Pacific region. In Oceania it is recorded from Tonga, Samoa, New Guinea and New Hebrides. There is no record of *L. macrolepis* from Australian waters. It is common in most parts of the Indo-Pacific including China, Hong Kong, Taiwan, the Philippines, Thailand, Borneo, Korea, Indonesia, Sumatra and Java.

Habitat: Schools in shallow coastal waters, estuaries and harbours. Brackish water species, form large schools during spawning at sea. Feed on small algae, diatoms, benthic polychaetes, crustaceans, molluscs, organic matter and detritus.

Diamond mullet ***Liza alata*** Steindachner, 1892

D1 IV; D2 9; A III, 9; pectoral fins 1 spur 15–17; LSS 29–34; TRS 10–12; GR 20–22/50–60; pyloric caeca 5; body depth 22.3–31.7% and head length 25–31.1% of SL.

Distribution: In Oceania it is reported from Tonga and New Guinea. In Australia it is abundant in the tropical waters of North Queensland, northern Territory and North western Australia.

Habitat: Inhabits coastal waters and estuaries, sometimes ascending rivers into freshwater. Prefers slow moving waters and turbid muddy lagoons. Feeds on microalgae, detritus, terrestrial plant material and aquatic insects.

Otomebora mullet ***Liza melinoptera*** Valenciennes, 1836

D1 IV; D2 9; A III, 9; pectoral fins 1 spur 14–15; LSS 28–31; TRS 9–10; GR 40–45/47–65; pyloric caeca 5; body depth 25.7–32.3% and head length 26.9–32.2% of SL.

Distribution: In Oceania it is reported from the Santa Cruz Islands, Solomon Islands, Vanuatu, New Caledonia, Tonga Islands, New Guinea, Fiji and Marquesas Islands. In the Indo-Pacific it is recorded from Taiwan, Korea, the Philippines, Malaysia, the Indonesian archipelago, Ambon, Timor, Sumatra, Ceram, Celebes, and Moluccas Islands. In Australia it is recorded from tropical Australia (North Queensland, northern Territory, and North western Australia).

Habitat: Schools in shallow coastal waters, enters mangroves, estuaries and rivers. Feeds on plant detritus, microalgae, benthic organisms and organic matter in sand and mud. Juveniles feed on planktonic algae.

Genus ***Crenimugil*** Schultz, 1946

Fringelip mullet ***Crenimugil crenilabis*** Forskål, 1775

D1 IV; D2 9; A III, 9; pectoral fins 1 spur 15–17; LSS 36–40; TRS 11–13; GR 35–42/45–58; pyloric caeca 7–10; body depth 24.5–31% and head length 24.8–31.8% of SL.

Distribution: In Oceania it is recorded from New Guinea, Cocos-Keeling Islands, Santa Cruz Islands, New Caledonia, New Hebrides, Samoa, Guam and Tahiti. It is also reported in the tropical Indo-Pacific from the Indonesian archipelago, Ryukyu Islands, and Miyazaki in southern Japan, to Marianas Islands, Marshall Islands, Phoenix Islands, Christmas Islands, and perhaps extending to the Hawaiian Islands in the north. In Australia it is reported from the Coral Sea, Queensland, Great Barrier Reef, Lizard Island, Herald Group Islands, One Tree Island, Heron Island, down to Lord Howe Island in the south.

Habitat: Found in coastal waters. Common in tide pools and reef flats. Forms large schools in coral reef areas. Feeds on diatoms and algae on dead corals.

Half fringelip mullet ***Crenimugil heterocheilus*** Bleeker, 1855a

D1 IV; D2 9; A III, 9; pectoral fins 1 spur 16–17; LSS 36–41; TRS 11–13; GR 40–45/38–56; pyloric caeca 5–9; body depth 27.3–29.6% and head length 29.9–32% of SL.

Distribution: In Oceania it is reported from the Indonesian archipelago and New Guinea, extending to New Hebrides and Samoa. It is reported from Australia in Queensland waters, including Moreton Bay.

Habitat: Found in coastal waters, ascending rivers into freshwater, penetrating far upstream.

Genus ***Moolgarda*** Whitley 1945 (***Valamugil*** Smith, 1948)

Bluetail mullet ***Moolgarda (Valamugil) buchanani*** Bleeker, 1853

D1 IV; D2 9; A III, 9; pectoral fins 1 spur 14–17; LSS 32–36; TRS 10–13; GR 31–41/34–55; pyloric caeca 6–9; body depth 29.1–33.6% and head length 25.1–30.9% of SL.

Distribution: In Oceania it is reported from parts of the Indonesian archipelago to New Guinea, Melanesia and Micronesia. It is abundant in tropical Australia, in North Queensland, northern territory and western Australia.

Habitat: Inhabits shallow coastal waters. Young enter estuaries and lagoons. Feeds on algae, diatoms, detritus, and crustaceans. Usually schooling in coral reef areas.

Bluespot mullet ***Moolgarda (Valamugil) seheli*** Forskål, 1775

D1 IV; D2 9; A III, 9; pectoral fins 1 spur 16–18; LSS 36–40; TRS 12–14; GR 20–45/34–70; pyloric caeca 7–9; body depth 28.9–32.2% and head length 25.3–30.0% of SL.

Distribution: In Oceania it is reported from parts of the Indonesian Archipelago, Sumatra, Java, Tonga, Samoa, New Guinea, and New Caledonia to Marquesas Islands. This species is common in the tropical waters of Australia in Queensland, northern territory and western Australia.

Habitat: Inhabits shallow coastal waters. Young enter estuaries and lagoons. Forms schools in coral reef areas. Feeds on algae, diatoms, detritus and crustaceans.

Longarm mullet ***Moolgarda* (*Valamugil*) *cunnesius*** Valenciennes, 1836

D1 IV; D2 9; A III, 9; pectoral fins 1 spur 14–16; LSS 32–36; TRS 10–13; GR 20–45/33–45; pyloric caeca 5–8; body depth 25.9–32.9% and head length 25.8–31.6% of SL.

Distribution: In Oceania it is reported from the Indonesian archipelago, and New Guinea. It is abundant in tropical Australia: Queensland, northern territory and western Australia.

Habitat: Inhabits shallow coastal waters. Young enter estuaries and lagoons. Forms schools in coral reef areas. Feeds on algae, diatoms, detritus and crustaceans.

Kanda mullet ***Moolgarda* (*Valamugil*) *engeli*** Bleeker, 1858b

D1 IV; D2 9; A III, 9; pectoral fins 1 spur 14–17; LSS 32–36; TRS 10–12; GR 20–32/36–52; pyloric caeca 4–7; body depth 26.5–30.9% and head length 26.2–32.8% of SL.

Distribution: In Oceania it is reported from the Indonesian Archipelago, islands of Melanesia, Micronesia, Polynesia, French Polynesia, Marquesas and Tuamotu Islands. It is also present in the tropical waters of North western Australia.

Habitat: Inhabits coastal waters, shallow lagoons, protected inlets, and over sandy to muddy areas of reef flats. Juveniles may enter rivers. Feeds on small algae, diatoms and other organic matter in sand and mud.

Longfinned mullet ***Moolgarda* (*Valamugil*) *perusii*** Valenciennes, 1836

D1 IV; D2 9; A III, 9; pectoral fins 1 spur 14–15; LSS 31–34; TRS 9–10; GR 20–32/40–50; pyloric caeca 6–7; body depth 22.4–25.1% and head length 23.6–26.1% of SL.

Distribution: In Oceania it is reported from the Indonesian Archipelago, islands of Melanesia, Micronesia, Polynesia, French Polynesia, Marquesas and Tuamotu Islands. It is also present in the tropical waters of North western Australia.

Habitat: Inhabits coastal waters and estuaries. Schooling over mudflats. Feeds on algae, diatoms, detritus and crustaceans in sand and mud.

Speigler's mullet ***Moolgarda* (*Valamugil*) *speigleri*** Bleeker, 1858

D1 IV; D2 9; A III, 9; pectoral fins 1 spur 15–16; LSS 37–41; TRS 10–11; GR 20–29/34–45; pyloric caeca 4; body depth 24.1–29.5% and head length 24.9–28.1% of SL.

Distribution: In Oceania it is reported from the Indonesian Archipelago and New Guinea. It is not present or reported from Australian waters.

Habitat: Schools in shallow coastal waters, estuaries, and backwaters. Feeds on small algae, diatoms and other organic matter taken with sand and mud. Fry feed on copepods and floating algae.

Table 5.1. Key to the species of Mugilidae occurring in Australia and Oceania.

1a.	Lips thick, upper and lower jaws posteriorly terminating in fleshy lobes; lower jaw bordered ventrally by a ridge of numerous lamellae; teeth on jaws proximal type (directly attached to jaw bones), or a combination of proximal type and distal type (borne on distal extreme of long and closely packed fibrous strands, derived from jaw bones and supported by lip tissue)	***[Cestraeus]* → 2**
1b.	Lips thin or thick, simple, and without posterior fleshy lobes; lower jaw without lamellae on lateral border; teeth on jaws proximal or distal type	**→ 3**
2a.	Medial lobes on lower jaw not reaching corner of mouth; teeth in upper jaw multicuspid and close-packed; vomer edentate; dentary produced into an elongate cylindrical arm; snout is pointed and the head profile is moderately convex both in juveniles and adults	***Cestraeus oxyrhynchus***
2b.	Medial lobes on lower jaw reaching corner of mouth; teeth in upper jaw bicuspid or unicuspid; vomer toothed; dentary produced into a short and robust cylinder; in young fish snout is pointed and the head profile is moderately convex, but in adults the snout is blunt or rounded and head is sharply convex	***Cestraeus plicatilis***
3a.	Scales in lateral series 54 or more; several rows of proximal teeth directly attached to jaw bones	***Aldrichetta forsteri***
3b.	Scales in lateral series less than 54; teeth (distal type) borne on edges of lips, often minute or absent	**→ 4**
4a.	Second pharyngobranchial toothed; lips thin; adipose eyefold absent, maxilla exposed	***[Myxus, Trachystoma]* → 5**
4b.	Second pharyngobranchial edentate; lips thick or thin; adipose eyefold present or absent; maxilla exposed or hidden.	**→ 6**
5a.	Scales cycloid; upper lip with one row of spatulate teeth with strangulate tips; maxilla relatively robust in lateral view, gradually narrowing towards distal end and moderately curved down over premaxilla; teeth present on vomer, palatine, endopterygoid, flat tongue and basihyal; opercle broad, roughly triangular with dorsal border slightly higher than the apex; 42–46 scales in lateral series	***Myxus elongatus***
5b.	Scales ctenoid; upper lip with one row of fine ciliform teeth; maxilla stocky and robust, with wide horizontal arm, and narrow sigmoid posterior end which is moderately curved down over premaxilla; opercle deeper than long, with dorsal border slightly higher than apex; 47–52 scales in lateral series; iris yellow orange to pink	***Trachystoma petardi***
6a.	Anal fin with more than nine rays and 10 pterygiophores; second dorsal fin with nine rays or more, and nine pterygiophores; two pyloric caeca	**[*Liza* "*Gracilimugil*", *Neomyxus*] → 7**
6b.	Anal fin with nine rays or less, and 10 pterygiophores or less; second dorsal fin with nine rays and nine pterygiophores; pyloric caeca variable	**→ 8**
7a.	Scales ctenoid; anal fin with three spines and 10-11 rays, and 11-12 pterygiophores in adults; second dorsal fin with nine-10 rays (including the first spine-like ray), and nine pterygiophores in adults; lips thin, each with one row of minute setiform teeth, palatine, tongue, endopterygoid and basihyal toothed but vomer edentate; posterior end of maxilla visible posteroventral to corner of closed mouth; 34–37 scales in lateral series	***Liza argentea* (*Gracilimugil argenteus*)**
7b.	Scales cycloid; anal fin with two spines and 10–11 rays, and 11 pterygiophores in adults; second dorsal fin with 10 rays (including the first spine-like ray), and 10 pterygiophores in adults; lips thick, lower lip folded down, lips bearing two-three rows of tricuspid teeth; tongue and endopterygoid toothed, but vomer, palatine and basihyal edentate; maxilla hidden; 45–48 scales in lateral series	***Neomyxus leuciscus***
8a.	Anal fin with three spins, eight rays and nine pterygiophores in adults	**→ 9**
8b.	Anal fin with three spines, nine rays and 10 pterygiophores in adults	**[*Rhinomugil, Ellochelon*] → 10**
9a.	Caudal fin truncate, with posterior margin nearly straight; adipose eyefold absent or moderately developed; teeth on endopterygoid, tongue and basihyal present, palatine edentate and vomer variably toothed; pyloric caeca branched and more than 15; basisphenoid present; urohyal anterodorsal process absent or reduced; urohyal posterior border truncate and simple; predorsal pterygiophores not joined; pelvic basipterygia broad and separated from each other; anterior pterygiophore of first dorsal fin straight or emarginate; 26–31 scales in lateral series	**11**
9b.	Caudal fin forked, adipose eyefold well-developed; endopterygoid toothed, but vomer, palatine, tongue and basihyal edentate; pyloric caeca two and simple; basisphenoid absent; urohyal anterodorsal process distinct; urohyal posteroventral border produced into a pair of protruding processes; predorsal pterygiophores joined; pelvic basipterygia articulated together; anterior pterygiophore of first dorsal fin deeply curved; 38–43 scales in lateral series	***Mugil cephalus***

Table 5.1. contd....

Table 5.1. contd.

10a.	Head flattened dorsally and concave between eyes, nostrils set low on snout, and eyes positioned dorsolateral and high on head; snout projecting beyond mouth and upper lip inferior to it; adipose eyefold moderately developed; vomer edentate; pelvic basipterygia broad and separated from each other, but each basipterygium has a medial alar process overlapping the contralateral structure; angular, metapterygoid, hyomandibula, opercle, preopercle, dermosphenotic, nasal, frontal and parasphenoid with distinct morphology; circumorbital bones short and robust; postzygapophysis of the second abdominal vertebra elongate; haemal bridge present in vertebra 11; 28–31 scales in lateral series . . . . **_Rhinomugil nasutus_**
10b.	Head broad, dorsally flattened and slightly convex between eyes; nostrils set high on snout and eyes positioned lateral on head; upper lip not inferior to snout; adipose eyefold absent; pectoral fins black (lower part yellowish in adults); vomer toothed; pelvic basipterygia broad and widely separated from each other; postzygapophysis of the second abdominal vertebra forming a pair of short blunt processes, expanding ventrolaterally; haemal bridges present in vertebra 10 and 11; scales in lateral series 26–28 . . . . **_Ellochelon vaigiensis_**
11a.	Scales ctenoid without membranous digitated hind margin; pectoral axillary scale absent or rudimentary; posterior end of maxilla either visible posteroventral to corner of closed mouth, or completely or partially concealed when the mouth is closed, anteroventral processes of basioccipital absent; posterolateral processes of basioccipital intermediate. . . . . **[_Oedalechilus, Crenimugil, Moolgarda "Valamugil"_] → 12**
11b.	Scales ctenoid with membranous digitated hind margin; pectoral axillary scale present and distinct; maxilla slender and weakly curved down posteriorly; posterior end of maxilla completely concealed posteroventral to corner of closed mouth; anteroventral and posteroventral processes of basioccipital present and developed . . . . **[_Liza, Liza "Paramugil"_] → 13**
12a.	Preorbital deeply notched; lips with distinct rugose fringe; upper lip thick, split into longitudinal lobes, lips deeply folded into preorbital notch at the corner of mouth; posterolateral border of frontal and proximal border of sphenotic are joined; postzygapophysis of the second abdominal vertebra absent; haemal bridges present in vertebra 10 and 11; three pyloric caeca. . . . . **_Oedalechilus labiosus_**
12b.	Preorbital moderately notched or only slightly kinked; lips thin, simple, without lobes or rugose fringe and not folded at the corner of mouth, posterolateral border of frontal and proximal border of sphenotic not joined; postzygapophysis of the second abdominal vertebra present; haemal bridges in vertebra 10 and 11 present or absent; pyloric caeca more than three. . . . . **[_Crenimugil, Moolgarda "Valamugil"_] → 19**
13a.	Maxilla more-or-less stocky, and sigmoid, posterior tip distinctly curved down and visible posteroventral to corner of closed mouth; angle of dentary symphysis obtuse; preorbital kinked at anterior serrate margin, broad and squarish at posteroventral border; postzygapophysis of the second abdominal vertebra spinous or elongate, expanding posterodorsally; haemal bridges absent in abdominal vertebrae; anterior process of proximal ribs not developed; intercalar developed; articulation of intercalar and ventral arm of posttemporal complicated; pyloric caeca five-six . . . . **[_Liza_] → 14**
13b.	Maxilla slender and narrow in lateral aspect, posterior end sharply curved down and partially or completely concealed posteroventral to corner of closed mouth; angle of dentary symphysis broadly obtuse; preorbital distinctly notched anteriorly and squarish at posteroventral border; angular roughly elongate and trapezoid shape; palatine elongate with expanded anterior process; haemal bridges present in vertebra 10 and 11; postzygapophysis of the second abdominal vertebra either elongate, or reduced to a pair of short and blunt processes; anterior process of proximal ribs developed; intercalar reduced; articulation of intercalar and ventral arm of posttemporal simple; pyloric caeca seven–26. . . . . **[_Liza "Paramugil"_] → 18**
14a.	Adipose eyefold moderately developed, covering most of iris. **15**
14b.	Adipose eyefold poorly developed, as a rim around eye. **16**
15a.	Body elongate, relatively slender; head pointed and depressed dorsally, and more-or-less bulging laterally, head length greater than body depth; eyes relatively small; parasphenoid nearly straight in lateral aspect; dermosphenoid long with anterior process; ethmoid reduced. . . . . **_Liza tade_**
15b.	Body moderately robust, head broad, dorsally flattened and bluntly pointed, head length equal or slightly bigger than body depth; snout shorter than eye diameter; parasphenoid convex in lateral aspect; ethmoid developed . . . . **_Liza subviridis_**

Table 5.1. contd....

Table 5.1. contd.

16a.	Body moderately robust, adipose eyefold poorly developed as a rim around eye; upper lip with one row of very close-set, small peg-like unicuspid teeth at the edge and two–four inner rows of more wide-set and irregularly scattered smaller teeth; lower lip with one row of small ciliform teeth; teeth on vomer, palatine, endopterygoid, tongue and basihyal; caudal fin forked; dentary coronoid process weakly curved; 32–35 scales in lateral series . . . . ***Liza macrolepis***
16b.	Body elongate or moderately robust; upper lip with one row of small, curved unicuspid or setiform teeth; palatine edentate; 28–32 scales in lateral series; caudal fin forked or emarginate. **19**
17a.	Body elongate, moderately deep and compressed at caudal peduncle; upper lip with one row of small, curved and close-set unicuspid teeth; vomer, endopterygoid, tongue and basihyal toothed but palatine edentate; second dorsal and anal fins emarginate; caudal fin emarginate with dark margin; scales on upper part of flanks with dark margins, giving reticulated appearance and dark horizontal streaks on upper rows of scales; origin of pelvic fins midway between origin of pectoral fin and origin of first dorsal fin; neural postzygapophysis of the second abdominal vertebra spinous, medium size and extended posterodorsally; dentary coronoid process concave; 29–32 scales in lateral series . . . . ***Liza alata***
17b.	Body moderately robust and well compressed at caudal peduncle; maximum body depth at origin of first dorsal fin 26–33% standard length; upper lip with one row of close-set setiform teeth and one inner row of well-spaced coliform teeth; caudal fin emarginate and dusky; vomer and palatine edentate, but endopterygoid, tongue and basihyal dentigerous; neural postzygapophysis of the second abdominal vertebra elongate; dentary coronoid process weakly concave; 28–31 scales in lateral series. . . . . ***Liza melinoptera***
18a.	24–26 scales in lateral series; seven–15 pyloric caeca; vomerine teeth absent; urohyal anterior process sharp; postzygapophysis of the 2nd abdominal vertebra elongate; pterotic with a lateral crest; posterolateral process of basioccipital rudimentary; more than five epipleural ribs . . . . ***Liza parmata* (*Paramugil parmatus*)**
18b.	28–32 scales in lateral series; 18–26 pyloric caeca; vomerine teeth present; urohyal anterior process blunt; postzygapophysis of the 2nd abdominal vertebra short and blunt, expanded ventrolaterally posterolateral process of basioccipital intermediate; five epipleural ribs. . . . . ***Liza georgii* (*Paramugil georgii*)**
19a.	Upper lip more-or-less thick, lower quarter or third part of lip bearing enlarged crenulations or papillae; lower lip thin either simple or fringed with a row of fine crenulations; adipose eyefold absent; angular deltoid with dorsal process; anterior pterygiophore of the first dorsal fin emarginate. . . . . ***Crenimugil* → 20**
19b.	Lips thin and simple, without crenulations or papillae; adipose eyefold present or absent; angular deltoid without dorsal process; anterior pterygiophore of the first dorsal fin straight; pharyngobranchial organ with large denticulate area, broad sulcus and no valves. . . . . ***Moolgarda* (*Valamugil*) → 21**
20a.	Upper lip very thick, with several rows of papillae on the lower part; lower lip thin, folded out and downwards, with fine crenate edge and one–two rows of fine papillae on its inner surface. . . . . ***Crenimugil crenilabis***
20b.	Upper lip moderately thick with two–five rows of papillae medial to its lower surface; lower lip thin, directed forwards and not papillate or crenate. . . . . ***Crenimugil heterocheilus***
21a.	Adipose eyefold absent; number of scales around caudal peduncle 19–20; origin of second dorsal fin at vertical through origin of anal fin or just behind it; second dorsal and anal fins falcate; pectoral fins moderately falcate; caudal fin deeply emarginate to forked; number of pyloric caeca 6–9. . . . . **22**
21b.	Adipose eyefold present; number of scales around caudal peduncle 16; origin of second dorsal fin at vertical behind anterior fourth or more of the anal fin; second dorsal and anal fins emarginate to moderately falcate; caudal fin emarginate or forked; number of pyloric caeca four–eight. **23**
22a.	Number of scales in lateral series 32–36 (usually 35) . . . . ***Moolgarda buchanani* (*Valamugil buchanani*)**
22b.	Number of scales in lateral series 37–38. . . . . ***Moolgarda seheli* (*Valamugil seheli*)**
23a.	Origin of second dorsal fin nearer to snout tip than to base of caudal fin. **24**
23b.	Origin of second dorsal fin nearer to base of caudal fin than to snout tip. **25**

Table 5.1. contd....

Table 5.1. contd.

24a.	Body moderately robust; origin of second dorsal fin at vertical posterior to about 1/3 anal fin base; both fins emarginate to slightly falcate; pectoral fin considerably long and falcate (22–29% SL), distinctly greater than length of head minus snout, and often equals head length (75–101% head length); caudal fin moderately forked; 32–36 (usually 35) scales in lateral series; 5–8 pyloric caeca. . . . . ***Moolgarda cunnesius* (*Valamugil cunnesius*)**
24b.	Body moderately robust and fusiform; origin of second dorsal fin at vertical posterior to 1/3 to 1/2 anal fin base, both fins emarginate, and scaled anterobasally; pectoral fin 21–25% standard length, and greater than length of head minus snout; caudal fin forked; 37–41 scales in lateral series; four pyloric caeca. . . . . ***Moolgarda speigleri* (*Valamugil speigleri*)**
25a.	Body robust, more-or-less fusiform; predorsal scales extending nearly up to tip of snout; origin of second dorsal fin at vertical posterior to 1/4 to 1/3 anal fin base, both fins well emarginate, and scaled anterobasally; caudal fin emarginate; 31–36 scales (usually 32) in lateral series; 10–12 scales in transverse series; four–seven pyloric caeca . . . . ***Moolgarda engeli* (*Valamugil engeli*)**
25b.	Body elongate and compressed; predorsal scales extending to just behind posterior nostrils; origin of second dorsal fin at vertical posterior to 1/3 to 1/2 anal fin base, both fins emarginate, and scaled anterobasally; caudal fin deeply emarginate; 31–34 scales in lateral series; nine-10 scales in transverse series; six–seven pyloric caeca . . . . ***Moolgarda perusii* (*Valamugil perusii*)**

Notes:

The names in square brackets [...] correspond to valid generic names in Eschmeyer (2015).
The names in quotation marks "..." within a square bracket [..."..."] correspond to their preceding generic name which is considered as valid by Eschmeyer (2015).
The names in normal brackets (...) correspond to their preceding species name which is considered as valid by Eschmeyer (2015).

References

Berra, T.M. 1981. An Atlas of Distribution of the Freshwater Fish Families of the World. University of Nebraska Press, London. 197pp.

Chubb, C.F., I.C. Potter, C.J. Grant, R.C.J. Lenanton and J. Wallace. 1981. Age structure, growth rates and movement of sea mullet *Mugil cephalus* L., and yellow eye mullet, *Aldrichetta forsteri* (Valenciennes) in the Swan Avon River system, Western Australia. Aust. J. Mar. Freshw. Res. 32: 605–628.

Durand, J.D. 2015. Implications of molecular phylogeny for the taxonomy of Mugilidae. *In*: D. Crosetti and S.J.M. Blaber (eds.). Biology, Ecology and Culture of Grey Mullet (Mugilidae). CRC Press, Boca Raton, USA (this book).

Durand, J.D., K.N. Shen, W.J. Chen, B.W. Jamandre, H. Blel, K. Diop, M. Nirchio, F.J. Garcia de León, A.K. Whitfield, C.W. Chang and P. Borsa. 2012a. Systematics of the grey mullets (Teleostei: Mugiliformes: Mugilidae): molecular phylogenetic evidence challenges two centuries of morphology-based taxonomic studies. Mol. Phylogenet. Evol. 64: 73–92.

Durand, J.D., W.J. Chen, K.N. Shen, C. Fu and P. Borsa. 2012b. Genus-level taxonomic changes implied by the mitochondrial phylogeny of grey mullets (Teleostei: Mugilidae). Comptes Rendus Biologies 335: 687–697.

Eschmeyer, W.N. (ed.). 2015. Catalog of Fishes: Genera, Species, References. (http://researcharchive.calacademy.org/research/ichthyology/catalog/fishcatmain.asp). Electronic version accessed on 15/01/2015.

Ghasemzadeh, J. 1998. Phylogeny and systematics of Indo-Pacific mullets (Teleostei: Mugilidae) with special reference to the mullets of Australia. Ph.D. dissertation, Macquarie University, Sydney.

Ghasemzadeh, J., W. Ivantsoff and Aarn. 2004. Historical overview of mugilid systematics with description of *Paramugil* (Teleostei: Mugiliformes: Mugilidae), new genus. Aqua. J. Ichthyol. Aquat. Biol. 8: 9–22.

González-Castro, M. and J. Ghasemzadeh. 2015. Morphology and morphometry based taxonomy of Mugilidae. *In*: D. Crosetti and S.J.M. Blaber (eds.). Biology, Ecology and Culture of Grey Mullet (Mugilidae). CRC Press, Boca Raton, USA (this book).

Grant, C.J. and A.V. Spain. 1975a. Reproduction growth and size allometry of *Mugil cephalus* Linnaeus (Pisces: Mugilidae) from North Queensland inshore waters. Aust. J. Zool. 23: 181–201.

Grant, C.J. and A.V. Spain. 1975b. Reproduction growth and size allometry of *Valamugil seheli* (Forskål) (Pisces: Mugilidae) from North Queensland inshore waters. Aust. J. Zool. 23: 463–474.

Grant, C.J. and A.V. Spain. 1975c. Reproduction growth and size allometry of *Liza vaigiensis* (Quay & Gaimard) (Pisces: Mugilidae) from North Queensland inshore waters. Aust. J. Zool. 23: 475–485.

Grant, C.J. and A.V. Spain. 1977. Variation of the body shape of three species of Australian Mullets (Pisces: Mugilidae) during the course of development. Aust. J. Mar. Freshw. Res. 28: 723–738.
Grant, C.J., A.V. Spain and P.N. Jones. 1977. Studies of sexual dimorphism and other variations in nine species of Australian Mullets (Pisces: Mugilidae) Aust. J. Zool. 25: 615–630.
Harris, J.A. 1968. The yellow-eye mullet. Age structure, growth rate and spawning cycle of a population of yellow eye mullet *Aldrichetta forsteri* (Cuv. and Val.) from the Coorong Lagoon, South Australia. Trans. R. Soc. South Australia 92: 37–50.
Harrison, I.J. and H. Senou. 1999. Order Mugiliformes. pp. 2069–2790. *In*: K.E. Carpenter and V.H. Niem (eds.). The Living Marine Resources of the Western Central Pacific, FAO Species Identification Guide for Fisheries Purposes. FAO, Rome.
Langi, S.A., T.F. Latu and S. Tulua. 1990. Preliminary study of the biology of mullets (Pisces: Mugilidae) from Nuku'Alofa, Tonga. Pap. Fish Sci. Pac. Isl. 1: 37–42.
Lenanton, R.C.J., I.C. Potter, N.R. Lonergan and P.J. Chrystal. 1984. Age structure and changes in abundance of three important species of teleost in a euphrotic estuary (Pisces: Teleostei) J. Zool. Lond. 203: 311–327.
Macleay, W. 1880. On the Mugilidae of Australia. Proc. Linn. Soc. N.S.W. 4: 410–27.
Macleay, W. 1884a. On a new and remarkable fish of the family Mugilidae from the interior of New Guinea. Proc. Linn. Soc. N.S.W. 8: 2–6.
Macleay, W. 1884b. Notes on the collection of fishes from the Burdekin and Mary Rivers, Queensland. Proc. Linn. Soc. N.S.W. 8: 199–212.
Merrick, J.R. and G.E. Schmida. 1984. Australian Freshwater Fishes. Biology and Management. Griffin Press Limited, 409pp.
Munro, I.S.R. 1956. Handbook of Australian Fishes. CSIRO Publication, Canberra, Australia. 172pp.
Munro, I.S.R. 1967. The Fishes of the New Guinea. Dept. of Agriculture, Stock and Fisheries, Port Moresby. 650pp.
Ogilby, J.D. 1887. On a new genus and species of Australian Mugilidae. Proc. Zool. Soc. Lond. 1887: 614–616.
Rossi, A.R., S. Livi and D. Crosetti. 2015. Genetics of Mugilidae. *In*: D. Crosetti and S.J.M. Blaber (eds.). Biology, Ecology and Culture of Grey Mullet (Mugilidae). CRC Press, Boca Raton, USA (this book).
Spain, A.V., C.J. Grant and D.F. Sinclair. 1980. Phenotypic affinities of 11 species of Australian mullet (Pisces: Mugilidae) Aust. J. Mar. Freshwater Res. 31: 69–83.
Thomson, J.M. 1954. The Mugilidae of Australia and adjacent seas. Salisbury South, Australia Aust. J. Mar. Freshwater Res. Salisbury South, Australia 5: 70–131.
Thomson, J.M. 1963. Synopsis of Biological data on the grey mullet *Mugil cephalus* Linnaeus 1758. Aust. CSIRO Div. Fish. Oceanogr. Fish Synop. 1: 1–75.
Weber, M. and L.F. de Beaufort. 1922. The fishes of the Indo-Australian Archipelago. (Leiden). Vol. 4: 1–410.
Whitley, G.P. 1930. Five new generic names for Australian fishes. Aust. Zool. 6: 250–251.
Whitley, G.P. 1941. Ichthyological notes and illustrations. Aust. Zool. 10: 1–52.
Whitley, G.P. 1945. New sharks and fishes from Western Australia Part 2. Aust. Zool. 11: 1–42.
Whitley, G.P. 1948. New sharks and fishes from Western Australia Part 4. Aust. Zool. 11: 259–276.

CHAPTER 6

Biogeography and Distribution of Mugilidae in the Western, Central and Southern Regions of Africa

Jean-Dominique Durand[1,*] and *Alan K. Whitfield*[2]

Introduction

The African continent has a shoreline of 26,000 km that spans from latitude 37° 21'N to 34° 51'S. This large latitudinal range, as well as the presence of important oceanographic and topographic features, is the main factor responsible for the delimitation of four marine regions: the temperate north-eastern Atlantic region, the tropical East Atlantic region, the temperate southern African region and the western Indo-Pacific region (Spalding et al. 2007). There are diverse oceanographic regimes present in African coastal waters, e.g., the cold Canary Current is responsible for major tropical upwelling off the coast of Morocco and Mauritania, the cold Benguela Current upwelling off the Atlantic coast of Namibia and South Africa, the warm Agulhas Current influencing the Indian Ocean coast of south eastern Africa, and the warm Somali Current flowing past Somalia and Oman.

The presence of upwelling regimes in particular limit connectivity of tropical marine species by creating cold water barriers to coastal movements and migrations (Maree et al. 2000). In addition, upwelling favours fish larval retention (Moyano et al. 2014) and would thus be responsible for promoting regional diversity differences. The most striking difference observable on a continental scale is the large asymmetry in coral reef distribution. While coral reefs are large and extensive along the eastern African coast, they are nearly absent on the western side, primarily due to Atlantic upwellings and strong cold coastal currents that reduce water temperatures in these areas (Nybakken 1997). Similarly, sea-surface temperature plays a major role in marine fish distribution, with the observed geographic range boundaries of most species being closely linked to their potential latitudinal ranges and thermal tolerance (Sunday et al. 2012).

Beyond physical factors and species physiological tolerance, the distribution range of a species is also shaped by its evolutionary history. During the Pleistocene era, glaciations deeply impacted coastal marine life and distribution through both sea level lowering and sea surface temperature changes. In western Africa,

[1] IRD, Center for Marine Biodiversity, Exploitation and Conservation, Université Montpellier 2 CC 093, Place Eugène Bataillon, 34095 Montpellier Cedex 5, France.
Email: jean-dominique.durand@ird.fr
[2] South African Institute for Aquatic Biodiversity (SAIAB), Private Bag 1015, Grahamstown 6140, South Africa.
Email: a.whitfield@saiab.ac.za
* Corresponding author

the lower temperature and greater aridity periodically experienced during the Plio-Pleistocene restricted forest areas (Schefub et al. 2003) as well as potential refuge for aquatic species (Lévêque 1997). During the Quaternary, forest refuges occupied core areas located in Sierra Leone–Liberia, East Côte d'Ivoire–West Ghana, and Cameroon–Gabon (Maley 1991, Nichol 1999). A good concordance was found between the location of these core areas and the distribution of the genetic diversity of estuary-associated fishes such as the black-chinned tilapia, *Sarotherodon melanotheron* (Falk et al. 2003) or the bonga shad, *Ethmalosa fimbriata* (Durand et al. 2005, Durand et al. 2013a), thus attesting to the impact of Pleistocene climatic changes on the population demography of these species. In South Africa, the last sea level regression drastically reduced the size and number of suitable habitats due to the very narrow continental shelf on the west and the east coasts. This led to the isolation and divergence of populations of marine species such as *Clinus superciliosus* and *C. cottoides* (von der Heyden et al. 2011). Earlier, during the Pliocene-Pleistocene transition, establishment of the present day characteristics of the Benguela Current and the intensification of the upwelling regime would have been the initial isolation mechanism for some marine species previously present from Angola to South Africa (Henriques et al. 2012, Henriques et al. 2014).

From the above it is apparent that the geographic distribution and genetic structure of a species are windows on the factors and evolutionary processes that significantly act upon dispersal opportunities and determine the demographic characteristics. However, the accuracy of any inferences relating to modern day distributions relies deeply on the quality of historic and present geographic and oceanographic data, as well as taxonomy. Hence there is an urgent need to provide an up-to-date taxonomic review of African Mugilidae especially in light of the recent molecular phylogeny results (Durand et al. 2012a,b, Fig. 6.1). The Mugilidae diversity in the western, central and southern regions of Africa consists of 31 mitochondrial lineages corresponding to 26 morphological species and five putative species, all belonging to 10 genera according the taxonomic revision based on the mitochondrial phylogeny (Durand et al. 2012b). Genus names hereafter follow the nomenclature of Eschmeyer (2014) and in brackets the genus name considered as valid, when different, by Durand et al. (2012b).

Distribution of Mugilidae Species in Western, Central and Southern Regions of Africa

Endemic Species of West and Central Africa

Liza (Chelon) bandialensis Diouf, 1991. The diassanga mullet occurs in Senegal, Gambia and Guinea Bissau (Trape et al. 2012, Fig. 6.2A). It is phylogenetically closely related to the South African species *Liza (Chelon) tricuspidens* (Fig. 6.1) with no genetic heterogeneity recorded between individuals collected in Senegal (Saloum Estuary) and Guinea Bissau (Bijagos Islands) (Durand et al. 2012a). It is generally a rare species, with larval and juvenile stages unrecorded (Trape et al. 2009).

Chelon bispinosus, Bowdich, 1825. The Cape Verde mullet is endemic to the Cape Verde Islands (Thomson 1997, Wirth et al. 2013, Fig. 6.2B) and has a close phylogenetic relationship with the north eastern Atlantic species, *C. labrosus* (Fig. 6.1).

Mugil bananensis Pellegrin, 1927. The banana mullet is distributed in coastal waters from Senegal to Angola including the Cape Verde Islands (Albaret 1992, Wirth et al. 2013, Fig. 6.2C). No genetic divergence has been documented between individuals collected in Senegal and Cote d'Ivoire (Durand et al. 2012a, Fig. 6.2C).

Mugil curema Valenciennes, 1836. Based on museum records and observations present in the GBIF biodiversity database, the white mullet occurs from Senegal to the Congo, including the islands of Cape Verde, Sao Tome and Principe (Fig. 6.2D). Outside of Africa, *M. curema* is present in American Pacific and Atlantic coastal waters. This distributional range encompasses several biogeographic areas spanning two continents and recent phylogenetic investigations demonstrate that *M. curema* is cryptically diverse (Fraga et al. 2007, Heras et al. 2009, Heras et al. 2009, Durand et al. 2012a). Between two and four evolutionary lineages characterize this so-called single species depending on the sampling regime over the

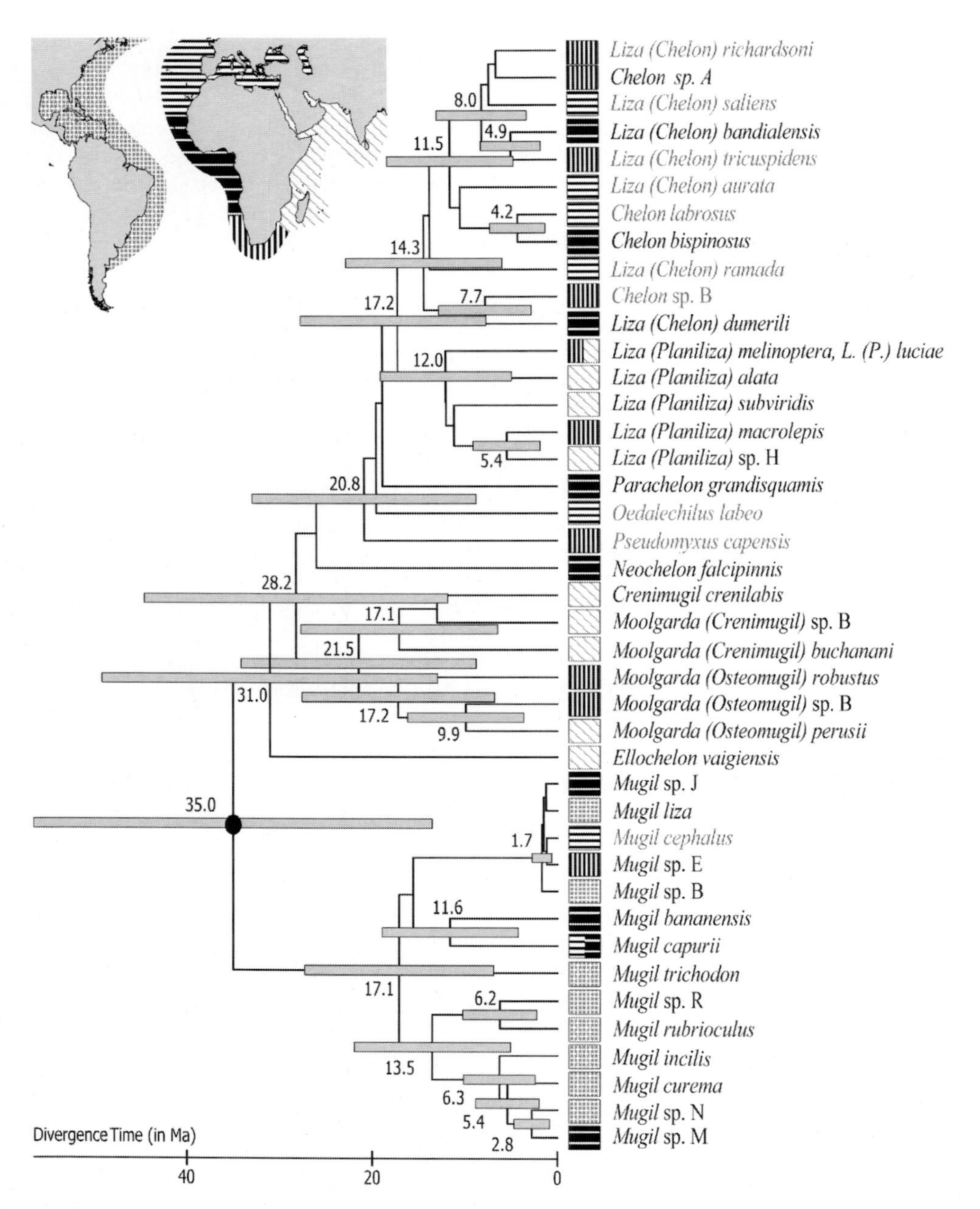

Figure 6.1. Maximum Likelihood Tree (timetree) generated using the RelTime method (Tamura et al. 2012) based on mitochondrial Cytochrome *b* and RNA 16S sequences (1282 positions in the final dataset).

The black dot on the dendrogram corresponds to the ancestor of the Mugilidae and was calibrated using the oldest known grey mullet fossil, namely *Mugil princeps* (30–40 My). Divergence times for all branching points in the topology were calculated using the Maximum Likelihood method based on the General Time Reversible model (Nei and Kumar 2000). Grey bars around the nodes (only those supported by a bootstrap value > 70%, with the exception of the *Planiliza/ Chelon* divergence) represent 95% confidence intervals and were computed using the method described in Tamura et al. (2013). The estimated log likelihood value for the topology is 15,274.05. The tree is drawn to scale, with branch lengths measured according to the relative number of substitutions per site. Evolutionary analyses were conducted in MEGA6 (Tamura et al. 2013). The distributional range of mugilid species in the known biogeographic provinces (shaded area on the map) is depicted in the first column of boxes adjacent to the dendrogram. Species names in grey corresponds to temperate species, in black name tropical species. Ma = Million years.

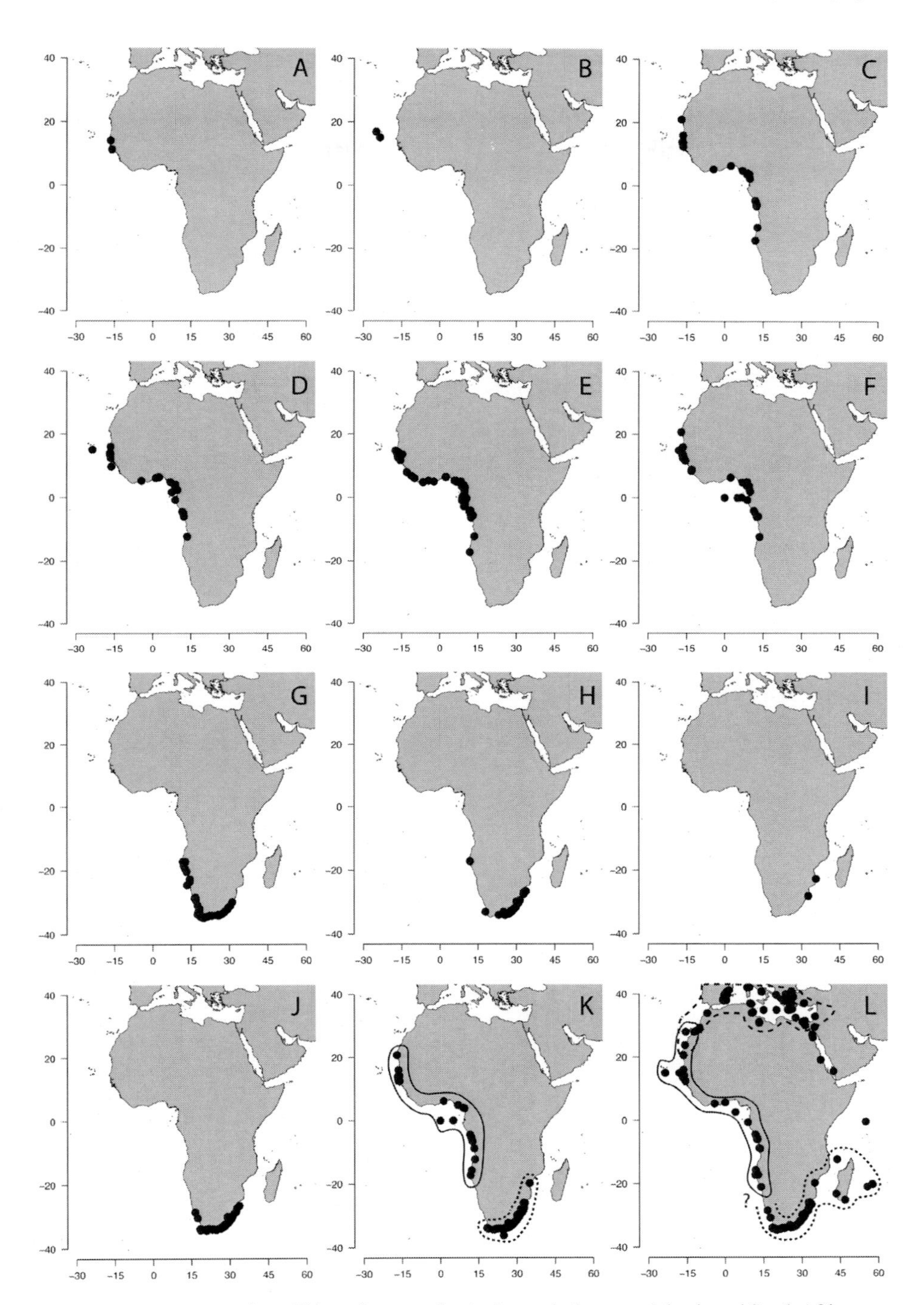

Figure 6.2. Distributional range of mugilid species occurring in the tropical eastern Atlantic and South Africa.

Each dot corresponds to a record in the GBIF data bank (http://www.gbif.org) plus some specimens barcoded by Durand et al. (2012a). (A) *Liza* (*Chelon*) *bandialensis*; (B) *Chelon bispinosus*; (C) *Mugil bananensis*; (D) *Mugil curema*; (E) *Neochelon falcipinnis*; (F) *Parachelon grandisquamis*; (G) *Liza* (*Chelon*) *richardsonii*; (H) *Liza* (*Chelon*) *tricuspidens*; (I) *Chelon* sp. A, J. *Paramyxus capensis*; (K) *Liza* (*Chelon*) *dumerili ssl* (*L.* (*C.*) *dumerili* = continuous line, *Liza* (*Chelon*) sp. B = dotted line), L. *Mugil cephalus ssl* (*Mugil* sp. J = continuous line; *M. cephalus* = discontinuous line; *Mugil* sp. E = dotted line). In the absence of exact distributional data, lines show the probable distributional range of cryptic species considering the distribution of evolutionary lineages (Durand et al. 2012a).

distribution range (Fraga et al. 2007, Heras et al. 2008, Heras et al. 2009, Durand et al. 2012a). All African *M. curema* belong to a unique evolutionary lineage, endemic to Africa. As the American *M. curema* was suggested to be a complex of cryptic species (Heras et al. 2009), which is congruent with chromosomatic formulae identified in West Atlantic specimens (Nirchio and Cequea 1998, Durand et al. 2012a), it is probable that the allopatric evolutionary lineage present in Africa is also a distinct species. Historically, *M. metzelaari* Chabanaud 1926 has been described in West Africa but later synonymized with *M. curema* (Thomson 1997). Pending taxonomic revision, this lineage has been provisionally designated as *Mugil* sp. M (Durand and Borsa 2015).

Neochelon falcipinis, Valenciennes, 1836. Distribution range of the sicklefin mullet spans from North Senegal to Congo or Angola according to Thomson (1997) and GBIF records (Fig. 6.2E), respectively. No significant genetic divergence was observed between samples collected in Senegal and Togo (Durand et al. 2012a). *N. falcipinis* is the type species of the monotypic genus *Neochelon* (Durand et al. 2012b).

Parachelon grandisquamis, Valenciennes, 1836. Based on museum records and observations present in the GBIF biodiversity database, distribution of the large scaled mullet occurred in estuarine and coastal areas from Mauritania to Angola, including Sao Tome and Principe (Fig. 6.2F). In contrast, Thomson (1997) described a northern range limit of the Senegal Estuary and a southern limit of the Niger Estuary. Based on records in the GBIF database, the widespread presence of this species in Mauritania is doubtful since only one record from this country exists, which is not the case for its documented presence south of Nigeria. This *P. grandisquamis* is the type species of the monotypic genus *Parachelon* (Durand et al. 2012b).

Endemic Species of South Africa

Liza (Chelon) richardsonii Smith, 1846. The South African mullet has been recorded from southern Angola (Cunene River) to St. Lucia (Heemstra and Heemstra 2004, Fig. 6.2G). It is abundant in temperate South African estuaries and the adjacent coastal zone but is rarely recorded in subtropical eastern Cape and KwaZulu-Natal waters (Smith and Smith 1986, Whitfield 1998).

Liza (Chelon) tricuspidens Smith, 1935. The striped mullet is frequently recorded from False Bay in the western Cape Province to Kosi Bay in KwaZulu-Natal, and is absent from cool-temperate estuaries on the South African west coast (Smith and Smith 1986, Whitfield 1998, Heemstra and Heemstra 2004). However, information from the GBIF database suggests that this species is present in Angola (Fig. 6.2H) and until genetic studies have been conducted it remains uncertain as to whether this is an allopatric population of the South African species or not.

***Chelon* sp. A.** This species has not yet been described but was characterized genetically in the phylogenetic study of Durand et al. (2012a) (Fig. 6.2I). The analyzed specimen (SAIAB-78131) was collected at low tide from a Cape Vidal rock pool in KwaZulu-Natal. The distributional range of this species is probably not restricted to northern KwaZulu-Natal as the DNA sequence (cytochrome oxidase I) is identical to an individual (ADC08 Smith 222.5 #3) identified as *Liza* (*Planiliza*) *macrolepis* collected at Pomene, Mozambique. The new species has probably been misidentified with *L.* (*P.*) *macrolepis*, and may therefore have a distribution range, at least in the South eastern Indian Ocean, which matches that of *L.* (*P.*) *macrolepis*. Pending taxonomic revision this lineage has been provisionally designated as *Chelon* sp. A (Durand and Borsa 2015).

Pseudomyxus capensis Valenciennes, 1836. The freshwater mullet is known from False Bay to Kosi Bay (Whitfield 1998, Heemstra and Heemstra 2004). However, the distribution range needs to be expanded on the western coast as there are two specimens conserved in the SAIAB collection, collected in the Orange and Spoeg Rivers mouths (SAIAB-43602 and SAIAB-11792) (Fig. 6.2J). *P. myxus* is the type species of the monotypic genus *Pseudomyxus* (Durand et al. 2012b).

Species Apparently Present in West, Central and Southern Africa

Liza (Chelon) dumerili Steindachner, 1869. The grooved mullet is present along African shorelines from Mauritania to Namibia and from the Western Cape Province (False Bay) in South Africa to Mozambique (Smith and Smith 1986, Albaret 1992, Gushchin and Fall 2012, Fig. 6.2K). It is also listed in fish inventory of Sao Tome Island in the Gulf of Guinea (Afonso et al. 1999). Phylogenetic reconstruction indicated cryptic genetic diversity for this species consisting of two evolutionary lineages (Durand et al. 2012a). The distributional range of these lineages appears to be allopatric with one occurring in western Africa (Senegal, Guinea Bissau and Togo) and the other in South Africa and southern Mozambique. The level of genetic divergence and the disjunct distributional range suggest that these lineages may correspond to two sister species. Because the type species has been described from a specimen collected in St. Louis, Senegal, *L.* (*C.*) *dumerili* is allocated to the western African lineage while the southern African lineage may be named *L.* (*C.*) *natalensis*, Castelnau 1861 or *L.* (*C.*) *canaliculatus*, Smith 1983. To date, the latter two species described from South Africa were synonymized with *L.* (*C.*) *dumerili* by Smith and Smith (1986) and Thomson (1997). Pending taxonomic revision, this latter lineage has been provisionally designated as *Chelon* sp. B by Durand and Borsa (2015). Consequently, *L.* (*C.*) *dumerili* distribution would be from St Louis (Senegal) to Cunene (Namibia) while that of *Chelon* sp. B would range from the western Cape (False Bay) in South Africa to Mozambique (Fig. 6.2K).

Mugil cephalus, Linnaeus, 1758. The flathead mullet is apparently present in all tropical and temperate waters of the world between 42° N and 42° S (Thomson 1966, Thomson 1997). However, a closer inspection of mugilid material from museum collections indicates that the overall distributional range of this species is probably more restricted, especially in tropical waters (Gilbert 1993). Along the African Indian Ocean shoreline, from Somalia to Tanzania, no record of the species exists, thus confirming that the distribution range of this species is probably more complex than generally thought (GBIF data, Fig. 6.2L). Furthermore, phylogeographic and population genetic studies have demonstrated that this species actually consists of several evolutionary lineages (Crosetti et al. 1994, Rossi et al. 1998a, Rossi et al. 1998b, Livi et al. 2011, Durand et al. 2012a). In Africa, from the Mediterranean Sea to the South West Indian Ocean, three evolutionary parapatric lineages have been highlighted genetically (Durand et al. 2012a, Durand et al. 2013b). It is probable that these lineages are cryptic species as it has been demonstrated that individuals belonging to different lineages in the north western Pacific are reproductively isolated (Shen et al. 2011). Because the type locality of *M. cephalus* Linnaeus 1758 is from Europe where only one evolutionary lineage is present (Durand et al. 2013b), the distributional range of *M. cephalus* would be restricted in Africa to the Mediterranean Sea and the Atlantic Moroccan shoreline north of Dakhla (Durand et al. 2012a). The lineage observed along the West African coast, from Knifiss in Morocco to the Cunene Estuary in Angola (Durand et al. 2012a), would correspond to *Mugil ashanteensis* Bleeker 1863 which was synonymized with *M. cephalus* by Thomson (1997). Pending taxonomic revision, this latter lineage was provisionally designated as *Mugil* sp. J by Durand and Borsa (2015). The third lineage recorded from South Africa, Réunion Island, Mauritius and Mayotte (Durand et al. 2012a, pers. comm.) would indicate a third African species that could correspond to *M. borbonicus* Valenciennes 1836 considering the type locality and the principle of priority. Pending taxonomic revision, this lineage was provisionally designated as *Mugil* sp. E by Durand and Borsa (2015).

Indo-Pacific Species Present in South Africa

The following species, with rare exceptions, are typical Indo-Pacific taxa with a distributional range limit extending to South Africa.

Crenimugil crenilabis Forsskål, 1775. The fringelip mullet is an Indo-Pacific species that is present in Africa from Somalia to the Eastern Cape Province in South Africa (Smith and Smith 1986, Fig. 6.3A). Low genetic heterogeneity has been recorded using individuals collected over its distribution range (Seychelles, Maldives, Saipan, Chesterfield, French Polynesia), thus confirming its single species taxonomic status (Durand et al. 2012a).

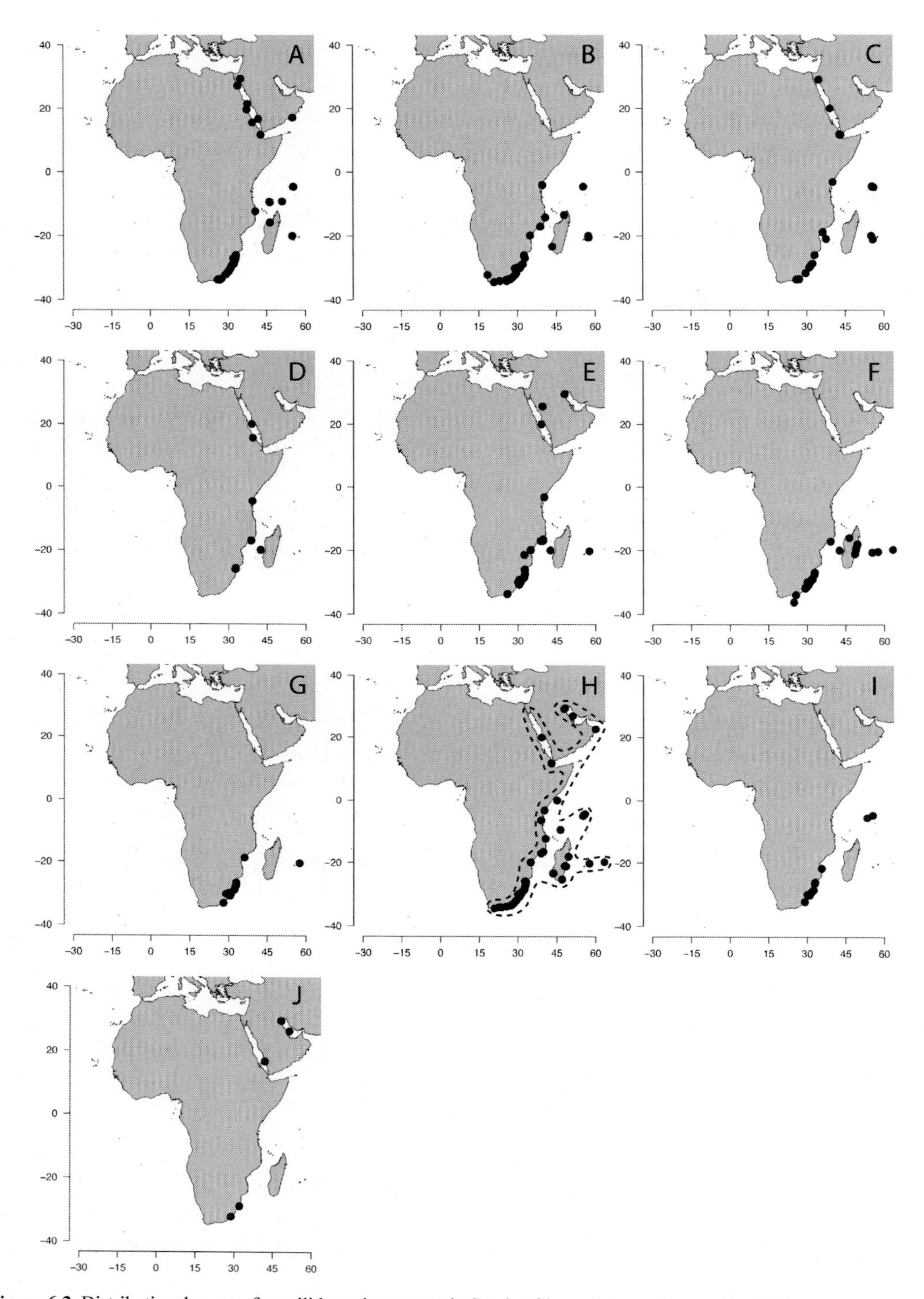

Figure 6.3. Distributional range of mugilid species present in South Africa and the eastern coast of Africa.

Each dot corresponds to a record in the GBIF data bank (http://www.gbif.org) plus some specimens that were barcoded by Durand et al. (2012a). (A) *Crenimugil crenilabis*; (B) *Moolgarda* (*Crenimugil*) *buchanani*; (C) *Moolgarda* (*Crenimugil*) *seheli ssl*; (D) *Ellochelon vaigiensis*; (E) *Moolgarda* (*Osteomugil*) sp. B; (F) *Moolgarda* (*Osteomugil*) *robustus*; (G) *Liza* (*Planiliza*) *alata*; (H) *Liza* (*Planiliza*) *macrolepis* (dotted line indicate the probable distribution range of *L.* (*P.*) *macrolepis sst*); (I) *Liza* (*Planiliza*) *melinoptera* (including the synonym *P. luciae*); (J) *Liza* (*Planiliza*) *subviridis*.

Moolgarda (Crenimugil) buchanani Bleeker, 1853. The bluetail mullet is regularly recorded in Africa from Somalia to the Breë Estuary in South Africa (Sommer et al. 1996, Whitfield 1998) but is also sometimes present on the South African Atlantic coast as a specimen was collected in Lamberts Bay (Western Cape Province) (GBIF dataset, Fig. 6.3B). Low genetic heterogeneity has been recorded using individuals over its distribution range (South Africa, Kenya, New Caledonia), which confirms its taxonomic status (Durand et al. 2012a).

Moolgarda (Crenimugil) seheli Forsskål, 1775. The bluespot mullet is an Indo-Pacific species present in East Africa from the Red Sea, where the type species has been described, to the Eastern Cape Province in South Africa (Thomson and Luther 1984, Smith and Smith 1986, Fig. 6.3C). Molecular phylogenetic investigation revealed a major level of cryptic diversity (Durand et al. 2012a). Among the three lineages discovered to date, two represent an Indo-Pacific distribution (Durand et al. 2012a). Pending taxonomical revision, these lineages were provisionally designated as ***Moolgarda* (*Crenimugil*)** sp. A, B and C by Durand and Borsa (2015), with the former two being present along the eastern and south eastern African shoreline.

Ellochelon vaigiensis Quoy and Gaimard, 1825. The squaretail mullet is present along eastern and south eastern African shores from Somalia to KwaZulu-Natal in South Africa (Thomson and Luther 1984, Smith and Smith 1986). This species is probably rare in South Africa as no specimen from this country is listed in the SAIAB museum collection (GBIF, Fig. 6.3D). Phylogenetic investigation has also revealed a high cryptic diversity, but the limited number of samples prevents any conclusions being reached on this species complex (Durand et al. 2012a).

Moolgarda (Osteomugil) cunnesius Valenciennes, 1836. The longarm mullet is depicted as an Indo-Pacific species present in Africa from Somalia to the Eastern Cape Province in South Africa (Thomson and Luther 1984, Smith and Smith 1986, Fig. 6.3E). However, phylogenetic reconstruction has demonstrated that this species is not monophyletic and consists of three allopatric cryptic species (Durand et al. 2012a). Because the type locality of *M.* (*O.*) *cunnesius* is in Molucca, the lineage sampled in Taiwan and in Vietnam (geographically closest to Molucca) was provisionally retained as the actual *M.* (*O.*) *cunnesius*. The two other lineages, one from eastern South Africa and the other one from Western Australia, were assigned provisional names *Moolgarda* (*Osteomugil*) sp. B and *Moolgarda* (*Osteomugil*) sp. A, respectively (Durand and Borsa 2015). To date actual specimens of *Osteomugil* sp. B have only been documented from the Eastern Cape and KwaZulu-Natal provinces (DNA barcode JQ060498 in Genbank). It is probable that the distributional range of *Osteomugil* sp. B is limited to eastern and south eastern African waters.

Moolgarda (Osteomugil) robustus Günther, 1861. The robust mullet is endemic to the south western Indian Ocean and has been recorded from the Eastern Cape Province in South Africa to Inhambane Province in Mozambique, as well as from Madagascar and Reunion Island (Smith and Smith 1986, pers obs., Fig. 6.3F). Of all the *Moolgarda* (*Osteomugil*) species, *M.* (*O.*) *robustus* is the most divergent, with no genetic intraspecific heterogeneity recorded among samples collected in South Africa and Reunion Island (Durand et al. 2012a, Xia et al. submitted).

Liza (Planiliza) alata, Steindachner, 1892. The diamond mullet is a western Indo-Pacific species from Madagascar in the north, which is the type locality, along the southern Mozambique coast to Algoa Bay in South Africa (Thomson and Luther 1984, Smith and Smith 1986, Fig. 6.3G). It was previously thought to occur in the north of Australia, Tonga and North western Pacific but a recent phylogenetic investigation demonstrated that the Australian population belonged to a distinct species (Durand et al. 2012a). *Liza* (*Planiza*) *ordensis* that was described in Australia and later synonymized by Thomson (1997) may need to be resurrected. Larson et al. (2013) supported the above suggestion on basis of eye colour difference observed between the South African *L.* (*P.*) *alata* and the Australian *L.* (*P.*) *ordensis*. Consequently, it is probable that *L.* (*P.*) *alata* will be restricted to South eastern Africa, even though the phylogenetic relationship with the north western Pacific *L.* (*P.*) *alata* has not been evaluated.

Liza (Planiliza) macrolepis Smith, 1849. The large-scale mullet is an Indo-Pacific species present from Somalia to Knysna in South Africa, including populations in Madagascar and the Seychelles (Thomson and Luther 1984, Smith and Smith 1986, Fig. 6.3H). A recent phylogenetic investigation highlighted two clades (Durand et al. 2012a), with all *L.* (*P.*) *macrolepis* sampled in the eastern Indian Ocean (Oman, Seychelles and South Africa) belonging to the same lineage, with other specimens from the Indo-Pacific region (Sri Lanka, Maldives, Japan, Taiwan, Fiji and New Caledonia) belonging to a different lineage. Large genetic divergence was recorded between these sister lineages, which suggests that there are actually two species, leading Durand and Borsa (2015) to propose that *macrolepis* be retained for the eastern Indian Ocean lineage and provisionally designated the second lineage as *Liza* (*Planiliza*) sp. H.

Liza (Planiliza) melinoptera Valenciennes, 1836. The otomebora mullet is an Indo-Pacific species present in eastern and south eastern Africa from Somalia in the north to the Eastern Cape Province in South Africa (Thomson and Luther 1984, Fig. 6.3I). According to Senou (1997) this species is a senior synonym of *Liza luciae* Penrith and Penrith 1967 which is in agreement with phylogenetic reconstruction based on polymorphism of three mitochondrial genes (Fig. 6.1).

Liza (Planiliza) subviridis Valenciennes, 1836. The greenback mullet is an Indo-Pacific species that is not mentioned as occurring in East Africa or South Africa by major reference works (Thomson and Luther 1984, Smith and Smith 1986, Whitfield 1998, Heemstra and Heemstra 2004). However, in the past decade several specimens have been collected in South Africa, thus attesting its presence in south eastern Africa (GBIF data, Mbande et al. 2005).

North Eastern Atlantic Species Present in Western Africa

Among the mugilids present in the north eastern Atlantic, two have their southern distribution extending into western Africa. *Liza* (*Chelon*) *aurata* and *Chelon labrosus* are both abundant in the north eastern Atlantic and appear on the fish checklist for Cape Verde (Wirth et al. 2013, Fig. 6.4A and B). However, the presence of these two north eastern Atlantic species in western Africa remains extremely limited.

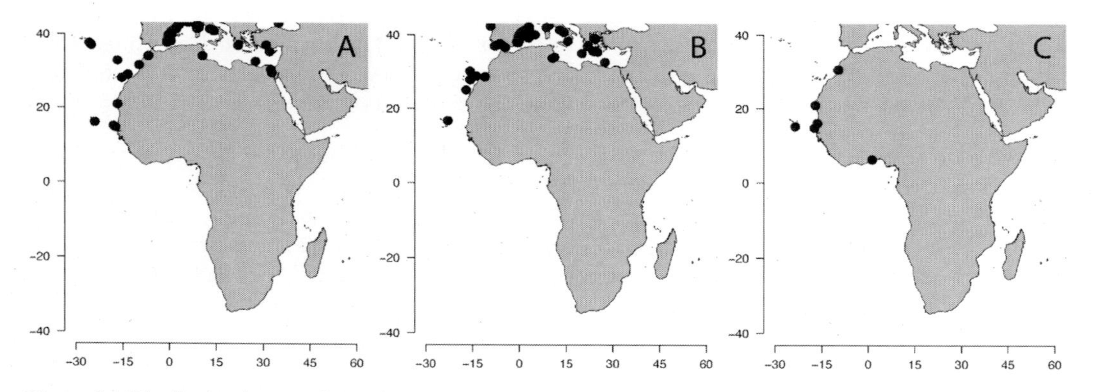

Figure 6.4. Distributional range of mugilid species present in the North eastern and eastern tropical Atlantic.

Each dot corresponds to a record in the GBIF data bank (http://www.gbif.org) plus some specimens that were barcoded by Durand et al. (2012a). A. *Liza* (*Chelon*) *aurata*; B. *Chelon labrosus*; C. *Mugil capurii*.

Mugil capurii, Perugia, 1892. The distribution of this species encompasses two biogeographic provinces as it has been recorded from northern Morocco (north eastern Atlantic) to Guinea Bissau (south eastern Atlantic) (Ben-Tuvia 1986, Sanches 1991, Thomson 1997, Froese and Pauly 2014, Fig. 6.4C). It was recently recorded in Togo which represents a major southward extension in its distribution

(Trape and Durand 2011). No cryptic genetic diversity has been highlighted over its distributional range and its closest relative is the West African species *M. bananensis* (Durand et al. 2012a, Fig. 6.1).

Drivers of Mugilid Biogeography in Africa

Distributional Range and Present Day Biogeographic Barriers

With one exception (*M. capurii*), most African mullet species are confined to either tropical or temperate waters. Even species presenting an apparent broad latitudinal distribution, such as *Mugil cephalus* or *Liza* (*Chelon*) *dumerili*, consist of evolutionary lineages with a geographic distribution limited to one biogeographic region. Indeed, it has been suggested that both these lineages are part of a species complex which reduces latitudinal distribution even further (Durand et al. 2012a, Whitfield et al. 2012, Durand et al. 2013b).

As mentioned earlier, marine biogeographic boundaries in Africa correspond to intense upwelling areas where temperature is probably too low to allow dispersion of tropical species. Many of the tropical fish taxa found in subtropical south eastern African estuaries occur mainly in the more northern systems within this region, with decreasing numbers being recorded in a south-westerly direction (Maree et al. 2000, Harrison 2005). Although some of these tropical taxa have been recorded from estuaries in the warm temperate zone of southern Africa, they are usually only present sporadically and in small numbers. Similarly, the temperate South African mullet *Liza* (*Chelon*) *richardsonii* is seldom recorded in subtropical estuaries but is known to extend its distribution into warmer estuaries during winter and then retreat southwards in the following summer (Branch and Grindley 1979). If sea surface temperature is a major physiological barrier to the latitudinal dispersion of tropical mugilids, then their presence or absence in estuarine ecosystems along the western Sahara and Namibia (where upwelling is strong), may also constitute a strong dispersal barrier for these species.

Most mugilid species present in Africa are strongly associated with estuaries and make intensive use of these systems as nursery areas and adult feeding grounds (Blaber 1977). Thus the absence of suitable estuarine ecosystems for the juvenile life stages on certain sections of the African coastline (e.g., Namibia) may limit mullet abundance and reduce population connectivity along such coasts.

Despite these ecological and physiological barriers, isolation of temperate and tropical mugilids is only partial, with mixing zones also present. In sub-Saharan Africa this zone is limited to the Cape Verde Archipelago and a stretch of continental shore that ranges from the western Sahara to the Banc d'Arguin National Park in northern Mauritania. By comparison, in South Africa the marine mixing zone is much broader and includes most of the eastern coast of South Africa (Fig. 6.3). This contrasting situation is probably directly linked to the oceanographic features that distinguish South Africa from the sub-Saharan Africa. In South Africa, the warm Agulhas Current flows southward down the Mozambique Channel towards the Eastern and Western Cape provinces, thus facilitating dispersion of tropical species in a southward direction (Harrison 2005). Sporadic marine upwelling along the Eastern Cape coast, especially during summer (Goschen et al. 2012), can restrict the southward extension of tropical mugilids but if they gain access to estuaries between such events, they will be protected from sharp declines in water temperature (James et al. 2008). In sub-Saharan Africa the cold Canary Current flows southward as far as Senegal and will prevent, at least seasonally, the northward dispersion of tropical species.

Mugilid Phylogenetic Relationships and their African Evolutionary History

African coastal waters were an important diversification centre for the Mugilidae. Among the four subfamilies (see Chapter 2—Durand 2015), the Cheloninae subfamily *sensu* Xia et al. (submitted) consists, with one exception, of genera exclusively recorded in African or north eastern Atlantic waters, e.g., *Neochelon*, *Oedalechilus*, *Pseudomyxus*, *Liza* (*Chelon*), *Chelon* and *Parachelon*. The exception is the specious Indo-Pacific genus *Planiliza* (Fig. 2.1 in Chapter 2) which is closely related to the East Atlantic genera *Chelon* and *Parachelon*. The time-calibrated phylogeny (Fig. 6.1) shows that the genetic divergence of African genera *Chelon* and *Parachelon* with the Indo-Pacific genus *Planiliza* probably dates back to the middle Miocene Climate Transition (17 to 11 Ma). This is consistent with two important vicariant events, namely the closure of the eastern Tethys Gateway and loss of the marine connection between the Mediterranean and the Indian Ocean (Hüsing et al. 2009). The above events were followed by the *Chelon* (*sensu* Durand et al. 2012b) diversification, mainly in temperate areas (in the Mediterranean Sea and in South African waters), which contrasts with the genus *Liza* (*Planiliza*) and other mugilid genera that have maximum species diversity in tropical latitudes (Fig. 6.1). The youngest vicariant event influencing the genus *Chelon* (*sensu* Durand et al. 2012b) probably corresponded to the formation of the two permanent upwellings off West Africa and off the western southern African coast. This view is supported by the existence of three sibling species associated with either temperate or tropical environments, e.g., *Chelon labrosus/Chelon bispinosus; Liza* (*Chelon*) *bandialensis/Liza* (*Chelon*) *tricuspidens; Liza* (*Chelon*) *dumerili/ Chelon* sp. B. For these last two sibling species, the divergence date estimated using the RelTime method (Tamura et al. 2012) is congruent with the Benguela Current system formation that occurred between 12 to 3 Mya (Shannon 1985).

African mugilid diversity was also enriched by waves of colonization from the western Atlantic region and the Indo-Pacific. The origin of the genus *Mugil* was probably located in American waters, considering both species and the phylogenetic diversity within this biogeographic province. This implies dispersion events across the Mid-Atlantic Ridge (MAR) that would have been difficult due to the distance which would have to be covered without access to benthic food resources (Helfman et al. 1997). However, this dispersal event probably occurred at least twice during the evolutionary history of the genus *Mugil*. The ancestors of the African *Mugil*, *M. capurii* and *M. bananensis*, reached African waters during the early Miocene based on time-calibrated phylogeny, while the level of divergence among species of the *M. curema* and *M. cephalus* complexes suggest a much more recent connection that dates back to the Pliocene (Fig. 6.1).

The MAR should represent a strong barrier for reef and coastal fishes, yet intercontinental dispersion has been suggested for a number of taxa such as parrotfishes (Robertson et al. 2006) and seahorses (Boehm et al. 2013). Joyeux et al. (2001) proposed that the larval stage duration directly influences range expansion across large oceanic dispersal barriers. However, Luiz et al. (2012) demonstrated that beyond larval-dispersal potential, large body size, wide latitudinal-range and rafting ability are better predicators of the mid-Atlantic crossing. Rafting by tropical reef fishes may be common (Robertson et al. 2004, Teske et al. 2005, Rocha et al. 2008).

If rafting was possible from West Africa to South America due to the Equatorial Current, then rafting from the Caribbean to the north eastern Atlantic would also be possible using the Gulf Stream (Boehm et al. 2013). Consequently it can be assumed that the settlement of the genus *Mugil* in eastern Atlantic waters occurred due to rafting in the Gulf Stream, a scenario in agreement with *Hippocampus hippocampus* that crossed the MAR from the Caribbean Sea 3.35 Ma (Boehm et al. 2013), a date consistent with the divergence observed among *M. cephalus* and *M. curema* species complexes (Fig. 6.1).

Finally, dispersion from the Indo-Pacific region into East African waters would have facilitated the arrival of *Ellochelon*, *Crenimugil sensu* Durand et al. (2012b) and *Moolgarda* (*Osteomugil*) species. Settlement in southern African waters of the latter genus is certainly ancient, as suggested by the presence in the south eastern Indian Ocean of endemic species [*Moolgarda* (*Osteomugil*) *robustus* and *Moolgarda* (*Osteomugil*) sp. B] and their high level of divergence when compared to other species of the genus *Osteomugil*. Dispersion of these tropical genera over large distance was certainly easier than crossing the MAR as the Indian Ocean coastline is continuous from Asia to Africa and mostly tropical or subtropical in terms of water temperatures. Nevertheless, some potential restrictions do exist such as in the form of the Indo-Pacific Barrier on the western border of the Indian Ocean and the mid-Indian Barrier (Randall 1998, Bellwood and Wainwright 2002). This latter biogeographic barrier is vindicated by a number of studies on coastal fish (Bay et al. 2004, Rocha et al. 2007) and matches the distributional border of two sibling species belonging to the genus *Liza* (*Planiliza*), namely *Liza (Planiliza) macrolepis* and *Liza* (*Planiza*) sp. H.

References

Afonso, P., F.M. Porteiro, R.S. Santos, J.P. Barreiros, J. Worms and P. Wirtz. 1999. Coastal marine fishes of Sao Tomé Island (Gulf of Guinea). Arquipel. Life Mar. Sci. 17A: 65–92.

Albaret, J.J. 1992. Mugilidae. pp. 780–788. *In*: C. Lévêque, D. Paugy and G.G. Teugels (eds.). The Fresh and Brackish Water Fishes of West Africa (Vol. 2). Collection Faune Tropicale 27.

Bay, L.K., J.H. Choat, L. van Herwerden and D.R. Robertson. 2004. High genetic diversities and complex genetic structure in an Indo-Pacific tropical reef fish (*Chlorurus sordidus*): Evidence of an unstable evolutionary past? Mar. Biol. 144: 757–767.

Bellwood, D.R. and P.C. Wainwright. 2002. The history and biogeography of fishes on coral reefs. pp. 5–32. *In*: P.F. Sale (ed.). Coral Reef Fishes: Dynamics and Diversity in a Complex Ecosystem. Academic Press, San Diego.

Ben-Tuvia, A. 1986. Mugilidae. pp. 1197–1204. *In*: P.J.P. Whitehead, M.L. Blauchot, J.C. Hureau, J. Nielsen and E. Tortonese (eds.). Fishes of North-eastern Atlantic and the Mediterranean. UNESCO, Paris.

Blaber, S.J.M. 1977. The feeding ecology and relative abundance of mullet (Mugilidae) in Natal and Pondoland estuaries. Biol. J. Linn. Soc. 9: 259–275.

Boehm, J.T., L. Woodall, P.R. Teske, S.A. Lourie, C. Baldwin, J. Waldman and M. Hickerson. 2013. Marine dispersal and barriers drive Atlantic seahorse diversification. J. Biogeogr. 40: 1839–1849.

Branch, G.M. and J.R. Grindley. 1979. Ecology of southern African estuaries. Part XI. Mngazana: A mangrove estuary in the Transkei. S. Afr. J. Zool. 14: 149–170.

Crosetti, D., W.S. Nelson and J. Avise. 1994. Pronounced genetic structure of mitochondrial DNA among populations of the circumglobally distributed grey mullet (*Mugil cephalus* Linnaeus). J. Fish Biol. 44: 47–58.

Durand, J.D. 2015. Implications of molecular phylogeny for the taxonomy of Mugilidae. *In*: Crosetti, D. and S.J.M. Blaber (eds.). Biology, Ecology and Culture of Grey Mullet (Mugilidae). CRC Press, Boca Raton, USA (this book).

Durand, J.-D., M. Tine, J. Panfili, O.T. Thiaw and R. Lae. 2005. Impact of glaciations and geographic distance on the genetic structure of a tropical estuarine fish, *Ethmalosa fimbriata* (Clupeidae, S. Bowdich 1825). Mol. Phylogenet. Evol. 36: 277–287.

Durand, J.D., K.N. Shen, W.J. Chen, B.W. Jamandre, H. Blel, K. Diop, M. Nirchio, F.J. Garcia de Leon, A.K. Whitfield, C.W. Chang and P. Borsa. 2012a. Systematics of the grey mullets (Teleostei: Mugiliformes: Mugilidae): Molecular phylogenetic evidence challenges two centuries of morphology-based taxonomy. Molecular and Phylogenetics Evolution 64: 73–92.

Durand, J.D., W.J. Chen, K.N. Shen, C. Fu and P. Borsa. 2012b. Genus-level taxonomic changes implied by the mitochondrial phylogeny of grey mullets (Teleostei: Mugilidae). C. R. Biologies 335: 687–697.

Durand, J.D., B. Guinand, J.J. Dodson and F. Lecomte. 2013a. Pelagic life and depth: coastal physical features in West Africa shape the genetic structure of the Bonga shad, *Ethmalosa fimbriata*. PloS One 8: 1–14. doi:10.1371/journal.pone.0077483.

Durand, J.D., H. Blel, K.N. Shen, E.T. Koutrakis and B. Guinand. 2013b. Population genetic structure of *Mugil cephalus* in the Mediterranean and Black Seas: A single mitochondrial clade and many nuclear barriers. Mar. Ecol. Prog. Ser. 474: 243–261.

Durand, J.D. and P. Borsa. 2015. Mitochondrial phylogeny of grey mullets (Acanthopterygii: Mugilidae) suggests high proportion of cryptic species. C. R. Biologies 338: 266–277.

Eschmeyer, W.N. (ed.). 2014. Catalog of Fishes: Genera, Species, References. Available at: http://research.calacademy.org/research/ichthyology/catalog/fishcatmain.asp. Electronic version accessed 01-09-2014.

Falk, T.M., G.G. Teugels, E.K. Abban, W. Villwock and L. Renwrantz. 2003. Phylogeographic patterns in populations of the black-chinned tilapia complex (Teleostei, Cichlidae) from coastal areas in West Africa: Support for the refuge zone theory. Mol. Phylogenet. Evol. 27: 81–92.

Fraga, E., H. Schneider, M. Nirchio, E. Santa-Brigida, L.F. Rodrigues-Filho and I. Sampaio. 2007. Molecular phylogenetic analyses of mullets (Mugilidae: Mugiliformes) based on two mitochondrial genes. J. Appl. Ichthyol. 23: 598–604.

Froese, R. and D. Pauly. 2014. FishBase. World Wide Web electronic publication. Available at: http//www.fishbase.org/. Electronic version accessed 10-2014.

Gilbert, C.R. 1993. Geographic distribution of the striped mullet (*Mugil cephalus* Linnaeus) in the Atlantic and eastern Pacific oceans. Fla. Sci. 56: 204–210.

Goschen, W.S., E.H. Schumann, K.S. Bernard, S.E. Bailey and S.H.P. Deyzel. 2012. Upwelling and ocean structures off Algoa Bay and the south-east coast of South Africa. Afr. J. Marine Sci. 34: 525–536.

Gushchin, A.V. and K.O.M. Fall. 2012. Ichthyofauna of littoral of the Gulf Arguin, Mauritania. J. Ichthyol. 52: 160–171.

Harrison, T.D. 2005. Ichthyofauna of South African estuaries in relation to the zoogeography of the region. Smithiana Bulletin 6: 1–27.

Heemstra, P.C. and E. Heemstra. 2004. Coastal Fishes of Southern Africa. NISC, Grahamstown and SAIAB, Grahamstown.

Helfman, G.S., B.B. Collette and D.E. Facey. 1997. The Diversity of Fishes. Blackwell Science, Malden, MA.

Henriques, R., W.M. Potts, W.H.H. Sauer and P.W. Shaw. 2012. Evidence of deep genetic divergence between populations of an important recreational fishery species, *Lichia amia* L. 1758, around southern Africa. Afr. J. Marine Sci. 34: 585–591.

Henriques, R., W.M. Potts, C.V. Santos, W.H.H. Sauer and P.W. Shaw. 2014. Population connectivity and phylogeography of a coastal fish, *Atractoscion aequidens* (Sciaenidae), across the Benguela current region: evidence of an ancient vicariant event. PLoS One 9(2): e87907.

Heras, S., M.I. Roldan and M. Gonzalez Castro. 2009. Molecular phylogeny of Mugilidae fishes revised. Rev. Fish Biol. Fish. 19: 217–231.

Hüsing, S.K., W.J. Zachariasse, D.J.J. van Hinsbergen, W. Krijgsman, M. Inceöz, M. Harzhauser, O. Mandic and A. Kroh. 2009. Oligocene Miocene basin evolution in SE Anatolia, Turkey: constraints on the closure of the eastern Tethys gateway. pp. 107–132. *In*: D.J.J. van Hinsbergen, M.A. Edwards and R. Govers (eds.). Collision and Collapse at the Africa–Arabia–Eurasia Subduction Zone. Special Publications, 311, The Geological Society, London.

James, N.C., A.K. Whitfield and P.D. Cowley. 2008. Preliminary indications of climate-induced change in a warm-temperate South African estuarine fish community. J. Fish Biol. 72: 1855–1863.

Joyeux, J.C., S.R. Floeter, C.E.L. Ferreira and J.L. Gasparini. 2001. Biogeography of tropical reef fishes: The South Atlantic puzzle. J. Biogeogr. 28: 831–841.

Larson, H.K., R.E.X.S. Williams and M.P. Hammer. 2013. An annotated checklist of the fishes of the Northern Territory, Australia. Zootaxa 3696: 1–293.

Lévêque, C. 1997. Biodiversity Dynamics and Conservation of the Fresh Water Fish of Tropical Africa. Cambridge University, Cambridge.

Livi, S., L. Sola and D. Crosetti. 2011. Phylogeographic relationships among worldwide populations of the cosmopolitan marine species, the striped grey mullet (*Mugil cephalus*), investigated by partial cytochrome b gene sequences. Biochem. Syst. Ecol. 39: 121–131.

Maley, J. 1991. The African rain forest vegetation and paleoenvironments during Late Quaternary. Clim. Change 19: 79–98.

Maree, R.C., A.K. Whitfield and A.J. Booth. 2000. Effect of water temperature on the biogeography of South African estuarine fish species associated with the subtropical/warm temperate subtraction zone. S. Afr. J. Sci. 96: 184–188.

Mbande, S., A.K. Whitfield and P.D. Cowley. 2005. The ichthyofaunal composition of the Mngazi and Mngazana estuaries: a comparative study. Smithiana Bull. 4: 1–20.

Moyano, M., J.M. Rodriguez, V.M. Benitez-Barrios and S. Hernandez-Leon. 2014. Larval fish distribution and retention in the Canary Current system during the weak upwelling season. Fish. Oceanogr. 23: 191–209.

Nei, M. and S. Kumar. 2000. Molecular Evolution and Phylogenetics. Oxford University Press, New York.

Nichol, J.E. 1999. Geomorphological evidence and Pleistocene refugia in Africa. Geogr. J. 165: 79–89.

Nirchio, M. and H. Cequea. 1998. Karyology of *Mugil liza* and *M. curema* from Venezuela. Bol. Invest. Mar. Cost. 27: 45–50.

Nybakken, J. 1997. Marine Biology: An Ecological Approach. Longman, New York.

Randall, J.E. 1998. Zoogeography of shore fishes of the Indo-Pacific region. Zool. Stud. 37: 227–268.

Robertson, D.R., J.S. Grove and J.E. McCosker. 2004. Tropical transpacific shore fishes. Pac. Sci. 58: 507–565.

Robertson, D.R., F. Karg, R. Leao, L. de Moura, B.C. Victor and G. Bernardi. 2006. Mechanisms of speciation and faunal enrichment in Atlantic parrotfishes. Mol. Phylogenet. Evol. 40: 795–807.

Rocha, L.A., M.T. Craig and B.W. Bowen. 2007. Phylogeography and the conservation of coral reef fishes. Coral Reefs 26: 501–512.

Rocha, L.A., C.R. Rocha, D.R. Robertson and B.W. Bowen. 2008. Comparative phylogeography of Atlantic reef fishes indicates both origin and accumulation of diversity in the Caribbean. BMC Evol. Biol. 8: 157.

Rossi, A.R., M. Capula, D. Crosetti, L. Sola and D.E. Campton. 1998a. Allozyme variation in global populations of striped mullet, *Mugil cephalus* (Pisces: Mugilidae). Mar. Biol. 131: 203–212.

Rossi, A.R., M. Capula, D. Crosetti, D.E. Campton and L. Sola. 1998b. Genetic divergence and phylogenetic inferences in five species of Mugilidae (Pisces: Perciformes). Mar. Biol. 131: 213–218.

Sanches, J.G. 1991. Catalogo dos principais peixes marinhos da Republica de Guine-Bissau. INIP, Lisboa.

Schefub, E., S. Schouten, J.H.F. Jansen and J.S.S. Damste. 2003. African vegetation controlled by tropical sea surface temperatures in the mid-Pleistocene period. Nature 422: 418–421.

Senou, H. 1997. Redescription of a mullet, *Chelon melinopterus* (Perciformes: Mugilidae). Bull. Kanagawa prefect. Mus. 26: 51–56.

Shannon, L.V. 1985. The Benguela ecosystem. Part I. Evolution of the Benguela, physical features and processes. Oceanogr. Mar. Biol. Ann. Rev. 23: 105–182.

Shen, K.N., B.W. Jamandre, C.C. Hsu, W.N. Tzeng and J.D. Durand. 2011. Plio-Pleistocene sea level and temperature fluctuations in the northwestern Pacific promoted speciation in the globally-distributed flathead mullet *Mugil cephalus*. BMC Evol. Biol. 11: 83.

Smith, M.M. and J.L.B. Smith. 1986. Mugilidae. pp. 714–720. *In*: M.M. Smith and P.C. Heemstra (eds.). Smiths' Sea Fishes. Springer-Verlag, Berlin.

Sommer, C., W. Schneider and J.M. Poutiers. 1996. The Living Marine Resources of Somalia. FAO Species Identification Guide for Fishery Purposes. Food and Agriculture Organization of the United Nations, Rome.

Spalding, M.D., H.E. Fox, G.R. Allen, N. Davidson, Z.A. Ferdana, M. Finlayson, B.S. Halpern, M. Jorgue, A.L. Lombana, S.A. Lourie, K.D. Martin, E. McManus, J. Molnar, C.A. Recchia and J. Robertson. 2007. Marine ecoregions of the world: a bioregionalization of coastal and shelf areas. BioScience 57: 573–583.

Sunday, J.M., A.E. Bates and N.K. Dulvy. 2012. Thermal tolerance and the global redistribution of animals. Nat. Clim. Change 2: 686–690.

Tamura, K., F.U. Battistuzzi, P. Billing-Ross, O. Murillo, A. Filipski and S. Kumar. 2012. Estimating divergence times in large molecular phylogenies. Proc. Natl. Acad. Sci. U.S.A. 109: 19333–19338.

Tamura, K., G. Stecher, D. Peterson, A. Filipski and S. Kumar. 2013. MEGA6: Molecular Evolutionary Genetics Analysis version 6.0. Mol. Biol. Evol. 30: 2725–2729.

Teske, P.R., H. Hamilton, P.J. Palsbøll, C.K. Choo, H. Gabr, S.A. Lourie, M. Santos, A. Sreepada, M.I. Cherry and C.A. Matthee. 2005. Molecular evidence for long-distance colonization in an Indo-Pacific seahorse lineage. Marine Ecology Progress Series 286: 249–260.

Thomson, J.M. 1966. The grey mullets. Oceanogr. Mar. Biol. Ann. Rev. 4: 301–355.

Thomson, J.M. 1997. The Mugilidae of the world. Memoirs of the Queensland Museum 43: 457–562.

Thomson, J.M. and G. Luther. 1984. Mugilidae (mullets) Page. Var. *In*: W. Fischer and G. Bianchi (eds.). FAO Species Identification Sheets for Fishery Purposes. Western Indian Ocean (Fishing Area 51). Vol. 3. Food and Agriculture Organization of the United Nations, Rome.

Trape, S. and J.D. Durand. 2011. First record of *Mugil capurii* (Mugilidae: Perciformes) in the Gulf of Guinea. J. Fish Biol. 78: 937–40.

Trape, S., J.D. Durand, F. Guilhaumon, L. Vigliola and J. Panfili. 2009. Recruitment patterns of young-of-the-year mugilid fishes in a West African estuary impacted by climate change. Estuar. Coast. Mar. Sci. 85: 357–367.

Trape, S., I.J. Harrison, P.S. Diouf and J.D. Durand. 2012. Redescription of *Liza bandialensis* (Teleostei: Mugilidae) with an identification key to mullet species of Eastern Central Atlantic. C. R. Biologies 335: 120–128.

von der Heyden, S., R.C.K. Bowie, K. Prochazka, P. Bloomer, N.L. Crane and G. Bernardi. 2011. Phylogeographic patterns and cryptic speciation across oceanographic barriers in South African intertidal fishes. J. Evol. Biol. 24: 2505–2519.

Whitfield, A.K. 1998. Biology and Ecology of Fishes in Southern African Estuaries. Ichthyological Monographs of the J. L. B. Smith Institute of Ichthyology, Grahamstown.

Whitfield, A.K., J. Panfili and J.D. Durand. 2012. A global review of the cosmopolitan flathead mullet *Mugil cephalus* Linnaeus 1758 (Teleostei: Mugilidae), with emphasis on the biology, ecology, genetics and fisheries aspects of this apparent species complex. Rev. Fish Biol. Fish. 22: 641–681.

Wirth, P., A. Brito, J.M. Falcon, R. Freitas, R. Fricke, V. Monteiro, F. Reiner and O. Tariche. 2013. The coastal fishes of the Cape Verde Islands—New records and an annotated check-list. Spixiana 1: 113–142.

Xia, R., J.D. Durand and C. Fu. Submitted. Multilocus resolution of the Mugilidae phylogeny (Teleostei: Mugiliformes): Implication on the family's taxonomy.

CHAPTER 7

Biogeography and Distribution of Mugilidae in the Mediterranean and the Black Sea, and North-East Atlantic

Cemal Turan

Geological and Geomorphological History of the Mediterranean and the Black Sea

The Mediterranean Sea is a remnant of the Tethys Ocean, a wedge-shaped, eastward-open equatorial sea that indented Pangea during the Triassic (Fig. 7.1) (Dutch 1998). The Tethys Sea connected, through an uninterrupted equatorial belt, the newly born ocean to the older Indo-Pacific Ocean after the opening of the Atlantic Ocean during the Cretaceous. During that time, the Tethys Sea harboured a highly diverse warm-water fauna from the tropical Indo-West Pacific Ocean (Dutch 1998, Bianchi and Morri 2000). In the Miocene orogeny, plate collisions sealed off the eastern Mediterranean Sea, trapping small remnants of ocean floor in the Black and Caspian Seas. At that time, the Isthmus of Suez was formed, separating the Mediterranean Sea from the Indo-Pacific Ocean. The connection between the Mediterranean Sea and the Atlantic Ocean was also closed at the end of the Miocene, and the Mediterranean Sea became a closed and isolated sea. River inflow was not enough to maintain the level of the Mediterranean, which dried out. Dehydration of the closed Mediterranean Sea might have driven the Tethyan deep water fauna to extinction (Bouchet and Taviani 1992), but at least part of the shallow-water biota may have survived through the Neogene (Stanley 1990, Bellan-Santini et al. 1992, Myers 1996, Bianchi and Morri 2000). After the re-opening of the Straits of Gibraltar at the dawn of the Pliocene (6 million years ago), the Mediterranean Sea was re-populated by species of Atlantic origin and the Mediterranean Sea biota became that of an Atlantic province (Briggs 1974, Briggs and Bowen 2012, Bianchi and Morri 2000, Spalding et al. 2007). The word Mediterranean derives from the Latin *mediterraneus*, meaning 'in the middle of earth' or 'between lands' (*medius*, 'middle, between' + *terra*, 'land, earth'). This is on account of the sea's intermediary position between the continents of Africa and Europe.

A great change in the Mediterranean fauna occurred after the opening of the Suez Canal in 1869. The Suez Canal connects two major water bodies, the Red Sea and the Mediterranean, which differ fundamentally, both faunistically and hydrographically. The Canal, which is 162.5 km long, 200–300 m

Molecular Ecology and Fisheries Genetics Laboratory, Faculty of Marine Sciences and Technology, Iskenderun Technical University, Iskenderun, Hatay, Turkey.
Email: turancemal@yahoo.com

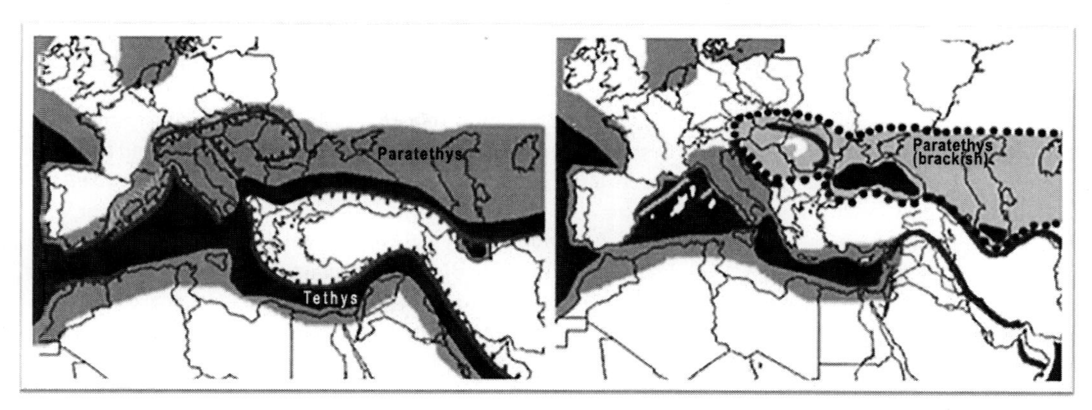

Figure 7.1. The Mediterranean region 20 (left) and 10 (right) million years ago respectively. Present coastlines are shown for reference only. Deep water is dark, shallow seas are grey, and brackish water is grey circled with black dots (modified from Dutch 1998).

wide and 10–15 m deep, crosses Lake Timsah and the Bitter Lakes on its way to the city of Suez and the Gulf of Suez. The main abiotic difference between the two regions is the temperature regime, which is stable in the tropical Red Sea, but experiences wide fluctuations in the subtropical Mediterranean. The fauna of the Red Sea is of tropical Indo-Pacific origin, while that of the Mediterranean is mainly of temperate Atlantic origin. The Red Sea and Mediterranean Sea were exposed to invasions of organisms from each other, known as the 'lessepsian migration', named after Ferdinand de Lesseps, the French diplomat in charge of the canal construction. The minor migration in the opposite direction is known as 'anti-lessepsian migration'. However, the great majority of migrational movement has been northwards from the Red Sea to the Mediterranean Sea (Por 1978, Ozturk and Turan 2013).

Biodiversity in the Mediterranean Sea, the Black Sea and the North-East Atlantic Ocean

The eastern Atlantic region comprises the British Isles, Lusitania Province, Morocco and eastwards through the Mediterranean Sea, Black Sea, Caspian and Aral provinces. There are no recent evaluations of endemisms in these three provinces (Briggs and Bowen 2012). The Lusitania Province includes the offshore islands of the Canaries, Azores and Madeira. The endemism in this province is concentrated within the Mediterranean itself, where 28% of marine species are endemic (IUCN 2010). The Straits of Gibraltar are often considered to be a natural barrier between the Mediterranean and the Atlantic segments of the Lusitania province. Although population genetic studies of fishes, molluscs, crustaceans and marine mammals show some genetic separations, there is no consistent pattern of evolutionary partitions at the Straits (Paternello et al. 2007, Briggs and Bowen 2012).

Mediterranean marine biodiversity has great cultural and economic importance for the Mediterranean countries. A rough estimate of more than 8,500 species of macroscopic marine organisms live in the Mediterranean Sea, corresponding to between 4 and 18% of the world's marine species. The high biodiversity of the Mediterranean Sea may be explained by historical, palaeogeographic and ecological reasons. Current Mediterranean biodiversity is undergoing rapid alteration under the combined pressures of climate change and human impact, but protection measures, either for species or ecosystems, are still scarce (Bianchi and Morri 2000, Turan et al. 2009).

The biota of the Black Sea reflects situations that may be found at higher latitudes: few species with high biomasses at high latitudes in contrast to many species with low biomasses at low latitudes (Ormond et al. 1997). Therefore, there is low species diversity relative to the high habitat diversity in the Black Sea (Turan 2014). The process of the Black Sea ichthyofauna formation continued for about 8,000 years after the last joining of the Black Sea with the world ocean. Over 200 species and subspecies, including occasional freshwater and marine fish reported only once and not occurred and reported elsewhere, live in the Black Sea (Svetovidov 1964, Oven 1993, Boltachev 2003, Boltachev et al. 2009). A considerable decrease in the numbers of many native Black Sea fish has been observed; some have not occurred for several decades, largely resulting from human activities such as chronic pollution, in particular eutrophication that causes fish kills, overfishing and poaching, physical destruction of spawning grounds and feeding areas, and invasion of alien species. On the other hand, newcomers have penetrated into the Black Sea via migration from the Mediterranean Sea, such as blue crab *Calonictis sapidus*, blue whiting *Micromesistius poutassou*, or with ship ballast waters, such as veined rapa whelk *Rapana venosa*, and warty comb jelly *Minemiopsis leidyi* (Shiganova 1998, Boltachev 2006, Yaglioglu et al. 2013). The Black Sea is connected to the Mediterranean Sea via the Turkish Straits, comprising the Istanbul Strait, the Sea of Marmara and the Çanakkale Strait. Global climate change, mediterranization of the Black Sea and human transport in ballast waters removed geographical barriers and changed the geographical distribution of species, associated with changes in biological and genetic diversity (Turan et al. 2009).

Mullet Species in the Mediterranean Sea, the Black Sea and the North-East Atlantic Ocean

The family Mugilidae is distributed worldwide and includes mainly coastal marine species that are present in all tropical, subtropical and temperate seas (Nelson 1984). Mugilid species are generally considered to be ecologically important and are a major food resource for human populations in certain parts of the world (Whitfield et al. 2012). In total, the family Mugilidae includes 20 genera and 71 species in the world (Eschmeyer and Fong 2014, Eschmeyer 2014). Eight species of Mugilidae, belonging to four genera, *Mugil, Liza, Chelon* and *Oedalechilus*, inhabit the Mediterranean, the Black Sea, and the North-East Atlantic Ocean (Nelson 1984, Turan 2007): *Mugil cephalus*, *Chelon labrosus, Oedalechilus labeo, Liza aurata, Liza ramada*, *Liza saliens, Liza carinata, Liza haematocheila* (Nelson 2006, Turan 2007). *Liza carinata* is a lessepsian immigrant species and recently invaded the eastern Mediterranean from the Red Sea through the Suez Canal (Ben-Tuvia 1975). Although native to the western North Pacific, *L. haematocheila* was introduced into the Azov Sea and the Black sea during the early 1980s (Zaitsev 1991).

Mullets are the target of commercial capture fisheries and are mainly caught by gill nets, seines and hooks. They are also cultured in some regions of the Mediterranean and the Black Sea, mainly in extensive ponds or confined coastal lagoons.

The taxonomic classification of the family Mugilidae is mainly based on morphological and genetic characters (see Chapter 1—González-Castro and Ghasemzadeh 2015 and Chapter 2—Durand 2015 for details and discussion), but unfortunately only a few morphological characters are suitable for use in a key to unambiguously establish the relationships among species (Stiassny 1993). Some taxonomically significant diagnostic characters of the Mediterranean, Black Sea and North-East Atlantic mullet species are given in Table 7.1.

Table 7.1. Descriptive taxonomic characters commonly used to distinguish Mediterranean mullets. First dorsal fin rays (DFR1), second dorsal fin rays (DFR2), ventral fin rays (VFR), anal fin rays (AFR), pectoral fin rays (PFR), pyloric caeca (PC) (Turan et al. 2011).

Species	DFR 1	DFR 2	VFR	AFR	PFR	PC
Chelon labrosus	IV	I 8	I 5	III 8–9	17	6–7
Oedalechilus labeo	IV	I 8	I 5	III 8–10	16–17	6
Mugil cephalus	IV	I 8	I 5	III 8–9	17	2
Liza haematocheila	IV	I 8–9	I 5	III 8–9	16	4–5
Liza aurata	IV	I 8	I 5	III 8–9	16	7–8
Liza carinata	IV	I 7	I 5	III 7	15	5
Liza ramada	IV	I 7–8	I 5	III 8–9	16–17	6–8
Liza saliens	IV	I 7	I 5	III 7–8	16	7–9

Endemic mullets

The Mediterranean endemic mullets include four genera (*Chelon, Liza, Mugil and Oedalechilus*) and six species: thicklip mullet *C. labrosus,* golden mullet *L. aurata,* thinlip mullet *L. ramada,* sharpnose mullet *L saliens,* flathead mullet *M. cephalus,* boxlip mullet *O. labeo* (Nelson 1984, Whitehead 1984, Harrison 1995, Turan 2007).

Chelon labrosus

The thicklip mullet *Chelon labrosus* is found inshore in schools, frequently entering brackish lagoons and freshwater (Billard 1997). *Chelon labrosus* females are larger than males (Kottelat and Freyhof 2007). The economic importance of *C. labrosus* is variable in different countries in the Mediterranean region. It is distributed along the Mediterranean Sea and South-western Black Sea, North-eastern Atlantic Coasts, including the British Isles, North Sea, Norwegian Sea, Barents Sea, Baltic Sea, Bay of Biscay, the offshore islands of the Canaries, Azores and Madeira, and coasts of West Africa (Fig. 7.2) (Nelson 1984, Whitehead 1984, Turan 2007, Froese and Pauly 2014). According to Froese and Pauly (2014), *C. labrosus* is absent in the southern part of Turkey (Mugla and Antalya Bay) although its presence is reported in this area (Sumer and Balik 2007, Sumer and Teksam 2013).

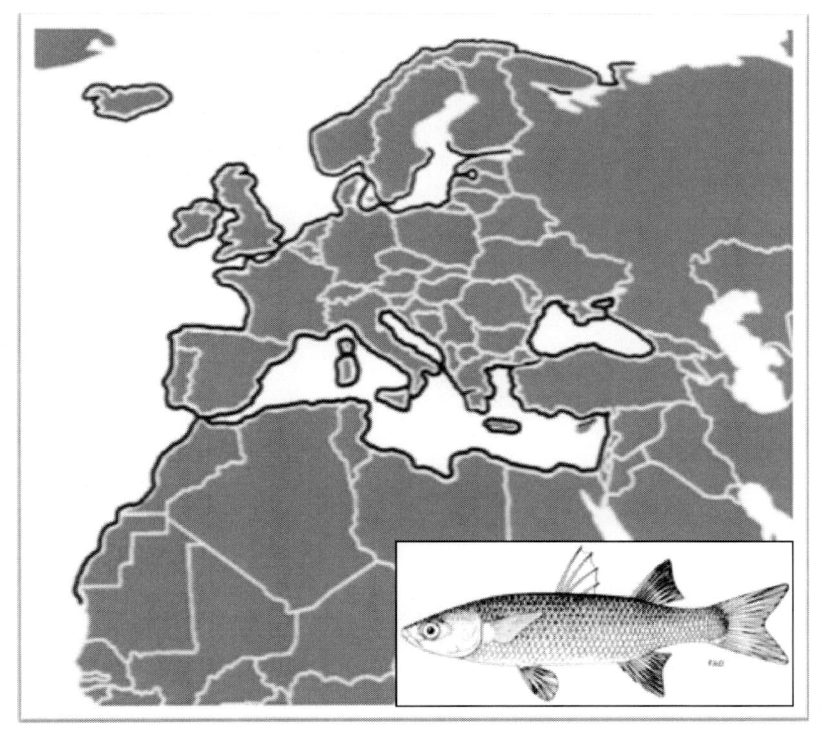

Figure 7.2. Distribution of *Chelon labrosus* in the Mediterranean, Black Sea and North-East Atlantic Ocean.

Liza aurata

The golden mullet *Liza aurata* is a neritic species, usually living in schools, entering lagoons and lower estuaries, but rarely freshwater (Thomson 1986, 1990). Juveniles move to coastal lagoons and estuaries in winter and especially in spring (Kottelat and Freyhof 2007). *Liza aurata* is found in the Mediterranean and the Black Sea, North-eastern Atlantic coasts, including the British Isles, North Sea, Norwegian Sea, Barents Sea, Bay of Biscay, the offshore islands of the Canaries, Azores and Madeira, and coasts of West Africa (Fig. 7.3) (Nelson 1984, Whitehead 1984, Turan 2007, Froese and Pauly 2014). Fazli et al. (2008) report its presence in the Caspian Sea.

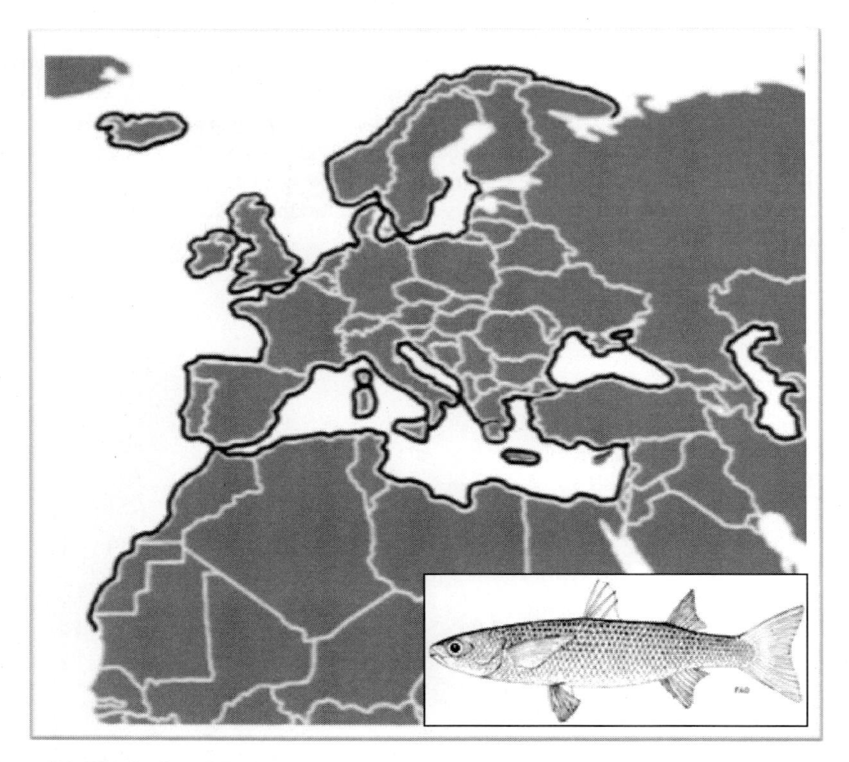

Figure 7.3. Distribution of *Liza aurata* in the Mediterranean, Black Sea and North-East Atlantic Ocean.

Liza ramada

The thinlip mullet *Liza ramada* is found in nearshore waters, and enters lagoons and the lower reaches of rivers in schools (Rochard and Elie 1994). Juveniles colonize the littoral zone and estuaries (Rochard and Elie 1994). It is found in the Mediterranean and the Black Sea, North-eastern Atlantic coasts, including the British Isles, North Sea, Norwegian Sea, Barents Sea, Bay of Biscay, the offshore islands of the Canaries, Azores and Madeira, and coasts of West Africa down to Mauritania (Fig. 7.4) (Nelson 1984, Whitehead 1984, Turan 2007, Froese and Pauly 2014).

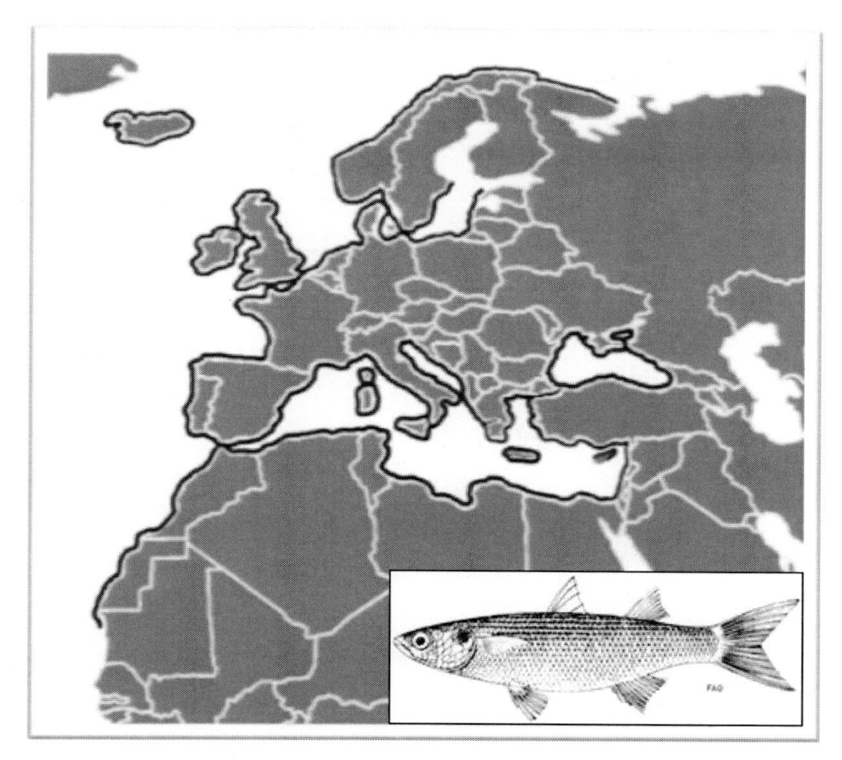

Figure 7.4. Distribution of *Liza ramada* in the Mediterranean, Black Sea and North-East Atlantic Ocean.

Liza saliens

The sharpnose mullet *Liza saliens* inhabits coastal waters and sometimes lagoons and estuaries (Thomson 1986, Kottelat and Freyhof 2007). It occurs in the Mediterranean Sea and the Black Sea, North-eastern Atlantic coasts, including the British Isles, Bay of Biscay, the offshore islands of the Canaries, Azores and Madeira, and coasts of West Africa down to the western Sahara (Fig. 7.5) (Nelson 1984, Whitehead 1984, Baltz 1991, Turan 2007, Froese and Pauly 2014). Patimar (2008) also reported its presence in the Caspian Sea.

Mugil cephalus

The flathead mullet *M. cephalus* inhabits inshore marine waters, estuaries, coastal lagoons, and rivers in the Mediterranean region. Taxonomically, *M. cephalus* differs from the other mullet species by pyloric caeca number (Table 7.1) and by the presence of an adipose eyelid. *Mugil cephalus* is traditionally cultured in coastal lagoons, also for the production of mullet roe, an expensive delicatessen. Culture is still based on fry captured from the wild, as despite several successful experimental trials of induced spawning (Chapter 13), no commercial mass propagation is carried out. Flathead grey mullet are distributed all along the Mediterranean and Black Sea coasts, North-eastern Atlantic Coasts, including the British Isles Coasts, North Sea, Norwegian Sea, Barents Sea, Bay of Biscay, the offshore islands of the Canaries, Azores and Madeira, and coasts of West Africa (Froese and Pauly 2014) (Fig. 7.6).

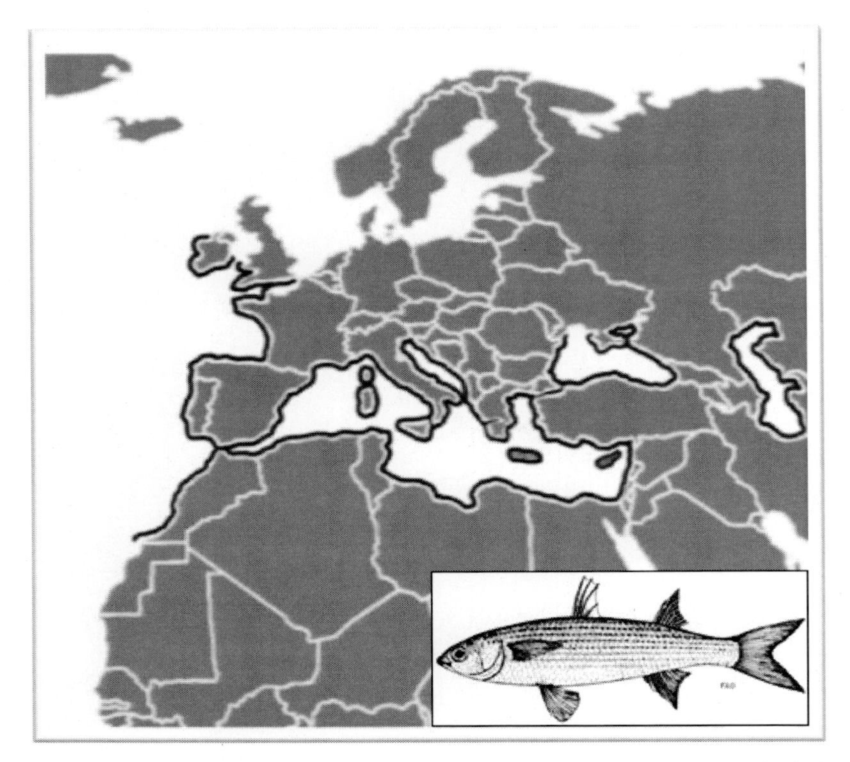

Figure 7.5. Distribution of *Liza saliens* in the Mediterranean, Black Sea and North-East Atlantic Ocean.

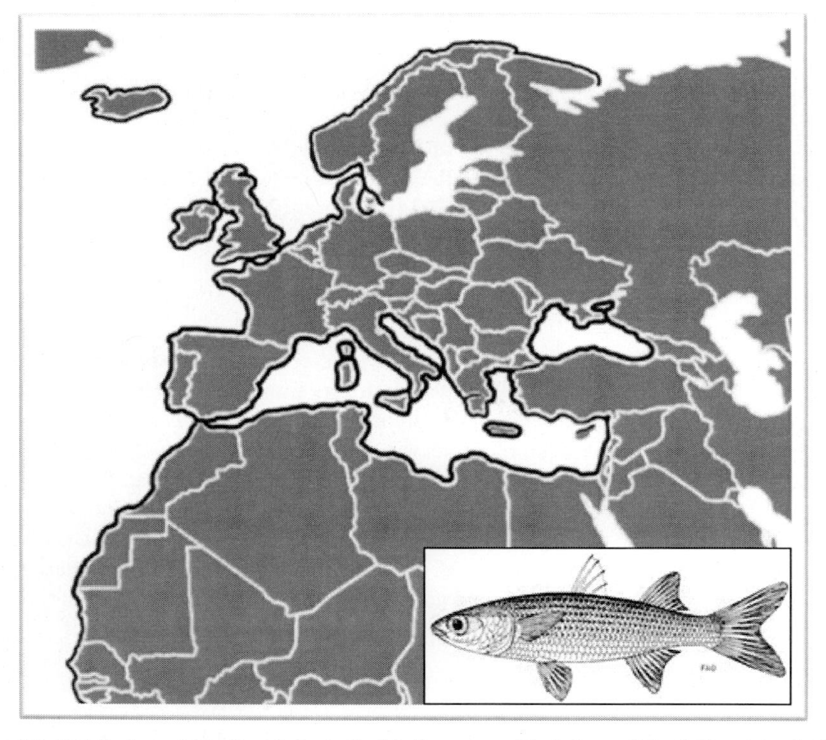

Figure 7.6. Distribution of *Mugil cephalus* in the Mediterranean, Black Sea and North-East Atlantic Ocean.

Oedalechilus labeo

The boxlip mullet *Oedalechilus labeo* inhabits inshore waters, at the mouths of rivers and sewage effluents, but does not enter brackish or freshwater areas (Trewavas 1979). *Oedalechilus labeo* is a target of fisheries in many countries in the Mediterranean. It is distributed throughout the Mediterranean Sea, but it is not found in the Black Sea and the North-East Atlantic (Fig. 7.7) (Nelson 1984, Whitehead 1984, Turan 2007, Froese and Pauly 2014), which may be related to water temperature.

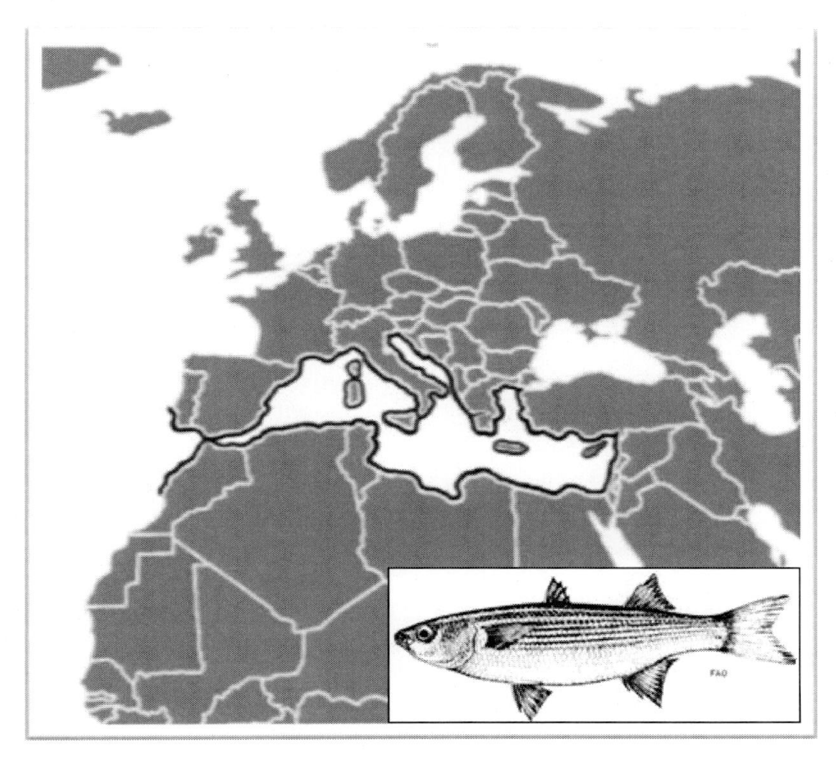

Figure 7.7. Distribution of *Oedalechilus labeo* in the Mediterranean, Black Sea and North-East Atlantic Ocean.

Exotic mullets

The exotic species of the family Mugilidae in the Mediterranean and Black Sea include the lessepsian keeled mullet *Liza carinata* and the introduced redlip mullet *Liza haematocheila* (Whitehead 1984, Turan 2007, Kottelat and Freyhof 2007).

Liza carinata

The keeled mullet *Liza carinata* is native to the Indian Ocean, and has recently spread to the South-eastern Mediterranean from the Red Sea through the Suez Canal as a lessepsian immigrant species (Fig. 7.8) (Thomson 1997). It was recorded first off Port Said, Egypt (Norman 1929, Halım and Rızkalla 2011), and later extended northwards to Iskenderun Bay, Turkey (Kosswig 1956, Turan 2007, Golani and Bogorodsky 2010). *Liza carinata* has also been found along the North African coast of the Mediterranean to Libya (Shakman and Kinzelbach 2007) (Fig. 7.8). It occurs mainly in marine coastal waters, but is also found

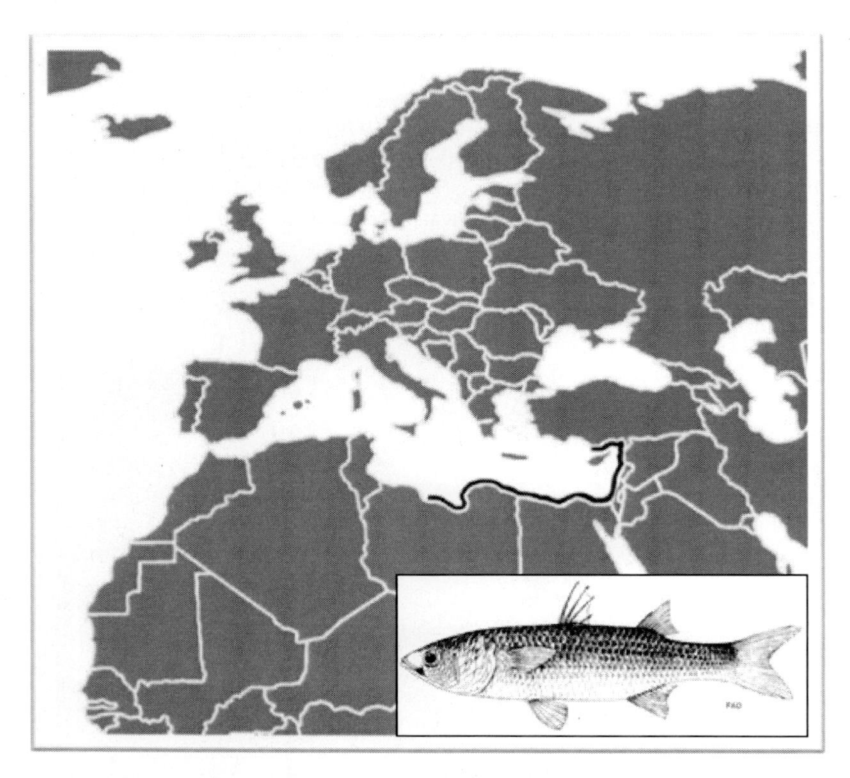

Figure 7.8. Distribution of *Liza carinata* in the Mediterranean Sea.

in inlets, lagoons and estuaries of rivers (Masuda et al. 1984). *Liza carinata* is a pelagic and euryhaline species that moves into lagoons and estuaries with thick aquatic macrophyte vegetation in spring and back into deep coastal waters in winter. Spawning takes place between August and October. It feeds on benthic algae and small molluscs, and tolerates wide temperature and salinity variations (Torcu and Mater 2000). Because of its small size, *L. carinata* has low economic value, and is normally not a target species for capture fisheries, though it is appreciated locally in Egypt as grilled fish.

Liza haematocheila

The redlip mullet (or haarder) *Liza haematocheila* is the valid name of *Mugil soiuy*, often still found in literature for the species (Kottelat and Freyhof 2007).

Liza haematocheila is native to the Amur River estuary and brackish waters of the Sea of Japan and western North Pacific (Zaitsev 1991). It has been introduced for aquaculture purposes to the Azov and the Black Seas where the local mullets had suffered from extreme water temperatures and salinities (Zaitsev 1991, Starushenko and Kazansky 1996). Following eight (1972–1980) and six years (1978–1984) of translocations into lagoons of the Black and Azov Seas respectively, together with three years (1984, 1987 and 1989) of releases of hatchery produced fry, a self-reproducing population was established (Starushenko and Kazansky 1996). This species was reported to be very abundant along the South Crimean coast. Therefore *L. haematocheila* can be considered as an alien invasive species from the Pacific.

Some of these specimens crossed the Bosphorus, the Sea of Marmara and the Dardanelles and appeared soon after in the Mediterranean (e.g., the Gulf of Smyrna in Turkey (Kaya et al. 1998)) and the Thracian Sea in Greece (Koutrakis and Economidis 2000). Thereafter, it spread progressively and has been recorded in several other localities in the northern Aegean Sea (mainly in the Strymon and Nestos estuaries), but there is no clear evidence that it has established a self-sustaining population (Minos et al. 2010).

At present, *L. haematocheila* is the most abundant mullet in the Black Sea, but it is rare in the Aegean Sea (Fig. 7.9) (Kaya et al. 1998, Minos et al. 2010). The successful expansion of *L. haematocheila* corresponds to a sharp decline of native species of Mugilidae (Kottelat and Freyhof 2007). *Liza haematocheila* is a euryhaline species and inhabits shallow coastal waters as well as freshwater regions of rivers (Nakabo 2002). It has high ecological plasticity and adaptability to waters of varying salt content and ion composition, so it can reproduce and the eggs can be fertilized at salinities ranging from 3 to 45‰ (Matishov and Luzhnyak 2007). *Liza haematocheila* has high economic value, and today is a main target of capture fisheries in the Black Sea, but its culture has not developed.

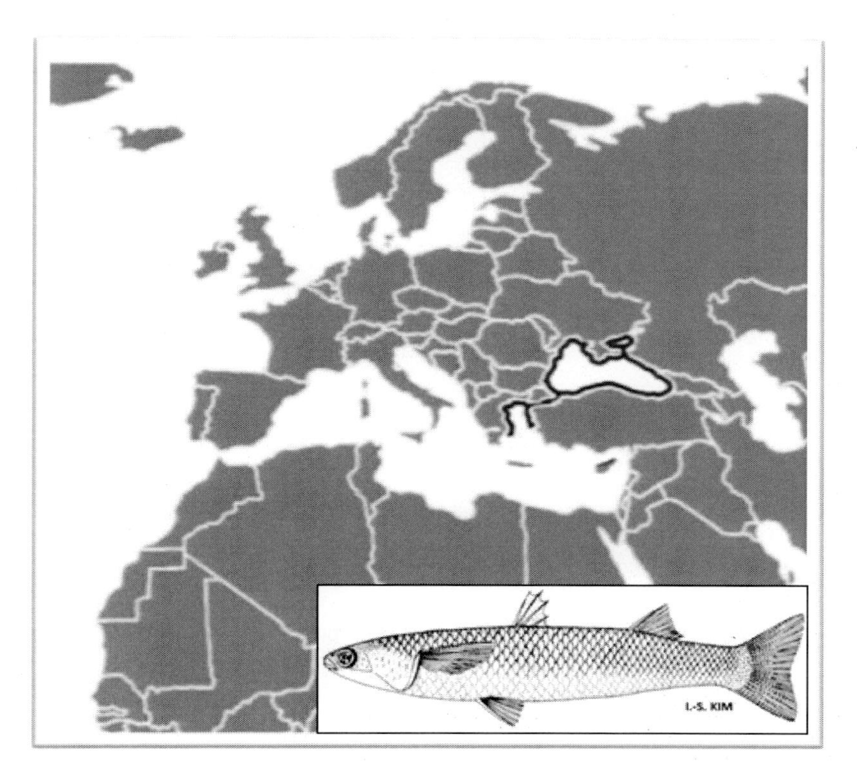

Figure 7.9. Distribution of *Liza haematocheila* in the Mediterranean and Black Sea.

Acknowledgements

Thanks to S. Karan, A. Uyan, S.A. Dogdu for their kind help.

References

Baltz, D.M. 1991. Introduced fishes in marine systems and inland seas. Biol. Cons. 56: 151–177.

Bellan-Santini, D., G. Fredj and G. Bellan. 1992. Mise au point sur les connaissances concernant le benthos profond méditerranéen. Oebalia 17: 21–36.

Ben-Tuvia, A. 1975. Comparison of the fish fauna in the Bardawil Lagoon and the Bitter Lakes. Rapp. Comm. Int. Mer. Mediterr. 23: 125–126.

Bianchi, N.C. and C. Morri. 2000. Marine biodiversity of the Mediterranean Sea: situation, problems and prospects for future research. Mar. Pollut. Bull. 40: 367–376.

Billard, R. 1997. Les Poissons d'Eau Douce des Rivières de France. Identification, Inventaire et Répartition des 83 Espèces. Lausanne, Delachaux & Niestlé. 192p.

Boltachev, A.R. 2003. Diversity of the food fishes. pp. 409–424. *In*: V.N. Eremeev and A.V. Gaevskaya (eds.). Modern Condition of Biological Diversity in Near-Shore Zone of Crimea (the Black sea sector). Ecosi-Gidrofyzika, Sevastopol (in Russian).
Boltachev, A.R. 2006. Trawl fishery and its effect on the bottom biocenoses in the Black Sea. Russian Mar. Ecol. J. 5: 45–56.
Boltachev, A.R., E.P. Karpova and O.N. Danilyuk. 2009. Findings of new and rare fish species in the coastal zone of the Crimea (the Black Sea). Vopr Ikhtyologii 49: 277–291.
Bouchet, P. and M. Taviani. 1992. The Mediterranean deep-sea fauna: pseudopopulations of Atlantic species. Deep-Sea Res. 39: 169–184.
Briggs, J.C. 1974. Marine Zoogeography. McGraw-Hill, New York.
Briggs, J.C. and B.W. Bowen. 2012. A realignment of marine biogeographic provinces with particular reference to fish distributions. J. Biogeogr. 39: 12–30.
Durand, J.D. 2015. Implications of molecular phylogeny for the taxonomy of Mugilidae. *In*: D. Crosetti and S.J.M. Blaber (eds.). Biology, Ecology and Culture of Grey Mullet (Mugilidae). CRC Press, Boca Raton, USA (this book).
Dutch, S. 1998. Closure of the Tethys. Continental Drift and Plate Tectonics, 22 December 1998. Available at http://www.uwgb.edu/dutchs/platetec/closteth.htm. Electronic version accessed 28/11/2014.
Eschmeyer, W.N. (ed.). 2014. Catalog of Fishes: Genera, Species, References. (http://research.calacademy.org/research/ichthyology/catalog/fishcatmain.asp). Electronic version accessed 27/11/2014.
Eschmeyer, W.N. and J.D. Fong. 2014. Catalog of Fishes: Species by Family/Subfamily. (http://research.calacademy.org/research/ichthyology/catalog/SpeciesByFamily.asp). Electronic version accessed 27/11/2014.
Fazli, H., A.A. Janbaz, H. Taleshian and F. Bagherzadeh. 2008. Maturity and fecundity of golden grey mullet (*Liza aurata* Risso 1810) in Iranian waters of the Caspian Sea. J. Appl. Ichthyol. 24: 610–613.
Froese, R. and D. Pauly (eds.). 2014. FishBase. World Wide Web electronic publication. www.fishbase.org, version 08-2014.
Golani, D. and S.V. Bogorodsky. 2010. The fishes of the Red Sea—reappraisal and updated checklist. Zootaxa 2463: 1–135.
González-Castro, M. and J. Ghasemzadeh. 2015. Morphology and morphometry based taxonomy of Mugilidae. *In*: Crosetti, D. and S.J.M. Blaber (eds.). Biology, Ecology and Culture of Grey Mullet (Mugilidae). CRC Press, Boca Raton, USA (this book).
Halım, Y. and S. Rızkalla. 2011. Aliens in Egyptian Mediterranean waters. A check-list of Erythrean fish with new records. Mediterr. Mar. Sci. 12: 479–490.
Harrison, I.J. 1995. Mugilidae. Lisas. pp. 1293–1298. *In*: W. Fischer, F. Krupp, W. Schneider, C. Sommer, K.E. Carpenter and V. Niem (eds.). Guia FAO para Identification de Especies para lo Fines de la Pesca. Pacifico Centro-Oriental. 3 Vols. FAO, Rome.
IUCN. 2010. IUCN Red List of Threatened Species. Version 2010.2. <http://www.iucnredlist.org> 26th July 2010.
Kaya, M., S. Mater and A.Y. Korkut. 1998. A new grey mullet species "*Mugil so-iuy* Basilewsky" (Teleostei: Mugilidae) from the Aegean Coast of Turkey. Turk. J. Zool. 22: 303–306.
Kosswig, C. 1956. Beitrag zur Faunengeschichte des Mittelmeeres. Pubbl. Staz. Zool. Napoli 28: 78.
Kottelat, M. and J. Freyhof. 2007. Handbook of European Freshwater Fishes. Publications Kottelat, Cornol, Switzerland. 646p.
Koutrakis, E.M. and P.S. Economidis. 2000. First record in the Mediterranean (North Aegean Sea, Greece) of the Pacific mullet *Mugil soiuy* Basilewsky 1855 (Mugilidae). Cybium 24: 299–302.
Masuda, H., K. Amaoka, C. Araga, T. Uyeno and T. Yoshino (eds.). 1984. The Fishes of the Japanese Archipelago. Tokai University Press, Tokyo, Japan. 438pp.
Matishov, G.G. and V.A. Luzhnyak. 2007. Extension of the spawning area of the Far Eastern mullet *Liza haematocheilus* (Temminck and Schlegel 1845) acclimated in the Sea of Azov and Black Sea basin: recent data on reproduction ecology. Doklady Biological Sciences 414: 221–222.
Minos, G., A. Imsiridou and P.S. Economidis. 2010. *Liza haematocheilus* (Pisces: Mugilidae) in Northern Aegean Sea. pp. 313–332. *In*: D. Golani and B. Appelbaum-Golani (eds.). Fish Invasions in the Mediterranean Sea: Change and Renewal. Pensoft Publishers, Sofia-Moscow.
Myers, A.A. 1996. Species and generic gamma-scale diversity in shallow-water marine Amphipoda with particular reference to the Mediterranean. J. Mar. Biol. Assoc. U.K. 76: 195–202.
Nakabo, T. 2002. Fishes of Japan with Pictorial Keys to the Species, English Edition I. Tokai University Press, Japan. 866p.
Nelson, J.S. 1984. Fishes of the World. John Wiley and Sons, New York. 523pp.
Nelson, J.S. 2006. Fishes of the World, 4th Edition. John Wiley and Sons, Inc. Hoboken, New Jersey, USA. 601p.
Norman, J.R. 1929. The Teleostean fishes of the family Chiasmodontidae. J. Nat. Hist. 3: 529–544.
Ormond, R.F.G., J.D. Gage and M.V. Angel. 1997. Marine Biodiversity: Patterns and Processes. Cambridge University Press, New York. 449pp.
Oven, L.S. 1993. Ichthyofauna of the Black Sea Bays under Anthropogenic Impact. Naukova Dumka Publishing House, Kiev, Ukraine. 144pp.
Ozturk, B. and C. Turan. 2013. Alien species in the Turkish seas. pp. 92–130. *In*: A. Tokac, A.C. Gucu and B. Ozturk (eds.). The State of the Turkish Fisheries. Turkish Marine Research Foundation Publication, Istanbul, Turkey.
Paternello, T., A.M.J. Volckaert and R. Castilho. 2007. Pillars of Hercules: is the Atlantic Mediterranean transition a phylogeographic break? Mol. Ecol. 16: 4426–4444.
Patimar, R. 2008. Some biological aspects of the sharpnose mullet *Liza saliens* (Risso 1810) in Gorgan Bay-Miankaleh wildlife refuge (the Southeast Caspian Sea). Turk. J. Fish. Aquat. Sc. 8: 225–232.
Por, F.D. 1978. Lessepsian Migration: The Influx of Red Sea Biota into the Mediterranean by Way of the Suez Canal Berlin. 228p.

Rochard, E. and P. Elie. 1994. La macrofaune aquatique de l'estuaire de la Gironde. Contribution au livre blanc de l'Agence de l'Eau Adour Garonne. pp. 1–56. *In*: J.-L. Mauvais and J.-F. Guillaud (eds.). Etat des connaissances sur l'estuaire de la Gironde. Agence de l'Eau Adour-Garonne, Editions Bergeret, Bordeaux, France.

Shakman, E.A. and R. Kinzelbach. 2007. Distribution and characterization of lessepsian fishes along the coast of Libya. Acta Ichthyol. Piscat. 37: 7–15.

Shiganova, T.A. 1998. Invasion of the Black Sea by the ctenophore *Mnemiopsis leidyi* and recent changes in pelagic community structure. Fish. Oceanogr. 7: 305–310.

Spalding, M.D., H.E. Fox, G.R. Allen, N. Davidson, Z.A. Ferdana, M. Finlayson, B.S. Halpern, M.A. Jorge, A. Lombana, S.A. Lourie, K.D. Martin, E. McManus, J. Molnar, C.A. Recchia and J. Robertson. 2007. Marine ecoregions of the world: a bioregionalization of coastal and shelf areas. BioScience 57: 573–583.

Stanley, D.J. 1990. Med desert theory is drying up. Oceanus 33: 14–23.

Starushenko, L.I. and A.B. Kazansky. 1996. Introduction of mullet harder (*Mugil so-iuy* Basilewsky) into the Black Sea and the Sea of Azov. Stud. Rev. Gen. Fish. Counc. Mediterr. 67: 29p.

Stiassny, M.L. 1993. What are grey mullets? Bull. Mar. Sci. 52: 197–219.

Sumer, C. and I. Balik. 2007. Comparison of two lagoons situated in the west and east Mediterranean coast of Turkey in terms of the catch per unit area and catch composition. Turkish J. Aquat. Life 5-8: 87–92.

Sumer, C. and I. Teksam. 2013. Catch efficiency and composition of Beymelek Lagoon. Anadol. J. Agri. Sci. 28: 47–51.

Svetovidov, A.N. 1964. The Fishes of the Black Sea. Nauka Publ. Moscow-Leningrad. 550p.

Thomson, J.M. 1986. Mugilidae. pp. 344–349. *In*: J. Daget, J.-P. Gosse and D.F.E. Thys van den Audenaerde (eds.). Check-list of the Freshwater Fishes of Africa (CLOFFA). ISNB, Brussels, MRAC; Tervuren; and ORSTOM, Paris. Vol. 2.

Thomson, J.M. 1990. Mugilidae. pp. 855–859. *In*: J.C. Quero, J.C. Hureau, C. Karrer, A. Post and L. Saldanha (eds.). Check-list of the Fishes of the Eastern Tropical Atlantic (CLOFETA). JNICT, Lisbon; SEI, Paris; and UNESCO, Paris: Vol. 2.

Thomson, J.M. 1997. The Mugilidae of the world. Memoirs of the Queensland Museum 41: 457–562.

Torcu, H. and S. Mater. 2000. Lessepsian fishes spreading along the coast of the Mediterranean and the Southern Aegean Sea of Turkey. Turk. J. Zool. 24: 139–148.

Trewavas, E. 1979. Sciaena nibe Jordan and Thompson 1911 (Pisces): Proposed conservation of the species name n/:beby use of the Plenary Powers Z.N. (S.)2226. Bull. Zool. Nomencl. 36: 155–157.

Turan, C. 2007. Atlas and Systematics of Marine Bony Fishes of Turkey. Nobel Publishing, Adana, Turkey. 549p.

Turan, C. 2014. Genetic studies on the Black Sea marine biota. pp. 457–468. *In*: E. Duzgunes, B. Ozturk and B. Zengin (eds.). Turkish Fisheries in the Black Sea. Turkish Marine Research Foundation Publication. Istanbul, Turkey.

Turan, C., F. Boero, A. Boltachev, E. Düzgüneş, Y.P. Ilyin, A. Kıdeys, D. Micu, J.D. Milliman, G. Minicheva, P. Moschella, T. Oğuz, B. Öztürk, H.O. Portner, T. Shiganova, A. Shivarov, E. Yakushev and F. Briand. 2009. Climate forcing and its impacts on the Black Sea marine biota. Executive summary of CIESM Workshop Monograph, Monaco. 39–152p.

Turan, C., M. Gürlek, D. Ergüden, D. Yağlıoğlu and B. Öztürk. 2011. Systematic status of nine mullet species (Mugilidae) in the Mediterranean Sea. Turk. J. Fish. Aquat. Sc. 11: 315–321.

Whitehead, P.J.P. 1984. Clupeidae. pp. 268–281. *In*: P.J.P. Whitehead, M.L. Bauchot, J.C. Hureau, J. Nielsen and E. Tortonese (eds.). Fishes of the North-Eastern Atlantic and the Mediterranean. UNESCO, Paris.

Whitfield, A.K., M. Elliott, A. Basset, S.J.M. Blaber and R.J. West. 2012. Paradigms in estuarine ecology–a review of the Remane diagram with a suggested revised model for estuaries. Estuar. Cost. Shelf Sci. 97: 78–90.

Yaglioglu, D., C. Turan and T. Ogreden. 2013. First record of blue crab *Callinectes sapidus* (Rathbun 1896) (Crustacea, Brachyura, Portunidae) from the Turkish Black Sea coast. Journal of the Black Sea/Mediterranean Environment 20: 13–17.

Zaitsev, Y.P. 1991. Eutrophication in the Black Sea. Paper Presented at International Workshop on the Black Sea. Focus on the Western Black Sea Shelf. 30 September–04 October 1995. Varna, Bulgaria.

CHAPTER 8

Musculoskeletal Anatomy of the Flathead Grey Mullet *Mugil cephalus*

Javad Ghasemzadeh

Introduction

The Mugilidae, commonly known as grey mullets, is a speciose family of teleostean fishes, which includes 20 genera and 81 species (Eschmeyer and Fong 2015). Mullets are small to medium-sized fishes occurring in various coastal aquatic habitats of the world's tropical, subtropical and temperate regions in all continents (Harrison and Howes 1991). Most species are euryhaline, inhabiting coastal marine waters, hypersaline to brackish-water lagoons, estuaries and freshwater (González-Castro 2007).

Mullets occupy a comparatively low position in the food web, and are thus relatively efficient secondary producers of protein. They are situated at the base of the food pyramid and, by virtue of their consumption of particulate organic matter, detritus and benthic microalgae, they are able to 'telescope' the food chain and make high quality fish protein available to top predators (Whitfield et al. 2012). Their larvae are usually planktonic feeders in the offshore marine environment (Brownell 1979), surf zone (Inoue et al. 2005) and when they first enter estuaries (Gisbert et al. 1996). In coastal and estuarine nursery habitats the fry and juveniles initially feed on small invertebrates that undergo vertical migrations within the water column and later feed mainly on benthic organisms and plant material by browsing and sifting the bottom sediments and detritus (Blaber and Whitfield 1977). The larger juvenile and adult mullets feed mainly on detritus and benthic microalgae (especially diatoms), together with foraminiferans, filamentous algae, protists, meiofauna and small invertebrates (Thomson 1963, Blaber 1976, Lawson and Jimoh 2010, Whitfield et al. 2012, Chapter 9—Cardona 2015).

Mullets are commercially exploited in all regions where they occur, constituting an important part of the human diet (Menezes 1983, Chapter 16—Crosetti 2015). They are widely used in both capture fisheries and aquaculture (Bacheler et al. 2005, Thomson 1963). Depending on the availability of different taxa within each region, the harvested mugilids will vary in species composition, but *M. cephalus* is often an important species in the catch (Katselis et al. 2003, Chaoui et al. 2006). Adults are targeted mainly by small-scale fisheries, while fry and juveniles are captured for aquaculture in certain areas (Whitfield et al. 2012).

Faculty of Marine Sciences, Chabahar Maritime University, Iran.
Email: jghasemz@yahoo.com.au; ghasemzadeh@cmu.ac.ir

Until the end of the 20th century, the identification and taxonomy of mullets mainly relied on external morphology, meristics, morphometrics and the structure of some internal organs (Chapter 1—González-Castro and Ghasemzadeh 2015). Characters which have been used by different authors include dentition (Ebeling 1957, 1961, Thomson 1975, Farrugio 1977), scales (Cockerell 1913, Jacot 1920, Pillay 1951, Thomson 1981, Chervinski 1984, Liu and Shen 1991), number of pyloric caeca (Perlmutter et al. 1957, Hotta and Tung 1966, Luther 1977), the alimentary tract (Thomson 1966), intestinal convolution (Hotta 1955), otoliths (Morovic 1953, González-Castro et al. 2009), morphology of the cephalic lateral line canals (Song 1981), osteology (Ishiyama 1951, Hotta and Tung 1966, Sunny 1971, Kobelkowsky and Resendez 1972, Luther 1977, Mohsin 1978, Senou 1988, Ghasemzadeh 1998, Ghasemzadeh et al. 2004), pharyngobranchial organ (Harrison and Howes 1991), and dentition, pigmentation and melanophore patterns in identification of fry and juveniles (Van Der Elst and Wallace 1976, Cambrony 1984, Reay and Cornell 1988, Serventi et al. 1996). A high degree of similarity in external morphology and a wide range of variability in taxonomic features have created a lot of confusion in correct identification, taxonomy and evolutionary relationships of the members of this family.

The external appearance of mullets is remarkably uniform. The body is elongate, sub-cylindrical, and more-or-less compressed posteriorly. The head is broad and horizontally compressed in most genera (convex in *Cestraeus* and *Joturus*), and almost completely covered by scales. The lateral line is absent. The scales are moderate-to-large, usually ctenoid, with one or more longitudinal rows of striae (there is a crenulated membranous caudal margin in *Valamugil* and *Crenimugil*). The eye is fairly large, laterally-placed (in *Rhinomugil*, dorsolaterally-placed) and visible from below. The eye is often covered by adipose eyefold tissue. There are two nostrils anterior to each eye. The mouth is small-to-moderate, terminal or inferior. The teeth are relatively small, hidden or absent from the lips, and variably-developed (in different genera) on the palatine, vomer, pterygoids, tongue and basihyal. The premaxilla is protractile. The maxilla is almost, or entirely, hidden beneath the preorbital. The two dorsal fins are widely-separated and well set back, the first dorsal being roughly at the mid-point between the snout and tail-base (placed further anteriorly in *Joturus*). The first fin has four spines, and an elongate scale at its base. The second dorsal fin has nine–11 segmented rays, the first one being short, slender and unbranched, and is often segmented near its tip in adults, and so is frequently mistaken for a spine. The anal fin is usually set slightly in advance of the second dorsal fin, and has two–three spines and eight–12 soft rays. The pectoral fins are inserted high on the body, with or without an elongate axillary scale, a short spine, and 14–20 rays. The pelvic fins are subabdominal, each with one spine, five rays, a long axillary scale, and an elongate cuneiform interpelvic scaly flange. The caudal fin is symmetrical, and markedly forked in the majority of mullets, but truncate or emarginate in a few species including *Ellochelon vaigiensis* and *Rhinomugil nasutus*. All fins except the first dorsal are mostly scaly. The number of scales of the midlateral line is 24–64, and in transverse series eight–16. The branchial basket is adapted for filter feeding with a specialized pharyngobranchial organ and pharyngeal sulcus, which vary in size and shape between species. The gill rakers are variable in size, slender and very numerous. The stomach has a muscular gizzard (except in *Agonostomus* and *Cestraeus*) with a variable number of pyloric caecae. The intestine is mainly elongate and elaborately convoluted. The number of vertebrae is 24–26. The body is normally dark olive, dark blue, dark brown, greenish or greyish dorsally. The flanks are silvery, in some species with a pale yellowish tinge, and often have more-or-less distinct dark stripes on some transverse rows of scales. The ventral parts of the body are silvery, or yellowish. The pectoral, dorsal, anal and caudal fins are usually dusky. The pelvic fins are pale yellow, and often have dusky margins.

Molecular genetics, protein or enzymes electrophoresis and mitochondrial DNA sequence analysis have provided additional characters which are useful in population structure and phylogenetic studies (Peterson and Shehadeh 1975, Rosenblatt and Waples 1986, Campton and Mahmoudi 1991, Menezes et al. 1992, Crosetti et al. 1993, 1994, Lee et al. 1995, Rossi et al. 1996, 1998a,b, 2004, Caldara et al. 1996, Rocha-Olivares et al. 2000, Papasotiropoulos et al. 2001, 2002, 2007, Murgia et al. 2002). Several Polymerase Chain Reaction (PCR) methods of genotype analysis have been developed for fish identification. Among them, the analysis of Restriction Fragment Length Polymorphism (RFLP) of PCR-amplified mitochondrial DNA (mtDNA) fragments has recently been widely used (Chen et al. 2003, Chen and Mayden 2009, 2010, Miya et al. 2003, Turan et al. 2005, Gornung et al. 2007, Imsiridou et al. 2007, Semina et al. 2007, Aurelle et al. 2008,

Ke et al. 2009, Heras et al. 2009, Jamandre et al. 2009, Liu et al. 2009, 2010, Erguden et al. 2010, Livi et al. 2011, Durand et al. 2012a,b, Chapter 2—Durand 2015).

In this chapter we describe in detail the skeletal bones and the muscular anatomy of neurocranium, suspensorium, pectoral and pelvic girdles of the flathead grey mullet *Mugil cephalus* as the most common and first-described species of the family Mugilidae. Specimens for examination were borrowed from the British Museum of Natural History, London (BMNH); Museum National d' History Naturelle, Paris (MNHN); Rijksmuseum van Natuurlijke Historie, Leiden (RMNH); Australian Museum, Sydney (AMS); Queensland Museum, Brisbane (QM); National Museum of Victoria, Melbourne (NMV); Western Australian Museum, Perth (WAM) and Northern Territory Museum and Art Galleries, Darwin (NTM). Specimens of *M. cephalus* for osteological examination were cleared and stained following the method of Taylor and Van Dyke (1985). Specimens were dissected under an Olympus SZM stereomicroscope and drawings were made by the author with the aid of a camera lucida attachment. All of the drawings and illustrations presented in this chapter are based on the specimen (AMS I.31251-010 SL = 59 mm. locality: Rainbow Creek beach, New South Wales, Australia). Nomenclature of bones is based on Patterson (1964), Patten (1987) and Rojo (1990), and terminology of muscles follows Winterbottom (1974).

Muscoulskeletal Anatomy of the Flathead Grey Mullet *Mugil cephalus*

Neurocranium

In Mugilidae the neurocranium has an elongate anteroposterior axis, the lateral axis is moderately elongate, and the vertical axis is small. The dorsal surface is broad, smooth and roughly quadrangular (Fig. 8.1A). The anterior border of the neurocranium is limited by the maxillary condyles of the vomer. The dorsal surface is formed by the frontals, parietals and supraoccipital. On each side of the dorsal surface, behind the lateral process of the frontal, a shelf is formed by union of the borders of the frontal and pterotic. The subfrontal fossa is formed under this shelf, and the sphenotics are visible dorsal to these shelves (Fig. 8.1A). The border of the orbit is limited anteriorly by the margin of the lateral process of the lateral ethmoid and posteriorly by the postorbital process of sphenotic (Fig. 8.1A). Although the parietals lie flush with the frontals and the supraoccipital dorsally, there is a midlateral ridge which continues toward the epioccipital. (Fig. 8.1A). This ridge, along with an indistinct ridge on the pterotic, forms the temporal fossa in which the epaxial muscles is inserted. The pterotic joins with the lateral part of the parietal and with the epioccipital along the floor of this fossa (Fig. 8.1A). In the posterior part of the neurocranium, the occipital fossa is formed by the posterior part of the supraoccipital and epioccipitals. The vertical plate of the supraoccipital crest divides the right and left occipital fossa from each other, and the exoccipitals delimit the floor of these fossae (Figs. 8.1A, 8.1C). The epaxial muscles insert into these fossae. The posterior alar crest of the epioccipitals as well as some bony projections in these fossae, provide insertion points for the epaxial muscles. A unique feature of the mugilid neurocranium is the posterior extension of the supraoccipital crest and epioccipital alar processes (Figs. 8.1A, B, and C).

In ventral view the orbital cavities are observed in the anteromedial part of the neurocranium (Figs. 8.1B and C). These cavities are separated by the parasphenoid, and occupy less than half of the total length of the neurocranium. A ventral fossa is formed at the juncture of the sphenotic, prootic and pterotic. In lateral view the ventral crest of the neurocranium, corresponds to the ventromedial ridge on the parasphenoid (Fig. 8.1C).

The sphenotics' postorbital processes form the lateral extremes of the neurocranium, and the parasphenoid forms its ventral border (Fig. 8.2A). A large foramen magnum is formed by exoccipitals and basioccipital. The occipital fossae, which are partly obscured by the wing-shaped processes of the epioccipitals, are dorsal to the exoccipitals. The intercalars extend posteroventrally to the epioccipitals. The posterior part of neurocranium is delimited by the minor occipital condyles of the exoccipitals and the major occipital condyle of the basioccipital, which join with the first vertebra.

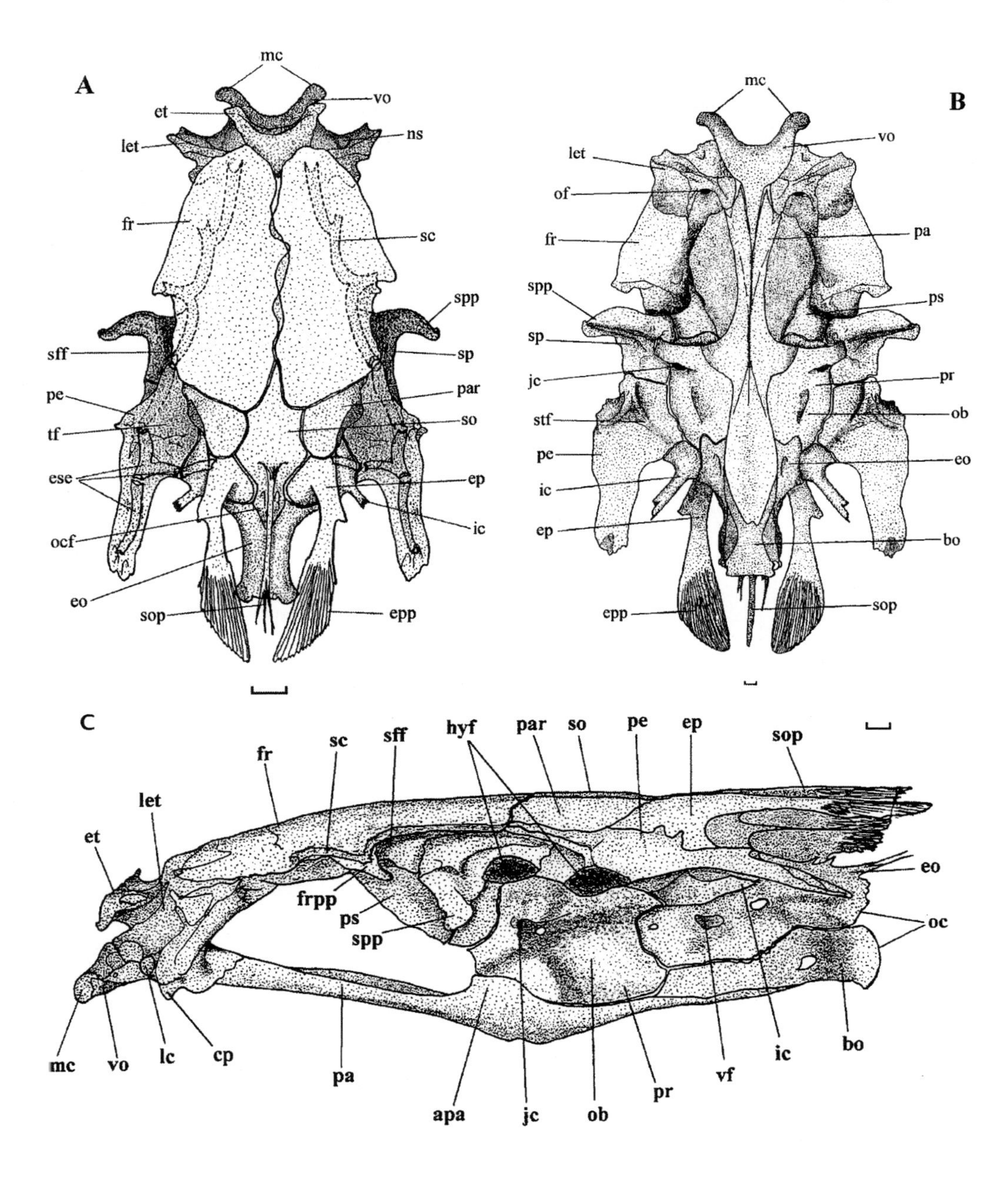

Figure 8.1. Neurocranium, *Mugil cephalus* AMS I.31251-010, 59 mm SL. (A) Dorsal view (nasals removed). (B) Ventral view. (C) Lateral view (nasals removed). Scale bar: 1 mm.

Abbreviations: apa: scending arm of parasphenoid, **bo:** basioccipital, **cp:** palatine condyle, **eo:** exoccipital, **ep:** epioccipital, **epp:** epioccipital process, **ese:** extrascapulars, **et:** ethmoid, **fm:** foramen magnum, **fr:** frontal, **frpp:** frontal postorbital process, **hyf:** hyomandibular facet, **ic:** intercalar, **jc:** jagular canal, **let:** lateral ethmoid, **mc:** maxillary condyle, **ns:** nasal sac, **ob:** otic bulla, **oc:** occipital condyle, **ocf:** occipital fossa, **of:** olfactory nerve foramen, **pa:** parasphenoid, **par:** parietal, **pc:** preorbital condyle, **pe:** pterotic, **pr:** prootic, **ps:** pterosphenoid, **sc:** sensory canal, **sff:** subfrontal fossa, **so:** supraoccipital, **sop:** supraoccipital process, **sp:** sphenotic, **spp:** sphenotic postorbital process, **stf:** subtemporal fossa, **tf:** temporal fossa, **vf:** vagus foramen, **vo:** vomer.

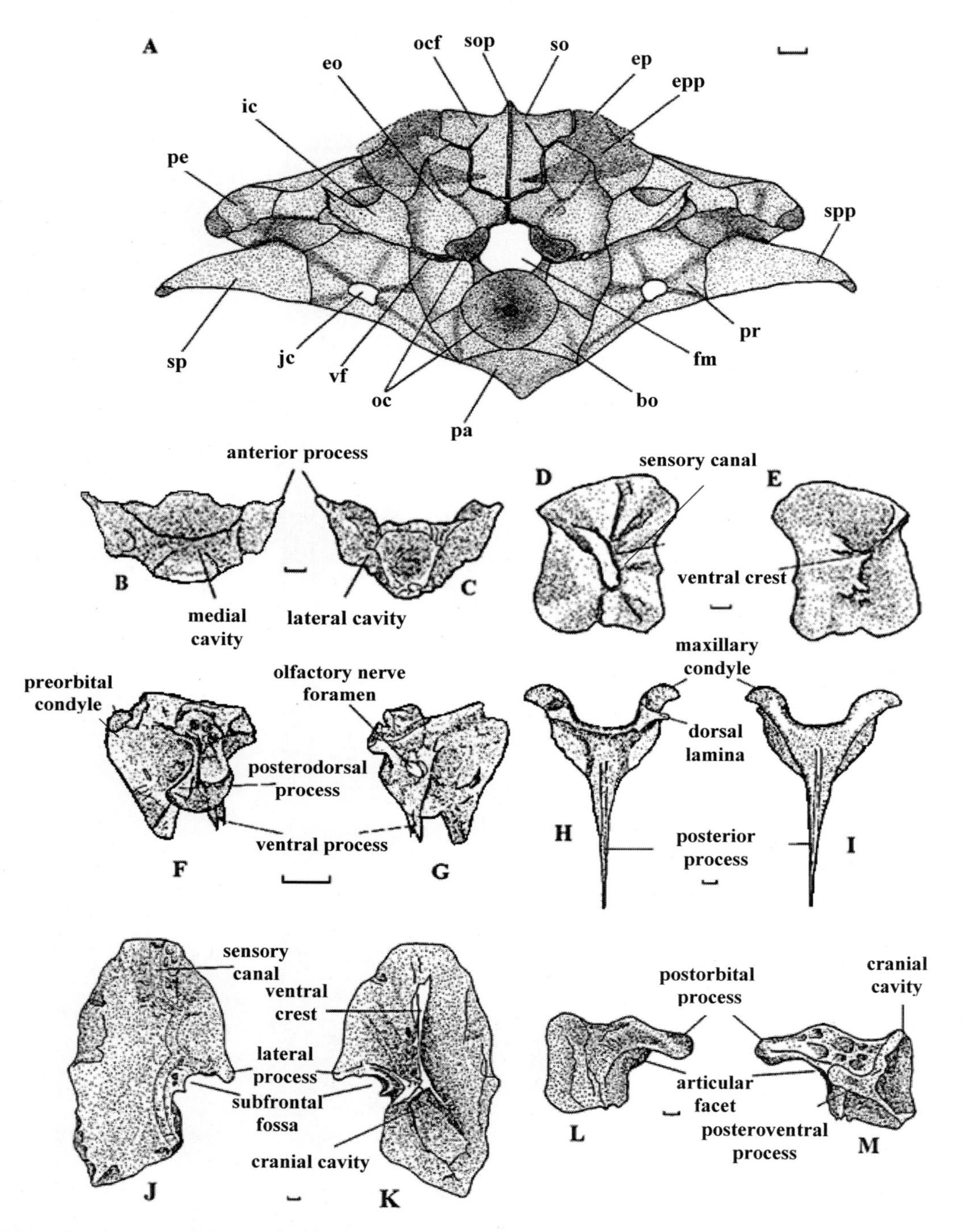

Figure 8.2. (A) Neurocranium, *Mugil cephalus* AMS I.31251-010, 59 mm SL. Posterior view (Abbreviations as in Fig. 8.1); (B) Ethmoid, dorsal view; (C) Ethmoid, ventral view; (D) Nasal (dorsal view, left bone); (E) Nasal (ventral view, left bone); (F) Lateral ethmoid (dorsal view, left bone); (G) Lateral ethmoid (ventral view); (H) Vomer (dorsal view); (I) Vomer (ventral view); (J) Frontal (dorsal view, right bone); (K) Frontal (ventral view); (L) Sphenotic (dorsal view, right bone); (M) Sphenotic (ventral view). Scale bar: 1 mm.

Ethmoid: The ethmoid, a bone of prechondral origin, is situated medial to the lateral ethmoids and dorsal to the vomer (Fig. 8.1A). It is irregular and approximately semilunar in shape, with two anterior processes, forming a shallow cavity which is occupied by the ascending processes of the premaxillae (Fig. 8.2B). In *M. cephalus* the dorsal surface of the ethmoid has a posteromedial cavity and two small lateral crests, which form two small lateral depressions. The ventral surface has a medial circular cavity and two lateral cavities formed by the circular cavity borders and dorsal body (Fig. 8.2C). The structure of the posteroventral body of this bone is porous and spongy. The ethmoid joins ventrally with the dorsomedial border of the vomer and laterally with the lateral ethmoids. The nasals partially overlie the ethmoid, and join with the anterior processes of the ethmoid through their ventral ridges.

Lateral ethmoids: The lateral ethmoids are a pair of prechondral origin bones with an irregular and complicated structure. They are well separated from each other, and located lateral to the ethmoid and vomer (Figs. 8.1A-B). A lateral condyle strongly characterizes each lateral ethmoid with its respective preorbital (Figs. 8.1C, 8.2F). Dorsally, irregular laminar structures present porous cavities which are overlaid by the frontal. An oblique anterodorsal lamina forms two funnel-shaped nasal sacs on the lateral aspect of each lateral ethmoid. The anterior nasal sac is wider and restricted to the space between the preorbital laterally, and the nasal and lateral ethmoid medially. This sac leads to a relatively large foramen, through which pass the olfactory nerves from the orbit to the nasal sac (Fig. 8.2G). The posterior nasal sac is considerably smaller, and limited to a small space between the nasal and preorbital bones and located dorso-laterally to the anterior nasal sac. The lateral ethmoid has two posterodorsal small processes which join with the frontal. Ventrally it has a depression and a small condyle for linking with the palatine. It also joins with the parasphenoid through a posteroventral process (Fig. 8.1B). The lateral border of the lateral ethmoid is serrated in adult fish. Its ventral surface is convex, with a laminar shelf originating from the olfactory nerve foramen.

Nasals: The nasals are a pair of irregular quadrangular dermal bones, covering the anterior processes of the ethmoids and lateral ethmoids. Each nasal has a dorsal depression bearing the anterior part of the supraorbital branch of the sensory canal system (Figs. 8.2D-E). This sensory canal extends obliquely towards the joining point of the nasal and preorbital, and dorsally is covered by a laminar shelf (in young fish), transforming to an irregular and partly squamate ridge in adults. The nasal is connected to the extreme anterior border of the frontal posteriorly, and via its anterolateral apex with the preorbital and the anterior process of palatine (Figs. 8.5A, 8.8A). Ventrally it has a small process which is produced as a high ridge extending anteriorly. A strong ligament attaches the anteromedial part of this ridge to the lateral process of the ethmoid (Fig. 8.5B).

Vomer: The vomer is the anterior-most ventral bone of the skull (Figs. 8.1A-B-C). Its anterior border has a pair of edentulous maxillary condyles directed anterolaterally, forming a deep curve medially (Figs. 8.2H-I). This cavity unites with the median cavity of the ethmoid, providing a space for housing the ascending processes of the premaxillae and rostral cartilage (Fig. 8.8A). The maxillary condyles of the vomer are connected to the maxillae via a cylindrical submaxillary menisci (Figs. 8.8A, 8.5B). The vomer has a long and sharp pointed posterior process which lies flush within the ventral cavity of the parasphenoid (Fig. 8.1B). Dorsally, the medial border is raised and produced as two bilateral laminar expansions, which join with the ventral margin of the lateral ethmoids. The ventral margin of the ethmoid inserts behind the raised dorsomedial border of the vomerine cavity.

Frontal: The frontals are a pair of large, broad, and flat bones forming the major area of the neurocranium roof. They are located dorsally flush with the parietals and supraoccipital (Fig. 8.1A). The dorsal surface of the frontal is smooth, except for a roguse region anteriorly, corresponding with the prominent ridge of the supraorbital sensory canal, and giving the appearance of being divided (Fig. 8.2J). Its anterior border is curved and joined with the nasal dorsally, and the ethmoid and lateral ethmoid ventrally. Medially the frontal is produced as a lateral process which bears a branch of the supraorbital sensory canal. Behind this process a large cavity (subfrontal fossa) is formed, which is complemented with the sphenotic and pterotic bones. The opercular muscles originate from this fossa. The postorbital process of the sphenotic is visible through this fossa. The posterior border of the frontals joins with the anterior margins of the supraoccipital and parietals (Fig. 8.1A).

The ventral surface of the frontal bears a medial crest, which is the most wide anteriorly (Fig. 8.2K). Posteriorly this crest is higher with a lateral expansion, which joins with the pterosphenoid and sphenotic, thus forming part of cranial cavity. The lateral ethmoid links to the ventral surface of the frontal through the anterior protrusion of this crest and a small lateral eminence.

Sphenotics: Each sphenotic has a quadrangular dorsal surface with a thick postorbital process projecting laterally. Although this bone is partly covered by the frontal and pterotic, it is visible dorsally and ventrally (Fig. 8.1A-B). The m. *levator arcus palatini* originates from the posterior side of this process, coursing ventrally over the hyomandibular facet and inserting on the preopercle horizontal arm (Fig. 8.5B). Ventrolaterally, the sphenotic has a prominent ovoid articular facet which receives the anterior condyle of hyomandibula (Fig. 8.2L-M). Ventrally it has a cavity which is part of the brain case. The anterior border joins the frontal dorsally, and pterosphenoid ventrally. The posterior border joins with the pterotic and prootic bones.

Pterosphenoids: These are a pair of almost trapezoid shape bones forming the posterolateral margin of the orbit and part of the wall and floor of the cranial cavity (Figs. 8.1B-C). Each pterosphenoid is linked with its respective frontal in the dorsolateral margin, and posteroventrally with its respective sphenotic and prootic. The anterior border is free. The ventral surface of the pterosphenoid is plain, but there is a small fold on dorsal surface (Figs. 8.3C-D).

In *M. cephalus* the pterosphenoid of one side does not join its opposite partner, but in some mugilid species, which have a basisphenoid, the basisphenoid either has a weak connection with both pterosphenoids, or strongly linked with them.

Parietals: The parietals are a pair of split-level bones which form part of the cranial roof (Fig. 8.1A). The midlateral border of each parietal is curved inwardly, forming a shelf which continues to the epioccipital, and a shallow depression which extends outward and joins with the pterotic, forming the temporal fossa for insertion of epaxial muscles. (Figs. 8.1A, 8.3B). A small crest gives a rather corrugated appearance to the ventral side of this bone (Fig. 8.3A). The parietal also overlaps portions of the supraoccipital, pterotics and epioccipitals. It is linked with the frontal anteriorly, epioccipital posteriorly, pterotic at its lateral border, and supraoccipital at its internal border (Fig. 8.1A).

Parasphenoid: The parasphenoid lies along the base of the neurocranium, median to the interorbital plane (Figs. 8.1B-C, 8.2A). The parasphenoid comprises anterior and posterior arms, leaving between them an angle corresponding with the most inferior part of the neurocranium. The anterior arm is long and broad anteriorly (Fig. 8.3E-F). It has a V-shaped ventral fossa which receives the posterior arm of the vomer. Two anterolateral borders are interposed between the vomer and lateral ethmoids (Fig. 8.1B). From the apex of the ventral fossa, a crest rises towards the middle, with a ventral depression, disappearing in the posterior portion (Fig. 8.3F-G). The alar (ascending) processes of the parasphenoid are directed dorsolaterally, joining with anteroventral surface of the prootics through sutures. There is a foramen between each alar process and prootic. The posterior arm of the parasphenoid is shorter, with an ovoid and slightly bifurcated fossa distally, which covers a large part of the basioccipital (Fig. 8.1B).

Prootics: The prootics are two relatively large bones of the otic region forming the anterior floor of the cranial vault. A horizontal and another oblique depression are present on the external surface of this bone, forming the external boundaries of the otic bulla. A foramen is present at the vertices of these cavities. The anteroventral border of prootic also represents a small depression (Figs. 8.1C, 8.3H). There is a large jugular canal ventral to the articular facet of the prootic and sphenotic (Figs. 8.1B-C, 8.4A, 8.3H-I). The lateral wall of this canal is reduced to a narrow strut. The posterior opening of the jugular canal is confluent with the exit of the hyomandibular nerve. The internal surface of the prootic is very irregular. An inferior horizontal lamina joins the contralateral bone, forming the roof of the anterior myodome and floor of the

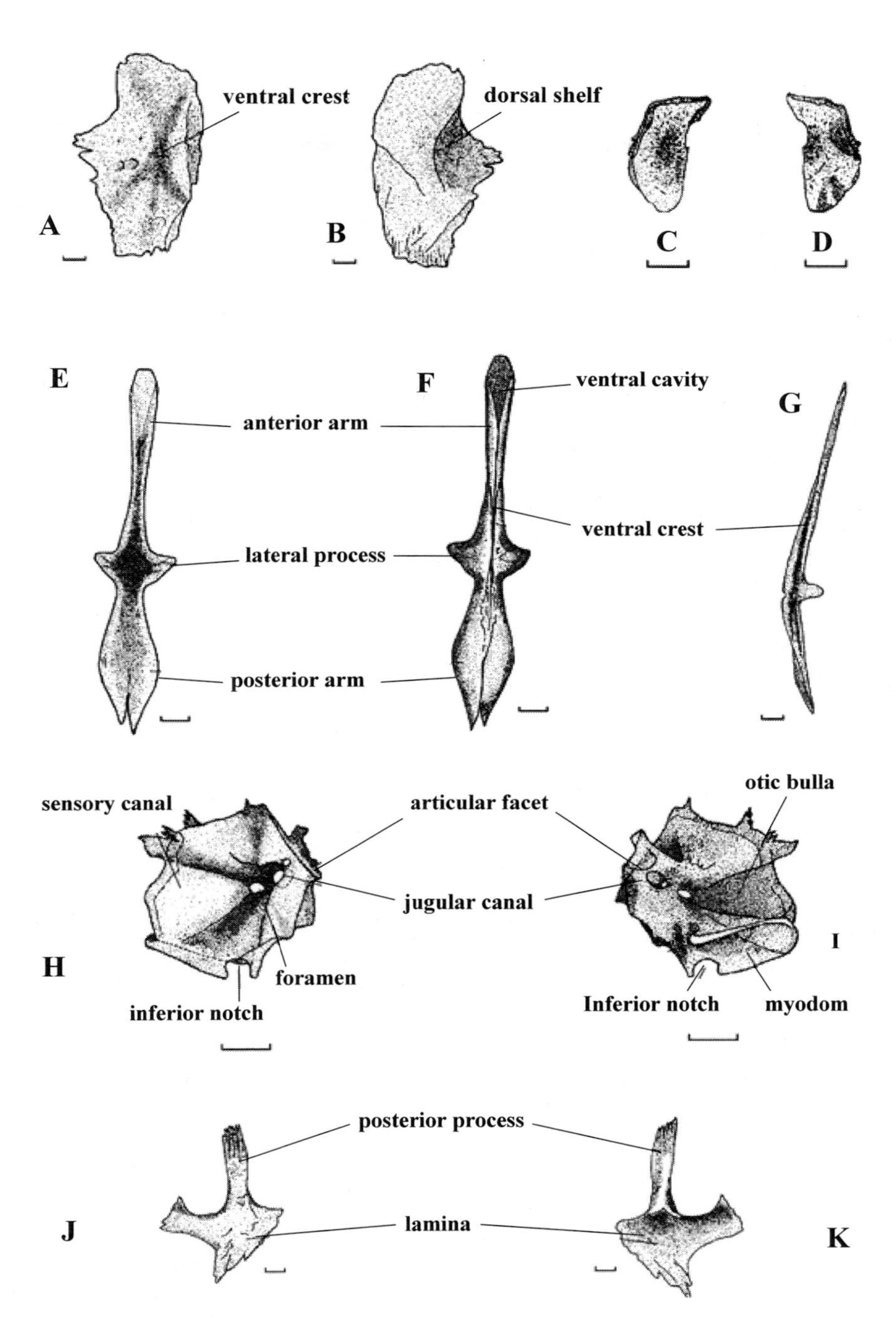

Figure 8.3. (A) Parietal (ventral view, right bone); (B) Parietal (dorsal view); (C) Pterosphenoid (lateral view, right bone); (D) Pterosphenoid (medial view); (E) Parasphenoid (ventral view); (F) Parasphenoid (dorsal view); (G) Parasphenoid (lateral view); (H) Prootic (external view, right bone); (I) Prootic (internal view); (J) Intercalar (external view, right bone); (K) Intercalar (internal view); *Mugil cephalus* AMS I.31251-010 SL = 59 mm. Scale bar: 1 mm.

cranial cavity (Fig. 8.3I). Over these laminae, the wall represents some irregular folds and two foraminae. These foraminae are close together, at the level of the jugular canal, the larger anterior foramina opening into the jugular canal.

The otic bullae are located above the horizontal laminae (Figs. 8.3I, 8.1C). They are relatively large and bulbous, extending postero-ventrally and laterally to the posterior myodome, into the basioccipital. A pair of oblong otoliths are housed in each otic bulla. The prootic joins with the sphenotic and pterotic dorsally, with pterosphenoid anteriorly, parasphenoid ventrally, and posteriorly with the basioccipital and exoccipital. It also has a tight suture with the intercalar.

Intercalars: The intercalars are a pair of small bones which are not directly involved in the formation of the cranial cavity. Each intercalar comprises a small lamina with two small lateral expansions. It has a flat external surface and a convex internal facet which overlies the exterior surface of the prootic and pterotic anteriorly, and the epioccipital and exoccipital posteriorly. A process emerges from this lamina posterolaterally, joining with the ventral ramus of the posttemporal (Figs. 8.1A-B, 8.3J-K).

Pterotics: The pterotics are a pair of roughly paddle or oar shaped bones, with a wider and thick anterior body. Each pterotic has a small anterior process diffusing between respective frontal and sphenotic, and a posterior process which is flat, and moderately convex dorsally (Figs. 8.4E-F). The latter process has a posterodorsal depression which joins with the dorsal arm of the posttemporal. The anterolateral border of the pterotic has an elevated crest which is linked with the frontal, forming the subfrontal fossa over the sphenotic bone (Figs. 8.1A-C). The anterior thick portion is concave ventrally, contributing to the formation of the subtemporal fossa. In lateral view, the pterotic has an oval facet which joins with the posterodorsal condyle of hyomandibula (Fig. 8.1C). The supratemporal sensory canal continues along the external border of the pterotic as a closed tube. The extrascapular sensory elements, consisting of three flat tubular bones, overlie the pterotic and part of the epioccipital (Fig. 8.1A). Dorsally the medial border of the pterotic joins with the parietal and epioccipital, anteriorly it converges with the sphenotic, and ventromedially with the prootic and intercalar (Fig. 8.1B).

Epioccipitals: The epioccipitals are situated at the posterolateral edges of cranium (Fig. 8.1A). The anterior part is almost quadrangular with a biconcave ventral facet, and the posterior part is produced as a long process which is divided into two rami. One small lateral ramus joins the dorsal arm of the posttemporal, and a medial ramus with a flat feather-like body which extends as fine intermuscular projection dorsally (Figs. 8.4C-D).

A short lamina divides the concave ventral facet of the epioccipital into two cavities. The lateral cavity joins the respective cavity of the pterotic (forming the posttemporal fossa), and a medial cavity forming part of the brain case. The medial border of the epioccipital joins with the supraoccipital, and the ventral border connects with the exoccipital and intercalar. The epioccipital links with the parietal in anterior border and laterally connects with the pterotic, and prootic.

Supraoccipital: The supraoccipital forms the posteromedial portion of the cranial roof (Fig. 8.1A). It comprises a horizontal lamina with a curved anterior border, two mediolateral earflap-shaped processes, and a medial crest extending posteriorly. A lateral oblique shelf extends posteroventrally at each side of this crest, which is produced as a relatively broad process anterolaterally, terminating in two irregular filamentous processes posteriorly (Fig. 8.4A). The lateral border of these shelves joins with the epioccipitals. These shelves extend over the exoccipitals, forming the occipital fossae, and providing an insertion area for the occipital epaxial muscles. The medial crest is laminar and thin; its dorsal border is at the level of the cranial roof (Fig. 8.1C). The proximal base of this crest is broad and interposed between the articulation sutures of the exoccipitals, but the distal part is thin and free, and divided into filamentous projections. The medial crest penetrates between the bifurcate neural plate of the first vertebra, to which it appears to be conjoined. The ventral surface of supraoccipital is moderately concave and forms part of the skull roof (Fig. 8.4B). The anterior margin of the supraoccipital is overlapped by the posterior margin of frontals. The lateral border and earflap process of each side is overlaid by its respective parietal, joining with it.

Exoccipitals: This pair of bones is located ventrolaterally on the basioccipital, forming a canal which is part of the spinal cord canal (Fig. 8.1C). The posterior opening of this canal is the foramen magnum (Fig. 8.2A). The median articulation of exoccipitals is ventral to the supraoccipital crest. The minor

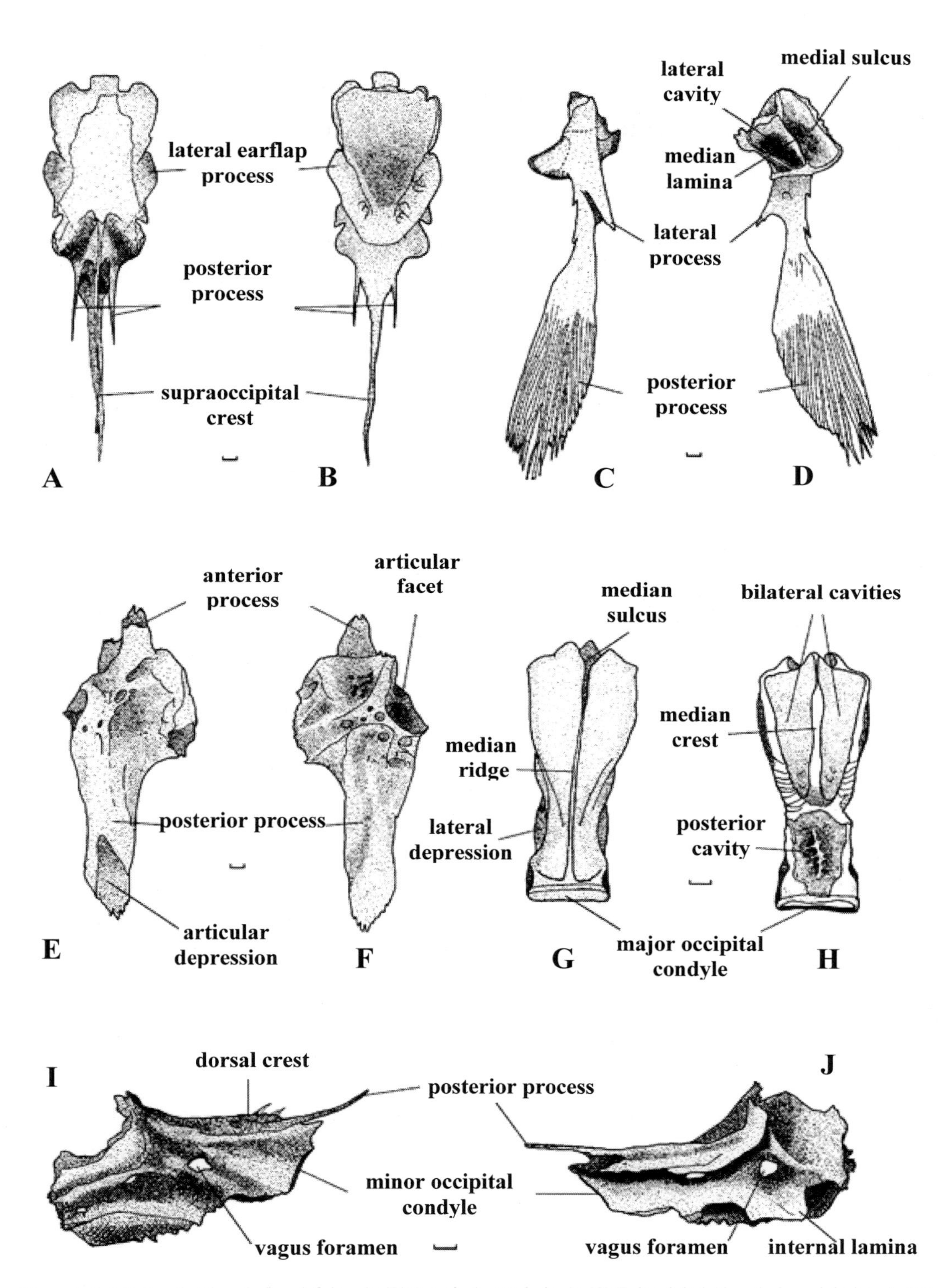

Figure 8.4. (A) Prootic (dorsal view, left bone); (B) Prootic (ventral view); (C) Epioccipital (dorsal view, right bone); (D) Epioccipital (ventral view); (E) Supraoccipital (dorsal view); (F) Supraoccipital (ventral view); (G) Basioccipital (external view); (H) Basioccipital (internal view) (I) Exoccipital (lateral view, left bone); (J) Exoccipital (medial view). *Mugil cephalus* AMS I.31251-010 SL = 59 mm. Scale bar: 1 mm.

occipital condyles are located at the posterior end of the exoccipitals and laterally to the foramen magnum, joining with the first vertebra (Fig. 8.2A). There is an anterodorsal cavity in the internal surface of each exoccipital which unites with the medial cavity of its respective epioccipital, forming part of the cranial vault. Ventral to this cavity, there is a large ovoid foramen for passage of the vagus nerve (Figs. 8.1C, 8.4I-J). Another foramen is present posteriorly. The dorsal surface of the exoccipital has a lateral crest produced as two spinous projections medially and a filamentous branched process posteriorly (Fig. 8.4I). These facets, with their ancillary projections, constitute the floor of occipital fossae, providing insertion points for epaxial muscles. There is a lamina in the internal facet of each exoccipital originating from the posterior part of otic cavity, joining to the medial crest of the basioccipital, and the lamina of the opposite exoccipital (Fig. 8.4J).

The exoccipitals join dorsally with the supraoccipital and epioccipitals. They are linked anterodorsally with their respective intercalar and pterotic under the intercalar lamina, and anteroventrally with their respective prootic.

Basioccipital: The basioccipital forms the posterior floor of the cranial vault, and is located ventral to the exoccipitals (Fig. 8.1C). It is partially covered by the parasphenoid in ventral part. The posterior part forms the major occipital condyle, with a round articular facet which contacts the first vertebral centrum (Figs. 8.1C-8.2A), together with two minor occipital condyles of the exoccipitals. The dorsal (internal) surface of the basioccipital is divided into two parts. The anterior part is deep and subdivided into two large and anteriorly wide cavities by a median crest, forming the posterior part of otic bullae. The posterior part is less deep and almost quadrangular, forming the floor of the foramen magnum (Fig. 8.4H). The ventral (external) facet is anteriorly divided into two flat surfaces by a median conical sulcus, forming the posterior myodome. There is also a low, median ridge on the ventral surface, which continues towards the apex of the median sulcus, and a pair of bilateral depressions close to posterior end (Fig. 8.4G). A pair of ligaments (Baudelot's ligament) originates bilaterally from these depressions, extending laterally and inserting on the anteroventral border of the dorsal cleithral process, just above the anterior spine. The basioccipital joins with exoccipitals dorsally, parasphenoid ventrally, prootics anteriorly, and the first abdominal vertebra posteriorly (Fig. 8.1C).

Orbit and circumorbital series

The orbital cavity is formed dorsally by the lateral ethmoid, frontal and sphenotic, and ventromedially by the endopterygoid. The circumorbital series are associated with the infraorbital branch of the sensory canal system (Fig. 8.5A). These series consist of preorbital, dermosphenotic, and four accessory infraorbital bones in *M. cephalus* (four–six in other mullets). The preorbital demarcates the anterolateral angle of the orbit, the accessory infraorbitals rim the ventrolateral margin, and the dermosphenoid demarcates the posterolateral angle (Fig. 8.5A). Each orbit is bound anteriorly and posteriorly by a pair of separate and semilunar sclerotic bones (Fig. 8.6F).

Preorbitals: The preorbitals are a pair of triangular bones, situated obliquely in front of the orbits (Fig. 8.5A). The anterior axis is elongate and denticulate. The internal surface is smooth and moderately concave (Fig. 8.6C). The external surface exhibits two longitudinal crests, the first one forming the posterior border of the preorbital. It is usually enlarged to form a shelf extending below the nasal bone (Figs. 8.6B-C). The apex of the anterior angle of the preorbital joins with the anterolateral margin of the nasal and anterodorsal process of the palatine (Fig. 8.5A). The second crest rests above the tubular sensory canal, which is open anteriorly. On the internal surface of the preorbital, there is an articular facet which is closely linked with the preorbital condyle of the lateral ethmoid.

Dermosphenotic: The dermosphenotic is a laminar elongate tubiform bone with a small elevated dorsal process (Figs. 8.6D-E). It is located on the upper posterior part of the orbit (Fig. 8.5A), joining proximally with the frontal lateral process, and (in some specimens) there is a weak posterior connection with the inferior surface of the sphenotic process. A sensory tube runs along this bone, connected proximally to the supraorbital sensory canal, and distally to the adjacent infraorbital.

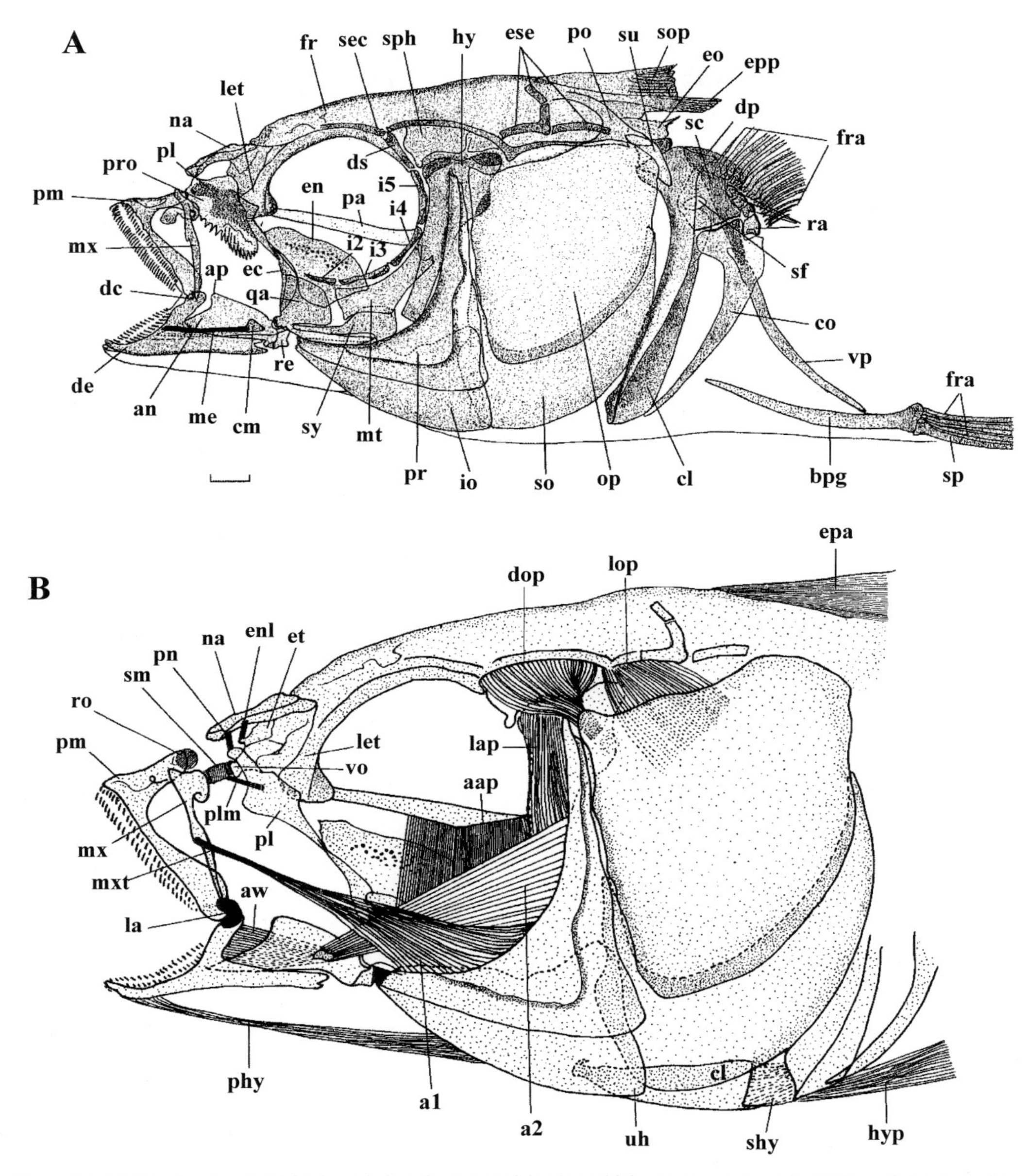

Figure 8.5. (A) Head and pectoral fin, lateral view (jaw opened, hyobranchial apparatus removed). (B) Superficial musculature and ligaments of the head, lateral view (hyoid bar removed), dashed lines indicate obscured structures. Scale bar: 1 mm.

Abbreviations: a1, a2, aw: divisions of *adductor mandibulae*, **aap:** *adductor arcus palatine*, **an:** angular, **ap:** angular dorsal process, **bpg:** basipterygium, **cl:** cleithrum, **cm:** coronomeckelian bone, **co:** coracoid, **dc:** dentary coronoid process, **de:** dentary, **dop:** *dilator operculi*, **dp:** dorsal postcleithrum, **ds:** dermosphenotic, **ec:** ectopterygoid, **en:** endopterygoid, **enl:** ethmoid-nasal ligament, **eo:** exoccipital, **epa:** epaxial muscle, **epp:** epioccipital process, **ese:** extrascapulars, **et:** ethmoid, **fr:** frontal, **fra:** fin rays, **hy:** hyomandibula, **hyp:** hypaxial muscle, **i:** infraorbitals, **io:** interopercle, **la:** labial ligament, **lap:** *levator arcus palatini*, **let:** lateral ethmoid, **lop:** *levator operculi*, **me:** mandibular (Meckel's cartilage), **mt:** metapterygoid, **mx:** maxilla, **mxt:** maxillary tendon, **na:** nasal, **op:** opercle, **pa:** parasphenoid, **pl:** palatine, **phy:** *protractor hyoideus*, **plm:** palatine-maxilla ligament, **pm:** premaxilla, **pn:** palatine-nasal ligament, **po:** posttemporal, **pr:** preopercle, **pro:** preorbital, **qu:** quadrate, **ra:** proximal radials, **re:** retroarticular; **ro:** rostral cartilage, **sc:** scapula, **sec:** sensory canal, **sf:** scapular foramen, **shy:** *sternohyoideus*, **sm:** submaxillary meniscus, **so:** subopercle, **sop:** supraoccipital process, **sp:** spine, **sph:** sphenotic, **su:** supracleithrum, **sy:** symplectic, **uh:** urohyal, **vo:** vomer, **vp:** ventral postcleithrum. *Mugil cephalus* AMS I.31251-010 SL = 59 mm.

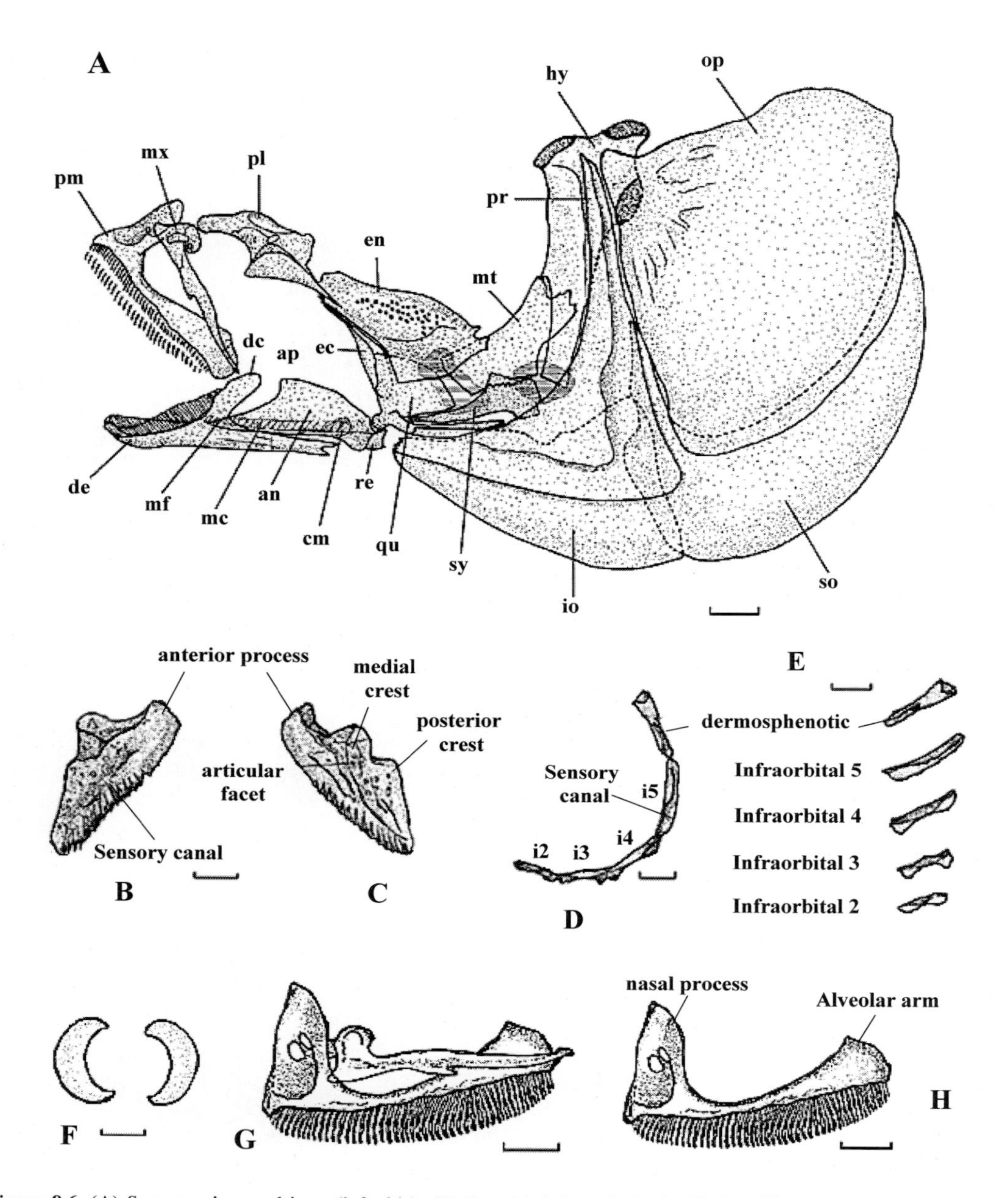

Figure 8.6. (A) Suspensorium and jaws (left side); (B) Preorbital (lateral view); (C) Preorbital (medial view); (D & E) Dermosphenoid and lnfraorbital bones (left side); (F) Sclerotic bones; (G) Premaxilla and maxilla (lateral view); (H) Premaxilla (lateral view). Scale bar: 1 mm. Dense horizontal lines: cartilage. Dashed lines: obscured bones.

Abbreviations: an: angular, **ap:** angular process, **cm:** coronomeckelian bone, **dc:** dentary coronoid process, **de:** dentary, **ec:** ectopterygoid, **en:** endopterygoid, **hy:** hyomandibula, **i:** infraorbitals, **io:** interopercle, **mc:** mandibular (Meckel's cartilage), **mt:** metapterygoid, **mx:** maxilla, **op:** opercle, **pl:** palatine, **pm:** premaxilla, **pr:** preopercle, **qu:** quadrate, **re:** retroarticular, **so:** subopercle, **sy:** symplectic. *Mugil cephalus* AMS I.31251-010 SL = 59 mm.

Accessory infraorbitals: Infraorbitals 2–4 are of similar shape, increasing in size sequentially. Their extremes are wider than their medial part, with the exception of infraorbital 4, which is pointed anteriorly (Figs. 8.5A, 8.6D-E). Infraorbital 5 is a chevron-shaped laminar bone with a ventrally-open sensory canal which is connected to the dermosphenotic dorsally, and infraorbital 4 ventrally.

Sclerotics: The sclerotics are two laminar semilunate bones, each with a concave face and another convex surface. They are equal in size, situated one in front and the other behind the orbital globe leaving a ventral opening for the passage of the orbital nerves and muscles (Fig. 8.6F).

Jaws

Premaxilla: The premaxilla has an elongate alveolar arm with two rows of setiform teeth at its inferior border, which is elevated as a ridge, and an ascending process originating from the anterior part, which is directed dorsally (Fig. 8.6G-H). In *Mugil cephalus* the jaw teeth are of distal-type and borne on the distal extreme of flexible and closely packed fibrous strands, emerging from the exposed edges of jaw bones. The alveolar arm is narrow proximally, broadening gradually towards its distal end which is thicker, sloped down and moderately curved medially. There is a shallow sulcus on the distal part of the face of the alveolar arm in which lies the maxillary horizontal arm. The ascending process has a medial depression leading into a ventral foramen and continuing internally as a short canal. There is a median ligamentous symphysis between the external borders of the ascending processes of the two premaxillae. Both ascending processes are dorsal to the ovoid rostral cartilage, to which they have a ligamentous attachment (Figs. 8.6G-H).

Maxilla: The maxilla is located dorsolateral to the alveolar arm of the premaxilla. It is thin, almost as long as the premaxilla, and comprises two parts, a horizontal arm and another vertical or ascending arm. The horizontal arm lies laterally to the premaxillary alveolar arm and terminates distally in a truncate uncinate process. Proximally this arm produces a dorsal spherical body from which the ascending arm emerges posterior to the premaxillary ascending process. There is another small anterior process anteroventral to the spherical body which is in line with the horizontal arm (Figs. 8.6G, 8.7A-C). A shallow sulcus extends between the spherical body and this small process. The horizontal arm is narrow and straight proximally, widening gradually until two-thirds of its length, then narrowing sharply towards the distal end which is slightly curved medially so it lies flush with the premaxilla. There is a ventral process midway along this arm, which is connected to a tendon attached to section A1 of m. *adductor mandibulae*, which originates from the horizontal border of the interopercle (Fig. 8.5B). The ascending arm of each maxilla is joined anteriorly to the rostral cartilage which is located medially behind the tips of the ascending processes of the premaxilla. This cartilage can rotate forward about this joint, facilitating protrusion of the upper jaw (Figs. 8.7H, 8.8A). An intermaxillary ligament originates from the spherical body of one side, running over the rostral cartilage and connecting to its contralateral, thus forming a ring through which the premaxillae slide during protrusion (Figs. 8.7H). Posteriorly, the spherical body of each maxilla is connected to the maxillary condyle of the respective lateral process of the vomer via ligaments which run through a cylindrical cartilaginous submaxillary meniscus (Figs. 8.8A, 8.5B). The anterior edge of the maxilla is bound to the nasal by a thick dermal tract. Other connections to the premaxilla and cranium, are via a system of flexible ligaments and connective tissue, providing considerable versatility of jaw movements.

Dentary: The dentary is the anterior most bone of the lower jaw, having a horizontal and another ventral arm (Figs. 8.5A, 8.6A). The horizontal arm slightly curves in toward the dentary symphysis, and bears one or two rows of setiform and distal-type teeth. Posteriorly it bends inwards, forming a rounded edentulous coronoid process at the posterodorsal margin (Fig. 8.7D-E). The posterior end of the premaxilla and maxilla fit into this process. The ventral arm extends obliquely at an acute angle from the horizontal arm. At this angle an anteriorly-directed conical fossa (angular fossa) is formed in the lateral face of the dentary. The anterior process of angular extends along the medial face of the dentary as far as the base of the angular fossa. The mandibular (Meckel's) cartilage which originates from the posterior angle of the angular and ventral to the coronomeckelian bone, also runs on the medial surface of the angular and dentary to its extreme length which is inserted deep to the anteromedian margin of the angular fossa. The mandibular sensory canal runs along the ventral border of the angular and dentary. The shorter section, on the angular, may be closed or open, while in the dentary there is a notch at the posterior extremity of the ventral arm, which is connected to the mandibular sensory canal and as a tube, runs forward until it opens on the anterolateral

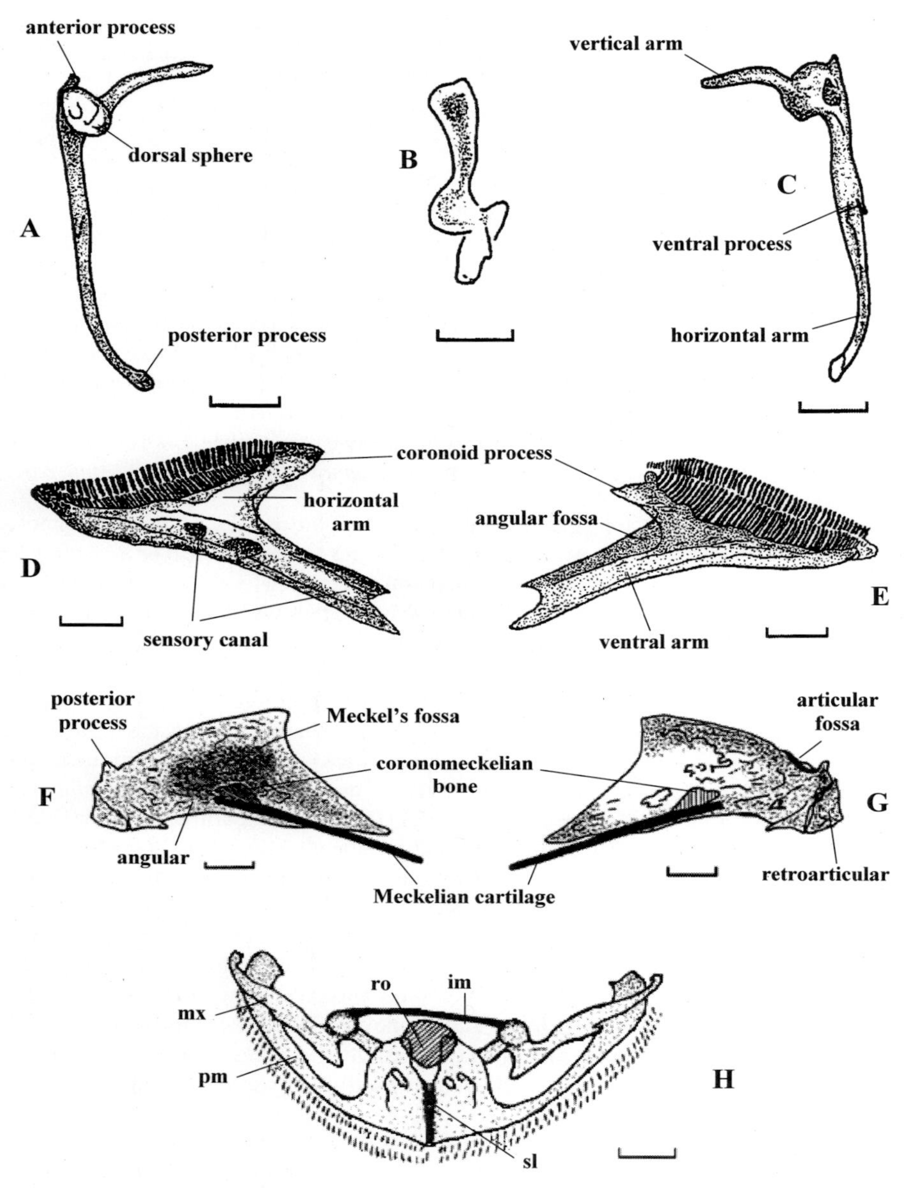

Figure 8.7. (A) Maxilla (dorsal view); (B) Lateral view of maxillary vertical (ascending) arm; (C) Maxilla (ventral view); (D) Dentary (lateral view); (E) Dentary (medial view); (F) Angular (lateral view); (G) Angular (medial view). All bones from the left side of jaw; (H) Dorsal view of premaxilla and maxilla, and their mode of attachment. *Mugil cephalus* AMS I.31251-010 SL = 59 mm.

Abbreviations: im: intermaxillary ligament, **mx:** maxilla, **pm:** premaxilla, **ro:** rostral cartilage, **sl:** symphysial ligaments.

surface of the dentary. The horizontal arm meets anteriorly with the contralateral dentary, forming the lower jaw symphysial knob. A fold of skin and connective tissue joins the posterolateral face of the dentary and its coronoid process to the distal end of the premaxilla, forming the labial ligament (*sensu* Forey 1975). In *M. cephalus* and all other mullets, the length of maxilla nearly equals the length of the alveolar arm of the premaxilla, and the labial ligament is attached to both bones (Fig. 8.5B).

Angular: The angular is triangular, and firmly attached to the dentary (Figs. 8.5A, 8.6A, 8.7F-G). It has a convex lateral surface, and a concave medial surface (Meckel's fossa). The greater part of the angular penetrates the angular fossa of the dentary and fills it. Within the Meckel's fossa the small coronomeckelian bone joins with a small crest, and the mandibular (Meckel's) cartilage starts ventral to it, running forward to its extreme at the base of the angular fossa of dentary. The tendon of section W of m. *adductor mandibulae* originates from the coronomeckelian bone, running forward in Meckel's fossa and attaches to the dentary (Fig. 8.5B).

The posterodorsal corner of the angular has an articular facet which receives the condyle of the quadrate. This facet has a small thin posterior process which limits the opening of the mouth. A short ligament joins this process to the ventral side of the quadrate.

Retroarticular: The retroarticular is a small and almost triangular bone, which is firmly attached to the posterior surface of the angular. A short thick ligament joins the corner of the retroarticular to the anterior tip of the interopercle (Fig. 8.5B).

Suspensorium

The suspensorium comprises the palatine, quadrate, ectopterygoid, endopterygoid, metapterygoid, symplectic, hyomandibular and preopercle.

Palatine: In *M. cephalus* the palatine is almost sigmoidal, with an edentulous expanded vertical plate which is divided longitudinally into dorsal and ventral alae (wings) by ridges on both medial (more prominent) and lateral surfaces (Figs. 8.8D-E). The ridges taper posteriorly to become a sharp posterior process which joins with the ectopterygoid and endopterygoid (Fig. 8.6A). Anterior to the palatine body a thick and rounded process emerges anterolaterally which joins the anterior tip of the preorbital and anterolateral corner of nasal via short ligaments, leaving a deep fossa ventral to its head (Figs. 8.5A-B). The palatine is also connected to the latral angle of the submaxillary meniscus by the palatomaxillary ligament which originates from its dorsomedial facet, thus contacting both the vomer and maxilla (Figs. 8.8A, 8.5B). There is a small ethmoidal process (Fig. 8.8D), posterior to the anterior process of the palatine, which is directed dorsolaterally. A short ethmopalatine ligament originating from the anterior angle of the lateral ethmoid attaches to this process (Fig. 8.8A). The palatine has an anterior ventral fossa which is delimited by its ventral ridge.

Endopterygoid: The endopterygoid is a laminar ovoid bone forming part of the orbital floor and the roof of the buccal cavity. Its plane is tilted medially about 45° relative to the direction of the palatoquadrate assemblage and it curves slightly medially to fit beneath the orbit. An ovoid dentigerous patch is present over its anteroventral surface (Fig. 8.8H). The dorsal border of endopterygoid is free, forming the ventromedial rim of the orbit. Ventrally, it is produced as a small lamina which is extended over the quadrate plate (Figs. 8.5A, 8.6A). Its posterior border also slightly overlaps the anterior section of the metapterygoid. The anteroventral border joins with the ectopterygoid, while its anterolateral border is produced as small digitiform projections, attached to a thick ligament which originates from the ventrolateral facet of the vomer, posterior to the vomerine lateral process (Fig. 8.8H).

Ectopterygoid: The ectopterygoid is an elongate thin bone located on the anterior margin of palatoquadrate assemblage. It has a dorsal spinous process emerging dorsolaterally with an obtuse angle from the ventral part, joining the posterior process of the palatine (Figs. 8.5A, 8.6A, 8.8G). Posteroventrally, the ectopterygoid overlaps the anterior border of quadrate while its internal border is joined to the endopterygoid (Fig. 8.6A).

Metapterygoid: The metapterygoid is a plain distorted quadrangle bone which contributes to the wall of the pharyngeal cavity (Figs. 8.5A, 8.6A, 8.8F). It has a posterodorsal broad process, which overlies on the ventral surface of the hyomandibular arm (Fig. 8.6A). The posteroventral border has a cartilaginous contact with the anterior margin of the hyomandibula in young fish, which becomes ossified in adults. Its anterodorsal process sutures with the anterior process of hyomandibula. The anterior border of the metapterygoid is

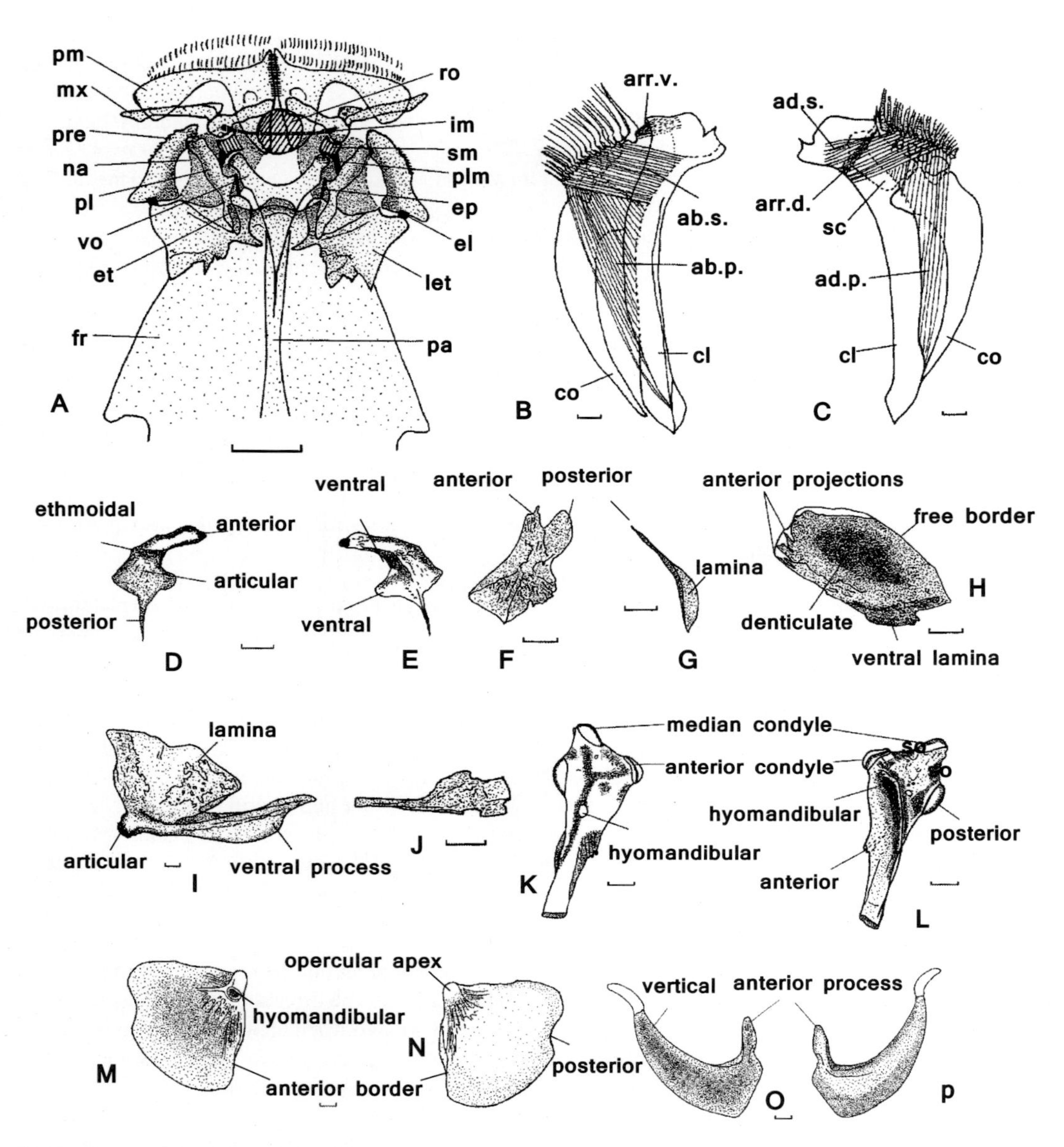

Figure 8.8. (A) Ventral view of ligaments of the upper jaw (premaxilla and maxilla protruded); (B) Superficial musculature of pectoral girdle (lateral view, right side); (C) Superficial musculature of pectoral girdle (medial view, right side); (D) Palatine (lateral view); (E) Palatine (medial view); (F) Metapterygoid (lateral view); (G) Ectopterygoid (lateral view); (H) Endopterygoid (lateral view); (I) Quadrate (lateral view); (J) Symplectic (lateral view); (K) Hyomandibula (lateral view); (L) Hyomandibula (medial view); (M) Opercle (lateral view); (N) Opercle (medial view); (O) Subopercle (lateral view); (P) Subopercle (medial view). All illustrated bones are from left side. Dashed lines indicate obscured structures. Scale bar: 1 mm.

Abbreviations: ab.p.: abductor profundus, **ab.s.:** abductor superficialis, **ad.p.:** adductor profundus **ad.s.:** adductor superficialis, **arr.d.:** arrector dorsalis **arr.v.:** arrector ventralis, **cl:** cleithrum, **co:** coracoid, **et:** ethmoid, **el:** ethmo-preorbital ligament, **ep:** ethmo-palatine ligament, **fr:** frontal, **im:** intermaxillary ligament, **let:** lateral ethmoid, **mx:** maxilla, **na:** nasal, **pa:** parasphenoid, **pl:** palatine, **plm:** palatine-maxilla ligament, **pm:** premaxilla, **pre:** preorbital, **ro:** rostral cartilage, **sc:** scapula, **sl:** symphysial ligament, **sm:** submaxillary meniscus, **vo:** vomer.

free dorsally, but is overlapped by the endopterygoid ventrally. Its ventral margin interdigitates with the symplectic ventrally, and has a cartilaginous contact with the quadrate lamina.

Quadrate: The quadrate is an important bone in articulation and movement of the mandibule. Its body is a triangular plate which terminates ventrally as a broad articular condyle (Fig. 8.8I). This condyle fits into the articular facet of the angular (Figs. 8.5A, 8.6A). A long grooved process emerges behind the condyle, terminating in a sharp point. The anterior process of the symplectic extends forward to insinuate itself between the quadrate lamina and the process. The anterior part of the quadrate plate is partially overlaid on the ectopterygoid lamina, while dorsally it is overlapped by the ventral lamina of the endopterygoid. Posteriorly the plate is connected to the metapterygoid via a narrow strip of cartilage (in young fish) which ossifies in adults (Fig. 8.6A).

Symplectic: The symplectic is an elongate bone with an irregular border. It is expanded posteriorly, and narrow anteriorly (Figs. 8.5A, 8.6A, 8.8J), penetrating into a groove between the lamina and ventral process of quadrate (Fig. 8.6A). The posterior broad portion has a slightly denticulate dorsal border interdigitating with the ventral border of the metapterygoid. The posterior and posteroventral border of symplectic are connected to the hyomandibular and interhyal ligament via a cartilage (Fig. 8.6A). The quadrate and symplectic both have broad ventral facets forming a strong articulation with the preopercle.

Hyomandibula: The hyomandibula is a deltoid-shaped bone which has an important function in the attachment of the suspensorium through condyles joining with the cranium. It has a broad and stout dorsal head, narrowing gradually towards its truncate vertical arm (Figs. 8.8K-L). Dorsally, it has two condyles which are widely-separated by a shallow notch. The anterior condyle joins with the sphenotic facet, and the posterior condyle fits in the pterotic facet. A prominent posterior opercular condyle joins with the anterodorsal facet of the opercle (Fig. 8.6A). The lateral surface of the hyomandibula has a very prominent crest (Fig. 8.8L), beginning under the first dorsal condyle, diverging posteriorly towards the opercular condyle, and then tapering ventrally down the arm providing a posterior shelf for insertion of the anterodorsal part of the preopercle (Fig. 8.6A). There is a deep depression anterior to the base of this ridge extending rostrally to become a flat plate above the ventral part of the hyomandibula. This surface bears a ventrally-directed process at its anterior border, joining the anterodorsal process of the metapterygoid. The distal border of the hyomandibular arm is connected to the symplectic and interhyal ligament via a cartilage. The ventral surface of hyomandibula is more or less flat or slightly concave dorsomedially with a distinct foramen located anteroventrally to the opercular condyle for passage of the hyomandibular nerves (Fig. 8.8K).

Two muscles are associated with the hyomandibula. The m. *adductor arcus palatini* forms the floor of the orbit between the skull and the palatal arch. It inserts into the hyomandibula, metapterygoid and posterior part of the endopterygoid, and extends towards the parasphenoid and pterosphenoid (Fig. 8.5B). The m. *levator arcus palatini* occupies the area at the rear of the orbit between the skull and the palatal arch. It originates from the edge of the sphenotic. Its fibres insert into the dorsal face of the hyomandibula and metapterygoid, and its ventrolateral region is overlaid by fibres of the m. *adductor mandibulae* (Fig. 8.5B).

Preopercle: The preopercle is a broad lunate bone with a long vertical arm and short horizontal arm (Figs. 8.5A-B, 8.6A, and 8.9A-B). The central part of this bone is narrow and thick, while the surrounding peripheral part is wide, thin and semi-transparent. The anterior border of the vertical arm is inserted in the posterior hyomandibular crest, while its posterior part lies superficial to the opercle and subopercle. A deep open canal starts from the dorsal tip of vertical arm, running medially down as a right angular trough towards the tip of the horizontal arm. This canal is connected to the supracranial sensory canal system and ventrally joins the mandibular sensory canal of the lower jaw (angular and dentary). The preopercle bone is considerably thicker anterior to this canal compared with the posterior part which is laminar, thin and transparent. The anterodorsal border of horizontal arm lies medial to the quadrate, while the thin laminar portion of the ventral margin overlaps part of the interopercle and subopercle (Fig. 8.5A). The ventral surface is concave and smooth, and attached to the opercle and subopercle through ligaments and connective tissue.

Opercular Series

Opercle: The opercle is the largest bone of opercular series with a roughly triangular shape (Figs. 8.5A-B, 8.6A, 8.8L-M). The lateral surface of this bone is slightly convex, resulting in concavity of the medial surface. The anterior border is the thickest part and joins with the preopercle and subopercle anterior rim. The opercular apex is thick and placed anterodorsally, with a deep medial articular socket joining with the opercular condyle of the hyomandibula.

There are some short ventral struts supporting the articular socket, and contributing to the formation of a deep cavity above the articular socket. The dorsal border of the opercle is thinner and curved inward dorsomedially, to form the roof to the opercular chamber, with the posterior angle lying over the posttemporal apex. The opercular posterior and ventral borders are thin, lying over the dorsal edge of the subopercle. The posterior border has a small notch posteromedially.

The m. *dilator operculi* is almost deltoid, originating from subfrontal fossa over the sphenotic bone (Fig. 8.5B), passing over the dorsal part of the palatine and m. *levator arcus palatini*, and inserted by a bunch of strong tendons to the ventral face of the opercular apex just above the articular facet. The m. *levator operculi* passes between the lateral skull wall and opercle, behind m. *dilator operculi*. In *M. cephalus* this muscle originates from the posterior part of the sphenotic and lateral border of the pterotic in the subfrontal fossa and above its hyomandibular facet, and inserts both inside the anterodorsal border of the opercle, and the internal surface of the opercle over some prominent short osseous struts supporting the articular facet. The angle of insertion of m. *levator operculi* varies with the slope of the medially inverted upper edge of the opercle, which can be horizontal to nearly vertical.

Subopercle: The subopercle is a laminar, thin, sickle shape bone (Figs. 8.5A-B, 8.6A, 8.8O-P). It has a pointed anterior arm which is overlapped by the opercle and preopercle, and a broad posterior arm curving upward, following the curvature of the opercle, and terminating in a very thin posterior protrusion. This arm is overlapped by the ventral border of the opercle.

Interopercle: The interopercle is relatively deltoid and placed ventral to the preopercle (Figs. 8.5A-B, 8.6A, 8.9C-D). Anteriorly, it is connected to the retroarticular by a series of short ligaments (Fig. 8.5B), and there is an articular facet on the dorsomedial border for articulation with the epihyal. There is a shallow notch above this articular facet, and the border anterior to it curves inwardly and connects firmly to the ventral margins of the quadrate and symplectic. A large part of the interopercle lateral surface is covered by the preopercle. The posterior end of interopercle partially covers the anterior end of subopercle, which is visible externally. The ventral edge of the interopercle connects with the branchiostegal membrane, and the branchiostegal rays lie along the medial side of the interopercle and subopercle.

Hyoid bar

The hyoid bar consists of the interhyals, epihyals, ceratohyals, hypohyals, basihyal, glossohyal, urohyal, and branchiostegals (Figs. 8.9E-F).

Basihyal: The basihyal is the anterior most median endochondral bone of the basibranchial series. It is located anteriorly to the urohyal, joining both branches of the hyoid arch and forming the skeleton of the tongue. In *M. cephalus* the basihyal is a moderately elongated bone with both ends expanded laterally, giving a biconcave appearance to its medial part (Fig. 8.9S). A semilunar, cartilaginous glossohyal joins with its anterior tip, while its posterior end is pointed medially and produced into two small articular facets which rest on the dorsal surface of the dorsal hypohyals (Fig. 8.9E).

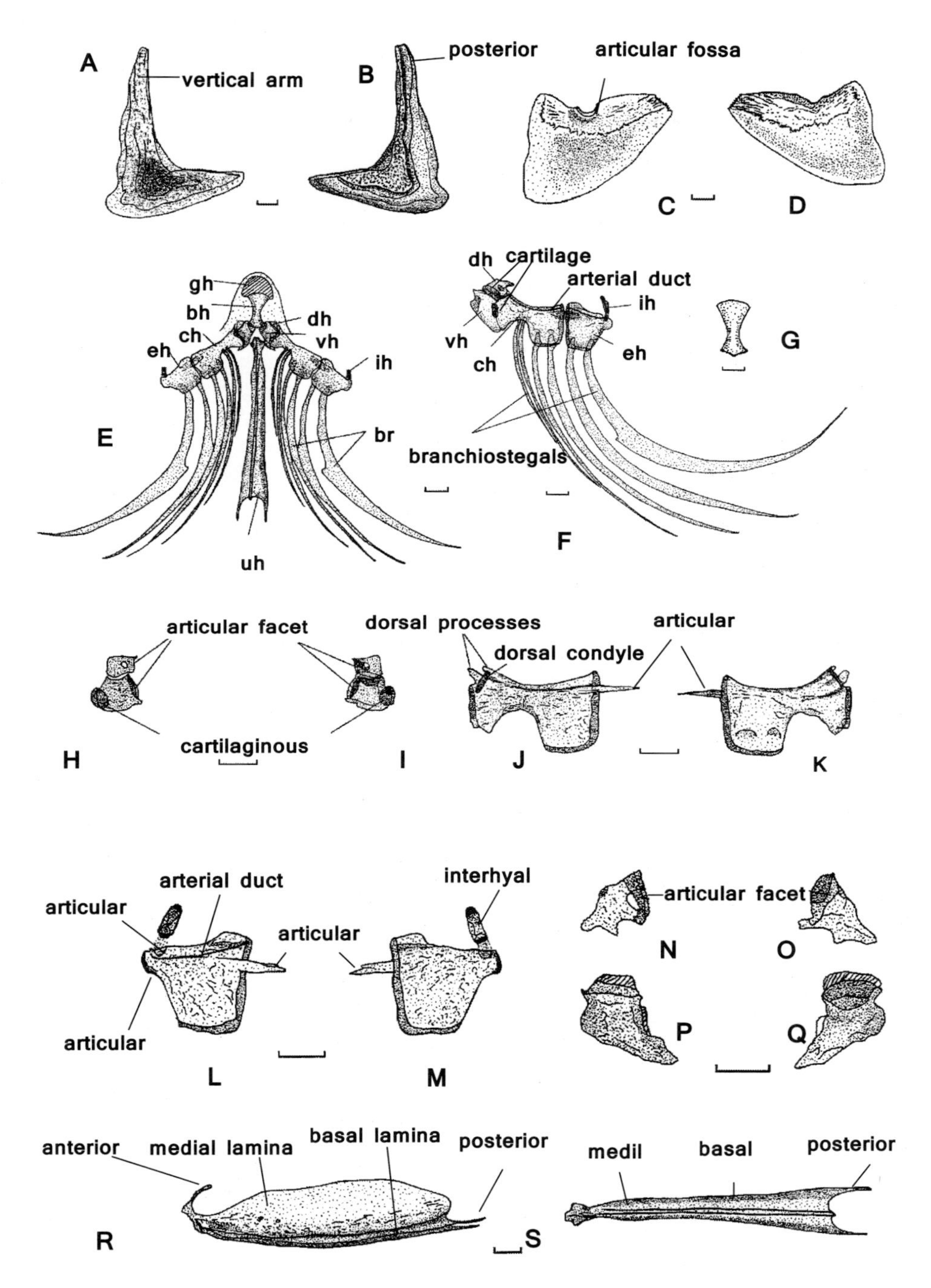

Figure 8.9. (A) Preopercle (LV); (B) Preopercle (MV); (C) Interopercle (LV); (D) Interopercle (MV); (E) Hyoid apparatus (dorsal view); (F) Hyoid bar (LV); (G) Hypohyals (LV); (H) Hypohyals (MV); (I) Ceratohyal (LV); (J) Ceratohyal (MV); (K) Epihyal (LV); (L) Epihyal (MV); (M) Dorsal hypohyal (LV); (N) Dorsal hypohyal (MV); (O) Ventral hypohyal (LV); (P) Ventral hypohyal (MV); (Q) Urohyal (LV); (R) Urohyal (MV); (S) Basihyal. All illustrated bones are from the left side. Scale bar: 1 mm.

Abbreviations: bh: basihyal, **br:** branchiostegal rays, **ch:** ceratohyal, **dh:** dorsal hypohyal, **eh:** epihyal, **gh:** glossohyal, **ih:** interhyal, **LV:** lateral view, **MV:** medial view, **uh:** urohyal, **vh:** ventral hypohyal.

Hypohyals: The hypohyals consist of two separate bones, one located above the other (Figs. 8.9E-F). The dorsal hypohyal is smaller and somewhat triangular in shape with a small projection at the anterodorsal tip. It has a dorsal articular surface which is weakly connected with the basihyal (Fig. 8.9E). It is attached to the ventral hypohyal and anterodorsal border of the ceratohyal via a distinct cartilage (Fig. 8.9F). Anteromedially, it has a foramen, leading into a duct for the hyoid artery, which runs through the ceratohyal and epihyal (Figs. 8.9F-H-M). The ventral hypohyal is rectangular, and smooth externally (Figs. 8.9G-O). It has an anteroventral articular facet which is connected with its opposite pair, and also the lateral facet of the urohyal through strong ligaments. Dorsally, it is attached to the dorsal hypohyal, and posteriorly to the anterior margin of the ceratohyal via cartilage (Figs. 8.9E-H).

Ceratohyal: The ceratohyal is the largest bone of the hyoid arch. It is narrow at its midlength and expanded at its ends, forming an irregular dumbbell-shape (Figs. 8.9E-F-I-J). The anterior border joins with the hypohyals through the cartilage, and has two anterodorsal processes, between which lies the duct for the hyoid artery (Figs. 8.9F-I). The posterior border joins the epihyal through interdigitating splint-like projections spanning the two bones (Figs. 8.9E-F).

Four branchiostegal rays are attached to the ventral edge of the ceratohyal. Two anterior small branchiostegals attach to the medial face of the narrow middle part of the ceratohyal. Two larger branchiostegals join into two shallow depressions on the lateral aspect of the posteroventral part of the ceratohyal (Figs. 8.9E-F). Allis (1903) considered the epihyal and ceratohyal to be a single bone having two centres of ossification.

Epihyal: The epihyal is a roughly trapezoid bone which is located between the interhyal and ceratohyal (Figs. 8.9E-F). The arterial duct continues from the ceratohyal along its dorsal margin. The posterior border of the epihyal is expanded posterodorsally and has a small condyle which attaches to the fossa of the interopercle (Figs. 8.9L-K). The anterior margin is straight and joins with the ceratohyal. There are two ovoid depressions on the lateral aspect of the ventral edge of the epihyal for articulation of the last two branchiostegal rays.

Interhyal: The interhyal is a small trabecular bone which is connected ventrally to the posterodorsal margin of the epihyal, forming an obtuse angle with the main axis of the hyoid bar (Fig. 8.9L). Its dorsal end attaches to the cartilaginous depression under the hyomandibula and symplectic via short ligaments. This attachment suspends the hyoid bar from the hyomandibula allowing a small amount of anteroposterior movement relative to the suspensorium.

Urohyal: The urohyal is an elongate anteriorly narrow, and posteriorly broad laminar bone, which is embedded in the muscles of the isthmus. It is made up of a vertical lamina and two basal lateral lamina which are located horizontally opposite to each other, forming an obtuse angle with the vertical lamina (Figs. 8.9Q-R). The posterior end of the lateral laminae extend backward, and are produced as two thin processes. The urohyal also has an anterodorsal process projecting posteriorly and attached to the posteroventral side of the basihyal ligamentously. The urohyal has two bilateral, anteroventral, flat articular facets which are attached to the ventral hypohyals ligamentously (Fig. 8.9E). The posterior border of the vertical lamina is thickened, and is connected with the coracoidal symphysis of the pectoral girdle.

Branchiostegals: There are six branchiostegal rays bilaterally, increasing in size from the anterior ray to the posterior ray (Figs. 8.9E-F). The proximal part of each ray is curved and short, bearing an articular facet. The rest of the bone is flat, with a thicker dorsal border. The four smaller branchiostegal rays are attached to the medial side of the ceratohyal. The two larger ones are attached to the lateral surface of the ventral edge of epihyal. All branchiostegal rays follow the curvature of the opercular complex.

Branchial arch

The branchial arch comprises the basibranchials, hypobranchials, ceratobranchials, epibranchials and pharyngobranchials (Figs. 8.10A-B).

Basibranchials: There are three ossified basibranchials which are medially-placed at the base of branchial basket. The first basibranchial is conical, embedded in a patch of thick ligaments, joining with the basihyal anterodorsally, and cartilaginously connected to the second basibranchial posteriorly.

The second basibranchial is elongate, rectangular and biconcave anterolaterally. It is cartilaginously connected to the first basibranchial anteriorly, to the third basibranchial posteriorly and to the first pair of hypobranchials bilaterally.

The third basibranchial is almost spear-shaped, and longer than the others. It is cartilaginously connected to the second basibranchial anteriorly, and has two small anterolateral depressions for articulation with the second pair of hypobranchials. The third pair of hypobranchials also join with it posterolaterally.

Hypobranchials: The first hypobranchial is irregularly-shaped, almost triangular dorsally, and rectangular ventrally. It has a notch in its anterolateral border, and an anteroventral small protrusion, forming a moderate concavity in its posteromedial border. It joins with the lateral side of the second basibranchial anteriorly, and with the first ceratobranchial posteriorly.

The second hypobranchial is rectangular, with a wide and curved anterior portion. It joins anteriorly with the anterolateral margin of the third basibranchial and posteriorly with the second ceratobranchial.

The third hypobranchial is triangular and has an anteriorly-directed process which is connected to the posterior border of the third basibranchial. Posteriorly, it joins with the third ceratobranchial.

In small specimens of *M. cephalus* (49.5–60 mm SL), remnants of the fourth basibranchial and hypobranchials were present as alcianophilic cartilaginous nodules (Fig. 8.10A). This is similar to the fourth hypobranchial observed in *Scomber* (Allis 1903), *Otolithus ruber* (Dharmarajan 1936) and *Cynoscion* (Mohsin 1973).

Ceratobranchials: The first four pairs of ceratobranchials are of similar shape, increasing in size progressively from the first to the last. They are slender, rod-like, slightly bent and markedly different from the 5th ceratobranchials. These bones are edentulous, and covered by gill rakers. The anterior end of the first, second and third ceratobranchials unite with their respective hypobranchials. The fourth ceratobranchial is broad at both ends, uniting anteroventrally with the fifth ceratobranchial through a cartilaginous process. The first four ceratobranchials posteriorly join with their respective epibranchials. The fifth ceratobranchial (lower pharyngeal) is broad and petal-shaped, with a thick anterior border, and thin laminar posterior part.

Each ceratobranchial has a lateral canal for passage of the branchial artery. The canal of the last two ceratobranchials is deeper and wider in its lower extreme.

Epibranchials: There are four pairs of epibranchials, forming the dorsal part of the first four branchial arches, and located close to the base of the neurocranium. The fifth arch lacks an epibranchial. These bones represent complex shapes and articulations. They are connected posteriorly with their respective pharyngobranchial and laterally to their ceratobranchials.

The first epibranchial is trabecular and follows the curvature of the branchial arches. In dorsomedian facet it has a process and a concavity (Figs. 8.10A-B, 8.11D).

The second epibranchial is scythe-shaped, formed by a slightly curved lamina with a long process arising from its anterior angle, and a shorter process from the opposite angle, parallel to the first process, leaving a curvature between them (Figs. 8.10A-B, 8.11E-F). The first process is bent and joins with the second ceratobranchial, while the second process is projected dorsally to the base of neurocranium. The anterior border of the second epibranchial joins with the second pharyngobranchial, and its ventral border is linked with the third pharyngobranchial.

The third epibranchial consists of a large curved ventral wing, and a short, straight ventral wing. A deep constriction separates the wings. A thin process emerges from the juncture of these two wings, which joins with the third ceratobranchial (Fig. 8.11G). The articulation with the condyle of the pharyngeal suspensorium is achieved through a facet located at the extremity of the larger wing (Fig. 8.11H).

The fourth epibranchial is formed by two perpendicular arms. One of the arms joins with the pharyngeal suspensorium. The distal end of the second arm terminates in a curved articular facet and a small process joining the fourth ceratobranchial (Figs. 8.10A-B, 8.11I-J).

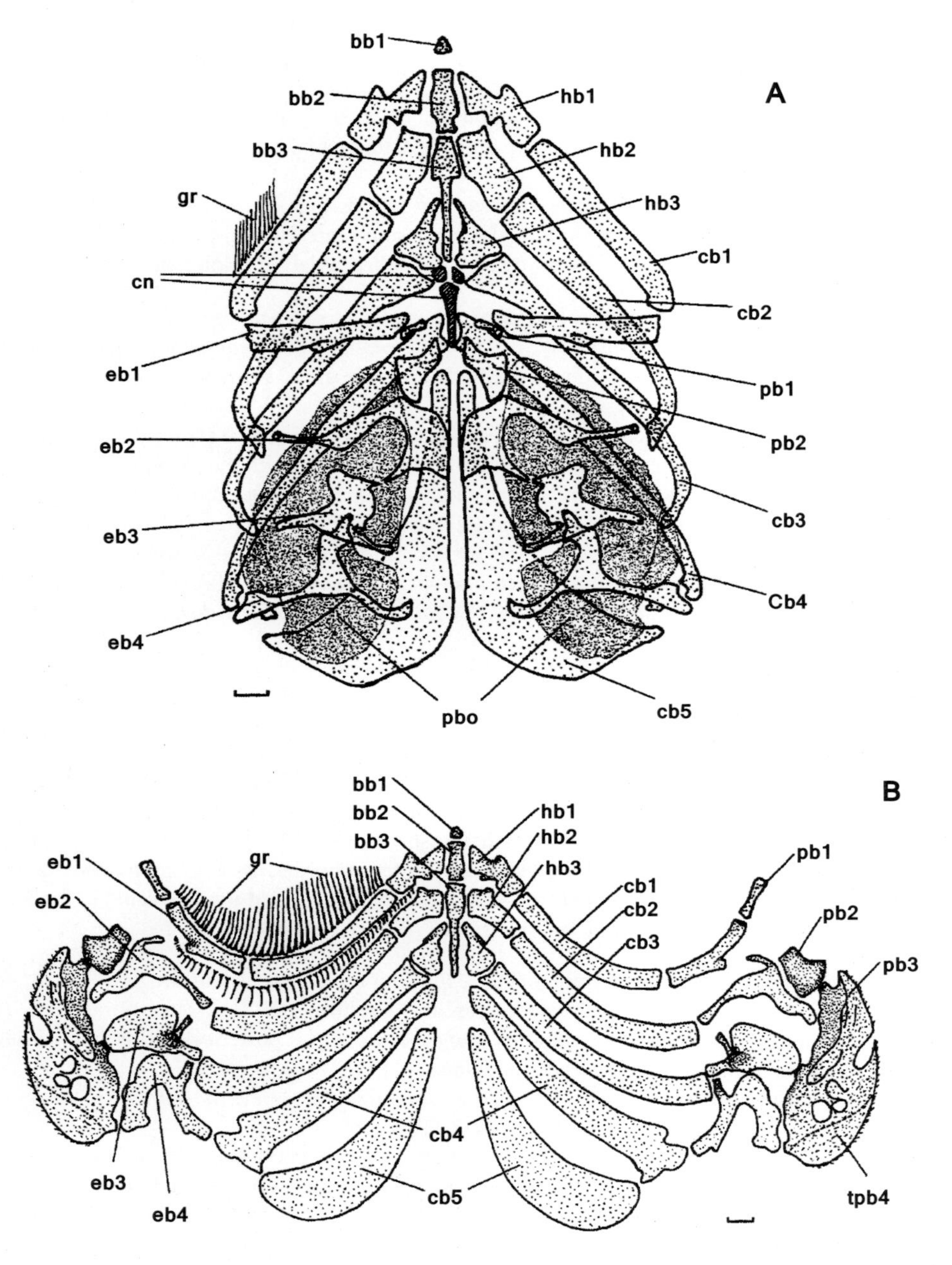

Figure 8.10. (A) Branchial apparatus (dorsal view), hyoid bar, gill rakers, pb4, pb4 and articular cartilages not shown. (B) Branchial apparatus (dorsal view), lower elements of both sides folded out (to show ventral surface). Dense parallel lines: cartilage; densely stippled areas: obscured pharyngobranchial organ folded inward; Scale bar: 1 mm.

Abbreviations: bb: basibranchial, **cb:** ceratobranchial, **cn:** cartilaginous noduls, **eb:** epibranchials, **gr:** gill rakers, **hb:** hypobranchials, **pb:** pharyngobranchial, **pbo:** pharyngobranchial organ, **tpb4:** toothplate of pharyngobranchial 4.

The first pharyngobranchial bone (first pharyngeal suspensory element) is a small trabecular bone, attached to the proximal end of the first epibranchial, and connected ligamentously to the prootic (Figs. 8.10A-B, 8.11K).

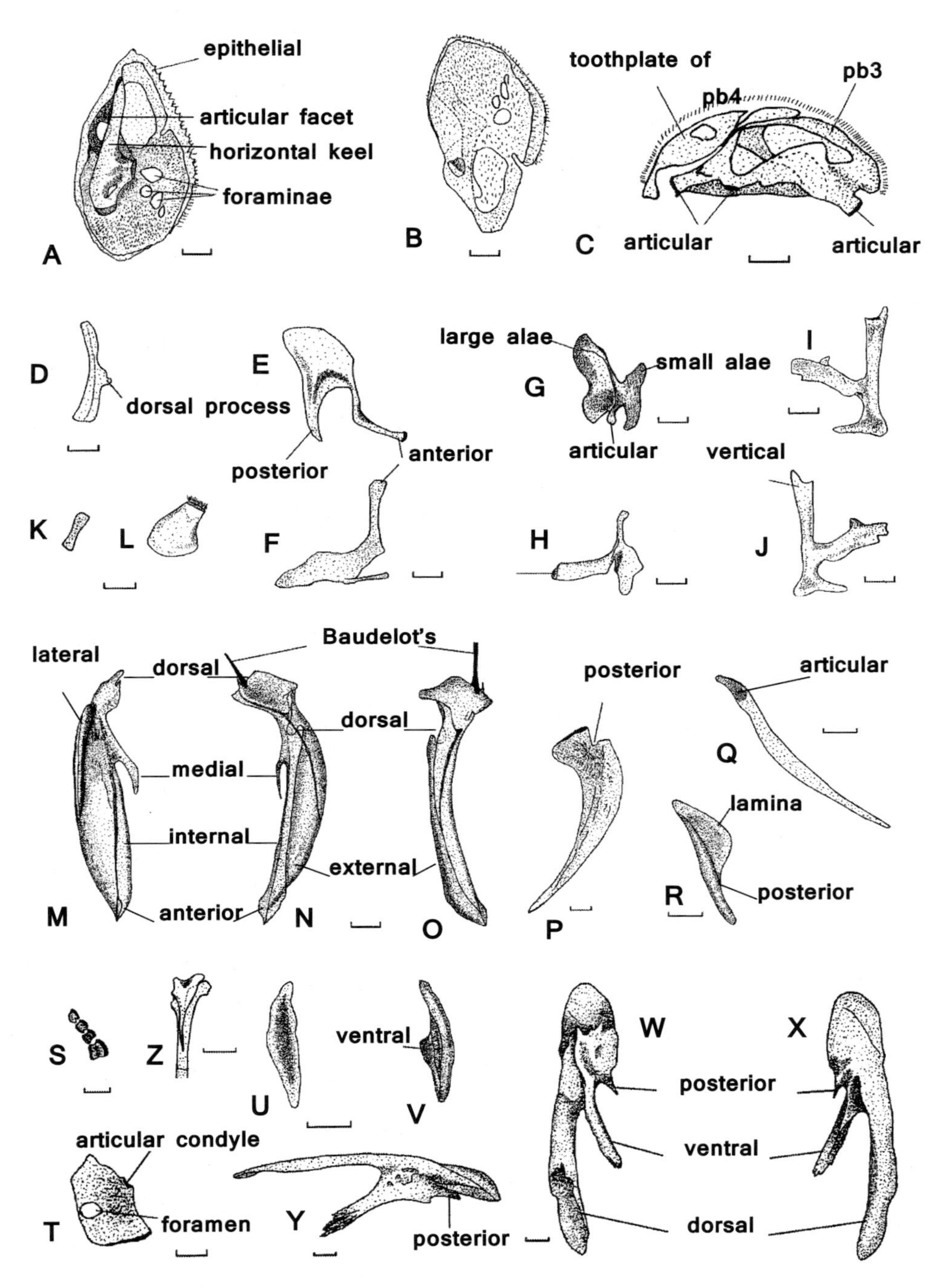

Figure 8.11. (A-L) Pharyngeal suspensory elements. (A) Pharyngobranchial organ (DV); (B) Pharyngobranchial organ (VV); (C) Pharyngobranchial organ (LV); (D) 1st epibranchial (LV); (E) 2nd epibranchial (DV); (F) 2nd epibranchial (LV); (G) (DV); (H) 3rd epibranchial (LV); (I) 4th epibranchial (DV); (J) 4th epibranchial (LV); (K) 1st pharyngobranchial; (L) 2nd pharyngobranchial; (M) Cleithrum, (AV); (N) Cleithrum (MV); (O) Cleithrum (PLV); (P) Coracoid (LB); (Q) Ventral postcleithrum (LB); (R) Dorsal postcleithrum (LB); (S) Proximal radials (LBs); (T) Scapula (LB); (U) Supracleithrum (DV, LB); (V) Supracleithrum (LV, LB);. (W) Posttemporal (DV, LB); (X) Posttemporal (VV, LB); (Y) Posttemporal (LV, LB); (Z) Pectoral fin spur. Scale bar: 1 mm.

Abbreviations: AV: anterior view, **DV:** dorsal view, **LV:** lateral view, **MV:** medial view, **LB:** left bone, **PLV:** posterolateral view. **VV:** ventral view.

The second pharyngobranchial (second pharyngeal suspensory element) is a piriform bone (Figs. 8.10A-B, 8.11L). It joins with the pharyngeal suspensorium at its base, and laterally with the second epibranchial. Its dorsal extremity is truncate and attached to the external surface of the parasphenoid.

Pharyngobranchial organ: Capanna et al. (1974) studied the structure of the pharyngobranchial organ (PBO) of the Mediterranean mullet, and presented an account of its anatomy, histology, dentition and possible complex filtering function (for feeding on small benthic particles). Capanna et al. (1974) published photographic images of the skeletal components of the PBO but did not identify individual bones.

Harrison and Howes (1991) reviewed the PBO of mullet, and gave a detailed and concise account of its structure, associated musculature and dentition, ontongeny, possible function, and its taxonomic utility among the genera of mullets.

The skeleton of the PBO is formed by epibranchials 2–4 and pharyngobranchials 2–4. The second pharyngobranchial attaches to an articular facet of the third phayngobranchial. The third pharyngobranchial is enlarged and constitutes the major part of the PBO, whereas the fourth pharyngobranchial is a small cartilaginous element inserted at the articulation of the 4th epibranchial with the 4th toothplate (Figs. 8.11A-C). The main body of the third pharyngobranchial is a fenestrated base-plate which is produced into an anterodorsally-directed keel, and terminates in an articular facet. The base of the second epibranchial attaches to this keel dorsally. Posteriorly, the base-plate of the third pharyngobranchial is deep and broad, and represents two articular processes. The posterior articular process is directed outward, and joins with the base of the fourth epibranchial. The medial articular process is vertical and connects to the base of the third epibranchial (Fig. 8.10B).

The external surface of the pharyngobranchial organ is covered by flexible ciliform distal-type teeth which develop on elongate, poorly-mineralized shafts of bone, supported by epithelial tissue of the mucosa covering the toothplate.

Lower pharyngeal (fifth ceratobranchial): Although it follows the same curvature as the other ceratobranchials, it is noticeably broader proximally, widening toward its distal end. Its anterior border is thick, bearing an arterial canal, while its posterior part is laminar and thin and covered with gill rakers (Figs. 8.10A-B).

Pectoral fin

The pectoral fin consists of the posttemporal, supracleithrum, cleithrum, postcleithra, scapula, coracoid, proximal radials (actinosts), spine and rays (Fig. 8.5A).

Posttemporal: This forked bone facilitates attachment of the pectoral girdle to the neurocranium (Fig. 8.5A). It has a dorsally flat round margin body which is produced into two rami. An elongate, flat and anteriorly-directed dorsal ramus joins with dorsal part of the epioccipital (Figs. 8.11W-X-Y). A trabecular ventral ramus is directed anteroventrally to interdigitate with the intercalar posterior process. The ventral ramus is elongated posteriorly, and produced as a small process, providing a shelf for attachment to the supracleithrum anterior part. Lateral to the cranium, the posttemporal border expands anterolaterally, overlapping the distal end of the posterior process of the pterotic. In *M. cephalus* and other mugilids studied, the posttemporal lacks sensory canals.

Supracleithrum: The supracleithrum is a small rhomboidal bone, with a relatively flat dorsal surface and a medially-elevated ventral crest (Figs. 8.11U-V). Its anterodorsal surface is connected to the ventral surface of the posttemporal, and its wedge-shape ventral crest fits between the anterior spinous process and anterodorsal surface of the cleithrum. This bone also lacks sensory canals.

Cleithrum: The cleithrum is the longest bone of the pectoral girdle (Figs. 8.11M-N-O). It is moderately curved anterodorsally, and folded into two lateral laminae, with a flat medial ridge terminating in a pointed anteroventral keel which joins with its contralateral partner via ligaments at the cleithral symphysis. This symphysis joins anteriorly with the posterior border of the urohyal through strong ligaments. A very small dorsal spinous process is present on the anterodorsal corner of the cleithrum. Baudelot's ligament, originating from the basioccipital, inserts behind the cleithrum dorsal spine.

The external lamina is broader and has an exterior lateral crest which originates lateral to its dorsal border, and continues ventrally to two-thirds of the cleithrum length. It projects slightly forward at its dorsal apex, providing a space for insertion and articulation of the supracleithrum. A shallow shelf under this crest gives origin to m. *abductor superficialis*, which continues as a series of discrete tendons attaching to the distal anterior bases of the lateral hemitrichia of the pectoral fin rays (Fig. 8.8B). The external lamina is curved backwards dorsally. The dorsomedial margin of the external lamina is connected to the anterodorsal surface of the dorsal postcleithrum. The internal lamina has a prominent medial process and a deep dorsal notch, which overlays the scapular foramen. Posteroventrally, the internal lamina articulates with the ventral end of the coracoid. The extremes of both cleithra are joined together at their ventral borders and anterior keels, forming the cleithral symphysis.

Other muscles of pectoral fin include the *arrector ventralis*, which originates from the ventral part of the cleithrum, and inserts onto the anterolateral base of the pectoral fin spine via a strong tendon. The m. *abductor profundus* is sheet-like, deep to m. *abductor superficialis*, originating from the lateral face of the cleithrum and coracoid, a small portion also originating from the scapular facet. The muscle sheet continues as a series of tendons inserting into the posteroventral flanges of the lateral hemitrichia of the pectoral fin rays. In all mullets, m. *abductor profundus* is extensive, and a large portion is exposed (where not overlain by m. *abductor superficialis*).

The m. *adductor superficialis* originates from dorsomedial face of the cleithrum, and the scapula by means of tendons inserted into the posteromedial faces of the fin rays (Fig. 8.8C). The m. *adductor profundus* is deeper than the m. *adductor superficialis*, originating from the posteroventral portion of cleithrum and the coracoid margin. This muscle continues as a series of tendons which insert into the posteroventral bases of the medial hemitrichium of all, except the first, fin rays (Fig. 8.8C). The m. *arrector dorsalis* originates from dorsomedial part of the posterior face of the cleithrum, and inserts into the base of the marginal (first) ray.

Coracoid: The coracoid is a flat arcuate bone which is dorsally expanded, gradually narrowing towards its articulation with the posteroventral surface of the internal lamina of cleithrum, leaving an elongate triangular space between it and the cleithrum (Figs. 8.5A, 8.11P). Its dorsal margin joins with the ventral border of the scapula and part of the internal lamina of the cleithrum. It has a posterodorsal notch housing the fourth and half of the third proximal radials (Fig. 8.5A).

Scapula: The scapula is mostly an irregular five-sided bone with round corners, which is larger in the vertical axis than horizontal. It is curved dorsomedially to fit into the anterior depression of the internal lamina of the cleithrum (Figs. 8.5A, 8.11T). The scapula has an ovoid foramen in its anteromedian part, corresponding with the notch of the internal lamina of the cleithrum. Almost half of the ventral surface of the scapula is attached to this lamina. Its ventral border joins with the coracoid.

The pectoral fin has a small, unsegmented marginal spur which is formed by one or fusion of both elements of the marginal ray hemitrichia, appearing as a short spine (Fig. 8.11Z). This spur has a complex articular base, and is closely applied to the first pectoral ray, together directly joining with the scapular condyle. Four proximal radials (actinosts) support the rest of pectoral rays. Posterior to its articular condyle, the scapula holds two proximal radials, and part of the third one.

All pectoral fin rays of acanthomorphs are composed of two halves; a lateral and a medial hemitrichium (Geerlink 1979, Gosline 1980). The base of each hemitrichium enlarges to act as a supporting surface for the fin muscles.

Proximal radials (actinosts): These four proximal radials have an irregular biconcave shape (Figs. 8.5A, 8.11S). They increase in size from the smallest (located posterior to the scapular condyle) towards the last (most ventral) element. The second and part of the third proximal radials are supported by the posterior margin of scapula, while the fourth is posterior to the coracoid. The posterior border of the proximal radials supports the fin rays through discoid cartilage and ligaments.

Dorsal postcleithrum: The dorsal postcleithrum is a flat bone with an expanded laminar dorsal part and a constricted ventral part (Fig. 8.11R). Its anterior border is very thick, while the posterior part is thin and laminar. The anterodorsal surface of this bone joins directly with the anteroventral margin of the cleithrum. Its posterior end has a ventral depression which joins with the anterior tip of the ventral postcleithrum (Fig. 8.5A).

Ventral postcleithrum: This bone is longer and thinner than the dorsal postcleithrum, and is pointed at both ends (Fig. 8.11Q). Its anterior portion is broader and curved medially, with an anterodorsal concavity for articulation with the posterior end of the dorsal postcleithrum. The ventral extremity of this bone joins with the lateral process of the pelvic basipterygium through a strong ligament.

Pelvic girdle

The pelvic girdle is formed by contralateral symmetrical basipterygia which are placed within the abdominal muscles (Fig. 8.5A). Each basipterygium is triangular, with its apex directed anteriorly. From the middle part, the basipterygium is inclined dorsally and moderately angled away from the ventral body wall. The pelvic girdle is not directly connected with the pectoral girdle, and connection is via the abdominal muscles to the cleithral symphysis. Each basipterygium has a thick base which is produced into a posterior process. The rest of the bone is laminar, with a lateral process at the lateral border, attaching to the ventral postcleithrum via a strong ligament (Figs. 8.5A, 8.12E-F).

Each pelvic fin has one spine and five rays, all supported directly by the basipterygium; there are no proximal radials or pterygiophores. The spine has a complex articulation with the basipterygium condyle, whilst the rays are directly supported by this bone through strong ligaments. Ventrally, behind the articular condyle, a concavity is pierced by a foramen opening into the sensory canal which runs laterally along the lateral border of basipterygium. The two basipterygia join to each other at their medial thick posterior border, and a well-developed interpelvic ligament firmly links them together (Figs. 8.12E-F). Ventral to this articulation, are a pair of elongate and thin processes, which are directed anteriorly, and deeply embedded within the pelvic musculature.

The m. *infracarinalis* runs from the cleithral symphysis to the caudal fin. A pair of tendons from the m. *infracarinalis anterior* attach to the mid-ventral cleithral processes. (Figs. 8.12E-F). The m. *abductor superficialis* originates from the median region of the ventral face of basipterygium, and anastomoses with the m. *infracarinalis anterior*, thus forming a more-or-less continuous muscle sheath (see also Stiassny 1993). The posterior processes of the basipterygia also connect to the tendinous m. *infracarinalis medialis*. Such tendons are present in many teleosts. Their presence in mugilids, atherinids and sphyraenids suggests that the pelvic girdle was attached to the cleithrum in some primitive ancestor (Gosline 1962).

The m. *abductor profundus pelvicus* originates from the ventral face of the basipterygium and inserts into the bases of the soft fin rays. The m. *arrector ventralis pelvicus* originates from the ventrolateral face of each basipterygium, and inserts into the ventral face of the respective pelvic spine (Fig. 8.12E). The m. *adductor profundus pelvicus* lies immediately above the dorsal surface of the basipterygium, and inserts into the anterolateral bases of the fin spine and rays. The m. *adductor superficialis pelvicus* lies dorsal to m. *adductor profundus pelvicus*, but is hardly distinguishable from the latter, inserting on the fin spine and all fin rays. The m. *arrector dorsalis pelvicus* originates from the dorsolateral face of each basipterygium, and inserts into the dorsolateral face of the respective pelvic spine.

Vertebral column

The vertebral column comprises 24 vertebrae. The last is modified into an urostyle to support the caudal fin structures (Fig. 8.13). No variation in the total number of vertebrae was observed in any of the species studied.

The 11 abdominal (precaudal) vertebrae lack a haemal spine. Transverse parapophyses, which are large and laterally-directed are present on all of the abdominal vertebrae. From the second vertebra, the parapophyses decrease in size gradually, and become inclined more ventrally in the posterior abdomen until

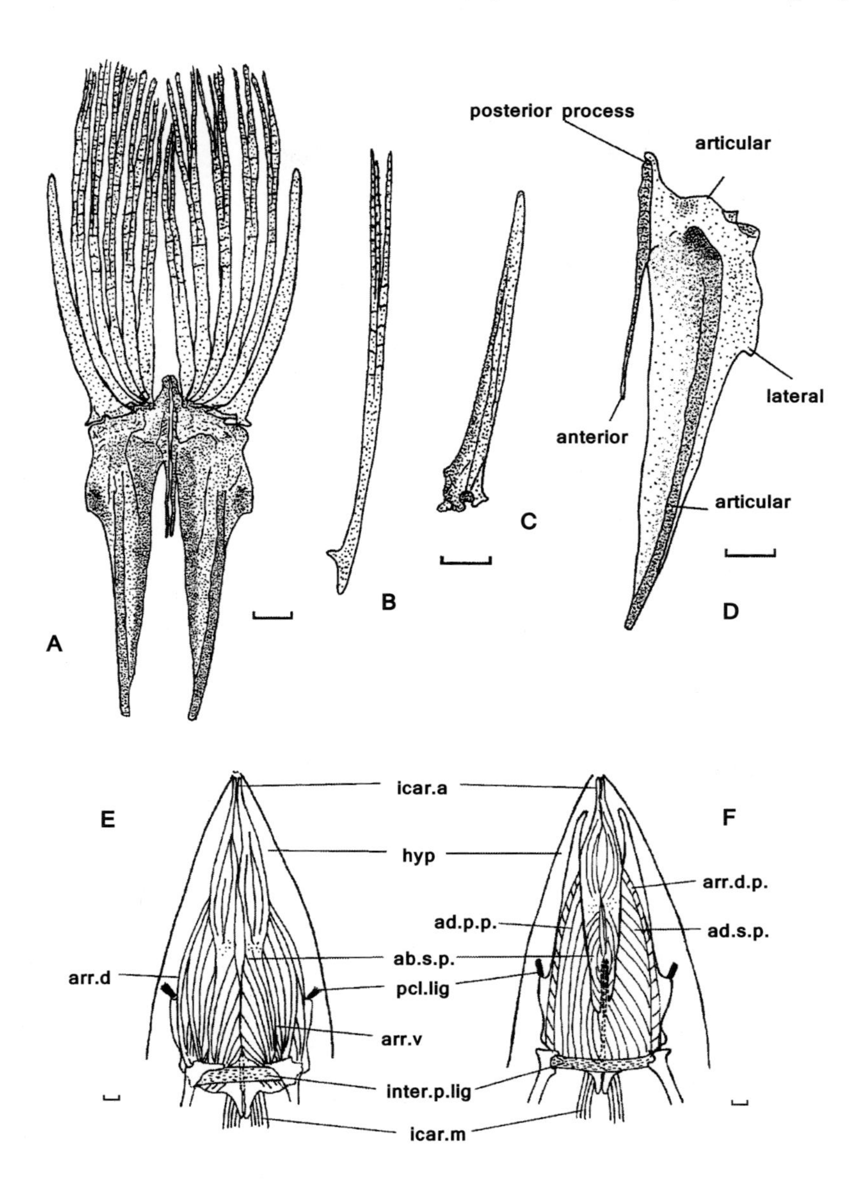

Figure 8.12. (A) Pelvic girdle (dorsal view); (B) Pelvic fin ray; (C) Pelvic fin spine; (D) Basipterygium (ventral view); (E) Ventral view of the musculature of pelvic girdle. *Abductor profundus pelvicus* is concealed by *abductor superficialis pelvicus* and *arrector ventralis pelvicus*; (F) Dorsal view of the musculuture of pelvic girdle. Left *adductor superficialis pelvicus* removed. Scale bar: 1 mm.

Abbreviations: ab.s.p.: *abductor superficialis pelvicus*, **ad.p.p.:** *adductor profundus pelvicus*, **ad.s.p.:** *adductor superficialis pelvicus*, **arr.d.p.:** *arrector dorsalis pelvicus*, **arr.v.p.:** *arrector ventralis pelvicus*, **bpt:** *basipterygium*, **hyp:** *hypaxialis*, **icar.a:** *infracarinalis anterior*, **icar.m:** *infracarinalis medialis*, **inter.p.lig:** inter-pelvic ligament, **pcl.lig:** postcleithrum ligament.

they unite to form the first complete haemal arch and haemal spine of the 12th (first caudal) vertebra. In the 10th and 11th vertebrae of *M. cephalus*, ossified haemal bridges are formed between the parapophyses, which are not true haemal arches.

The body or centrum of each vertebra is cylindrical and amphicoelous. The space between the centra of adjacent vertebrae is filled with notochordal tissue. Neural arches arise bilaterally from the dorsolateral

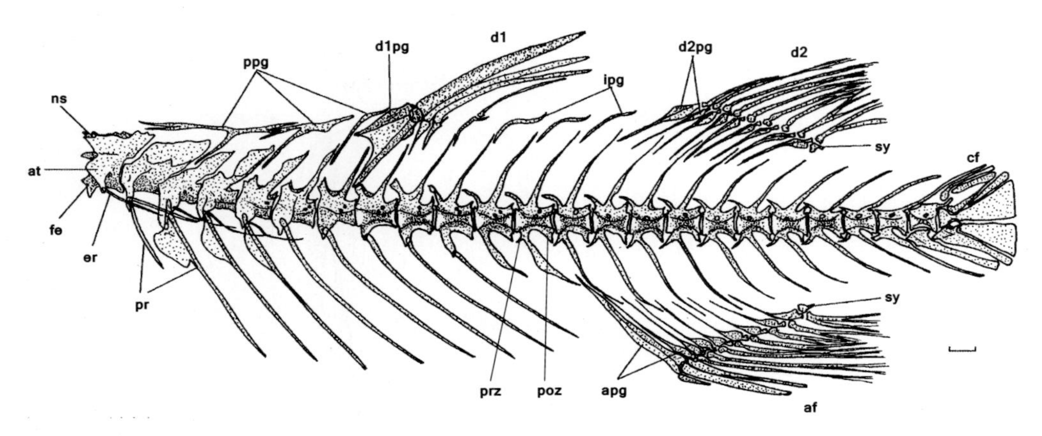

Figure 8.13. Axial skeleton.

Abbreviations: af: anal fin, **apg:** anal fin pterygiophore, **at:** first vertebrae, **cf:** caudal fin, **d:** dorsal fin, **d1pg:** proximal pterygiophore of the 1st dorsal fin, **d2pg:** pterygiophores of the 2nd dorsal fin, **er:** epipleural rib, **fe:** facet for exoccipital condyle, **ipg:** interdorsal pterygiophore, **ns:** neural spine (neural plate), **ppg:** predorsal pterygiophore, **poz:** postzygapophysis, **pr:** preural rib, **prz:** prezygapophysis, **sy:** fin stay, **tp:** transverse parapophysis. *Mugil cephalus* AMS I.31251-010 SL = 59 mm. Scale bar: 1 mm.

region of each centrum, uniting dorsally to form a long posteriorly-directed neural spine. The neural spines of the first five abdominal vertebrae are short and anteroposteriorly-expanded as neural plates, of varying shape and size.

The centrum of the first abdominal vertebra has a slightly-compressed horizontal axis, and two ventrolateral ridges. The anterodorsal margin of the centrum is produced into a pair of anteriorly-directed ovoid articular facets for articulation with the corresponding minor exoccipital condyles. The anterior discoidal facet joins with the major basioccipital condyle. The neural plate is very broad, and posteriorly attaches to the anterior border of the second neural plate, leaving no intervening lacuna. The first neural plate is divided into two laminae anteriorly, which penetrates between the supraoccipital crest. The parapophyses are long and broad, emerging from the dorsolateral part of the centrum with a curved posteriorly-directed border. There is no distinct zygapophysis in the first vertebra. The neural postzygapophyses are poorly developed.

The centrum of the second vertebra is longer than that of the first, with a constriction in the middle and two ventrolateral ridges present. The neural plate is similar to that of the first vertebra, but is undivided and is anteroposteriorly-expanded. The parapophyses are narrower but longer compared with those of the first vertebra, emerging from the anterior dorsolateral part of the centrum. There are no conspicuous neural prezygapophyses. The neural postzygapophyses are produced into a pair of uncinate processes which are directed anteroventrally from either side of the neural arch.

The centrum of the third vertebra is similar to that of the second vertebra, with a central constriction and two ventrolateral ridges. It has a neural plate considerably thinner than the preceding ones, broad at the base and pointed at the tip. The parapophyses, arising from the midlateral part of the centrum, are the broadest of all vertebrae. This vertebra also has no conspicuous neural prezygapophyses, but its postzygapophyses are poorly developed.

The neural plate of the fourth vertebra is enlarged and considerably elevated. Narrowing of the neural plate starts in this vertebra, becoming spinous in the eighth vertebra. The parapophyses, emerging from the ventrolateral region of the centrae, start to become narrow and short from the fourth vertebra. Fourth vertebral neural prezygapophyses and postzygapophyses are most distinct. The neural arches are enclosed in the second, third and fourth vertebrae due to the expansion of neural plates.

In the fifth vertebra, the neural plate is more dilated and parapophyses are more reduced. The neural prezygapophyses and postzygapophyses are more developed and the neural arch is more open. This trend continues until the eighth vertebra where the neural spine is thin and short, the dorsal opening of the neural arch is wide, the parapophyses are short and more ventrally-inclined, and neural prezygapophyses and postzygapophyses are well-developed. In the eighth vertebra, the haemal postzygapophyses make their first appearance.

In the ninth and the 10th vertebrae, the neural spine is longer and the parapophyses diverge further ventrally. A thin transverse haemal bridge, between the parapophyses, is present on the 10th vertebra. In the 11th (last abdominal) vertebra, the parapophyses are stouter, with a thicker haemal bridge. In the 12th (first caudal) vertebra a complete haemal arch is present.

Neural prezygapophyses first appear in the fourth abdominal vertebra. Each prezygapophysis ankyloses with its contralateral partner and is directed anterodorsally, with a relatively broad dorsal margin, forming a second dorsal arch in front of each neural arch. From the ninth or 10th vertebra these arches become uncinate and narrow, joining with each other anterodorsally until the 21st vertebra, in which they are not conjoined. Dorsal postzygapophyses are present from the third vertebra. They are uncinate, separated from each other, and posterodorsally projected, contacting the next vertebral prezygapophyses, or overlapping the edge of following vertebra.

The 13 caudal vertebra have a single median haemal spine or arch, of variable length. The spine is present until the penultimate vertebra, in which the haemal spine is displaced by the autogenous parhypural. With the exception of the first (largest) and last three (modified to support elements of the caudal fin), the caudal vertebrae have a somewhat uniform structure.

The first caudal vertebra has a haemal spine bent proximally, projecting posteriorly, its tip almost attains the haemal spine of the 13th vertebra. The first pterygiophore of the anal fin inserts anterior to the gap between these two haemal spines (Fig. 8.13). The haemal prezygapophyses are present but small in the first caudal vertebra, developing as larger spines in the following vertebrae. They are projected anteroventrally, where they meet or overlap the spinous postzygapophyses of the preceding vertebra which project posteroventrally. The haemal pre- and post-zygapophyses decrease in size from the 19th vertebra, till they are absent from the 23rd vertebra. The size of the neural and haemal spines and arches of the caudal vertebrae gradually decrease anteroposteriorly. The neural and haemal spines of the 22nd vertebra are longer and stouter than those of the preceding vertebrae. These elongated strong spines help in supporting the caudal fin. The 23rd (penultimate) vertebra has an autogenous, stout haemal spine. The neural arches meet dorsally but the neural spine is modified into a low-lying crest over the neural arch. This crest is posterodorsally produced into a short spine. The last vertebra is profoundly modified and incorporated into the caudal fin skeleton. The centrum is posteriorly produced as a laminar urostyle.

Caudal skeleton

The caudal skeleton is formed by modification of the terminal vertebrae of the axial column (Fig. 8.14). The dorsal and ventral borders are limited by the strong neural and haemal spines of the 22nd vertebra. The penultimate vertebra (pu2) has a considerably-reduced neural spine located between the epurals and uroneural. Its haemal spine is autogenous, thick and robust, with a small anterior crest.

The last vertebra is modified as an urostyle consisting of terminal half centrum anteriorly and a triangular lamina (resulting from fusion of hypurals 3 and 4) posteriorly. Hollister (1937) recognized this lamina as the upper hypural plate, but hypural 5 is also included in this plate. Hypural 5 is conical, forming the dorsal margin of the upper hypural plate and supporting two caudal rays at its posterior border.

A pair of uroneurals are closely connected together and located between hypural 5 and epural 1. Each uroneural is expanded at the base dorsal to the terminal half-centrum, tapering distally to a spinous process dorsal to hypural 5. Two epurals are located between the uroneurals and the neural spine of the second preterminal vertebra. Epural 2 is relatively uncinate and lies dorsal the trabecular epural 1, above the uroneurals.

There is a gap between the upper and lower hypural plates. Hypurals 1-2 are fused to form the lower hypural plate, articulating anteriorly with the terminal half-centrum. The autogenous parhypural which is the modified haemal spine of urostyle, is located ventral to the lower hypural plate. It has a large basal process (hypurapophysis) on either side, which is directed posterolaterally below the horizontal plane through the vertebral centrae.

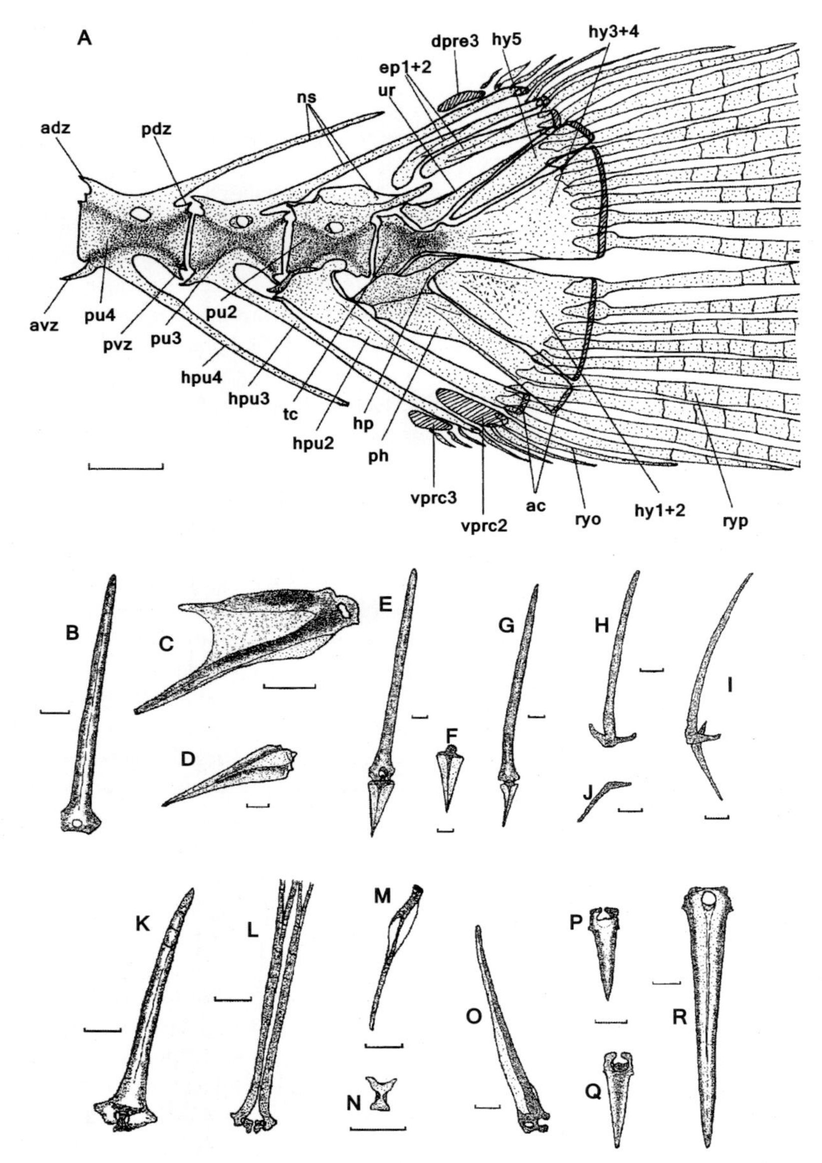

Figure 8.14. (A) Caudal skeleton of *Mugil cephalus* AMS I.31251-010 SL = 59 mm. Scale bar: 1 mm. Dashed area: Cartilage.

Abbreviations: ac: accessory cartilage, **adz:** anterior dorsal zygapophysis, **avz:** anterior ventral zygapophysis, **bp:** basal process of parhypural, **dprc3:** third dorsal preural radial cartilage, **ep:** epurals, **hpu:** haemal spine of the preural vertebrae, **hy:** hypural, **ns:** neural spine, **ph:** parhypural, **pdz:** posterior dorsal zygapophysis, **pu:** preural vertebrate, **pvz:** posterior ventral zygapophysis, **ryo:** procarrent ray, **ryp:** principal ray, **tc:** terminal half centrium, **ur:** uroneural, **vprc2:** second ventral preural radial cartilage, **vprc3:** third ventral preural radial cartilage.

(B-R) First dorsal fin, second dorsal fin and anal fin elements.

(B) First spine of the first dorsal fin. (C) Lateral view of the pterygiophore of the first spine of the first dorsal fin. (D) Dorsal view of the pterygiophore of the first spine of the first dorsal fin; (E) Second spine and pterygiophore of the first dorsal fin; (F) Pterygiophore of the second spine of the first dorsal fin; (G) Third spine and pterygiophore of the first dorsal fin; (H) Fourth spine of the first dorsal fin; (I) Fourth spine and pterygiophore of the first dorsal fin; (J) Pterygiophore of the fourth spine of the first dorsal fin; (K) First ray and distal radial of the second dorsal fin; (L) Dorsal view of a fin ray and distal radials of the second dorsal fin; (M) Pterygiophore of the first ray of the second dorsal fin; (N) Distal radial of the first ray of the second dorsal fin; (O) First pterygiophore of the anal fin; (P) Dorsal view of the first spine of the anal fin; (Q) Ventral view of the second spine of the anal fin; (R) Dorsal view of the second spine of the anal fin.

The preural dorsal cartilage is located distally, posterior to the neural spine of the second preterminal vertebra (pu3), supporting the last dorsal procurrent ray. The preural ventral cartilage is located distally, posterior to the haemal spine of the second preterminal vertebrae (pu3), giving support to the last two ventral procurrent rays. A second larger preural ventral cartilage is also situated distally, posterior to the autogenous spine of penultimate vertebra (pu2), supporting three procurrent rays.

In young fishes, strips of accessory cartilage are attached to the posterior border of caudal fin elements, providing extra support for fin rays. The distal edges of the caudal complex, except the neural spine of the penultimate vertebrae (pu2), support the principal and procurrent rays of the caudal fin. The caudal principal rays originate on the posterior margins of the hypural plates. There are 8 + 7 (upper + lower) principal rays on the hypural plates; the first two dorsal rays of the upper plate and the last ventral ray of the lower plate are unbranched, and the other 12 rays are branched.

Pleural ribs and epineurals

The abdominal vertebrae (v2–v10) support large pleural ribs, which protect the viscera. The epineurals are located between the dorsal and ventral epaxial muscles, and are smaller and less numerous than the ribs (Fig. 8.13).

Pleural ribs: There are nine pairs of pleural ribs. The first and the last abdominal vertebrae do not support pleural ribs. The first pleural rib is short, stout, and united with the first epineural rib forming a V-shaped bone which is connected to the anterolateral side of parapophysis of the second abdominal vertebra. The other ribs are long and curved with an anterior discoidal articular facet contacting the anterolateral border of parapophyses of the corresponding vertebra. There is an anterior curved laminar expansion close to the articular facet of the ribs which is well-developed in the second rib and become gradually smaller in subsequent ribs, until it disappears in the last rib. The ribs do not meet ventrally.

Epineurals: There are five pairs of epineurals, contacting the tip of the parapophysis of the first five vertebrae. The first epineural is cylindrical, thickened proximally, and flattened towards its distal end. It joins with the dorsolateral angle of the first vertebral parapophysis, and is directed posterolaterally towards the internal facet of the dorsal postcleithrum.

The second epineural is fused to the first rib (as described above). It is slightly narrower than the first epineural. The third-to-fifth epineurals are similar, decreasing in size posteriorly. Dorsally they are attached to the articular facet of the second-to-fourth rib respectively. The epineurals extend posteroventrally, between the dorsal and ventral abdominal muscles, parallel to the axial skeleton.

First dorsal fin (Figs. 8.14B, C, D, E, F, G, H, I and J)

The first dorsal fin consists of four spines. Each spine is supported by a single basal pterygiophore formed by the fusion of proximal and medial elements, distal radials being absent. The first pterygiophore of this fin is quite large and stout, with a triangular dorsal aspect, and notched triangular lateral aspect (Fig. 8.14C). It is formed by a medial vertical plate which is produced as an elongate spinous process anteroventrally, and another much shorter process anterodorsally, forming a curved notch in its anterior border. This plate is broad posterodorsally. At its base it is expanded into two bilateral triangular laminae. Posteriorly, this pterygiophore is produced into a vertical bony ring which is interlocked in the anterior horizontal foramen of the anteriormost robust spine. This pterygiophore is placed between the seventh and eighth vertebrae and its anteroventral spinous tip deeply penetrates into the space between these two vertebrae (Fig. 8.13). The triangular pterygiophore of the second spine is considerably smaller than the first one, and placed immediately after it (Figs. 8.13, 8.14D). A similar mechanism holds it in place.

The third spine and its pterygiophore lack the articular ring and foramen, and are firmly joined via strong ligaments (Fig. 8.14G). The first three spines are placed very close to each other, and their pterygiophores are attached together via strong ligaments. The second and third pterygiophores are intercalated between the spines of the eighth and ninth abdominal vertebrae. The fourth spine is well-separated from the first three, and has a bilaterally-expanded base (Figs. 8.13, 8.14H, I and J). It is connected to a simple sigmoidal pterygiophore located behind the spine of the ninth abdominal vertebra. All spines of the first dorsal fin are in the horizontal axis of the body, except the second spine which is slightly slanted from this line.

Second dorsal fin (Figs. 8.14K, L, M and N)

There are nine rays in the second dorsal fin of *M. cephalus*. The anterior most ray of this fin is frequently mistaken for a spine (Fig. 8.14K). In fact it is a short and slender ray which is often unbranched and only segmented near its tip in adults (it is branched in observed cleared and stained specimens ca. 30 mm SL).

The rays of the second dorsal fin are supported by nine pterygiophores and one fin stay (Fig. 8.13). The first and second pterygiophores are intercalated between the 13th and 14th neural spines. The last two pterygiophores are placed between the 16th and 17th neural spines. The first ray is strongly-supported by the first pterygiophore (Fig. 8.13). This ray has an ossified biconcave distal radial with two opposed lateral uncinate processes which is fused to the spine's ventrally-open foramen, and hold it firmly. Each of the rest of pterygiophores have a pair of bilateral ovoid distal radials which join with two hemitrichia of their corresponding fin ray. Distal radials of each pterygiophore are located in the dorsal depression of the following pterygiophore, thus giving more support to the neighbouring rays, as occurs in percoids (Smith and Bailey 1962). The last ray of the second dorsal and anal fins is composed of two rays sharing the same pterygiophore. The last pterygiophore articulates posteriorly with a triangular fin stay.

There are three interdorsal pterygiophores between the two dorsal fins. They are sigmoidal, and their dorsal portion is laterally-flattened. They are inserted posterior to the tip of the 10th–12th neural spines respectively.

Anal fin (Figs. 8.14O, P, Q and R)

In adult *M. cephalus*, this fin has three spines and eight rays, which connect to nine pterygiophores and one fin stay. The first anal pterygiophore is very different from the following pterygiophores. It is fused with the following pterygiophore, forming a long, triangular plate, modified to hold the first and second anal spines (Fig. 8.14O). It has an anterior foramen at its base in which are placed two ventrolateral hooks of the first anal spine. Posterior to this foramen is a bluntly-curved hook which is firmly linked with the anterior foramen of the second anal spine (Fig. 8.14R). Posteriorly, this pterygiophore also gives support to the third spine. The anal fin rays join with their respective pterygiophores via distal radials, as described for the second fin rays. The first anal pterygiophore is intercalated between the first and second haemal spines, and anteriorly is connected to the ventral tip of the first haemal spine (Fig. 8.3). The last anal pterygiophore is placed between the fifth and sixth haemal spines, and is followed posteriorly by a fin stay.

Predorsal pterygiophores (supraneurals)

There are three narrow predorsal pterygiophores in medial plane, between the supraoccipital crest and the first dorsal fin (Fig. 8.13). The first and second predorsal pterygiophores are T-shaped, and formed by a horizontal ramus joined about half-way to an anteroventrally directed ventral ramus. The horizontal ramus of the first predorsal pterygiophores has bifurcate anterior and posterior branches. The anterior branch extends over the entire neural plate of the second abdominal vertebra, and posterior part of the neural plate of the first abdominal vertebra. The posterior branch courses over the neural plates of the third and fourth abdominal vertebrae. The ventral ramus is placed between the neural plates of the second and third abdominal vertebrae.

In the second predorsal pterygiophores only the anterior branch of the horizontal ramus is bifurcate, interlocking with the posterior branch of the first predorsal pterygiophore, over the fourth neural plate. The posterior branch extends over the fifth and sixth neural plates. Its ventral ramus intercalates between the fourth and fifth neural plates.

The third predorsal pterygiophore is simple and almost sigmoidal. It has a short horizontal part which is placed over the first pterygiophore of the first dorsal fin and an anterior oblique part extending anteroventrally behind the sixth neural spine.

Sensory canals

Sensory canals are developed on the nasal, frontal, pterotic, preopercle, infraorbitals, angular and dentary. Additionally, extrascapulars are present. There is no lateral line in Mugilidae. All sensory canals are closed except those of the preopercle, nasal, suborbital and infraorbital bones.

In *M. cephalus*, the infraorbital series comprises six bones (Fig. 8.5A). The preorbital (first infraorbital) is located anterior to the orbit. The obliquely-oriented preorbital canal is unbranched, with a complete posterior wall and relatively large anterior opening (Fig. 8.6B). There is a noticeable distance between preorbital and the second infraorbital which is the smallest of these series. This bone is followed posteriorly by the third, fourth and fifth infraorbitals, which progressively increase in size. The sensory canal of these bones forms an open ventral trough connected to the ventral opening of the dermosphenotic canal.

The dermosphenotic is the posterior-most infraorbital, forming the posterior border of the orbit, anterolateral to the sphenotic. It has a closed sensory canal which is dorsally connected to the postorbital opening of the supraorbital sensory canal. The nasal has an obliquely-extended sensory canal, which is the continuation of the supraorbital sensory canal, terminating in a pore dorsal to the articulation with the preorbital.

The frontal supraorbital canal has a relatively large post-nasal pore in the anterior rugose area of the frontal. The canal curves posterolaterally to form a pore, followed by another infraorbital pore, and continues as the infraorbital canal, which is in turn connected to the temporal canal. The temporal canal extends along the anterior border of pterotic and connects with the anterior L-shaped extrascapular bone.

The extrascapular elements are an apomorphy of the mugilid cranium, consisting of three distinct and well-developed tubular bones located dorsal to the temporal fossa (Fig. 8.1A), and connecting the pterotic sensory canal to the post-cranial nervous system.

The preopercle has a posteriorly-open sensory canal which is connected to the supratemporal sensory canal. This canal runs between the medial and lateral plates of preopercle to join the mandibular sensory canal of the angular and dentary.

References

Allis, E.P. 1903. The skull and the cranial and first spinal muscles and nerves in *Scomber scomber*. J. Morph. 18: 45–328.

Aurelle, D.R., M. Barthelemy, J.P. Quignard, M. Trabelsiand and E. Faure. 2008. Molecular phylogeny of Mugilidae (Teleostei: Perciformes). Open Mar. Biol. J. 2: 29–37.

Bacheler, N.M., R.A. Wong and J.A. Buckel. 2005. Movements and mortality rates of striped mullet in North Carolina. N. Am. J. Fish Manag. 25: 361–373.

Blaber, S.J.M. 1976. The food and feeding ecology of Mugilidae in the St. Lucia lake system. Biol. J. Linn. Soc. 8: 267–277.

Blaber, S.J.M. and A.K. Whitfield. 1977. The feeding ecology of juvenile mullet (Mugilidae) in south east African estuaries. Biol. J. Linn. Soc. 9: 277–284.

Brownell, C.L. 1979. Stages in the early development of 40 marine fish species with pelagic eggs from the Cape of Good Hope. Ichthy. Bull. J.L.B. Smith Inst. Ichthyol. 40: 1–84.

Caldara, F., L. Bargelloni, L. Ostellari, E. Penzo, L. Colombo and T. Patarnello. 1996. Molecular phylogeny of grey mullets based on mitochondrial DNA sequence analysis: evidence of a differential rate of evolution at the intrafamily level. Mol. Phylogenet. Evol. 6: 416–424.

Cambrony, M. 1984. Identification et périodicité du recrutement des juvéniles de Mugilidae dans les étangs littoraux du Languedoc-Roussillon. Vie Milieu 34: 221–227.

Campton, D.E. and B. Mahmoudi. 1991. Allozyme variation and population structure of striped mullet (*Mugil cephalus*) in Florida. Copeia 1991: 485–492.

Capanna, E., S. Cataudella and G. Monaco. 1974. The pharyngeal structure of Mediterranean Mugilidae. Monitore Zoologica. Italiano (NS) 8: 29–46.

Cardona, L. 2015. Food and feeding of Mugilidae. *In*: D. Crosetti and S.J.M. Blaber (eds.). Biology, Ecology and Culture of Grey Mullet (Mugilidae). CRC Press, Boca Raton, USA (this book).

Chaoui, L., M.H. Kara, E. Faure and J.P. Quignard. 2006. The fish fauna of Mellah Lagoon (north-east Algeria): diversity, production and commercial catches analysis. Cybium 30: 123–132.

Chen, W.J. and R.L. Mayden. 2009. Molecular systematics of the Cyprinoidea (Teleostei: Cypriniformes), the world's largest clade of freshwater fishes: further evidence from six nuclear genes. Mol. Phylogenet. Evol. 52: 544–549.

Chen, W.J. and R.L. Mayden. 2010. A phylogenomic perspective on the new era of ichthyology. BioScience 60: 421–432.

Chen, W.J., C. Bonillo and G. Lecointre. 2003. Repeatability of clades as a criterion of reliability: a case study for molecular phylogeny of acanthomorph (teleostei) with larger number of taxa. Mol. Phylogenet. Evol. 26: 262–288.

Chervinski, J. 1984. Using scales for identification of four Mugilidae species. Aquaculture 38: 79–81.

Cockerell, T.D.A. 1913. The scales of some Queensland fishes. Mem. Qld. Mus. 2: 51–9.

Crosetti, D. 2015. Current state of capture fisheries and culture of Mugilidae. *In*: D. Crosetti and S.J.M. Blaber (eds.). Biology, Ecology and Culture of Grey Mullet (Mugilidae). CRC Press, Boca Raton, USA (this book).

Crosetti, D., J.C. Avise, F. Placidi, A.R. Rossi and L. Sola. 1993. Geographic variability in the grey mullet *Mugil cephalus*: preliminary results of mt DNA and chromosome analyses. Aquaculture 111: 95–101.

Crosetti, D., W.S. Nelson and J.C. Avise. 1994. Pronounced genetic structure of mitochondrial DNA among populations of the circumglobally distributed grey mullet (*Mugil cephalus*). J. Fish Biol. 44: 47–58.

Dharmarajan, M. 1936. The anatomy of *Otolithus ruber* (Bl. & Schn) Part 1. The Endoskeleton. Royal Asia Soc. Beng. 2: 1–71.

Durand, J.D., K.N. Shen, W.J. Chen, B.W. Jamandre, H. Blel, K. Diop, M. Nirchio, F.J. García de León, A.K. Whitfield, C.W. Chang and P. Borsa. 2012a. Systematics of the grey mullet (Teleostei: Mugiliformes: Mugilidae): molecular phylogenetic evidence challenges two centuries of morphology-based taxonomic studies. Mol. Phylogen. Evol. 64: 73–92.

Durand, J.D., W.J. Chen, K.N. Shen, C. Fu and P. Borsa. 2012b. Genus-level taxonomic changes implied by the mitochondrial phylogeny of grey mullets (Teleostei: Mugilidae). Comp. Rend. Biol. 335: 687–697.

Ebeling, A.W. 1957. The dentition of eastern Pacific mullets, with special reference to adaptation and taxonomy. Copeia 3: 173–185.

Ebeling, A.W. 1961. *Mugil galapagensis*, a new mullet from the Galapagos Islands, with notes on related species and a key to the Mugilidae of the eastern Pacific. Copeia 3: 295–304.

Erguden, D., M. Gurlek, D. Yaglioglu and C. Turan. 2010. Genetic identification and taxonomic relationship of Mediterranean mugilid species based on mitochondrial 16S rDNA sequence data. J. Anim. Vet. Adv. 9: 336–341.

Eschmeyer, W.N. and J.D. Fong. 2015. Catalog of Fishes: Species by Family/Subfamily. (http://researcharchive.calacademy.org/research/ichthyology/catalog/SpeciesByFamily.asp). Electronic version accessed on 15/01/2015.

Farrugio, H. 1977. Clés commentées pour la détermination des adultes et des alevins de Mugilidae de Tnnisie. Cybium 2: 57–73.

Geerlink, P.J. 1979. The anatomy of the pectoral fin in *Sarotherodon niloticus* Trewavas (Cichlidae) Neth. J. Zool. 29: 9–32.

Ghasemzadeh, J. 1998. Phylogeny and systematics of Indo-Pacific mullets (Teleostei: Mugilidae) with special reference to the mullets of Australia. Ph.D. dissertation, Macquarie University, Sydney.

Ghasemzadeh, J., W. Ivantsoff and Aarn. 2004. Historical overview of mugilid systematics with description of *Paramugil* (Teleostei: Mugiliformes: Mugilidae), new genus. Aqua. J. Ichthyol. Aquat. Biol. 8: 9–22.

Gisbert, E., L. Cardona and F. Castello. 1996. Resource partitioning among planktivorous fish larvae and fry in a Mediterranean coastal lagoon. East Coast. Shelf Sci. 43: 723–735.

González-Castro, M. 2007. Los peces representantes de la Familia Mugilidae en Argentina. Ph.D. thesis. Universidad Nacional de Mar del Plata, Argentina.

González-Castro, M. and J. Ghasemzadeh. 2015. Morphology and morphometry based taxonomy of Mugilidae. *In*: D. Crosetti and S.J.M. Blaber (eds.). Biology, Ecology and Culture of Grey Mullet (Mugilidae). CRC Press, Boca Raton, USA (this book).

González-Castro, M., V. Abachian and R.G. Perrotta. 2009. Age and growth of the striped mullet, *Mugil platanus* (Actinopterygii: Mugilidae), in a southwestern Atlantic coastal lagoon (37°32′ S –57°19′ W): a proposal for a life-history model. J. Appl. Ichthyol. 25: 61–66.

Gornung, E., P. Colangelo and F. Annesi. 2007. 5S ribosomal RNA genes in six species of Mediterranean grey mullets: genomic organization and phylogenetic inference. Genome 50: 787–795.

Gosline, W.A. 1962. Systematic position and relationships of the percesocine fishes. Pac. Sci. 16: 207–217.

Gosline, W.A. 1980. The evolution of some structural systems with reference to the interrelationships of modern lower teleostean fish groups. Jap. J. Ich. 27: 1–27.

Harrison, I.J. and G.J. Howes. 1991. The pharyngobranchial organ of mugilid fishes; its structure, variability, ontogeny, possible function and taxonomic utility. Bull. Br. Mus. Nat. Hist. (Zool.) 57: 111–132.

Heras, S., M.I. Roldan and M. Gonzalez Castro. 2009. Molecular phylogeny of Mugilidae fishes revised, Rev. Fish Biol. Fisheries 19: 217–231.

Hollister, G. 1937. Caudal skeleton of Bermuda shallow water fishes. II. Order Percamorphi, suborder Percesoces: Atherinidae, Mugilidae, Sphyraenidae. Zoologica 22(3): 265–279.

Hotta, H. 1955. On the mature mugilid fish from Kabashima, Nagasaki Pref., Japan, with additional notes on the intestinal convolution of Mugilidae. Jpn. J. Ichthyol. 4: 162–169.

Hotta, H. and I.S. Tung. 1966. Identification of fishes of the family Mugilidae based on the pyloric caeca and the position on inserted first interneural spine. Jpn. J. Ichthyol. 14: 62–66 (In Japanese, English summary and key).

Imsiridou, A., G. Minos, V. Katsares, N. Karaiskou and A. Tsiora. 2007. Genetic identification and phylogenetic inferences in different Mugilidae species using 5S rDNA markers. Aquaculture Res. 38: 1370–1379.

Inoue, T., Y. Suda and M. Sano. 2005. Food habits of fishes in the surf zone of a sandy beach at Sanrimatsubara, Fukuoka Prefecture, Japan. Ichthyol. Res. 52: 9–14.

Ishiyama, R. 1951. Revision of the Japanese mugilid fishes, especially based upon the osteological characters of the cranium. Jap. J. Ichthyol. 1: 238–250.

Jamandre, B.W., J.D. Durand and W.N. Tzeng. 2009. Phylogeography of the flathead mullet *Mugil cephalus* in the Northwest Pacific inferred from the mtDNA control region. J. Fish Biol. 75: 393–407.

Jacot, A.P. 1920. Age, growth and scale characters of the mullet *Mugil cephalus* and *Mugil curema*. Trans. Amer. Micr. Soc. 39: 199–229.

Katselis, G., C. Koutsikopoulos, E. Dimitriou and Y. Rogdakis. 2003. Spatial patterns and temporal trends in the fishery landing of the Messolonghi-Etoliko lagoon system (western Greek coast). Sci. Mar. 67: 501–511.

Ke, H.M., W.W. Lin and H.W. Kao. 2009. Genetic diversity and differentiation of grey mullet (*Mugil cephalus*) in the coastal waters of Taiwan. Zool. Sci. 26: 421–428.

Kobelkowsky, A.D. and A.M. Resendez. 1972. Estudio comparative del endosqueleto de *Mugil cephalus* y *Mugil curema* (Pisces: Perciformes) An Inst. Biol. Univ. Nac. Auton. Mex. Ser. Cienc. Mar. Limnol. 43: 33–81.

Lawson, E.O. and A.A. Jimoh. 2010. Aspects of the biology of grey mullet, *Mugil cephalus*, in Lagos Lagoon, Nigeria. AACL Bioflux 3: 181–193.

Lee, S.C., J.T. Chang and Y.Y. Tsu. 1995. Genetic relationships of four Taiwan mullets (Pisces: Perciformes: Mugilidae) J. Fish Biol. 46: 159–162.

Liu, C.H. and S.C. Shen. 1991. Lepidology of the Mugilid fishes. J. Taiwan Mus. 44: 321–357.

Liu, J.Y., C.L. Brown and T.B. Yang. 2009. Population genetic structure and historical demography of grey mullet, *Mugil cephalus*, along the coast of China, inferred by analysis of the mitochondrial control region. Biochem. Syst. Ecol. 37: 556–566.

Liu, J.Y., C.L. Brown and T.B. Yang. 2010. Phylogenetic relationships of mullets (Mugilidae) in China Seas based on partial sequences of two mitochondrial genes. Biochem. Syst. Ecol. 38: 647–655.

Livi, S., L. Sola and D. Crosetti. 2011. Phylogeographic relationships among worldwide populations of the cosmopolitan marine species, the striped grey mullet (*Mugil cephalus*), investigated by partial cytochrome b gene sequences. Biochem. Syst. Ecol. 39: 121–131.

Luther, G. 1977. New characters for consideration in the taxonomic appraisal of grey mullets. J. Mar. Biol. Assoc. India 19: 1–9.

Menezes, N.A. 1983. Guia prático para conhecimento e identificaçao das tainhas e paratis (Pisces: Mugilidae) do litoral Brasileiro. Revta. Bras. Zool., S. Paulo 2: 1–12.

Menezes, M.R., M. Martins and S. Naik. 1992. Interspecific genetic divergence in grey mullets from the Goa region. Aquaculture 105: 117–129.

Miya, M., H. Takeshima, H. Endo, N.B. Ishiguro, J.G. Inoue, T. Mukai, T.P. Satoh, M. Yamaguchi, A. Kawaguchi, K. Mabuchi, S.M. Shirai and M. Nishida. 2003. Major patterns of higher teleostean phylogenies: a new perspective based on 100 complete mitochondrial DNA sequences. Mol. Phylogenet. Evol. 26: 121–138.

Mohsin, A.K.M. 1973. Comparative osteology of the weakfishes (*Cynoscion*) of the Atlantic and Gulf Coasts of the United States. (Pisces: Sciaenidae) Unpubilshed Ph.D. dissertation. Texas A. & M. Univ. 1–148.

Mohsin, A.K.M. 1978. The osteology of the grey mullet, Liza subviridis (Valenciennes). The branchiocranium, vertebral column, and girdles. Malayan Nat. J. 31(4): 202–218.

Morovic, D. 1953. Sur la determination des muges adriatiques d'apres la forme de l'otolith sagitta. Notes Inst. Oceanogr. Split 9: 1–7.

Murgia, R., G. Tola, S.N. Archer, S. Vallerga and J. Hirano. 2002. Genetic identification of grey mullet species (Mugilidae) by analysis of mitochondrial DNA sequence. Application to identify the origin of processed ovary products (bottarga). Mar. Biotechnol. 4: 119–126.

Papasotiropoulos, V., E. Klossa-Kilia, G. Kilias and S. Alahiotis. 2001. Genetic divergence and phylogenetic relationships in grey mullets (Teleostei: Mugilidae) using allozyme data. Biochem. Genet. 39: 155–168.

Papasotiropoulos, V., E. Klossa-Kilia, G. Kilias and S. Alahiotis. 2002. Genetic divergence and phylogenetic relationships in grey mullets (Teleostei: Mugilidae) based on PCR–RFLP analysis of mtDNA segments. Biochem. Genet. 40: 71–86.

Papasotiropoulos, V., E. Klossa-Kilia, G. Kilias and S. Alahiotis and G. Kilias. 2007. Molecular phylogeny of grey mullets (Teleostei: Mugilidae) in Greece: evidence from sequence analysis of mtDNA segments. Biochem. Genet. 45: 623–636.

Patten, J.M. 1987. Osteology, relationships and classification of hardyheads of the subfamily Atherinidae (Pisces: Atherinidae) Unpublished M.Sc. thesis, Macquarie University, Sydney.

Patterson, C. 1964. A review of Mesozoic acanthopterygian fishes, with special reference to those of the English Chalk. Phil. Trans. R. Soc. (ser. B) 247: 213–482.

Perlmutter, A., L. Bogard and Y. Pruginin. 1957. Use of the estuarine and sea fish of the family Mugilidae (grey mullets) for pond culture in Israel. Proc. Fish Counc. Mediterr., FAO 4: 289–304.

Peterson, G.L. and Z.H. Shehadeh. 1975. Subpopulation of the Hawaiian stripped mullet *Mugil cephalus*: analysis of variations of nuclear eye-lens protein electropherograms and nuclear eye-lens weights. Mar. Biol. 11: 52–60.

Pillay, T.V.R. 1951. Structure and development of the scales of five species of grey mullets of Bengal. Proc. Nat. Inst. Sci. India 17: 413–24.

Reay, P.J. and V. Cornell. 1988. Identification of grey mullet (Teleostei: Mugilidae) Juveniles from British waters. J. Fish Biol. 32: 95–99.

Rocha-Olivares, A., N.M. Garber and K.C. Stuck. 2000. High genetic diversity, large inter-oceanic divergence and historical demography of the striped mullet. J. Fish Biol. 57: 1134–1149.

Rojo, A.L. 1990. Dictionary of Evolutionary Fish Osteology. CRC Press, Boca Raton, Florida. 273pp.
Rosenblatt, R.H. and R.S. Waples. 1986. A genetic comparison of allopatric populations of shore fish species from the eastern and central Pacific Ocean: dispersal or vicariance? Copeia 1986: 275–284.
Rossi, A.R., D. Crosetti, E. Gornung and L. Sola. 1996. Cytogenetic analysis of global populations of *Mugil cephalus* (striped mullet) by different staining techniques and fluorescent *in situ* hybridization. Heredity 76: 77–82.
Rossi, A.R., M. Capula, D. Crosetti, L. Sola and D.E. Campton. 1998a. Allozyme variation in global populations of striped mullet, *Mugil cephalus* (Pisces: Mugilidae). Mar. Biol. 131: 203–212.
Rossi, A.R., M. Capula, D. Crosetti, D.E. Campton and L. Sola. 1998b. Genetic divergence and phylogenetic inferences in five species of Mugilidae (Pisces: Perciformes). Mar. Biol. 131: 213–218.
Rossi, A.R., A. Ungaro, S. De Innocentiis, D. Crosetti and L. Sola. 2004. Phylogenetic analysis of Mediterranean mugilids by allozymes and 16S mt-rRNA genes investigation: are the Mediterranean species of *Liza* monophyletic? Biochem. Genet. 42: 301–315.
Semina, A.V., N.E. Polyakova and V.A. Brykov. 2007. Analysis of mitochondrial DNA: taxonomic phylogenetic relationships in two fish taxa (Pisces: Mugilidae and Cyprinidae). Biochemistry (Moscow) 72: 1349–1355.
Senou, H. 1988. Phylogenetic Interrelationships of the Mullets (Pisces: Mugilidae) Tokyo University. Ph.D. Thesis (In Japanese).
Serventi, M., I.J. Harrison, P. Torriecelli and G. Gandolfi. 1996. The use of pigmentation and morphological characters to identify Italian mullet fry. J. Fish Biol. 49: 1163–1173.
Smith, C.L. and R.M. Bailey. 1962. The subocular shelf of fishes. Jour. Morphol. 110: 1–17.
Song, J.K. 1981. Chinese mugilid fishes and morphology of their cephalic lateral line canals. Sinozoologia 1: 9–22 (in Chinese with English abstract and key).
Stiassny, M.L.J. 1993. What are grey mullets? Bull. Mar. Sci. 52(1): 197–219.
Sunny, K.G. 1971. Morphology of the vertebral column of *Mugil macrolepis* (Smith) Bull. Dep. Mar. Oceanogr. Univ. Cochin 5: 101–108.
Taylor, W. and G. Van Dyke. 1985. Revised procedures for staining and clearing small fishes and other vertebrates for bone and cartilage study. Cybium 9: 107–119.
Thomson, J.M. 1963. Synopsis of biological data on the grey mullet *Mugil cephalus* Linnaeus 1758. Aust. CSIRO Div. Fish Oceanogr. Fish Synop. No. 1. Variable page numbers.
Thomson, J.M. 1966. The grey Mullets. Annu. Rev. Oceangr. Mar. Biol. 4: 301–335.
Thomson, J.M. 1975. The dentition of grey mullets. Aquaculture 5: 108.
Thomson, J.M. 1981. The taxonomy of grey mullets. pp. 1–16. *In*: O.H. Oren (ed.). Aquaculture of Grey Mullets. IBP, 26, Cambridge University Press, Cambridge.
Turan, C., M. Caliskan and H. Kucuktas. 2005. Phylogenetic relationships of nine mullet species (Mugilidae) in the Mediterranean Sea. Hydrobiologia 532: 45–51.
Van Der Elst, R.P. and J.H. Wallace. 1976. Identification of the juvenile mullet of the east coast of South Africa. J. fish Biol. 9: 371–374.
Whitfield, A.K., J. Panfili and J.D. Durand. 2012. A global review of the cosmopolitan flathead mullet *Mugil cephalus* Linnaeus 1758 (Teleostei: Mugilidae), with emphasis on the biology, genetics, ecology and fisheries aspects of this apparent species complex. Rev. Fish Biol. Fish 22: 641–681.
Winterbottom, R. 1974. A descriptive synonymy of the striated muscles of the Teleostei. Proc. Acad. Nat. Sci. Phil. 125: 225–317.

CHAPTER 9

Food and Feeding of Mugilidae

Luis Cardona

Introduction

Grey mullets have often been described as mud-eaters, iliophagous, detritus feeders, deposit feeders and interface feeders (Brusle 1981) because the diet of most of these species is based on the organic matter present in the sediment, although they can also exploit benthic invertebrates, green filamentous macroalgae, plankton and other suspended organic matter. No other group of fish relies so much on microphytobenthos, which represents a major component of the diet of most grey mullet species. Most marine herbivorous fishes consume primarily fleshy macroalgae and seagrasses (parrot fishes, sea breams and surgeon fishes), turf algae (damselfishes) or encrusting coralline algae (some parrot fishes), but not the microscopic algae growing on the sediment. Some freshwater cyprinids and armoured catfishes also consume significant amounts of detritus and benthic filamentous and microscopic benthic algae, but scrape them from rocks and boulders instead of filtering from the water column and the sediment. Only the Indo-Pacific milkfish *Chanos chanos* and some tilapias, native to African freshwater and estuarine ecosystems, have diets close to those of grey mullets and share with them the ability to behave alternatively as deposit feeders and suspension feeders (Odum 1970, Whitfield and Blaber 1978). All these species are relatively large and forage primarily on microscopic algae, thus bypassing steps in the food web and telescoping primary production to fisheries yield (Hiatt 1944, Odum 1970).

When feeding on sediment, grey mullets angle their heads downwards, protract the premaxillaries and scrape the surface of the sediment as they take a mouthful of sediment and associated food material (Odum 1970, King 1988). After working the material between the pharyngeal bones they reject some particles through their gills and mouth in a characteristic 'coughing and spitting' action (Thomson 1954). No other group of fish forages in this way and grey mullets exploit a unique trophic niche thanks to a highly sophisticated feeding apparatus, including toothed lips for scraping microbial films, pharyngeal teeth to remove large sediment particles from the oral cavity, densely packed gill rakers to retain fine particles, a two-chambered stomach with a powerful gizzard to break cell walls and a very long intestine to digest plant material (Ghazzawi 1935, Al-Hussaini 1947, Thomson 1954, Ebeling 1957, Capanna et al. 1974, Harrison and Howes 1991, Drake et al. 1984). These structures develop when young grey mullets are just a few centimetres long, which lead to a major ontogenetic dietary shift, from zooplanktivorous larvae to sedimentivorous juveniles and adults.

IRBio and Department of Animal Biology, Faculty of Biology, University of Barcelona, Avenida Diagonal 643, 08028 Barcelona, Spain.
Email: luis.cardona@ub.edu

The specialized feeding mode of grey mullets restrict most of them to ecosystems where the sediment is enriched with organic matter, like sheltered coastal habitats, estuaries and the lower reaches of rivers. In these habitats, the organic matter contents of the upper layer of the sediment ranges between 10 and 250 mg g^{-1} dry weight, as compared with less than 10 mg g^{-1} dry weight in most marine sediments (Blaber 1977, Odum 1970, Odum 1988, Mallo et al. 1993, Cardona et al. 2001). Nevertheless, a few species have adapted to colonize other habitats, such as the snouted mullet *Chaenomugil proboscideus* from the rocky shores of Central America and the bobo mullet *Joturus pichardi* and the mountain mullet *Agonostomus monticola* from fast flowing rivers in Central America and the Caribbean. This has been made possible by the abandonment of sedimentivory and the modification of the feeding organs to allow diets based on macroalgae, cyanobacteria films and benthic invertebrates.

The Feeding Apparatus

The mouth is terminal or sub-terminal in most grey mullet species (Thomson 1966, Drake et al. 1984), but in the genera *Agonostomus*, *Rhinomugil* and *Joturus* is inferior, being overhung by a fleshy snout (Thomson 1966). The premaxilla is protrusible in all species because the internal hook of the maxilla forces the premaxilla outwards (Thomson 1954). Lips are papillated in *Chaenomugil*, *Neomyxus*, *Oedalechilus*, and *Chelon*, although the latter is probably polyphyletic (Durand et al. 2012a,b).

Teeth are weak and present on the premaxillaries and mandibles of all species (Fig. 9.1). Furthermore, they may occur on the tongue (e.g., the yellow-eye mullet *Aldrichetta forsteri*, the snouted mullet *Chaenomugil proboscideus*, the bobo mullet *Joturus pichardi*, and the leaping mullet *Liza saliens*), the vomer and palatines (e.g., the mountain mullet *Agonostomus monticola*, the bobo mullet *Joturus pichardi,* the flat-tail mullet *Liza aregentea*, and the sand grey mullet *Myxus elongatus*) and the pterygoid bone (e.g., the yellow-eye mullet *Aldrichetta forsteri*, the thicklip mullet *Chelon labrosus*, and the leaping mullet *Liza saliens*) (Thomson 1954, Ebeling 1957, Drake et al. 1984, Correa Polo et al. 2012).

Grey mullet teeth are not anchored to the bone, but attached by fibrous strands. Teeth on the upper jaw are below the premaxillary bone and usually occur in two rows or bands. The primary teeth are external and are usually arranged in a uniserial row of large teeth (Ebeling 1957, Drake et al. 1984), although they form a compact band in species such as the snouted mullet *Chaenomugil proboscideus* (Ebeling 1957). Most species have a secondary band of smaller teeth internal to the primary row, composed of smaller teeth with a variable arrangement (Ebeling 1957, Drake et al. 1984). A toothless area of gum separates both groups of teeth. The mandibles usually contain only an external row of primary teeth, but an internal row

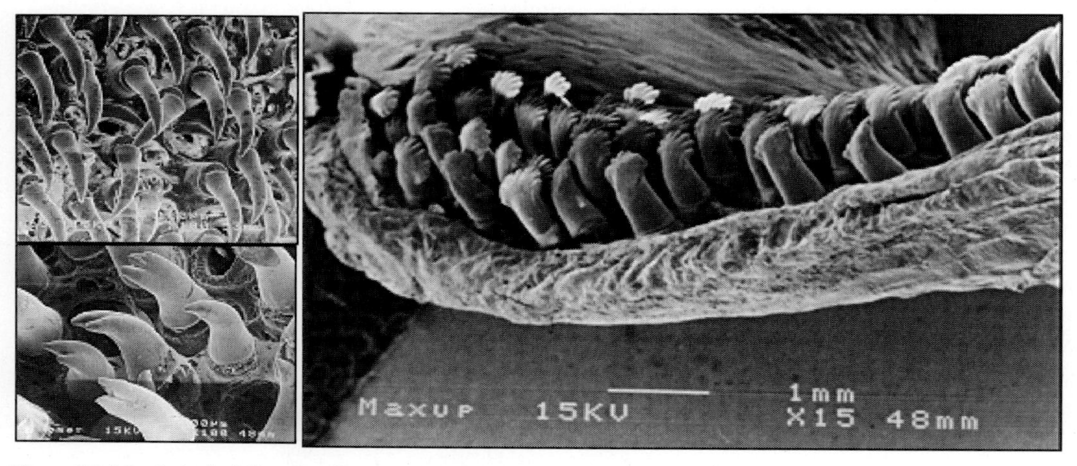

Figure 9.1. Morphological diversity of grey mullet teeth. Grey mullet teeth are usually unifid (top left), but bifid (bottom left), trifid or even serrated teeth (top right) exist. The primary teeth are external and are usually arranged in a uniserial row of large teeth, but several rows exist in the dentary bone of the bobo mullet *Joturus pichardi* (right). Reproduced with permission from INVEMAR; from Correa Polo et al. (2012).

of secondary teeth exists in some species (Ebeling 1957, Drake et al. 1984). Mandibular teeth are usually smaller than those on the premaxillary bone (Ebeling 1957, Drake et al. 1984). Teeth grow continuously and teeth size is correlated with body size, although teeth are replaced throughout the life of the individual (Ebeling 1957). The number of primary teeth in the upper jaw also increases with body size (Drake et al. 1984).

The tips of teeth in both jaws are usually yellowish and can be unifid, bifid, trifid or serrated and range in size from the minute, almost microscopic secondary teeth of most species to the bifid, trifid and serrated primary teeth found respectively in the snouted mullet *Chaenomugil proboscideus*, the striped mullet *Liza tricuspidens* and the bobo mullet *Joturus pichardi* (Ebeling 1957, Marais 1980, Drake et al. 1984, Correa Polo et al. 2012). Unifid, minute teeth are associated with sedimentivory (Ebeling 1957, Marais 1980), whereas large, relatively complex teeth with more than two tips are associated with rock scraping or the consumption of macroalgae.

Although the snouted mullet *Chaenomugil proboscideus* (bifid teeth) and the bobo mullet *Joturus pichardi* (serrated teeth) belong to the same lineage according to molecular phylogenies, they are more closely related to typical sedimentivores in the genus *Mugil* than to the striped mullet *Liza tricuspidens* (trifid teeth) (Durand et al. 2012a,b). Accordingly, teeth with more than one tip have probably evolved independently in at least two lineages as an adaptation to non-sedimentivorous diets. This hypothesis receives further support from the sedimentivory and the unifid teeth of the two most basal genera of grey mullets, *Rhinomugil* and *Sicamugil* (Johar and Singh 1981, Munshi et al. 1991, Durand et al. 2012b).

The presence of teeth on the tongue and the palate is thought to favour the capture of animal prey (Drake et al. 1984) and are present in those species that rely heavily on invertebrates, such as the yellow-eye mullet *Aldrichetta forsteri* and the mountain mullet *Agonostomus monticola* (Thomson 1954, Cruz 1987, Edgar and Shaw 1995, Aiken 1998, Phillip 1993, Torres-Navarro and Lyons 1999).

The posterior roof of the pharynx of most grey mullets has a filtering apparatus formed by toothed pads (Fig. 9.2), dorsal to the gill rakers (Ghazzawi 1935, Al-Hussaini 1947, Pillay 1953, Thomson 1954, Ebeling 1957, Capanna et al. 1974, Ching 1977, Drake et al. 1984, Harrison and Howes 1991, Guinea and Fernández 1992). This pharyngobranchial organ is missing from *Agonostomus* and *Joturus* and

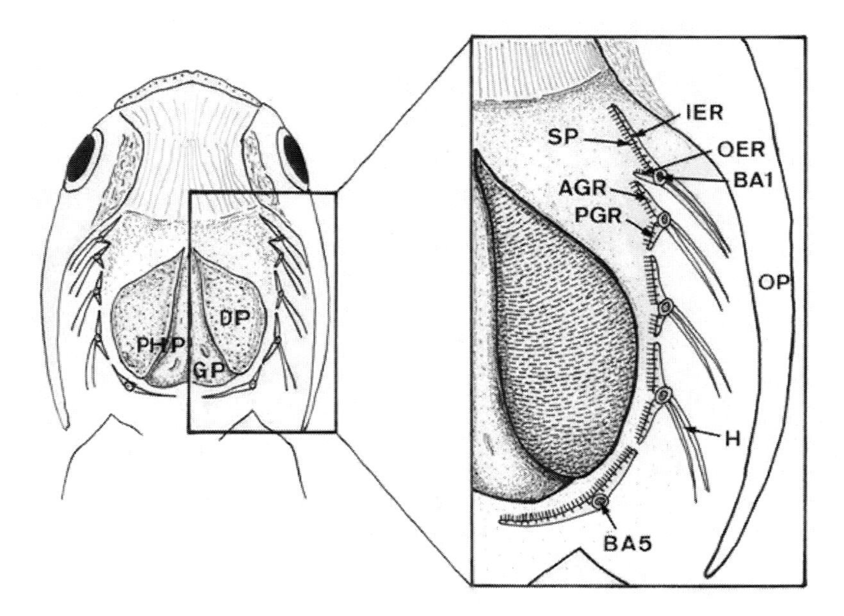

Figure 9.2. Position of the toothed pads of the pharyngobranchial organ and the branchial arches within the pharyngeal cavity. OP, Operculum; PHP, pharyngeal pads; DP, denticulated part; GP, gustatory part; SP, secondary processes; BA1, first branchial arch; BA5, fifth branchial arch; H, holobranchs, ARG, anterior gill raker; PGR, posterior gill raker; IER, inner edge of the gill raker; OER, outer edge of the gill raker. Reproduced with permission from John Wiley and Sons; from Guinea and Fernández (1992).

is poorly developed in *Cestraeus* and *Aldrichetta*, which led Harrison and Howes (1991) to consider them the most basal genera of grey mullet. However, molecular phylogenies identify *Sicamugil* and *Rhinomugil*, both with a well developed pharyngobranchial organ, basal to all other mullet (Durand et al. 2012a,b). Furthermore, the clade *Agonostomus* + *Joturus* is the system group of *Mugil*, and *Aldrichetta* is the sister group of *Liza argentea* (Durand et al. 2012a,b). If so, the loss of or simplification of the pharyngobranchial organ observed in *Agonostomus, Joturus, Cestraeus* and *Aldrichetta* is a derived character associated with diets different from sedimentivory (Thomson 1954, Cruz 1987, Edgar and Shaw 1995, Aiken 1998, Philip 1993, Torres-Navarro and Lyons 1999).

Pharyngeal teeth resemble those of the jaws, but are slender and more curved. The largest pharyngeal teeth are at the anterior edges of the pads (proximal-type teeth) and become progressively smaller towards the posterior edge (secondary type teeth) (Ghazzawi 1935, Capanna et al. 1974, Harrison and Howes 1991). The posterior region of the pad has been proposed to have a sensorial role (Capanna et al. 1974, Guinea and Fernández 1992) and is devoid of teeth in most species, although in large specimens of the yellow-eye mullet *Aldrichetta forsteri* the entire pad is studded with teeth (Thomson 1954). The spacing between pharyngeal teeth ranges from 120 μm in the leaping grey mullet *Liza saliens* to 58 μm in the thinlip mullet *Liza ramada* and spacing increases with body length (Drake et al. 1984, Guinea and Fernández 1992).

The gill rakers of the fifth pharyngeal arch form concavities on the floor of the pharynx to fit the toothed pads and the pharyngeal teeth span the space between the ventral surface of the dorsal pads and the dorsal surface of the gill rakers (Al-Hussaini 1947, Thomson 1954, Ebeling 1957, Capanna et al. 1974, Drake et al. 1984, Harrison and Howes 1991). The combined action of the pharyngobranchial organ with the gill rakers of the fifth pharyngeal arch is thought to allow grey mullets to reject the coarser particles in the sediment (Fig. 9.3). Nevertheless, primarily sedimentovorous species can swallow invertebrates as large as the larvae of chironomid midges, snails, polychaetes and amphipods (e.g., Bishop and Miglarese 1978, Hickling 1970, Drake et al. 1984, Cardona 2001, Dankwa et al. 2005), thus indicating that they have some control on the filtering action of the pharyngobranchial organ.

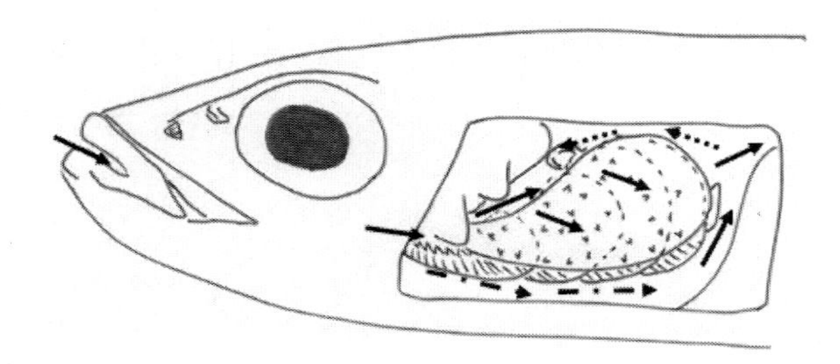

Figure 9.3. Diagrammatic representation of hypothesized water flow in the pharynx of grey mullet. Food-laden water (thick arrow), filtered and recirculated water (dashed arrow), and mucus (dashed/dotted arrow) around PBO. The surrounding gill arches are indicated by broken lines. Reproduced with permission from The Natural History Museum; from Harrison and Howes (1991).

The gill rakers (Fig. 9.4) form a second sieve thought to prevent smaller particles from escaping through the gill openings (Thomson 1954, Thomson 1966, Drake et al. 1984, Guinea and Fernández 1992). The average density of gill rakers along the branchial arches is five–seven gill rakers mm^{-1} (Drake et al. 1984), but the actual number of gill rakers on each branchial arch varies across species (Thomson 1954, Ching 1977, Drake et al. 1984, Guinea and Fernández 1992) and the spacing between gill rakers increases with body length and decreases from the first to the fifth branchial arch (Drake et al. 1984, Guinea and Fernández 1992). Furthermore, the morphology of the gill raker varies within and among branchial arches (Drake et al. 1984, Harrison and Howes 1991).

The mesh size of the branchial filter is further reduced by the existence of two series of microbranchial spines on the inner side of each gill raker (Drake et al. 1984, Guinea and Fernández 1992) although they are absent from the gill rakers of the first and second branchial arches of some species (Ching 1977). The average distance between microbranchial spines is 10 μm in the otomebora mullet *Chelon melinopterus*

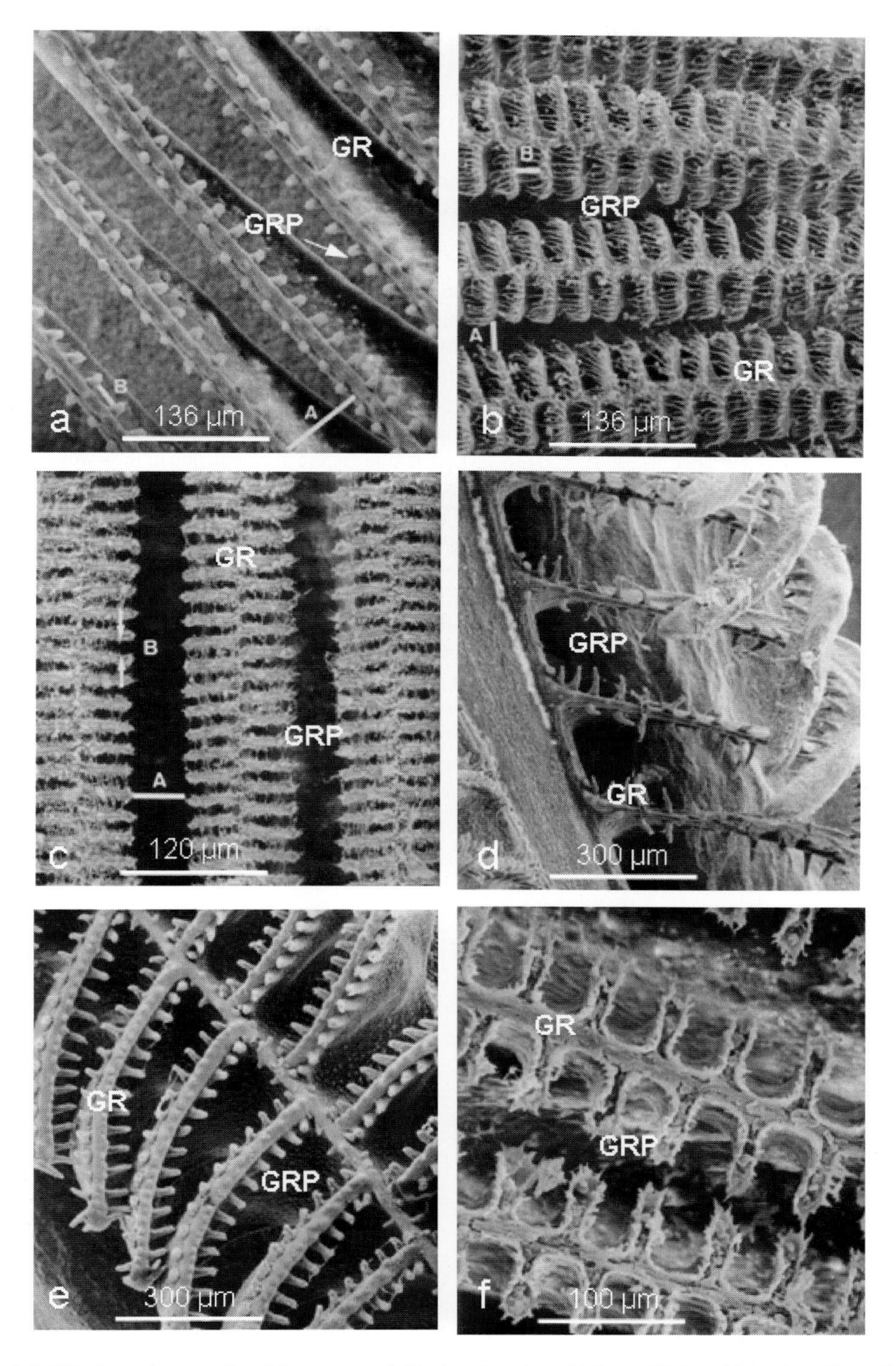

Figure 9.4. Gill rakers of grey mullets. Micrographs of gill rakers from first, third and fifth branchial arches of juvenile *Liza aurata* (a, b and c) and *Liza saliens* (d, e and f). For the first and third branchial arches, the images correspond to the anterior gill raker row. Posterior gill raker row can be partially seen in (d) and (e). GR, gill raker; GRP, gill raker secondary process; A, distance between gill rakers; B, distance between secondary processes. The figures above scale bars are expressed in µm. Reproduced with permission from John Wiley and Sons; from Guinea and Fernández (1992).

(Ching 1977), 24 μm in the thinlip mullet *Liza ramada* (Guinea and Fernández 1992) and 65 μm in the leaping mullet *Liza saliens* (Guinea and Fernández 1992). As with the spacing between gill rakers, the spacing between microbranchial spines of the same gill raker increases with body length (Drake et al. 1984, Guinea and Fernández 1992).

Gill rakers sometimes occur in the stomach of several species of grey mullets (Blaber 1975, Das 1977, Drake et al. 1984). This is thought to be an artifact caused by gill netting, which results in the abrasion, shedding and swallowing of the gill-raker filaments (Blaber 1975) or may be the result of regular moulting (Drake et al. 1984).

The oral cavity is lined by a stratified epithelium with abundant taste buds and mucous cells (Thomson 1954) and a fold of tissue from the posterior edge of the upper and lower pharyngeals forms a kind of valve at the rear of the pharynx so the mouth ends in a slit (Thomson 1966). The oesophagus is a straight, short, uniform muscular tube with a stratified epithelium (Ghazzawi 1935, Al-Hussaini 1947, Thomson 1954). The stomach of the grey mullet is divided into a thin-walled cardiac and a thick-walled pyloric stomach (Fig. 9.4). The cardiac stomach is a long pointed pouch leading off the dorsal side of the passage from the oesophagus to the pyloric stomach. The epithelium of the blind end of the cardiac stomach is columnar and devoid of gastric glands, which abound at the proximal end of the cardiac stomach (Ghazzawi 1935, Al-Hussaini 1947). There are several longitudinal folds running along the lumen of the cardiac stomach, one of them operating as a valve (Ghazzawi 1935). These folds allow the expansion of the cardiac stomach to store the food for later processing in the pyloric stomach, often called the gizzard (Ghazzawi 1935, Thomson 1954, Hickling 1970). The wall of the pyloric stomach has an extremely well developed muscular layer made of rings of smooth, non-striated muscle fibres (Ghazzawi 1935). The lumen is lined with a single type of columnar epithelial cells responsible for the production of a matrix that surrounds food particles and protects epithelial cells from abrasion (Ghazzawi 1935, Hickling 1970). Sand is thought to assist the gizzard in grinding plant material, a hypothesis supported by the poor growth of mullet raised with a microalgae-based pellet fed without access to sand (Sánchez et al. 1993).

There is a number of caeca in the ventral part of the pyloric stomach (Fig. 9.5), from two in the genus *Mugil* up to 10 in some *Liza* species and 22 in *Paramugil georgii*, with some individual variability in those species with more than two caeca (Ghazzawi 1935, Thomson 1954, Drake et al. 1984). The histology of

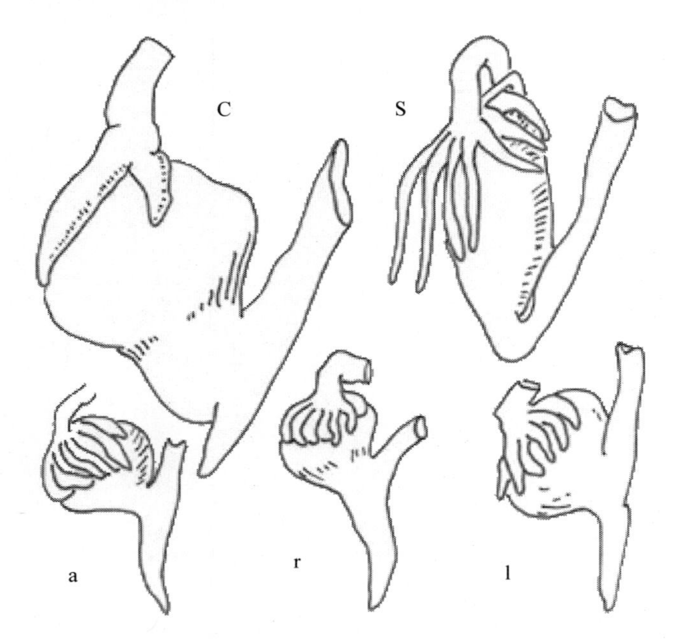

Figure 9.5. Stomach morphology in grey mullets. Golden mullet *Liza aurata* (a), flathead mullet *Mugil cephalus* (c), thicklip mullet *Chelon labrosus* (l), leaping *mullet Liza saliens* (s) and thinlip mullet *Liza ramada* (r). Note the variability in the number and size of pyloric caeca. Reproduced with permission from INSTM; from Heldt (1948).

the pyloric caeca is like that of the intestine and consists of a thin muscular wall, with an outer longitudinal muscle layer and an inner circular muscle layer, and a highly folded mucosa made of columnar epithelial cells and mucus-secreting goblet cells (Ghazzawi 1935, Albertini-Berhaut 1987). Enzymatic activity is usually higher in the caeca than in the intestine (Trellu et al. 1978, Establier et al. 1985) and hence enzyme production is its major role.

The intestine is arranged in several convolutions and its length is always larger than that of the body (Ghazzawi 1935, Thomson 1954, Albertini-Berhaut 1987). The whole intestine is wrapped in a peritoneal connective tissue which acts as a fat storage. The histology of the intestine is like that of the pyloric caeca, but folding is less complex (Ghazzawi 1935). There are no major external regions in the intestine, although the duodenum has a greater diameter and lacks mucus-secreting goblet cells (Ghazzawi 1935, Albertini-Berhaut 1987). Enzymatic activity is much higher in the intestine than in the stomach (Trellu et al. 1978). Proteins are absorbed along the whole intestine in adults, whereas lipid absorption takes place essentially in the foregut, as revealed by the presence of lipid droplets (Albertini-Berhaut 1988).

The gut of adult grey mullets is 1.5–4.6 times longer than the total body length (Table 9.1), which traditionally has been considered an adaptation to digest detritus and plant material (Hickling 1970, Odum 1970, Drake et al. 1984, Albertini-Berhaut 1987). This hypothesis is supported by the straight gut of larvae (Albertini-Berhaut 1987) and the much shorter gut of carnivorous fry as compared with that of sedimentivorous late juveniles and adults (Hickling 1970, Albertini-Berhaut 1987). Furthermore, the grey mullet reared in captivity on an animal based diet have shorter guts than wild specimens of the same species (Drake et al. 1984) and there is a large variability in the relative length of guts among populations

Table 9.1. Mean gut length relative to body length in several species of grey mullet.

Species	Mean relative gut length	Reference
Chelon labrosus	3.6	Drake et al. 1984
Chelon labrosus	3.9	Hickling 1970
Liza aurata	3.7	Al-Hussaini 1947
Liza aurata	3.9	Drake et al. 1984
Liza aurata	3.8	Albertini-Berhaut 1987
Liza dumerilii	1.8	Dankwa et al. 2005
Liza falcipinnis	4.0	Dankwa et al. 2005
Liza grandisquamis	2.0	Dankwa et al. 2005
Liza ramada	3.4	Albertini-Berhaut 1987
Liza ramada	2.9	Drake et al. 1984
Liza saliens	1.5	Drake et al. 1984
Liza saliens	1.5	Albertini-Berhaut 1987
Liza vaigiensis	3.2	Wijeyaratne and Costa 1990
Mugil bananensis	3.8	Dankwa et al. 2005
Mugil cephalus	4.6	Dankwa et al. 2005
Mugil cephalus	3.5	Drake et al. 1984
Mugil cephalus	5.5	Odum 1970
Mugil cephalus	3.3	Odum 1970
Mugil cephalus	3.2	Hiatt 1944
Mugil curema	4.0	Dankwa et al. 2005
Valamugil buchanani	3.8	Wijeyaratne and Costa 1990
Valamugil cunnesius	2.8	Wijeyaratne and Costa 1988

of some species thought to be connected to the relative importance of detritus and algae in the diet (Odum 1970, Collins 1981, Drake et al. 1984).

Other factors are however, probably more important than the relative length of gut when comparing the capacity of different species to assimilate plant material. For instance, the gut of the leaping mullet *Liza saliens* is only 1.5 times longer than total body length and accordingly has been consider the most carnivorous of the species inhabiting Mediterranean coastal lagoons (Drake et al. 1984, Albertini-Berhaut 1987). However, stable isotope analysis has revealed that this species derives 68% of the assimilated nutrients from plant resources (Koussoroplis et al. 2011). Conversely, the gut of the thicklip mullet *Chelon labrosus* is 3.6–3.9 times longer than total body length (Hickling 1970, Drake et al. 1984) and this species derives only 45% of the assimilated nutrients from plant material (Koussoroplis et al. 2011). This is probably because the activity of the enzyme α-amylase, responsible for the hydrolysis of starch into sugars, is much higher in the intestine of the leaping mullet *Liza saliens* than in that of the thicklip mullet *Chelon labrosus* (Establier et al. 1985) and hence has a higher capacity to assimilate plant material. Accordingly, enzymatic activity may be more relevant for the assimilation of plant material than the relative length of gut. Unfortunately, very little research has been conducted on the enzymatic activity of the gut of grey mullets. Gut pH may also play role in the capacity to assimilate plant material, as it ranges from 3.5 in the sicklefin mullet *Liza falcipinnis*, to 7.8 in the grooved mullet *Liza dumerili* and 8.5 in the flathead mullet *Mugil cephalus* and the white mullet *Mugil curema* from western Africa (Payne 1978) and they differ in trophic level (Faye et al. 2011). Again, insufficient information has been published to reach any conclusion.

The relative length of gut may also increase with body size in some species (Odum 1970, Hickling 1970, Collins 1981), but this trend is not consistent across populations (Odum 1970, Drake et al. 1984, Blay 1995, Dankwa et al. 2005). Furthermore, diet seldom changes with body size in juveniles and adult grey mullets (Hickling 1970, Blaber 1976), although an increased consumption of plant material associated with the lengthening of the gut has been reported in the mountain mullet *Agonostomus monticola* (Cotta-Ribeiro and Molina-Ureña 2009, Torres-Navarro and Lyons 1999, Cotta-Ribeiro and Molina-Ureña 2009).

Methods for Studying the Diet of Grey Mullets

Ideally, dietary studies are based on the identification to the species level of the prey items occurring into the stomach contents of the predator and the assessment of the relative importance of each prey species to the diet according to its weight. However, this is seldom possible when studying the diet of grey mullets, because of the large amount of inorganic particles present in the stomach contents, the wide diversity of size of the prey consumed, and the impossibility of gravimetric studies when tiny prey are considered.

The stomach content of grey mullets is usually a fine, compact paste, without macroscopic details. When inspected under a dissecting microscope at 10x or 20x, only the largest, coloured prey such as the red larvae of chironomid midges and green filamentous algae are obvious amidst a multitude of sand grains, mud aggregates and fine detritus. Transparent prey such as copepods and nematodes are hardly visible in this setting, and single-celled alga other than the largest diatoms pass unnoticed. On the other hand, inspecting the sample at higher magnification (40x or 100x) will reveal a high diversity and abundance of single-celled algae and perhaps some copepods, but hardly any larger prey. Thus, there is not a single method that allows the simultaneous detection and identification of all the prey species occurring in the stomach contents of grey mullets.

To cope with this problem, samples are first mixed with 70% ethanol (1:3 sample ethanol ratio) and stirred. A liquid sample is collected immediately after stirring and observed at 40x–100x for the identification of microscopic algae. Each individual cell is identified to the lowest taxonomic level and counted. The sample is stirred a second time and another liquid sample is collected and observed under the microscope. The process is repeated several times and the accumulated number of species counted. When the rarefaction curve stabilizes, a complete description of the microalgal assemblage in the sample has been achieved. At that point, the sediment sample preserved in ethanol is stained with Bengal Rose to make evident the transparent invertebrates, although a small sub-sample of the stomach content in ethanol may be preserved unstained for further work with microalgae. The stained sample is then examined under a dissecting microscope at 10x or 20x and all the invertebrates are counted. The presence of filamentous algae, sand and detritus is recorded. If the Petri dish has an underlying lattice, the cover of sand, detritus and each species of macroalgae can be assessed easily as percentage cover.

After which, the researcher has a list of microalgae species, a list of invertebrates and a list of macroalgae. The relative importance of each microalgal, invertebrate and macroalgal species can be computed, but only for that particular assemblage, so only the frequency of occurrence can be calculated (Albertini-Berhaut 1973, Payne 1976, Blaber 1976, 1977, Drake et al. 1984, King 1988, Cardona 2001). Frequencies of occurrence are non-additive, although some researchers manage to publish frequency data in this way (Blanco et al. 2003), but can be used to calculate trophic overlap and niche breadth using the Morisitia-Horn index and a modification of Gladfelter-Johnson index respectively (Cardona 1991, Cardona 2001).

Stable Isotopes and Fatty Acids

Microscopic analysis of stomach contents produces a biased image of grey mullet diet, as heterotrophic microbes are seldom detected and never quantified accurately. Moriarty (1976) was the first to assess the contribution of bacteria to the diet of grey mullets using taxa specific chemical markers, an approach that has increased in popularity during the past two decades due to the improvement of analytical methods, cost reduction and increasing availability in a number of laboratories.

The rational for stable isotope analysis is that the isotopic composition of an animal depends on that of its food sources, though a small difference usually exists between the isotopic values of prey and predator (De Niro and Epstein 1978, 1981, Post 2002, Martínez del Rio et al. 2009). The ratio $^{13}C/^{12}C$ (expressed as $\delta^{13}C$) varies between different types of primary producers, thus allowing discrimination between terrestrial and marsh plants, macroalgae, submerged vascular plants and phytoplankton. On the other hand, the ratio $^{15}N/^{14}N$ (expressed as $\delta^{15}N$) increases with each trophic level due to the preferential excretion of the lighter isotope and is, therefore, used to assess trophic level. Accordingly, stable isotopes of carbon and nitrogen are the most widely used, but the stable isotope ratios of sulphur and other elements are also useful.

Stable isotope analysis can be performed on any tissue, but white dorsal muscle is commonly selected (Carlier et al. 2007, Lin et al. 2007, Quan et al. 2007, Vizzini and Mazzola 2008, Lebreton et al. 2011, Faye et al. 2011, Giarrizzo et al. 2011). Nevertheless, turnover rate varies dramatically among tissues, from a few weeks in the liver to several months in muscle (Buchheister and Latour 2010), so tissue selection has to be based on the relevant time-window to be investigated. This is critical in species like grey mullets, which move between habitats with contrasting stable isotope baselines, as inadequate tissue selection may result in major mistakes when interpreting the stable isotope ratios in the muscle of recent immigrants. Lipids are usually depleted in ^{13}C in comparison with protein, because of a strong negative enzymatic selection during lipid synthesis. Accordingly, lipids have to be removed from the samples prior to analysis or stable isotope ratios have to be corrected mathematically (Post et al. 2007).

Stable isotope ratios can be used simply as habitat tracers (Fig. 9.6), but also to infer diet composition by means of mixing models, algorithms that use a mass balance approach to calculate the contribution of dietary sources to the nutrients actually assimilated by the predator (Parnell et al. 2010). Mixing models produce reliable results only when the fractionating factor from diet to predator is known and if large differences exist between the isotope ratios of the considered sources. Fractionation factors are tissue specific and have to be assessed experimentally. Furthermore, mixing models are extremely sensitive to the stable isotope ratios of the potential food sources included into the analysis, so accurate identification of potential prey by means of stomach contents analysis is still required.

Fatty acids are a second major group of chemical markers used in dietary studies. They are synthesized by plants and bacteria and transferred to consumers with only slight modifications. As a consequence, the fatty acid profiles of consumers integrate dietary information during several months and inform about the major sources of lipids in the diet. Odd fatty acids with 15 and 17 C are bacterial markers, including cyanobacteria, the EPA/DHA ratio is informative about the relative importance of diatoms over dinoflagellates, and fatty acids with more than 24 carbons are markers of terrestrial vegetation

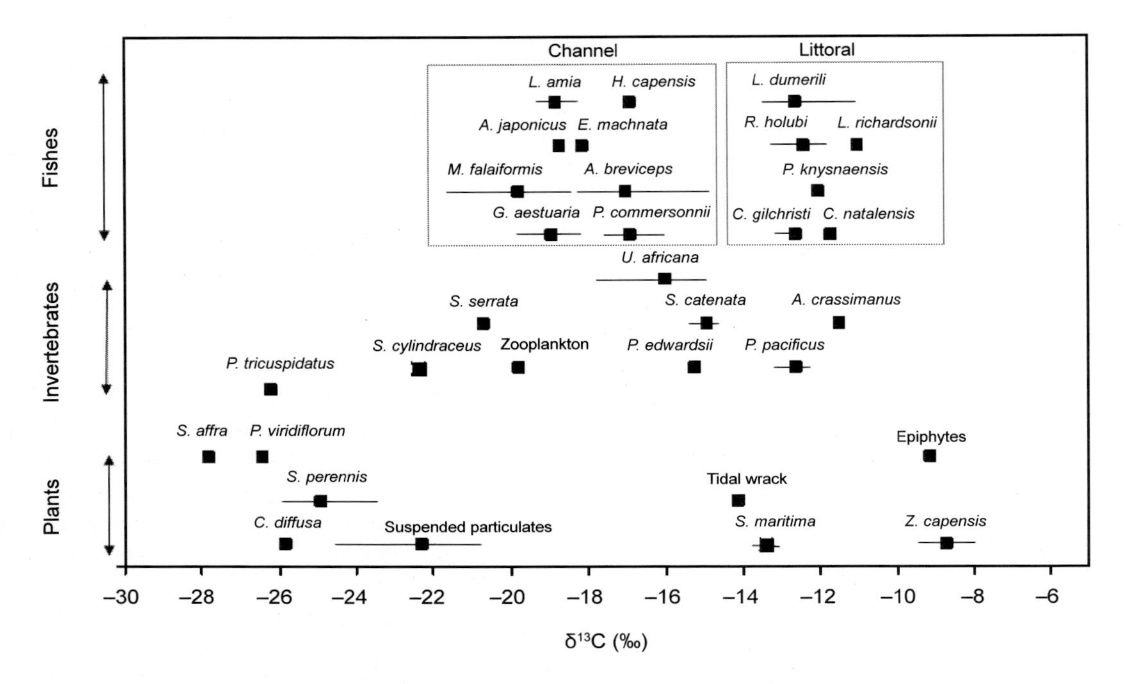

Figure 9.6. Identification of primary food sources for grey mullets through stable isotope analysis. The $\delta^{13}C$ values of the fishes inhabiting the Kariega estuary (South Africa) reveal that the grooved mullet *Liza dumerili* and the South African mullet *Liza richardsoni* forage primarily in shallow littoral areas and avoid the channels. Furthermore, they differ in diet, with the South African mullet *Liza richardsoni* more reliant on epiphytes and the seagrass *Zostera capensis* than the grooved mullet *Liza dumerili*. Reproduced with permission from Elsevier; from Paterson and Whitfield (1997).

(Sargent 1997, Li et al. 1998, Patil et al. 2007). Furthermore, quantitative fatty acid analysis, or QFASA, allows estimating predator diets if the fatty acid profiles of potential dietary sources and the calibration coefficients are known (Iverson et al. 2004).

Fry Diet and Ontogenic Dietary Shifts

The mouth of the larva of the flathead mullet *Mugil cephalus* opens three days after hatching (Nash et al. 1974), that of the thinlip mullet *Liza ramada* four or five days after hatching (Mousa 2010) and that of the thicklip mullet *Chelon labrosus* five to six days after hatching (Cataudella et al. 1988, Zouiten et al. 2008). The mouth is terminal and adapted to capture swimming prey in larvae and postlarvae, the gut is straight, the walls of the pyloric stomach are not thickened and pyloric caeca are absent (Drake et al. 1984). The pharyngobranchial organ starts to develop in 10 mm SL larvae (Harrison and Howes 1991) and teeth and the pharyngobranchial organ are totally formed in 40 mm SL fish (Harrison and Howes 1991). Gut folding develops as postlarvae grow larger than 30 mm TL (Thomson 1966, Albertini-Berhaut 1987). Trypsin and amylase activity are detected from hatching, but the diversity and the activity of digestive enzymes increases dramatically with size (Trellu et al. 1978, Zouiten et al. 2008).

Nothing is known about the diet of larvae in the wild, but in captivity they consume diatoms and zooplankton (Nash et al. 1974). Postlarvae measuring 10–30 mm TL, usually referred to as fry, are primarily zooplanktivores in all the species studied to date, including the abu mullet *Liza abu* (Ahmad and Hussain 1982), the thicklip mullet *Chelon labrosus* (Albertini-Berhaut 1973, Tosi and Torricelli 1988, Gisbert et al. 1995, 1996), the largescale mullet *Liza macrolepis* (Luther 1962), the thinlip mullet *Liza ramada* (Albertini-Berhaut 1973, Ferrari and Chieregato 1981, Tosi and Torricelli 1988, Gisbert et al. 1995, 1996), the golden mullet *Liza aurata* (Albertini-Berhaut 1973, Ferrari and Chieregato 1981, Gisbert et al. 1995, 1996), the leaping mullet *Liza saliens* (Albertini-Berhaut 1973, Ferrari and Chieregato 1981, Tosi and Torricelli 1988, Gisbert et al. 1995, 1996), the greenback mullet *Liza subviridis* (Chan and Chua 1979) and the flathead mullet *Mugil cephalus* (Zismann et al. 1975, Tossi and Torricelli 1988, Eggold and Motta 1992, Gisbert et al. 1995, 1996).

Harpacticoid and cyclopoid copepods are the main prey of grey mullet fry in brackish environments, whereas cyclopoid copepods, cladocerans and rotifers prevail in oligohaline and freshwater sites (Fig. 9.7). Most prey range from 700 to 1000 μm in length (Fig. 9.8).

Fry of white mullet *Mugil curema* and flathead mullet *Mugil cephalus* (20–40 mm SL) may consume marine snow, but prefer zooplankton if available and do not grow well with a marine snow diet (Larson and Shanks 1996). Similarly, the growth rate of the fry of the thin lip mullet *Liza ramada* 25 mm TL increases with the protein content of the diet (El-Sayed 1991), but fish measuring 50 mm SL require only 25% of protein in the diet and growth is not improved with a higher protein contents (Papaparaskeva-Papoutsoglou and Alexis 1986). Likewise, the proportion of proteins in the stomach contents of the flathead mullet *Mugil cephalus*, grooved mullet *Liza dumerili*, South African mullet *Liza richardsoni* and *Liza tricuspidens* inhabiting South African estuaries decreases with body length from 30 mm TL fry to late juveniles > 15 cm TL (Marais and Erasmus 1977).

Young mullet shift to a benthic diet when 20–30 mm TL, although the exact size of the ontogenic dietary shifts is species specific (Luther 1962, Albertini-Berhaut 1973, Zismann et al. 1975, Ferrari and Chieregato 1981). The amount of plant material and detritus increases steadily and the fish larger than 40–50 mm TL have an adult diet (Luther 1962, Ferrari and Chieregato 1981, Eggold and Motta 1992). The fry of the flathead mullet *Mugil cephalus* from Sri Lanka start consuming sand grains when 20 mm TL and sand occurs in the stomach all specimens larger than 50 mm TL (De Silva and Wijeyaratne 1977). The consumption of plant material follows a similar pattern and diatoms occur in the stomachs of specimens larger than 20 mm TL from Sri Lanka and Spain (De Silva and Wijeyaratne 1977, Drake et al. 1984, Gisbert

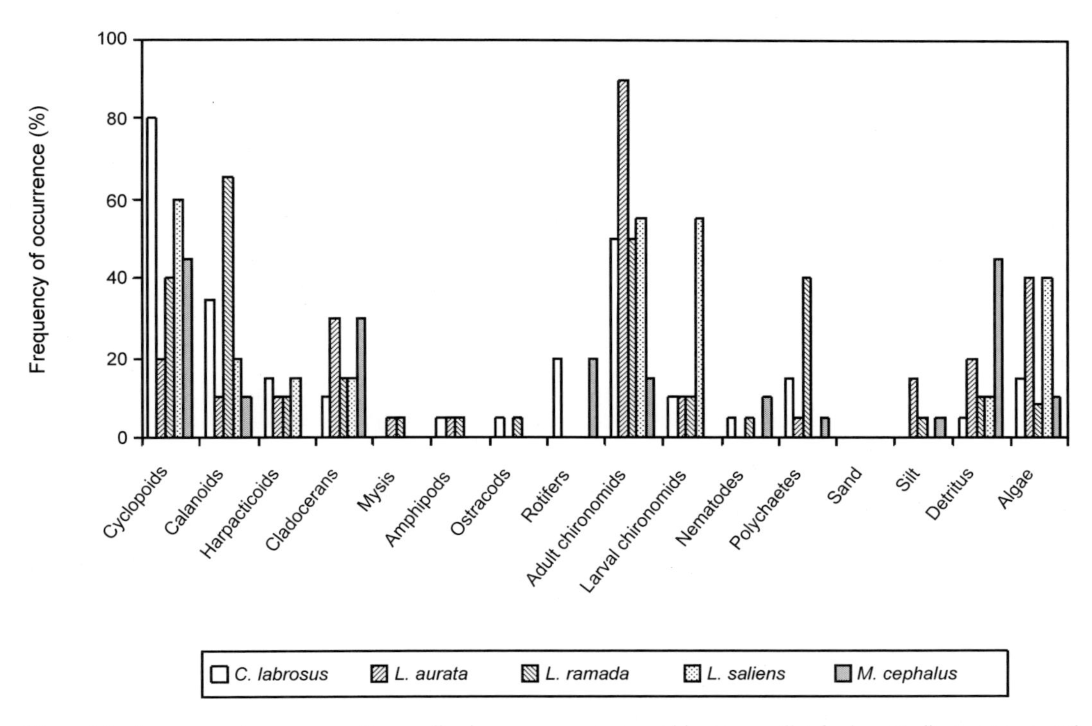

Figure 9.7. Frequency of occurrence of generalized prey groups consumed by grey mullet fry in a Mediterranean coastal lagoon. Note the high prevalence of cyclopoid and harpacticoid copepods in all the species and the absence of sand or silt in the stomach contents, both characteristic of fry diet in brackish lagoons. Consumption of adult chironomids was high in that particular locality, but is not a common characteristic. Reproduced with permission from John Wiley and Sons; from Gisbert et al. (1995).

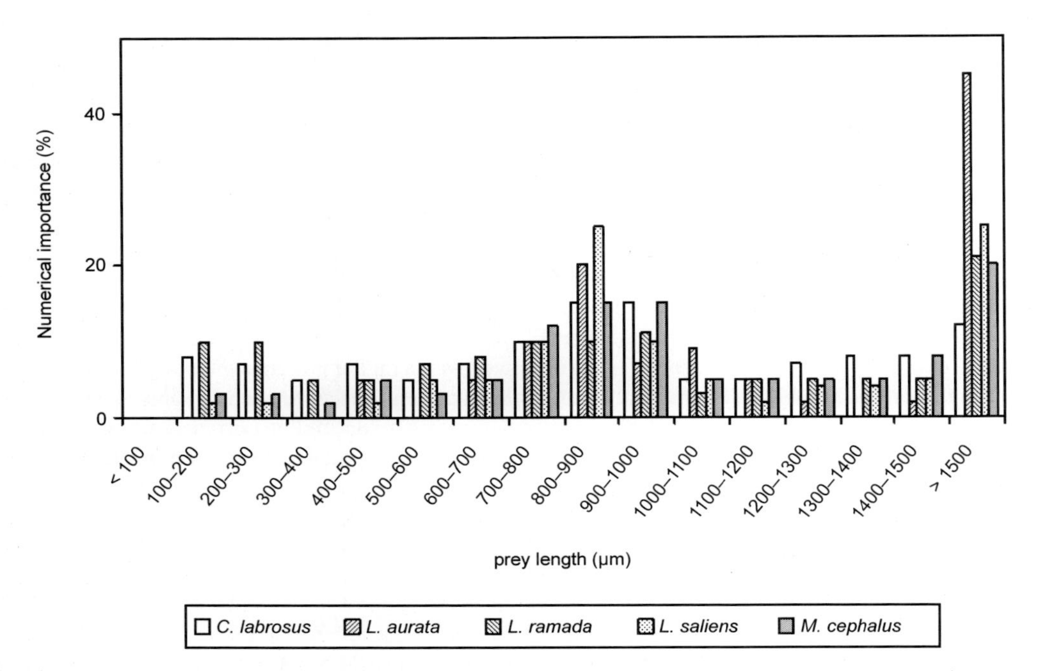

Figure 9.8. Size distribution of animal prey found in the stomachs of mullet fry from a Mediterranean coastal lagoon. Note that most species capture primarily prey ranging 700–1000 µm in length, but there are relevant differences in the total range. Bar patterns denote different sampling months. Reproduced with permission from Elsevier; from Gisbert et al. (1996).

et al. 1995, 1996). The shift from a planktonic to a benthic diet is demonstrated in the leaping grey mullet *Liza saliens* by changes in the fatty acid profile, as fatty acids associated with dinoflagellates (22:6n3) prevail over those associated with diatoms (20:5n3) in fry shorter than 30 mm TL and the opposite is true in larger fry and adults (Koussoroplis et al. 2010). Fry of the squaretail mullet *Ellochelon vaigiensis* < 20 mm consume primarily calanoid and cyclopoid copepods and a little sand, detritus and microalgae. Proportions change as fish grow, without any obvious break-point, and the stomach contents of fingerlings longer than 8 cm are dominated by sand, detritus, microalgae and polychaetes, with just a few copepods (Hajisamae et al. 2004).

Juvenile and Adult Diets

The morphology of the feeding apparatus and the gut of juvenile grey mullets larger than 5–8 cm TL is similar to that of the adults (Ebeling 1957, Albertini-Berhaut 1987) when they have completed the shift from the zooplanktophagous diet typical of early developmental stages to the sedimentivorous diet typical of adults (Albertini-Berhaut 1973). After that shift, the stomach contents of juvenile and adult grey mullets is usually a mixture of sand, detritus, microphytobenthos, green macroalgae, infauna, zooplankton and benthic macrofauna and no major dietary changes are usually recorded during the life of most grey mullet species (Blaber 1976, 1977, Hajisamae et al. 2004, Platell et al. 2006), although in some species, juveniles shorter than 9 cm do not consume filamentous green algae (Thomson 1954). This is in sharp contrast to the situation in most bony fishes, whose diets change throughout life. This is because mouth gape diameter is a major factor determining the diet of most bony fishes (Karpouzi and Stergiou 2003), but not for grey mullets. Actually, grey mullets depart from the general equations describing the relationship between body size and prey size or trophic level for bony fishes, as they usually consume prey much smaller and at a lower trophic level than expected for fishes their size (Edgar and Shaw 1995, Akin and Winemiller 2008).

Grey mullets may pump-filter phytoplankton, scrape microalgae and filamentous algae from rocks and submerged macrophytes, and capture invertebrates, but sedimentivory is the primary feeding mode of most species in most situations. Grey mullets usually forage in unvegetated areas (Blaber et al. 1989, Thomas and Connolly 2001), as dense stands of submerged macrophytes may limit their capacity to access the sediment. Furthermore, grey mullet abundance usually increases after the collapse of macrophyte beds (Whitfield 1986, Sheppard et al. 2011). Certainly, submerged macrophytes may be a major source of carbon for estuarine grey mullets (Paterson and Whitfield 1997, Abrantes and Sheaves 2008), but microphytobenthos is often a much better food source than the detritus from vascular plants (see below) and hence the replacement of macrophyte beds by bare sediment is profitable for most grey mullet species. Nevertheless, the white mullet *Mugil curema* is more abundant in seagrass beds than in unvegetated areas in the estuaries of the Gulf of Mexico (Castillo-Rivera et al. 2002) and the same is true for the tiger mullet *Liza argentea* in Australian salt marshes (Thomas and Connolly 2001).

Sources of Organic Matter in the Sediment

The organic matter present in the upper layer of sediments is a mixture of detritus, microphytobenthos and infauna (meiofauna and macrofauna). Detritus usually represents the bulk of sediment organic matter, and detritus availability and consumption by grey mullets may increase during the rainy season as a result of increased freshwater runoff and the collapse of submerged macrophytes (Luther 1962, Payne 1976, Blaber 1977, Cardona 2001). However, the nutritional quality of detritus is highly variable and often low. Conversely, microphytobenthos and infauna are usually scarcer, but often more relevant as food sources for grey mullets.

Detritus from vascular plants (terrestrial plant, marsh plants, mangroves and seagrasses) is the primary source of organic matter for species like the so-iny mullet *Liza haematocheila* according to stable isotope analysis (Quan et al. 2007) and the relative abundance of long chain fatty acids (Köse et al. 2010). On the other hand, detritus from vascular plants can be hard to digest for most species, including the white mullet *Mugil curema*, the largescale mullet *Liza macrolepis*, the thinlip mullet *Liza ramada*, the golden mullet *Liza aurata* and the thicklip grey mullet *Chelon labrosus* (Lin et al. 2007, Giarrizzo et al. 2011, Koussoroplis et al. 2010, 2011, Lebreton et al. 2011). This is because detritus is rich in cellulose and other complex carbohydrates and grey mullet lack the enzymes required to digest them and there is no experimental evidence that the microbial flora of the gut may play a role in the digestion of complex carbohydrates, as suggested (Odum 1970). Accordingly, the flathead mullet *Mugil cephalus* has been reported to prefer microphytobenthos to detritus (Odum 1970).

Grey mullets can however, exploit the microbial flora growing on decaying detritus, and bacteria represent 15–30% of the organic carbon in the stomach contents of the flathead mullet *Mugil cephalus* from seagrass beds in Australia (Moriarty 1976). Recent research using fatty acids has confirmed the relevance of bacteria in the diet of this species, as proven by the high levels of fatty acids with an odd number of carbons observed in individuals from the North-western Atlantic (Recks and Seaborn 2008), the Black Sea (Özogul and Özogul 2007) and the Mediterranean Sea (Şengör et al. 2003). Conversely, the scarcity of long chain fatty acids in the profile of the same samples confirms that macrophyte detritus is not a relevant food source itself (Şengör et al. 2003, Özogul and Özogul 2007, Recks and Seaborn 2008). The so-iny mullet *Liza haematocheila* introduced into the Black Sea is also enriched in fatty acids with an odd number of carbons and hence is highly dependent on bacteria as a source of lipids (Köse et al. 2010). Conversely, bacteria had only a minor role as a lipid source for the thinlip grey mullet *Liza saliens* from Mediterranean coastal lagoons (Koussoroplis et al. 2010).

Microphytobenthos is the second major component of the organic matter present in the sediment and is usually dominated by diatoms and cyanobacteria. Most grey mullet species digest diatoms easily (Odum 1970, Payne 1976), but cyanobacteria are more difficult to digest (Odum 1970). Stable isotope analysis has revealed that microphytobenthos is the major source of assimilated carbon and nitrogen for species such as the largescale mullet *Liza macrolepis* (Lin et al. 2007), the bluetail mullet *Valamugil buchanani* (Abrantes and Sheaves 2008) and the white mullet *Mugil curema* (Giarrizzo et al. 2011). A high EPA/DHA index reveals the importance of diatoms as a source of lipids for the flathead mullet

Mugil cephalus in the North-western Atlantic (Recks and Seaborn 2008) and the Black Sea (Özogul and Özogul 2007), but not in a coastal lagoon in the Aegean Sea (Şengör et al. 2003). Likewise, diatoms are also more relevant than dinoflagellates and bacteria as lipid sources for the thinlip grey mullet *Liza saliens* from Mediterranean coastal lagoons, a species highly reliant on sediment organic matter (Koussoroplis et al. 2010, 2011).

Benthic animals are the third major component of the organic matter present in the sediment and can be sub-divided into meiofauna, macrofauna and epifauna. The meiofauna includes those species that live among the particles of the substratum (ostracods, harpacticoid copepods and nematods), the infauna includes those species living in burrows (polychaetes, the larvae of chironomid midges, some amphipods and clams) and the epifauna includes species living on the surface (snails and some amphipods). All these groups are commonly reported from the stomach contents of most grey mullet species (Fig. 9.9), which accordingly are often reported as omnivores in the literature (e.g., Fagade and Olaniyon 1973, Blaber 1976, 1977, Drake et al. 1984, Cardona 2001). Consuming as much animal material as possible was hypothesized to be advantageous for grey mullets due to higher assimilation efficiency (Blaber 1976, 1977). However, careful studies comparing the abundance of potential prey in the sediment and stomach contents has demonstrated that the otembora mullet *Chelon melinopterus*, consistently avoided infauna, but strongly select detritus and microphytobenthos (Ching 1977).

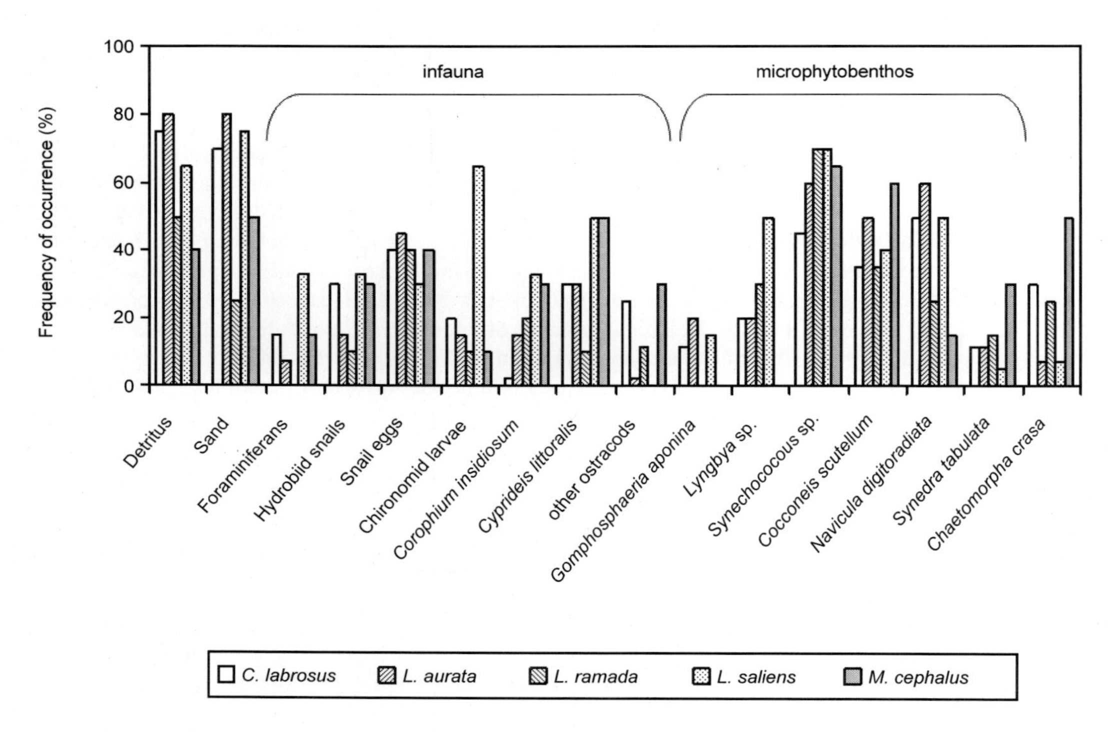

Figure 9.9. Frequency of occurrence of prey groups consumed by juvenile and adult grey mullet in a Mediterranean coastal lagoon. From top to bottom, thicklip mullet *Chelon labrosus* (a), golden mullet *Liza aurata* (b), thinlip mullet *Liza ramada* (c), leaping mullet *Liza saliens* (d) and flathead mullet *Mugil cephalus* (e). Note the high prevalence of detritus, sand, cyanobacteria (*Lyngbya* sp. and *Synechococcus* sp.), diatoms (*Cocconeis scutellum* and *Navicula digitoradiata*) and ostracods (*Cyprideis littoralis*) in the diet of the five species. Snails and their eggs, amphipods (*Corophium insidiosum*) and the larvae of chironomid midges were consumed frequently only by some species. Bars correspond to different seasons. Reproduced with permission from John Wiley and Sons; from Cardona (2001).

Sediment Processing and Particle Selection

Earlier researchers thought of grey mullets as selective feeders with a great capacity to concentrate fine particles from the sediment and hence increase the organic matter content of their diets (Al-Hussaini 1947, Pillay 1953, Thomson 1954, 1966, Odum 1968a, 1970). However, the degree of sediment processing and the capacity of grey mullets to concentrate fine inorganic particles are highly variable, as demonstrated by the diversity of granulometric profiles reported for sympatric species (Masson and Marais 1975, Blaber 1976, 1977, Blaber and Whitfield 1977, Mariani et al. 1987, Osorio Dualiby 1988, Dankwa et al. 2005). This is because of the large differences in the mesh size of the pharyngeal filter of grey mullets (Ching 1977, Drake et al. 1984, Harrison and Howes 1991, Guinea and Fernández 1992) and also because grey mullets may adopt different processing strategies depending on the species and the environment.

Sometimes, no major differences exist in the granulometric profiles and the prey assemblages of the sediment and those of grey mullet stomach contents (Fig. 9.10), as reported for the flathead mullet *Mugil cephalus* in one Florida site (Eggold and Motta 1992) and in two Mediterranean sites (Mariani et al. 1987). This is probably the result of bulky sedimentivory, i.e., the swallowing of mouthfuls of sediment after minimal processing, except for the removal of the coarser inorganic particles. Alternatively, the average diameter of the particles in the stomach contents is smaller (Fig. 9.10) and the animal prey are scarcer than in the sediment available, as reported for the flathead mullet *Mugil cephalus* at several sites along the east coast of North America (Odum 1968a, 1970), the otembora mullet *Chelon melinopterus* in Malaysian mangroves (Ching 1977) and the thinlip mullet *Liza ramada* in the Mediterranean Sea and the adjoining Atlantic Ocean (Mariani et al. 1987, Almeida 2003). This is probably the result of extensive processing of the sediment, which results in the concentration of those particles smaller than the spacing between pharyngeal teeth and larger than the spacing between the microbranchial spines of the gill rakers. In this scenario, animal prey are scarce because they probably escape when the larger particles are voided. The processing strategy of most grey mullet species is probably intermediate between these two extremes.

The capacity of each species to retain and concentrate the finest particles in the sediment depends on the spacing between the microbranchial spines of the gill rakers, although grey mullets may retain some particles smaller than the mesh size of the branchial filter (Eggold and Motta 1992, Cardona 1996). This is probably because of the presence of mucus, which reduces the functional mesh size (Ruberstein and Kohel 1977). The otembora mullet *Chelon melinopterus* is one of the best concentrators of fine particles, as the distance between microbranchial spines is 10 μm and the average diameter of the particles from the stomach contents is 14 μm (Ching 1977). Likewise, the distance between microbranchial spines in the thinlip mullet *Liza ramada* is 24 μm (Guinea and Fernández 1992) and the threshold for positive selectivity of microalgae is 20 μm (Cardona 1996). Accordingly, it can concentrate silt and fine sand and the granulometric profile of the stomach content is usually enriched with particles ranging from 50 to 100 μm in diameter as compared with the sediment available (Mariani et al. 1987, Almeida 2003). Conversely, the distance between the microbranchial spines of the leaping grey mullet *Liza saliens* is 65 μm (Guinea and Fernández 1992) and the stomach content is depleted of particles less than 63 μm and enriched with particles ranging from 250 to 1000 μm when compared to the available sediment (Mariani et al. 1987).

The spacing between the elements of the branchial filter increases allometrically with body length (Drake et al. 1984, Guinea and Fernández 1992, Eggold and Motta 1992), which results in a reduced capacity to retain the finest particles and concentrate fine particulated organic matter as grey mullet grow. This is shown well by the proportion of particles less than 63 μm in the stomach contents of the grooved mullet *Liza dumerili* in the Swartkops estuary (South Africa): 78.6% for fishes between 6 and 10 cm TL, 57.5% for fish from 10 to 15 cm TL and 50.4% for fish larger than 15 cm TL (Masson and Marais 1975). Accordingly, the energy density of the stomach contents of fish less than 15 cm TL was 5.3 times higher than that of the sediment, but only 2.5 times higher for fish larger than 15 cm TL (Marais 1980).

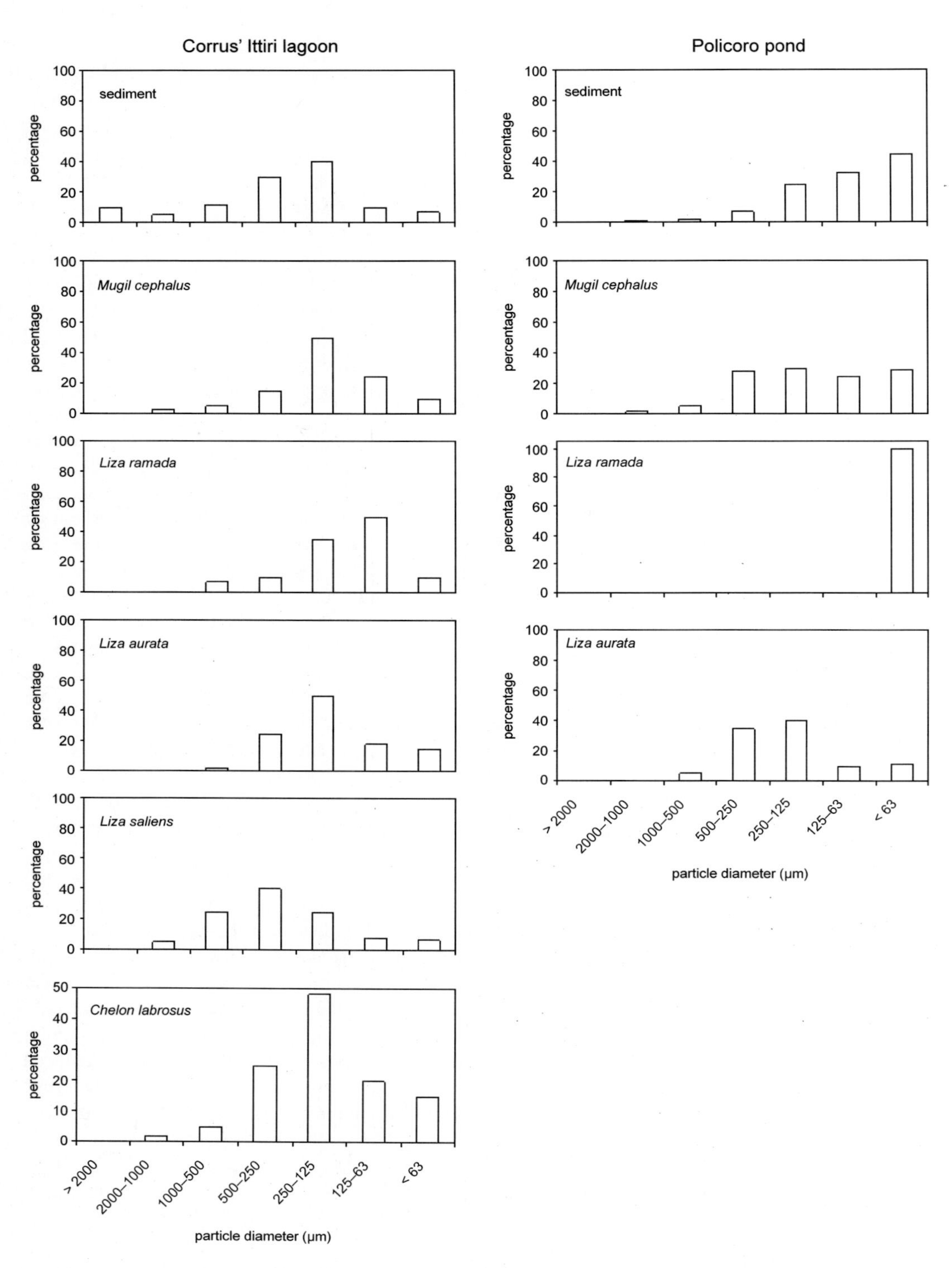

Figure 9.10. Size-frequency distribution of inorganic particles in the sediment (top panel) and the intestine of grey mullets for two Mediterranean sites. Reproduced with permission from Elsevier; from Mariani et al. (1987).

Concentrating the finest particles significantly increases the amount of organic matter and the energy density of the diet. For instance, particles less than 125 μm represent 95% of the stomach contents of the flathead mullet *Mugil cephalus* and 65% of that of the grooved mullet *Liza dumerili* in the Swartkops estuary, South Africa (Masson and Marais 1975), which results in a much higher energy density in stomach contents than that of the sediment (Marais 1980). Similar results have been reported for the flathead mullet *Mugil cephalus* in North America (Odum 1968a, 1970) and Australia (Moriarty 1976) and for the grooved mullet *Liza dumerili* and the sicklefin mullet *Liza falcipinnis Liza falcipinnis* in western Africa (Payne 1976).

Nevertheless, the capacity to concentrate the finest particles varies dramatically between species and so the chemical composition of the food. In South African estuaries, the grooved mullet *Liza dumerili* ingest the coarser sand (Blaber 1977, Marais 1980) and hence their stomach contents had a much lower carbohydrate and protein and higher age ash than those of the South African mullet *Liza richardsoni* and the flathead mullet *Mugil cephalus*. The latter two species have similar granulometric profiles (Blaber 1977, Marais 1980) and also similar chemical composition of the stomach contents (Fig. 9.11). In the Mediterranean Sea, the stomach contents of the thinlip mullet *Liza ramada* have a higher organic content and energy density than that of sympatric leaping grey *Liza saliens* (Cardona 2001), which is in agreement with the coarser branchial filter of the latter (Drake et al. 1984, Guinea and Fernández 1992) and its poor capacity to concentrate fine particles (Mariani et al. 1987). However, the spacing between the pharyngeal teeth and between the branchial processes of the golden mullet *Liza aurata* and those of the thinlip mullet *Liza ramada* are rather similar (Guinea and Fernández 1992), but the diet of the former contains less organic matter (Cardona 2001) and coarser particles than that of the latter when living in sympatry (Mariani et al. 1987).

Nevertheless, the organic matter content of the diet of sedimentivorous grey mullets is usually less than 250 mg g^{-1} dry matter (Odum 1968a, Hickling 1970, Ching 1977, Marais and Erasmus 1977, Marais 1980, Cardona 2001) and the energy density of the diet ranges from 0.2 to 6.5 kJ g^{-1} dry matter (Marais 1980, Cardona 1999a, 2001). It should be noted that the actual amount of energy available for grey mullets is much lower, as carbohydrates may represent half the organic matter in the diet (Marais and Erasmus 1977, Marais 1980, Cardona 1999a, 2001) and are hard to digest for most species, even if sand grains abrade plant tissue (Payne 1978). As a result, the assimilation efficiency of the organic matter in the diet is only 50% (Hickling 1970, Odum 1970, Payne 1976).

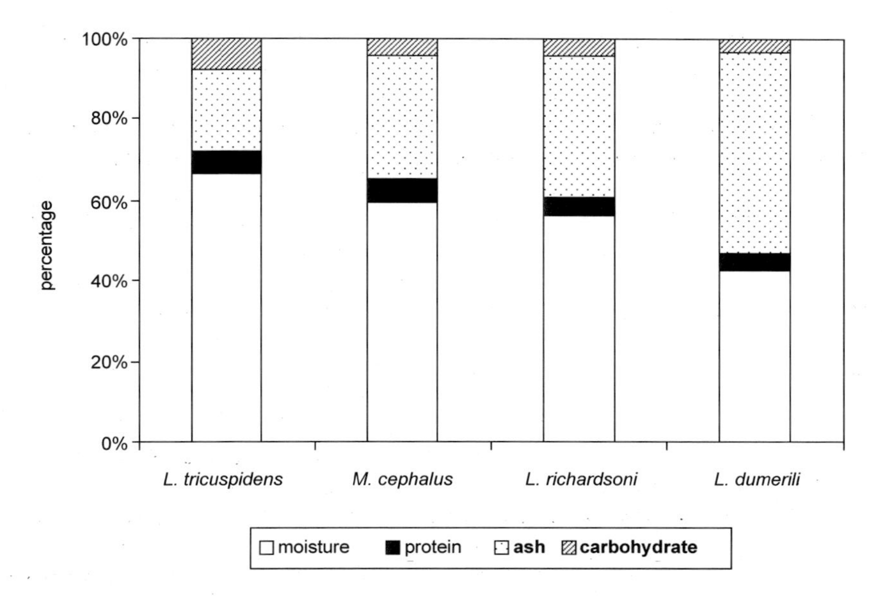

Figure 9.11. Chemical composition of the stomach contents of four grey mullet species from the Swartkops estuary (South Africa). Note that the species ingesting the coarser sand (*Liza dumerili*) also has the lowest carbohydrate and protein and the highest age ash. Reproduced with permission from Elsevier; from Marais and Erasmus (1977).

Furthermore, the granulometric profile of the sediment has a dramatic influence on the capacity of grey mullets to concentrate the fine particles and hence the average diameter of the inorganic particles in the stomach contents of any species varies locally (Masson and Marais 1975, Blaber 1976, 1977, Blaber and Whitfield 1977, Mariani et al. 1987, Osorio Dualiby 1988, Dankwa et al. 2005). The flathead mullet *Mugil cephalus* and the golden mullet *Liza aurata* concentrate on particles smaller than 125 μm when foraging on fine sand and particles in the range 125–500 μm when foraging on finer sediment (Mariani et al. 1987). Conversely, the thinlip mullet *Liza ramada* concentrate on particles smaller than 125 μm at both sites (Mariani et al. 1987, Almeida 2003). However, species ranking based on average particle size is rather constant and, in South African estuaries, *Valamugil buchanani*, *Valamugil robustus*, *Valamugil seheli* and *Valamugil cunnesius* always consume more fine sediment than *Mugil cephalus*, *Liza macrolepis* and *Liza richardsoni* which in turn consume finer sediment than *Liza dumerili* and *Liza tricuspidens* (Blaber 1977, Marais 1980).

Manipulative experiments have demonstrated the existence of complex relationships between grey mullet biomass, the organic matter content of the sediment and the abundance of infauna. Sometimes, the sediment of enclosures stocked with grey mullets has a lower content of organic matter than that of fish-less enclosures, but this is not always true (Cardona 1996, Torras et al. 2000, Cardona et al. 2001). This is because the infauna usually consumes much finer particles of organic matter than those retained by grey mullets, which in turn may reduce the density of infauna when stocked at a high density. Nevertheless, when stocked at a density closer to that of natural habitats, grey mullets often did not reduce the abundance of infauna (Service et al. 1992, Cardona et al. 2001) and the organic matter content of the sediment is lower than that of fish-less controls because of the synergistic effects of infauna and grey mullets.

Other Feeding Modes

There is not a single grey mullet species that relies exclusively on plankton, but the South African mullet *Liza richardsoni* may directly exploit surf diatoms (Romer and McLachlan 1986) and planktonic diatoms represent the bulk of the diet of the thinlip mullet *Liza ramada* in eutrophic freshwater ecosystems in western Europe and the Mediterranean (Cardona and Castello 1994, Almeida 2003). This species can be seen swimming with the mouth open, filtering water in Mediterranean lagoons and harbours, can capture many species of phytoplankton and stable isotope analysis indicates that up to 35% of the nutrients assimilated in these environments may derive from suspended particulate organic matter (Koussoroplis et al. 2011). *Mugil cephalus* is also capable of filtering phytoplankton and zooplankton when dense phytoplankton blooms occur (Odum 1968b, 1970, Cardona and Castello 1994, Blanco et al. 2003). Electivity experiments have revealed that the flathead mullet *Mugil cephalus* and the thinlip mullet *Liza ramada* behave as pump filters which select prey by size when foraging on plankton (Fig. 9.12).

Controlled experiments have demonstrated the capacity of the flathead mullet *Mugil cephalus* and the leaping grey mullet *Liza saliens* to reduce the density of planktonic rotifers, cladocerans and cyclopoid copepods (Cardona et al. 1996, Torras et al. 2000, Cardona et al. 2001), although rotifers and cladocerans are seldom reported from the stomach contents of juveniles and adults (Cardona 2001, Almeida 2003). Other species have been reported to consume zooplankton, but in the absence of detailed taxonomic information this can be a misleading term including a mixture of true zooplankton and meiofauna.

There are, however, two species whose stomach contents are dominated by animal prey. The mountain mullet *Agonostomus monticola* inhabits turbulent areas in streams and rivers in Central America and the West Indies. Insects and plant material usually have a similar volumetric contribution to the stomach contents of adult fishes (> 12 cm TL), but in some localities insects may represent the bulk of the diet (Cruz 1987, Aiken 1998, Phillip 1993, Torres-Navarro and Lyons 1999). Plant material includes filamentous algae, leaves, fruits and seeds, and sediment is always very scarce, thus suggesting that animal prey are captured individually and not mixed with sediment. Furthermore, the relevance of plant material may vary seasonally, as increased water turbidity may reduce the availability of algae in some streams during the rainy season, when only insects are usually consumed (Phillip 1993, Torres-Navarro and Lyons 1999).

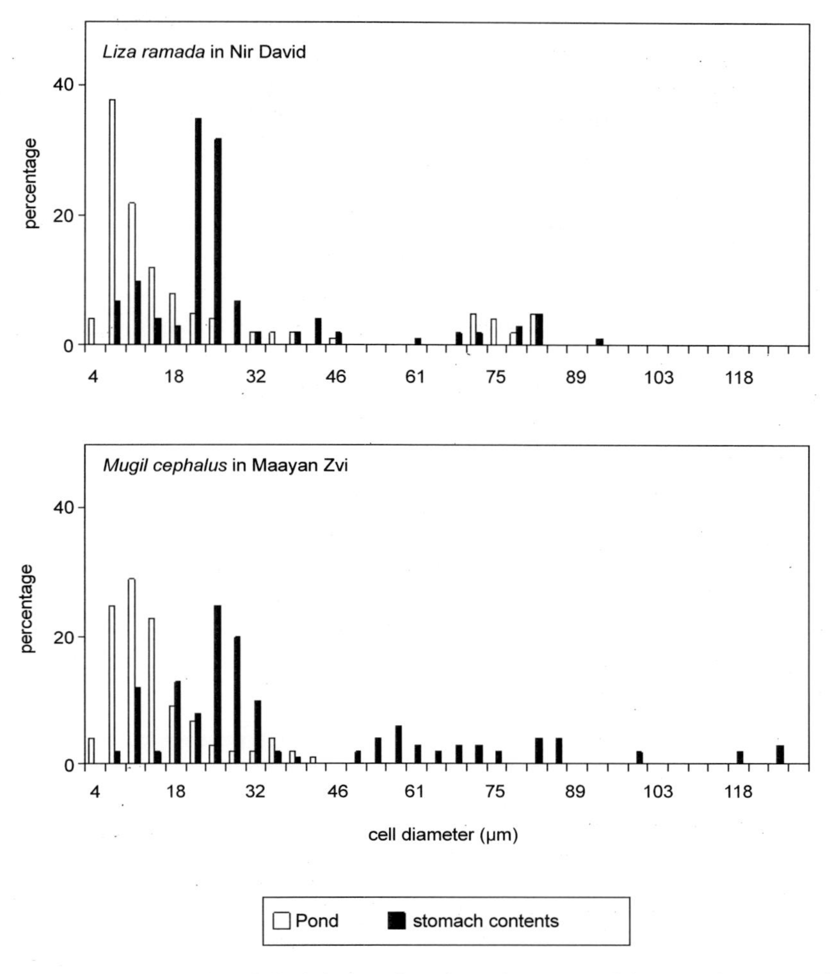

Figure 9.12. Size-frequency distribution of phytoplankton from the environment and the stomach contents of grey mullets from semi-intensive fish ponds in Israel. Note that only cells larger than approximately 20 µm, are proportionally more abundant in the stomach contents than in the environment and that grey mullet cannot retain the smallest cells. Reproduced with permission from The Israeli Journal of Aquaculture - Bamidgeh; from Cardona et al. (1996).

Interestingly, the mountain mullet is the only species in the family with teeth on the vomer and palatines, which might play a role in the handling of large invertebrates, and without a pharyngobranchial organ (Harrison and Howes 1991).

Information about the diet of the yellow-eyed mullet *Aldrichetta forsteri* is scant, but invertebrates and plant material make a similar contribution to stomach volume in Australia. Invertebrates are represented by benthic molluscs, amphipods and harpacticoid copepods (Thomson 1954, Edgar and Shaw 1995). Interestingly, the yellow-eyed mullet *Aldrichetta forsteri* has a simplified pharyngobranchial organ (Harrison and Howes 1991).

Most grey mullet species can scrape hard substrata occasionally to forage on attached microbial films and green macroalgae and the flathead mullet *Mugil cephalus* has been reported to scrape epiphytes from submerged macrophytes (Odum 1970, Collins 1981). However, rock scraping is the main feeding mode of the snouted mullet *Chaenomugil proboscideus*, the bobo mullet *Joturus pichardi* and probably the striped mullet *Liza tricuspidens*, although they differ in diet. The snouted mullet *Chaenomugil proboscideus* inhabits the rocky coasts of western Central America and North-western Colombia and consumes primarily macroalgae (Castellanos-Galindo and Giraldo 2008). The large, bifid, scoop-like teeth of the snouted mullet *Chaenomugil proboscideus* probably play a role in the scraping of macroalgae, but the role of the teeth on the tongue remain unknown and have been interpreted as plesiomorphic (Ebeling 1957). The denticulate region of the pharyngobranchial organ of the snouted mullet *Chaenomugil proboscideus* is highly reduced, as it is in the poorly known boxlip mullet *Oedalechilus labeo*, another inhabitant of marine rocky shores.

The bobo mullet *Joturus pichardi* lives in large rivers in Central America flowing to the Caribbean and forages primarily on the cyanobacteria film growing on submerged boulders, although there is an increase in the consumption of animals, mainly insects and freshwater prawns, with size (Cruz 1987). The large, serrated, scoop-like primary teeth of the bobo mullet *Joturus pichardi* probably play a role in rock scraping, whereas the simple teeth on the palate and tongue may help in handling large animal prey, such as large freshwater prawns.

The striped mullet *Liza tricuspidens* has been reported to consume primarily green algae in coastal lagoons (Blaber 1976) and sandy estuaries (Masson and Marais 1975) and green and red seaweeds in rocky, open estuaries (Blaber 1977). It can also graze on seagrasses (Masson and Marais 1975). Nevertheless, it also consumes macrodetritus and coarse sand (Masson and Marais 1975, Blaber 1977, Marais 1980), which may assist in grinding plant material. Whether it scrapes up the algae or forages on detached fragments transported by currents is unknown, although the large, trifid, scoop-like teeth (Marais 1980), suggest that rock scraping is the primary feeding mode.

The diamond mullet *Liza alata* consumes primarily filamentous green algae and macrodetritus in South African estuaries (Blaber 1976, 1977). The pinkeye mullet *Trachystoma petardi* is another poorly known species that may forage primarily on filamentous green algae (Thomson 1954).

Description of Adult Diets

Dietary information on the most basal species is scant, but corsula *Rhinomugil corsula* from Indian rivers feeds on detritus, cyanobacteria and macroalgae (Munshi et al. 1991), whereas the yellowtail mullet *Sicamugil cascasia* has been described as relying primarily on benthic microalgae (Johar and Singh 1981).

All *Mugil* species are primarily estuarine and have very similar diets, based on fine particulate organic matter filtered from the sediment (Ibáñez-Aguirre 1993, Sánchez Rueda 2002, Dankwa et al. 2005). The banana mullet *Mugil bananensis* and the white mullet *Mugil curema* usually behave as sediment filterers and their stomach contents are dominated by fine sand, microphytobenthos and detritus, mixed with some filamentous green macroalgae and invertebrates such as copepods and foraminiferans (Payne 1976, Osorio Dualiby 1988, Ibáñez Aguirre 1993, Blay 1995, Sánchez Rueda 2002, Dankwa et al. 2005). The parassi mullet *Mugil incilis* and the Lebranche mullet *Mugil liza* have similar diets (Osorio Dualiby 1988).

The flathead mullet *Mugil cephalus* may behave as a sediment filterer and feed mainly on detritus and microphytobenthos (Odum 1968a, 1970, Das 1977). However, it may also behave as a bulk sedimentivore, also swallowing foraminiferans and other protists, meiofauna and small invertebrates, and as a grazer foraging on filamentous green algae (Thomson 1953, Thomson 1954, Luther 1962, Blaber 1976, 1977, Payne 1976, Marais 1980, Collins 1981, Tandel et al. 1986, Eggold and Motta 1992, Cardona 2001, Blanco et al. 2003). Blaber (1977) hypothesized that the flathead mullet *Mugil cephalus* from South Africa may prefer animal prey to plant material when available and large numbers of polychaetes have been reported to be also consumed in North America (Bishop and Miglarese 1978), India (Tandel et al. 1986) and western Africa (Dankwa et al. 2005). Stable isotope data indicate that the flathead mullet *Mugil cephalus* from the Mediterranean Sea may have a trophic level much higher than that of other sympatric grey mullets (Carlier et al. 2007). The relationship between the large variability in the feeding habits reported for this species and the high regional variability in the anatomy of the pharyngobranchial organ (Harrison and Howes 1991) remains to be explored.

Species in the genus *Liza* are more diverse in diet, but most of them are best classified as bulky sedimentivores. The abu mullet *Liza abu*, from freshwater marshes in Iraq, consumes primarily detritus, diatoms and macroalgae (Ahmad and Hussain 1982, Hussain et al. 2009). Klunzinger's mullet *Liza klunzingeri* from freshwater marshes in southern Iraq consumes primarily detritus and diatoms, but not macroalgae (Hussain et al. 2009). The largescaled mullet *Liza grandisquamis*, the largescale mullet *Liza macrolepis*, the grooved mullet *Liza dumerili* and the sicklefin mullet *Liza falcipinnis* from African estuaries consume primarily fine sand, detritus, microphytobenthos, and meiofauna, although the sicklefin mullet *Liza falcipinnis* and the largescale mullet *Liza macrolepis* may also consume filamentous green algae (Fagade and Olaniyon 1973, Blaber 1976, 1977, King 1988, Dankwa et al. 2005). The largescale mullet *Liza macrolepis* has a similar diet in India (Luther 1962) and Taiwan (Lin et al. 2007). However, stable isotope analyses have revealed that most of the assimilated organic matter by the largescale mullet *Liza macrolepis* in Taiwan comes from microphytobenthos (Lin et al. 2007), whereas invertebrates are the major source of assimilated organic matter for the sicklefin mullet *Liza falcipinnis* in western Africa (Faye et al. 2011). The flat-tail mullet *Liza argentea* consume primarily diatoms, green filamentous algae and fine detritus (Thomson 1954).

In the Mediterranean Sea, the thinlip mullet *Liza ramada*, the golden mullet *Liza aurata* and the leaping mullet *Liza saliens* consume primarily fine sand, detritus, microphytobenthos and meiofauna, although filamentous green algae can be locally abundant and the larvae of chironomid midges are usually a major prey (Drake et al. 1984, Cardona 2001). Stable isotope ratios suggest that *Liza aurata* from Atlantic and Mediterranean coastal lagoons rely primarily on meiofauna, seston and suspended organic matter, with the actual contributions varying locally (Cardona 2001, Blanco et al. 2003, Carlier et al. 2007, Lebreton et al. 2011, Koussoroplis et al. 2011). Diatoms dominated the diet of the thinlip mullet *Liza ramada* in European estuaries and Mediterranean coastal lagoons (Cardona 2001, Laffaille et al. 2002, Almeida 2003, Kasimoglu and Yilmaz 2012), but stable isotope ratios suggest that they rely primarily on meiofauna, seston and suspended organic matter, with the actual contributions varying locally (Lebreton et al. 2011, Koussoroplis et al. 2011, França et al. 2011, Vinagre et al. 2011).

The squaretail mullet *Liza vaigiensis* consumes detritus, microphytobenthos and polychaetes in coastal lagoons in Sri Lanka (Wijeyaratne and Costa 1990) and Singapore (Hajisamae et al. 2004). The stomach contents of the greenback mullet *Liza subviridis* are dominated by diatoms, detritus and filamentous green algae in proportions that vary locally (Chan and Chua 1979, Hussain et al. 2009). However, at least in Australia, the greenback mullet *Liza subviridis* derives most assimilated organic matter from animal prey (Abrantes and Sheaves 2008).

The goldspot mullet *Chelon parsia* has been reported to forage primarily on detritus, although filamentous green algae and diatoms are also consumed (Surendra Babu and Neelakantan 1983). The otembora mullet *Chelon melinopterus* from Malaysian mangroves rich in benthic invertebrates consumes primarily detritus, sand and microphytobenthos (Ching 1977). On the other hand, the thicklip mullet *Chelon labrosus* from European and Mediterranean estuaries consumes a wide diversity of prey, from detritus and diatoms to amphipods, snails and larvae of chironomid midges (Hickling 1970, Cardona 2001). Stable isotope ratios reveal that most of the assimilated nutrients derive from the meiofauna, seston and suspended organic matter, with the actual contributions varying locally (Carlier et al. 2007, Koussoroplis et al. 2011).

The bluetail mullet *Valamugil buchanani* consumes detritus, microphytobenthos and polychaetes in coastal lagoons in Sri Lanka (Wijeyaratne and Costa 1990) and detritus, microphytobenthos and green filamentous algae in South Africa, where animal prey are seldom consumed (Blaber 1976, 1977). Actually, the bluetail mullet *Valamugil buchanani* has a trophic level of 2 in Australian estuaries, consistent with a diet based on macrophytobenthos (Abrantes and Sheaves 2008). The long-finned mullet *Valamugil cunnesius* consumes polychaetes and detritus (Wijeyaratne and Costa 1988) and diatoms and some detritus in South Africa (Blaber 1976, 1977). The bluespot mullet *Valamugil seheli* consumes primarily diatoms, but not detritus in South Africa (Blaber 1976). The robust mullet *Valamugil robustus* consumes primarily diatoms, filamentous green algae and detritus in South Africa (Blaber 1976).

The silver mullet *Paramugil georgii* consumes copepods, diatoms, filamentous green algae and amphipods but little sand and detritus (Thomson 1954).

The freshwater mullet *Myxus capensis* consumes diatoms, filamentous green algae and detritus in South Africa, together with a few animal prey (Blaber 1976, 1977). The sand mullet *Myxus elongatus* consumes detritus, filamentous green algae and infauna in Australia (Thomson 1954).

The Trophic Level of Grey Mullets

Meiofauna (usually harpacticoid copepods and ostracods) and infauna (polychaetes, larvae of chironomid midges, snails and clams) are often consumed by estuarine grey mullets (Hickling 1970, Blaber 1977, Bishop and Miglarese 1978, Tandel et al. 1986, Wijeyaratne and Costa 1990, Cardona 2001, Dankwa et al. 2005). Traditionally, these animal prey were considered minor dietary items, as the volume of detritus and microphytobenthos is usually much larger. Nevertheless, traditional methods are usually unable to assess the actual contribution of animal prey to the diet of grey mullets. Recent studies using stable isotope analysis have challenged that point of view, as the $\delta^{15}N$ values of some species are consistent with primarily carnivorous diets (Table 9.2). Furthermore, the actual relevance of animal prey to the diet of grey mullets varies not only between species, but also geographically within most species (Table 9.2).

The white mullet *Mugil curema* and the parassi mullet *Mugil incilis* from Brazilian mangroves (Giarrizzo et al. 2011), the white mullet *M. curema* in the Greater Caribbean (Vaslet et al. 2012), Speigler's mullet *Valamugil speigleri* from mangroves in Thailand (Thimdee et al. 2008) and the bluetail mullet *Valamugil buchanani* from mangroves in Australia (Abrantes and Sheaves 2008) are true herbivores, with a trophic level of two or slightly higher. Conversely, the banana mullet *Mugil bananensis* and the white mullet *Mugil curema* from estuaries in western Africa have trophic levels close to 2.7 and hence are best classified as omnivores (Faye et al. 2011).

Table 9.2. Trophic level of grey mullets according to stable isotope ratios.

Species	Trophic level	Reference
Chelon labrosus	3.0	Carlier et al. 2007
Liza aurata	3.0	Carlier et al. 2007
Liza aurata	3.0	Lebreton et al. 2011
Liza dumerilii	3.1	Faye et al. 2011
Liza falcipinnis	3.0	Faye et al. 2011
Liza haematocheila	2.6	Quan et al. 2009
Liza haematocheila	3.1	Quan et al. 2010
Liza macrolepis	2.0	Lin et al. 2007
Liza ramada	3.0	Lebreton et al. 2011
Liza ramada	2.7	França et al. 2011
Liza ramada	4.0	França et al. 2011
Liza ramada	3.0	Vinagre et al. 2011
Liza ramada	3.5	Vinagre et al. 2011
Liza subviridis	3.0	Abrantes and Sheaves 2008
Mugil bananensis	2.8	Faye et al. 2011
Mugil cephalus	2.5	Akin and Winemiller 2008
Mugil cephalus	3.5	Fry et al. 1999
Mugil cephalus	3.5	Carlier et al. 2007
Mugil curema	2.0	Giarrizzo et al. 2011
Mugil curema	2.0	Vaslet et al. 2011
Mugil curema	2.7	Faye et al. 2011
Mugil incilis	2.0	Giarrizzo et al. 2011
Valamugil buchanani	2.0	Abrantes and Sheaves 2008
Valamugil speigleri	2.1	Thimdee et al. 2008

Liza species usually have more carnivorous diets than *Mugil* species from the same habitats (Faye et al. 2011), except in the Mediterranean (Carlier et al. 2007). Stable isotope ratios indicate omnivorous diets, with a strong reliance on harpacticoid copepods, for the thinlip mullet *Liza ramada*, the golden grey mullet *Liza aurata*, the leaping grey mullet *Liza saliens* and the thicklip mullet *Chelon labrosus* from Mediterranean coastal lagoons (Carlier et al. 2007, Koussoroplis et al. 2011) and Portuguese estuaries (França et al. 2011, Vinagre et al. 2011). The reliance on animal prey is even higher for the thinlip mullet *Liza ramada* and the golden mullet *Liza aurata* from Atlantic salt marshes (Lebreton et al. 2011), the grooved mullet *Liza dumerili* and the sicklefin mullet *Liza falcipinnis* from western Africa (Faye et al. 2011) and the flathead mullet *Mugil cephalus* from Mediterranean coastal lagoons (Carlier et al. 2007), best classified as carnivores. However, it should be noted that none of these species captures individual animal prey, but swallow them mixed with sand, detritus and microphytobenthos.

When available, data from stable isotope analysis reveal that most of the organic matter assimilated by bulk sedimentivores comes from animal prey. Accordingly, they are at the same trophic level than other fishes preying on invertebrates and one trophic level above true sediment filterers (Fig. 9.13). This is because most of the bulk sedimentivores have widely spaced pharyngeal teeth and gill rakers and short guts, like the grooved mullet *Liza dumerili*. Only species with long guts, like the bluetail mullet *Valamugil buchanani,* may behave in certain situations as sediment filterers and hence rely primarily on detritus and plant material, but this is not necessarily true, as shown by the thinlip mullet *Liza ramada*, a species with a long gut and finely packed gill rakers, but a high trophic level according to stable isotope ratios. This is probably because of a higher assimilation rate of animal matter due to a limited capacity to digest plant tissues in most grey mullet species. Studies on enzymatic activity are urgently needed to shed additional light of this issue.

Resource Partitioning and Competition

Dietary analyses have often revealed a high similarity in the diet of sympatric species of grey mullets at the planktophagous fry stage (Gisbert et al. 1996) and the juvenile/adult sedimentivorous stage (Fagade and Olaniyon 1973, Blaber 1976, 1977, Ibáñez Aguirre 1993, Cardona 2001, Sánchez Rueda 2002, Dankwa et al. 2005). The trophic similarity in the diet of sympatric species of *Mugil* revealed by dietary studies is usually supported by stable isotopes (Faye et al. 2011, Giarrizzo et al. 2011), but in some areas, sympatric species differ in their stable isotope ratios, thus revealing different diets. For instance, the bluetail mullet *Valamugil buchanani* from estuaries in tropical Australia is depleted in ^{15}N and enriched in ^{13}C as compared with the greenback mullet *Liza subviridis* from the same sites (Fig. 9.13). Accordingly, the bluetail mullet *Valamugil buchanani* is thought to rely primarily on microphytobenthos and detritus, whereas the greenback mullet *Liza subviridis* is an omnivore or carnivore relying on invertebrates thriving on detritus from mangroves and marsh plants (Abrantes and Sheaves 2008). Likewise, the South African mullet *Liza richardsonii* and the grooved mullet *Liza dumerili* from South African estuaries differ in their δ^{13}C values, suggesting that the former is more dependent on the detritus from seagrasses and the latter on saltmarsh plants (Paterson and Whitfield 1997), but differences are highly idiosyncratic. For instance, Carlier et al. (2007) suggested similar trophic levels for the golden mullet *Liza aurata* and the thicklip mullet *Chelon labrosus* in a coastal lagoon in France, although they differed in their δ^{13}C values perhaps due to the higher reliance of the thicklip mullet *Chelon labrosus* on seagrass detritus. Conversely, the thicklip mullet *Chelon labrosus* is at a higher trophic level than the golden mullet *Liza aurata* in the Greek coastal lagoon studied by Koussoroplis et al. (2011), where no difference existed in δ^{13}C.

Interspecific competition was also thought to be reduced by particle size preferences (Blaber 1976, 1977, Marais 1980). For instance, in Colombia mean particle size is 163 μm for the white mullet *Mugil curema*, 229 μm in the parassi mullet *Mugil incilis* and 401 μm in Lebranche mullet *Mugil liza* (Osorio Dualiby 1988). In western Africa, mean particle size is 100 μm for the largescaled mullet *L. grandisquamis*, 100 μm for the banana mullet *Mugil bananensis*, 120 μm for the sicklefin mullet *Liza falcipinnis*, 135 μm for the white mullet *Mugil curema,* 200 μm for the flathead mullet *Mugil cephalus* and 220 μm for the grooved mullet *Liza dumerili* (Dankwa et al. 2005).

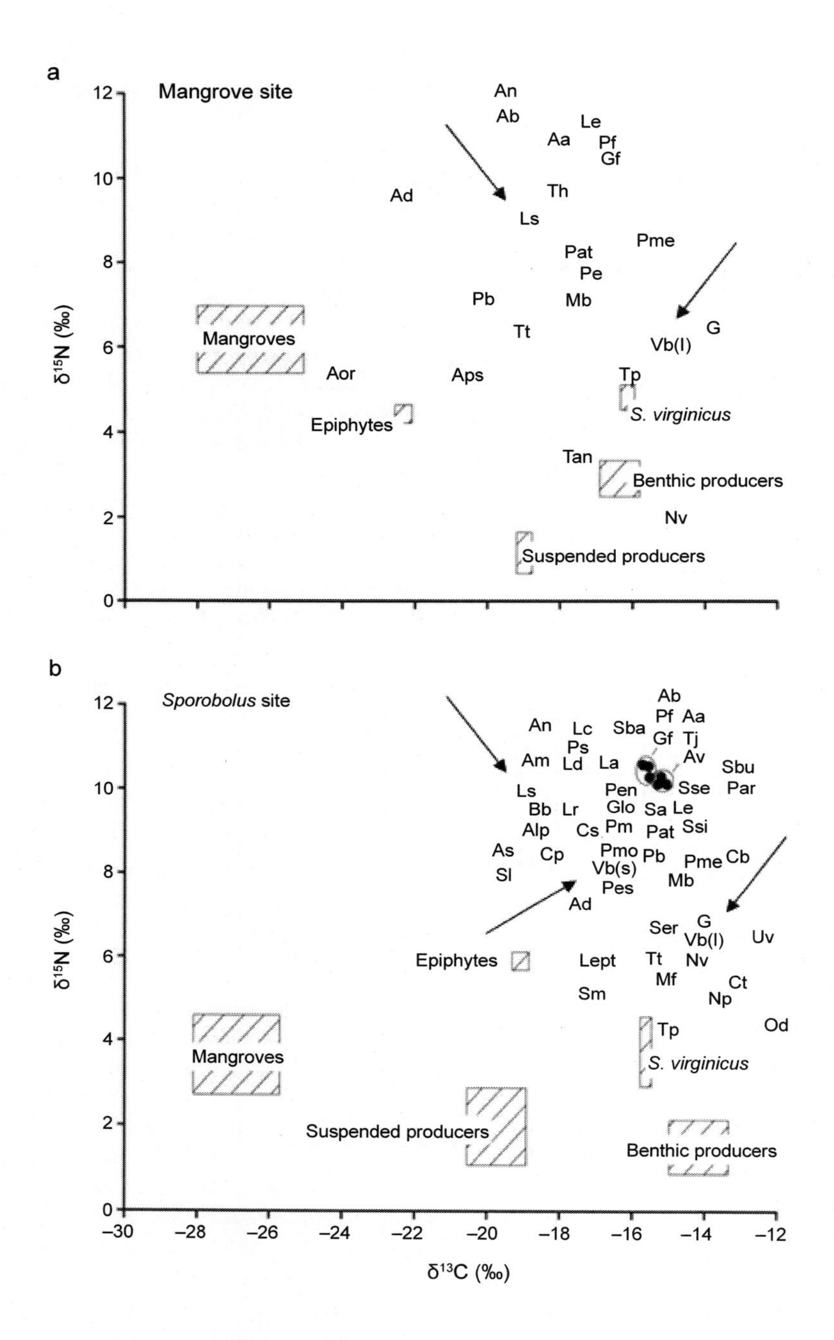

Figure 9.13. Assessment of trophic level through stable isotope analysis. Stable isotope ratios reveal that juvenile bluetail mullet *Valamugil buchanani* (Vb(l)) from tropical Queensland (Australia) rely primarily on microphytobenthos and detritus, whereas juvenile greenback mullet *Liza subviridis* (Ls) forage one trophic level above and cluster with invertebrate predators. The trophic level of the zooplanktophagous fry of the bluetail mullet *Valamugil buchanani* (Vb(s)) is also higher than that of the juveniles of the same species, but lower than that of juvenile greenback mullet *Liza subviridis*. Reproduced with permission from Elsevier; from Abrantes and Sheaves (2008).

In any case, resource similarity may lead to competition only if resources are limiting. Grey mullets have been suggested not to be food limited at the sedimentivorous stage (Odum 1970, Cardona 2001), but they could be during early life (Gisbert et al. 1995, 1996). Accordingly, competition at the fry stage has been suggested as one of the major drivers of the structure of grey mullet assemblages in Mediterranean estuaries (Cardona et al. 2008).

Seasonal and Diel Feeding Rhythms

Grey mullet usually forage during light hours, both at the zooplanktophagous fry stage (De Silva and Wijeyaratne 1977, Albertini-Berhaut 1979, Torricelli et al. 1981, 1988, Gisbert et al. 1997) and the sedimentivorous juvenile/adult stage (Blaber 1976, Marais 1980, Collins 1981, King 1988, Cardona 1999a). Fry are probably visual predators, but there is no obvious reason why sedimentivorous juveniles and adults have to restrict foraging to light hours, particularly in turbid estuarine habitats. The flathead mullet *Mugil cephalus* and the leaping mullet *Liza saliens* at the sedimentivorous stage may forage at night in some situations (Blaber 1976, Cardona 1999a) and the grooved mullet *Liza dumerili* have been reported to forage continuously, although feeding intensity may decrease at night (Blaber 1976, Marais 1980).

The influence of tides on the feeding rhythm of juvenile and adult grey mullet at the sedimentivorous stage is highly variable, even within the same species. Available evidence suggests that the flathead mullet *Mugil cephalus* and the thinlip mullet *Liza ramada* synchronize their feeding rhythms with tidal cycles only where extensive mud flats become available at high tide (Odum 1970, Almeida et al. 1993, Laffaille et al. 2002). Otherwise, they forage independently of tide height (Collins 1981, Almeida 2003). This is probably because there is little benefit in waiting till the next high tide unless it guarantees access to a huge amount of trophic resources. Conversely, the sicklefin mullet *Liza falcipinnis* and the grooved mullet *Liza dumerili* from western Africa have been reported to forage primarily at low tide (Dankwa et al. 2005).

The metabolic rate of grey mullets is highly sensitive to water temperature, which results in seasonal changes in their energy budgets (Marais 1978, Guinea and Fernández 1991). Furthermore, food availability and quality may also vary seasonally because of increased amount of detritus in the sediment at the end of the rainy season in tropical regions (Blaber 1976, 1977, Payne 1976, Das 1977) or after the collapse of macrophyte beds in early autumn in Mediterranean lagoons (Cardona 1999a, 2001). Accordingly, grey mullet modify seasonally stomach fullness and the length of the foraging period (Fig. 9.14) to match the energy requirements imposed by variable water temperature with changes in energy availability derived from changes in food abundance and quality (Cardona 1999a).

Juvenile and adult thicklip mullet *Chelon labrosus*, thinlip mullet *Liza ramada* and golden mullet *Liza saliens* eat year round in the Mediterranean and western Europe, but feeding intensity declines dramatically in winter (Hickling 1970, Cardona 2001, Almeida 2003), when low water temperature dramatically reduces the metabolic rate. Juvenile leaping mullet *Liza saliens* are more sensitive to low water temperatures than any other European species of grey mullet and stop feeding in captivity when water temperature drops below 10ºC (Guinea and Fernández 1991). Accordingly, they fast during winter in the wild (Cardona 1999a), which probably explains why they fail to colonize cold temperate areas with long winters (Cardona 1999b). On the other hand, the daily ration of juvenile leaping mullet *Liza saliens* peaks during summer, not only because of high water temperatures, but also because of the low energy density of the food (Cardona 1999a).

At least the flathead mullet *Mugil cephalus* and the leaping mullet *Liza saliens* stop feeding during the spawning migration (Odum 1970, Cardona 1999b). This is probably because the ripe gonads of both sexes are bulky and fill most of the visceral cavity during the spawning season. Leaping mullet *Liza saliens* spawn during the summer months in the Mediterranean and hence adults fast twice annually, in summer and winter (Cardona 1999b).

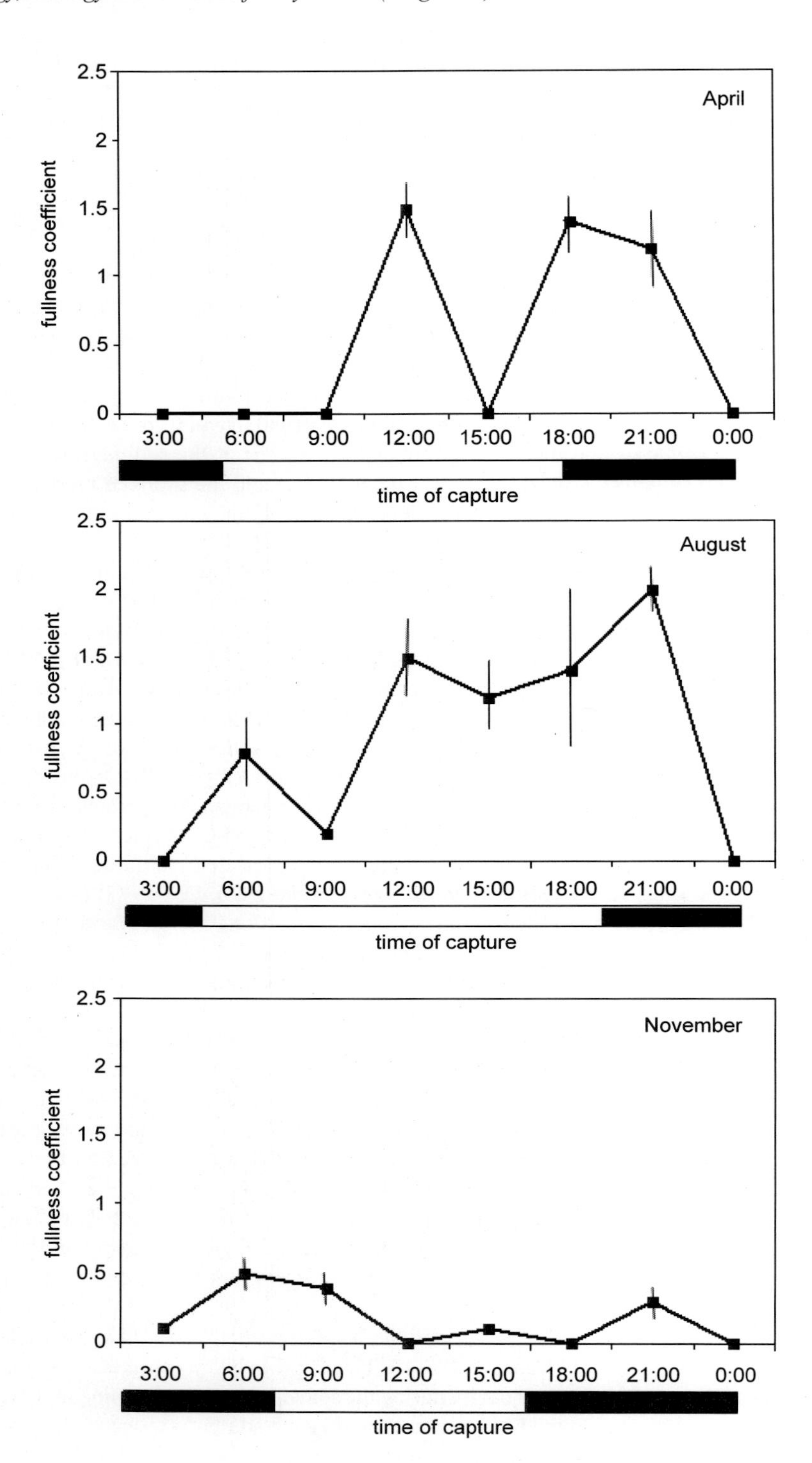

Figure 9.14. Diel feeding rhythm of juvenile leaping mullets *Liza saliens* in a Mediterranean coastal lagoon, shown as changes in the stomach fullness coefficient. Vertical bars show the standard deviation. Changes in the stomach fullness and the length of the foraging period resulted in daily rations ranging from 6% of body weight in August to 1.4% of body weight in November. Juvenile leaping mullets *Liza saliens* fast in winter. Reproduced with permission from Elsevier; from Cardona (1999a).

References

Abrantes, K. and M. Sheaves. 2008. Incorporation of terrestrial wetland material into aquatic food webs in a tropical estuarine wetland. Est. Coast. Shelf Sci. 80: 401–412.

Ahmad, T.A. and N.A. Hussain. 1982. Observations on the food of young *Liza abu* Heckel from Salihyia River Basrah, Iraq. Iraqi J. Mar. Sci. 11: 79–88.

Aiken, K.A. 1998. Reproduction, diet and population structure of the mountain mullet, *Agonostomus monticola*, in Jamaica, West Indies. Env. Biol. Fish 53: 347–352.

Akin, S. and K.O. Winemiller. 2008. Body size and trophic position in a temperate estuarine food web. Acta Oecologica 33: 144–153.

Albertini-Berhaut, J. 1973. Biologie des stades juvéniles de Téleostéens Mugilidae *Mugil auratus* Risso 1810, *Mugil capito* Cuvier 1892 et *Mugil saliens* Risso 1810. Aquaculture 2: 251–266.

Albertini-Berhaut, J. 1979. Rythme alimentaire chez les jeunes *Mugil capito* (Téléostééns Mugilidae) dans le Golfe de Marseille. Tehys 9: 79–82.

Albertini-Berhaut, J. 1987. L'intestin chez les Mugilidae (Posisson: Téléostéens) à différentes étapes de leur croissance. I. Aspects morphologiques et histologiques. J. Appl. Ichthyol. 3: 1–12.

Albertini-Berhaut, J. 1988. L'intestin chez les Mugilidae (Posisson: Téléostéens) à différentes étapes de leur croissance. II. Aspects ultrastructuraux et cytophysiologiques. J. Appl. Ichthyol. 4: 65–78.

Al-Hussaini, A.H. 1947. The feeding habits and the morphology of the alimentary tract of some teleosts living in the neighbourhood of the marine biological stations, Ghardaqa, Re Sea. Publ. Mar. Biol. Sta. Ghardaqa, Fouad I Univ. 5: 1–61.

Almeida, P.R. 2003. Feeding ecology of *Liza ramada* (Risso 1810) (Pisces: Mugilidae) in a south-western estuary of Portugal. Est. Coas. Shelf Sci. 57: 313–323.

Almeida, P.R., F. Moreira, J.L. Costa, C.A. Assis and M.J. Costa. 1993. The feeding strategies of *Liza ramada* (Risso 1826) in fresh and brackish water in the River Tagus, Portugal. J. Fish Biol. 42: 95–107.

Bishop, J.M. and J.V. Miglarese. 1978. Carnivorous feeding in adult striped mullet. Copeia 1978: 705–707.

Blaber, S.J.M. 1975. The shedding of gill-raker filaments in grey mullet (Mugilidae). S. Afr. J. Sci. 71: 377.

Blaber, S.J.M. 1976. The food and feeding ecology of Mugilidae in the St. Lucia lake system. Biol. J. Linn. Soc. 8: 267–277.

Blaber, S.J.M. 1977. The feeding ecology and relative abundance of mullet (Mugilidae) in Natal and Pondoland estuaries. Biol. J. Linn. Soc. 9: 259–275.

Blaber, S.J.M. and A.K. Whitfield. 1977. The feeding ecology of juvenile Mugilidae in south east African estuaries. Biol. J. Lin. Soc. 9: 277–284.

Blaber, S.J.M., D.T. Brewer and J.P. Salini. 1989. Species composition and biomasses of fishes in different habitats of a tropical northern Australian estuary: their occurrence in the adjoining sea and estuarine dependence. Est. Coast. Shelf Sci. 129: 509–531.

Blanco, S., S. Romo, M.J. Villena and S. Martínez. 2003. Fish communities and food web interactions in some shallow Mediterranean lakes. Hydrobiologia 506-509: 473–480.

Blay, J., Jr. 1995. Food and feeding habits of four species of juvenile mullets (Mugilidae) in a tidal lagoon in Ghana. J. Fish Biol. 46: 134–141.

Brusle, J. 1981. Food and feeding in grey mullet. pp. 185–217. *In*: O.H. Oren (ed.). Aquaculture of Grey Mullets. Cambridge University Press, Cambridge.

Buchheister, A. and R.J. Latour. 2010. Turnover and fractionation of carbon and nitrogen stable isotopes in tissues of a migratory, coastal predator, summer flounder (*Paralichthys dentatus*). Can. J. Fish Aquatic Scie. 67: 445–461.

Capanna, E., S. Cataudella and G. Monaco. 1974. The pharyngeal structure of Mediterranean Mugilidae. Monitore Zoologico Italiano–Italian J. Zool. 8: 29–46.

Cardona, L. 1991. Measurement of trophic niche breadth using occurrence frequencies. J. Fish Biol. 39: 901–903.

Cardona, L. 1996. Microoalge selection by mullets (*Mugil cephalus* and *Liza ramada*) in Israeli semi-intensive fish ponds. Israeli J. Aquaculture Bamidgeh 48: 165–173.

Cardona, L. 1999a. Seasonal changes in the food quality, diel feeding rhythm and growth rate of juveniles leaping grey mullet *Liza saliens*. Aquat. Living Resour. 12: 263–270.

Cardona, L. 1999b. Age and growth of leaping grey mullet (*Liza saliens*) (Risso 1810) in Minorca (Balearic Islands). Sci. Mar. 63: 93–99.

Cardona, L. 2001. Non-competitive coexistence between Mediterranean grey mullet: evidence from seasonal changes in food availability, niche breadth and trophic overlap. J. Fish Biol. 59: 729–744.

Cardona, L. and F. Castello. 1994. Relative importance of plankton and benthos as food sources for *Liza ramada* and *Mugil cephalus* in Israeli semi-intensive fish ponds. Israeli J. Aquaculture Bamidgeh 46: 197–202.

Cardona, L., X. Torras, E. Gisbert and F. Castelló. 1996. The effect of striped grey mullet (*Mugil cephalus* L.) on freshwater ecosystem. Israeli J. Aquaculture Bamidgeh 48: 179–185.

Cardona, L., P. Royo and X. Toras. 2001. Effects of leaping grey mullet *Liza saliens* (Osteichthyes, Mugilidae) in the macrophyte beds of oligohaline Mediterranean coastal lagoons. Hydrobiologia 462: 233–240.

Cardona, L., B. Hereu and X. Torras. 2008. Juvenile bottlenecks and salinity shape grey mullet assemblages in Mediterranean estuaries. Est. Coast. Shelf Sci. 77: 623–632.

Carlier, A., P. Riera, J.-M.- Amouroux, J.-Y. Bodiou, K. Escoubeyrou, M. Desmalades, J. Caparros and A. Grémare. 2007. A seasonal survey of the food web in the Lapalme Lagoon (northwestern Mediterranean) assessed by carbon and nitrogen stable isotope analysis. Est. Coast. Shelf Sci. 73: 299–315.

Castellanos-Galindo, G.A. and A. Giraldo. 2008. Food resource use in a tropical eastern Pacific tidepool fish assemblage. Mar. Biol. 53: 1023–1035.

Castillo-Rivera, M., J.A. Zavala-Hurtado and R. Zárate. 2002. Exploration of spatial and temporal patterns of fish diversity and composition in a tropical estuarine system of Mexico. Rev. Fish Biol. Fish 12: 167–177.

Cataudella, S., F. Massa, M. Rampacci and D. Crosetti. 1988. Artificial reproduction and larval rearing of the thick lipped mullet (*Chelon labrosus*). J. Appl. Ichthyol. 4: 130–139.

Chan, E.H. and T.E. Chua. 1979. The food and feeding habits of greenback grey mullet, *Liza subviridis* (Valenciennes), from different habitats and at various stages of growth. J. Fish Biol. 15: 165–171.

Ching, C.V. 1977. Studies on the small grey mullet *Liza melinoptera* (Valenciennes). J. Fish Biol. 11: 293–308.

Collins, M.R. 1981. The feeding periodicity of striped mullet, *Mugil cephalus* L., in two Florida habitats. J. Fish Biol. 19: 307–315.

Correa Polo, F., P. Eslava Eljaiek, C. Martínez and J.C. Narváez Barandica. 2012. Descripción de la morfología dental y del hábito alimentario del besote *Joturus pichardi* (Mugiliformes: Mugilidae). Bol. Invest. Mar. Cost. 41: 463–470.

Cotta-Ribeiro, T. and H. Molina-Ureña. 2009. Ontogenic changes in the feeding habits of the fishes *Agonostomus monticola* (Mugilidae) and *Brycon behreae* (Characidae), Térraba River, Costa. Rica. Rev. Biol. Trop. 57: 285–290.

Cruz, G.A. 1987. Reproductive biology and feeding habits of cuyamel, *Joturus pichardi* and tepemechín, *Agonostomus monticola* (Pisces: Mugilidae) from Río Plátano, Mosquitia, Honduras. Bull. Mar. Sci. 40: 63–72.

Dankwa, H.R., J. Blay, Jr. and K. Yankson. 2005. Food and feeding habits of grey mullets (Pisces: Mugilidae) in two estuaries in Ghana. W. Afr. J. Appl. Ecol. 8: 1–13.

Das, H.P. 1977. Food of the grey mullet *Mugil cephalus* (L.) from the Goa region. Mahasagar 10: 35–43.

De Niro, M.J. and S. Epstein. 1978. Influence of diet on the distribution of carbon isotopes in animals. Geochim. Cosmochim. Acta 42: 495–506.

De Niro, M.J. and S. Epstein. 1981. Influence of diet on the distribution of nitrogen isotopes in animals. Geochim. Cosmochim. Acta 45: 341–351.

De Silva, S.S. and M.J.S. Wijeyaratne. 1977. Studies on the biology of young grey mullet, *Mugil cephalus* L. II. Food and feeding. Aquaculture 12: 157–167.

Drake, P., A.M. Arias and L. Gállego. 1984. Biología de los mugílidos (Osteichthyes, Mugilidae) en los esteros de las salinas de San Fernando (Cádiz). III. Hábitos alimentarios y su relación con la morfometría del aparato digetivo. Inv. Pesq. 48: 337–367.

Durand, J.-D., W.-J. Chen, K.-N. Shen, C. Fu and P. Borsa. 2012a. Genus-level taxonomic changes implied by the mitochondrial phylogeny of grey mullets (Teleostei: Mugilidae). C.R. Biologies 335: 687–697.

Durand, J.-D., K.-N. Shen, W.J. Chen, B.W. Jamandre, H. Blel, K. Diop, M. Nirchio, F.J. Garcia de León, A.K. Whitfield, C.-W. Chang and P. Borsa. 2012b. Systematics of the grey mullets (Teleostei: Mugiliformes: Mugilidae): molecular phylogenetic evidence challenges two centuries of morphology-based taxonomy. Mol. Phyl. Evol. 64: 73–92.

Ebeling, A.W. 1957. The dentition of eastern Pacific mullets, with special reference to adaptation and taxonomy. Copeia 1957: 173–185.

Edgar, G.J. and C. Shaw. 1995. The production and trophic ecology of shallow-water fish assemblages in southern Australia. II. Diets of fishes and trophic relationships between fishes and benthos at Western Port, Victoria. J. Exp. Mar. Biol. Ecol. 194: 83–106.

Eggold, B.T. and P.J. Motta. 1992. Ontogenetic dietary shifts and morpholocial correlates in striped mullet, *Mugil cephalus*. Environ. Biol. Fish 34: 139–158.

El-Sayed, A.-F.M. 1991. Protein requirements for optimum growth of *Liza ramada* fry (Mugilidae) at different water salinities. Aquat. Living Resour. 4: 117–123.

Establier, R., J. Blasco, M. Gutiérrez, M.C. Sarasquete and E. Bravo. 1985. Enzimas en organismos marinos:III. Actividad α-amiásica en diversos órganos de mugílidos. Inv. Pesq. 49: 255–259.

Fagade, S.O. and C.I.O. Olaniyon. 1973. The food and feeding interrelationship of the fishes of Lagos lagoon. J. Fish Biol. 5: 205–227.

Faye, D., L. Tito de Morais, J. Raffray, O. Sadio, O. Thiom Thiaw and F. Le Loc'h. 2011. Structure and seasonal variability of fish food webs in an estuarine tropical marine protected area (Senegal): Evidence from stable isotope analysis. Est. Coast. Shelf Sci. 92: 607–617.

Ferrari, I. and A.R. Chieregato. 1981. Feeding habits of juvenile stages of *Sparus aurata* L., *Dicentrachus labrax* L., and mugilids in brackish environments of the Po river delta. Aquaculture 25: 243–257.

França, S., R.P. Vasconcelos, S. Tanner, C. Máguas, M.J. Costa and H.N. Cabral. 2011. Assessing food web dynamics and relative importance of organic matter sources for fish species in two Portuguese estuaries: A stable isotope approach. Mar. Environ. Res. 72: 204–215.

Fry, B., P.L. Mumford, F. Tam, D.D. Fox, G.L. Warren, K.E. Havens and A.D. Steinman. 1999. Trophic position and individual feeding histories of fish from Lake Okeechobee, Florida. Can. J. Fish. Aquat. Sci. 56: 590–600.

Ghazzawi, F.M. 1935. The pharynx and intestinal tract of the Egyptian mullets *Mugil cephalus* and *Mugil capito*. Part II. On the morphology and histology of the alimentary canal in *Mugil capito* (Tobar). Notes and Memoirs of the Fisheries Research Directorate, Cairo. 31pp.

Giarrizzo, T., R. Schwamborn and U. Saint-Paul. 2011. Utilization of carbon sources in a northern Brazilian mangrove ecosystem. Est. Coast. Shelf Sci. 95: 447–457.

Gisbert, E., L. Cardona and F. Castelló. 1995. Competition between mullet fry. J. Fish Biol. 47: 414–420.

Gisbert, E., L. Cardona and F. Castelló. 1996. Resource partitioning among planktivorous fish larvae and fry in a Mediterranean coastal lagoon. Est. Coast. Shelf Sci. 43: 723–735.

Gisbert, E., L. Cardona and F. Castelló. 1997. Diel feeding rhythm of grey mullet fry in northeastern Spain. Vie Milieu 47: 47–51.

Guinea, J. and F. Fernández. 1991. The effect of SDA, temperature and daily rhythm on the energy metabolism of the mullet *Mugil saliens*. Aquaculture 97: 353–364.

Guinea, J. and F. Fernández. 1992. Morphological and biometrical study of the gill rakers in four species of mullet. J. Fish Bio. 41: 381–397.

Hajisamae, S., L.M. Chou and S. Ibrahim. 2004. Feeding habits and trophic relationships of fishes utilizing an impacted coastal habitat, Singapore. Hydrobiologia 520: 61–71.

Harrison, I.J. and G.J. Howes. 1991. The pharyngobranchial organ of mugilid fishes; its structure, variability, ontogeny, possible function and taxonomic utility. Bull. Brit. Mus. Nat. Hist. (Zool.) 57: 111–132.

Heldt, H. 1948. Contribution à l'étude de la biologie des muges des lacs tunisiens. Bull. Stat. Océanogr. Salammbô 41: 1–35.

Hiatt, R.W. 1944. Food-chains and the food cycle in Hawaiian fish ponds. Part I. The food and feeding habits of mullet (*Mugil cephalus*), milkfish (*Chanos chanos*), and the ten-pounder (*Elops machnata*). Trans. Am. Fish Soc. 74: 250–261.

Hickling, C.F. 1970. A contribution to the natural history of English grey mullet (Pisces: Mugilidae). J. Mar. Biol. Ass. UK 50: 609–633.

Hussain, N.A., H.A. Saod and E.J. Al-Shami. 2009. Specialization, competition and diet overlap of fish assemblages in the recently restored southern Iraqi marshes. Marsh Bull. 4: 21–35.

Ibáñez-Aguirre, A.L. 1993. Coexistence of *Mugil cephalus* and *M. curema* in a coastal lagoon in the Gulf of Mexico. J. Fish Biol. 42: 959–961.

Iverson, S.J., W. Chrisfield, D. Bowen and W. Blanchard. 2004. Quantitative fatty acid signature analysis: a new method of estimating predator diets. Ecol. Mon. 74: 211–235.

Johar, M.A.O. and N.K. Singh. 1981. The food and feeding habits of *Sicamugil cascasia* (Ham.). Biol. Bull. India 3: 191–192.

Karpouzi, V.S. and K.I. Stergiou. 2003. The relationships between mouth size and shape and body length for 18 species of marine fishes and their trophic implications. J. Fish Biol. 62: 1353–1365.

Kasimoglu, C. and F. Yilmaz. 2012. Feeding habits of the thin-lipped grey mullet, *Liza ramada*, in Goekova Bay in the southern Aegean Sea. Zool. Mid. East 56: 55–61.

King, R.P. 1988. Observations on *Liza falcipinnis* (Valenciennes 1836) in the Bonny River, Nigeria. Rev. Hydrobiol. Trop. 21: 63–70.

Köse, S., S. Koral, Y. Özogu and B. Tufan. 2010. Fatty acid profile and proximate composition of Pacific mullet (*Mugil so-iuy*) caught in the Black Sea. Int. J. Food Sci. Tech. 45: 1594–1602.

Koussoroplis, A.-M., A. Bec, M.-E. Perga, E. Koutrakis, C. Desvilettes and G. Bourdier. 2010. Nutritional importance of minor dietary sources for leaping grey mullet *Liza saliens* (Mugilidae) during settlement: insights from fatty acid $\delta^{13}C$ analisis. Mar. Ecol. Prog. Ser. 404: 207–217.

Koussoroplis, A.-M., A. Bec, M.-E. Perga, E. Koutrakis, G. Bourdier and C. Desvilettes. 2011. Fatty acid transfer in the food web of a coastal Mediterranean lagoon: evidence for high arachidonic acid retention in fish. Est. Coast. Shelf Sci. 91: 50–461.

Laffaille, P., E. Feunteun, C. Lefebvre, A. Radureau, G. Sagan and J.-C. Lefeuvre. 2002. Can thin-lipped mullet directly exploit the primary and detritic production of European macrotidal Salt Marshes? Est. Coas. Shelf Sci. 54: 729–736.

Larson, E.T. and A.L. Shanks. 1996. Consumption of marine snow by two species of juvenile mullet and its contribution to their growth. Mar. Ecol. Prog. Ser. 130: 19–28.

Lebreton, B., P. Richard, E.P. Parlier, G. Guillou and G.F. Blanchard. 2011. Trophic ecology of mullets during their spring migration in a European saltmarsh: a stable isotope study. Est. Coast. Shelf Sci. 91: 502–510.

Li, R., A. Yokota, J. Sugiyama, M. Watanabe, M. Hiroki and M.M. Watanabe. 1998. Chemotaxonomy of planktonic cyanobacteria based on non-polar and 3-hydroxy fatty acid composition. Phycol. Res. 46: 21–28.

Lin, H., W.-Y. Kao and Y.-T. Wang. 2007. Analyses of stomach contents and stable isotopes reveal food sources of estuarine detritivorous fish in tropical/subtropical Taiwan. Est. Coast. Shelf Sci. 73: 527–537.

Luther, G. 1962. The food habits of *Liza macrolepis* (Smith) and *Mugil cephalus* Linnaeus (Mugilidae). Ind. J. Fish 9: 604–626.

Mallo, S., F. Vallespinós, S. Ferrer and D. Vaqué. 1993. Microbial activities in estuarine sediments (Ebro Delta, Spain) influenced by organic matter influx. Sci. Mar. 57: 31–40.

Marais, J.F.K. 1978. Routine oxygen consumption of *Mugil cephalus*, *Liza dumerili* and *Liza richardsoni* at different temperatures and salinties. Mar. Biol. 50: 9–16.

Marais, J.F.K. 1980. Aspects of food intake, food selection, and alimentary canal morphology of *Mugil cephalus* (Linnaeus 1958), *Liza tricuspidens* (Smith 1935), *L. richardsoni* (Smith 1846), and *L. dumerili* (Steindachner 1869). J. Exp. Mar. Biol. Ecol. 44: 193–209.

Marais, J.F.K. and T. Erasmus. 1977. Chemical compositon of alimentary canan contents of mullet (Teleostei: Mugilicade) caught in the Swartkops estuary near Port Elizabeth, South Africa. Aquaculture 10: 263–273.

Mariani, A., S. Panella, G. Monaco and S. Cataudella. 1987. Size analysis of inorganic particles in the alimentary tracts of Mediterranean mullet species suitable for aquaculture. Aquaculture 62: 123–129.

Martínez del Rio, C., N. Wolf, S.A. Carleton and L.Z. Gannez. 2009. Isotopic ecology ten years after a call for more laboratory experiments. Biol. Rev. 84: 91–111.

Masson, H. and J.F.K. Marais. 1975. Stomach content analysis of mullet from the Swartkops estuary. Zoologia Africana 10: 193–207.

Moriarty, D.J.W. 1976. Quantitative studies on bacteria and algae in the food of the mullet *Mugil cephalus* L. and the prawn Metapenaeus bennettae (Racek & Dall.). J. Exp. Mar. Biol. Ecol. 22: 131–143.

Mousa, M.A. 2010. Induced spawning and embryonic development of *Liza ramada* reared in freshwater ponds. Anim. Repr. Sci. 199: 115–122.

Munshi, J.S., J. Munshi and K. Singh. 1991. Graphic representation of the food and feeding habits of certain animals of River Ganga: Mode of feeding and dimension of food items. J. Fresh. Biol. 3: 45–58.

Nash, C.E., C.-M. Kuo and S. McConnel. 1974. Operational procedures for rearing larvae of the grey mullet (*Mugil cephalus* L.). Aquaculture 3: 15–24.

Odum, W.E. 1968a. The ecological significance of the fine particulate selection by the striped mullet *Mugil cephalus*. Limnol. Oceanogr. 13: 92–97.

Odum, W.E. 1968b. Mullet grazing on a dinoflagellate bloom. Chesapeake Sci. 9: 202–204.

Odum, W.E. 1970. Utilization of the direct grazing and plant detritus food chains by the striped mullet *Mugil cephalus*. pp. 220–240. *In*: J.H. Steele (ed.). Marine Food Chains. Oliver & Boyd, London.

Odum, W.E. 1988. Comparative ecology of tidal freshwater and salt marshes. An. Rev. Ecol. Syst. 19: 147–176.

Osorio Dualiby, D. 1988. Ecología trófica de *Mugil curema*, *M. incilis* y *M. liza* (Pisces: Mugilidae) en la Ciénaga Grande de Santa Marta, Caribe colombiano. Análisis cualitativo y cuantitativo. An. Inst. Inv. Mar. Punta de Betín, Colombia 18: 113–126.

Özogul, Y. and F. Özogul. 2007. Fatty acid profiles of commercially important fish species from the Mediterranean, Aegean and Black Seas. Food Chem. 100: 1634–1638.

Papaparaskeva-Papoutsoglou, E. and M.N. Alexis. 1986. Protein requirements of young mullet, *Mugil capito*. Aquaculture 52: 105–115.

Parnell, A., R. Inger, S. Bearhop and A.L. Jackson. 2010. Source partitioning using stable isotopes: coping with too much variation. PLoS ONE 5(3): e9672.

Paterson, A.W. and A.K. Whitfield. 1997. Stable carbon isotope study of the food-web in a freshwater-deprived South African estuary, with particular emphasis on the ichthyofauna. Est. Coast. Shelf Sci. 45: 705–715.

Patil, V., T. Kallqvist, E. Olsen, G. Vogt and H.R. Gislerød. 2007. Fatty acid composition of 12 microalgae for possible use in aquaculture feed. Aquacult. Int. 15: 1–9.

Payne, A.I. 1976. The relative abundance and feeding habits of the grey mullet species occurring in an estuary in Siere Leone, West Africa. Mar. Biol. 35: 277–286.

Payne, A.I. 1978. Gut pH and digestive strategies in estuarine grey mullet (Mugilidae) and tilapia (Cichlidae). J. Fish Biol. 13: 627–629.

Phillip, D.A.T. 1993. Reproduction and feeding of the mountain mullet, *Agonostomus monticola*, in Trinidad, West Indies. Env. Biol. Fishes 37: 47–55.

Pillay, V.T. 1953. Studies on the food, feeding habits and alimentary tract of the grey mullet, *Mugil tade* Forsskål. Proc. Nat. Inst. Sci. India 19: 777–827.

Platell, M.E., P.A. Orr and I.C. Potter. 2006. Inter- and intra-specific portioning of food resources by six large and abundant fish species in a seasonally open estuary. J. Fish Biol. 69: 243–262.

Post, D. 2002. Using stable isotopes to estimate trophic position: models, methods, and assumptions. Ecology 83: 703–718.

Post, D.M., C.A. Layman, D.A. Arrington, G. Takimoto, J. Quattrochi and C.G. Montaña. 2007. Getting to the fat of the matter: models, methods and assumptions for dealing with lipids in stable isotope analyses. Oecologia 152: 179–189.

Quan, W., C. Fu, B. Jin, Y. Luo, B. Li, J. Chen and J. Wu. 2007. Tidal marshes as energy sources for commercially important nektonic organisms: stable isotope analysis. Mar. Ecol. Prog. Ser. 352: 89–99.

Quan, W., L. Shi and Y. Chen. 2010. Stable isotopes in aquatic food web of an artificial lagoon in the Hangzhou Bay, China. Chin. J. Oceanol. Limnol. 28: 489–497.

Quan, W.M., D.Q. Huang, T.J. Chu, Q. Sheng, C.Z. Fu, J.K. Chen and J.H. Wu. 2009. Trophic relationships in the Yangtze River estuarine salt marshes: preliminary investigation from δ13C and δ15N analysis. Acta Ocean. Sin. 28: 50–58.

Recks, M.A. and G.T. Seaborn. 2008. Variation in fatty acid composition among nine forage species from a southeastern US estuarine and nearshore coastal ecosistema. Fish Physiol. Biochem. 34: 275–287.

Romer, G.S. and A. McLachlan. 1986. Mullet grazing on surf diatom accumulations. J. Fish Biol. 28: 93–104.

Ruberstein, D.I. and M.A.R. Kohel. 1977. The mechanisms of filter-feeding: some theoretical considerations. Am. Nat. 111: 981–994.

Sánchez, A., L. Cardona and F. Castelló. 1993. Crecimiento de alevines de *Liza ramada* (Osteichthyes, Mugilidae) en agua dulce: efectos de la alimentación. Actas IV Conres Nac. Acuicult. 91–96.

Sánchez Rueda, P. 2002. Stomach content analysis of *Mugil cephalus* and *Mugil curema* (Mugiliformes: Mugilidae) with emphasis on diatoms in the Tamiahua lagoon, Mexico. Rev. Biol. Trop. 50: 245–252.

Sargent, J.R. 1997. Fish oils and human diet. Brit. J. Nutr. 78(Suppl. 1): S5–S13.

Şengör, G.F., Ö. Özden, N. Erkan, M. Tüter and H.A. Aksoy. 2003. Fatty acid compositions of flathead grey mullet (*Mugil cephalus* L. 1758) fillet, raw and beeswaxed caviar oils. Tur. J. Fish Aquat. Sci. 3: 93–96.

Service, S., R.J. Feller, B.C. Coull and R. Woods. 1992. Predation effect of three fish species and a shrimp on macrobenthos and meiobenthos in microcosmos. Est. Coast. Shelf Sci. 34: 277–293.

Sheppard, J.N., N.C. James, A.K. Whitfield and P.D. Cowle. 2011. What role do beds of submerged macrophytes play in structuring estuarine fish assemblages? Lessons from a warm-temperate South African estuary. Est. Coast. Shelf Sci. 95: 145–155.

Surendra Babu, K. and B. Neelakantan. 1983. Biology of *Liza parsia* in the Kali estuary, Karwar. Mahasagar-Bull. Nat. Inst. Ocean. 16: 381–389.

Tandel, S.S., R.P. Athalye, K.S. Gokhale and B.N. Bandodkar. 1986. On the seasonal changes in food habit of *Mugil cephalus* in the Thana Creek. Indian J. Fish 33: 270–276.

Thimdee, W., G. Deein, N. Nakayama, Y. Suzuki and K. Matsunaga. 2008. $\delta^{13}C$ and $\delta^{15}N$ indicators of fish and shrimp community diet and trophic structure in a mangrove ecosystem in Thailand. Wetl. Ecol. Manag. 16: 463–470.

Thomas, B. and R.M. Connolly. 2001. Fish use of subtropical saltmarshes in Queensland, Australia: relationships with vegetation, water depth and distance onto the marsh. Mar. Ecol. Prog. Ser. 209: 275–288.

Thomson, J.M. 1953. Growth and habits of the sea mullet, *Mugil dobula* Gunther, in Western Australia. Aust. J. Mar. Freshw. Res. 2: 193–225.

Thomson, J.M. 1954. The organs of feeding and the food of some Australian mullet. Aust. J. Mar. Freshw. Res. 5: 469–485.

Thomson, J.M. 1966. The grey mullets. Oceanogr. Mar. Biol. Ann. Rev. 4: 301–335.

Torras, X., L. Cardona and E. Gisbert. 2000. Cascading effects of flathead grey mullet *Mugil cephalus* on the ecosystem of eutrophic freshwater microcosmos. Hydrobiologia 429: 49–57.

Torres-Navarro, C.I. and J. Lyons. 1999. Diet of *Agonostomus monticola* (Pisces: Mugilidae) in the Río Ayuquila, Sierra de Manantlán Biosphere Reserve, México. Rev. Biol. Trop. 47: 1087–1092.

Torricelli, P., P. Tongiorgi and P. Almansi. 1981. Migration of grey mullet fry into the Arno river: seasonal appearance, daily activity, and feeding rhythms. Fish Res. 1: 219–234.

Torricelli, P., P. Tongiorgi and G. Gandolfi. 1988. Feeding habits of mullet fry in the Arno River (Tyrrhenian coast). I. Daily feeding cycle. Boll. Zool. 3: 161–169.

Tosi, L. and P. Torricelli. 1988. Feeding habits of mullet fry in the Arno River (Tyrrhenian coast). II. The die. Boll. Zool. 3: 171–177.

Trellu, J., J. Albertini-Berhaut and H.J. Ceccaldi. 1978. Caractérisation de quelques activités enzymatiques digestives chez *Mugil capito* en relation avec la taille. Biochem. Syst. Ecol. 6: 255–259.

Vaslet, A., D.L. Phillips, C. France, I.C. Feller and C.C. Baldwin. 2011. The relative importance of mangroves and seagrass beds as feeding areas for resident and transient fishes among different mangrove habitats in Florida and Belize: evidence from dietary and stable-isotope analyses. J. Exp. Mar. Biol. Ecol. 434-435: 81–93.

Vinagre, C., J. Salgado, H.N. Cabral and M.J. Costa. 2011. Food web structure and habitat connectivity in fish estuarine nurseries—impact of river flow. Est. Coast. 34: 663–674.

Vizzini, S. and A. Mazzola. 2008. The fate of organic matter sources in coastal environments: a comparison of three Mediterranean lagoons. Hydrobiologia 611: 67–79.

Whitfield, A.K. 1986. Fish community structure response to major habitat changes within the littoral zone of an estuarine coastal lake. Env. Biol. Fish 17: 41–51.

Whitfield, A.K. and S.J.M. Blaber. 1978. Resource segregation among iliophagous fish in Lake St. Lucia, Zululand. Env. Biol. Fish 3: 293–296.

Wijeyaratne, M.J.S. and H.H. Costa. 1988. The food, fecundity and gonadal maturity of *Valamugil cunnesius* (Pisces: Mugilidae) in the Negombo Lagoon, Sri Lanka. Indian J. Fish 35: 71–77.

Wijeyaratne, M.J.S. and H.H. Costa. 1990. Food and feeding of two species of grey mullets *Valamugil buchanani* (Bleeker) and *Liza vaigiensis* Quoy and Gaimard inhabiting brackishwater environments in Sri Lanka. Indian Fish 37: 211–219.

Zismann, L., V. Berdugo and B. Kimor. 1975. The food and feeding habits of early stages of grey mullets in the Haifa Bay region. Aquaculture 6: 59–75.

Zouiten, D., I. Ben Khemis, R. Besbes and C. Cahu. 2008. Ontogeny of the digestive tract of thick lipped grey mullet (*Chelon labrosus*) larvae reared in "mesocosms". Aquaculture 279: 166–172.

CHAPTER 10

Age and Growth of Mugilidae

Ana L. Ibáñez

Introduction

Almost 35 years ago Quignard and Farrugio (1981) reviewed the subject of grey mullet age and growth. Since then, quite a number of studies have been carried out, particularly on commercially important species, and there is now much information in the Fishbase data base (Froese and Pauly 2014) and other specialized data bases (ISI Web of Knowledge, Scopus, Biological Abstract). A review showed that some areas like the Mediterranean Sea and the North-eastern Atlantic have been intensely studied, while information is very scarce for the rest of the world. Many studies deal with age and growth, but concentrate on a few species such as thicklip grey mullet *Chelon labrosus*, three species of the genus *Liza* (golden grey mullet *Liza aurata*, thinlip grey mullet *L. ramada* and leaping mullet *L. saliens*) and of course the species with the widest geographical distribution, flathead grey mullet *Mugil cephalus*. Recent studies have dealt with the mullets of the American coasts, mainly the eastern American coasts, as well as those of western Africa. Given the present availability of studies and publications, this chapter analyzes the information on age, absolute growth and relative growth, taking into account the geographical distribution and the methods used to record the age, sex and habitat, among other aspects.

Each section summarizes the main points of the information and the method of analysis used, as each study was carried out with a different number of data, depending on availability. Apart from the information in Fishbase (Froese and Pauly 2014) and in Quignard and Farrugio (1981), approximately 60 papers published after 1981 were reviewed, providing an extensive list of publications. The marked variability with which the information is presented resulted in some missing data. These were obtained from other sources or were deduced from the study area, as was the case for salinity values. In some cases however, it was not possible to provide complete information.

The information presented here provides some particularly interesting data, such as the fact that the recording of the size at each age differs between the Mediterranean and the NE Atlantic. This was validated by records obtained for the commercial species of the genus *Liza* and the species *Chelon labrosus*. It was also noteworthy that the condition index '*b*' of the size-weight relationship does not vary among species, geographical areas or sexes and it barely reaches isometry, which may respond to most of the collected specimens being either young or slim young adults. A reduction in the estimation of L_∞ for the Mugilidae family during the last three decades is very obvious, and is considered a warning sign for the fisheries that

Metropolitan Autonomous University, Hydrobiology Department, Av. San Rafael Atlixco 186, Col. Vicentina, México, D.F. 09340 México.
Email: ana@xanum.uam.mx

are based on this family. Also, significant differences were observed in the size values recorded during the first year of life between brackish and marine waters, though no differences were recorded for subsequent ages. These results stand out among others. The greatest and most discussed problem in the review of studies focused on age, that determined the first growth mark, for which various authors have recorded different dates of appearance. It is considered necessary to start a phase of comparative and/or experimental analyses in the field, designed to gain knowledge by using hard structures (scales, otoliths, vertebrae), particularly in the case of the first years of life, in order to make significant progress with this problem.

Considerations on Growth Categories

Growth is understood as the increase in length or weight generated by physiological processes, and it is directly related to age. Thus a population of fish includes specimens of both sexes, recently born small fish and very old fish. The curve that is best adjusted to the growth curve is the type S or sigmoid curve, which illustrates the absolute growth of an individual or population throughout its life. Thus, the parameters of the von Bertalanffy equation are used here to analyze absolute growth.

$$\mathrm{Tl} = L_{\infty}[1 - e^{(-k(t-t_0))}]$$

where: Tl is the length at age 't', L_{∞} is the average length a fish could acquire in the case of growing to a very old age (in fact, infinite), k is the growth coefficient, and t_0 is the hypothetical 'age' a fish should be when its length is zero, if it had always grown according to the equation (t_0 generally has a negative value as at age zero—that is, at the moment of birth—the fish already have a certain length).

Also, fish present different proportions in different parts of the body at different ages, with growth being proportional with respect to size. This type of growth is called relative growth and is defined as the increase in length or weight during a certain period of time with respect to the length or weight at the start of that period. It is expressed numerically or in percentages.

The differential growth of the different parts of the body of an organism with respect to its total growth is known as relative growth. Huxley (1932) was the first author to establish the basis for the study of relative growth or allometry in his book 'Problems of Relative Growth'.

Relative growth may be given between a linear and a volume measurement as the relationship between the length or size (as a linear measurement) and the weight, in which case, in order for the body proportions to remain equal, has the form:

$$W = a.Tl^3$$

where: W is the body weight, Tl is the total body length of a fish at age 't', a is the constant or intercept, and 3 is the slope of the relationship called 'b' or condition index. When growth is isometric, that is, when the size and weight increase in the same proportion, the value of the slope is equal to 3. If b is smaller or bigger than 3 however, relative growth does not occur in the same proportion.

Knowledge of relative growth is of great interest in the biological sciences as it has several practical applications, such as to relate the proportions of the different body structures to the ontogenetic habits of a species associated with movement and the capture of food, and to detect specimen sex, subspecies and populations. It may also be used as a condition index (Safran 1992) and, from the point of view of the Chaos theory, as a fractal. This text focuses on the application of this allometry law in the first sense.

In general, a condition index 'b' smaller than 3 indicates the sample is of slim individuals in which length increases more than weight, whereas when b is bigger than 3 the individuals in the sample are robust. The interpretation of the condition index provides information on the general shape of the population under study. Thus, $b < 3.0$ indicates a decrease in condition or elongation in form with an increase in length, and $b > 3.0$ indicates an increase in condition or increase in height or width with an increase in length, and the larger the distance from 3.0, the larger the change in condition or form.

In addition, the relative growth between two linear measurements such as the cephalic length (Cl) and the size or total length (Tl) is compared, although in this case '*b*' has values near one when both linear measurements increase in the same proportion.

$$Cl = a\ .\ Tl^{1}$$

which may be made linear as: $\log x_i = \log a + \beta \log x_1$.

In the case of two linear variables, $b < 1$ indicates that the dependent variable increases less than the size, whereas $b > 1$ indicates that it increases more than the size. Thus, during fish ontogeny, different parts of the body grow at different rates according to the species and the population. When $b = 1$ the organism grows evenly, as if it were made bigger in a photocopier, that is to say, it changes 'proportionally'.

Availability of Information for the Exploration of Growth Data

A literature review was carried out, with at least half of the information obtained from the Fishbase data base (Froese and Pauly 2014) and the previous review of Quignard and Farrugio (1981). Thus this study was based on three data bases, two to analyze absolute growth: (1) Length-age data base and (2) von Bertalanffy parameters data base, and one to analyze relative growth: (3) Length-weight relationship data base.

In order to be able to compare the information, lengths were converted from fork length and standard length to total length (Tl) in agreement with Quignard and Farrugio (1981) for *Mugil cephalus*, *Chelon labrosus*, *L. ramada*, *L. aurata*, *L. saliens* and *Moolgarda cunnesius*. In the case of *M. curema*, the conversion of Santana et al. (2009) was used. The original size values were used for the other species as no regressions were available for the various sizes.

In most cases the information was initially examined for the whole Mugilidae family, after which it was analyzed for the five species (thicklip grey mullet *Chelon labrosus*, golden grey mullet *Liza aurata*, thinlip grey mullet *L. ramada*, leaping mullet *L. saliens* and flathead grey mullet *Mugil cephalus*) for which the availability of data was sufficient to allow statistical analyses. ANOVAs were used when the homoscedasticity and normality of the parameters were satisfied. Otherwise, non-parametric analyses were used.

In order to carry out a comparative analysis, geographical areas were classified as: Pacific and Atlantic oceans divided into north, south, east and west, the Gulf of Mexico and Caribbean Sea, Mediterranean Sea, Indian Ocean and South China Sea. The information on salinity was classified as: freshwater (0–5 psu), brackish (5–30 psu) and marine (> 30 psu). Authors' data on salinity were taken as valid, but when no data were found, salinity values were assigned according to the geographical location of the study after a review of the available information on the area. For example, the Caspian Sea was classified as a brackish environment.

The methods by which age was recorded were catalogued as: indirect size frequency method, direct scale reading methods, and reading of otoliths and spines. The reading of operculi or other hard parts, as well as of marks, was included as 'others'.

Five species are the most frequently studied, as was mentioned in the introduction. In order to avoid excluding the other species of the family however, information is provided for some of these. Each section where information is analyzed presents its characteristics, the methods that were used and the details of the analyses that were applied.

Absolute Growth

Length-age data base

This includes the size for the age groups L1 to L6, the method by which the age was calculated, the species, the geographical area of the study and the salinity of the environment classified as freshwater, brackish and marine. This information applies to an N = 110 distributed among eight species. However, the analysis of only five species is presented here (thicklip grey mullet *Chelon labrosus*, golden grey mullet *Liza aurata*, thinlip grey mullet *L. ramada*, leaping mullet *L. saliens* and flathead grey mullet *Mugil cephalus*) as it is for these that there is the necessary or at least sufficient information to carry out comparisons.

von Bertalanffy parameters data base

The available information includes the parameters of the von Bertalanffy function: t_0, L_∞ and k, which are referenced to the geographical area, the method by which the age of each sex was calculated and the salinity of the environment.

ϕ' was estimated in order to compare the growth performance of species with similar shapes (Pauly and Munro 1984). According to Pauly and Munro (1984), ϕ' represents and quantifies the energetics of a given habitat or niche as it is directly related to growth performance, and hence to metabolism and food consumption. This indicator helps compare growth. Notwithstanding that information is available for 24 species of the family, it is again only for the same five species (thicklip grey mullet *Chelon labrosus*, golden grey mullet *Liza aurata*, thinlip grey mullet *L. ramada*, leaping mullet *L. saliens* and flathead grey mullet *Mugil cephalus*) that there is sufficient information to carry out a comparative analysis among species, geographical areas, sex, and method by which the age is calculated.

The analysis uses descriptive statistics expressed through averages for the central tendency, and data dispersion was described as the standard deviation. An inferential analysis was carried out with parametric statistical tests (MANOVAs and ANOVAs) and a Mann-Whitney's U non-parametric test. In most cases, the information was analyzed initially for the whole Mugilidae family, followed by the analysis of the five species for which the availability of data was sufficient for statistical analyses.

Relative Growth

Length-weight relationship data base

This includes the parameters of the relationship, the intercept '*a*' and the slope '*b*', the geographical area, the sex and the salinity of the environment. Of the two parameters of the length-weight relationship, only '*b*' is used as the measurement unit does not affect this exponent. As it does affect the estimation of '*a*', it is best to use only the most reliable information.

Morphometric relationships

Studies that are important in this section are only commented on, as the limited information on morphometric growth makes it impossible to carry out statistical analyses with the information that is available at present.

Results of the Absolute Growth Analysis

Length-Age Examination

The length of the first six age groups was compared using a data base of 104 values obtained directly from authors' reports. Most of the information came from the study of the five species *Liza aurata*, *L. ramada*, *L. saliens*, *Chelon labrosus* and *Mugil cephalus* with 41, 22, 8, 10 and 23 reports of length at age for each species. Few reports of size for each age exist for the Mugilidae family, and only the species of greatest commercial importance have been studied. Also, most of the information corresponds to the geographical area of the Mediterranean and NE Atlantic (54.8 and 30.8% respectively). The information for the Gulf of Mexico and Caribbean Sea, the Indian Ocean, the Caspian Sea, the SW Pacific and the SE Pacific corresponds to 4.8, 3.8, 3.8, 1 and 1% respectively.

In order to analyze the similarities and differences among the sizes of each age group of the five species for which there is available information, box plot diagrams are presented for the first six age groups of *Mugil cephalus*, *Chelon labrosus*, *L. ramada*, *L. aurata* and *L. saliens* (Fig. 10.1). Notwithstanding that the information is unequal for the five species, it is clear that greater sizes have been recorded for thicklip

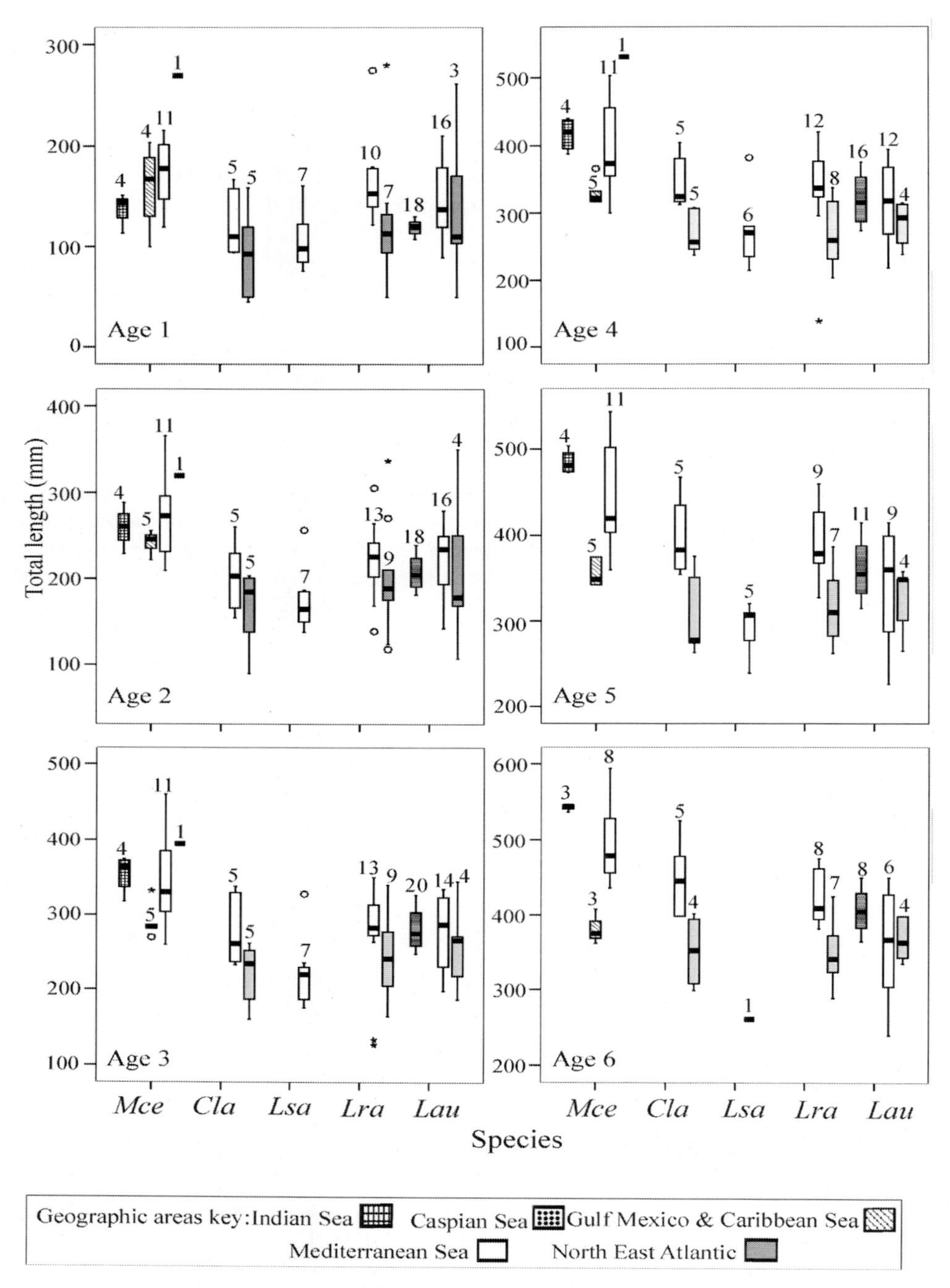

Figure 10.1. Boxplot showing total length by age for different geographic areas by species. Mce = *Mugil cephalus*, Cla = *Chelon labrosus*, Lsa = *Liza saliens*, Lra = *Liza ramada* and Lau = *Liza aurata*. Numbers are sample size.

grey mullet *Chelon labrosus*, thinlip grey mullet *L. ramada* and golden grey mullet *L. aurata* collected from the Mediterranean (white boxes), in contrast with the sizes recorded for specimens from the NE Atlantic (grey boxes). This may be verified for the six age groups that are presented. Similarly, the sizes reported for *M. cephalus* from the Mediterranean have been markedly greater than those of the Gulf of

Mexico and Caribbean Sea, while those of Australia have been greater than those of the Mediterranean with respect to the sizes of ages three to six. Differences in the determination of the total length (Tl) per age were assessed by MANOVAs, with Tl per age as the dependent variable and the species, method and environment as the grouping factor. MANOVA tests were run to assess the effect of the species (flathead grey mullet Mce = *Mugil cephalus*, thicklip grey mullet Cla = *Chelon labrosus*, leaping mullet Lsa = *Liza saliens*, thinlip grey mullet Lra = *L. ramada*, golden grey mullet Lau = *L. aurata*) on Total length (Tl) per age (from age one to age six), the method used (Length frequency (Lfr) and counting of Scales (Scl) or Otoliths (Oth)) and salinity (Freshwater, Brackish Water and Sea Water). The results indicated significant interspecific differences ($P = 0.008$) for the size of each age group, and significant differences among methods ($P = 0.007$) and salinities ($P = 0.003$) (Table 10.1).

Table 10.1. Manova test to assess the effect of species, method and salinity (fixed factors) on total length by age, from age one to age six (dependent variables). *P*-value of significance test.

Effect	**Wilks'λ**	**Fs**	***P***
Intercept	0.025	164.219	< 0.000
Species	0.212	2.061	0.008
Salinity	0.341	2.968	0.003
Method	0.371	2.663	0.007
All interactions			> 0.05

Significant differences among methods were recorded for the sizes of the first age group (one-way ANOVA) $P_{Age1} = 0.006$, with no significant differences for the other sizes ($P_{Age2} = 0.158$, $P_{Age3} = 0.735$, $P_{Age4} = 0.387$, $P_{Age5} = 0.709$, $P_{Age6} = 0.906$). In the case of salinity, significant differences were recorded only for the first age group $P_{Age1} = 0.034$ ($P_{Age2} = 0.053$, $P_{Age3} = 0.072$, $P_{Age4} = 0.938$, $P_{Age5} = 0.991$, $P_{Age6} = 0.875$). The lengths of each age group, from age one to age nine, may be seen in Table 10.2, and the average values and standard deviations for the different environments are presented in Table 10.3. Reports for freshwater environments are very scarce, as is the case for *Chelon labrosus*, *Liza saliens* and *L. ramada*, or non-existent as in the case of *L. aurata* and *M. cephalus*. In short, the size of each age group varies among species, and in the first age group it varies with respect to salinity and the method by which they were calculated.

Differences in age group sizes between the Mediterranean, with greater sizes, and the more northern Atlantic coasts have been previously recorded by different authors. When comparing mullet populations living in the Atlantic (Arné 1938, Hickling 1970, Reay 1987, Arruda et al. 1991) with those from the Mediterranean (Farrugio and Quignard 1974, Modrusan et al. 1988, Sinovcic et al. 1986) and the Bay of Cadiz (Drake et al. 1984), the latter ones have greater mean lengths per age group (Moura and Serrano-Gordo 2000). Hickling (1970) also recognized that the growth rate in the Mediterranean (Erman 1961, Morovic 1960), Morocco (Rossignol 1951) and the rather special environment of fish ponds (Le Dantec 1955) was greater than in Ireland or England, as could be expected due to the higher temperatures and longer growing seasons of the more southerly habitats. This is supported by reports that this species attains mean lengths of 181–191 mm during the same period in more southern waters (Wimpenny 1932, Arné 1938). Moreover, *M. cephalus* may acquire average total lengths of 178–220 mm in just under a year (Chubb et al. 1981), equal to those of three-year old *L. ramada* specimens in British waters (Hickling 1970).

Some studies have considered that there are no significant differences between the age read using scales or otoliths (Bentzel 1968, Snovsky and Shapiro 2000). Others have proposed the use of scales (Ghadirnejad 1996 in Fazli et al. 2008) as they consider otoliths to be very thick and the counting of marks difficult. Yet others have suggested that although using scales to estimate age is reliable (Bok 1983 in Ellender et al. 2012), they underestimate age, particularly in older specimens (Ellender et al. 2012). This is possible since juveniles are born naked (that is, without scales), for example *Chelon labrosus* pre-larvae are poorly developed at the moment of hatching, the juvenile stage is rapidly reached as the nostrils and homocercal tail appear as early as day 22 (p. h.), and by day 60 (p. h.) the fish are entirely covered by scales

Table 10.2. Mean total length ± SD (minimum and maximum values) at ages by species (mm).

	Tl 1	Tl 2	Tl 3	Tl 4	Tl 5	Tl 6	Tl 7	Tl 8	Tl 9
M. cephalus	168 ± 42(100–270)	265 ± 43(210–366)	336 ± 55(260–460)	393 ± 67(300–531)	432 ± 68(342–543)	481 ± 71(362–594)	540 ± 101(380–637)	559 ± 144(396–670)	
N =	20	21	21	21	20	14	5	3	
C. labrosus	109 ± 43(45–167)	183 ± 49(90–261)	250 ± 55(161–338)	309 ± 54(237–405)	354 ± 68(263–468)	406 ± 72(299–526)	437 ± 77(350–574)	474 ± 97(381–612)	485 ± 119(411–622)
N =	10	10	10	10	10	9	7	5	3
L. saliens	107 ± 30(76–161)	176 ± 40(138–257)	223 ± 52(176–328)	275 ± 58(214–382)	291 ± 33(240–320)	262	271		
N =	7	7	7	6	5	1	1		
L. ramada	149 ± 59(50–280)	215 ± 55(118–337)	265 ± 65(127–350)	313 ± 70(138–420)	367 ± 55(263–460)	394 ± 53(289–475)	374 ± 69(311–448)	359 ± 32(337–382)	379 ± 32(357–402)
N =	17	21	21	19	15	14	3	2	2
L. aurata	142 ± 53(50–262)	215 ± 54(107–350)	269 ± 48(187–345)	305 ± 50(217–395)	340 ± 51(226–415)	373 ± 59(240–450)	414 ± 61(324–480)	432 ± 71(345–535)	
N =	37	38	38	32	24	18	13	8	
M. cephalus	168 ± 42(100–270)	265 ± 43(210–366)	336 ± 55(260–460)	393 ± 67(300–531)	432 ± 68(342–543)	481 ± 71(362–594)	540 ± 101(380–637)	559 ± 144(396–670)	
N =	20	21	21	21	20	14	5	3	
M. curema	208 ± 25(179–223)	244 ± 10(233–252)	273 ± 6(267–278)	298 ± 11(285–305)	320 ± 16(302–332)	337 ± 28(317–357)	358 ± 37(331–384)	372 ± 41(343–401)	424
N =	3	3	3	3	3	2	2	2	1

Table 10.3. Total length at age (from age one to age nine) by species and salinity (mm).

Species	Salinity	Tl 1	Tl 2	Tl 3	Tl 4	Tl 5	Tl 6	Tl 7	Tl 8	Tl 9
C. labrosus	Freshwater	95 ± 1(94–95)	161 ± 8(155–166)	235 ± 3(233–237)	318 ± 8(312–324)	372 ± 16(361–383)	422 ± 33(399–445)	456 ± 44(425–487)	523	
	N =	2	2	2	2	2	2	2	1	
	Brackish water	130 ± 33(93–158)	205 ± 19(185–230)	272 ± 41(234–330)	315 ± 51(256–381)	360 ± 66(275–435)	426 ± 45(399–478)	437	467	
	N =	4	4	4	4	4	3	1	1	
	Marine water	96 ± 59(45–167)	173 ± 74(90–261)	235 ± 79(161–338)	299 ± 77(237–405)	340 ± 94(263–468)	382 ± 103(299–526)	427 ± 105(350–574)	460 ± 132(381–612)	485 ± 119(411–622)
	N =	4	4	4	4	4	4	4	3	3
L. saliens	Freshwater	90	145	176						
	N =	1	1	1						
	Brackish water	90 ± 12(81–98)	147 ± 12(138–155)	188 ± 2(186–189)	225 ± 15(214–235)	258 ± 26(240–277)	262	271		
	N =	2	2	2	2	2	1	1		
	Marine water	121 ± 35(76–161)	198 ± 40(165–257)	252 ± 51(220–328)	301 ± 54(270–382)	312 ± 7(307–320)				
	N =	4	4	4	4	3				
L. ramada	Freshwater	159	220 ± 72(169–271)	219±120(134-303)	332	357	378			
	N =	1	2	2	1	1	1			
	Brackish water	156 ± 72(79–280)	221 ± 61(124–337)	271 ± 57(165–350)	309 ± 69(203–420)	360 ± 61(263–460)	382 ± 54(289–475)	337 ± 37(311–363)	359 ± 32(337–382)	380 ± 32(357–402)
	N =	9	11	11	10	9	8	2	2	2
	Marine water	138 ± 45(50–180)	206 ± 49(118–242)	268 ± 69(127–332)	316 ± 81(138–377)	383 ± 54(294–427)	418 ± 54(331–462)	448		
	N =	7	8	8	8	5	5	1		
L. aurata	Brackish water	160 ± 58(105–262)	230 ± 56(165–350)	272 ± 51(198–345)	303 ± 53(217–392)	344 ± 62(226–415)	382 ± 75(240–450)	442 ± 31(408–480)	465 ± 23(444–490)	
	N =	16	18	16	12	7	6	4	3	
	Marine water	128 ± 45(50–211)	200 ± 50(107–280)	267 ± 46(187–335)	306 ± 49(229–395)	338 ± 48(265–415)	369 ± 52(301–450)	401 ± 68(324–480)	413 ± 85(345–535)	
	N =	21	20	22	20	17	12	9	5	
M. cephalus	Brackish water	176 ± 48(100–270)	267 ± 51(216–366)	333 ± 63(270–460)	391 ± 75(317–531)	415 ± 66(342–540)	444 ± 65(362–562)	380	396	
	N =	11	12	12	12	11	8	1	1	
	Marine water	160 ± 33(114–210)	261 ± 32(210–313)	341 ± 47(260–410)	396 ± 58(300–460)	452 ± 67(342–543)	530 ± 45(470–594)	580 ± 53(510–637)	641 ± 42(611–670)	
	N =	9	9	9	9	9	6	4	2	
M. curema	Brackish water	208 ± 25(179–223)	244 ± 10(233–252)	273 ± 6(267–278)	298 ± 11(285–305)	320 ± 16(302–332)	337 ± 28(317–357)	358 ± 37(331–384)	372 ± 41(343–401)	424
	N =	3	3	3	3	3	2	2	2	1

(Khemis et al. 2006). It is thus not possible to follow the life history of an organism for approximately two months. According to Smith and Deguara (2003), a comparison of results suggests that the use of scales may underestimate age in some cases. Scales or otoliths may be unreliably employed in both species to estimate age in both sexes. However, the best fit to the growth equation is obtained using the age-length determination through the analysis of otoliths (Ibáñez-Aguirre and Gallardo-Cabello 1996), where results have shown that only for the first age group are different sizes recorded following different methods to determine age. This may be related to the interpretation of the first growth mark. This is discussed later.

Age and Growth of Grey Mullet

Von Bertalanffy Parameters. Phi (ϕ') Comparison

The Phi (ϕ') parameter was analyzed using 357 data, most from Quignard and Farrugio (1981, 34.7%) and Fishbase (32.7%), with the other data reported by authors who on occasion listed their von Bertalanffy parameter values and compared them with those of other areas and/or species. Of the 357 data, those of *Aldrichetta forsteri* reported by Curtis and Shima (2005) were eliminated as they were extreme cases, regarding either the minimum or the maximum values. The final number of data in the data base was then 343, with 160 reports for the years 1932 to 1980, making up 46.6% of the data available for this analysis. Most of the studies (53.4%) were carried out over the last 30 years, from 1980 to the present, with 28.3% generated during the 1990s.

The genera *Liza* and *Mugil* represent 45.8 and 32.1% (together 77.9%) of the data, while *Chelon* and *Moolgarda* represent 12.5 and 2.0% respectively. Most of the information corresponds to the geographical area of the Mediterranean and NE Atlantic (49.2 and 12.4% respectively). For Australia, the Gulf of Mexico and Caribbean Sea, and the Indian Ocean, the percentages are 9.8, 9.3 and 7.6 respectively. The data base of the von Bertalanffy growth parameters includes 24 species, although there is little information for most of them (Table 10.4). As in the previous case, the most frequently studied species are *Mugil cephalus*, *Liza aurata*, *L. ramada*, *Chelon labrosus* and *L. saliens* (with 25.8, 15.4, 14.0, 12.1 and 11.0% respectively). The other species have been recorded very few times, and there are some species (*Moolgarda buchanani*, *M. cunnesius*, *M. seheli* and *Mugil trichodon*) for which there are only one to three reports.

Figure 10.2A presents the frequency distribution of ϕ' for the Mugilidae family, with values ranging from 1.82 for *L. carinata* in Pakistan (Abbas 2001) to 3.47 for *M. cephalus* in India (Jhingran and Misra 1962). The average values of ϕ' for the best studied species are shown in Fig. 10.2B, where *M. cephalus* presents the best growth performance with an average of 2.996 (± 0.208) and *L. saliens* presents the lowest average of 2.540 (± 0.174). As significant differences were recorded for the ϕ' of the five studied species ($P < 0.0001$), comparisons (ANOVAs) of the ϕ' values for the different variables were carried out for each species. Regarding salinity, only *L. ramada* showed significant differences in growth performance, as there is significantly less ($P = 0.017$) information on ϕ' for freshwater areas. However, once those four data were eliminated, no differences in performance were recorded (that is, in ϕ') between brackish and marine waters for the other species (Fig. 10.3A). Regarding the methods used, *L. aurata* had significantly different values, with estimations of ϕ' using otoliths to record age providing markedly smaller values (Fig. 10.3B), as in the case of *M. cephalus* ($P = 0.003$ and Mann-Whitney $U = 22.323$, $P < 0.001$, respectively). In the case of the other species, ϕ' did not vary in relation to the method used. With respect to the geographical area, the value of ϕ' calculated for *M. cephalus* was much smaller and significantly different for the NE Pacific, and the Gulf of Mexico and Caribbean Sea (Mann-Whitney $U = 12.000$, $P < 0.0001$). None of the other four species recorded a significant difference in ϕ'. Between the sexes, the only species that showed significant differences in ϕ' was *M. cephalus*, for which the value for the males was smaller than that for the females ($P = 0.002$) (Figs. 10.3C and D). It is evident that *M. cephalus* has a greater variability for each variable, basically because of its wide distribution. As a result of this, many different authors have dealt with this species and its different geographical areas, which in turn has produced a greater variance. This variation also could be caused by the presence of several species identified as flathead grey mullet since there is molecular evidence that *M. cephalus* is a complex of cryptic species (Durand et al. 2012, Chapter 2—Durand 2015).

Table 10.4. Number of growth studies available by geographic area and species.

Species	Aus	CasS	GMCS	IndO	MedS	NEPac	NWPac	SEPac	NEAtl	NWAtl	SEAtl	SWAtl	Total
Aldrichetta forsteri	19												19
Chelon labrosus		2			28				13				43
Ellochelon vaigiensis	1												1
Liza abu					2								2
Liza aurata		10			34				11				55
Liza carinata				1									1
Liza grandisquamis											1		1
Liza haematocheila					3								3
Liza macrolepis				1									1
Liza parsia				3									3
Liza ramada					37				13				50
Liza saliens		7			27				5				39
Liza subviridis				1			1						2
Liza tade				2									2
Moolgarda buchanani				1									1
Moolgarda cunnesius				3									3
Moolgarda seheli	1			2									3
Mugil cephalus	14	2	15	13	40	2	4		2				92
Mugil curema			5			4		3		2		2	16
Mugil hospes			4										4
Mugil incilis			2										2
Mugil liza			5									2	7
Mugil trichodon			2										2
Oedalechilus labeo					4								4
Total	35	21	33	27	175	6	5	3	44	2	1	4	356
Percentage	9.8	5.9	9.3	7.6	49.2	1.7	1.4	0.8	12.4	0.6	0.3	1.1	100

Key: Aus = Australia; CasS = Caspian Sea; GMCS = Gulf of Mexico and Caribbean Sea; IndO = Indian Ocean; MedS = Mediterranean Sea; NEPac, NWPac and SEPac are North East, North West and South East Pacific Ocean respectively; NEAtl, NWAtl, SEAtl and SWAtl are North East, North West, South East and South West Atlantic Ocean respectively.

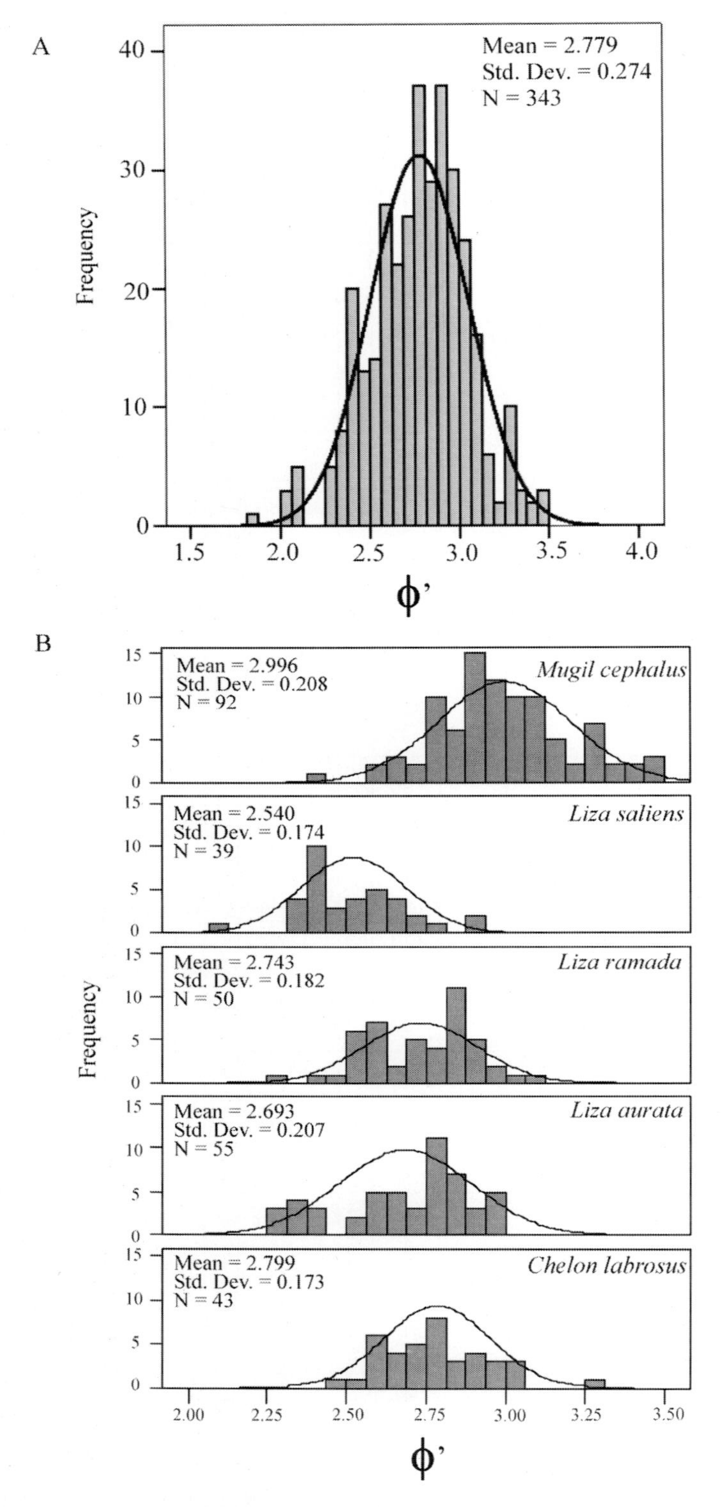

Figure 10.2. φ' frequency distribution by Mugilidae family (A) and by species (B). The information for this assessment came from the localities reported in Table 10.4.

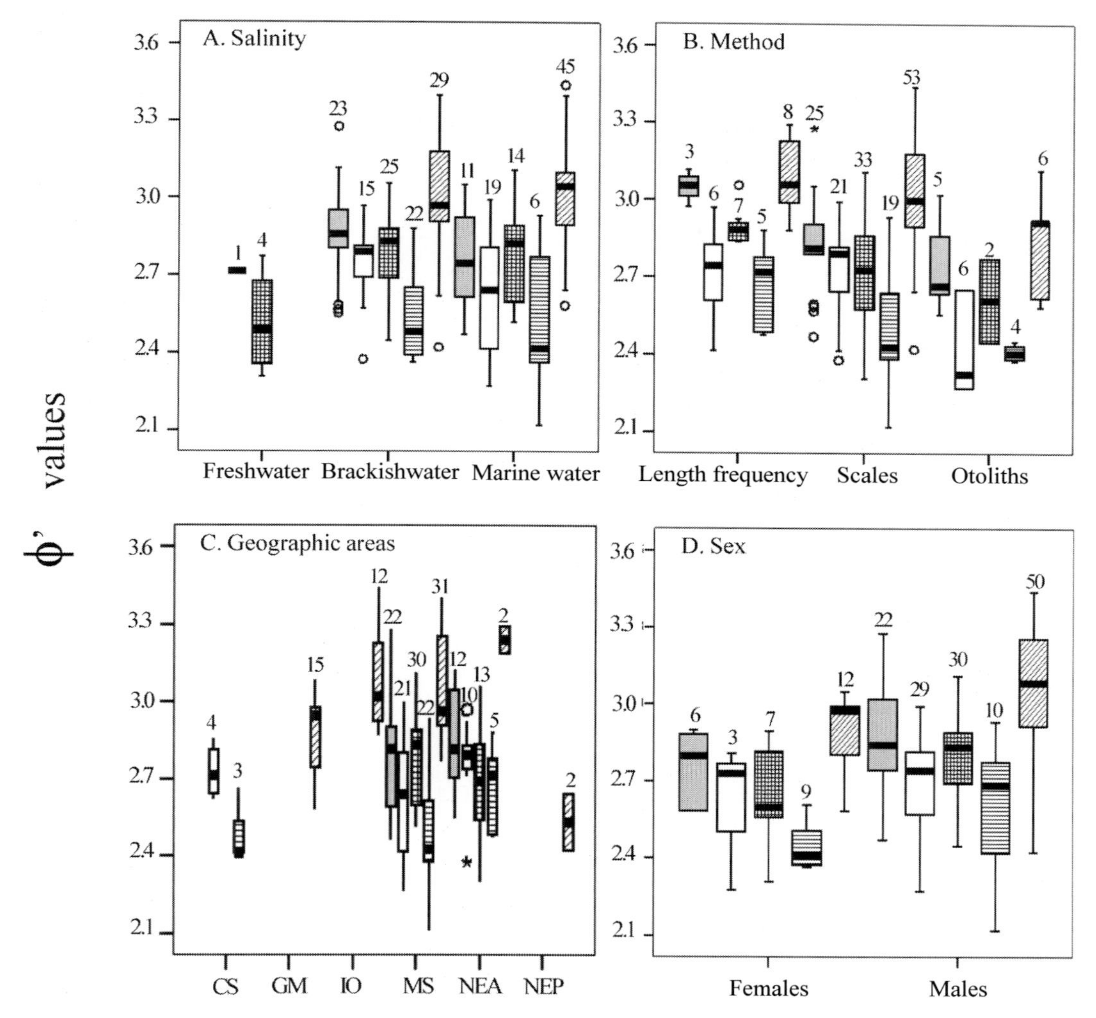

Figure 10.3. Boxplot showing φ' by Salinity (A), Method (B), Geographic areas (C) and Sex (D). *Chelon labrosus* (grey), *Liza aurata* (white), *Liza ramada* (grid), *Liza saliens* (horizontal lines) and *Mugil cephalus* (diagonal lines). CS = Caspian Sea, GM = Gulf of Mexico, IO = Indian Ocean, MS = Mediterranean Sea, NEA = North East Atlantic, NEP = North East Pacific. Numbers are sample size.

Apart from the five mullet species mentioned here, there is information on φ' for another 19, most of which (16) have one to four reports, and only *Aldrichetta forsteri*, *Mugil curema* and *M. liza* have 19, 16 and 8 citations in the literature (Table 10.5). On average, the species with greater φ' values are *Liza haematocheila* (three reports) and *Mugil liza* (eight reports) and those with smaller φ' values are *Mugil trichodon* (two reports) and *Oedalechilus labeo* (four reports). The minimum φ' values were reported for *Aldrichetta forsteri*, *Mugil trichodon* and *O. labeo* and the maximum values were for *A. forsteri*, *Mugil curema* and *Liza haematocheila*. Except for *A. forsteri*, the method most used to determine age has been the reading of scales, followed by reading of otoliths and size frequency (with 52.0, 13.5 and 11.8% respectively). Other methods such as the reading of spines, fin rays and marks, mentioned as 'others', make up 22.7%. Most studies on growth have taken place in marine and brackish environments (54.4 and 43.3% respectively), while those of freshwater environments are very scarce (2.3%). Of the studies, 36.1% focused on the growth of the different sexes, 41.2% dealt with whole populations, and 22.7% did not specify the objective of the study.

Table 10.5. ϕ' characteristics of different species of Mugilidae.

Species	ϕ' ± STD	Mín-Máx	N	Salinity	Methods	Geographic areas	Sex
Aldrichetta forsteri	2.361 ± 1.738	0.510–5.009	19	Sw^{19}	Scl^{5}, Otol^{14}	IO^{3}, NW-Pac^{16}	F^{9}/M^{9}, Mix^{1}
Chelon labrosus	2.799 ± 0.173	2.469–3.276	43	Bw^{25},Sw^{18}	Lfr^{3}, Scl^{25}, Otol^{5}, Ot^{2}, NAI^{8}	CS^{2}, MED^{28}, NEAtl^{13}	F^{8}/M^{7}, Mix^{15}, NAI^{13}
Ellochelon vaigiensis	3.042		1	Fw^{1}	Scl^{1}	NW-Pac^{1}	NAI^{1}
Liza abu	2.205 ± 0.146	2.102–2.308	2	Sw^{2}	Otol^{2}	MED^{2}	F^{1}/M^{1}
Liza aurata	2.693 ± 0.207	2.270–2.993	55	Fw^{1},Bw^{23},Sw^{31}	Lfr^{6}, Scl^{21}, Otol^{6}, Ot^{5}, NAI^{17}	CS^{10}, MED^{34}, NEAtl^{11}	F^{5}/M^{5}, Mix^{32}, NAI^{13}
Liza carinata	1.816		1	Fw^{1}	Scl^{1}	IO^{1}	Mix^{1}
Liza grandisquamis	2.753		1	Bw^{1}	Lfr^{1}	SE-Atl^{1}	Mix^{1}
Liza haematocheila	3.089 ± 0.054	3.027–3.128	3	Bw^{3}	Scl^{3}	MED^{3}	F^{1}/M^{1}, Mix^{1}
Liza macrolepis	2.705		1	Sw^{1}	Lfr^{1}	IO^{1}	Mix^{1}
Liza parsia	2.523 ± 0.074	2.479–2.608	3	Bw^{2},Sw^{1}	Scl^{3}	IO^{3}	Mix^{3}
Liza ramada	2.743 ± 0.182	2.305–3.107	50	Fw^{4},Bw^{25},Sw^{18}, NAI^{3}	Lfr^{7}, Scl^{34}, Otol^{2}, NAI^{7}	MED^{37}, NEAtl^{13}	F^{8}/M^{8}, Mix^{21}, NAI^{13}
Liza saliens	2.540 ± 0.174	2.117–2.929	39	Bw^{28},Sw^{11}	Lfr^{5}, Scl^{19}, Otol^{4}, NAI^{11}	CS^{7}, MED^{27}, NEAtl^{5}	F^{12}/M^{12}, Mix^{11}, NAI^{4}
Liza subviridis	2.819 ± 0.149	2.713–2.924	2	Sw^{2}	Lfr^{2}	IO^{1}, NW-Pac^{1}	Mix^{2}
Liza tade	2.804 ± 0.202	2.662–2.947	2	Bw^{1}, Sw^{1}	Lfr^{1}, Scl^{1}	IO^{2}	Mix^{2}
Moolgarda buchanani	2.735		1	Sw^{1}	Lfr^{1}	IO^{1}	Mix^{1}
Moolgarda cunnesius	2.435 ± 0.042	2.401–2.482	3	Bw^{1}, Sw^{2}	Lfr^{1}, Scl^{1},NAI^{1}	IO^{3}	Mix^{3}
Moolgarda seheli	2.965 ± 0.085	2.910–3.063	3	Fw^{1}, Sw^{2}	Scl^{3}	IO^{2}, NW-Pac^{1}	F^{1}/M^{1}, NAI^{1}
Mugil cephalus	2.996 ± 0.208	2.421–3.468	92	Bw^{31},Sw^{60}, NAI^{1}	Lfr^{8}, Scl^{53}, Otol^{7}, Ot^{6}, NAI^{18}	IO^{13}, CS^{2}, GMC^{15}, MED^{40}, NEAtl^{2}, NEPac^{2},NW-Pac^{18}	F^{15}/M^{15}, Mix^{31}, NAI^{31}
Mugil curema	2.756 ± 0.314	2.415–3.313	16	Bw^{8}, Sw^{8}	Lfr^{2}, Scl^{5}, Otol^{4}, Spi^{3},NAI^{2}	SE-Pac^{3}, GM&C^{5}, SW-Atl^{2}, NE-Pac^{4}, NW-Atl^{2}	F^{2}/M^{2}, Mix^{12}
Mugil hospes	2.588 ± 0.240	2.410–2.921	4	Sw^{4}	Otol^{1}, Spi^{2}, NAI^{1}	GM&C^{4}	F^{1}/M^{1}, Mix^{2}
Mugil incilis	2.267 ± 0.000		2	Bw^{1}, Sw^{1}	Lfr^{2}	GM&C^{2}	Mix^{2}
Mugil liza	2.983 ± 0.074	2.878–3.121	8	Bw^{2}, Sw^{6}	Lfr^{1}, Otol^{4}, Spi^{2}, NAI^{1}	GM&C^{5}, SW-Atl^{3}	F^{1}/M^{1}, Mix^{6}
Mugil trichodon	2.041 ± 0.021	2.027–2.056	2	Sw^{2}	Spi^{1}, NAI^{1}	GM&C^{2}	Mix^{2}
Oedalechilus labeo	2.090 ± 0.034	2.039–2.111	4	Sw^{4}	Scl^{4}	MED^{4}	F^{1}/M^{1}, Mix^{2}

Clues: Salinity: Sw: Salt water, Fw = Freshwater, Bw = Brackish water. Methods: Lfr = Length frequency, Scl = Scales, Otol = Otolith, Spi = Spine, Ot = Others (as opercular bones, tag, back-calculated), NAI = Not available information. Geographic areas: CS = Caspian Sea; IO = Indian Ocean, NW-Pac = North West Pacific, MED = Mediterranean Sea, SE-Atl = South East Atlantic, SE-Pac = South East Pacific, GM&C = Gulf of Mexico and Caribbean Sea, SW-Atl = South West Atlantic, NE-Pac = North East Pacific, NW-Atl = North West Atlantic. Sex: F = female, M = male, Mix = both sexes. Superscripts indicate frequency of reports.

Mugil cephalus is the species with the best growth performance, as indicated by the frequency distribution of ϕ'. Alvarez-Lajonchere (1981) estimated growth in decreasing order for *M. liza*, *M. curema*, *M. hospes* and *M. trichodon*. A comparison of these results with those of Thomson (1966) places *M. liza* among those with the quickest growth, similar to *M. cephalus*. *Mugil curema* and *M. hospes* present an intermediate growth rate, while *M. trichodon* grows slowly. According to De Silva and Silva (1979), the striped mullet *Mugil cephalus* Linnaeus 1758 has a worldwide distribution in coastal waters, lagoons and estuaries, between 42ºN and 42ºS, while Menezes et al. (2010) suggested that individuals from the NW Atlantic identified as *M. cephalus* may represent a population of *M. liza* in that region. It is to be noted that *M. liza* also presents a greater growth performance than the other *Mugil* species along the coasts of America, since it may be its vicarious species.

Another phenomenon that is found constantly in the literature is that of mullet growth speed. It is generally accepted that all mullets have a very rapid growth rate during the first two years of life, after which growth declines (Brusle 1981). The annual increase in growth during the succeeding ages is greatest during the first year, but decreases markedly throughout the remaining years of life (Abbas 2001). After three to five years of age, the decrease in growth recorded for *C. labrosus*, *L. ramada* and *L. saliens* is probably associated with the age at first maturity of these species (Abbas 2001). Koutrakis and Sinis (1994) considered that grey mullets can be considered fast growing, moderately long-lived species that reach about 75% of their maximum size in their first three to four years of life. According to the von Bertalanffy growth equation, the growth of *O. labeo* is intensive during the first four years of life, with growth slowing down considerably after five years of age (Matic-Skoko et al. 2012). In *Liza haematocheilus*, growth appears to be continuous throughout life, but is rapid during the early stages and declines gradually in older fish (Okumus and Bascinar 1997). A possible reason for the quick growth during the first years of life may be the high predation mullets face in coastal and estuarine environments where they spend their early life stages, but this requires further study.

L_∞ for the Mugilidae Family

The L_∞ values reported for the 24 species of Mugilidae vary from 17.3 cm for *Liza carinata* to 194 cm for *Mugil cephalus*. The average value for the family is 53.4 ± 25.0 cm, with a mean of 51.0 cm. From the 353 data used in the analysis in Fig. 10.4, half were below 51 cm, with a mean of 36.5 ± 9.2 cm, and two thirds of the distribution had values below 57.7 cm, with a mean of 40.8 ± 11.0 cm. As has been mentioned throughout the chapter, few species have been well studied and the mean values are thus strongly influenced by sample size. This is to say that a biased mean value is produced as there are some species with few studies. Figure 10.4 shows the amount of available data for each of the 24 species shown, as well as their median and distribution values.

According to Kraiem et al. (2001), the values given for the asymptotic maximum length L_∞ of *Liza ramada* specimens in European populations are, for example, 31.6 cm for males and 47 cm for females in France (Djabali et al. 1993) and 31.6 cm for males and 45.9 cm for females in Greece (Koutrakis and Sinis 1994). *Liza ramada* males range from 25.6 to 48.0 cm in Tl and 162 to 936 g in weight, while the females range from 27.5 to 58.0 cm in Tl and 190 to 2350 g in weight (Glamuzina et al. 2007).

Mullet length varies between males and females. Earlier authors found that the females of *Mugil saliens* (El Zarka and El Sedfy 1970), *M. capito* (El Maghraby et al. 1974), *M. cephalus* (Brulhet 1975, Cech and Wohlschlag 1975, Quignard and Farrugio 1981, Ibáñez-Aguirre et al. 1999) and *M. curema* (Ibáñez-Aguirre and Gallardo-Cabello 1996) grew faster or slightly faster than the males (Abbas 2001). Most authors have agreed that *Liza subviridis* females are slightly longer than the males of all ages, and also live longer (Al-Daham and Wahab 1991). At least some authors have recognized that the females are dominant only in the older age groups (Patimar 2008). The literature review revealed that only Koutrakis and Sinis (1994) considered that growth in length was the same for the males and females of *C. labrosus*, *L. ramada* and *L. saliens*, recognizing that the females live longer and predominate in the older age groups.

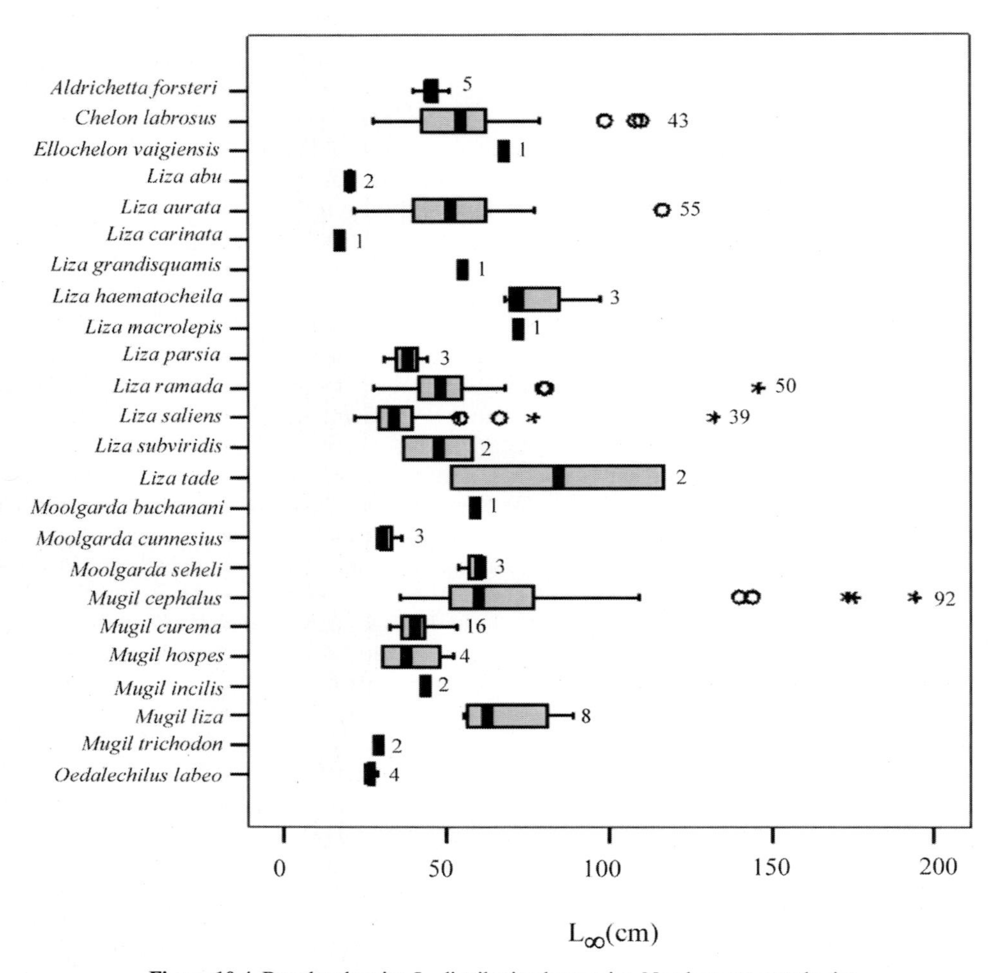

Figure 10.4. Boxplot showing L_∞ distribution by species. Numbers are sample size.

L_∞ Variation in Time

The infinite size L_∞ was analyzed using 357 data, most from Quignard and Farrugio (1981, 33.3%) and Fishbase (31.4%), with the other data reported by authors who on occasion listed their von Bertalanffy parameter values and compared them with those of other areas and/or species. Of the 357 data, those of species infrequently reported were eliminated, resulting in 279 data distributed among the five already mentioned most studied species.

Despite it being well known that the growth parameters L_∞ and k are correlated, L_∞ was used instead of the maximum L, as there is little information in the literature on this last parameter, or it is available mainly for juveniles. Thus, although L_∞ is not the ideal indicator to compare maximum size among populations and/or species, according to Sparre and Venema (1997) it is interpreted as the average size of a very old fish (in a strict sense, infinitely old) or the asymptotic length, for which reason it may be used as an indicator of the decrease in size or the 'deterioration' in the size of a species through time. In order to carry out this comparison, the averages of the L_∞ values were obtained for each of the decades under study, from the 1930s, for which there is little information, at the present. The results obtained for the five most studied species, *Chelon labrosus*, *Liza aurata*, *L. ramada*, *L. saliens* and *Mugil cephalus*, are presented.

Most of the values (83.5%) have been calculated from the 1960s to the present. Notwithstanding that the information is unequal for the five species, a reduction in the value of L_∞ seems to be common (Fig. 10.5), particularly for *C. labrosus* (grey box plot), while for *L. ramada* the L_∞ value has remained around 50 cm. Most of the extreme cases presented in the figure correspond to specimens of *M. cephalus* of India and Australia of the 1940s to 1960s (Thakur 1967, Thomson 1951, Kesteven 1942).

Different estimations with high extreme values have been recorded from the 1930s to 1960s, particularly for *M. cephalus*. However, infinite size values began to decrease in the 1970s and became more homogeneous among species, despite each species having a different growth rate (Fig. 10.5). Distributions from the 1970s to the present clearly lie between 25 and 75 cm, while those from the 1940s to the 1960s lie between 50 and 100 cm.

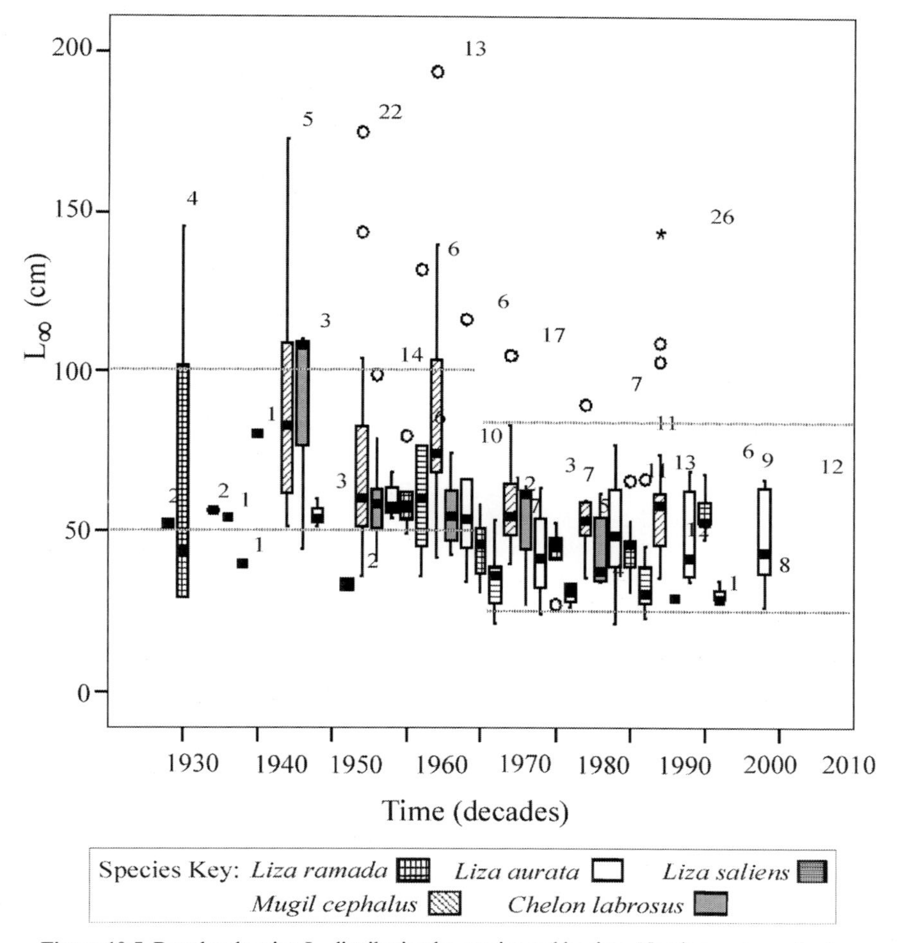

Figure 10.5. Boxplot showing L_∞ distribution by species and by time. Numbers are sample size.

Growth of Transplanted Grey Mullets

When growth rates of the redlip mullet (*Liza haematocheila*) collected from the Black Sea were compared with those of native mullet species living in the Black Sea and the Mediterranean, the redlip mullet exhibited a better growth. In addition, Kazanskji and Stashenko (1980) reported that the growth rate of redlip mullet collected from the new environment was two to three times greater than that recorded in their native waters. This shows that, in spite of strong seasonal variations in water temperature and primary production, the Black Sea appears to favour this microphagous species (Okumus and Bascinar 1997).

According to Patimar (2008), the three mullet species *Liza aurata*, *L. saliens* and *Mugil cephalus* were first transplanted by Soviet authorities from the Black Sea to the Caspian Sea between 1930 and 1934 (Dmitriev 1964, Berg 1965). The first two species are now common along the Iranian coast of the Caspian Sea and reach maturity earlier than those that live in the Black Sea (Beliaeva et al. 1989).

Relative Growth Analysis

Length-Weight (L-W) Relationship. Condition Index 'b'

The review of parameter b was carried out with 343 data, most from Fishbase (50.7%) and Quignard and Farrugio (1981, 23.6%). The other data were collected from various studies that have been carried out on each species and/or from tables that condense published information (25.7%). The genera *Liza* and *Mugil* represent 58.5 and 23.3% of the data, whereas *Chelon* and *Moolgarda* represent 11.0 and 3.6%. Also, most of the information corresponds to the geographical areas of the Mediterranean and NE Atlantic (54.3 and 17.6% respectively), the Indian Ocean, the SW Pacific and the Gulf of Mexico and Caribbean Sea (8.4, 5.7 and 5.1% respectively). There are 33 species listed in the condition index data base. The most frequently studied species are *Liza ramada*, *L. aurata*, *Mugil cephalus*, *Chelon labrosus*, *L. saliens* and *M. curema* (19.4, 16.7, 14.9, 11.0, 9.0 and 7.5% respectively). Other species have been studied very little, and there are species such as *Joturus pichardi*, *Liza lauvergnii*, *Ellochelon vaigiensis*, *Mugil hospes*, *M. incilis* and *Moolgarda robustus* for which there are only one or two condition index determinations, mostly because they are rare species.

Figure 10.6A presents the distribution of frequencies for the Mugilidae family. A normal distribution is observed with an average of 2.95 ± 0.25 and some extreme cases, to the left with a value of $b = 1.75$ and to the right with two values of $b = 4.05$, the first case corresponding to *M. cephalus* juveniles of the Gulf of Mexico (Broadhead 1953) and the second to two studies of *L. aurata* specimens from the coast of Tunisia (Fehri-Bedoui and Gharbi 2005) and the coasts of Great Britain (Hickling 1970). The distribution of frequencies of the condition index 'b' for the Mugilidae family presents a leptokurtic shape (Fig. 10.6A), with no significant differences in 'b' among species ($P = 0.808$) or among geographical areas ($P = 0.085$). Of the five species with the most data, *C. labrosus* has the highest condition index average value, followed by *M. cephalus*, *Liza ramada* and *L. aurata*, with *L. saliens* presenting the lowest condition index value (Fig. 10.6B). The average value for the family, as well as for four of the five studied species, barely reaches the isometric value of this condition index. The 'b' index is highly sensitive to data sampling and, in the case of fishery studies, the data range is generally small. These results indicate that the specimens are not very robust, since four of the five species have mostly negative allometric values or a barely isometric growth, with positive allometric growth only for *C. labrosus*. In view of the fact that the specimens considered in studies on age and growth are captured commercially, it is possible to state that most individuals were thin and were represented mainly by slim adults or young-adults (less robust), for which results provided negative allometric values or a barely isometric growth. These results together with those that show that L_∞ values have decreased over the last decades, are a warning call that may indicate the presence of overfishing.

According to Patimar (2008), variations in the b exponent may be caused by environmental conditions and state of nutrition that vary among sites, and exert a local selective pressure on the populations. Different b values between neighbouring sites suggest a different fish condition or fitness. According to Matic-Skoko et al. (2012), numerous authors have provided values of the b parameter for other mugilids, indicating that this parameter is strongly affected by region, species, sex and sample size (Dulcic and Kraljevic 1996, Moura and Serrano-Gordo 2000, Fehri-Bedoui and Gharbi 2005, Ilkyaz et al. 2006).

Variability in L-W estimates can be attributed to a number of other factors including the season, habitat, gonad maturity, sex, diet and stomach fullness, health, preservation technique and differences in the observed length range of the collected specimens (Froese 2006). According to Kennedy and Fitzmaurice (1969), adult bodies vary greatly according to sex, spawning condition and season. Since the gonads of a gravid female may represent a quarter of its total weight, there is a big length-weight ratio difference between nearly ripe and recently spent females.

A

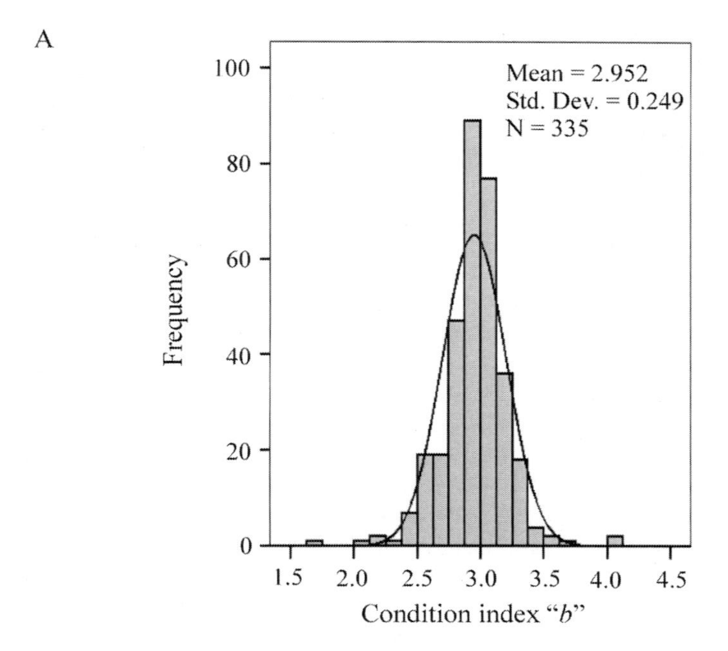

B

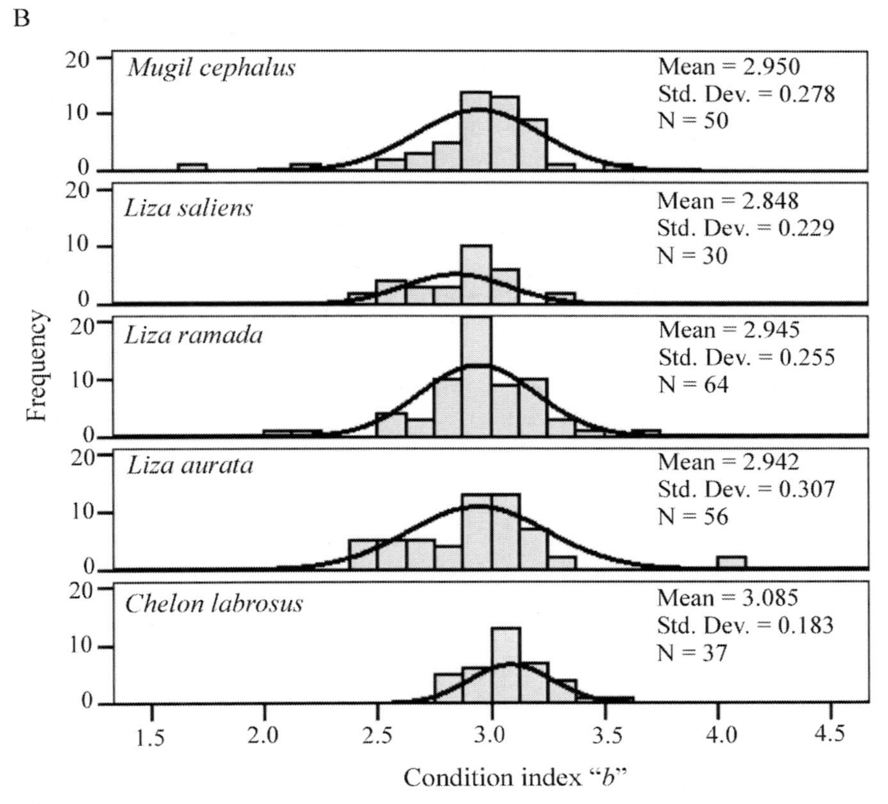

Figure 10.6. Condition index '*b*' frequency distribution by Mugilidae family (A) by species (B). The information on *M. cephalus* came from different localities around the world, on the other species from the Mediterranean and North East Atlantic regions.

Albertini-Berhaut (1975) also described a change in the length-weight relationship of *L. ramada* and *L. aurata*, at 67 mm and 72 mm respectively, probably also caused by a difference in diet. The different length at which the change takes place is attributed to the smaller range of lengths (15 to 108 mm) used by this author (Koutrakis and Sinis 1994).

Length-Length Relationship

Studies on the relative growth of the body parts of fish are very scarce for the Mugilidae. In fact, there are only five studies that focus on the allometric growth of various body parts. Apparently Grant and Spain (1975) were the first to carry out this study on *Mugil cephalus* in North Queensland, Australia, followed by Drake and Arias (1984) in Cádiz, Spain and later Pérez-García and Ibáñez-Aguirre (1992), Ibáñez-Aguirre and Lleonart (1996) and Ibáñez et al. (2006) along the Mexican coasts. Grant and Spain (1975), Ibáñez-Aguirre and Lleonart (1996) and Ibáñez-Aguirre et al. (2006) analyzed the relative growth of both sexes of *Mugil cephalus*. Although these studies did not record exactly the same measurements, all were based on the first. Eight morphometric measurements recorded in these studies may be compared: Sl = standard length; Cl = cephalic length (distance from the snout tip to the posterior extreme edge of the operculum); PD1 = distance from the snout tip to the anterior base to the erect spinous dorsal fin, and PD2 = to the second dorsal fin; Al = anal length (distance from the snout tip to the anterior edge of the anal aperture); MaxH = Maximum height or depth at the first dorsal (depth of the body, perpendicularly through the anterior insertion of the first dorsal fin); MaxT = Maximum thickness at the pectoral fins (width of the body between the pectoral fin bases); and MinH = Height of the caudal peduncle (Fig. 10.7).

Comparing the type of allometry, it may be observed that most studies in general agree that the cephalic length (Cl) decreases as the size increases (Table 10.6). The three species of the genus *Liza* had a similar relative growth in 50% of the measurements, and had a greater frequency of negative and positive allometries (87.5%) than the two species of the genus *Mugil*, in which isometric growth was predominant (58.0%). Table 10.6 shows the variations in the allometry results of the populations and sexes, possibly resulting from the number of individuals in the sample, as the slope of the relationship between a variable (dependent) and the size (the independent variable), or the allometric factor, is highly sensitive to sample size (Froese 2006).

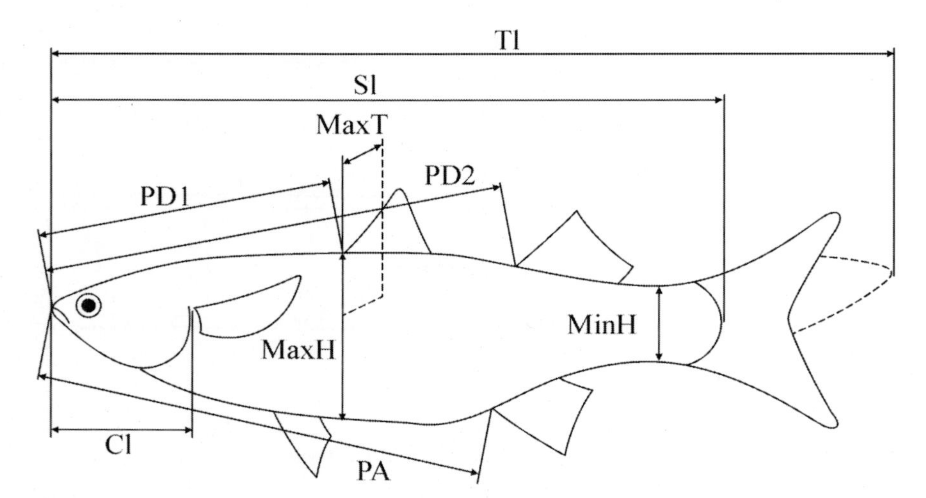

Figure 10.7. Morphometric measurements of grey mullets. Tl = Total length; Sl = standard length; Cl = cephalic length (distance from the snout tip to the posterior extremity of the operculum); PD1 = distance from the tip of the snout to the anterior base to the erect spinous dorsal fin; and to the second dorsal fin (PD2); PA = anal length (distance from the snout tip to the anterior limit of the anal aperture. MaxH = Maximum height or depth at first dorsal (depth of the body, perpendicularly through the anterior insertion of the first dorsal fin); MaxT = Maximum thickness at pectoral fins (width of the body between the pectoral fin bases); MinH = Depth of the caudal peduncle (from Drake and Arias 1984, Fig. 1). Copyright © by the Editor in Chief of Scientia Marina. Reprinted by permission of Scientia Marina.

Table 10.6. Relative growth of five species of grey mullet according to Grant and Spain (1975), Drake and Arias (1984), Pérez-García and Ibáñez-Aguirre (1992) and Ibáñez-Aguirre and Lleonart (1996). F = Female number; M = Male numbers. Female/Male N of individuals. Key for variables: Sl = standard length; Cl = cephalic length; PD1 = distance from the snout tip to the anterior base to the erect spinous dorsal fin, and PD2 = to the second dorsal fin; Al = anal length; MaxH = Maximum Height or depth at the first dorsal; MaxT = Maximum Thickness at the pectoral fins and MinH = Height of the caudal peduncle.

Species	Sl	Cl	PD1	PD2	PA	MaxH	MaxT	MinH	N (F/M)	Geogr.area	Author
Mugil cephalus	=/=	=/=	=/=	+/=	=/=	=/=	=/=	=/=	36/32	NE Australia	Grant and Spain 1975
	-	-	-	-	-	-	+	-	441	NE Atlantic	Drake and Arias 1984
	=	+	=	=	+	=	-	=	140	Gulf of Mex	Pérez-García and Ibáñez-Aguirre 1992
	=/=	–/–	+/=	+/–	+/–	=/+	=/+	–/+	126/123	Gulf of Mex	Ibáñez-Aguirre and Lleonart 1996
Mugil curema	-	=	=	=	=	=	=	-	204	Gulf of Mex	Pérez-García and Ibáñez-Aguirre 1992
Mugil curema	=/=	–/–	=/=	=/=	=/=	=/=	+/=	=/+	179/130	Gulf of Mex	Ibáñez-Aguirre and Lleonart 1996
Mugil curema	=/=	–/–	=/=	=/–	+/+	=/–	=/=	–/–	368/232	NE Pacific	Ibáñez-Aguirre et al. 2006
Liza ramada	-	-	-	+	+	=	+	+	828	NE Atlantic	Drake and Arias 1984
Liza aurata	-	-	-	-	-	+	+	=	790	NE Atlantic	Drake and Arias 1984
Liza saliens	-	-	-	-	=	+	+	+	538	NE Atlantic	Drake and Arias 1984
Chelon labrosus	-	-	-	+	+	=	+	=	359	NE Atlantic	Drake and Arias 1984

In addition to the bivariate analyses, Ibáñez-Aguirre and Lleonart (1996) carried out multivariate analyses that indicated that the males and females of *Mugil cephalus* formed two clearly separated groups, while those of *Mugil curema* showed no important differences. The separation of the groups due to a difference in shape and not in size was made clear through a correspondence analysis (reciprocal averaging) and a canonical analysis of the populations' normalized data. The morphometric variables that differentiated the sexes were the Minimum Height (MinH) and the Maximum Thickness (MaxT).

Ibáñez-Aguirre et al. (2006) compared the shape of *Mugil curema* specimens of the Pacific and Atlantic oceans and found no significant differences between the two populations. Only the eye diameter and the body width (greatly influenced by sexual maturity) showed differences among populations. However, Atlantic specimens showed the greatest morphological variability, which may be due to the possibility of more than one population being present.

Relative and Compensatory Growth

The proportional growth of the different parts of the body allows the use of structures such as scales or otoliths to calculate relationships and proportions that are useful in the analysis of growth. As already noted, this growth responds to, among other causes, complex changes in the environment. It is thus possible that specimens or organisms that co-habit in one area have similarities in growth. As a consequence, relative growth has been useful in discriminating fishery stocks. In the case of the mugilids, not only has the relative growth of the different parts of the body been used to discriminate species, as well as the sex of the same species (Ibáñez and Lleonart 1996), but the shape of the scales has also been used to discriminate species, genera and populations (Ibáñez et al. 2007). The change in the shape of the scales along the fish body in relation to the type of swimming has also been analyzed (Ibáñez et al. 2009). Variations in growth from slow to rapid, that occur at some stages in periods of compensatory growth, may cause morphological changes in scale shape, since growth ridges are laid down on the surface of the scales (bony surface layer). As a result, it is reasonable to assume that factors related to the cycle, location or environment variability in the growth characteristics of populations can control scale morphology. Therefore, morphological characters, together with morphometric dimensions and meristic counts, provide information to discriminate stocks, reflecting the place or season where the population developed (Ibáñez et al. 2007). Variability in the hydro-geomorphology or limnology of waters, the productivity and the abundance of fish stocks probably account for the difference in scale morphology observed, and explain why scale shape can be used to discriminate between stocks (Ibáñez et al. 2012a).

Main Issues Discussed in Age and Growth Studies

Age: First Mark

The recording of the first ring is one of the greatest controversies in the determination of mugilid age, and it is here that significant differences were observed regarding age determination both by different methods and salinity. Most authors agree that in most mugilid species the first ring is formed between 12 and 18 months after spawning (Kesteven 1953, Kennedy and Fitzmaurice 1969, Almeida et al. 1995, Hotos 2003, Abdallah et al. 2012). With respect to the second ring, Almeida et al. (1995) recorded it for *Liza ramada* between 24 and 30 months.

The problem of reading the first ring is associated with the recognition of the first annulus in older specimens, because of the opaque appearance of the scale's central region, as is recorded in the literature. This has been reported especially for *M. cephalus* (Erman 1959, Hendricks 1961, Kennedy and Fitzmaurice 1969, Grant and Spain 1975, Libosvarsky 1976, Ibañez-Aguirre and Gallardo-Cabello 1996). Cardona (1999a) mentioned that the reading of otoliths is difficult because the sagittae of older specimens often become too opaque (Erman 1959, Kennedy and Fitzmaurice 1969, Ibáñez-Aguirre et al. 1996). Kaya et al. (2000) suggested reading thick and opaque otolith sagittae of older fish after processing with sandpaper. The reading of bones is also difficult (Hickling 1970).

Among the studies that have helped clarify the formation of the first annulus is that of Hsu and Tzeng (2009) on *M. cephalus*. These authors studied the periodicity of age marks on scales and otoliths of this species using cultured fish of known ages and comparing them with wild fish. They found that multiple concentric increases were deposited around the core region of the otolith before the first annulus formed. This indicates that the multiple concentric rings around the core region of otoliths are not annuli (Hsu and Tzeng 2009). The possible causes of these deposits could be several life-history events during the early life stage, such as the change from larval to juvenile stage and environmental changes that may leave a mark on the otoliths (Chang et al. 2000), and confuse validation of the age mark. Ibáñez-Aguirre and Gallardo-Cabello (1996) recorded an incomplete mark in an 82 mm long *M. cephalus*, considered a 'juvenile mark', that might correspond to the change in habitat from a marine birth place to a lagoon. This hyaline band is incorporated into the nucleus as the scales and otoliths grow. After the nucleus is formed, new material is deposited unevenly. A greater number of bands are formed at the back than at the front, resulting in a confused accretion pattern. However, carrying out an adequate treatment for the reading of the otoliths, considering that the material that is concentrated near the core is not rings (according to Hsu and Tzeng 2009), makes it possible to obtain a better reading of this structure. The recommendation of Hsu and Tzeng (2009) to frontally section otoliths along the anterior-posterior axis and to transversely section them along the dorsal-ventral axis allows an easier reading than along other otolith axes. Several authors have mentioned that this is because the annuli in otoliths are clearer along the long axis than the short axis. Also, the different axes of the otoliths have different growth rates. This is unfortunately, more time consuming and expensive.

Thus, it is necessary to progress to a stage at which doubts can be eliminated when carrying out comparative field studies or experimental studies. In order to be able to shed light on the subject of ring formation, either in the laboratory or with approaches or hypotheses in field studies in relation to geographical areas or environmental differences it is necessary to answer questions such as: at what time is the first slow-growing ring formed; which are the factors that most affect its development; and which cases show the incomplete rings or the so-called juvenile ring?

Age: Explanation of Marks

Most authors agree that growth marks are caused by low temperatures during the winter or follow metabolic changes that include gonad development and the storing of reserves for spawning migrations under poor feeding or fasting conditions (Alvarez-Lajonchere 1981). Spawning is considered the event that triggers a period of slow growth and the formation of the hyaline band in adults of many species (Morales-Nin 1987).

Most references support the formation of only one annulus per year during the cold months in most mugilid species of temperate and subtropical regions (Thomson 1951, Erman 1959, Farrugio and Quignard 1974, Libosvarsky 1976, Cambrony 1983, Koutrakis et al. 1994). Hotos and Katselis (2011) and Abdallah et al. (2012) observed that one translucent opaque zone was deposited on the otolith in February or January of each year in *Liza aurata*. A similar pattern has been described for *Liza aurata* (Arruda et al. 1991), *Chelon labrosus*, *Mugil cephalus* (Farrugio 1975, Ellender et al. 2012), *M. liza* (González-Castro et al. 2009) and *Oedalechilus labeo* (Matic-Skoko et al. 2012) for the winter. Also, one annulus has been reported for *M. cephalus* and *M. curema* in subtropical regions, as well as a strong correlation between the slow band formation and the spawning migration period (Díaz-Pardo and Hernández-Vásquez 1980, Ibañez-Aguirre and Gallardo-Cabello 1996). Erman (1959) recorded a fast growth from May to September and a slow growth after October for *Mugil cephalus* from the Bosphorus and the Sea of Marmara. Thomson (1951), in Australia, recorded that the growth of the species practically stopped in mid-winter. Cech and Wohlschlag (1975) differed from the conventional concept and reported that age marks on otoliths of *M. cephalus* from coastal waters of Texas, USA were deposited twice a year, in summer and winter. Alvarez-Lajonchere (1981) recorded two marks per year for local Cuban mugilids. Marks in *M. curema* are formed in mid-summer and at the end of the year coinciding with the spawning seasons, in *M. liza* from mid-summer to the end of the year coinciding with a low productivity in coastal lagoons, an explosive gonad development and spawning, and in *M. hospes* at mid-summer coinciding also with the reproductive season. A probable origin of scale mark formation in tropical Mugilidae was suggested by Pillay (1954) and Thakur (1967).

They mentioned that monsoon rains caused a major disruption of benthic flora and fauna of littoral waters and reduced the feeding intensity, which was predominantly by benthic detrital scavengers. Odum (1970) considered it more likely that a lower feeding intensity in *M. cephalus* was caused by the turbidity and turbulence of inshore waters during the monsoon season, as the detritus-rich upper benthic sediments were mostly in suspension and made feeding difficult.

Winter temperatures in temperate climates retard fish growth and hence scale accretion, causing scale breaks. Thomson (1951) stated that annual ring formation in *M. cephalus* took place at temperatures of 16 to 18°C. Kennedy and Fitzmaurice (1969) stated that temperature is related to feeding. At water temperatures of 8 to 9°C, mullets fed only spasmodically and sparingly, and at temperatures below 8°C they ceased feeding and became inactive. These authors found no trace of food in either juveniles or adults in March (of 1968), though food was present in both juveniles and adults in April. According to Cardona (1999b), mullets of all ages gained weight only at water temperatures above 20°C. However, the growth of both adults and immature fish decreased in mid-summer when the growth of juveniles peaked. This difference was not a consequence of reproductive cost, as the growth rate of immature specimens decreased markedly, despite the fact that they did not spawn.

According to Hsu and Tzeng (2009), annuli of both scales and otoliths are deposited in winter. This is when annuli in calcified structures are usually formed, during the slow-growing period, in fish in low temperature and poor-nutrition environments (Bilton and Robbins 1971). Some authors have mentioned that intrinsic factors such as ontogenetic changes, metabolic rate, maturation and population density, as well as extrinsic factors such as the migratory environment, food availability and temperature, all have the potential to influence the growth of fish and the subsequent deposition rate of annuli (Panfili et al. 2002). As a result of this, there is still controversy surrounding the reasons behind the increases and decreases in mullet growth.

Figure 10.8 shows an example of annuli microstructure in a transverse section. This is a *M. cephalus* specimen with four annuli. It is considered to be four years of age since one annulus is formed per year. As previously noted, the growth rate in mullets is in general very rapid during the first years of life, and

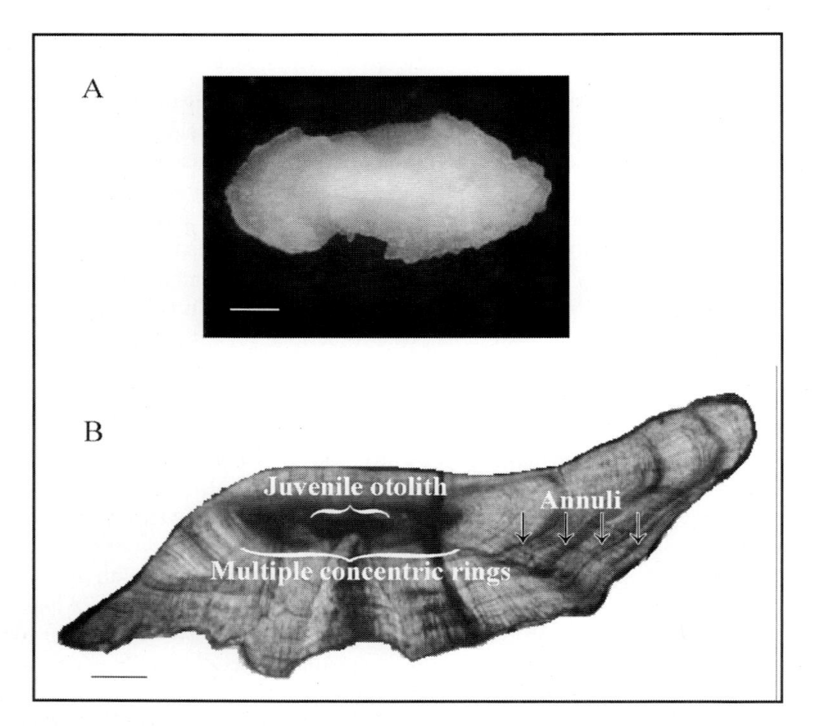

Figure 10.8. Comparison of a specimen of *Mugil cephalus* from Tamiahua Lagoon, Mexico. Tl = 31.7 cm, four years old (Photo: C.C. Hsu). (a) whole otolith, (b) microstructure of annuli in a transverse section. Scale bar: A = 1 mm; B = 300 μm.

becomes very slow afterwards. In this particular case, a 31.7 cm long specimen, based on back-calculation, would have had an average approximate size of 15.8 cm at one year of age, meaning that in one year it reached half the size it would reach at four years of age, and at two years of age it would have been approximately 20.4 cm long. Thus before the formation of the first ring, the deposition of organic matrix fibres and carbonate crystals results in growth increases and confirms the rapid growth of the fish. This deposition is considerable and decreases after the formation of the second ring and, in consequence, the growth rate decreases markedly in agreement with the rate of deposition of material in the otolith. Specimens of approximately three or four years of age clearly show each ring, although the marks that are formed afterwards are difficult to read as the growth rate becomes very slow.

Figure 10.8 also clearly shows the shadow of the otolith in its juvenile stage, with the later formation of multiple concentric rings around the otolith core that, as Hsu and Tzeng (2009) stated, are not rings (annuli) and must be marks due to changes in the environment, from sea water to brackish water, that are accompanied by changes in feeding, temperature and other environmental parameters.

Age: Other Methods as Alternatives for Age Determination

Various authors have proposed alternatives to better determine age. Cardona (1999a) proposed an alternative method based on Petersen's procedure of cohort identification using a length-frequency plot. Njoku and Ezeibekwe (1996) suggested the analysis of a large number of length-frequency data for tropical species, as has been suggested by different authors (Bagenal and Tesch 1978, Pauly 1983), considering that several studies have not been able to easily determine mullet age from hard parts (particularly *Mugil cephalus*: Brulhet 1975, De Silva 1980). This may be especially useful when well defined size classes exist, as a consequence of a short spawning season. Indeed, this method has been successfully used for several grey mullet species, including the leaping grey mullet (Thomson 1951, Thomson 1957, El Zarka and El Din 1964, Kennedy and Fitzmaurice 1969, El Zarka et al. 1970, Drake et al. 1984, Claridge and Potter 1985, Ibañez-Aguirre et al. 1995). Paton et al. (1994) proposed a new method to determine age in mullets based on the thoracic spines. Sr:Ca ratios in otoliths have also been used to reconstruct the past salinity history of mullets throughout their migration from freshwater habitats to the sea. Since *Mugil cephalus* and *M. curema* migrate yearly from estuaries to the sea to spawn, Sr:Ca ratios have been used to calculate the age of these species. This method is generally accompanied by the reading of the annual rings (Chang et al. 2004, Ibáñez et al. 2012b).

Causes of Growth Variations

According to Sudha et al. (1992), the greater growth rate of fish living in sea water is due to the lower expenditure of energy required to maintain an osmotic gradient. Sudha et al. (1992) stated that this could be considered as the major factor that contributes to the different growth rates exhibited by juveniles and adults in two different backwaters of South India.

Matic-Skoko et al. (2012) stated that the growth of euryhaline species is often affected by salinity, as the energy used for osmoregulation is then not available for growth (Brett 1979). Moreover, Cardona (2006) recorded that the distribution of some mugilids was dramatically affected by salinity. As *O. labeo* inhabits stenohaline environments with relatively constant salinity values (approximately 37 psu) in open Adriatic waters (Grbec et al. 1998), it is presumed that growth is not primary affected by salinity, but rather by a low availability of food. In particular, the food supply for detritivorous species such as *O. labeo*, is expected to be smaller in oligotrophic, stenohaline rocky coves than in the eutrophic, soft bottom estuaries inhabited by other species.

Age and growth in length and weight recorded for *Liza aurata* in the lagoon system of Messolonghi-Etoliko and in the neighbouring sea waters of the Gulf of Patraikos (Greece) were not significantly different. This may be explained by the fact that the relatively high salinity of the lagoon may negate its high trophic advantage and/or alternatively by a scenario that is based on the seasonal migrations of the species between the sea and the lagoon (Hotos and Katselis 2011).

The *L. ramada* of the Tagus (Spain) has been reported to be heavier at any given length than many other species, and only that from the Rhóne, at 340 mm Tl or longer, was heavier than that from the study area, suggesting that *L. ramada* in freshwater habitats are in good condition (Oliveira and Ferreira 1997). *Liza ramada* is considered to be the grey mullet that travels the farthest upstream to freshwater feeding areas (Sauriau et al. 1994), and studies of species with similar life histories have shown that the adults tend to return to good feeding grounds every year (Oliveira and Ferreira 1997). Several authors have mentioned that the abundance of *L. ramada* far upstream in the Tagus and the better condition of the fish, indicate the presence of sites with very favourable growing conditions. This, plus high food availability, can partly explain the increasing *L. ramada* populations in the study area.

In contrast, there is no clear relationship between the rate of increase in length and salinity in the natural environment, as De Silva and Perera (1976) observed. Regarding population density, negative correlation coefficients have always been obtained for the number of individuals per surface area and the final specimen size, although these coefficients have not always been statistically significant (Drake et al. 1984).

Hotos and Katselis (2011) mentioned that different growth rates of golden grey mullet *Liza aurata* in different locations are probably due to local differences in important factors that influence growth, such as water temperature. This is because grey mullets spend most of their time in shallow inshore waters where the water temperature is influenced more by local conditions (with noticeable fluctuations from location to location) than by the more stable temperature of the open sea (Kennedy and Fitzmaurice 1969). However, the different growth rates of the golden grey mullet in different locations may be related to other factors apart from temperature, such as food availability and/or density dependent relationships (El Zarka et al. 1970, Drake et al. 1984).

In marine areas with significant temperature variations, grey mullets grow in length during the spring and summer, followed by a period of slow growth during the following five or six months. A comparison of growth curves for monthly mean length and temperature shows that most of that annual growth takes place when temperature is highest (> 14ºC). Kennedy and Fitzmaurice (1969) kept mullets over the winter in an aquarium tank in an unheated laboratory and related the cessation of growth to a reduced metabolic rate caused by low temperatures, as well as to a partial winter fast. Moreover, juveniles at water temperatures above 10ºC fed actively, at 8–9ºC only spasmodically and sparingly, and at temperatures below 8ºC they ceased feeding and became inactive. Cardona (1999a) recorded that *L. saliens* did not grow at water temperatures below 20ºC, and found a significant correlation between the growth rate in weight of juveniles and water temperature.

Hickling (1970) stated that grey mullet tend to exit estuaries and shallow waters in cold weather, and may suffer heavy mortalities if they do not. The presence of 'full' stomachs appears to be greatest in the late summer and autumn. Wimpenny (1932) found that thin-lipped grey mullet in Egypt fed most intensively in October just before the spawning migration, and recorded an accumulation of fat reserves in the mesentery. When the fish were being examined, the presence or absence of abundant intestinal fat was noted. According to Hickling (1970), a slow growth rate is the result of poor quality food. However, when given nutritious food, the fish are capable of growing fast (Yashouv 1966).

Other reasons for differences in growth have been stated by McDonough and Wenner (2003), who suggested that environmental factors that may affect growth may include the temperature and the photoperiod, which could vary for larvae spawned at different times during the spawning season. In agreement with this, fish that spawned during the middle of the spawning season (December to February for *M. cephalus* along the Carolinas, USA) could have some advantage over fish that spawned either earlier or later in the season. In other words, fish that arrived earlier would spend more time in cold water and grow more slowly. Another consideration suggested by McDonough and Wenner (2003) was that the fish could have arrived from a wide range of geographic areas (where spawning occurred) at different times of the spawning season, and food resources offshore could have been better for some groups of larvae than for others.

Differences in growth values could also be generated by the various methods used by researchers to read fish age, and this was reflected in the results obtained (Kraljevic and Dulcic 1996). However, as Ilkyaz et al. (2006) suggested, it is possible that the variations in the population parameters of the golden grey mullet represent epigenetic responses to the different environmental conditions (temperature, food, geographic location) and nutrient levels in the study areas.

As is known, growth is a complex process in which not only one, but probably several different factors and variables, have an effect. However, comparative analyses could shed light on this subject. In summary, studies on the age and growth of mugilids from a variety of areas have revealed differences in growth rates that could be explained by the worldwide distribution of this group, which provides it with different survival strategies (Ibáñez-Aguirre et al. 1999, Kraiem et al. 2001, Matic-Skoko et al. 2012).

Conclusions and Suggestions

A great variety of results has been presented, both from the studies presented and from the references. The length of each age group varied among geographical areas and different salinities, and among the methods used to read the age of the first age group. The growth performance ϕ' showed differences among the methods, sexes, geographical areas and salinities for a few species (*L. ramada*, *L. aurata* and *M. cephalus*). It is yet to be decided whether these differences in the case of *M. cephalus* are due to the species being the most studied, or because it has the widest geographical distribution, or also because of taxonomic differences given that *M. cephalus* has been considered as a complex of species (see Chapter 2—Durand 2015). There is still not enough evidence to establish categorical conclusions, but it has been possible to observe that mullet growth is more homogeneous than previously recorded. As has previously been suggested and is clearly presented here, the length of each age group varies among environments, as was recorded for specimens of the Mediterranean Sea and the northern Atlantic. The Mediterranean is characterized by higher temperatures and salinity. Literature reports have accepted that growth decreases in low temperatures, and it is believed that this is in response to mullets eating less or not eating at temperatures below 8ºC, but there are no studies that can confirm these statements.

The decrease in growth and the seasons when growth marks form are subjects that occupy most discussions in studies of mullet age and growth. One of the greatest controversies regarding age is the formation of the first ring or growth mark. Notwithstanding that there is agreement that this mark appears between one year and one and a half years of age, and that most authors associate it with a decrease in temperature, there is still controversy on the subject. Most authors agree that both the scales and the otoliths are adequate structures to determine age. Present opinion suggests that it would be advisable to carry out experimental research to provide greater knowledge on the way that material is deposited on hard structures during the juvenile phases, as otoliths become very opaque and scales grow irregularly at the core. The use of a size frequency analysis, as previously mentioned, is convenient when the sample size is very large and spawning takes place over a short period of time (around four months), and makes it possible to clearly differentiate cohorts. When this is not the case, it is not convenient to use this method.

The formation of a mark per year is a generalized finding recorded in the literature, except in the case of tropical environments in Cuba and subtropical environments in Texas. Most authors agree that marks form when the temperature decreases, when the mullets stop feeding or reduce the amount they eat, but there are no studies that deal directly with this. Other studies have related low growth to the low productivity of the winter season and turbidity during the monsoons in tropical regions, but it is necessary to analyze this further as there are no hard data on this subject. The relationship between the formation of a double annual mark and a double spawning is however, documented.

Mugil cephalus is the species with the greatest growth rates, compared with the species of the genus *Liza*: *L. ramada*, *L. saliens* and *L. aurata*, and *Chelon labrosus*. Also, and according to studies carried out in tropical areas of the eastern American continent, *Mugil liza*, which replaces *M. cephalus* along the South American coasts, is also the species that has the highest growth rates among those present in the area: *M. curema*, *M. hospes* and *M. trichodon*. The most recorded reasons for high mullet growth rates are temperature, availability of food, salinity and photoperiod, but a great vein of experimental or comparative studies could shed light on this subject.

As is known, mullet species that have been transplanted from their original habitats have shown good adaptation to their new environments. This has been analyzed for *Liza haematocheila* in the Black Sea, and particularly for *Liza aurata* and *Liza saliens* in the Caspian Sea.

Relative growth has proved its utility for weight and size potential relationships, as in the potential relationships of different linear measurements. The condition index 'b' of the length-weight relationship presented an isometric value with a tendency to a negative allometry, which means that the specimens Mugilidae that have been studied are slim, and this may be related to the fact that young adults are ever more better represented in the fisheries of these species, than robust adults which are less frequent in the catches. Also, the L_∞ size has decreased in value over the last decades, as may be seen when comparing 1930–1960 with 1970–2000. In consequence, and considering that L_∞ and '*b*' are low, overfishing has occurred.

It would also be very useful if studies and comparative analyses were carried out that would help answer questions and solve hypotheses as, for example, whether a decrease in feeding rate is sufficient to reduce the formation of material on hard structures or to form a low growth mark? ... or up to what point the temperature of the different species affects the rate of formation of deposited material on hard structures and the formation of low growth marks. It would seem worthwhile to examine different groups of specimens under varying the environmental conditions, and to analyze the results. Many more questions can be formulated.

Acknowledgements

I would like to thank to Professors Manuel Gallardo-Cabello and Jordi Lleonart who guided me in my PhD studies in the biology of mullets.

References

Abbas, G. 2001. Age, growth and mortality of the mullet, *Liza carinata* (Pisces: Mugilidae) in the backwaters of Bhanbhore, Sindh (Pakistan: Northern Arabian Sea). Pakistan. J. Zool. 33: 1–5.

Abdallah, Ch., M. Ghorbel, G. Hajjej and O. Jarboui. 2012. Age and growth of grey mullet *Liza aurata* in Tunisian south coast. Cah. Biol. Mar. 53: 461–468.

Albertini-Berhaut, J. 1975. Biologie des stades juvéniles de Téléosteéens Mugilidae *Mugil auratus* Risso 1810, *Mugil capito* Cuvier 1829 ex *Mugil saliens* Risso 1810. III—Croissance linéaire et pondérale de *Mugil capito* dans le Golfe de Marseille. Aquaculture 5: 179–97.

Alvarez-Lajonchere, L.S. 1981. Determinación de la edad y el crecimiento de *Mugil curema*, *M. hospes*, *M. trichodon* y *M. liza*. Rev. Inv. Mar. 2: 142–162.

Al-Daham, N.K. and N.K. Wahab. 1991. Age growth and reproduction of the greenback mullet *Liza subviridis* Valenciennes in an estuary in southern Iraq. J. Fish Biol. 28: 81–88.

Almeida, P.R., F.M. Moreira, I.M. Domingos, J.L. Costa, C.A. Assis and M.J. Costa. 1995. Age and growth of *Liza ramada* (Risso 1826) in the River Tagus, Portugal. Sci. Mar. 59: 143–147.

Arné, P. 1938. Contribution á l'étude de la biologie des muges du Golfe de Gascogne. Rapp. P.-V. Comm. Int. Explor. Mediterr. 11: 77–115.

Arruda, L.M., J.N. Azevedo and A.I. Neto. 1991. Age and growth of the grey mullet Pisces Mugilidae in Ria de Aveiro Portugal. Sci. Mar. 55: 497–504.

Bagenal, T.B. and F.W. Tesch. 1978. Age and growth. pp. 101–136. *In*: T.B. Bagenal (ed.). Methods for Assessment of Fish Production in Fresh Waters. IBP Handbook No 3, Blackwell Publications, Oxford.

Beliaeva, V.N., A.D. Vlasenko and V.P. Ivanov. 1989. Caspian Sea: Ichthyofauna and Fisheries Resources. Nauk Press, Russian Academy of Science, Moscow.

Bentzel, H. 1968. Methods for determining age in the grey mullets. Fish Fishbrd. 3: 25–27 (in Hebrew).

Berg, L.S. 1965. Freshwater fishes of the U.S.S.R. and adjacent countries. Israel Program for Scientific Translations Ltd. (Russian version published in 1949), Jerusalem.

Bilton, H.T. and G.L. Robbins. 1971. Effects of feeding level on circulus formation on scales of young sockeye salmon (*Oncorhynchus nerka*). J. Fish Res. Bd. Canada 28: 861–868.

Brett, J.R. 1979. Environmental facts and growth. pp. 1–89. *In*: W.S. Hoar and D.J. Randall (eds.). Fish Physiology. Academic Press, New York.

Broadhead, G.C. 1953. Investigations of the black mullet *Mugil cephalus* L. in northwest Florida. Florida St. Board Conserv. Mar. Lab. Tech. Ser. 7: 1–33.

Brulhet, J. 1975. Observations on the biology of *Mugil cephalus ashenteensis* and the possibility of its aquaculture on the Mauritanian coast. Aquaculture 5: 271–81.

Brusle, J. 1981. Food and feeding in grey mullet. pp. 185–217. *In*: O. Oren (ed.). Aquaculture of Grey Mullets. Cambridge University Press, Cambridge.

Cambrony, M. 1983. Recrutement et biologie des stades juvéniles de Mugilidae (Poissons-Téléostéens) dans trois milieux lagunaires Roussillon et du Narbonnais (Salses-Leucate, Lapalme Bourdigou). Thesis, Docteur de 3ème cycle. University of Perpignan, Perpignan, France.

Cardona, L. 1999a. Age and growth of leaping grey mullet (*Liza saliens* (Risso 1810)) in Minorca (Balearic Islands). Sci. Mar. 63: 93–99.

Cardona, L. 1999b. Seasonal changes in the food quality, dial feeding rhythm and growth rate of juvenile leaping grey mullet *Liza saliens*. Aquat. Living Resour. 12: 263–270.

Cardona, L. 2006. Habitat selection by grey mullets (Osteichthyes, Mugilidae) in Mediterranean estuaries: the role of salinity. Sci. Mar. 70: 443–455.

Cech, J.J. and D.E. Wohlschlag. 1975. Summer growth depression in the striped mullet, *Mugil cephalus*. Contrib. Mar. Sci. 19: 92–100.

Chang, C.W., W.N. Tzeng and Y.C. Lee. 2000. Recruitment and hatching dates of grey mullet (*Mugil cephalus* L.) juveniles in the Tanshui estuary of northwest Taiwan. Zool. Stud. 39: 99–106.

Chang, C.W., Y. Iizuka and W.N. Tzeng. 2004. Migratory environmental history of the grey mullet *Mugil cephalus* as revealed by otoliths Sr:Ca ratios. Mar. Ecol. Prog. Ser. 269: 277–288.

Chubb, C.F., I.C. Potter, C.J. Grant, R.C.J. Lenanton and J. Wallace. 1981. Age structure growth rates and movements of sea mullet *Mugil cephalus* and yellow-eye mullet *Aldrichetta forsteri* in the swan Avon River system Western Australia. Aust. J. Mar. Fresh. Res. 32: 605–628.

Claridge, P.N. and I.C. Potter. 1985. Distribution abundance with size composition of mullet populations in the Severn Estuary and Bristol Channel England UK. J. Mar. Biol. Ass. UK 65: 325–336.

Curtis, T.D. and J.S. Shima. 2005. Geographic and sex-specific variation in growth of yellow-eyed mullet, *Aldrichetta forsteri*, from estuaries around New Zealand. New Zeal. J. Mar. Fresh. Res. 39: 1277–1285.

De Silva, S. 1980. Biology of juvenile grey mullet: a short review. Aquaculture 19: 21–36.

De Silva, S.S. and P.A.B. Perera. 1976. Studies on the young grey mullet, *Mugil cephalus* L. I. Effects of salinity on food intake, growth and food conversion. Aquaculture 7: 327–338.

De Silva, S.S. and E.I.L. Silva. 1979. Biology of young grey mullet, *Mugil cephalus* L., populations in a coastal lagoon in Sri Lanka. J. Fish Biol. 15: 9–20.

Díaz-Pardo, E. and S. Hernández-Vázquez. 1980. Crecimiento, reproducción y hábitos alimenticios de la lisa *Mugil cephalus* en la Laguna de San Andres, Tamps. An. Esc. Nal. Cienc. Biol., Mexico 23: 109–127.

Djabali, F., A. Mehailia, M. Koudil and B. Brahim. 1993. Empirical equations for the estimation of natural mortality in Mediterranean teleosts. Naga ICLARM 16: 35–39.

Dmitriev, A.N. 1964. Mullet in the Iranian waters of Caspian. Priroda 12: 74–75 (In Russian).

Drake, P. and A.M. Arias. 1984. Biology of mullets Osteichthyes, Mugilidae in the fish ponds of San-Fernando Cadiz Spain II. Crecimiento relativo. Inv. Pesq. 48: 157–173.

Drake, P., A.M. Arias and R.B. Rodríguez. 1984. Biology of mullets Osteichthyes, Mugilidae in the fish ponds of San-Fernando Cadiz Spain I. Growth in length and weight. Inv. Pesq. 48: 139–156.

Dulcic, J. and M. Kraljevic. 1996. Weight-length relationships for fish species in the eastern Adriatic (Croatian waters). Fish Res. 28: 243–251.

Durand, J.D. 2015. Implications of molecular phylogeny for the taxonomy of Mugilidae. *In*: D. Crosetti and S. Blaber (eds.). Biology, Ecology and Culture of Grey Mullet (Mugilidae). CRC Press, Boca Raton, USA (this book).

Durand, J.D., K.N. Shen, W.J. Chen, B.W. Jamandre, H. Bled, K. Diop, M. Nirchio, F.J. García de León, A.K. Whitfield, C.-W. Chang and P. Borsa. 2012. Systematics of the grey mullet (Teleostei: Mugiliformes: Mugilidae): molecular phylogenetic evidence challenges two centuries of morphology-based taxonomic studies. Mol. Phylogenet. Evol. 64: 73–92.

Ellender, B.R., G.C. Taylor and O.L.F. Weyl. 2012. Validation of growth zone deposition rate in otoliths and scales of flathead mullet *Mugil cephalus* and freshwater mullet *Myxus capensis* from fish of known age. Afr. J. Mar. Sci. 34: 455–458.

El Maghraby, A.M., M.T. Hashem and H.M. El Sedfy. 1974. Sexual, spawning migration and fecundity of *Mugil capito* (Cuv) in lake Borollus. Bull. Inst. Ocean. Fish Area 4: 33–56.

Erman, F. 1959. Observations on the biology of the common grey mullet (*Mugil cephalus*). Proc. Tech. Pap. Gen. Fish Counc. Mediterr. 5: 157–169.

Erman, F. 1961. On the biology of thicklipped grey mullet (*Mugil chelo* Cuv.) Rap. P.-V. Comm. Int. Explor. Sci. Mer. Mediterr. 16: 277–85.

El Zarka, S. and El Din. 1964. Acclimatization of *Mugil saliens* (Risso) in Lake Quarun U.A.R. Proc. Gen. Fish Comm. Mediterr. 7: 337–346.

El Zarka, S. and H.M. El Sedfy. 1970. The biology and fishery of *Mugil saliens* (Risso) in Lake Quarun, United Arab Republic, U.A.R. J. Oceanogr. Fish 1: 1–26.

El Zarka, S., A.M. El Maghraby and K. Abdel-Hamld. 1970. Studies on the distribution, growth and abundance of migratory fry and juveniles of mullet in a brackish coastal lake (Edku) in the United Arab Republic. Etud. Rev. FAO-CPM 46: 1–19.

Farrugio, H. 1975. Les muges (Poissons, Teléostéens) de Tunisie. Répartition et pêche. Contribution á leur étude systématique et biologique. Ph.D. Thesis, Univer. Montpellier II.

Farrugio, H. and J.P. Quignard. 1974. Biologie de *Mugil* (*Liza*) *ramada* Risso 1826 et de *Mugil* (*Chelon*) *labrosus* Risso 1826 (Poissons, Téléostéens, Mugilidés) du Lac de Tunis. Bull. Inst. Oceanogr. Peche Salammbo 2: 565–579.

Fazli, H., D. Ghaninejad, A.A. Janbaz and R. Daryanabard. 2008. Population ecology parameters and biomass of golden grey mullet (*Liza aurata*) in Iranian waters of the Caspian Sea. Fish Res. 93: 222–228.

Fehri-Bedoui, R. and H. Gharbi. 2005. Age and growth of *Liza aurata* (Mugilidae) along Tunisian coasts. Cybium 29: 119–126.

Froese, R. 2006. Cube law, condition factor and weight–length relationships: history, meta-analysis and recommendations. J. Appl. Ichthyol. 22: 241–253.

Froese, R. and D. Pauly (eds.). 2014. Fish Base. World Wide Web electronic publication. www.fishbase.org, Electronic version accessed 6/2014.

Glamuzina, B., J. Dulcic, A. Conides, I.V. Bartulovic, S. Matic-Skoko and C. Papaconstantinou. 2007. Some biological parameters of the thin-lipped mullet *Liza ramada* (Pisces: Mugilidae) in the Neretva River delta (Eastern Adriatic, Croatian coast). Vie et Milieu 57: 131–136.

González-Castro, M., V. Abachian and R.G. Perrotta. 2009. Age and growth of the striped mullet, *Mugil platanus* (Actinopterygii: Mugilidae), in a southwestern Atlantic coastal lagoon (37 degrees 32'S-57 degrees 19'W): a proposal for a life-history model. J. Appl. Ichthyol. 25: 61–66.

Grant, C.J. and A.V. Spain. 1975. Reproduction, growth and size allometry of *Mugil cephalus* Linnaeus (Pisces: Mugilidae) from North Queensland Inshore Waters. Aust. J. Zool. 23: 181–201.

Grbec, B., M. Morovic and M. Zore-Armanda. 1998. Some new observations on the long-term salinity changes in the Adriatic Sea. Acta Adriat. 39: 3–12.

Hendricks, L.J. 1961. The striped mullet *Mugil cephalus* L. Fish Bull. Calif. 113: 95–103.

Hickling, C.F. 1970. A contribution to the natural history of the English grey mullet (Pisces: Mugilidae). J. Mar. Biol. Ass. UK 50: 609–633.

Hotos, G.N. 2003. A study on the scales and age estimation of the grey golden mullet, *Liza aurata* (Risso 1810), in the lagoon of Messolonghi (W. Greece). J. Appl. Ichthyol. 19: 220–228.

Hotos, G.N. and G.N. Katselis. 2011. Age and growth of the golden grey mullet *Liza aurata* (Actinopterygii: Mugiliformes: Mugilidae), in the Messolonghi-Etoliko lagoon and the adjacent gulf of Patraikos, western Greece. Acta Ichthyol. Piscat. 41: 147–157.

Hsu, C.C. and W.N. Tzeng. 2009. Validation of annular deposition in scales and otoliths of flathead mullet *Mugil cephalus*. Zool. Stud. 48: 640–648.

Huxley, J.S. 1932. Problems of Relative Growth. The Dial Press, New York.

Ibáñez-Aguirre, A.L. and J. Lleonart. 1996. Relative growth and comparative morphometrics of *Mugil cephalus* L. and *M. curema* V. in the Gulf of Mexico. Sci. Mar. 60: 361–368.

Ibáñez-Aguirre, A.L. and M. Gallardo-Cabello. 1996. Age determination of the grey mullet *Mugil cephalus* L. and the white mullet *Mugil curema* V. (Pisces: Mugilidae) in Tamiahua Lagoon, Veracruz. Cien. Mar. 22: 329–345.

Ibáñez-Aguirre, A.L., M. Gallardo-Cabello and P. Sánchez-Rueda. 1995. Estimación de la edad de la lisa *Mugil cephalus* y la lebrancha, *M. curema,* por métodos indirectos. Hidrobiologica 5: 105–111.

Ibáñez-Aguirre, A.L., M. Gallardo-Cabello and X. Chiappa-Carrara. 1999. Growth analysis of striped mullet, *Mugil cephalus*, and white mullet, *M. curema* (Pisces: Mugilidae) in the Gulf of Mexico. Fish Bull. 97: 861–872.

Ibáñez, A.L., E. Cabral-Solís, M. Gallardo-Cabello and E. Espino-Barr. 2006. Comparative morphometrics of two populations of *Mugil curema* in the Atlantic and Pacific coasts. Sci. Mar. 70: 139–145.

Ibáñez, A.L., I.G. Cowx and P. O'Higgins. 2007. Geometric morphometric analysis of fish scales for identifying genera, species and local populations within the Mugilidae. Can. J. Fish Aquat. Sci. 64: 1091–1100.

Ibáñez, A.L., I.G. Cowx and P. O'Higgins. 2009. Variation in elasmoid fish scale patterns is informative with regard to taxon and swimming mode. Zool. J. Linn. Soc.-Lon. 155: 834–844.

Ibáñez, A.L., E. Pacheco-Almanzar and I.G. Cowx. 2012a. Does compensatory growth modify fish scale shape? Env. Biol. Fish 94: 477–482.

Ibáñez, A.L., C.C. Hsu, C.W. Chang, C.H. Wang, Y. Iizuka and W.N. Tzeng. 2012b. Diversity of migratory environmental history of striped mullet *Mugil cephalus* and white mullet *M. curema* in the Mexican waters as indicated by otolith Sr: Ca ratios. Cien. Mar. 38: 73–87.

Ilkyaz, A.T., K. Firat, S. Saka and H.T. Kinacigil. 2006. Age, growth, and sex ratio of golden grey mullet, *Liza aurata* (Risso 1810) in Homa Lagoon (Izmir Bay, Aegean Sea). Turk. J. Zool. 30: 279–284.

Jhingran, V.G. and N.K. Misra. 1962. Further fish-tagging experiments in Chilka Lake (1959) with special reference to *Mugil cephalus* Linnaeus. Indian J. Fish 9: 129–197.

Kaya, M., M. Bilecenoglu and O. Ozaydin. 2000. Growth characteristics of the leaping mullet (*Liza saliens* Risso 1810) in Homa Lagoon, Aegean Sea. Isr. J. Aquacult.-Bamid. 52: 159–166.

Kazanskji, B.N. and L.I. Stashenko. 1980. Acclimatization of Pacific mullet in the Black Sea Basin. Biol. Morya 6: 46–50.

Kennedy, M. and P. Fitzmaurice. 1969. Age and growth of thick-lipped grey mullet *Crenimugil labrosus* in Irish waters. J. Mar. Biol. Ass. UK 49: 683–699.

Kesteven, G.L. 1942. Studies in the biology of Australian mullet. I. Account of the fishery and preliminary statement of the biology of *Mugil dobula* Gunter. Bull. Aust. CSIRO Melb. 157: 1–99.

Kesteven, G.L. 1953. Further results of tagging sea mullet, *Mugil cephalus* L., on the eastern Australian coast. Aust. J. Mar. Freshwater Res. 4: 251–306.

Khemis, I.B., D. Zouiten, R. Besbes and F. Kamoun. 2006. Larval rearing and weaning of thick lipped grey mullet (*Chelon labrosus*) in mesocosm with semi-extensive technology. Aquaculture 259: 190–201.

Koutrakis, E.T. and A.I. Sinis. 1994. Growth analysis of grey mullets (Pisces: Mugilidae) as related to age and site. Isr. J. Zool. 40: 37–53.

Koutrakis, E.T., A.I. Sinis and P.S. Economidis. 1994. Seasonal occurrence, abundance and size distribution of gray mullet fry (Pisces: Mugilidae), in the Porto-Lagos lagoon and Lake Vistonis (Aegean Sea, Greece). Isr. J. Aquacult.-Bamid. 46: 182–196.

Kraiem, M.M., C.B. Hamza, M. Ramdani, A.A. Fathi, H.M.A. Abdelzaher and R.J. Flower. 2001. Some observations on the age and growth of thin-lipped grey mullet, *Liza ramada* Risso 1826 (Pisces: Mugilidae) in three North African wetland lakes: Merja Zerga (Morocco), Garaat Ichkeul (Tunisia) and Edku Lake (Egypt). Aquat. Ecol. 35: 335–345.

Kraljevic, M. and J. Dulcic. 1996. Age, growth and mortality of the golden grey mullet *Liza aurata* (Risso 1810) in the eastern Adriatic. Arch. Fish Mar. Res. 44: 69–80.

Le Dantec, J. 1955. Quelques observations sur la biologie des muges des réservoirs de Certes á Audenge. Rev. Trav. Off. (Sci. Tech.) Peches. Marit. 19: 93–97.

Libosvarsky, J. 1976. Lepidological note on grey mullet (*Mugil capito*) from Egypt. Zool. Listy 25: 73–79.

Matic-Skoko, S., J. Ferri, M. Kraljevic and A. Pallaoro. 2012. Age estimation and specific growth pattern of boxlip mullet, *Oedalechilus labeo* (Cuvier 1829) (Osteichthyes, Mugilidae), in the eastern Adriatic Sea. J. Appl. Ichthyol. 28: 182–188.

McDonough, C.J. and C.A. Wenner. 2003. Growth, recruitment and abundance of juvenile striped mullet (*Mugil cephalus*) in South Carolina estuaries. Fish Bull. 101: 343–357.

Menezes, N.A., C. De Oliveira and M. Nirchio. 2010. An old taxonomic dilemma: the identity of the western south Atlantic lebranche mullet (Teleostei: Perciformes: Mugilidae). Zootaxa 2519: 59–68.

Modrusan, Z.E. Teskeredzic and S. Jukic. 1988. Biology and ecology of Mugilidae species on the eastern Adriatic coast (Sibenik bay). FAO Fish Rpt. 394: 159–167.

Morales-Nin, B. 1987. Influence of environmental factors on microstructure of otoliths of three demersal fish species caught off Namibia. pp. 255–262. *In*: A.I.L. Payne, J.A. Gulland and K.H. Brinkd (eds.). The Benguela and Comparable Ecosystems. S. Afr. J. Mar. Sci. 5.

Morovic, D. 1960. Contribution à la connaissance de la croissance annuelle du muge *Mugil cephalus* (L.) et de *M. chelo* Cuv. dans l'étang Pantan (Dalmatie). Rapport préliminaire. Rap. P.-V. Comm. Int. Explor. Sci. Mer. Méditerr. 15: 115–8.

Moura, I.M. and L. Serrano-Gordo. 2000. Abundance, age, growth and reproduction of grey mullets in Obidos lagoon, Portugal. Bull. Mar. Sci. 67: 677–686.

Njoku, D.C. and I.O. Ezeibekwe. 1996. Age composition and growth of the large-scaled mullet, *Liza grandisquamis* (Pisces: Mugilidae), Valenciennes 1836 on the New Calabar Estuary, off the Nigerian coast. Fish Res. 26: 67–73.

Odum, W.E. 1970. Utilization of the direct grazing and plant detritus food chains by the striped mullet *Mugil cephalus*. pp. 222–240. *In*: J.H. Steele (ed.). Marine Food Chains. Oliver and Boyd, London.

Okumus, I. and N. Bascinar. 1997. Population structure, growth and reproduction of introduced Pacific mullet, *Mugil soiuy*, in the Black Sea. Fish Res. 33: 131–137.

Oliveira, J.M. and M.T. Ferreira. 1997. Abundance, size composition and growth of a thin-lipped grey mullet, *Liza ramada* (Pisces: Mugilidae) population in an Iberian River. Folia Zool. 46: 375–384.

Panfili, J., H. de Pontual, H. Troadec and P.J. Wright. 2002. Manual of fish sclerochronology. pp. 19–27. *In*: J. Panfili and H. de Pontual (eds.). Manual of Fish Sclerochronology. Ifremer-IRD coedition, Brest.

Patimar, R. 2008. Some biological aspects of the sharpnose mullet *Liza saliens* (Risso 1810) in Gorgan Bay-Miankaleh wildlife refuge (the Southeast Caspian Sea). Turk. J. Fish Aquat. Sci. 8: 225–232.

Paton, D., L. Cardona and E. Gisbert. 1994. Comparative growth in relation with the age assessed by skeletochronology in two mullet fishes of genus *Liza*. Anim. Prod. 59: 303–307.

Pauly, D. 1983. Some simple methods for the assessment of tropical fish stocks. FAO Fisheries Tech. Paper No. 234, 52p.

Pauly, D. and J.L. Munro. 1984. Once more on the comparison of growth in fish and invertebrates. ICLARM Fishbyte 2: 21.

Pérez-García, M. and A.L. Ibáñez-Aguirre. 1992. Morfometría de los peces *Mugil cephalus* y *M. curema* (Mugiliformes-Mugilidae) en Veracruz, México. Rev. Biol. Trop. 40: 335–339.

Pillay, S.R. 1954. The biology of the grey mullet *Mugil tade* Forsk with observation on its fishery in Bengal. Proc. Natl. Inst. Sci. India 20: 187–217.

Quignard, J.P. and H. Farrugio. 1981. Age and growth of grey mullet. pp. 155–218. *In*: O.H. Oren (ed.). Aquaculture of Grey Mullets. Cambridge University Press, Cambridge.

Reay, P.J. 1987. A British population of the grey mullet, *Liza aurata* (Teleostei: Mugilidae). J. Mar. Biol. Ass. UK 67: 1–10.

Rossignol, M. 1951. *Mugil chelo* Cuvier. Notes sur les muges des côtes marocaines. Ann. biol. Cons. Int. Expl. Mer. Copenhagen 8: 89–90.

Safran, P. 1992. Theoretical analysis of the weight length relationship in fish juveniles. Mar. Biol. 112: 545–551.

Santana, F., E. Morize, J. Clavier and R. Lessa. 2009. Otolith micro and macrostructure analysis to improve accuracy of growth parameter estimation for White mullet *Mugil curema*. Aquat. Biol. 7: 199–206.

Sauriau, P.G., J.P. Robin and J. Marchand. 1994. Effects of the excessive organic enrichment of the Loire estuary on the downstream migratory patterns of the amphihaline grey mullet *Liza ramada* (Pisces: Mugilidae). pp. 349–356. *In*: K.R. Dyer and R.J. Orth (eds.). Changes in Fluxes in Estuaries: Implications from Science to Management. Olsen and Olsen, Fredensborg.

Sinovcic, G., V. Alegría-Hernández, J. Jug-Duja, S. Jukic, I. Kacic, S. Regner and M. Tonkovic. 1986. Contribution to the knowledge of ecology of grey mullet *Liza ramada* (Risso 1826) from the Middle Adriatic (Sibenik area). Acta Adriat. 27: 147–162.

Smith, K.A. and K. Deguara. 2003. Formation and annual periodicity of opaque zones in sagittal otoliths of *Mugil cephalus* (Pisces: Mugilidae). Mar. Fresh. Res. 54: 57–67.

Snovsky, G. and J. Shapiro. 2000. The age and growth of the grey mullet, *Liza ramada*, in Lake Kinneret, Israel. J. Aquaculture in the Tropics 15: 91–95.

Sparre, P. and S.C. Venema. 1997. Introduction to tropical fish stock assessment. Part 1. Manual. FAO Fisheries Technical Paper. No. 306. 1, Rev. 2. Rome.

Sudha, S., C.M. Aravindan and N.K. Balasubramanian. 1992. Comparative study of length-weight relationship of wild stock of juveniles and adults of the grey mullet, *Liza tade* (Forsskal) from two backwater systems. J. Anim. Morph. Physiol. 39: 137–144.

Thakur, N.K. 1967. Studies on the age and growth of *Mugil cephalus* L. from the Mahanadi estuarine system. Proc. Natl. Inst. Sci. India (B)33: 128–43.

Thomson, I.M. 1951. Growth and habits of the sea mullet, *Mugil dobula* Günther in Western Australia. Aust. J. Mar. Freshwater Res. 2: 193–225.

Thomson, I.M. 1957. Interpretation of the scales of the yellow-eyed mullet *Aldrichetta forsteri* (Cuvier and Valenciennes) (Mugilidae). Aust. J. Mar. Freshwater Res. 8: 14–28.

Thomson, I.M. 1966. The grey mullet. Oceanogr. Mar. Biol. 4: 301–35.

Wimpenny, R.S. 1932. Observations on the size age and growth of two Egyptian mullets, *Mugil cephalus* (Linn.), the bouri and *M. capito* (Cuv.) the tobar. Rep. Fish Serv. Egypt, Coastguards and Fisheries Service, Ministry of Finance, Cairo. 53p.

Yashouv, A. 1966. Breeding and growth of grey mullet (*Mugil cephalus* L.). Bamidgeh 18: 3–13.

CHAPTER 11

Sexuality and Reproduction of Mugilidae

Mariano González-Castro[1,*] and *George Minos*[2]

Introduction

Mullets are gonochoristic or bisexual fish: this implies that individuals are either males or females. Normally, adult females are bigger than males, but there is no sexual dimorphism and thus it is impossible to externally distinguish between sexes. With respect to reproductive mechanisms, the species of Mugilidae are oviparous, which involves the release of gametes into the water (usually coastal marine waters), with external fertilization and development.

Reproductive System

Male: Anatomy of Testis

The testes are internal, longitudinal and paired organs. They are suspended by lengthwise mesenteries known as mesorchia, and lie lateral to the gas bladder. In Teleostei, there is no connection between the kidney and gonads at maturity. The sperm duct is new and originates from the testes (Helfman et al. 2009). Grier (1993) concluded that primitive osteichthyans have an anastomosing tubular testis, whereas derived teleosts, including atherinomorphs, have a lobular testis. The lobular testis can be divided into two types based on distribution and arrangement of spermatogonia. The mullet testis type can be defined as 'unrestricted lobular', which implies that spermatogonia occur along the entire length of the tubules, in contrast to atherinomorphs which are classified as 'restricted lobular', as spermatogonia are confined to the distal end of the tubules (Parenti and Grier 2004).

Female: Anatomy of Ovaries

The ovary of mullets is a hollow, paired organ, consisting of two ovarian lobes that are separated by a septum. Both lobes are joined near the urogenital pore. Numerous ovigerous folds project into the ovarian

[1] Laboratorio de Biotaxonomía Morfológica y molecular de peces, Instituto de Investigaciones Marinas y Costeras IIMyC-CONICET, UNMdP, Mar del Plata, Argentina.
Email: gocastro@mdp.edu.ar; gocastro@gmail.com
[2] Department of Aquaculture and Fisheries Technology, Alexander Technological Educational Institute of Thessaloniki, Nea Moudania, Chalkidiki, Greece.
Email: gminos@otenet.gr; gminos@aqua.teithe.gr
* Corresponding author

cavity. The lamellae consist of connective tissue lined by germinal epithelium, which contains nests of oogonia. Ovarian follicles develop along the lamellae and the mature oocytes are ovulated into the ovarian cavity (El-Halfawy et al. 2007).

Development and Microanatomy of Gonads

In gonochoristic species possessing purely ovarian or testicular tissues, two main types of gonadal differentiation can be seen. In most species gonad development proceeds from an indifferent gonad directly to ovary or testis. These species are called primary gonochorists or 'differentiated gonochorists' (Jakobsen et al. 2009).

In higher vertebrates, the gonad originates in an undifferentiated state and consists of a cortex, derived from stromal cells and a medulla from the peritoneal wall, which itself is drawn from the mesonephric blastema. During ovarian differentiation, the cortex proceeds through further development, but with simultaneous degeneration of the medulla; in the case of testicular differentiation, the developmental programme is reversed (Pandian 2012). However, the somatic component of fish gonads is of unitary origin, i.e., the somatic component of ovary and testis originates from the peritoneal wall alone (Nakamura 1978). Following the arrival of Primordial Germ Cells (PGCs), as guided by chemo-attractants and colonization of the gonadal anlage, the gonad now consists of both germ cells and gonadal supporting somatic cells. The somatic cell lineages then develop into granulose and thecal cells in the ovary and differentiate into Sertoli cells and Leydig cells in the testis. The granulose and the Sertoli cells are known as germ cells supporting cells and directly enclose the germ cells, whereas thecal and Leydic cells function as sex steroid hormone producing cells (Pandian 2012).

In a study related to ovary development of the flathead mullet *Mugil cephalus*, McDonough et al. (2005) showed that in undifferentiated juveniles the primordial gonad lobes are suspended by a mesentery connected dorsally to the peritoneum and attached ventrally to the intestines. The gonads in specimens less than 50 mm have lobes ranging from 70 to 100 μm in length. Lobes are made up of somatic cells and a peripheral germinal epithelium. The lobes are attached along their dorsomedial surface by loose fibrous connective tissue, known as stromal tissue. No defining male or female characteristics are present at this fish length. In specimens ranging from 50 to 100 μm, gonad lobes increase to 150 μm and appear more vascularized. The lobes are attached to the suspensory mesentery, which is attached to the peritoneum and a few deuterogonia are visible along the lateral periphery of each lobe. The remainder of the lobes contain somatic tissue. The individual germ cells are approximately 5 μm in size. In specimens ranging from 100 to 150 mm, the gonad lobes are obviously vascularized and attain a size of 200 to 300 μm. Early ductwork begins to become evident. Deuterogonia enlarge and form nests along the lateral and distal portions of the lobes. Somatic cells make up a large portion of each lobe and the stromal tissue is now more stalklike, attaching each lobe to the suspensory tissue. There were only four specimens in this size range that start to differentiate as males. Gonads destined to be males were identified by duct structures within the gonad lobe as well as by more elongated germ cell nests. These morphologically distinct features resulted in an early demonstration of the corradiating pattern of ducts and lobules seen in more advanced testes (McDonough et al. 2005).

The 150 to 200 mm size class individuals (*M. cephalus*) showed that 0.2% of females and 37.3% of males began initial differentiation, but the majority of all specimens (62.5%) remained undifferentiated. The undifferentiated gonads became larger, and lobe size was 600 to 800 μm. There was increased vascularization, particularly along the medial portion of the stroma. Germ-cell nests were now more organized, with four to eight cells visible in each. More than 83% of specimens > 200 mm had become sexually differentiated. In some cases, germ cell nests that were characteristic of female precursors could also be found in the centre portions of lobes adjacent to the characteristic male precursor lobule structures. The primary duct was now well formed; however there were still no definitive morphological characteristic that would enable sex determination (McDonough et al. 2005).

Male Differentiation

The initial differentiation of males is evident in the morphological features of the germ-cell tissue located along the peripheral portions of each lobe. The germ tissue begins to form elongated bands perpendicular to the edge of the lobe, whereas the somatic tissue begins to form fibrous bands originating along the edges of the primary duct. The primary duct is defined structurally at this point. With continued increase in fish length, lobes increase in size and vascularization. The germ tissue continues to elongate medially within the lobe in a corradiating pattern.

Somatic tissue continues to form band-like structures that eventually become secondary ductwork, and the germ-cell expands to form lobules. As the lobules become more developed, spermatogonia begin to line the lobules as part of the germinal epithelium. Sertoli cells are not visible because of the lack of resolution at this magnification (400 ×) level. Mitotic proliferation of spermatogonia causes lobular enlargement, although spermatogonia are very small at this stage (2–3 μm) (McDonough et al. 2005).

Female Differentiation

The first sign of female sexual differentiation is the organization of germ-cell tissue into round nests of eight–10 cells each. The germ-cell nests, which eventually give rise to oogonial nests, are first found along the lateral periphery of the lobe. There is evidence of early ovarian wall development, which consists of a single layer of cells forming the outer layer of the lobe, separate from the oogonial nests. Although some ductwork is present, there is no evidence of the formation of lamellae. With continued development, individual cells within the nests become more visible and the ovary wall becomes more evident. Stalks or buds of tissue are observed growing out of the base of the stroma on the dorsolateral surface. As development progresses, the ovarian wall attached to these stalks or tissue buds appears to grow over the dorsal surface of each ovarian lobe. The presence of both the ovary wall stalk buds and the rounded germ-cell nests located throughout the gonad lobe are diagnostic of female differentiation (McDonough et al. 2005).

Primary growth oocytes increase in number and begin to aggregate, forming distinct lamellae. The ovary wall continues to differentiate at this point but is only a few cell layers thick. There is still a great deal of stroma and somatic tissue left in the ovary, but it begins to form bands of fibrous tissue, resulting from the regression of stroma and somatic tissue (where present) as the lamellae continue to develop. Oogonia begin proliferating and differentiating into primary growth oocytes as folliculogenesis commences. The ovary wall, now becoming vascularized, begins to separate from the lamellae, opening a space that becomes the ovarian lumen. The ovary wall is made up of squamous cells on the inside layers and collagen and elastic tissue on the outer layers. The stroma and somatic cells continue to be reduced until they are primarily fibrous tissue from which the lamellae are suspended. Histological ovarian cross-sections change from the leaf or spade shape of the undifferentiated gonad to a more rounded one. Once ovarian differentiation is complete, the individual lamellae have oocytes within each and the stroma is reduced to suspensory tissue for the lamellae. The primary growth oocytes present in the lamellae remain small (80 to 100 μm) and relatively uniform in size (McDonough et al. 2005). At the initiation of reproductive development, the oocytes start to grow from the arrested prophase of the first meiotic division (Stenger 1959, Kuo et al. 1974).

Reproductive Biology

Although it is desirable to use data from both sexes when studying reproductive biology, there is a tendency to accept that females (the biological data obtained from) are biologically more informative than those obtained from males. This relates to the fact that analysis of histological slides of ovaries can accurately reveal both the geographic location and the period in which spawning occurs. Also, some important reproductive parameters such as fecundity are obviously estimated only from females. Lastly, the duration of the reproductive season of a population must be defined by the time between the beginning and the end of spawning events, and is measurable by microscopic observation of the maturity stage of only ovaries, because it is well known that males mature earlier, and finish their reproductive season after females.

Stages of Oocyte Development in Mullets

The process which involves the transformation of oogonia to oocytes is called oogenesis (Selman and Wallace 1989). In these cells the chromosomes are arrested at the diplotene stage of the first meiotic stage (MacMillan 2007). There is doubt whether in fish the oocytes develop from a stock of precursors already present in the ovary from puberty, as in mammals, or oogenesis (understood as the production of new oocytes) may occur in the adult ovary (Saborido-Rey 2002). MacMillan (2007) suggests that in most bony fishes, oogonial proliferation is found in the adult animal. There is a limited period of peak oogonial division in species with circumscribed annual breeding cycles; in species which do not have restricted breeding cycles, oogonial proliferation occurs either in waves or continuously throughout the year.

Oogonia

Oogonia are small cells, with a rounded large nucleus in the centre and lax chromatin, surrounded by scarce cytoplasm. In *Mugil liza* these cells are 4–8 μm in diameter, with only one nucleolus, and have only been observed fundamentally in virginal or immature specimens (GSI = 0.02) or in post-spawning females (GSI = 0.77) (Da Silva and Esper 1991, Albieri and Araújo 2010, González-Castro et al. 2011).

Primary Growth (previtellogenic stage)

Two phases can be observed in the Primary Growth (PW) stage in fish, based fundamentally on nucleolar morphology: the chromatin-nucleolus stage and the peri-nucleolus phase. The oocytes at the chromatin-nucleolus stage have scarce cytoplasm and a large nucleus, centrally located, which usually contain a single large basophilic nucleolus at the beginning of this phase. The oocyte itself is surrounded by a few squamous follicular cells (Wallace and Selman 1981, Saborido 2002, González-Castro 2007). Chromatin-nucleolus oocytes become arrested in late diplotene of meiotic prophase I, as the synaptonemal complexes dissociate within the oocyte nucleus (Selman and Wallace 1989).

Several studies on teleosts indicate that PW is independent of pituitary control (Khoo 1979). During this stage, oocytes increase in volume about 1,000 fold, and in mullets PW oocytes reach a diameter that varies from 50 to 200 μm, with polyhedral shapes and rounded-central nuclei (Da Silva and Esper 1991, Minos et al. 2010, González-Castro et al. 2011, Lemos et al. 2014). These cells commonly show one–nine prominent nucleoli and basophilic cytoplasm, as denoted for *M. liza* by González-Castro et al. (2011). Surprisingly, in the alien redlip *Liza haematocheila* (*L. haematocheilus*) primary growth oocytes exhibit only one giant nucleolus (Minos et al. 2010). As the oocyte grows, the definitive follicle forms. Oocytes of fishes undergo the early stages of prophase and become arrested in diplotene of the first meiotic division (MacMillan 2007).

At the peri-nucleolus phase, concomitantly with the growth of the oocytes, the nucleus increases in size to form the germinal vesicle. Multiple peripheral nucleoli appear, usually eight to 14 for *M. liza*, according to González-Castro (2007) and González-Castro et al. (2011), as a consequence of the amplification of ribosomal genes. At the end of this state, numerous microvilli extend on the surface of the oocytes while the precursor materials of the *zona radiata* begin to accumulate in spots (Wallace and Selman 1981, Saborido-Rey 2002).

Secondary Growth (Cortical Alveolus Stage)

The cortical alveolus stage is characterized by the initial appearance of three components: cortical alveoli, the *zona radiata* (also termed as chorion in a more advanced stage) and lipids. Cortical alveoli initially appear circumferentially at various depths in the cytoplasm. They are membrane bound, have a homogeneous appearance and generally stain for both protein and carbohydrates (PAS positive reaction). The size and structure is variable: some contain a single large granule within a homogeneous matrix while others contain only the homogeneous matrix. They appear to be formed by the granular reticulum endoplasmic and Golgi complex and contain a glycoprotein (polysialoglycoprotein with a molecular weight of 200,000

Daltons or above) synthesized within the oocytes (Tyler and Sumpter 1996, MacMillan 2007). Then, in this stage of development, the endogenous vitellogenesis starts. Glycoproteins are synthesized endogenously and incorporated into the cortical alveolus which grows in size. In *M. liza*, with diameters of 200 and 300 µm, the oocytes begin to lose their basophilic characteristics (González-Castro 2007). At fertilization, the cortical reaction takes place which involves the release of the contents of the cortical alveoli into the perivitelline space between oocytes and the vitelline envelope to prevent polyspermy and the penetration of pathogens. The zona radiata is an acellular vitelline envelope, which develops around the oocyte and continues to differentiate and increase in complexity throughout the oocyte growth (Tyler and Sumpter 1996, Patiño and Sullivan 2002).

Vitellogenesis

In this stage extraovarian proteins are sequestered, processed and packaged into oocytes. Basically, a hepatically derived plasma precursor called Vitellogenin (VTG) is the main precursor of yolk proteins (Tyler and Sumpter 1996). VTG are phosphoglycolipoproteins, members of the large lipid transfer protein superfamily, and play an important role as a source of nutrients during the early embryo development (Babin et al. 1999). Greeley et al. (1986) showed two large proteins to be the major yolk proteins present in mullet oocytes at the end of vitellogenic growth. As evidenced by the *de novo* appearance of these marker proteins, early vitellogenesis in the flathead mullet *M. cephalus* begins in oocytes that are between 160 and 180 µm in diameter. Late vitellogenic oocytes become competent to resume meiotic maturation and develop into a fertilizable egg in response to an *in vitro* steroid challenge, at 600 µm in diameter (Greeley et al. 1987).

Yolk proteins are accumulated in fluid-filled yolk spheres or globules. In those marine fishes that spawn pelagic eggs, the globules fuse centripetally, forming a continuous mass of fluid yolk during postvitellogenic maturation (Wallace and Selman 1985). Yolked oocytes of mullets usually have a range of 350–800 µm diameters (Silva and De Silva 1981, Su and Kawasaki 1995, Ergene 2000, Romagosa et al. 2000, Kendall and Gray 2008, Minos et al. 2010, González-Castro et al. 2011). Mousa (2010) showed that prespawning females of the thin-lipped grey mullet *Liza ramada* contained vitellogenic (tertiary yolk) oocytes, with a centrally located Germinal Vesicle (GV) and diameters from 600 to 650 µm.

During the vitellogenesis stage, the radiate zone surrounding the oocytes becomes evident. Follicular cells multiply and stratify, being notorious in histological slides. Thus, the granulosa cells (internal layer) and the theca cells (external layer, which includes fibroblasts, collagen fibres, capillaries and steroid-productions cells) can be identified. Both layers are separated by a basal membrane (González-Castro et al. 2007). Then, the oocytes, zona radiata and follicular cells are denominated as ovarian follicle. In *M. liza*, yolked oocytes present eosinophilic yolk protein granules throughout the cytoplasm. The zona radiata becomes thick, striated and highly eosinophilic. Although less frequent, lipid vesicles have been observed dispersed between the yolk granules (González-Castro et al. 2011, Lemos et al. 2014).

In an experimental study of oocyte growth in *M. cephalus* maturing at different salinities, Tamaru et al. (1994) demonstrated that the rate of oocyte growth from females maturing in freshwater was significantly slower than those matured in brackish and seawater. Moreover, only 30% of the females maintained in freshwater reached a state of maturity (i.e., oocytes ≥ 600 µm) at which spawning could be induced, contrasting with the 90% of mature females obtained for both brackish and seawater.

Maturation (Germinal Vesicle Migration; Hyalinization; Hydration)

The following events characterize the resumption of meiosis that occurs during oocyte maturation in fish: a) the germinal vesicle migrates towards the periphery (GVM) of the oocytes and the nuclear envelope dissociates (GVBD); b) the chromosomes condense and proceed to the first meiotic metaphase, followed by the elimination of the first polar body; and c) the remaining chromosomes then proceed to the second meiotic metaphase, where they arrest once again. Hydration of oocytes occurs just prior to spawning (Selman and Wallace 1989, Hsu et al. 2007). Hydration during the final phases of development can either be negligible or account for a considerable proportion of the final egg size, depending on the fish species.

A hydration phase during final maturation appears to be important in the production of pelagic (buoyant) mullet eggs (Wallace and Selman 1981).

Mullets undertake a reproductive migration from estuaries or coastal lagoons to the sea where they reach spawning areas (which, in general, are not well documented). Accordingly, scarce evidence relative to final oocyte maturation (hydration) and ovulation on wild populations has been published. Lemos et al. (2014), in a study of the migration and reproductive biology of *M. liza* in south Brazil, found females (42% of the specimens analyzed in the month of June) with hyaline oocytes. The presence of this oocyte stage was associated with high salinity and sea surface temperatures of 19–21ºC, and followed the seasonal northward displacement of these oceanographic conditions. Hsu et al. (2007) reported evidence for a few mature females of *M. cephalus* captured in the North-East of Taiwan, which had hydrated oocytes of 900–1,100 μm in diameter.

Postovulatory Follicles (POF)

The ovulatory process includes several changes in the oocytes and ovarian follicle prior to expulsion of the oocyte. Microvillar connections between the oocytes and the surrounding granulosa cells are broken (follicular separation), parts of the follicular wall decrease in thickness, and an opening forms in the follicle wall (follicular rupture) due to digestion of tissue by proteolytic enzymes and probably also by the internal pressure generated by the hydration process. The oocytes are released by contraction of smooth muscles in the follicular wall. This process is regulated in part by prostaglandins (Khan and Thomas 1999). Once the oocytes are ovulated, the follicular cells remain in the ovary and constitute the postovulatory follicles (POFs). As POFs are reabsorbed in a few days (usually within 72–96 hours), they can be considered as evidence of recent spawning. POFs have been reported in few species of Mugilidae (Da Silva and Esper 1991, Hsu et al. 2007, Lemos et al. 2014).

Atretic Follicles

Atresia is the oocyte degeneration that may occur at any stage of development. This process has been suggested to play a major role in determining the numbers of oocytes that are recruited into the successive stages of development, thereby affecting the number of developing oocytes that eventually form mature eggs. However, a wide analysis of the available data on atresia suggests that when the oocytes enter the cortical alveolus stage and during their development thereafter, atresia does not appear to play a significant role in controlling fecundity in normal ovarian physiology, and that its occurrence is largely a result of environmental stress (Tyler and Sumpter 1996 and citations therein).

McDonough et al. (2005) stated the histological criteria used to determine atretic stage in the flathead grey mullet (*M. cephalus*), based on ovarian atretic process described by Hunter and Macewicz (1985) and observational data of flathead grey mullet ovaries from this study. They found four types of atresias:

1) *Alpha atresia*: vitellogenic oocytes are present with distinct yolk globules, which are beginning to break down. The most developmentally advanced oocytes will undergo atresia first, followed by less developed oocytes. The oocyte will break down from the interior outward; the vitelline membrane and follicle layers are the last portion of the oocyte to decay. As the oocyte breaks down, a series of vacuoles of various sizes will appear within the oocytes.
2) *Beta atresia*: the oocytes continue to become reduced in size as they decay. The vacuoles that began to form during the alpha stage are now coalescing together to form one large vacuole within the oocyte. This gives the lamellae a distinct hollow matrix and just the outer layers of the oocyte and follicle are now left. This appears to be the shortest atretic phase.
3) *Gamma atresia*: the oocytes that were left in the hollow matrix during the beta stage now begin to shrink in size and the outer layers fold in on themselves as the oocyte collapses. The areas in and around the collapsed oocytes and lamellae become highly vascularized during this stage in order to facilitate rapid resorption of decaying cellular material. There will still be some vacuoles present within the collapsed oocytes, but they have become much smaller and there are much fewer of

them. This stage continues until most of the remaining oocytes that developed for spawning are no longer recognizable as oocytes.

4) *Delta atresia*: the remnants of old oocytes at this stage are identifiable only as decaying cellular material and will stain a distinct yellow-brown colour and are still present in approximately 30% or more of the material within the ovary. Undeveloped oocytes have a much more distinct presence, and are numerous within individual lamellae. The amount of vascularization seen in the gamma stage is reduced because most of the old material has been reabsorbed.

Ovaric Maturity Stages

The Macro and Microscopic Seven-Stage Scale

Four to seven ovarian stages have been traditionally described in Mugilidae (Kesteven 1942, Erman 1961, Ezzat 1965, Webb 1973, Ünlü et al. 2000, Ameur et al. 2003, Marin et al. 2003, Ibáñez-Aguirre and Gallardo-Cabello 2004, El-Halfawy et al. 2007, Fazli et al. 2008, Kendall and Gray 2008, González-Castro et al. 2011, Lemos et al. 2014). However, a seven-stage maturity ovarian scale is perhaps the most appropriate to use in mullets (see below, *Frequency distribution of oocyte diameters*), as it describes each one of the possible stages of gonad maturity, from the virginal to the adult, and therefore gives more accuracy in the results. The stages of ovarian maturity are macro-, but also microscopically described (according to González-Castro 2007 and González-Castro et al. 2011 for *M. liza*):

I. *virginal*: Very small ovaries, less than 1 g in weight, translucent with a thin tunic (ovarian wall). At the microscopic level only oogonia and incipient primary growth oocytes are observed.
II. *immature*: Swollen small ovaries, with weights between 1 and 3 g, with pink colour and thin tunic. At the microscopic level primary growth oocytes with compact and organized lamellar structure can be observed.
III. *incipient maturity*: ovaries reach weights of 10–20 g, and can reach the third part of the abdominal cavity. Colour from pale yellow to dark yellow. Ovarian arteries can be observed. Two stages of oocyte development occur at the microscopic level: primary growth oocytes and cortical alveoli oocytes, thus females in this stage should be considered as 'mature' for the length at first maturity estimation.
IV. *advanced maturity*: ovaries occupy half to three quarters of the abdominal cavity with weights between 30 and 280 g. Colour dark yellow to orange with prominent ovarian artery. Oocytes are discernible to the naked eye. Histology reveals primary growth (scarce) and yolked oocytes.
V. *spawning* (running ripe ovaries): ovaries occupy the entire abdominal cavity. Translucent (hyaline or hydrated) oocytes can be seen, are easily perceptible to the naked eye, and can be free (ovulated) in the lumen ovary.
VI. *spent*: flaccid ovaries, with a notable shrinking (25% of the abdominal cavity), highly vascularized (colour reddish due to capillaries rupture) and thick elastic gonad wall. Residual ovulated oocytes may be visible.
VII. *resting*: ovaries shows a thick tunic, more flaccid than in the juvenile stage, usually of reddish or greyish colour, with weights between 4 and 10 g. Primary growth oocytes are observed and, eventually, nests of oogonia.

Biological Parameters of Mullets: a Comparative Analysis

Frequency Distribution of Oocyte Diameters

The dynamic aspect of oocyte growth in mullets indicates they can be classified as group-synchronous *sensu* Wallace and Selman (1981). This implies that mature females of mullets have two batches of oocytes at the same time: a fairly synchronous group of larger oocytes (defined as a 'clutch') and a stock of smaller (previtellogenic) oocytes from which the clutch is recruited (González-Castro et al. 2011). This pattern has been widely reported for mullets (Chan and Chua 1980, Greeley et al. 1987, Wijeyaratne and Costa 1988,

Kim et al. 2004, González-Castro 2007, Kendall and Gray 2008, Albieri and Araujo 2010). Hence, mullets should be classified as total spawners or partial spawners with annual determinate fecundity, in accordance with the classification of Hunter et al. (1992). As a general rule mullets share a bimodal oocyte-distribution and a migratory behaviour, which suggest a conservative reproductive pattern from an evolutionary point of view (Greeley et al. 1987, Minos et al. 2010, González-Castro et al. 2011).

Estimation of Potential Fecundity

As usually fecundity estimations in mullets are based on ovaries at maturity stage IV (due to the difficulty in catching specimens with running ripe ovaries), it is more appropriate to use the term Potential Fecundity (PF) instead of absolute fecundity since some 'loss of oocytes' due to atresia cannot be disregarded. The PF should be estimated using ripe (or advanced maturity) ovaries, with no evidence of recent spawning (no postovulatory follicles) or atresia. The procedure basically consists of removing three pieces of ovary of approximately 0.1–0.2 g each, from the anterior, middle and posterior parts of the gonad. Later, after rehydration of each ovary sample, these samples should be weighed with an analytical balance (± 0.0001 g) and all the yolked oocytes counted. Consequently, PF can be estimated by the gravimetric method as the product between the mean number of yolked oocytes per ovary gram (Yo/g) and the Ovary weight (Ow) (PF = Yo/g x Ow). Another parameter which can be estimated is the relative fecundity (the number of yolked oocytes per unit weight of fish), which is a more comparable estimate among species than PF. Ibáñez-Aguirre and Gallardo-Cabello (2004) estimated the fecundity for *M. cephalus* and *Mugil curema* from Mexico, obtaining values ranging between 540,000–1,500,000 and 82,000–380,000 oocytes (potential fecundity) respectively (approximately a 3:1 relationship), while the mean relative fecundity was 1,680 (*M. cephalus*) and 1,064 (*M. curema*) (approximately 1.6:1).

The fecundity of all mullets generally increases with increase in size and age of the females. Increase in the growth rate of females and the accompanying increase in their fecundity are considered as an adaptive response to an improvement in the living conditions of populations. Moreover, estimations of fecundity vary greatly according to species, region, period and technical procedures employed (Brusle 1981, Alvarez-Lajonchere 1982, González-Castro et al. 2011).

Fecundity estimates for small to medium size species (i.e., *Liza abu, Liza parsia, Moolgarda cunnesius, M. curema*) usually vary between 12,000 to 400,000 oocytes/female, while for medium/large species (i.e., *M. cephalus, Liza aurata, M. liza*) from 500,000 to 3,000,000 (Wijeyaratne and Costa 1988, Su and Kawasaki 1995, Marin and Dodson 2000, Rheman et al. 2002, McDonough et al. 2003, Ibáñez-Aguirre and Gallardo-Cabello 2004, Patimar 2008, Eljaiek and Vesga 2011, Albieri and Araújo 2010, González-Castro et al. 2011, Şahinöz et al. 2011, Lemos et al. 2014).

Length at First Maturity (L_{50})

The L_{50} of a species is the length at which 50% of the specimens display maturing gonads (i.e., ovaries with oocytes in the cortical alveolus stage) (Vazzoler 1996). Normally, to estimate this biological parameter, the fish sample should be divided into 1 cm (0.5 for small species) length classes. Individuals must be classified as immature or mature, being considered as 'mature specimens' when ovaries are at the incipient maturity stage (stage III, with cortical alveoli stage oocytes), or of course, more advanced ovarian stages. The proportion of individuals assigned as being mature for each 1 cm length class must be calculated, and logistic curves should be fitted to these data, employing one of the methods available in the literature, i.e., maximum likelihood; non-linear least squares, non-linear regression (Saila et al. 1988, Beverton 1992, Roa et al. 1999). Finally, the L_{50} can be determined from the equation of the fitted logistic curve (González-Castro 2007, Fazli et al. 2008, Kendall and Gray 2008, González-Castro et al. 2011, Lemos et al. 2014).

As Brusle (1981) indicates, there is a rough correspondence between reported length (size) at maturity and the local mean sea temperature. It is normal that the youngest ages and smallest lengths at first maturity have been recorded from the warmer waters and the greatest length and age at first maturity from the coldest waters. Accordingly, it has been generally observed that sexual maturity occurs at a greater size and age in higher rather than lower latitudes. So, it seems *M. liza* offers a good example of these concepts, as apparently the L_{50} estimates for the species vary accordingly to latitude; for females, L_{50} was: 454 mm (at 37° S, González-Castro et al. 2011), 421.9 mm (for samples collected approximately from 32° to 26° S, Lemos et al. 2014); 412 mm (25° 31´S, Esper et al. 2000) and 350 mm (23° S, Albieri and Araújo 2010).

Sex differences of L_{50} estimates have been also reported in most of the studied species. Generally males mature earlier (in age and size) than females (Okumus and Bascinar 1997, Esper et al. 2000, McDonough et al. 2003, Kendall and Gray 2008, Eljaiek and Vesga 2011, González-Castro et al. 2011, Lemos et al. 2014). However, Oliveira et al. (2011) reported for *M. curema* in Northeastern Brazil higher L_{50} estimates for males than females (264 and 240 mm, respectively).

Gonadosomatic Index

During the progress of the stages of oocyte development the ovaries expand their volume and thereafter weight. The gonadosomatic index (GSI) is an efficient estimator of the physiological state of the gonads. Its monthly variation can be used with confidence to define the reproductive period of the species under study. GSI is related to maturity stages. Mean GSI values estimated for each maturity stage of females of *M. liza* showed significant differences as revealed by ANOVA (González-Castro et al. 2011). However, histological correlation is desirable, in order to compare the GSI values with oocyte growth stages. GSI can be defined as: $GSI = O_w \cdot 100/B_w$, where O_w is the ovary weight in grams and B_w is the body weight. An alternative formula to calculate GSI is to consider Bw as: total weight-Ow. This variation of the GSI formula magnifies the GSI values of mature specimens with respect to the immature ones.

Sample locations and timing are critical for analyzing monthly GSI values, based on an annual cycle. Most mullet species perform spawning migrations from the feeding grounds (coastal lagoons, lakes and estuaries) to the marine spawning grounds, so the highest GSI values should be found at sea. In this respect, the highest GSI (mean values) recorded for females of *M. liza* caught by the artisanal fleet operating inside the Los Patos lagoon, Brazil was 5.5%, and occurred in April; in contrast, peak GSI values for females caught by the industrial purse seine fleet operating at sea were much higher (11.5%), and occurred in June (Lemos et al. 2014).

Monthly average GSI values are usually quite different between sexes for mullets and usually, male values are lower than female ones (Okumus and Bascinar 1997, Hotos et al. 2000, El-Halfawy et al. 2007, González-Castro 2007, Albieri and Araújo 2010, Şahinöz et al. 2011, Lemos et al. 2014). The male gonads do not vary so much in weight in the spawning season, basically due to the fact that oocytes undergo a noticeable increase in volume and weight during development whereas spermatozoa do not.

The reproductive studies performed on mullets commonly show that GSI has a single annual peak and this can be correlated with a single spawning period per year (Table 11.1). However, two GSI peaks (correlated with advanced maturity ovaries) were described for *Liza parsia* (Rheman et al. 2002) of Bangladesh, *M. curema* of Brazil (Oliveira et al. 2011) and for *M. liza* of Argentina (González-Castro 2007, González-Castro et al. 2011), but not for *M. liza* of Brazil (Lemos et al. 2014). Similarly, the reproductive biology of *Liza argentea* occurring in two estuaries of South-eastern Australia was found to differ: GSI values and macroscopic staging of gonads identified the peak spawning period as occurring between March and November (Lake Macquarie) and January and April (St. Georges Basin). In contrast, peak spawning of *Myxus elongatus* was concentrated between January and March in both estuaries (Kendall and Gray 2008).

Gonadal Cycle

Monthly variation of gonad maturity stages on an annual basis provides information on the reproductive cycle of the adult stock of a mullet species. Therefore, accuracy in the classification of the gonad maturity stages is crucial and a histological approach is desirable.

McDonough et al. (2003) found immature and inactive males and females of *M. cephalus* (South Carolina estuaries) every month of the year. They point out that this fact could indicate that mature mullets do not spawn every year or that fish that remain in the estuary do not migrate offshore to spawn. They also suggest that the inactive (but mature) females found in the early part of the spawning season may not spawn until much later. Accordingly, González-Castro et al. (2011) highlight the presence in Mar Chiquita coastal

Table 11.1. Highest mean-gonadosomatic index (GSI), obtained for the species of Mugilidae in which only one GSI-Peak was reported.

Species	Region	Highest mean-GSI (%)		Month	Reference
		Females	Males		
Liza aurata	Messolonghi, W. Greece	5.32	2.78	September	Hotos et al. (2000)
-	Caspian Sea, Iran	6.4	2.7	October	Ghaninejad et al. (2010)
Liza ramada	Göksü Delta	16	-	November	Ergene (2000)
-	Lake Timsah, Egypt	12.4	-	November	El-Halfawy et al. (2007)
-	Adriatic Sea coast, Croatia	8	-	October	Bartulovic et al. (2011)
Liza saliens	Southeast Caspian Sea	5.9	2	June	Patimar (2008)
Liza klunzingeri	Arabian Gulf, Kuwait	7.5	5	November	Abou-Seedo and Dadzie (2004)
Valamugil cunnesius	Negombo lagoon, Sri Lanka	11.88	1.09	March	Wijeyaratne and Costa (1988)
Mugil cephalus	Taiwan	18	12	January	Su and Kawasaki (1995)
-	Jeju Island, Korea	6.97	-	December	Kim et al. (2004)
-	South Carolina estuaries, EEUU	17	11	November	McDonough et al. (2005)
Rhinomugil corsula	Meghna River, Bangladesh	5.5	-	July	Akter et al. (2012)

lagoon (Argentina) of not only juveniles, but also adult specimens of *M. liza* in a resting stage throughout the year, in co-occurrence with females in an advanced state of maturity (April–May, November–December). Chang et al. (2004) investigated the Sr:Ca ratios in otoliths of *M. cephalus* of Taiwan; they showed that the migratory environmental history of the mullet beyond the juvenile stage consists of two types. In Type 1 mullet, the high Sr:Ca ratios indicated that they migrated between estuary and offshore waters, but rarely entered the freshwater habitat. In Type 2 mullet, the low Sr:Ca ratios indicated that the mullet migrated to a freshwater habitat. Most mullet collected nearshore and offshore were of Type 1, while those collected from the estuaries were a mixture of Types 1 and 2. The mullet spawning stock consisted mainly of Type 1 fish. These facts could explain the co-occurrence of resting/advanced maturity stages.

Reproductive Physiology: the Molecular Basis of Fish Reproduction

The Hypothalamic–Hypophyseal System

Reproduction in fish is under neuroendocrine control, but the functions of hypothalamus, pituitary and gonads are particularly influenced by environmental factors such as photoperiod and water temperature. The hypothalamo-hypophysial system in fish is divided into three main areas: the hypothalamus, which is part of the diencephalon; the neurohypophysis, which derives from the ventral diencephalon and represents the neural compartment of the pituitary; and the adenohypophysis, which is the non-neuronal part of the gland.

There is no median eminence and no portal system in teleost fish, in which hypothalamic neurons terminate very close to adenohypophysial cells, reducing the diffusional distance, or making synaptoid contact with adenohypophysial cells. This means that hypothalamic control over the adenohypophysis can be exerted through direct action upon the secretory cells (Cerdá-Reverter and Canosa 2009).

Hypothalamic-Reproductive Hormones: Gonadotropin Releasing Hormone (GnRH)

The hypothalamic releasing hormones function to control the secretions of the pituitary hormones, mainly the anterior pituitary hormones, either positively or negatively.

In vertebrates, gonadotropin releasing hormone (GnRH) is a decapeptide. As many as 24 molecular isoforms of GnRH have been characterized so far and eight variants have been found in the teleost brain

(Kah et al. 2007). It is apparent that the appearance of GnRH and GnRH receptors plays an integral role in controlling the onset of gonadal growth and the seasonality of reproduction (Schulz and Goos 1999).

Hypophyseal-Reproductive Hormones: LH, FSH

Fish adenohypophysis synthesizes at least eight different hormones, of which two are directly related to reproduction in fishes: the follicle stimulating hormone (FSH/GTHI) and luteinizing hormone (LH/GTHII).

Reproduction in fish is controlled by the actions of Follicle Stimulating Hormone (FSH) and Luteinizing Hormone (LH). These belong to the glycoprotein hormone family and are heterodimeric glycoproteins composed of a common alpha subunit and a hormone specific β subunit. LH has been characterized in a number of teleost species whereas much less is known about FSH, as it has only been characterized relatively recently. Although the precise roles of LH and FSH are not known, they are produced in separate cells in the pituitary, exhibit distinct patterns of expression at different stages of the reproductive cycle and have been shown to act on the gonads to produce sex steroids and other gonadal factors which play important roles in the development and maturation of gametes (Devlin and Nagahama 2002). Based in large part on knowledge of their actions in salmonids, FSH has been implicated in the control of gametogenesis because it stimulates the production of 17b-estradiol and the incorporation of vitellogenin into the developing oocyte. In males, FSH stimulates Sertoli cell proliferation and maintenance of quantitatively normal spermatogenesis. LH is very low or undetectable during early stages of reproductive development, but high before maturation, when it is known to stimulate gonadal steroidogenesis and is involved in oocyte maturation, ovulation and spermiation.

Overview of Regulation of Reproductive Hormones: Inhibitory and Stimulatory Reproductive Pathways

The basic schedule of the neuro-endocrine regulation of reproductive function includes not only the environmental variables (such as temperature, photoperiod or water salinity) but also the hypothalamus-pituitary-gonadal axis and the liver. The sense organs are responsible for the reception of the environmental stimulus, and promote the synthesis of GnRH by the hypothalamus in the brain. GnRH neurons have direct contact with the pituitary gland, and the release of GnRH induces the synthesis and liberation of gonadotropins, hormones of which the main target organs are the gonads. Some oestrogens are produced by the gonads in response to gonadotropin stimulation. Particularly, oestradiol is produced by the ovary and after being targeted to the liver this steroid will stimulate the synthesis and liberation of vitellogenins. This glyco-phospho-lipo-protein will be captured by the ovary where it is responsible for vitellogenesis, a phase in which oocyte growth occurs by the uptake of this complex protein. Also, the gonadal steroids have a feedback effect over the hypothalamus and the adenohypophysis, thus controlling the synthesis and release of LH and FSH (Goos 1987, Van Der Kraak et al. 1998, Yaron et al. 2003, Levavi-Sivan et al. 2006). Both positive and negative feedback effects of testosterone and 17b-oestradiol have been demonstrated.

These actions are complex in that sex steroid feedback regulation of LH and FSH release involves actions at the level of the hypothalamus and pituitary. In addition to effects on LH and FSH synthesis within the pituitary, testosterone and 17b-oestradiol also affect the GnRH system and other neuroendocrine factors which control LH synthesis and release. Other steroids including the maturation-inducing steroid 17, 20b-dihydroxy-4-pregnen-3-one (17, 20bP) and the corticosteroid cortisol have been implicated in the regulation of LH and FSH release (Levavi-Sivan et al. 2006), but these effects are considered to be secondary to the dominant actions of testosterone and 17b-oestradiol.

Spawning Migration of Mullets: a Proposal for the Life-History Model for the Southern Population of *Mugil liza* as a Case Study

Recently, Fraga et al. (2007) and Heras et al. (2009) showed that *M. platanus* should be considered a junior synonym of *Mugil liza*. However, Mai et al. (2014) using microsatellite markers provided the first molecular evidence of the existence of distinct population clusters of *M. liza* along the South American Atlantic coast: one represented by Niteroi samples (Rio de Janeiro, Brazil) and the other including the southern localities of Brazil (Ubatuba, Laguna e Rio Grande) and Argentinean samples (Lavalle, Buenos Aires province). The identification of a possible barrier to gene flow in this region (Rio de Janeiro) provides a base to better understand *M. liza* life-history traits and to interpret the genetic variation inherent in the species complex in adaptive terms. Thus, as suggested by González-Castro et al. (2012), there may be at least two populations acting as different significant evolutionary units: one from RJ (Brazil) to Cuba, and a southern population from Sao Paulo State (Brazil) to Buenos Aires Province (Argentina).

Several works related to the reproductive biology/spawning migration of the southern population of *M. liza* has been published in the last 30 years (Vieira 1991, Vieira and Scalabrin 1991, Esper et al. 2000, Romagosa et al. 2000, Miranda et al. 2006, González-Castro 2007, Vieira et al. 2008, González-Castro et al. 2009a, González-Castro et al. 2009b, González-Castro et al. 2011, Garbin et al. 2013, Lemos et al. 2014). Most are complementary studies, which allow a proposal for the life-history model for this southern population. Based on Vieira and Scalabrin (1991), Vieira (1991) presents a hypothetical model for the life history of *M. liza* from Lagoa dos Patos (Brazil). The author relied mainly on the temporal–spatial distribution of larval–juvenile recruitment (measured by CPUE) and adult captures (during the reproductive migration), relating these data to different environmental variables (temperature, salinity, direction of coastal marine currents). González-Castro et al. (2011) (based on González-Castro 2007) proposed a hypothetical model for adult stocks of *M. liza* (Mar Chiquita coastal lagoon, Argentina), based not only on CPUE of adults and environmental data, but also on ovarian maturity stages, gonadosomatic index, the allometric growth coefficient b and border analyses of otoliths. Both models (Vieira 1991, González-Castro et al. 2011) basically agree and are complementary. Recently, the works of Lemos et al. (2014) and Garbin et al. (2013) give some additional information that supports the hypothetical life-history model of *M. liza*.

This model relates the specific growth with its life cycle as follows:

i) Grey mullets use the period between January and mid-April–May for growth and gonadal maturation. High gonadosomatic index (GSI) and allometric coefficient (b) values, opaque band otoliths (high protein deposition) and the presence of females with advanced ovarian maturation are strong evidence of gonadal maturation. Garbin et al. (2013) showed that mullets have highest condition factor in April and May. According to González-Castro et al. (2009b), shoals of maturing mullets were found in the nearby of the mouth of Mar Chiquita coastal lagoon (Argentina) in autumn (April–May).

ii) From May to the end of June, a reproductive migration (from the estuaries or coastal lagoons towards the sea) occurs. A decrease in feeding is therefore expected, and, accordingly, a decreasing growth rate. Vieira and Scalabrin (1991) indicated that the sudden drop of temperature between March and June triggers the reproductive migration of the flathead grey mullet, at least from Los Patos lagoon (Brazil). Accordingly, González-Castro (2007) and González-Castro et al. (2011) showed (for Mar Chiquita coastal lagoon, Argentina) a decrease of temperature below 20°C from March to August, with a sudden drop observed in May–June. Therefore this environmental variable could be the factor that starts the reproductive migration of southern populations of *M. liza* from coastal lagoons towards the sea.

iii) Accordingly, the gradual northward migration of these fish from Argentina (around 36° S) to Santa Catarina (around 26° S) appears to be correlated with preparation for spawning, as the GSI values for fish caught off Argentina (González-Castro et al. 2011) and those sampled from the artisanal catches of southern Brazil (Lemos et al. 2014) are lower in April (< 5.5) than those of fish sampled from industrial purse seine fishery catches in June (> 11). The peak-spawning in the southern population of *M. liza* appears to occur in June (with a range between May to August) in offshore waters in Santa Catarina and Paraná states, as indicated by the presence of hyaline oocytes (Lemos et al. 2014), and is closely related to the northward displacement of temperatures between 19 and 21°C along the coast. The broad geographical range of spawning locations along the coast of Santa Catarina suggests that *M. liza* spawning is more closely related to oceanographic conditions (salinity of 32–35, depth of *c.* 50 m and SSTs around 20°C) than to specific geographical locations. However, a small-secondary spawning event (probably in more southern latitudes than the main spawning) cannot be disregarded, as existence of females in advanced sexual maturity during November–December were observed uninterruptedly over several years (González-Castro 2007, González-Castro et al. 2011). There was also a strong recruitment of early juveniles in summer (Acha 1990, González-Castro, unpublished data).

iv) From August–September the reproductive events end and, probably, the mullet start feeding and migrate to estuaries and coastal lagoons. Eggs and larvae drift towards the surf zone by wind generated surface currents (Vieira and Scalabrin 1991). After reach approx. 20 mm TL, early juveniles gradually migrate to the bottom and begin to feed on benthic organisms and mineral particles (Vieira 1991). In the surf zone, they move with longshore currents that run southward most of the year resulting in passive transport towards the mouth of the lagoons or estuaries. Once close to the estuaries the bottom net upstream estuarine circulation moves recruits into estuaries, after which they actively spread into shallow waters (Vieira and Scalabrin 1991). In shallow waters and small coastal lagoons (i.e., Mar Chiquita lagoon) the winds and the tide would also be responsible for larval movement.

Lemos et al. (2014) observed the lowest condition factor values at the end of the spawning season (August). This state probably results from the energy expenditure required for migratory and reproductive processes. Values of the allometric coefficient b start to increase from September, and in November reach the values registered in (i) (González-Castro et al. 2009b). As well, the otolith opaque band rate starts to increase, exceeding (in December) the percentage of otoliths with hyaline bands. However, Garbin et al. (2013) showed a single peak for hyaline rings (slow growth) occurring from September to November (spring) and a single peak for opaque rings (fast growth) from June to August (winter).

Early Development (and Larval Development)

The present section aims to describe the eggs, embryonic and larval developmental stages together with the juvenile development of some common Mugilidae species. Specifically, it focuses on ontogeny, the changes that occur as the eggs develop, hatch into larvae, and finally transform into juveniles. A lot of published work is on *M. cephalus* and *Chelon labrosus*, two of the most commercial valuable species among the grey mullets in coastal extensive aquaculture systems.

The developmental stages of several grey mullet species have been described by many authors: *L. aurata* (Vodyanitskii and Kazanova 1954, Demir 1971); *Liza saliens* (Vodyanitskii and Kazanova 1954); *M. cephalus* (Sanzo 1930, 1936, Anderson 1958, Tang 1964, Yashouv and Berner-Samsonov 1970, Kuo et al. 1973a, Tung 1973, Liao 1975, Nash et al. 1974); *L. ramada* (Yashouv and Berner-Samsonov 1970); *Liza dumerili* (Benetti and Fagundes Netto 1980); *Liza macrolepis* (Sebastian and Nair 1975); *M. curema* (Anderson 1957, 1958); *Oedalechilus labeo* (Sanzo 1937).

The success of induced reproduction techniques has enabled researchers to describe the developmental stages of grey mullets. For *C. labrosus,* preliminary descriptions for embryonic and larval development were given by Sanzo (1936), Demir (1971) and Cassifour and Chambolle (1975) from eggs and larvae sampled at sea. The successful induced spawning of *C. labrosus* enabled Cataudella et al. (1988b) and Boglione et al. (1992) to describe embryonic and larval development until metamorphosis. They reported that although the prelarva is poorly developed at hatching, the juvenile stage is rapidly attained as nostrils and homocercal tail are observed as early as day 22 post hatching.

Various terminologies have been proposed to define the stages of development in fishes. For the ichthyoplanktonic developmental stages, Brownell (1979) proposed the following: egg, yolk sac, pre-flexion, flexion and post-flexion. More recently, Boglione et al. (1992) proposed the development to be divided only into four periods: embryonic, pre-larval, larval and juvenile, each one characterized by typical anatomical and physiological features. The stages proposed by Boglione et al. (1992) are adopted in this chapter.

The embryonic phase of development (or egg), is used to describe the development from spawning until hatching. At hatching, the fish becomes a larva. Most fish have a yolk sac for nourishment at hatching, and until it is absorbed they are termed yolk-sac larvae or pre-larvae. The larval period lasts until the number of fin rays reaches their adult complement when the juvenile stage begins. Squamation (the formation of scales) usually starts at this point in development also. The juvenile stage terminates when the fish first reaches sexual maturity (Miller and Kendal 2009).

Embryonic Stages—Development

Fertilization occurs when the sperm penetrates the chorion of the egg and begins the various stages leading to sperm-egg fusion. The ripe egg possesses a micropyle on the surface of the egg envelope at the animal pole, which allows passage of the sperm into the egg. The micropyle is funnel-shaped and only wide enough at its base for the passage of a single sperm. After sperm penetration, a plug forms at the base of the micropyle to prevent polyspermy (Miller and Kendal 2009).

The embryonic stage occurs inside the chorion and ends at hatching (Boglione et al. 1992). During incubation the embryo develops from a single cell into a complex organism. The embryonic period can be divided into three stages: early (fertilization to blastopore closure); middle (from blastopore closure to the time the tail begins to curve laterally away from the embryonic axis) and late (from the time the tail is curved away from the embryonic axis to hatching) (Rè and Meneses 2008, Miller and Kendal 2009).

The embryonic development of grey mullet eggs has been described in detail through the years. Here only the basic features of development as can be discerned by examining whole eggs with a dissecting microscope will be summarized. This level of detail is usually sufficient for early life-history studies dealing with field-caught pelagic eggs and on induced spawning procedures (Miller and Kendal 2009).

Morphology of Eggs (Unfertilized)

The morphology of eggs before fertilization was described for several Mugilidae by the early naturalists, most of whom made reference to the characteristic large oil globule. The availability of eggs at all stages of development during induced breeding procedures has resulted in full descriptions for *M. cephalus* by Sanzo (1930), Tang (1964), Yashouv (1969), Liao et al. (1972), Kuo et al. (1973a) and Tung (1973); by Sanzo for *C. labrosus* (1936) and *O. labeo* (1937); by Anderson (1957) for *M. curema*; by Perceva-Ostroumova (1951) and Dekhnik (1954) for *L. saliens* (Nash and Koningsberger 1981).

The unfertilized eggs of Mugilidae are pelagic, spherical, transparent and the chorion is smooth. The surface of the egg shell is smooth and unsculptured. The yolk appears unsegmented, the perivitelline space is narrow, and there is one (or more) large oil globule that makes the eggs extremely buoyant and merges as eggs develop. The eggs are not adhesive (Nash and Koningsberger 1981, De Sylva 1984).

Specifically, *M. cephalus* eggs in Taiwan are spherical with a diameter of 812 µm. An irregular sculpted pattern of ridges is found on the surface of the egg zona radiata (Fig. 11.1) but these patterns are not seen near the micropylar region. No pores or pore-traces are found in the egg envelopes. The surface of the micropylar region is rough and uneven. No microvilli or knobs are found. The micropylar canal is short

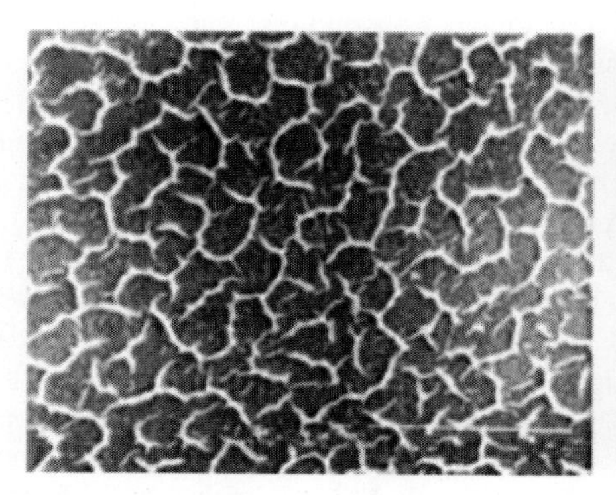

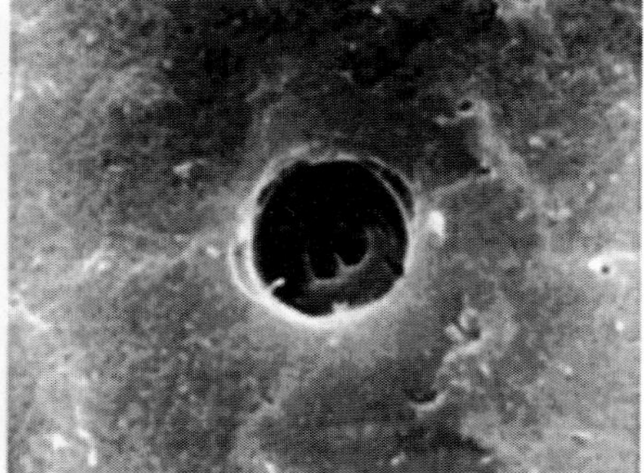

Figure 11.1. Envelope surface of *Mugil cephalus* egg (left) and micropyle of *Mugil cephalus* egg (right). (ia), internal aperture (after Li et al. 2000).

and funnel-shaped with three clockwise spiral ridges. The outer opening of micropylar canal is 3.55 μm in diameter and the diameter of the canal is 1.2 μm (Fig. 11.1) (Li et al. 2000).

The ripe unfertilized eggs of *M. cephalus* appear rounded, colourless and transparent, with a diameter of about 620 ± 30 μm at water temperature of 25 ± 1°C in Egypt. The egg membrane is smooth, not separated from the yolk and the cytoplasm is reduced to a thin layer covering the yolk; one oil globule was noticed in ripe ova (Fig. 11.2). After about 20 minutes from fertilization the egg membrane swells up and separates from the perivitelline space; the diameter of the eggs increases to 700 ± 30 μm (El-Gharabawy and Assem 2006).

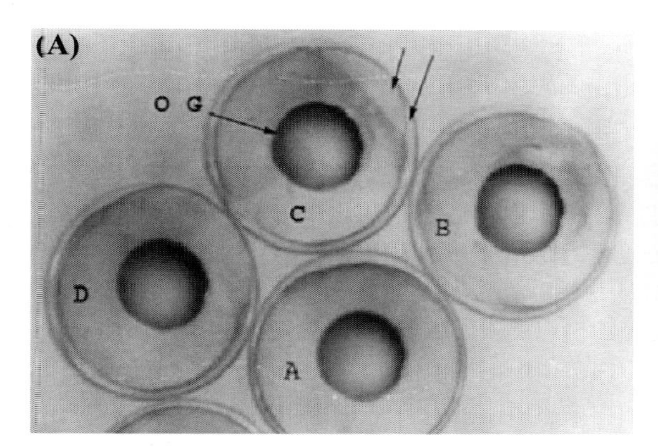

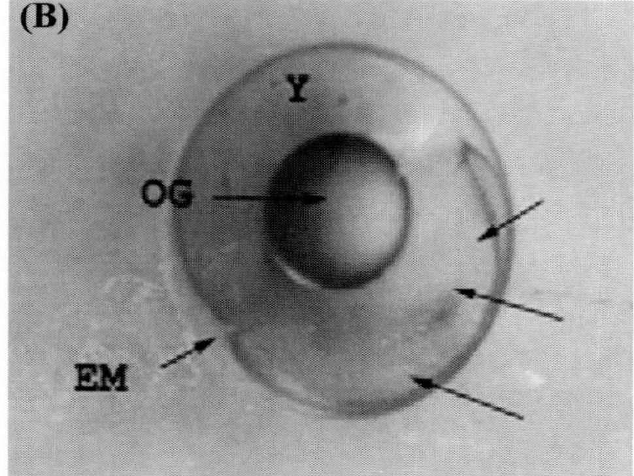

Figure 11.2. Eggs of *Mugil cephalus* (X31). **(A)** Unfertilized egg A), Fertilized egg B) Germinal disc stage C) Morula stage after 3.5 hours after fertilization (a.f.), D) Blastoderm consisted of multiple layered cellular plate (arrows) and blastula after about 4.25 hours from fertilization. Note large oil globule (OG); **(B)** Gastrula stage, after about 5 hours from fertilization, EM) egg membrane, OG) oil globule and the blastoderm layer (arrows) which spread over the yolk material (Y) (after El-Gharabawy and Assem 2006).

The egg diameter of *C. labrosus* at spawning have a diameter of 1.3–1.4 mm which is the highest among the Mugilidae, contain two–13 oil globules and number on average 648 · g^{-1} (Boglione et al. 1992, Cataudella et al. 1988b). These values correspond well with data from Sanzo (1936, 1937) and Yashouv and Berner-Samsonov (1970) on eggs collected in the wild.

Fertilized Egg Appearance

Shortly after fertilization, the cytoplasm becomes thickened at the animal pole of the egg where the nucleus occurs. Early development of fish eggs generally exhibits a meroblastic pattern of cleavage, in that the cells form only at the animal pole of the fertilized egg, and cleavage does not go through the entire yolk, as it does in holoblastic cleavage. The yolk concentrates at the vegetal pole (Miller and Kendal 2009).

The availability of fertilized eggs at all stages of development from induced breeding has resulted in full descriptions for *M. cephalus* by Sanzo (1930), Tang (1964), Yashouv (1969), Liao et al. (1972), Kuo et al. (1973a) and Tung (1973); Sanzo (1936) for *C. labrosus* and *O. labeo* (1937); Anderson (1957) for M. *curema;* Perceva-Ostroumova (1951) and Dekhnik (1954) for *L. saliens.*

Fertilized eggs of mullets generally are spherical and transparent. The surface of the egg shell is smooth and unsculptured and they are not adhesive. The yolk appears unsegmented and there is predominantly one large oil globule making them extremely buoyant (Nash and Shehadeh 1980). Liao (1975) mentioned that the fertilized grey mullet egg is buoyant and stays afloat near the surface of water with slight aeration, but settles down slowly in still water. Tang (1964) observed that the eggs were suspended in circulating water, but sank to the bottom in standing water, while Yashouv (1969) reported the sinking of mullet eggs towards the end of incubation. According to Kuo et al. (1973a,b), the majority of eggs which sink are unfertilized but the buoyancy of the eggs depends on several factors.

The fertilized eggs of *M. cephalus* are pelagic, spherical, transparent, non-adhesive, straw-coloured. The chorion is smooth with irregular delicate sculpturing or without marking, sculpturing or attachment mechanisms except hydrofuge membrane; with homogeneous unsegmented yolk and the perivitelline space is narrow (Martin and Drewry 1978, Fahay 2007).

Egg Diameter. The mean egg diameter of fertilized eggs of *M. cephalus* (Fig. 11.2) is 0.93 mm with a range of 0.88–0.99 mm (Kuo et al. 1973a,b, Nash et al. 1974, Tucker 1998, Fahay 2007). It seems that the egg diameter differs from different areas (see Table 11.2).

Table 11.2. Egg diameter of fertilized eggs of *Mugil cephalus* (Martin and Drewry 1978).

Egg diameter (mm)	Region/Country
0.60–0.72	Black Sea
0.66–0.72	Israel
0.72–0.81	Italy, Sicily, Corsica
0.8	India
0.65–1.08	Japan
0.9–0.96	Taiwan
0.88–0.98	Hawaii
0.81	Taiwan (Li et al. 2000)
0.89	Taiwan (Tung 1973)

According to Rè and Meneses (2008), within a single species there is little variation in egg characters (size, number and size of oil globules, chorion surface, yolk, pigmentation, and morphology of the developing embryo). Development time is highly related to temperature and is species-specific. The yolk is unsegmented, the shell is clear and smooth (Kuo et al. 1973a,b, Ditty et al. 2006). In Egypt, after five hours from fertilization, the diameter of egg of *M. cephalus* increased from 620 ± 30 μm to 820 ± 19 μm (El-Gharabawy and Assem 2006). The eggs of *M. curema* are pelagic and spherical with a diameter of 0.86–0.92 mm (Ditty et al. 2006, Fahay 2007).

Oil Globule. The oil globule diameter of the grey mullet species is about 35% of egg diameter (Ditty et al. 2006). Specifically, in all naturally spawned (eggs released spontaneously by the females) and fertilized eggs of *M. cephalus* a single large oil globule is present, 0.30–0.36 mm in diameter (mean: 0.33 mm) (Kuo et al. 1973a,b, Nash et al. 1974, Tucker 1998, Ditty et al. 2006) or 0.39 mm (Tung 1973). The oil globule is without colour, yellowish or light yellow (Martin and Drewry 1978). The oil globule diameter is 0.26–0.31 mm in the Black Sea, 0.28 mm uniform in Italy and Sicily, 0.37 mm in India, 0.36–0.40 in Taiwan, 0.33 mm uniform in Hawaii, 36.1–40% of egg diameter (Martin and Drewry 1978).

Yashouv and Berner-Samsonov (1970) reported that eggs of *M. cephalus* and *L. ramada* have one or more oil globules and that both types of eggs developed normally and hatch. During development they observed that the multiple oil globules merge into one since at hatching the larvae have one (or rarely two) oil globule located in the yolk sac.

Sometimes several globules are present in artificially removed eggs, these can develop normally and globules coalesce during development. The immature oocyte usually contains a great number of oil globules which gradually coalesce and decrease in number during the process of maturation. When eggs were removed from gravid females by manual pressure during artificial stripping multiple oil globules have been observed and the earlier the eggs were sampled by manual extrusion, the greater the frequency of multiple oil globules (Kuo et al. 1973a,b). The presence of multiple oil globules in manually extruded mature eggs probably indicates premature removal of eggs from the female. Eggs with more than four oil globules were fertilized, but failed to divide (Kuo et al. 1973a,b) and the survival of eggs which initially contained multiple oil globules was always low (Nash and Shehadeh 1980, Nash and Koningsberger 1981). Eggs with less than four oil globules developed normally and the globules tended to coalesce during the latter stages of embryonic development (Kuo et al. 1973a). The presence of multiple oil globules was also noted in *L. saliens* (Perceva-Ostroumova 1951) and *C. labrosus* (Sanzo 1936, Kuo et al. 1973a,b, Nash and Koningsberger 1981, Nash and Shehadeh 1980).

Sanzo (1936) described eggs of C. *labrosus* with one large and several smaller globules (Fig. 11.3), and Perceva-Ostroumova (1951) noted the same for *L. saliens*. The eggs of *M. curema* have a single oil globule, 0.03 mm in diameter (Fahay 2007).

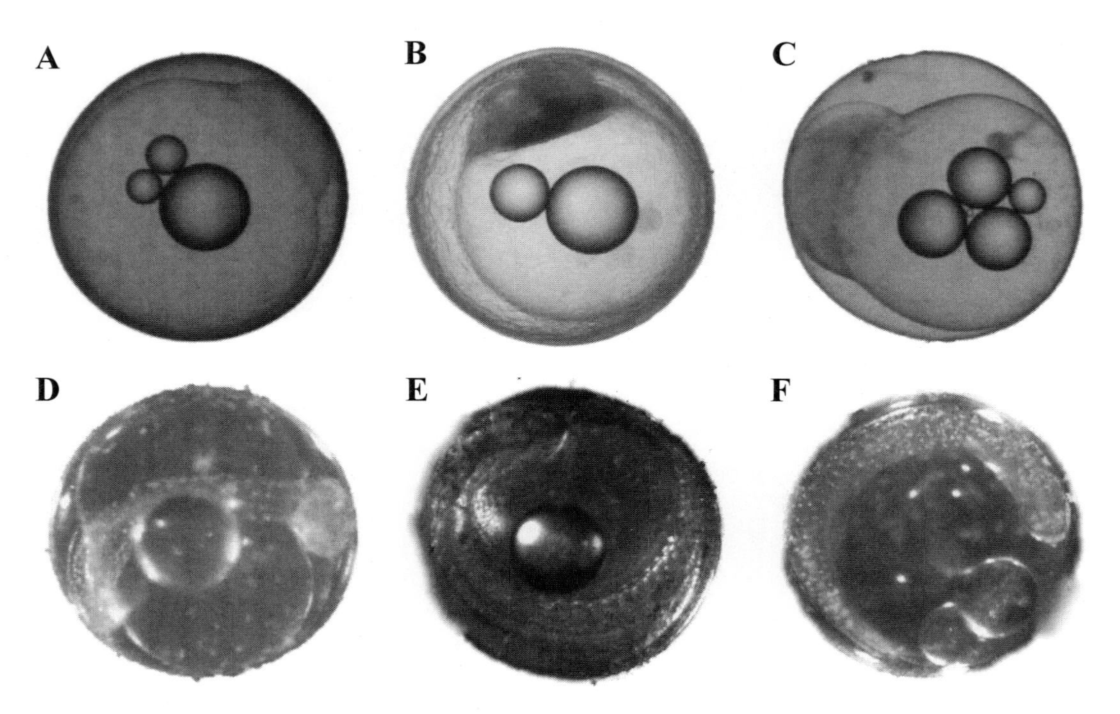

Figure 11.3. (A–C) Fertilized egg of *Chelon labrosus* in *Blastula* and *Gastrula* stage (X 63); (D–F) Developing embryo (X 50–63) (photos by G. Minos).

Germinal Disc Stage

Cell division at first proceeds in an orderly fashion with a single layer of two, four, eight, and 16 uniformly sized cells (*blastomeres*) forming the germinal disc (*blastodisc*) on the yolk. An acellular layer, the periblast, forms around the *blastodisc*, which metabolizes the yolk for the developing embryo. At first the periblast is continuous with the marginal blastomeres. The individual cells of the blastodisc decrease in size with each cell division (Miller and Kendal 2009).

After about 30 minutes from fertilization in *M. cephalus*, the protoplasm is gradually differentiated from yolk to form the blastodisc at the animal pole. The blastodisc is in the form of circular-shaped cap with diameter of 500 μm and height of 112 μm, the perivitelline space is visible only above the cytoplasm (Fig. 11.2) (El-Gharabawy and Assem 2006).

Cleavage (two, four, eight, 16, 32, 64, 128 cell stages) and Morula Stage

In *M. cephalus* the perivitelline space occurs 15 minutes after fertilization (Kuo et al. 1973a,b) when cleavage takes place for the blastodisc (two, four, eight and 64) and then irregular cleavage. Thereafter, the number of blastomeres becomes a multicellular cape (*morula* stage after about 3.5 hours from fertilization at water temperature of 25 ± 1°C). The mean total diameter of the egg is 715 ± 20 μm, the height of the blastodermal cap is about 176 μm and the oil globule is 300 μm (Fig. 11.2) (El-Gharabawy and Assem 2006).

The first meroblastic cleavage (2-cell stage) in *M. cephalus* takes place meridionally near the centre of the germ disc (Fig. 11.4) 50 minutes after fertilization (24ºC) to one hour 30 minutes (20–24.5ºC) variously equal to unequal (Kuo et al. 1973a,b, Martin and Drewry 1978). The perivitelline space is minute, appearing 15 to 40 minutes after fertilization (Martin and Drewry 1978).

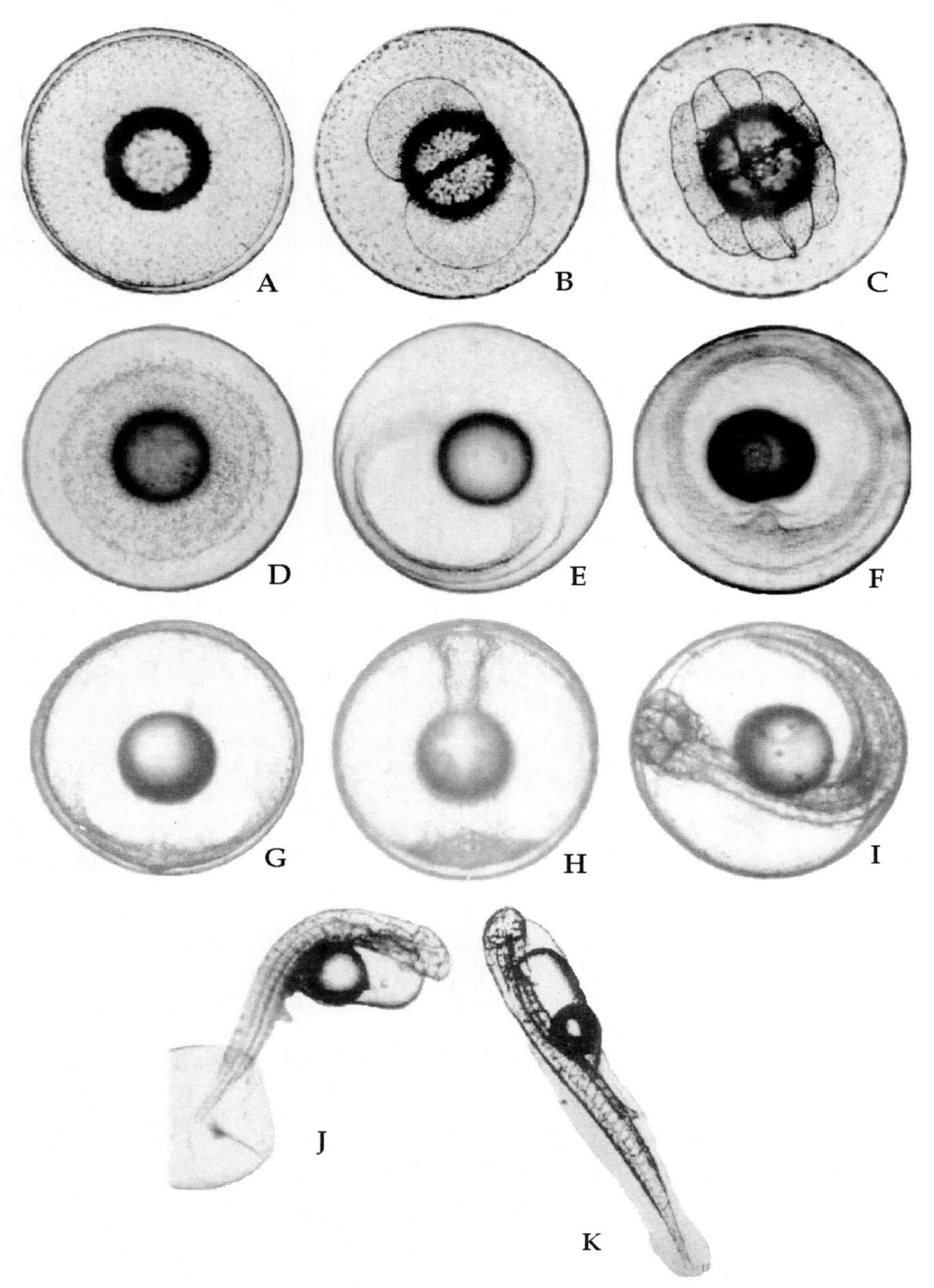

Figure 11.4. Embryonic development of *Mugil cephalus.* (A) Fertilized egg (X 60); (B) 2-cell stage (X 60); (C) 16-cell stage (X 56); (D) Blastula (X 56–57); (E) Gastrula (X 59–60); (F) Gastrula (X 59–60); (G) Late neurula, diameter 0.93 mm (X 57–58); (H) Developing embryo (X 54–60); (I) Developing embryo, note pigment on oil globule, diameter c.a. 0.93 mm (X 54–60); (J, K) Hatching (x 32) and newly hatched larval (X 36) (after Kuo et al. 1973a).

According to Kuo et al. (1973a) who described in detail the embryonic development of *M. cephalus* at 24°C, the second cleavage (four-cell stage) at right angles to the first occurs 65–70 minutes after fertilization. The third cleavage (eight-cell stage) occurs at right angles to the first division and parallel to the second, 85–90 minutes after fertilization. The third cleavage (16-cell stage) (Fig. 11.4) occurs parallel to the first division 95 minutes after fertilization when 16 blastomeres are in a single plane. The fourth cleavage (32-cell stage) occurs 110 minutes (one hour 50 minutes) after fertilization and the arrangement of blastomeres is irregular. Beyond the 32-cell stage, it is difficult to count the cells, and cell division is not as synchronous as earlier. In *M. cephalus*, the fifth cleavage (64-cell stage) occurs two hours 25 minutes after fertilization and the blastomeres are reduced in size as division continues (Kuo et al. 1973a). The sixth cleavage (128-cell stage) occurs two hours 40 minutes after fertilization while at three hours 25 minutes the blastodisc is well formed with the appearance of a flattened raspberry atop the yolk (the blastodermal cap or morula stage). At this point the individual cells are too small to be distinguished. The cells reach a maximum height of about 10 cells, and the egg usually floats with the blastodermal cap downward (Kuo et al. 1973a,b, Miller and Kendal 2009).

Blastula Stage

Following the blastodermal cap stage, the blastula stage occurs as the mass of cells begins to flatten and encircle the yolk (epiboly), and the edge of the cell mass becomes thickened (germ ring stage) (Fig. 11.3). At the same time, the blastomeres lift up in the centre of the mass, creating a central cavity called the blastocoel (Miller and Kendal 2009).

After about 4.25 hours from fertilization in *M. cephalus* at water temperature of 25 ± 1°C, the *blastoderm* is flattened out over the yolk to become circular; the height of the cellular cap is 224 μm. The mass of cytoplasm remains in the form of a syncytial layer adjacent to the yolk and called a periblast (Fig. 11.2) (El-Gharabawy and Assem 2006).

About five hours 15 minutes to five hours 40 minutes after fertilization in *M. cephalus,* blastulation (Fig. 11.4) is in progress, the perimeter of the blastodisc forms a thickened blastoderm and there is movement of the cells over the whole surface of the yolk in the epiboly process (Kuo et al. 1973a,b, El-Gharabawy and Assem 2006). The embryonic shield enlarges by addition of cells. At this stage, the egg diameter of *M. cephalus* in Egypt is about 820 ± 19 μm (Fig. 11.2) at water temperature of 25 ± 1°C (El-Gharabawy and Assem 2006).

Gastrula Stage

By the time the cell mass is about halfway around the yolk, the first indications of the embryo can be seen as a thickened line (neural keel) on top of the yolk perpendicular to the edge of the germ ring, indicating that gastrulation is occurring (Fig. 11.3). During the gastrula stage the single-layered blastoderm becomes a multilayered embryo (Miller and Kendal 2009) and is characterized by cell movement and recognition within the embryo (El-Gharabawy and Assem 2006). As the embryonic shield forms, there is a thickening of the caudal margin of the blastodermal cap. In this area cells invaginate to form a gastrula. The outer layer of cells becomes the skin of the adult and the inner layer the gut and mesoderm (Miller and Kendal 2009). *Gastrulation* (Figs. 11.5, 11.6) in *M. cephalus* starts seven hours 15 minutes to seven hours 35 minutes after fertilization when the blastocoele, germ ring and embryonic shield appear (Kuo et al. 1973a,b). Eleven hours 55 minutes after fertilization, the head fold extends to the vegetal pole and the embryonic shield expands. Invagination of blastomeres is complete 12 hours 45 minutes after fertilization (Kuo et al. 1973a,b).

Organogenesis Stage

About seven hours post fertilization in *M. cephalus*, the embryonic fold is observed as a thickening along the dorso-lateral margins of the yolk (Fig. 11.7), and consists of an aggregation of cells but they do not form any clear cut division or boundaries (El-Gharabawy and Assem 2006). At the age of about 13.5 ± 1 hours,

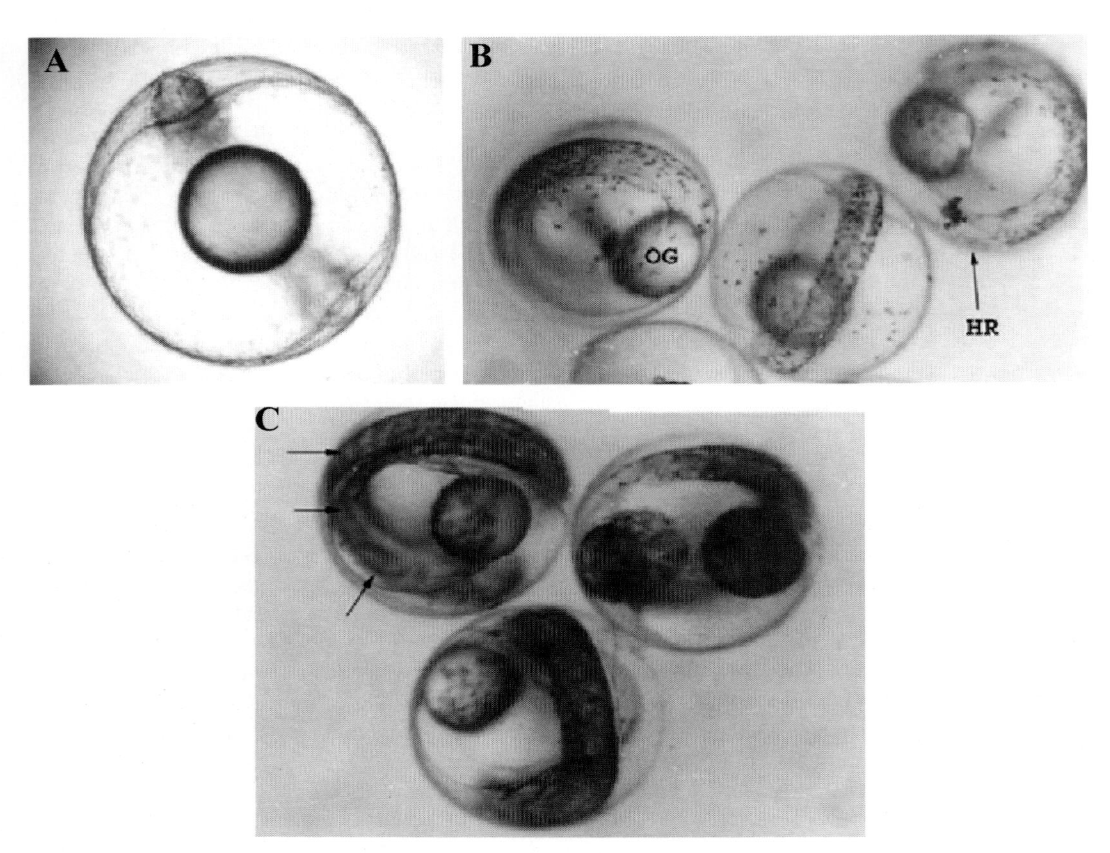

Figure 11.5. Embryo of *Mugil cephalus.* (A) At age 12 hours post fertilization (p.f.) (X31) (after Yousefian et al. 2009) and (B) at age 13.5 hours p.f. Head region (HR) started to be shaped, dark pigments cover almost all of the embryo (arrows) and on the oil globule (OG), X31 and (C) at age 13.5 hours p.f. Head region (HR) starts to be shaped; dark pigments cover almost all of the embryo (arrows) and on the oil globule (OG), X31 (after El-Gharabawy and Assem 2006).

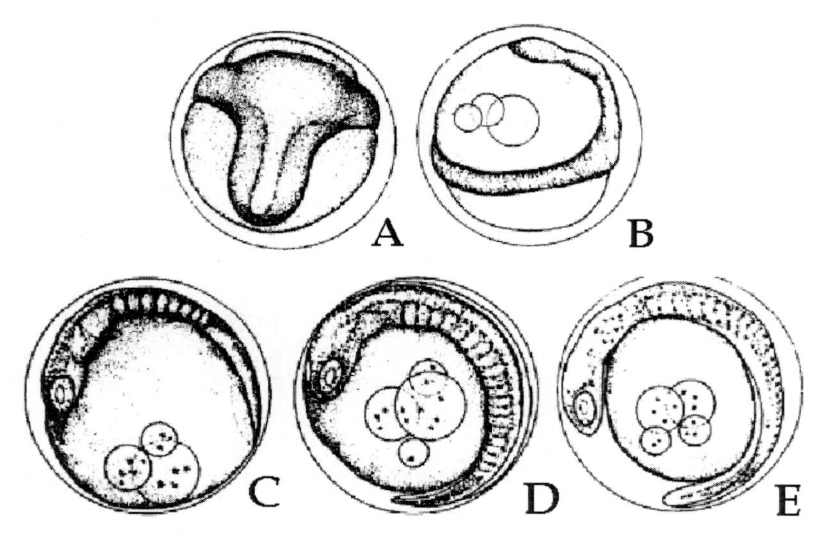

Figure 11.6. Embryonic development of *Chelon labrosus.* (A) 23 hours post fertilization (p.f.); (B) 1 day 4 hours p.f.; (C) 1 day 8 hours p.f.; (D) 2 days 17 hours p.f.; (E) 3 days p.f. (after Boglione et al. 1992).

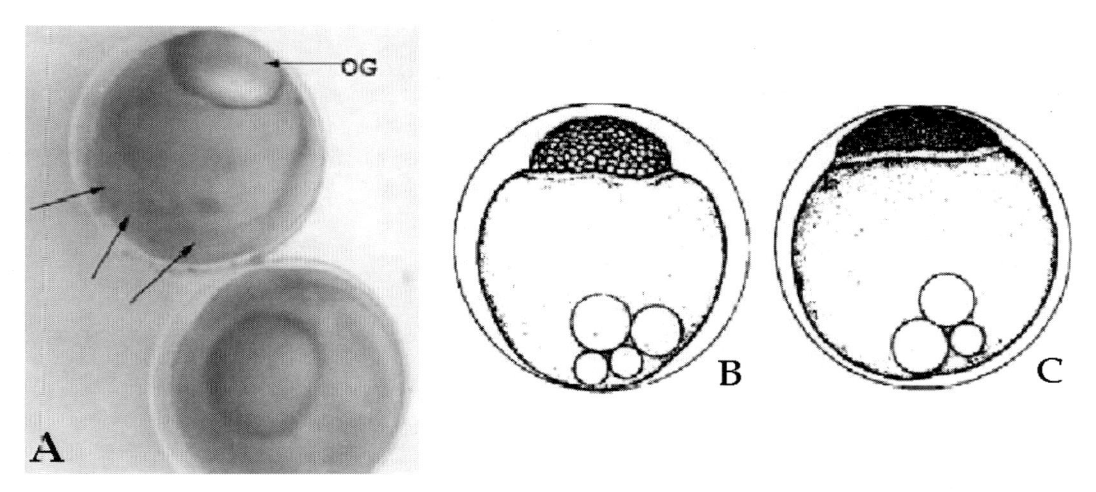

Figure 11.7. Embryonic Development. (A) Fold stage of *M. cephalus* 7 hours post fertilization (p.f.). Note beginning of organogenesis process as slight folding has not any clear boundaries (arrows), oil globule (OG), X31 (after El-Gharabawy and Assem 2006); *Chelon labrosus* (B) 7 hours 30 min (p.f.); (C) 10 hours p.f. (after Boglione et al. 1992).

the embryo begins to be noticeable. The head region starts to be shaped and the optic cup and tail region form. Dark pigments cover almost all of the embryo, yolk sac and the oil globule. The melanophores appear as stars in shape. The egg diameter is 870 (± 30) μm and oil globule is 350 μm in diameter (Fig. 11.7). The jerky striated muscle movement of the tail somites increase and the tail starts to bend to help and accelerate the hatching process, leading to the loosening of the egg membrane (El-Gharabawy and Assem 2006).

For *C. labrosus*, at the age of seven hours 30 minutes (T = 14–20°C), an early discoblastula appears (Fig. 11.4). At the age of about 10 hours, a late discoblastula appears when the blastomeres begin to migrate towards the vegetative pole (Fig. 11.4) (Boglione et al. 1992). At an age of 12 hours in *M. cephalus* embryo at water temperature 22–24°C in Iran (Fig. 11.5), the head region starts to shape and dark pigments cover almost the entire embryo (Yousefian et al. 2009).

Neurula. The neural groove (Fig. 11.4) appears 13 hours 30 minutes after fertilization while 14 hours 30 minutes after fertilization the optic vesicles are apparent and the tail is expanded (Kuo et al. 1973a,b). The embryo of *M. cephalus* 13 ± 1 hours post fertilization begins to be noticeable (Fig. 11.5).

Developing Embryo. The cells continue to grow around the yolk until only a small circle of yolk is not covered (the blastopore) and the embryo begins to take shape. Along the neural keel the neural tube and notochord are forming. Myomeres gradually form on either side of the notochord, the three main portions of the brain (forebrain, midbrain, and hindbrain) can be seen, and the optic vesicles appear as the blastopore closes (end of gastrulation). At the beginning of the middle stage of development, internal organs (liver, gut) can be seen and the heart forms and starts to beat when the embryo circles about halfway around the yolk. Auditory organs and eye lenses appear. Pigment often first appears on the embryo about this time. Myomeres continue to be added both anteriorly and posteriorly to those that first developed midway along the body, and the tailbud margin is defined (Fig. 11.3). Pectoral fin buds and the otic capsules appear as the tailbud lifts off the yolk (beginning of the late stage of development). Following this, the body lengthens, the various organs become better defined, and pigment is added. The embryo begins to move within the egg at this time (Fig. 11.3). Oxygen consumption increases as the embryo starts to move. Hatching occurs at various stages of development, depending on the species. Some fishes hatch when the embryo reaches full circle around the yolk, but other species hatch before or after this (Miller and Kendal 2009).

According to the descriptions of Kuo et al. (1973a,b) of the developing embryo of *M. cephalus*, the somatic segmentation begins 15 hours 30 minutes post fertilization (p.f.) and the optic vesicles begin to differentiate. The differentiation of olfactory lobes begins 20 hours 55 minutes to 21 hours 25 minutes (p.f.) while melanophores appear in the oil globule and dorsal aspect of the embryo. The tail is tapered and bent ventrally. Optic vesicles and olfactory lobes are well developed 25 hours (p.f.) and the finfold appears where the embryo is elongated with the tail near to the head; the heart begins to beat. Twenty seven hours (p.f.), the tail is free of the yolk sac and body movement begins while 30 hours 30 minutes to 31 hours (p.f.) the tip of the tail reaches the head while body movement is more frequent and the embryo position is changing in the egg. Thirty five hours 40 minutes (p.f.), wave-like body movement occurs regularly every three- five seconds with the tail movements very frequent.

In *C. labrosus*, at an age of 23 hours, the embryonic shield appears. The embryo (dorsal view) is elongating from the germ ring (Fig. 11.6). At an age of one day four hours, the germ ring moves towards the vegetal pole, engulfing it completely. The first two brain vesicles begin to be identifiable (Fig. 11.6). At an age of one day eight hours, three brain vesicles can be identified as well as the eye rudiment. Eight somites are present. Pigment appears on the body, among the somites and on the oil globules in the form of star-shaped melanophores (Fig. 11.6). At age of two days 17 hours, the embryo emerges from the yolk. The black pigmentation, due to melanophores, is extended wider along the trunk. The tail is rising from the yolk; 20 somites can be counted (Fig. 11.6). The two small olfactory placodes are recognizable. Their different cell types can be identified. In the frontal region of the head, two lateral line neuromasts are present, medial to the two olfactory placodes, as well as one neuromast on each side, near the branchial pore. The epidermal cells display microridges, although they do not yet show the typical pattern of the adult. At an age of three days, the head is rising from the yolk; 21 somites can be counted (Fig. 11.6) (Boglione et al. 1992).

Free larvae. The hatching (free larvae stage) of *M. cephalus* (Fig. 11.8) is evident 36–38 hours after fertilization at 24ºC and 48–50 hours at 22ºC (Kuo et al. 1973a,b). The larva breaks the egg membrane with tail movement and the head emerges first (Kuo et al. 1973a,b). Total length of newly hatched larvae is 2.65 ± 0.23 mm (mean ± standard deviation) (Kuo et al. 1973a,b) compared with 2.08–3.40 mm reported by Liao et al. (1972). These data are in agreement with previous reports of hatching time:

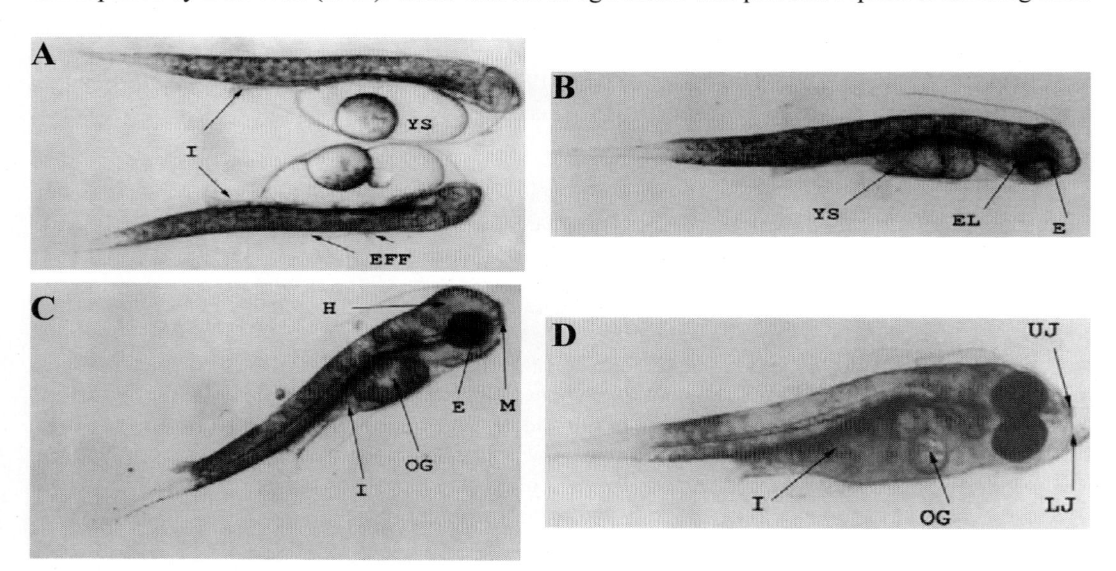

Figure 11.8. *Mugil cephalus* larvae. (A) 12 hours post hatching (p.h.). The yolk sac (YS) was still large, the dorsal, ventral and caudal fins were connected together to form the embryonic fin fold (EFF). The intestine (I) appeared as thin tube extended upper to the end of yolk sac, X31; (B) 24 hours post hatching (p.h.). A sharp decrease was observed in the yolk sac (YS), eye lens (EL) was formed and eye retinas (E) were darkly pigmented; (C) 36 hours post hatching (p.h.). Head (H) is elevated, eye (E) became darker and the mouth (M) is slightly opened, the intestine (I) extended downwards through the ventral fin fold, oil globule (OG) is still appear, X25; (D) 48 hours post hatching, the mouth parts are completely formed, upper jaw (UJ) and the lower jaw (LJ) could be easily seen, and the intestine (I) is function, while the oil globule (OG) is still present, X25 (after El-Gharabawy and Assem 2006).

34–38 hours at 23–24.5°C and 49–54 h at 22.5–23.7°C in salinities of 30.1–33.8‰ (Liao et al. 1972). Hatching time is prolonged further at lower temperatures: 60–65 hours at 21°C (Liao et al. 1971) and 60 hours at 18.2–19.5°C (Yang and Kim 1962).

Pre-larval stage (Yolk Sac Larvae or the Endotrophic Period)

Hubbs (1943) defined the terminology for the young stages of fishes and separated 'pre-larval', 'postlarval' and 'juvenile' stages on observed criteria. For most teleosts, the formation of the scales signifies the end of the postlarval stage. The Mugilidae, however, develop scales in the early postlarval stages (between 8–10 mm in length) and are soon well developed (12–14 mm in length). Hubbs' (1943) criteria do not therefore apply (Nash and Shehadeh 1980, Nash and Koningsberger 1981).

Roule (1917) divided the postlarval stage of Mugilidae into two successive periods. The first had rudimentary scales as the diagnostic criterion, followed by the second with true scales. Anderson (1958) regarded the formation of the third spine of the anal fin as signifying completion of the postlarval stage, and classed individuals as juveniles if the third anal ray had fused into a spine (Nash and Shehadeh 1980, Nash and Koningsberger 1981). Finally, Tung (1973) described five stages of larval development and morphology (Nash and Shehadeh 1980, Nash and Koningsberger 1981).

The pre-larval stage is characterized by nutritive contribution provided exclusively by the yolk sac and ends when the larva becomes capable of autonomously feeding, even though a yolk residue is still present (Fig. 11.9). Balon (1985) defines this period as eleuterembryonic. The pre-larva of *C. labrosus* develops rapidly during the final resorption of the vitelline sac and during the early feeding stages (Boglione et al. 1992).

Figure 11.9. (A–B) Pre-larvae of *Chelon labrosus* 1 to 2 days old (X 25); (C–D) Larvae 2 to 3 days old (X 40–50); (E–F) larvae 8 days old (X 25–50); (G–H) fry 25 to 30 days old (X 12,5–16) (photos by G. Minos).

Morphology

Some of the first descriptions on the morphology of the larvae of Mugilidae were made by Sanzo when he observed and illustrated the early stages of *C. labrosus* and *M. cephalus* (Sanzo 1936) and *O. labeo* (Sanzo 1937) considering that their prelarvae are poorly developed, measuring only 2.2–2.5 mm in length. The mouth was closed and there was no trace of a branchial skeleton. Characteristic of the larvae were the voluminous yolk sac and large oil globule often accompanied by smaller oil globules (Nash and Shehadeh 1980, Nash and Koningsberger 1981).

The first complete morphological descriptions of the larvae of Mugilidae were made by those workers involved in the induced spawning of the adults by hormone injection and describe mainly the development of *M. cephalus* (Tang 1964, Yashouv 1969, Yashouv and Berner-Samsonov 1970, Liao et al. 1972, Kuo et al. 1973a, Tung 1973). Thomson (1963) reviewed the work on *M. cephalus* and quoted in detail the embryonic development described by Sanzo (1936) and Anderson (1958). A full and accurate description of development and behaviour of *M. cephalus* was made by Liao (1975). Sebastian and Nair (1975) described the development of *L. macrolepis*.

Newly hatched larvae of *M. cephalus* vary in length between 2.2 and 3.5 mm. The oil globule (and any additional small globules) is situated in the posterior part of the yolk. Tung (1973) recorded 24 myotomes and described that the anterior half of the body is bent on the yolk sac. The larvae had five or six pairs of cupulae on the body side from eye to tail, and several pairs on the front of the head. Feeding began three–five days after hatching. Yashouv and Berner-Samsonov (1970) gave full descriptions to the eggs and larvae of five mullet species: *M. cephalus*, *L. ramada*, *L. saliens*, *C. labrosus* and *L. aurata*.

Hatching Stage

Hatching occurs once the embryo has reached a certain stage of development. At that point proteolytic enzymes in the perivitelline space begin to soften the envelope. Enzymes produced by the embryo digest proteins to form more soluble compounds. The envelope is digested from inside out, with the embryo using any products that are nutrients. As oxygen demand by the moving embryo exceeds the supply, the embryo moves more, helping it to escape the envelope. In large eggs the embryo tends to emerge tail-first; in small eggs the larva emerges head-first (Miller and Kendal 2009).

Environmental conditions for incubation. Egg development and hatching time are both temperature dependent. Nash et al. (1974) and Nash and Kuo (1975) report the survival of eggs of *M. cephalus* within broad ranges of temperature, salinity and dissolved oxygen. Minimal mortalities of eggs occurred at 22°C for normal sea water (32‰), and an effective temperature range for incubation was 11–24°C (Nash and Koningsberger 1981). Most workers prefer a working temperature range of 18–24°C for *M. cephalus*. Above 25°C incubation is inhibited although some eggs will hatch at 30°C; the mortality beyond 25°C is usually above 90% and often total. Optimal salinities for incubation are 30–32‰ under ambient temperature conditions (19.5–20.5°C), and significant decreases in egg survival occur with eggs incubated in mean oxygen concentrations below 5.0 ppm (Nash and Koningsberger 1981).

Incubation period. The duration of hatching tends to be greater at lower salinities and temperatures. The incubation time of the eggs of *M. cephalus* appears in Table 11.3. *L. ramada* egg developed and hatched within 36–44 hours at 22–32°C (Yashouv and Berner-Samsonov 1970). The incubation time of *M. macrolepis* was 23 hours at 26–29°C and 29–31‰ salinity (Sebastian and Nair 1975).

Hatching length. The length of newly hatched larvae of *M. cephalus* (W. Atlantic) is reported as 2.0–2.2 mm (Ditty et al. 2006); 2.65 mm (Kuo et al. 1973a,b); 1.97 ± 0.23 mm in total length and the yolk has an oval shape (0.7 mm x 0.34 mm) (El-Gharabawy and Assem 2006). The hatch size of the egg of *M. curema* is 1.6–1.8 mm (Ditty et al. 2006). The newly hatched larvae of *C. labrosus* (Fig. 11.9) measure 3.41 ± 0.48 mm (Zouiten et al. 2008) or 3.7 ± 0.1 mm (Ben Khemis et al. 2006) in total length at day 1 post hatching and 3.84 ± 0.31 mm (Zouiten et al. 2008) or 3.9 ± 0.1 mm (Ben Khemis et al. 2006) at day 5 post hatching when the mouth opens. At that stage most of the yolk is consumed, but an oil globule is still present. It

Table 11.3. Duration of hatching of *Mugil cephalus* at different salinities and temperatures.

Time (hours)	Temperature (°C)	Salinity (‰)	Region/Country	Reference
26 ± 4	25 ± 1	34	Egypt	El-Gharabawy and Assem 2006
38	23–24		W. Atlantic	Ditty et al. 2006
36–38	24		Taiwan	Kuo et al. 1973a
48–50	22		Taiwan	Kuo et al. 1973a
36–44	22–32		Israel	Yashouv and Berner-Samsonov 1970
34–38	24	32	Hawaii	Tucker 1998
34–38	23–24.5	30.1–33.8	Taiwan	Liao 1975
49–54	22.5–23.7	30.1–33.8	Taiwan	Liao 1975
48	22.5–23.5		India	Chaudhuri et al. 1977
59–64	20.0–24.0	24.39–35.29	Taiwan	Tang 1964

remains visible under the dissecting microscope until day 10 (p.h.). The swimbladder inflates between day 10 and day 15, but it is difficult to observe due to the dark pigmentation of the larvae (Ben Khemis et al. 2006). According to Cataudella et al. (1988a,b) and Boglione et al. (1992), in *C. labrosus* at ambient conditions of 15–18°C after incubation for three days, hatching was achieved and the larvae measured 4 mm in length.

Fertilized eggs from laboratory-reared redlip mullet *L. haematocheila* (*L. haematocheilus*) hatch after 80–95 hours, with incubation at 15.6–21.2°C and the mean total length of newly-hatched larvae is 3.62 mm (Yoshimatsu et al. 1992).

At an age of four hours post hatching (p.h.), the prelarvae of *C. labrosus* have an average total length of 4 mm; this value is in agreement with those of Sanzo (4.2 mm, 1936) and Cassifour and Chambolle (3.8 mm, 1975). The mouth is not yet formed and the branchial skeleton is still undeveloped. The optic vesicles are still unpigmented. The olfactory placodes are more extended and surrounded by a typical ring of melanophores. The heart is still in the form of a straight tube. The digestive tract is undifferentiated. The end part of the intestine is still blind. The embryo displays a continuous dorso-ventral unpaired fin. The pigmentation extends uniformly from the head along the trunk, except for the final section of the tail. There is a typical concentration of melanophores on the unpaired fin at the level of the 16th somite, while the remainder of the fin is unpigmented. 1–2 oil drops are still present (Figs. 11.6, 11.7) (Boglione et al. 1992).

At age one day (p.h.) the newly-hatched larvae of *M. cephalus* is about 2.56–3.52 mm TL having a large yolk and oil globule. The front part of the notochord is curved along the yolk sac and the curve degree is related to the duration of hatching. At low temperature, the duration is long and the curve is more distinct. Pigmentation is dependent on the individual and colourless on the eye. The mouth is not developed and the digestive tube is not well developed (Liao 1975).

At age one day 4 hours (p.h.) in *C. labrosus*, the caudal fin begins to separate from the continuous fin and its first rays can be distinguished (Fig. 11.9). The pigmentation extends to the yolk sac both dorsally and ventrally. Typical star-shaped melanophore clusters appear around the final tract of the intestine. The group of melanophores on the dorsal fin extends as far as the 18th myotome; 25 myotomes are present (Fig. 11.6) (Boglione et al. 1992).

At an age two days (p.h) in *M. cephalus* (2.64–3.28 mm TL), the formation of organs is in progress. More pigmentation is found on the eye and body. The total length is shorter than before and the mouth is under development. A bud of the pectoral fin appears. Nostrils are obvious (Liao 1975).

At an age of two days 22 hours (p.h.) in *C. labrosus*, the mouth is invaginated, but not yet open (Fig. 11.9). The pectoral fins are clearly identifiable. The pigmentation can be seen also at the level of the horizontal ribs and less evidently along the myosepta. Moreover, two rows of chromatophores, one dorsal and one ventral, are present on each side (Fig. 11.6) (Boglione et al. 1992).

Following the growth of *M. cephalus*, at age three–four days (p.h.) (3.11–3.53 mm TL), the mouth opens and it is able to take food. Upper and lower jaws become well developed. Yolk is diminished being

¼ of original size and oil globule is also reduced. At age five–seven days (p.h.) (3.06–3.40 mm TL), the digestive tube is well developed. There is formation of stomach, intestine, gall bladder, pancreas, gas bladder and continued reduction of the oil globule. At an age of eight days (p.h.) (3.35–3.80 mm TL), there is complete disappearance of the oil globule; formation of gill filaments and growth accelerates (Liao 1975).

Absorption of the oil globule is completed on the 10th day at 24°C and on the 15th day at 22°C (Kuo et al. 1973a,b).

At an age of 10–13 days (p.h.) in *M. cephalus* (3.45–5.10 mm TL), the finfold moves backward. Gill filaments are well developed and the body surface becomes dark in colour. There is formation of the hypural bone. This is the second critical period with a very low survival rate. At age 14–15 days (p.h.) (3.85–5.70 mm TL), there is formation of an urostyle. Seven–nine ray bases are found in each of the anal fin and second dorsal fin. Gill lamellae are formed on gill filaments at an age of 16–19 days (p.h.) (5.40–6.60 mm TL), and there are 17 soft rays in the caudal fin. Black spots are scattered over the whole body. Shiny silver white complexion appears from the gill cover along the ventral surface to the anus at age 20–21 days (6.00–7.65 mm TL). The colour appears to be brown to silver green. Four soft rays are found on the first dorsal fin. At age 22–24 days (8.20–10.9 mm TL), formation of complete 20 soft rays in the caudal fin, 11 rays in anal fin, six rays in pelvic fin, 15 rays in pectoral fin. Fin membrane of dorsal fin and pelvic fin almost entirely degenerated, and the scales have one–three circles of ridges. At age 25–28 days (8.80–15.0 mm TL), all scales and fin rays are well-formed. The appearance of teeth begin. Two nostrils are separated (Liao 1975).

Larval Stage

The larval stage begins at hatching and lasts until complete fin ray counts have been attained and squamation has begun. The yolk-sac larval stage starts at hatching and ends when the yolk sac is absorbed (Miller and Kendal 2009).

The larval stage is characterized by autonomous feeding and morphological changes begin to affect mostly the head region. Only later on are these changes followed by a transformation of the rest of the body. Hubbs (1943) defines as postlarval this stage which begins immediately on absorption of the yolk sac, and lasts as long as the structure and form are unlike that of the juvenile (Boglione et al. 1992).

Development (12 hours, 24 hours, 36 hours, 48 hours, p.h.)

After about 12 hours from hatching of *M. cephalus*, the yolk sac is still large; the dorsal, ventral and caudal fin are connected together to form the embryonic fin fold. The intestine appears as a thin tube extended upwards to the end of yolk sac (Fig. 11.5). The blood circulation is clear as a strong flow occurs in the heart. After about 24 hours from hatching a sharp decrease appears in the yolk sac diameter (0.26 mm). The body axis is straightening and the total length increases to 2.42 ± 0.24 mm. Eye lens are forming and the retina is darkly pigmented; eye diameter is about 0.15 mm (El-Gharabawy and Assem 2006).

After about 36 hours from hatching, head region is elevated and increases in width (0.37 mm) with more rounded configuration, the eye becomes darker with increasing pigmentation; also the mouth parts begin to be formed and the mouth is slightly open. The intestine extends downwards through the ventral fin fold increasing in both thickness and diameter, while the oil globule still appears at the abdominal region.

Yolk-Sac Stage (from Hatching to Complete Absorption of Yolk Sac)

Most fishes hatch as yolk-sac larvae, a stage that lasts until the yolk sac is absorbed and during which the yolk continues to supply nutrition to the larva (Fukuhara 1990, Johns and Howell 1980). The yolk sac is prominent and situated on the anterior ventral part of the body, occupying about half of the total length. In species that present a yolk-sac stage (generally with a short incubation period), larvae usually hatch without a functional mouth, open anus, eye pigment and differentiated fins (MedSudMed 2011). During development, the contents of the yolk sac and the oil globule (if present) are gradually absorbed (endogenous feeding). When the yolk is completely used, the development of the sensorial, circulatory, muscular and digestive organs is complete and the larva starts feeding actively on plankton (exogenous feeding). The availability of the right food organisms at this stage is thus a critical factor. In the yolk-sac stage, the end of the notochord (urostyle) is straight, but later, on its ventral side, a triangular thickening develops forming the rudiments of the hypural elements from which the first caudal fin rays will develop (MedSudMed 2011). Yolk-sac larvae are largely transparent and are about 2–6 mm long and are generally not well developed. The large yolk causes the larva to float upside down. Swimming usually consists of short, vertical bursts. The advantages of developing during this stage outside the egg include better oxygen supply and better mechanisms to get rid of metabolites. The larvae can increase in size and development without the restrictions of the egg, while still getting nourishment from the yolk. They can also practice feeding and obtain some nourishment from prey as well as from the yolk (Miller and Kendal 2009).

The larvae of cultured *C. labrosus* measures 3.7 ± 0.1 mm in total length at day one (p.h.) and 3.9 ± 0.1 mm at day five (p. h.) when the mouth opens. At this stage most of the yolk is consumed, but the oil globule is still present (Fig. 11.9). It remains visible under the dissecting microscope until day 10 (p.h.).

The larvae of *L. haematocheila* (*L. haematocheilus*) measures 3.02 mm; the anus is situated slightly posterior to the middle of body; the number of somites is 25 in total, with some variations (Yoshimatsu et al. 1992). Melanophores are present on the head, the edge of the eye, trunk and caudal regions, and on the tissue surrounding the inner side of the yolk-sac and oil globule. Xanthophores are present on the trunk and caudal regions and on the tissue surrounding the outer side of the yolk-sac and oil globule. The larvae have unpigmented eyes and no fins and the mouth has not yet formed (Yoshimatsu et al. 1992). The five-day yolk-sac larvae of *L. haematocheila* are 3.89 mm TL. They have open mouths, pigmented eyes, and fan-shaped pectoral fins. The intestine is convoluted, but no food organisms are visible in the gut (Yoshimatsu et al. 1992).

Internal-External Feeding Stage

After about 48 ± 12 hours post hatching, the mouth parts are completely formed, upper and lower jaws easily noticed. The length of the lower jaw slightly exceeds that of the upper jaw. Separation of the dorsal and anal fin folds from the caudal fin clearly appears with mesenchyme deposition. The notochord is clearly seen extending through the whole body. The swimming activity of the larvae increases as the enlargement of the swimbladder proceeds. At this stage the larvae can eat external food taking it in addition to internal food yolk material. The elimination of dark faeces and presence of large particles in the intestinal tract indicate that mixed feeding has begun (El-Gharabawy and Assem 2006).

The six-day yolk-sac larvae of *L. haematocheila* are 3.95 mm TL. The first inflation of the gas-bladder is observed in almost all individuals. Yolk absorption is nearly complete, but the oil globule is still large. Many individuals start to feed before the completion of yolk absorption. Individuals that have completed yolk absorption first appear on the 7th day, and all the larvae have completed it by the 9th day (Yoshimatsu et al. 1992).

External Feeding Stage (Three–Five Days)

Larvae of *M. cephalus* begin feeding three–five days after hatching (El-Gharabawy and Assem 2006). Such results are in agreement with Tung (1973) and Liao (1997) who concluded that in *M. cephalus* mouth opening and the development of upper and lower jaws allow the larvae to take food at three to four days. The time of eye completion is similar to that of mouth opening and the body axis becomes straight, which helps the larvae to identify their food as internal-external feeding starts. Kvenseth et al. (1996) observed the development of functional eyes at the time when larval halibut capture prey and when the digestive system appears histologically functional.

At an age of three days 21 hours post hatching (p.h.) in *C. labrosus*, the eye becomes pigmented. The macula, which appears differentiated, is recognizable in the histological preparations in cross section in the otacyst. The heart is folded over on itself. The different parts of the alimentary canal (oesophagus, stomach and intestine) begin to differentiate histologically. The base of the skull and the cartilaginous branchial arches, the pharyngobranchial, hypobranchial and epibranchial, can be identified. The swimbladder is activated. Rays of the pectoral and caudal fins are clearly recognizable (Fig. 11.9). Groups of chromatophores are now present also ventral to the mouth and on the dorsal surface of the head (Boglione et al. 1992).

On the fourth day after hatching in *M. cephalus*, the mouth is completely open and becomes functional, and the maxillaries are distinct. The foregut and hindgut are visible, the anus is open and the eye is completely pigmented. The body and the notochord are straight; the yolk sac decreases in diameter to 0.1 mm. Greatest body depth range from 0.37 to 0.38 mm, total length range from 2.55 to 2.75 mm, and pre-anal length range from 1.27 to 1.3 mm. The head length increases from 13.26 to 17.05% of the total body length. At the end of the sixth day after hatching the total length of larvae ranges between 2.68 and 2.83 mm, pre-anal length is about 1.35 ± 0.3 mm, head length varies between 0.53 to 0.58 mm and the height of the head is 0.37 mm (Fig. 11.8) (El-Gharabawy and Assem 2006).

Notochord Flexion and Stages of Larval Development. One of the fundamental events in the development of most fishes is the flexion of the notochord that accompanies the hypochordal development of the homocercal caudal fin (Miller and Kendal 2009). As the larva grows, the urostyle bends upwards, the hypural elements become defined and the caudal rays develop. The larval stage can be divided, based on the degree of flexion of the terminal section of the notochord, into three different developmental stages, a) 'preflexion', b) 'flexion', and c) 'postflexion' (referred to the notochord in the urostyle region) (Rè and Meneses 2008, Miller and Kendal 2009). These three phases lead to the last one that proceeds adulthood, which is defined as the juvenile phase (MedSudMed 2011).

a) *Preflexion stage (from complete yolk-sac absorption to start of notochord flexion).* This begins once both hatching and complete absorption of the yolk-sac has occurred and ends when the notochord starts to flex. The notochord is straight during the preflexion stage, and the larva is just starting to feed. Its sensory and locomotor powers are developing rapidly since the larva must have operational eyes, a movable jaw, and a functional gut to capture and utilize prey. The median finfold is usually continuous, not divided into separate fins. Larval pectoral fins are well developed and used for positioning. The pelvic fins are usually not developed yet. The larval pigment pattern is becoming established. The retina holds only cones, allowing vision in bright light only. The first red blood cells appear. Respiration is still cutaneous (Miller and Kendal 2009, MedSudMed 2011). Diagnostic characters for preflexion larvae are similar to those for other larval stages: meristics (number of myomeres, number of fin rays); body size and shape; fin development sequence; gut shape and length; pigmentation pattern; presence or absence of spines on the head; presence or absence of specialized larval characters (such as fin-ray ornamentation, stalked eyes, and a trailing gut) (MedSudMed 2011).
b) *Flexion stage (from the beginning of notochord flexion to completion of notochord flexion).* In many fishes is accompanied by rapid development of fin rays, change in body shape, change in locomotor ability, and feeding behaviour. This stage begins when the tip of the notochord bends dorsally concurrently with the development of the caudal-fin rays and the supporting skeletal elements ventral to it. This stage ends when the notochord tip has reached its final position and the flexion is complete at approximately 45 degrees from the notochord axis and the principal caudal-fin rays and supporting skeletal elements are in the adult longitudinal position (Miller and Kendal 2009, MedSudMed 2011). During this stage swimming changes from a burst and stop mode to a more continuous gliding mode. Swimbladder inflation may first occur during this stage, allowing the larva to be neutrally buoyant. Feeding becomes more proficient and the ability to avoid predators increases (Houde and Schekter 1980, Blaxter 1986). The digestive system develops and digestion improves (Govoni et al. 1986, Oozeki and Bailey 1995). Diel vertical migrations and other complex behaviours (e.g., schooling) may commence during the flexion stage. Rods begin to develop in the eyes. Larval pigment patterns are fully developed and larval specializations such as elongate fin rays and head spines may develop. Fin rays develop rapidly in most fins and they and other skeletal elements start to ossify (Dunn 1984). Cruising and burst swimming speeds of fish larvae increase as the larvae grow (Miller and Kendal 2009).

c) *Postflexion stage (from completion of notochord flexion to start of metamorphosis).* This stage starts when the notochord flexion is completed and ends at the onset of metamorphosis (transformation) when all of the fin rays have formed, and the juvenile stage begins. Diagnostic characters for this stage are similar to those for other larval stages, with additional osteological characters (such as timing and sequence of bone and cartilage development) that should be considered. During the Postflexion stage larvae continue to develop toward having all of their skeletal elements ossified. Respiration shifts from cutaneous to gills and the stomach differentiates in the gut. Rods further develop in the eyes, allowing vision in lower light (Miller and Kendal 2009, MedSudMed 2011).

Larval stage. At age of six days 8 hours post hatching (p.h.) for *C. labrosus*, maxillar and mandibular bones can be identified in the mouth region. The operculum almost completely covers the branchial pore. The yolk sac is reduced in size and will be almost completely resorbed after about nine days. The lateral line is made up of six–seven circumorbital neuromasts, plus two on the trunk (one dorsal, one ventral) and three medial ones on the tail. The melanophores on the residual yolk sac begin to grow denser and the ventral region is intensely pigmented. On the remainder of the body the pigmentation extends over the full height of the trunk, and is more intense around the intestine. The melanophore cluster in the continuous unpaired fin persists (Fig. 11.10) (Boglione et al. 1992).

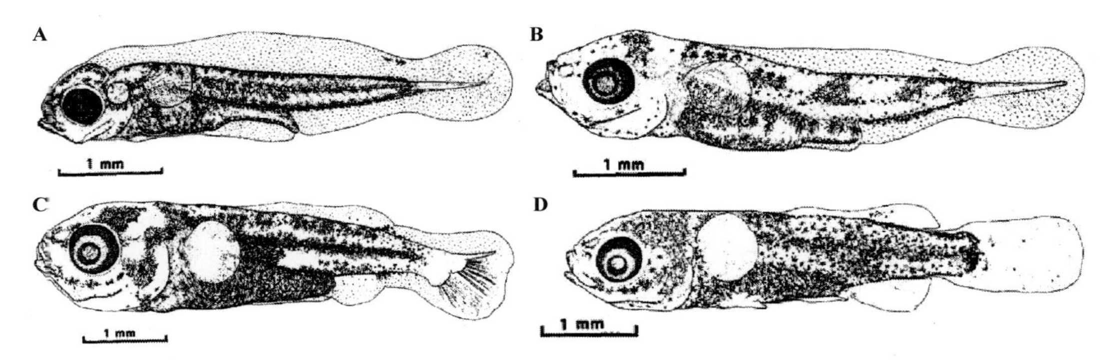

Figure 11.10. *Chelon labrosus* larvae. (A) 6 days and 8 hours post hatching (p.h.); (B) 12 days and 3 hours post hatching (p.h.); (C) 15 days and 22 hours post hatching (p.h.); (D) 22 days post hatching (p.h.) (after Boglione et al. 1992).

After eight days from hatching, larvae of *M. cephalus* were 2.8 ± 0.4 mm TL, with pigmented and functional swimbladder present. The oil globule decreased in size to reach 0.07 mm. At the 11th day after hatching the larvae average total body length was 3.1 ± 0.4 mm; the body depth was 0.5 mm, the head length was 0.59 mm, the eye diameter was 0.2 mm and the oil globule completely absorbed (El-Gharabawy and Assem 2006).

At an age of 12 days 3 hours after hatching (a.h.) in *C. labrosus*, the changes primarily affected the head (larval body, juvenile head). The olfactory placode began to deepen. The eye tissues were now fully differentiated. The lateral line of the body was now composed of six neuromasts: four medial, one dorsal and one ventral. The oesophagus was differentiated and also the differentiation of the stomach and the intestine was nearly complete. Liver, pancreas and spleen were present. Groups of melanophores cluster into patches latero-dorsally on the sides (Boglione et al. 1992).

The morphological development of the various parts of the larva of *C. labrosus* took place in a cephalo-caudal direction. At about 12 days after hatching the larva had a juvenile head and a larval body (Boglione et al. 1992). Although the prelarva was poorly developed at time of hatching, the thicklipped grey mullet very rapidly attained the juvenile stage, with the presence of anterior and posterior nostrils and the homocercal tail as early as day 22; by day 60 it was already entirely covered by scales (Boglione et al. 1992).

At an age of 15 days 22 hours post hatching (p.h.) for *C. labrosus*, dorsal, anal and caudal fin began to differentiate (Fig. 11.10) (Boglione et al. 1992).

The 13th day larvae of *L. haematocheila* were 5.72 mm TL. Oil globule absorption was complete. Xanthophores were no longer visible. A rudimentary caudal fin appeared at the lower tip of the notochord (Yoshimatsu et al. 1992). During the 15th day, larvae of *L. haematocheila* were 6.42 mm TL. The notochord started to flex upwards. The anlagen of the 1st and 2nd dorsal and anal fins appeared (Yoshimatsu et al. 1992).

At the 15th day after hatching, the larvae of *M. cephalus* were 3.39 ± 0.41 mm TL with a slow transformation of *M. cephalus* post larval stage to the young fish (El-Gharabawy and Assem 2006).

At an age of 22 days post hatching (p.h) in *C. labrosus*, dorsal and anal fins were separated from the caudal fin. The tail reached its definitive homocercal form. Ventral fins began to take shape (Fig. 11.10). The rims of the olfactory placode began to fold together, separating the two nostrils (anterior and posterior). There were two anterior teeth on the mandibular bone as well as on each half of the palatine organ. The microridges on the epidermal cells were arranged in the 'fingerprint' pattern typical of the adult (Boglione et al. 1992).

The 24 day larvae of *L. haematocheila* were 11.2 mm TL having developed fin rays and the larval fin divides at the caudal peduncle. Clear bands of melanophores are on the median line and the posterior part of the trunk. The 25 day larvae were 11.5 mm TL having fully developed fin rays except for the lower part of the pectoral fins (Yoshimatsu et al. 1992).

The digestive tract and digestion process undergo major developmental changes during the first weeks of life in fish (Walford and Lam 1993, Zambonino-Infante and Cahu 1994, 2001). Pancreas secretion function constitutes the first step of the maturation process of digestive function, and the second is the onset of brush border membrane enzymes in the intestine (Ma et al. 2005). The capacity of synthesizing amylase and thus to feed efficiently on vegetal sources seems to be expressed since early larval development stages in *C. labrosus* (Zouiten et al. 2008).

Scale development in *C. labrosus* shows by silvering of the fish, at the end of day 25, indicating that it also corresponds to the transition from larval to post-larval stage (Nash and Shehadeh 1980).

Swimbladder. In most teleosts, including Mugilidae, the swimbladder is completely closed in adults. The pneumatic duct is only found in larvae, and it conveys swallowed air to the developing swimbladder and empties the gas bladder by passing gas bubbles into the alimentary canal. Once the swimbladder is functional, the pneumatic duct slowly degenerates (Nash et al. 1977).

Larval and juvenile behaviour. In rearing conditions, larvae of *C. labrosus* first feed as soon as the mouth opens, on day 5 (p.h.). The larvae seem to prospect mainly the upper part of water column (more or less 70 to 80 cm) during the first feeding days. During the entire rearing period, they display regular prospecting swimming and seem to feed easily as indicated by numerous successful prey captures and their distended abdominal segment. Schooling was first observed on day 19 (p.h.) but concerned only the biggest individuals which showed a silver appearance limited to flanks. Schooling and flanks' silvering extended progressively and included the whole population by day 23 (p.h.), simultaneously with escapement reflex to moving shade (Ben Khemis et al. 2006).

Grazing on wall enclosures developed around day 40–45 (p.h.), after complete silvering of the fish. After days 49–50 (p.h.), a jumping behaviour was noted in case of stress (noise, moving shade, immersion of something). Around days 60–62 (p.h.), this behaviour became spontaneous and was observed in the absence of any stressing agent.

In the larvae of *C. labrosus*, the acquisition of the intestinal adult mode of digestion seems to occur between day 14 and day 20. The maturation of the digestive tract in *C. labrosus* larvae is relatively precocious especially if compared to European sea bass *Dicentrarchus labrax* larvae, which achieve this development around week four of life (Cahu and Zambonino-Infante 1994). The larvae of *C. labrosus* might hence support earlier co-feeding and weaning strategies which could be initiated as early as the second week of life (Ben Khemis et al. 2006).

Larval Development. Larval development is temperature dependent. Shelbourne (1964) demonstrated the need for optimum temperature control for the culture of marine flatfish. He also noted differences between the survival of larvae in natural conditions and those in the intensive hatchery environment where bacterial activity was potentially more dangerous. The reasoning of Shelbourne is particularly relevant to the culture of fish and shellfish in tropical and subtropical latitudes. There the ambient temperatures are highly conducive to bacterial growth, and the optimum rates for yolk utilization by the larvae are probably narrowly defined and close to a critical level (Nash and Koningsberger 1981).

Strict temperature control for the incubation of the eggs of the Mugilidae is important. Emergent and developing larvae up to metamorphosis tolerate an ever widening temperature range and their growth rate responds accordingly (Nash and Koningsberger 1981). Liao (1975) reported the successful culture of the larvae of *M. cephalus* in Taiwan over a number of preceding years within the ambient temperature range of 19–24°C. Minimum mortality of the larvae of *M. cephalus* was recorded between 18.9 and 25.3°C, although some larvae survived temperatures as low as 15.9°C and as high as 29.1°C (Nash and Koningsberger 1981). Sebastian and Nair (1975) operated within a higher range of 26–29°C for the culture of *M. macrolepis*.

Transformation Stage (from the beginning of metamorphosis to the complete development of fin-ray and the beginning of scale formation). Between the larval and juvenile stages, there is a transitional stage, which may be abrupt or prolonged and which, in many fishes, is accompanied by a change from planktonic habits to demersal or schooling pelagic habits. In some fishes migration to a 'nursery' ground occurs during or just before this stage. Morphologically, in the transformation stage larval characters are lost and the juvenile/adult characters are attained. The end of transformation stage is defined as the completion of fin-ray development and the onset of scales. Its duration is different from taxon to taxon (Miller and Kendal 2009, MedSudMed 2011). Two ontogenetic processes occur during this stage of transition between the larva and juvenile: (1) loss of specialized larval characters, and (2) attainment of juvenile-adult characters. Changes that occur during this stage include pigment pattern, body shape, fin migration, photophore formation, loss of specialized larval characters, eye migration, and scale formation (Miller and Kendal 2009, MedSudMed 2011).

In many, the morph resembles a herring-like fish and is apparently adapted for neustonic life. The dorsal aspect of the fish is dark green or blue and the lateral and ventral is silvery or white. The body tends to be herring shaped, the mouth is terminal and the fins are generally unpigmented. Such a stage is present in some codfishes (hakes [*Urophycis* spp.]), squirrelfishes (*Holocentrus* spp.) perch-like fishes (perciforms), goatfishes [mullids], and grey mullets [mugilids] (Miller and Kendal 2009).

Juvenile Stage

Juvenile individuals generally resemble small adults and Hubbs (1943) considers the juvenile to be the young stage similar to the adults in all essentials. The juvenile stage is morphologically defined by the complete fin-ray development, the scale formation and the complete ossification of the skeleton. This stage ends when the sexual maturation is completed (MedSudMed 2011).

Nash and Shehadeh (1980), however, regard the criterion of scale formation as not being applicable to Mugilidae, which develop scales in the early larval stages (at 8–10 mm in length) and the scales are soon well developed (at 12–14 mm in length). These authors report several criteria to define the completion of Hubbs' (1943) postlarval stage, as the fusion of the third anal ray into a spine (Anderson 1958) or the transition from postlarval rudimentary scales to juvenile true scales (Roule 1917). The transition from larva to juvenile in *C. labrosus* is characterized by the development or the modification of crucial functions: feeding (ingestion and digestion), propulsion (prospection, capture and escapement), respiration (from cutaneous to gill), etc. (Van Snik et al. 1997). According to Osse et al. (1997), not all systems can grow simultaneously and to realize the necessities for survival, fish larvae apparently spend the available energy with priority for most important functions (Ben Khemis et al. 2006).

According to Boglione et al. (1992), at 45–50 days from hatching *C. labrosus* has already a differentiated digestive apparatus, taste buds, teeth on maxillary arch, palate, palatine organ and hyoid copula, homocercal tail, well formed ventral, dorsal and anal fins, and anterior and posterior nostrils. Ten days later, its scale covering is complete and they suggest this evident morphological criterion as marking the start of the juvenile stage. Sixty days after hatching, juvenile *C. labrosus* resemble the adult, except for its size and sexual maturation. It is entirely covered by cycloid scales (except for the ventral region, where they are ctenoid). SEM examination reveals the presence of free lateral line neuromasts on the medial trunk scales. The olfactory mucosa is enclosed in a nasal cavity which opens towards the exterior through the anterior and posterior nostrils. Teeth are now present on the maxillary arch, the palate, palatine organ and hyoid copula. Taste buds that are present already at day 32, increased in number.

The larvae of *L. haematocheila* transformed to juveniles by the 30th day at 552ºC-days post hatching while the first juvenile appeared on the 25th day at 454ºC-days (degree days). Therefore, the transformation from larvae to juvenile took place in the range 12.0 to 14.0 mm. The 60 day juveniles and young adults of *L. haematocheila* is 36.4 mm having a third anal spine, regarded as a young adult (Yoshimatsu et al. 1992).

Young mullet, which first appear in small schools along the coasts and in the estuaries, measuring 18–28 mm in length and about 30–45 days old, are fully scaled. The transition stage from postlarvae to juveniles used by Anderson (1958) does not occur until the young are 33–45 mm in length or between 45 and 60 days old. Thomson (1963) regarded the transition complete at about 50 mm when the third anal spine formed from the anterior ray and the adipose eyelid started to form (Nash and Shehadeh 1980, Nash and Koningsberger 1981). The transformation of *M. cephalus* from pre-juvenile to juvenile involves change from predaceous to iliophagus (detritivore) feeding habits (Martin and Drewry 1978).

Also the use of the term 'fry' for young fish is commonly used to describe all the resources of postlarvae and juveniles collected along the coastlines and transferred to nursery ponds or confined coastal lagoons (Nash and Shehadeh 1980, Chapter 16—Crosetti 2015).

Pigmentation

Young mullet are very heavily pigmented and are often associated with heavily pigmented young of the goatfishes (Mullidae) in offshore habitats (Ditty et al. 2006).

Vertebrae/Myomers

Grey mullet larvae have 24 vertebrae (except mountain mullet, *Agonostomus,* which has 25) and lack (or have few and very small) preopercular spines (Ditty et al. 2006).

Fin Rays: Formation of Third Anal Spine

Young grey mullets have two anal spines until about 30–40 mm SL when the third anal element transforms into a third spine (Ditty et al. 2006) while Thomson (1963) reported that the third anal spine is formed at about 50 mm. Some authors (Anderson 1957, 1958, Leis and Trnski 1989) have regarded the formation of the third anal spine as a criterion for the juvenile stage of Mugilidae. Specifically, individuals of *L. haematocheila* form the third anal spine on the 40th day and the frequency of occurrence increases gradually after the 52nd day while all the individuals complete spine-formation on the 76th day at a water temperature of 15.1 to 24.7ºC (Yoshimatsu et al. 1992).

References

Abou-Seedo, F. and S. Dadzie. 2004. Reproductive cycle in the male and female grey mullet, *Liza klunzingeri* in the Kuwaiti waters of the Arabian Gulf. Cybium 28: 97–104.

Acha, E.M. 1990. Estudio anatómico-ecológico de la lisa (*Mugil liza*) durante su primer año de vida. F. Marít. 7: 37–43.

Akter, H., M.R. Islam and M. Belal Hossain. 2012. Fecundity and Gonadosomatic Index (GSI) of Corsula, *Rhinomugil corsula* Hamilton 1822 (Family: Mugilidae) from the Lower Meghna River Estuary, Bangladesh. Global Veterinaria 9: 129–132.

Albieri, R.J. and F.G. Araújo. 2010. Reproductive biology of the mullet *Mugil liza* (Teleostei: Mugilidae) in a tropical Brazilian bay. Zoologia 27: 331–340.

Alvarez-Lajonchere, L.S. 1982. The fecundity of mullets (Pisces: Mugilidae) from Cuban waters. J. Fish Biol. 21: 607–613.

Ameur, B., A. Bayed and T. Benazzou. 2003. Rôle de la communication de la lagune de Merja Zerga avec l'océan Atlantique dans la reproduction d'une population de *Mugil cephalus* L. (Poisson, Mugilidae). Bulletin de l'Institut Scientifique, Rabat, section Sciences de la Vie 25: 55–62.

Anderson, W.W. 1957. Early development, spawning, growth and occurrence of the silver mullet *(Mugil curema)* along the South Atlantic coast of the United States. Fishery Bull. Fish Wildl. Serv. US 57: 397–494.

Anderson, W.W. 1958. Larval development, growth and spawning of striped mullet (*Mugil cephalus*) along the South Atlantic coast of the United States. Fish Bull. US Fish Wild. Serv. 58: 501–519.

Babin, P.J., J. Bogerd, F.P. Kooiman, W.J. Van Marrewijk and D.J. Van der Horst. 1999. Apolipophorin II/I, apolipoprotein B, vitellogenin, and microsomal triglyceride transfer protein genes are derived from a common ancestor. J. Mol. Evol. 49: 150–160.

Balon, E.K. 1985. The theory of saltatory ontogeny and life history model revisited. pp. 13–28. *In*: E.K. Balon (ed.). Early Life Histories of Fishes: New developmental, Ecological and Evolutionary Perspectives. Junk Publ., Dordrecht.

Bartulovic, V., J. Dulcic, S. Matic-Skoko and B. Glamuzina. 2011. Reproductive cycles of *Mugil cephalus*, *Liza ramada* and *Liza aurata* (Teleostei: Mugilidae). J. Fish Biol. 78: 2067–2073.

Ben Khemis, I., D. Zouiten, R. Besbes and F. Kamoun. 2006. Larval rearing and weaning of thick lipped grey mullet (*Chelon labrosus*) in mesocosm with semi-extensive technology. Aquaculture 259: 190–201.

Benetti, D.D. and E.B. Fagundes Netto. 1980. Consideracoes sobre desova e alevinagem de tainha (*Mugil liza* Valenciennes 1836) em laboratorio. Inst. Pesq. Marinha Publicacao 135: 26p.

Beverton, R.J.H. 1992. Patterns of reproduction strategy parameters in some teleosts fishes. J. Fish Biol. 41: 137–160.

Blaxter, J.H.S. 1986. Development of sense organs and behavior of teleost larvae with special reference to feeding and predator avoidance. Trans. Am. Fish. Soc. 115: 98–114.

Boglione, C., B. Bertolini, M. Russiello and S. Cataudella. 1992. Embryonic and larval development of the thick-lipped mullet (*Chelon labrosus*) under controlled reproduction conditions. Aquaculture 101: 346–359.

Brownell, C.L. 1979. Stages in the early development of 40 marine fish species with pelagic eggs from the Cape of Good Hope. Ichthyol. Bull. J.L.B. Smith Institute of Ichthyology, Rhodes University, Grahamstown 40: 84pp.

Brusle, J. 1981. Sexuality and biology of reproduction in grey mullets. pp. 99–154. *In*: O.H. Oren (ed.). Aquaculture of Grey Mullets. Cambridge University Press, Cambridge.

Cahu, C. and J.L. Zambonino-Infante. 1994. Early weaning of sea bass (*Dicentrarchus labrax*) larvae with a compound diet: effect on digestive enzymes. Comp. Biochem. Physiol., A109: 213–222.

Cassifour, P. and P. Chambolle. 1975. Induction de la ponte par injection de progestérone chez *Crenimugil labrosus* (R.), Poisson teléosteen, en milieu saumâtre. J. Physiol. Paris 70: 565–570.

Cataudella, S., D. Crosetti, F. Massa and M. Rampacci. 1988a. The propagation of juvenile mullet (*Chelon labrosus*) in earthen ponds. Aquaculture 68: 321–323.

Cataudella, S., F. Massa, M. Rampacci and D. Crosetti. 1988b. Artificial reproduction and larval rearing of the thick lipped mullet (*Chelon labrosus*). J. Appl. Ichthyol. 4: 130–139.

Cerdá-Reverter, J.M. and L.F. Canosa. 2009. Neuroendocrine systems of the fish brain. *In*: A.P. Farrell and C.J. Brauner (eds.). Fish Physiology Series, Vol. 28. Honorary Editors: William S. Hoar and David J. Randall. Academic Press, Elsevier.

Chan, E.H. and T.E. Chua. 1980. Reproduction in the greenback grey mullet, *Liza subviridis* (Valenciennes 1836). J. Fish Biol. 16: 505–519.

Chang, C.W., Y. Iizuka and W.N. Tzeng. 2004. Migratory history of the grey mullet *Mugil cephalus* as revealed by otolith Sr:Ca ratios. Mar. Ecol. Progr. Ser. 269: 277–288.

Chaudhuri, H., J.V. Juario, J.H. Primavera, R. Mateo, R. Samson, E. Cruz, E. Jarabejo and J. Canto, Jr. 1977. Artificial fertilization of eggs and early development of the milkfish *Chanos chanos* (Forskal). pp. 21–38. *In*: Induced spawning, artificial fertilization of eggs and larval rearing of the milkfish *Chanos chanos* (Forskal) in the Philippines (Technical Report No. 3), Tigbauan, Iloilo, Philippines: Aquaculture Department, Southeast, Asian Fisheries Development Center.

Crosetti, D. 2015. Current state of grey mullet fisheries and culture. *In*: Crosetti, D. and S.J.M. Blaber (eds.). Biology, Ecology and Culture of Grey Mullets (Mugilidae). CRC Press, Boca Raton, USA (this book).

Da Silva, R.M.P. and M.L.P. Esper. 1991. Observaçoes sobre o desenvolvimento citomorfológico dos ovarios de tainha *Mugil platanus* (Günther) da Baía de Paranagua (Brasil). Acta Biol. Par. Curitiba 20: 15–39.

De Sylva, D.P. 1984. Mugiloidei: development and relationships. pp. 530–533. *In*: H.G. Moser, W.J. Richards, D.M. Cohen, M.P. Fahay, A.W. Kendall, Jr. and S.L. Richardson (eds.). Ontogeny and Systematics of Fishes. Am. Soc. Ichthyol. Herp., Spec. Publ. No. 1.

Dekhnik, T.V. 1954. Reproduction of the anchovy and mullet in the Black Sea. Tr. Vses. Nauchnolssled. Inst. Morsk. Rybn. Khor. Okeanogr. 28: 34–48 (In Russian).

Demir, N. 1971. On the occurrence of grey mullet postlarvae off Plymouth. J. Mar. Biol. Assoc. UK 51: 235–246.

Devlin, R.H. and Y. Nagahama. 2002. Sex determination and sex differentiation in fish: an overview of genetic, physiological, and environmental influences. Aquaculture 208: 191–364.

Ditty, G., T. Farooqi and R.F. Shaw. 2006. Mugilidae: Mullets. pp. 891–899. *In*: W. Richards (ed.). Early Stages of Atlantic Fishes. An Introduction Guide for the Western Central North Atlantic, Vol. 1. Taylor and Francis, CRC Press, Boca Raton, Florida.

Dunn, J.R. 1984. Developmental osteology. pp. 48–50. *In*: H.G. Moser, W.J. Richards, D.M. Cohen, M.P. Fahay, A.W. Kendall, Jr. and S.L. Richardson (eds.). Ontogeny and Systematics of Fishes. Am. Soc. Ichthyol. Herp., Spec. Publ. No. 1.

El-Halfawy, M.M., A.M. Ramadan and W.F. Mahmoud. 2007. Reproductive biology and histological studies of the grey mullet, *Liza ramada* (Risso 1826), in lake Timsah, suez Canal. Egypt. J. Aquat. Res. 33: 434–545.

El-Gharabawy, M.M. and S.S. Assem. 2006. Spawning induction in the Mediterranean grey mullet *Mugil cephalus* and larval developmental stages. Afr. J. Biotechnol. 5: 1836–1845.

Eljaiek, P.E. and R.D. Vesga. 2011. Reproducción de *Joturus pichardi* y *Agonostomus monticola* (Mugiliformes: Mugilidae) en ríos de la Sierra Nevada de Santa Marta, Colombia. Rev. Biol. Trop. Int. J. Trop. Biol. 59: 1717–1728.

Ergene, S. 2000. Reproduction characteristics of thinlip grey mullet, *Liza ramada* (Risso 1826) inhabiting Akgöl-Paradentz Lagoons (Göksü Delta). Turk. J. Zool. 24: 159–164.

Erman, F. 1961. On the biology of the thicklipped grey mullet (*Mugil chelo* Cuv.). Rap. P.-V. Comm. Int. Explor. Sci. Mer. Mediterr. 16: 277–85.

Esper, M.L.P., M.S. de Menezes and W. Esper. 2000. Escala de desnvolvimento gonadal e tamanho de primeira maduraçao de femeas de *Mugil platanus* Günther 1880 da Baía de Paranagua, Paraná, Brasil. Acta Biol. Par., Curitiba 29: 255–263.

Ezzat, A. 1965. Contribution à l'étude de la biologie de quelques Mugilidae de la région de l'étang de Berre et de Port de Bouc. Thesis, Marseille.

Fahay, M.P. 2007. Early Stages of Fishes in the Western North Atlantic Ocean (Davis Strait, Southern Greenland and Flemish Cap to Cape Hatteras). Volume Two. Scorpaeniformes through Tetraodontiformes. Northwest Atlantic Fisheries Organization, Dartmouth, NS, Canada.

Fazli, H., A.A. Janbaz, H. Taleshian and F. Bagherzadeh. 2008. Maturity and fecundity of golden grey mullet (*Liza aurata* Risso 1810) in Iranian waters of the Caspian Sea. J. Appl. Ichthyol. 24: 610–613.

Fraga, E., H. Schneider, M. Nirchio, E. Santa-Brigida, L.f. Rodrigues-Filho and I. Sampaio. 2007. Molecular phylogenetic analyses of mullets (Mugilidae: Mugiliformes) based on two mitochondrial genes. J. Appl. Ichthyol. 23: 598–604.

Fukuhara, O. 1990. Effects of temperature on yolk utilization, initial growth, and behavior of unfed marine fish-larvae. Mar. Biol. 106: 169–174.

Garbin, T., J.P. Castello and P.G. Kinas. 2013. Age, growth, and mortality of the mullet *Mugil liza* in Brazil's southern and southeastern coastal regions. Fish Res. 149: 61–68.

Ghaninejad, D., S. Abdolmalaki and Z.M. Kuliyev. 2010. Reproductive biology of the golden grey mullet, *Liza aurata* in the Iranian coastal waters of the Caspian Sea. Iran. J. Fish Sci. 9: 402–411.

González-Castro, M. 2007. Los peces representantes de la familia Mugilidae en Argentina. Ph.D. thesis. Universidad Nacional de Mar del Plata, Argentina.

González-Castro, M., J.M. Díaz de Astarloa, M.B. Cousseau, D.E. Figueroa, M.S. Delpiani, D. Bruno, J.M. Guzonni, G.E. Blasina and M.Y. Deli Antoni. 2009a. Fish composition in a south-western Atlantic temperate coastal lagoon: spatial–temporal variation and relationships with environmental variables. J. Mar. Biol. Assoc. UK 89: 593–604.

González-Castro, M., V. Abachian and R.G. Perrotta. 2009b. Age and growth of the striped mullet *Mugil platanus* (Actinopterygii: Mugilidae), in a southwestern Atlantic coastal lagoon (37°32's–57°19'w): a proposal for a life-history model. J. Appl. Ichthyol. 25: 61–66.

González-Castro, M., G. Macchi and M.B. Cousseau. 2011. Studies on reproduction of the mullet *Mugil platanus* Günther 1880 (Actinopterygii: Mugilidae) from the Mar Chiquita coastal lagoon, Argentina: similarities and differences with related species. Ital. J. Zool. 78: 343–353.

González-Castro, M., A.L. Ibáñez, S. Heras, M.I. Roldán and M.B. Cousseau. 2012. Assessment of lineal versus landmarks—based morphometry for discriminating species of Mugilidae (Actinopterygii). Zoolog. Stud. 51: 1515–1528.

Goos, H.J. 1987. Steroid feedback on pituitary gonadotropin secretion. pp. 16–20. *In*: D.R. Idler, L.W. Crim and J. Walsh (eds.). Proceedings of the Third International Symposium on Reproductive Physiology of Fish. St. John's University, St. John, NS, Canada.

Govoni, J.J., G.W. Boehlert and Y. Watanabe. 1986. The physiology of digestion in fish larvae. Environ. Biol. Fish 16: 59–77.

Greeley, M.S., Jr., D.R. Calder and R.A. Wallace. 1986. Changes in teleost yolk proteins during oocytes maturation: Correlation of yolk proteolysis with oocytes hydration. Comp. Biochem. Physiol. 84B: 1–9.

Greeley, M.S., Jr., D.R. Calder and R.A. Wallace. 1987. Oocyte growth and development in the striped mullet *Mugil cephalus* during seasonal ovarian recrudescence: Relationship to fecundity and size at maturity. Fishery Bulletin 85: 187–200.

Grier, H.J. 1993. Comparative organization of Sertoli cells including the Sertoli cell barrier. pp. 703–739. *In*: L.D. Russell and M.D. Griswold (eds.). The Sertoli Cell. Cache River Press, Clearwater.

Helfman, G.S., B.B. Collette, D.E. Facey and B.W. Bowen. 2009. The Diversity of Fishes. Biology, Evolution, and Ecology. Willey-Blackwell, West Sussex.

Heras, S., M. Roldán and M. González-Castro. 2009. Molecular phylogeny of Mugilidae fishes revised. Rev. Fish Biol. Fish 19: 217–231.

Hotos, G.N., D. Avramidou and I. Ondrias. 2000. Reproduction biology of *Liza aurata* (Risso 1810), (Pisces: Mugilidae) in the lagoon of Klisova (Messolonghi, W. Greece). Fisheries Research 47: 57–67.

Houde, E.D. and R.C. Schekter. 1980. Feeding by marine fish larvae: development and functional responses. Environ. Biol. Fish 5: 315–334.

Hsu, C.C., Y.S. Han and W.N. Tzeng. 2007. Evidence of flathead mullet *Mugil cephalus* L. spawning in waters Northeast of Taiwan. Zool. Stud. 46: 717–725.

Hubbs, C.L. 1943. Terminology of early stages of fishes. Copeia 4: 260.

Hunter, J.R. and B.J. Macewicz. 1985. Rates of atresia in the ovary of captive and wild Northern anchovy, *Engraulis mordax*. Fish Bull. 83: 119–136.

Hunter, J.R., B.J. Maciewicz, N.C.H. Lo and C.A. Krimbell. 1992. Fecundity, spawning and maturity of female Dover sole, *Microstomus pacificus*, with an evaluation of assumption and precision. Fish Bull. 90: 101–128.

Ibáñez-Aguirre, A.L. and M. Gallardo-Cabello. 2004. Reproduction of *Mugil cephalus* and *M. curema* (Pisces: Mugilidae) from a coastal lagoon in the Gulf of Mexico. Bull. Mar. Sci. 75: 37–49.

Jakobsen, T., M.J. Fogarty, B.A. Megrey and E. Moksness. 2009. Fish Reproductive Biology. Implications for Assessment and Management. Blackwell Publishing Ltd., Malaysia.

Johns, D.M. and W.H. Howell. 1980. Yolk utilization in summer flounder (*Paralichthys dentatus*) embryos and larvae reared at two temperatures. Mar. Ecol. Prog. Ser. 2: 1–8.

Kah, O., C. Lethimonier, G. Somoza, L.G. Guilgur, C. Vaillant and J.J. Lareyre. 2007. GnRH and GnRH receptors in metazoa: a historical, comparative, and evolutive perspective. Gen. Comp. Endocrinol. 153: 346–364.

Kendall, B.W. and C.A. Gray. 2008. Reproductive biology of two co-occurring mugilids, *Liza argentea* and *Myxus elongatus*, in south-eastern Australia. J. Fish Biol. 73: 963–979.

Kesteven, G.L. 1942. Studies in the biology of Australian mullet. I. Account of the fishery and preliminary statement of the biology of *Mugil dobula* Günther. Bull. Aust. CSIRO Melb. 157: 1–99.

Khan, I.A. and P. Thomas. 1999. Ovarian cycle, teleost fish. *In*: E. Knobil and J.D. Neill (eds.). Encyclopedia of Reproduction, Vol. 3. Academic Press, San Diego.

Khoo, K.H. 1979. The histochemistry and endocrine control of vitellogenesis in goldfish ovaries. Can J. Zool. 57: 617–626.

Kim, S.J., Y.D. Lee, I.K. Yeo, H.J. Baek, H.B. Kim, M. Nagae, K. Soyano and A. Hara. 2004. Reproductive cycle of the female grey mullet, *Mugil cephalus*, on the coast of Jeju Island, Korea. J. Environ. Toxicol. 19: 73–80.

Kuo, C.M., Z.H. Shehadeh and K.K. Milisen. 1973a. A preliminary report on the development, growth and survival of laboratory reared larvae of the grey mullet (*Mugil cephalus* L.). J. Fish Biol. 5: 459–470.

Kuo, C.M., Z.H. Shehadeh and C.E. Nash. 1973b. Induced spawning of captive grey mullet (*Mugil cephalus* L.) females by injection of human chorionic gonadotropin (HCG). Aquaculture 1: 429–432.

Kuo, C.M., C.E. Nash and Z.H. Shehadeh. 1974. The effects of temperature and photoperiod on ovarian development in captive grey mullet (*Mugil cephalus* L.). Aquaculture 3: 25–43.

Kvenseth, A.M., K. Pittman and J.V. Helvik. 1996. Eye development in Atlantic halibut (*Hippoglossus hippoglossus*): differentiation and development of the retina from early yolk sac stage through metamorphosis. Can. J. Fish Aquat. Sci. 53: 2524–2532.

Leis, I.C. and T. Trnski. 1989. The Larvae of Indo-Pacific Shorefishes. New South Wales University Press, Australia.

Lemos, V.M., A.S. Varela, Jr., P.R. Schiwingel, J.H. Muelbert and J.P. Vieira. 2014. Migration and reproductive biology of *Mugil liza* (Teleostei: Mugilidae) in south Brazil. J. Fish Biol. 85: 671–687.

Levavi-Sivan, B., J. Biran and E. Fireman. 2006. Sex steroids are involved in the regulation of gonadotropin-releasing hormone and dopamine D2 receptors in female tilapia pituitary. Biol. Reprod. 75: 642–650.

Li, Y.H., C.C. Wu and J.S. Yang. 2000. Comparative ultrastructural studies of the *Zona radiata* of marine fish eggs in three genera in Perciformes. J. Fish Biol. 56: 615–621.

Liao, I.C. 1975. Experiments on the induced breeding of the grey mullet in Taiwan from 1963–1973. Aquaculture 6: 31–58.

Liao, I.C. 1997. Larviculture of finfish and shellfish in Taiwan. J. Fish Soc. Taiwan 23: 349–369.

Liao, I.C., Y.J. Lu, T.L. Huang and M.C. Lin. 1971. Experiments on induced breeding of the grey mullet, *Mugil cephalus* Linnaeus. Aquaculture 1: 15–34.

Liao, I.C., C.S. Cheng, L.C. Tseng, M.Y. Lim, L.S. Hsieh and H.P. Chen. 1972. Preliminary report on the mass propagation of grey mullet, *Mugil cephalus* Linnaeus. Fishery Ser. Chin.-Am. jt Comm. Rur. Reconstr. 12: 1–14.

Ma, H., C. Cahu, J. Zambonino, H. Yu, Q. Duan, M.M. Le Gall and K. Mai. 2005. Activities of selected digestive enzymes during larval development of large yellow croaker (*Pseudosciaena crocea*). Aquaculture 245: 239–248.

MacMillan, D.B. 2007. Fish Histology. Female Reproductive Systems. Springer, The Netherlands.

Mai, A.C.G., C.I. Miño, L.F.F. Marins, C. Monteiro-Neto, L. Miranda, P.R. Schwingel, V.M. Lemos, M. González-Castro, J.P. Castello and J.P. Vieira. 2014. Microsatellite variation and genetic structuring in *Mugil liza* (Teleostei: Mugilidae) populations from Argentina and Brazil. Estuar. Coast. Shelf S. 149: 80–86.

Marin, E.B. and J.J. Dodson. 2000. Age, growth and fecundity of the silver mullet, *Mugil curema* (Pisces: Mugilidae), in coastal areas of Northeastern Venezuela. Rev. Biol. Trop. 48: 389–398.

Marin, E.B.J., A. Quintero, D. Bussiere and J.J. Dodson. 2003. Reproduction and recruitment of white mullet (*Mugil curema*) to a tropical lagoon (Margarita Island, Venezuela) as revealed by otolith microstructure. Fish Bull. 101: 809–821.

Martin, F.D. and G.E. Drewry. 1978. Development of fishes of the mid-Atlantic bight. An atlas of eggs, larvae and juvenile stages. Vol. VI. Stromatidae through Ogcocephalidae. Biological Services Program, Fish and Wildlife Service, US Department of Interior.

McDonough, C.J., W.A. Roumillat and C. Wenner. 2003. Fecundity and spawning season of striped mullet (*Mugil cephalus* L.) in South Carolina estuaries. Fish Bull. 101: 822–834.

McDonough, C.J., W.A. Roumillat and C. Wenner. 2005. Sexual differentiation and gonad development in striped mullet (*Mugil cephalus*) from South Carolina estuaries. Fish Bull. 103: 601–619.

MedSudMed, 2011. Identification sheets of early life stages of bony fish (Western Libya, Summer 2006). GCP/RER/ITA/ MSM-TD-18. MedSudMed Technical Documents No 18: 251pp.

Miller, B.S. and A.W. Kendal, Jr. 2009. Early life history of marine fishes. University of California Press, California.

Minos, G., L. Kokokiris and P.S. Economidis. 2010. Sexual maturity of the alien redlip mullet, *Liza haematocheilus* (Temminck and Schlegel 1845) in north Aegean Sea (Greece). J. Appl. Ichthyol. 26: 96–101.

Miranda, L.V., J.T. Mendonça and M.C. Cergole. 2006. Diagnóstico do estoque e orientacões para o ordenamento da pesca de *Mugil platanus*. pp. 38–48. *In*: C.L.D.B.R. Wongtschowski, A.O. Ávila-da-Silva and M.C. Cergole (eds.). Análise

das principais pescarias comerciais da região Sudeste-Sul do Brasil: dinâmica populacional das espécies em explotacão II. São Paulo.
Mousa, M.A. 2010. Induced spawning and embryonic development of *Liza ramada* reared in freshwater ponds. Anim. Reprod. Sci. 119: 115–122.
Nakamura, M. 1978. Morphological and experimental studies on sex differentiation of the gonad in several teleost fishes. Ph.D. Thesis, Hokkaido University, Hokkaido.
Nash, C.E. and C.M. Kuo. 1975. Hypotheses for problems impeding the mass propagation of grey mullet and other finfish. Aquaculture 6: 119–134.
Nash, C.E. and Z.H. Shehadeh. 1980. Review of the breeding and propagation techniques for grey mullet, *Mugil cephalus* L. ICLARM Studies and Reviews 3, International Center for Living Aquatic Resources Management, Manila, Philippines. 87pp.
Nash, C.E. and R.M. Koningsberger. 1981. Artificial propagation. pp. 265–312. *In*: O.H. Oren (ed.). Aquaculture of Grey Mullets. IBP 26 Cambridge University Press, Cambridge, UK.
Nash, C.E., C.M. Kuo and S.C. McConnell. 1974. Operational procedures for rearing larvae of the grey mullet (*Mugil cephalus* L.). Aquaculture 3: 15–24.
Nash, C.E., C.M. Kuo, W.D. Madden and C.L. Paulsen. 1977. Swim bladder inflation and survival of *Mugil cephalus* to 50 days. Aquaculture 12: 89–94.
Okumus, I. and N. Bascinar. 1997. Population structure, growth and reproduction of introduced Pacific mullet, *Mugil so-iuy,* in the Black Sea. Fish Res. 33: 131–137.
Oliveira, M.F.D., E.F. Dos Santos Costa, F.A. De Morais Freire, J.E. Lins De Oliveira and A.C. Luchiari. 2011. Some aspects of the biology of white mullet, *Mugil curema* (Osteichthyes, Mugilidae), in the northeastern region, Brazil. Pan-Americ. J. Aquat. Sci. 6: 138–147.
Oozeki, Y. and K.M. Bailey. 1995. Ontogenetic development of digestive enzyme activities in larval walleye pollock, *Theragra chalcogramma.* Mar. Biol. 122: 177–186.
Osse, J.W.M., J.G.M. Van den Boogaart, G.M.J. Van Snick and L. Van der Sluys. 1997. Priorities during early growth of fish larvae. Aquaculture 155: 249–258.
Pandian, T.J. 2012. Sex Determination in Fish. CRC Press, Boca Raton, Florida.
Parenti, L.R. and H.J. Grier. 2004. Evolution and phylogeny of gonad morphology in bony fishes. Integr. Comp. Biol. 44: 333–348.
Patimar, R. 2008. Some biological aspects of the sharpnose mullet *Liza saliens* (Risso 1810) in Gorgan Bay-Miankaleh Wildlife Refuge (the Southeast Caspian Sea). Tur. J. of Fish Aquat. Sci. 8: 225–232.
Patiño, R. and C.V. Sullivan. 2002. Ovarian follicle growth, maturation, and ovulation in teleost fish. Fish Physiol. Biochem. 26: 57–70.
Perceva-Ostroumova, T.A. 1951. Reproduction and development of grey mullets introduced in the Caspian Sea. Tr. Vses. Nauchno-lssled. Inst. Morsk. Rybn. Khoz. Okeanogr. 18: 123–134.
Rè, P. and I. Meneses. 2008. Early stages of marine fishes occurring in the Iberian Peninsula. IPIMAR/IMAR. 282pp. ISBN-978-972-9372-34-6.
Rheman, S., M.L. Islam, M.M.R. Shah, S. Mondal and M.J. Alam. 2002. Observation on the fecundity and gonadosomatic index (GSI) of grey mullet *Liza parsia* (Ham.). online J. Biol. Sci. 2: 690–693.
Roa, R., B. Ernst and F. Tapia. 1999. Estimation of size at sexual maturity: an evaluation of analytical and resampling procedures. Fish Bull. 97: 570–580.
Romagosa, E., E.F. Andrade-Talmelli, M.Y. Narahara and H.M. Godinho. 2000. Desova e fecundidade da tainha *Mugil platanus* (Teleostei: Mugilidae) na regiao estuarino-lagunar de Cananéia, Sao Paulo, Brasil (25° 01' S; 47° 57' W). Atlàntica, Río Grande 22: 5–12.
Roule, L. 1917. Sur le développement larvaire et postlarvaire des poissons du genre *Mugil*. C.R. Acad. Sci. Paris 164: 194–196.
Saborido-Rey, F. 2002. Ecología de la reproducción y potencial reproductivo en las poblaciones de peces marinos. Curso Doutoramento do bienio 2002–2004. Universidad de Vigo, España. 71pp.
Şahinöz, E., Z. Doğu, F. Aral, R. Şevik and H.H. Atar. 2011. Reproductive characteristics of mullet (*Liza abu* H., 1843) (Pisces: Mugilidae) in the Atatürk Dam Lake, Southeastern Turkey. Turk. J. Fish Aquat. Sci. 11: 07–13.
Saila, S.B., C.W. Recksiek and M.H. Prager. 1988. Fishery Science Application System. A Compendium of Microcomputer Programs and Manual of Operation. Elsevier, New York.
Sanzo, L. 1930. Uova e larve di *Mugil cephalus* Cuv., ottenute per fecondazione artificiale (nota preventiva). Mem. R. Corn. Talassogr. Ital. 179: 1–5.
Sanzo, L. 1936. Contributi alla conoscenza dello sviluppo embrionale e post embrionale nei mugilidi. I. Uova e larve di *Mugil cephalus Cuv.* ottenute per fecondazione artificiale. II. Uova e larve di *Mugil chelo* Cuv. Mem. R. Corn. Talassogr. Ital. 230: 1–11.
Sanzo, L. 1937. Uova e larve di *Mugil labeo* Cuv. Boll. Pesca Piscic. Idrobiol. 13: 506–510.
Schulz, R. and H.J. Goos. 1999. Puberty in male fish: Concepts and recent developments with special reference to the African catfish (*Clarias gariepinus*). Aquaculture 177: 5–12.
Sebastian, M.J. and V.A. Nair. 1975. The induced spawning of the grey mullet, *Mugil macrolepis* (Aguas) SMITH and the large-scale rearing of its larvae. Aquaculture 5: 41–52.
Selman, K. and R.A. Wallace. 1989. Cellular aspects of oocyte growth in Teleosts. Zoologic. Sci. 6: 211–231.

Shelbourne, J.E. 1964. The artificial propagation of marine fish Adv. Mar. Biol. 2: 1–83.
Silva, E.I.L. and S.S. De Silva. 1981. Aspects of the biology of grey mullet, *Mugil cephalus* L., adult populations of a coastal lagoon in Sri Lanka. J. Fish Biol. 19: 1–10.
Stenger, A.H. 1959. A study of the structure and development of certain reproductive tissues of *Mugil cephalus* Linnaeus. Zoologica 44: 53–70.
Su, W.C. and T. Kawasaki. 1995. Characteristics of the life history of grey mullet from Taiwanese Waters. Fish Sci. 61: 377–381.
Tamaru, C.S., C.S. Lee, C.D. Kelley, G. Miyamoto and A. Moriwake. 1994. Oocyte growth in the striped mullet *Mugil cephalus* L. maturing at different salinities. J. World Aquacult. Soc. 25: 109–115.
Tang, Y.A. 1964. Induced spawning of striped mullet by hormone injection. Jpn. J. Ichthyol. 12: 23–28.
Thomson, J.M. 1963. Synopsis of biological data on the grey mullet (*Mugil cephalus* Linnaeus 1758). Fisheries Synopsis No. 1. Division of Fisheries and Oceanography, CSIRO. Sydney, Australia.
Tucker, J.W., Jr. 1998. Marine Fish Culture. Kluwer, Boston.
Tung, I.H. 1973. On the egg development and larval stages of the grey mullet (*Mugil cephalus* Linnaeus). Rep. Inst. Fish Biol. Minist. Econ. Aff. Nat. Taiwan Univ. 3: 187–215.
Tyler, C.R. and J.P. Sumpter. 1996. Oocyte growth and development in teleosts. Rev. Fish Biol. Fish 6: 287–318.
Ünlü, E., K. Balci and N. Meriç. 2000. Aspects of the biology of *Liza abu* (Mugilidae) in the Tigris River (Turkey). Cybium 24: 27–43.
Van Der Kraak, G., J.P. Chang and D.M. Janz. 1998. Reproduction. pp. 465–488. *In*: D.H. Evans (ed.). The Physiology of Fishes. CRC Press, Boca Raton, Florida.
Van Snik, G.M.J., J.G.M. Van den Boogaart and J.W.M. Osse. 1997. Larval growth patterns in *Cyprinus carpio* and *Clarias gariepinus* with attention to finfold. J. Fish Biol. 50: 1339–1352.
Vazzoler, A.E. 1996. Biología da reproducao de peixes Teleósteos: teoria e prática. EDUEM; Sao Paulo.
Vieira, J.P. 1991. Juvenile mullets (Pisces: Mugilidae) in the Estuary of Lagoa dos Patos, RS, Brazil. Copeia 2: 409–418.
Vieira, J.P. and C. Scalabrin. 1991. Migraçao reprodutiva da "tainha" (*Mugil platanus* Günther 1880) no sul do Brasil. Atlântica, Río Grande 13: 131–141.
Vieira, J.P., A.M. García and A.M. Grimm. 2008. Evidences of El Niño effects on the mullet fishery of the Patos Lagoon Estuary. Bras. Arch. Biol. Technol. 51: 433–440.
Vodyanitskii, V.A. and I.I. Kazanova. 1954. The identification of pelagic eggs and larvae of Black Sea fishes. Tr. Vses. Nauchno-lssled. Inst. Morsk. Rybn. Khoz. Okeanogr. 28: 240–325.
Walford, J. and T.J. Lam. 1993. Development of digestive tract and proteolytic enzyme activity in seabass (*Lates calcarifer*) larvae and juveniles. Aquaculture 109: 187–205.
Wallace, R.A. and K. Selman. 1981. Cellular and dynamic aspects of oocyte growth in teleosts. Am. Zool. 21: 325–343.
Wallace, R.A. and K. Selman. 1985. Major protein changes during vitellogenesis and maturation of *Fundulus* oocytes. Dev. Biol. 110: 492–498.
Webb, B.F. 1973. Fish populations on the Avon-Heathcote estuary. II. Breeding and gonad maturity. New Zeal. J. Mar. Fresh. Res. 7: 45–66.
Wijeyaratne, M.J.S. and H.H. Costa. 1988. The food, fecundity and gonadal maturity of *Valamugil cunnesius* (Pisces: Mugilidae) in the Negombo lagoon, Sri Lanka. Indian J. Fish 35: 71–77.
Yang, W.T. and U.B. Kim. 1962. A preliminary report on the artificial culture of grey mullet in Korea. Indo-Paci. J. Fish Coun. 9: 62–70.
Yaron, Z., G. Gur, P. Melamed, H. Rosenfeld, A. Elizur and B. Levavi-Sivan. 2003. Regulation of fish gonadotropins. Int. Rev. Cytol. 225: 131–185.
Yashouv, A. 1969. Preliminary report on induced spawning of *M. cephalus* (L.) reared in captivity in freshwater ponds. Bamidgeh 21: 19–24.
Yashouv, A. and E. Berner-Samsonov. 1970. Contribution to the knowledge of eggs and early larval stages of mullets (Mugilidae) along the Israeli coast. Bamidgeh 22: 72–89.
Yoshimatsu, T., S. Matsui and C. Kitajima. 1992. Early development of laboratory-reared redlip mullet, *Liza haematocheila*. Aquaculture 105: 379–390.
Yousefian, M., M. Ghanei, R. Pourgolam and H.K.H. Rostami. 2009. Gonad development and hormonal induction in artificial propagation of grey mullet, *Mugil cephalus* L. Res. J. Fish Hydrobiol. 4: 35–40.
Zambonino-Infante, J.L. and C. Cahu. 1994. Development and response to a diet change of some digestive enzymes in sea bass (*Dicentrarchus labrax*) larvae. Fish Physiol. Biochem. 12: 399–408.
Zambonino-Infante, J.L. and C. Cahu. 2001. Ontogeny of gastrointestinal tract of marine fish larvae. Comp. Biochem. Physiol. C 130: 477–487.
Zouiten, D., I. Ben Khemis, R. Besbes and C. Cahu. 2008. Ontogeny of the digestive tract of thick lipped grey mullet (*Chelon labrosus*) larvae reared in «mesocosms». Aquaculture 279: 166–172.

CHAPTER 12

Biology and Ecology of Fry and Juveniles of Mugilidae

Emmanuil Koutrakis

Introduction

Mugilidae is a widely distributed family and its species occur mainly in coastal marine and brackish waters in all tropical and temperate seas (Thomson 1966). They are usually euryhaline fish and are more commonly found in coastal waters and lagoons than in freshwaters; some of them seem capable of spending a great part of their life in fresh or brackish water, others only for a few months each year (Quignard and Farrugio 1981), while juvenile mullets have been found to tolerate up to 78 psu for short periods (Trape et al. 2009). According to Blaber (1987), grey mullets maximize their tenement in estuaries and freshwater in order to take advantage of the rich resources and the shelter from predators provided therein. As Blaber (1987) noted, these two advantages are the reason why the marine phase is short and is mostly related to the larval stage and spawning period. The high abundance of mullets globally, and the high biomass that these species can attain are due to the aforementioned characteristics along with their foraging at the first level of the food web (Whitfield et al. 2012). At the same time, they are also a significant food source for higher trophic level piscivores (McDonough and Wenner 2003).

Their fishery is popular in many parts of the world and usually supports artisanal fisheries, but they have also been farmed for centuries. They are caught with specialized gears, such as fixed trap systems or special nets (see Chapter 16—Crosetti 2015); they have been consumed for centuries, have considerable commercial value and, together with mullet fry and products, contribute to the fishery economies of many countries (Koutrakis 1999). Indeed, the demand for the most valuable part of grey mullet, the ovaries of the adult female, has grown considerably in recent decades and elevated the status of grey mullet (Gautier and Hussenot 2005, Hung and Shaw 2006, Whitfield et al. 2012). The ovaries are processed with salt to make caviar-like Italian '*bottarga*', French '*poutargue*' or Greek '*avgotaraxo*' (produced mainly from the flathead grey mullet *Mugil cephalus*), while in Taiwan and Japan the testes of the male grey mullet and the stomach are mouth-watering ingredients for seafood cuisine, hence, the grey mullet is also referred to as 'Grey Gold' (Hung and Shaw 2006).

The first reference to grey mullets worldwide comes from the Mediterranean, since they were fairly well known to the Ancient Greeks and to the Romans and served as an important nutritional resource in Mediterranean Europe (Thompson 1947, Nash 1978). More than 2100 years before 1758, when Linneaus

Fisheries Research Institute, Hellenic Agricultural Organization, 640 07 Nea Peramos, Kavala, Greece.
Email: manosk@inale.gr

described *Mugil cephalus* in his 'Systema Naturae', Aristotle had described the life history of different species of grey mullets in his work 'The history of Animals' (second half of the 4th c. BC). He used the name '*kefalos*' which means 'big head' for *Mugil cephalus*, '*kestrefs*' for *Liza ramada*, '*myxinos*' for *L. aurata* and '*chelon*', focusing on the big lips, for *Chelon labrosus* (Thompson 1947, Koutrakis 1999). He also noted that older fish had bigger scales, different spawning seasons and that they migrate to rivers and lakes to feed. He described their fishing methods and he characterized them as 'noble' food, because they did not feed on flesh of other fish (Koutrakis 1999).

The Mugilidae family taxonomy and nomenclature has not yet concluded; according to the most recent revisions, 14 to 20 genera are recognized as valid (Durand et al. 2012a,b; see Chapter 2—Durand 2015). During the last 130 years, up to 281 nominal species and 45 genera have been proposed for inclusion in the Mugilidae (Thomson 1997, Durand et al. 2012a,b). Many of these have been recognized as synonyms in regional reviews over the past 120 years (Thomson 1997). Eschmeyer and Fong (2014) list 20 valid genera and 71 species, with two genera, *Liza* and *Mugil*, representing 40% of the species within the family. The most common species of the family is the striped or flathead mullet *Mugil cephalus* L., a cosmopolitan species, occurring in tropical, subtropical and temperate coastal waters in all the world's major oceans (Thomson 1966), occupying a wide variety of freshwater, estuarine, marine and also hyperhaline environments (Whitfield et al. 2012). For its worldwide distribution, the flathead mullet shows a great variability, which has recently also been attributed to the presence of several cryptic within a *M. cephalus* species complex (Durand et al. 2012a,b, Durand 2015). In contrast, there are species that are highly endemic, such as the robust mullet *Valamugil robustus* and the St. Lucia mullet *Liza luciae*, that occur in and adjacent to only a few estuaries of Natal and Mozambique (Blaber 1987), others that are stenohaline and live only in the sea, such as the boxlip mullet *Oedalechilus labeo* (Nelson 2006) and about 10 mugilid species that mainly inhabit freshwaters (Blaber 1987).

The first stages of young grey mullet, according to their development stage, are usually called post-larvae, fry or juveniles. Young fish are also called fingerlings, but this is used more by aquarists or in aquaculture. In this chapter, all young mullets over 30 mm (Total Length, TL), which corresponds to ~ 25 mm Standard Length (SL) are considered juveniles; the juvenile period begins when the fins are fully differentiated and most of the temporary organs are replaced by definitive organs (Balon 1975) (see also Chapter 10—Ibáñez 2015). Young fish smaller than 30 mm, which undergo the larval period (Balon 1975), are considered post-larvae (Bensam 1989) or fry. Fish exceeding 100 mm (SL) are considered as sub-adults for all species.

Especially with regard to fry, most studies that deal with the juvenile stages of Mugilidae have been carried out in countries where grey mullets have a commercial interest for fishing, such as in the western Atlantic in the U.S.A.: in North Carolina the striped mullet have been fished since 1880 and was the most abundant and important saltwater fish (NCDENR 2005); in the western South Atlantic (e.g., Brazil, Vieira 1991, Chaves et al. 2002; Mexico, Cabral-Solís et al. 2010); in South Africa (Wallace and van der Elst 1975, Payne 1976); Sri Lanka (Wijeyaratne and Costa 1987), or Australia (Thomson 1963, 1966) and New Zealand, where grey mullet provided an important food resource for pre-European Mäori (Paulin and Paoul 2006); but mainly for stocking inland waters and aquaculture ponds, since the culture methods are still of the extensive pattern, based on the capture of fry in the coastal areas (IUCN 2007, Mićković et al. 2010). In other parts of the world however, catches are exclusively used as bait, as in South Africa (Wallace and van der Elst 1975) and Hawaii, East Pacific (Nishimoto et al. 2007).

Stocking of juveniles is mainly practiced in the Mediterranean region and in South East Asia (China, Taiwan, Liao 1981, Hung and Shaw 2006; India, Thomson 1966, Rajyalakshmi and Chandra 1987; the Philippines, Thomson 1966) where collection of fry for farming is practiced traditionally, usually by beach seines in coastal waters, a tradition introduced in Taiwan by Chinese fish farmers who immigrated there during the 16th century (Tang 1975). In Hong Kong and in Bengal (India) mullet fry are reared together with Asiatic carps (Thomson 1966). In the Mediterranean, there are references for successful stocking in Egypt, which date back to 1920 (e.g., *M. cephalus* and *L. ramada* stocking Lake Mariut in Egypt, El-Zarka and Kamel 1965; Lake Quarun, Ishak et al. 1982) and according to Saleh (2008) stocking had been traditionally practiced in the 'hosha' system in the Nile Delta region in Egypt for centuries. Collection of fry for farming is practiced traditionally also in other Mediterranean countries, such as in Italy that has a long tradition and especially in the Venice lagoon, where it has been regulated since the time of the 'Venice Republic' ('*Vallicoltura*': Rossi 1981, Gandolfi et al. 1981, Torricelli et al. 1982), in Tunisia (Farrugio 1975, 1977, Chauvet 1984), in Israel (e.g., Lake Kinneret, Bar-Ilan 1975, Chervinsky 1982; Lake Tiberias, Pruginin et al. 1975) and in Greece (Kladas et al. 1989; Lake Volvi: Koutrakis 1994, Bobori et al. 1998). *Mugil cephalus* juveniles are also used for stock-enhancement with releases of cultured fry in some parts of the world, such as Hawaii (Kraul 1983, Leber et al. 1996, 1997, Nishimoto et al. 2007), where this species also has a long history. Hawaiian words depict the different development stages, from fingerlings to juveniles and the species was revered by Hawaiian royalty as a food fish harvested from fishponds (artificially entrapped) and later as a valuable commercial product in the early 1900s (Nishimoto et al. 2007). It is important to note that IUCN (2007) suggests that although the removal of grey mullet individuals from wild stocks for the purpose of stocking aquaculture farms does not appear to have a negative impact on them, it should, nevertheless, be performed in a sustainable way, since as noted by Perez-Ruzafa and Marcos (2012) the effectiveness of stock enhancement after such practices remains uncertain in all cases.

Research has not evenly covered all species within the family. Most of the research on Mugilidae fry and juveniles is targeted on the most common or commercial species, such as the flathead mullet *M. cephalus* worldwide, the thinlip grey mullet *Liza ramada*, the golden grey mullet *L. aurata*, the leaping mullet *L. saliens* and the thicklip grey mullet *Chelon labrosus* in the Mediterranean and in the eastern Atlantic. Research on other species is only occasional or included in more generic studies regarding the fish fauna of a particular area.

This chapter is dedicated to the biology and ecology of fry and juveniles of grey mullets around the world. In order to do so, the migration of fry and juveniles to coastal areas is described, including habitat and seasonal distribution, first appearance to the coastal areas and factors and mechanisms influencing this important phase in the life cycle of grey mullets. The morphology of juveniles is described, mainly related to the characteristics used to discriminate species, and a comparative description of the melanophore pattern on the ventral side of the head in five species of grey mullet is presented. Finally, the growth and development from the larval to the juvenile stage is reviewed.

Migration of Fry and Juveniles to the Coastal Areas

Migration is defined as a coming and going with the seasons on a regular basis (Harden Jones 1984). Fish migration can be considered to stem from the two following prerequisites: the search and selection of suitable areas/environments to migrate, and the adoption of the appropriate environmental conditions that will mark the beginning of the migration period (Dodson 1988). Lucas and Baras (2001) proposed three types of migration: reproductive (spawning), feeding and refuge migration.

All diadromous Mugilidae are catadromous (McDowall 1997, Zydlewski and Wilkie 2013); they are amongst the most flexible and variable of catadromous fishes, but all of them, whether catadromous, facultative marine wanderers, or entirely marine, spawn at sea (McDowall 1997), as do most estuarine fish species. Only a few small specialized forms have adapted their entire life cycles to the variable conditions of temperature, salinity and turbidity characteristic of estuaries (Wallace and van der Elst 1975). Thus, the following pattern of migration of grey mullets is generally acknowledged: before maturity, young grey mullet live predominantly in the coastal systems of rivers and lakes (salt, brackish or freshwater); as maturation starts, they move to the sea where they complete maturation and spawn (Brusle 1981) spherical pelagic eggs with an oil droplet (Nash and Kuo 1975, Torricelli et al. 1982). About one to three months after the beginning of the reproductive period (Thomson 1966, Nash and Koningsberger 1981), there is an onshore migration at the post-flexion larval stage, which is followed by temporary occupation of the surf zone as early juveniles (Powles 1981, Collins and Stender 1989, Ditty and Shaw 1996, Hettler et al. 1997, Strydom and d'Hotman 2005). Once near the coast, they move in schools to coastal lagoons or lakes, estuarine waters and sometimes reach freshwater habitats several kilometres away from the coast (205 km for *M. cephalus* in Evros River, Greece, pers. observation). Such migrations are thereafter undertaken each year, but permanent sea-living populations do exist, as Whitfield et al. (2012) suggest for some *M. cephalus* populations. In his classification system, Blaber (2002), based on how fish utilize estuaries and where they spawn, included grey mullet as a marine migrant species, even though as he noted, *M. cephalus* in some cases in tropical estuaries can be considered as an estuarine species. Similarly, Koutrakis et al. (2005) and Franco et al. (2008) classified grey mullets as marine migrant species, a category that, in terms of habitat use, dominates European estuarine fish assemblages (Franco et al. 2008).

The migration of Mugilidae fry and juveniles to coastal areas is usually called recruitment. According to Trape et al. (2009) the recruitment of juveniles into estuarine nurseries is the result of complex interactions between spawning success, hydrodynamic processes, and pre-recruitment mortality. And as Brusle (1981) states, the migration of fry into inland waters and their abundance in coastal areas and lakes depends on the success of spawning and the chances of survival of the different developmental stages.

Habitat and Seasonal Occurrence

All over the world, fry of different species of mullet migrate from the sea to coastal areas, bays, gulfs, estuaries, lakes and lagoons, which function as 'nursery habitats' during the fattening period (Brusle 1981), but can also act as shelter from predators (Wallace and van der Elst 1975). Estuaries have been important nursery areas for juvenile fishes since the Devonian Period, c. 360 million years ago, as revealed in southern African estuaries (Whitfield and Elliot 2002) and as Blaber (1987) concludes, an estuarine phase is obligatory, at least for the grey mullet juveniles.

Fry of various lengths with a seasonal periodicity move into different coastal and inland habitats all over the world. Mathieson et al. (2000) consider that mugilids characterize the typical European tidal marsh fish assemblages of the northern and southern Atlantic coast, such as the Westerschelde estuary in the Netherlands and the Bay of Cadiz in Spain, where mugilids are found mainly in tidal marsh sites. Unlike salt marshes in northern America, marshes in Europe constitute more marginal estuarine habitats, usually flooded for only short periods during high tides (Mathieson et al. 2000).

Fry and post-larvae enter riverine systems, such as the Severn estuary that opens into the Bristol Channel (U.K.) (Claridge et al. 1986); coastal lagoons and salt marsh creeks, in the northern (Cambrony 1984, Koutrakis et al. 1994, Katselis et al. 1994, Malavasi et al. 2004, Franco et al. 2006a), and southern Mediterranean (Zismann and Ben-Tuvia 1975, Vidy and Franc 1992), which could also be hypersaline, like the Bardawil lagoon in Egypt (38–45 psu) (Zismann and Ben-Tuvia 1975); canals or lake-sea connections, as in Egypt (El-Zarka and Kamel 1965, Rafail and Hamid 1974); freshwater rivers, as in Israel (Perlmutter et al. 1957, Bograd 1961) and the Senegal River in Africa (Sarr et al. 2013).

In South Africa they spread over large lake systems, such as the St. Lucia, where they prefer calm areas (Whitfield and Blaber 1978); they find shelter in eel grass 'beds' (*Zostera capensis*) on the east coast (Wallace and van der Elst 1975, Beckley 1983), freshwater rivers (Bok 1979), but also hypersaline shallow pools, creeks and backwaters in salinities that can exceed 65 psu (Wallace and van der Elst 1975). They can also be found, in an uncommon sub-Saharan estuarine ecosystem in West Africa, where the salinity gradient has been permanently inverted (47 to 78 psu) due to an increasing rainfall shortage (Trape et al. 2009). Moreover, juveniles use mangroves as nursery grounds in temperate tidal mangrove creeks like the flat-tail mullet *Liza argentea* in New South Wales, Australia (Bell et al. 1984); in the southern Egyptian Red Sea (Abu El-Regal and Ibrahim 2014); and all over southern Asia, southern Africa and central America where mangroves are also related to commercial fish catches (Nagelkerken et al. 2008). Unsworth et al. (2008) in Indonesia showed that habitats with seagrass beds and mangroves were found to be of great importance for juvenile fishes and when there are reefs in the broader area, a greater fish nursery function is provided.

Most species enter brackish environments, but not necessarily freshwater (Lucas and Baras 2001). Only a few species live a substantial part of their life-cycle in freshwater, such as the freshwater mullet *Myxus capensis* (see Chapter 13—Nordlie 2015). This South African grey mullet species enters freshwater in late winter and early spring as post-larvae and remains there, up to 120 km from the sea, for two–five years until nearly mature; then the fish cease feeding, migrate downstream to estuaries to complete gonad development and probably spawn in coastal areas (Bok 1979). In North, Central and South America, the mountain mullet *Agonostomus monticola* occurs upstream in rivers for long periods of its life-cycle; it was also thought that the spawning of the species occurred in fresh or brackish environments, with the eggs drifting towards the sea after their release, and with the young swimming back to the freshwater environment (Loftus et al. 1984). A similar life-history cycle may occur for the abu mullet *Liza abu* in South East Asia. Other species that have been recorded moving into freshwater environments include the pinkeye mullet *Myxus (Trachystoma) petardi* (Australia), the sicklefin mullet *Liza falcipinnis* and the grooved mullet *L. dumerili* (North Africa), the fairy mullet *Agonostomus telfairii* (Africa and western Indian Ocean), the robust mullet *Valamugil robustus* (Madagascar), the longarm mullet *V. (Moolgarda) cunnesius,* the goldspot mullet *Liza (Chelon) parsia* (India), and the bobo mullet *Joturus pichardi* (Central America) (Lucas and Baras 2001).

For all species the juveniles do not always occur in the same grounds as the adults. This probably happens because osmoregulatory ability increases through development (Zydlewski and Wilkie 2013); adults have stronger osmoregulatory ability (Lucas and Baras 2001) and are therefore capable of exploring more areas. Whereas juvenile *L. ramada* remain in moderately brackish water in the Loire River (France), adults reach freshwater as far as 300–350 km upstream (Sauriau et al. 1994). Similarly adults of the red mullet *Liza haematocheila* are frequently found invading mainly freshwater habitats in North Aegean (Greece), but no juveniles have been found, not in these environments nor in the coastal systems of the area (Koutrakis and Economidis 2000, Minos et al. 2010).

The seasonal occurrence of fry and juveniles of 25 grey mullet species, from around the world, is shown in Table 12.1. The seasons in which fry and juveniles of different species of grey mullets are found close to the coasts and enter inland waters varies between species, with no more than two or three species present at the same time in a particular area, all being able to utilize the available food resources (Koutrakis 2004). The juveniles of some species, such as *L. saliens,* have a prolonged feeding migration (Savchuk 1973, Koutrakis et al. 1994) probably because of the extended breeding period of these species.

In South East Asia, in the island of Taiwan, even though several grey mullet species occur, only *M. cephalus* is of economic importance (see Chapter 19—Liao et al. 2015). Tang (1975) reported that fry of this species appeared from September to April, but are collected mainly from November to December, while Chang et al. (2000), also in Taiwan, stated that *M. cephalus* juveniles appeared from November to March, as also reported by Rajyalakshmi and Chandra (1987) in India and as shown by Martin and Drewry (1978) for more countries in South East Asia (India, Hong Kong, Taiwan), Japan and Hawaii (Table 12.1). In the East Pacific, Major (1978) described recruitment of *M. cephalus* in Hawaii from the end of January to the end of June. In Moreton Bay, Australia recruitment of grey mullet fry was observed throughout the year; silver mullet *Mugil (Paramugil) georgii* starts from December to May, *M. cephalus* from July to November, while *L. argentea* is from October to November (Blaber and Blaber 1980). In West Africa grey mullet juvenile recruitment (< 25 mm) occurred for *M. cephalus* from December to May, *L. falcipinnis* from July to February, *M. bananensis* and *M. curema* from June to November and *L. dumerili* juveniles (< 15 mm) occurred in June and from October to January (Trape et al. 2009).

Off the South-eastern United States (northern west Atlantic) larvae of *M. cephalus* are more abundant in winter (January–March), while white mullet *M. curema* occur in spring (April–May) (Fahay 1975, Powles 1981, Collins and Stender 1989) and can be found in South Carolina estuaries from January through May (McDonough and Wenner 2003). Also Hettler et al. (1997) in South Carolina, U.S.A., described recruitment of larvae from offshore in the period from December to April. In the Gulf of Mexico, *M. cephalus* larvae were observed from October to March, with a peak during November–December and *M. curema* primarily from April through September, but most abundant during April–May; while *Agonostomus monticola* was only found once in the sea in August (Ditty and Shaw 1996) (Table 12.1). In a review of the seasonal occurrence of fry and juveniles of *M. cephalus* in estuaries and lagoons compiled by Martin and Drewry (1978), it was revealed that in west North Atlantic (U.S.A. coasts), recruitment started in November and lasted until June (winter to spring); the peak season however, started in December in the warmer South-East U.S.A. states around the Gulf of Mexico (Texas, Mississippi, Florida) and shifted towards warmer months heading further north, reaching peak abundance during May in Maryland. In all studies carried out south of the equator a shift toward the winter months of June–August, thus winter to spring, is also observed (Table 12.1).

In the southern West Atlantic, in Lagoa dos Patos (Brazil), a lagoon-type bar-built estuary, the species occurrence and abundance had two seasonal periods related to changes in water temperature and salinity. One period (summer and autumn) was associated with warmer and more saline waters when *M. curema,*

Table 12.1. Seasonal occurrence and length range of fry and juveniles of grey mullet species in different coastal areas around the world. Seasonal migration presented is limited to the length range reported for each study (e.g., when larger specimens are caught later, these months are not included); data (lengths, months) were extrapolated from published studies (text, figures). For comparative reasons, length was transformed to standard length (mm) for all studies.

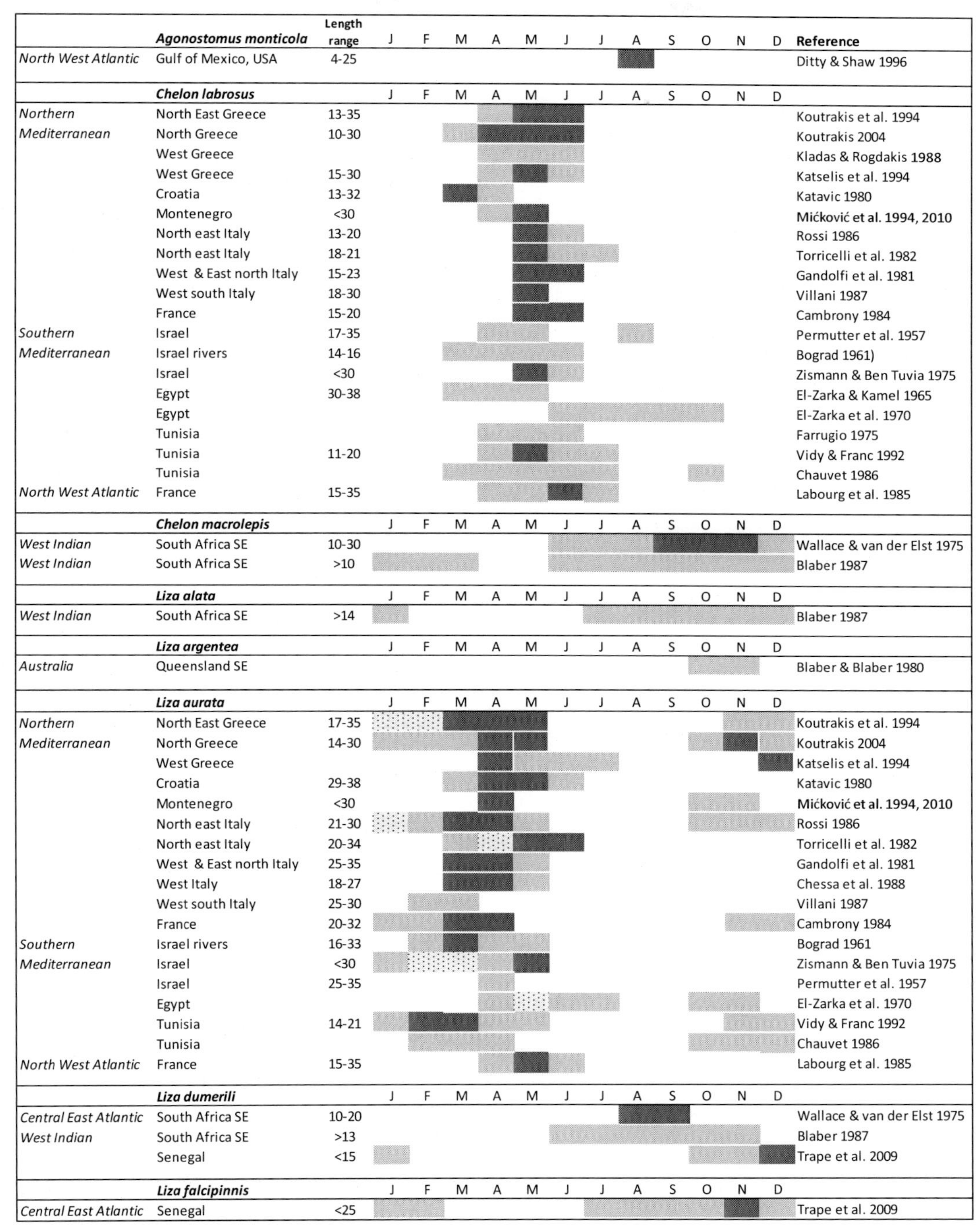

	Agonostomus monticola	**Length range**	J	F	M	A	M	J	J	A	S	O	N	D	**Reference**
North West Atlantic	Gulf of Mexico, USA	4-25													Ditty & Shaw 1996
	Chelon labrosus		J	F	M	A	M	J	J	A	S	O	N	D	
Northern Mediterranean	North East Greece	13-35													Koutrakis et al. 1994
	North Greece	10-30													Koutrakis 2004
	West Greece														Kladas & Rogdakis 1988
	West Greece	15-30													Katselis et al. 1994
	Croatia	13-32													Katavic 1980
	Montenegro	<30													Mićković et al. 1994, 2010
	North east Italy	13-20													Rossi 1986
	North east Italy	18-21													Torricelli et al. 1982
	West & East north Italy	15-23													Gandolfi et al. 1981
	West south Italy	18-30													Villani 1987
	France	15-20													Cambrony 1984
Southern Mediterranean	Israel	17-35													Permutter et al. 1957
	Israel rivers	14-16													Bograd 1961)
	Israel	<30													Zismann & Ben Tuvia 1975
	Egypt	30-38													El-Zarka & Kamel 1965
	Egypt														El-Zarka et al. 1970
	Tunisia														Farrugio 1975
	Tunisia	11-20													Vidy & Franc 1992
	Tunisia														Chauvet 1986
North West Atlantic	France	15-35													Labourg et al. 1985
	Chelon macrolepis		J	F	M	A	M	J	J	A	S	O	N	D	
West Indian	South Africa SE	10-30													Wallace & van der Elst 1975
West Indian	South Africa SE	>10													Blaber 1987
	Liza alata		J	F	M	A	M	J	J	A	S	O	N	D	
West Indian	South Africa SE	>14													Blaber 1987
	Liza argentea		J	F	M	A	M	J	J	A	S	O	N	D	
Australia	Queensland SE														Blaber & Blaber 1980
	Liza aurata		J	F	M	A	M	J	J	A	S	O	N	D	
Northern Mediterranean	North East Greece	17-35													Koutrakis et al. 1994
	North Greece	14-30													Koutrakis 2004
	West Greece														Katselis et al. 1994
	Croatia	29-38													Katavic 1980
	Montenegro	<30													Mićković et al. 1994, 2010
	North east Italy	21-30													Rossi 1986
	North east Italy	20-34													Torricelli et al. 1982
	West & East north Italy	25-35													Gandolfi et al. 1981
	West Italy	18-27													Chessa et al. 1988
	West south Italy	25-30													Villani 1987
	France	20-32													Cambrony 1984
Southern Mediterranean	Israel rivers	16-33													Bograd 1961
	Israel	<30													Zismann & Ben Tuvia 1975
	Israel	25-35													Permutter et al. 1957
	Egypt														El-Zarka et al. 1970
	Tunisia	14-21													Vidy & Franc 1992
	Tunisia														Chauvet 1986
North West Atlantic	France	15-35													Labourg et al. 1985
	Liza dumerili		J	F	M	A	M	J	J	A	S	O	N	D	
Central East Atlantic	South Africa SE	10-20													Wallace & van der Elst 1975
West Indian	South Africa SE	>13													Blaber 1987
	Senegal	<15													Trape et al. 2009
	Liza falcipinnis		J	F	M	A	M	J	J	A	S	O	N	D	
Central East Atlantic	Senegal	<25													Trape et al. 2009

Table 12.1. contd....

Legend: **Studies are presented from North to South and East to West.

Table 12.1. contd.

	Species / Location	Length range	J	F	M	A	M	J	J	A	S	O	N	D	Reference
	Liza grandisquamis														**Reference**
Central East Atlantic	Nigeria														Ezenwa et al. 1990
	Liza ramada		J	F	M	A	M	J	J	A	S	O	N	D	
Northern Mediterranean	North East Greece	15-35													Koutrakis et al. 1994
	North Greece	14-30													Koutrakis 2004
	West Greece	14-30													Katselis et al. 1994
	Croatia	23-36													Katavic 1980
	Croatia	16-21													Bartulovic et al. 2007
	Montenegro	<30													Mićković et al. 1994, 2010
	North east Italy	17-27													Rossi 1986
	North east Italy	14-28													Torricelli et al. 1982
	West & East Italy	17-23													Gandolfi et al. 1981
	West Italy	17-20													Chessa et al. 1988
	West south Italy	14-30													Villani 1987
	France	15-30													Albertini-Berhaut 1975
	France	18-30													Cambrony 1984
Southern Mediterranean	Israel	15-35													Permutter et al. 1957
	Israel rivers	8-13													Bograd 1961
	Israel	<30													Zismann & Ben Tuvia 1975
	Egypt	17-27													El-Zarka & Kamel 1965
	Egypt														El-Zarka et al. 1970
	Tunisia														Farrugio 1975
	Tunisia														Chauvet 1986
	Tunisia	12-13													Vidy and Franc 1992
North West Atlantic	France	15-35													Labourg et al. 1985
	Liza richardsoni		J	F	M	A	M	J	J	A	S	O	N	D	
West Indian	South Africa SE	>11													Blaber 1987
	Liza saliens		J	F	M	A	M	J	J	A	S	O	N	D	
Northern Mediterranean	North East Greece	9-25													Koutrakis et al. 1994
	North Greece	12-30													Koutrakis 2004
	West Greece														Kladas & Rogdakis 1988
	West Greece	13-30													Katselis et al. 1994
	Croatia	18-30													Katavic 1980
	Montenegro	<30													Mićković et al. 1994, 2010
	North east Italy	10-30													Rossi 1986
	North east Italy	12-32													Torricelli et al. 1982
	West & East north Italy	10-30													Gandolfi et al. 1981
	West south Italy	12-30													Villani 1987
	France	12-28													Cambrony 1984
Southern Mediterranean	Israel	15-35													Permutter et al. 1957
	Israel rivers	11-13													Bograd 1961
	Israel	<30													Zismann & Ben Tuvia 1975
	Egypt	16-25													El-Zarka et al. 1970
	Tunisia	10-17													Vidy & Franc 1992
	Tunisia														Chauvet 1986
	Liza tricuspidens		J	F	M	A	M	J	J	A	S	O	N	D	
West Indian	South Africa SE	>17													Blaber 1987
	Moolgarda cunnesius		J	F	M	A	M	J	J	A	S	O	N	D	
West Indian	South Africa SE	>11													Blaber 1987
	Mugil bananensis		J	F	M	A	M	J	J	A	S	O	N	D	
Central East Atlantic	Senegal	<25													Trape et al. 2009
	Mugil cephalus	**Length range**	J	F	M	A	M	J	J	A	S	O	N	D	**Reference**
Northern Mediterranean	North East Greece	20-25													Koutrakis et al. 1994
	North Greece	17-30													Koutrakis 2004
	West Greece	20-32													Kladas & Rogdakis 1988
	West Greece	23-30													Katselis et al. 1994
	Croatia	18-33													Katavic 1980
	Montenegro	<30													Mićković et al. 1994, 2010
	North east Italy	31-36													Rossi 1986
	North east Italy	18-24													Torricelli et al. 1982
	West & East north Italy	20-35													Gandolfi et al. 1981
	West Italy														Chessa et al. 1988

Legend: **Studies are presented from North to South and East to West.

Table 12.1. contd....

Table 12.1. contd.

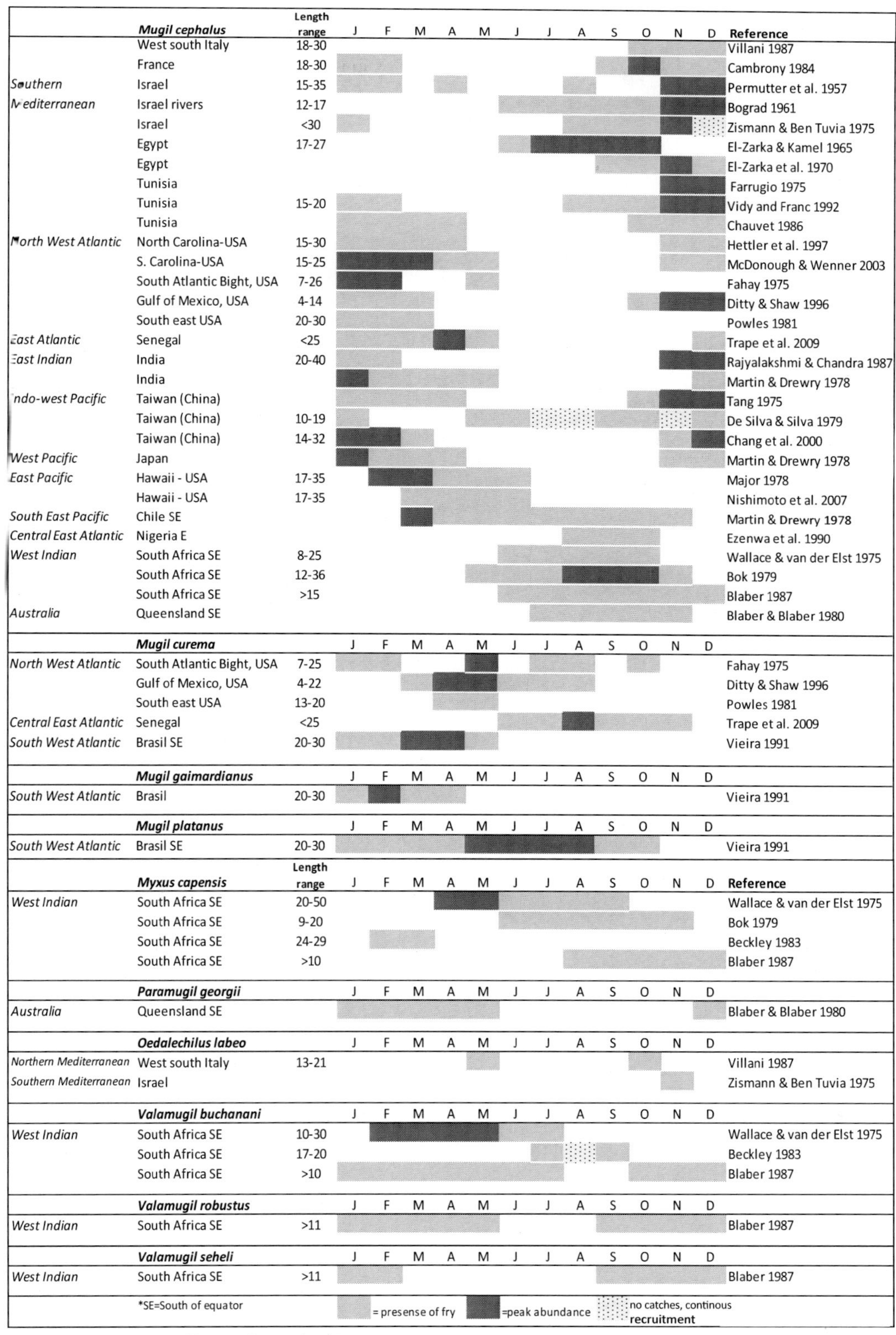

	Mugil cephalus	Length range	J	F	M	A	M	J	J	A	S	O	N	D	Reference
	West south Italy	18-30													Villani 1987
	France	18-30													Cambrony 1984
Southern	Israel	15-35													Permutter et al. 1957
Mediterranean	Israel rivers	12-17													Bograd 1961
	Israel	<30													Zismann & Ben Tuvia 1975
	Egypt	17-27													El-Zarka & Kamel 1965
	Egypt														El-Zarka et al. 1970
	Tunisia														Farrugio 1975
	Tunisia	15-20													Vidy and Franc 1992
	Tunisia														Chauvet 1986
North West Atlantic	North Carolina-USA	15-30													Hettler et al. 1997
	S. Carolina-USA	15-25													McDonough & Wenner 2003
	South Atlantic Bight, USA	7-26													Fahay 1975
	Gulf of Mexico, USA	4-14													Ditty & Shaw 1996
	South east USA	20-30													Powles 1981
East Atlantic	Senegal	<25													Trape et al. 2009
East Indian	India	20-40													Rajyalakshmi & Chandra 1987
	India														Martin & Drewry 1978
Indo-west Pacific	Taiwan (China)														Tang 1975
	Taiwan (China)	10-19													De Silva & Silva 1979
	Taiwan (China)	14-32													Chang et al. 2000
West Pacific	Japan														Martin & Drewry 1978
East Pacific	Hawaii - USA	17-35													Major 1978
	Hawaii - USA	17-35													Nishimoto et al. 2007
South East Pacific	Chile SE														Martin & Drewry 1978
Central East Atlantic	Nigeria E														Ezenwa et al. 1990
West Indian	South Africa SE	8-25													Wallace & van der Elst 1975
	South Africa SE	12-36													Bok 1979
	South Africa SE	>15													Blaber 1987
Australia	Queensland SE														Blaber & Blaber 1980
	Mugil curema		J	F	M	A	M	J	J	A	S	O	N	D	
North West Atlantic	South Atlantic Bight, USA	7-25													Fahay 1975
	Gulf of Mexico, USA	4-22													Ditty & Shaw 1996
	South east USA	13-20													Powles 1981
Central East Atlantic	Senegal	<25													Trape et al. 2009
South West Atlantic	Brasil SE	20-30													Vieira 1991
	Mugil gaimardianus		J	F	M	A	M	J	J	A	S	O	N	D	
South West Atlantic	Brasil	20-30													Vieira 1991
	Mugil platanus		J	F	M	A	M	J	J	A	S	O	N	D	
South West Atlantic	Brasil SE	20-30													Vieira 1991
	Myxus capensis	**Length range**	J	F	M	A	M	J	J	A	S	O	N	D	**Reference**
West Indian	South Africa SE	20-50													Wallace & van der Elst 1975
	South Africa SE	9-20													Bok 1979
	South Africa SE	24-29													Beckley 1983
	South Africa SE	>10													Blaber 1987
	Paramugil georgii		J	F	M	A	M	J	J	A	S	O	N	D	
Australia	Queensland SE														Blaber & Blaber 1980
	Oedalechilus labeo		J	F	M	A	M	J	J	A	S	O	N	D	
Northern Mediterranean	West south Italy	13-21													Villani 1987
Southern Mediterranean	Israel														Zismann & Ben Tuvia 1975
	Valamugil buchanani		J	F	M	A	M	J	J	A	S	O	N	D	
West Indian	South Africa SE	10-30													Wallace & van der Elst 1975
	South Africa SE	17-20													Beckley 1983
	South Africa SE	>10													Blaber 1987
	Valamugil robustus		J	F	M	A	M	J	J	A	S	O	N	D	
West Indian	South Africa SE	>11													Blaber 1987
	Valamugil seheli		J	F	M	A	M	J	J	A	S	O	N	D	
West Indian	South Africa SE	>11													Blaber 1987

*SE=South of equator; = presense of fry; =peak abundance; no catches, continous recruitment

Legend: **Studies are presented from North to South and East to West.

and redeye mullet *M. gaimardianus* were more abundant; from mid-winter through early summer, in a period associated with cold and less saline waters, *M. platanus* dominated the catches (Vieira 1991).

In the northern Mediterranean the inshore migration of juveniles exhibits two peaks throughout the year (Gandolfi 1978, Katavic 1980, Gandolfi et al. 1981, Torricelli et al. 1982, Cambrony 1984, Chauvet 1984, 1986, Rossi 1986, Villani 1987, Chessa et al. 1988, Kladas and Rogdakis 1988, Koutrakis et al. 1994, Katselis et al. 1994, Mićković et al. 1994, Koutrakis 2004, Mićković et al. 2010); the autumn–winter peak is characterized by the first appearance of *L. saliens* from August to September and *M. cephalus* later in October to December; the spring peak is characterized by higher abundances and the entrance first of *L. ramada* and later of *L. aurata* juveniles (February to May) and the first appearance of *C. labrosus* juveniles (May to June). Villani (1987) was the only author to observe recruitment of *Oedalechilus labeo* at a very small size (13–18 mm) in Lesina lagoon (South Italy) during May and October.

In the southern Mediterranean, as on the coast of Israel (Bograd 1961, Zismann and Ben Tuvia 1975), in Tunisia (Vidy and Franc 1992) or in Mex canal in Egypt (El-Zarka and Kamel 1965), a similar pattern is described, especially as regards the peak seasons. The recruitment of the three latter species on the Atlantic Coast of Europe to the salt marshes of Arcachon Bay in France was similar (Labourg et al. 1985), while in the Severn Estuary (South-West U.K., eastern Atlantic) only fry of *L. ramada* occurred from August to mid-January, with a higher abundance during October–November (Claridge et al. 1986).

The species' first appearance (Table 12.1), both in the northern and in the southern Mediterranean estuarine systems, followed the same order, with one–two-months variation (Katavic 1980, Torricelli et al. 1982, Cambrony 1984, Rossi 1986, Kladas and Rogdakis 1988, Vidy and Franc 1992, Koutrakis et al. 1994, Koutrakis 2004). This probably means that the fry of each species for both areas hatch at the same period, thus probably originating from single marine populations. There are however, cases that when compared with seasonal migrations from other areas, raise the possibility that two separate breeding stocks coexist. Evidence supporting this hypothesis comes from Egypt, where the first appearance of *M. cephalus* is recorded earlier than in other Mediterranean areas (El-Zarka and Kamel 1965), perhaps reflecting a mixing of Mediterranean and Red Sea stocks, introduced via the Suez Canal (Martin and Drewry 1978). Similar evidence comes from Whitfield et al. (2012) who reviewed recent genetic studies indicating that the flathead mullet may be a species complex (see Chapter 2—Durand 2015).

On the east coast of South Africa also two peaks of grey mullet juvenile abundance in the Zostera beds of the Swartkops Estuary were observed (Beckley 1983); *M. cephalus* juveniles were found in the estuaries from June to December, *Liza dumerili* from August to September and largescale mullet *L. (Chelon) macrolepis* from July to December (Wallace and van der Elst 1975). In West Africa, in a hypersaline estuary, they also show two peaks during the year; *M. cephalus* and *L. dumerili* recruit during the dry season, whereas *M. curema,* banana mullet *M. bananensis* and *L. falcipinnis* occur during the wet season (Trape et al. 2009).

Some authors suggest that the timing of grey mullet recruitment into estuaries and lagoons coincides with the onset of favourable conditions within the nursery area, such as the rainy season (Wallace 1975, Payne 1976, De Silva and Silva 1979), which ensures productive marine larval habitats (Blaber 1987, Whitfield et al. 2012). This is mentioned principally for *M. cephalus,* however there are cases such as the Mediterranean, where grey mullet juveniles reach the coast throughout the year, even in winter (e.g., *M. cephalus*) when climatic conditions are not favourable at all, but also less interspecific completion exists. Interaction between species may occur in West African estuaries where juveniles of *M. cephalus* and *L. dumerili* approach the coast during the dry season during hypersaline conditions (Trape et al. 2009). As Baldó and Drake (2002) showed in the Guadalquivir Estuary, in Spain, juvenile *L. ramada* and *L. saliens* enter the estuary during winter when there are low densities of their main prey (copepods), and feed on diverse foods in which insects play an important dietary role.

Size at First Appearance and Size Distribution of Juveniles

Every year the first appearance of the juveniles of each grey mullet species depends on the spawning period and the distance of the spawning grounds from the coast. Thus, different times of first inshore migration for the same species in different areas are not only the result of actual environmental changes or interannual environmental variability, but also due to the distance from spawning areas and the temperature of coastal waters, since floating eggs and larvae are drifted by sea currents (Rossi 1986). There are cases where post-larvae have been found far from the spawning grounds, as Reay (1992) proposed for the larvae of *M. cephalus* found in the Camel estuary, 500 km from the Bay of Biscay, where the species probably spawns. Similarly, Tully and O'Ceidigh (1989) have found *Liza* spp. post-larvae in the neuston of Galway Bay (Ireland) when no record of either fry or adult exists from the Irish coasts.

Grey mullet juveniles first enter estuaries when individuals are between 10 and 30 mm SL with most recruitment occurring at a size of about 15–25 mm SL. Differences in length between specimens of different species could be attributed to the different distances of the reproductive grounds of the various species, thus the different time fry need to approach the coast, giving the opportunity for rapid growth. Evidence that supports this hypothesis comes from a study of neustonic young off the South-eastern United States, where Powles (1981) found that standard lengths of larvae of *M. cephalus* and of *M. curema* were inversely related to the distance from shore; this relationship was strongest for *M. cephalus* (winter 1973: $SL = -0.13X + 30.56$; winter 1976: $SL = -0.111X + 20.48$; X = distance in km). Hettler et al. (1997) in South Carolina, U.S.A., suggested continued recruitment of new individuals from offshore, in the period from December to April, since no substantial change in length and weight of striped mullet over the recruitment season was observed.

Another hypothesis for the different lengths of species approaching the coast is the different growth rate at the larval and post-larval stage of the different species, due also to the different amount of food available in the different seasons in which fry are approaching coastal systems. Evidence that supports this hypothesis comes from studies in the northern Mediterranean (Koutrakis 2004), where *L. saliens* fry reaches the coast at the smallest length (9.3 mm, SL), followed by *C. labrosus* (10 mm), *L. ramada* (14 mm), *L. aurata* (16.9 mm) and *M. cephalus* (17 mm, SL). Indeed in the same area, *L. saliens* had the slowest growth, followed by *L. ramada* and *C. labrosus* (Koutrakis and Sinis 1994), while the faster growing species is *M. cephalus* (Chervinsky 1975) (see also Chapter 10—Ibáñez 2015).

Juveniles can be collected until about 50 mm length (SL); larger specimens disappear from the catches with experimental seine nets, probably as a result of a size-dependant change in habitat (De Silva and Silva 1979, Koutrakis 2004, Trape et al. 2009). These juvenile habitat shifts are probably associated with changes in feeding ecology that occur between 30 and 55 mm (Albertini-Berhaut 1974, De Silva 1980, Blaber and Whitfield 1977; see also Development and growth) and take place at a time when predation risk is reduced due to an increase in individual size and swimming capacities (Mićković et al. 2010). However, it should also be taken in consideration that the high swimming performances of these juveniles make them capable of avoiding net capture, since their schools are small and rapidly scattered (Bograd 1961, Trape et al. 2009, Mićković et al. 2010). Juveniles exceeding this length (60–70 mm, TL) have significant rates of survival when released for marine stock enhancement, as shown with cultured fry of *M. cephalus* in Hawaii (Leber et al. 1997) (see also Chapter 18—Leber et al. 2015).

On the western Atlantic coast of France, Labourg et al. (1985) found that *L. aurata*, *L. ramada* and *Chelon labrosus* juveniles enter tidal marshes in one, two and three cohorts respectively, at 42.9 mm, 23.6 mm and 21.8 mm (TL). El-Zarka and Kamel (1965) in Mex canal (Egypt) found *M. cephalus* fry at 17 mm, although most of the fry were 22–27 mm. Savchuk (1973) in the North-western part of the Black Sea found *M. cephalus* fry in July and August at 9 mm, but most of the fry were 15–23 mm and *L. saliens* from 16 mm, although most fry were 22–27 mm. In South Africa larvae of *M. cephalus* and *Myxus capensis* arrive in the coastal surf zone at about 8–25 mm (SL) (Strydom and d'Hotman 2005). Similar sized specimens of both species have been found in bays associated with rocky shores, 9–24 mm (SL) (Strydom 2008). In South-eastern Africa, 15 species of juvenile grey mullet reach the coasts throughout the year at a size of 9–14 mm, except for fringelip mullet *Crenimugil crenilabis* that was found at 28 mm (Blaber 1987). *Agonostomus monticola* off the Bahamas and South Atlantic Coast of U.S.A. remain at sea until they have reached a size of about 20 to 35 mm, while 22 mm is the smallest size at which this species has been taken from freshwater rivers (Anderson 1957b).

Motivation, Mechanisms and Factors Affecting Migration of Fry and Juveniles to the Coast

Many authors have discussed why grey mullet post-larvae migrate to the coastal areas, what is influencing larval orientation and which factors influence them while approaching the coast and during their distribution in the estuarine environments? As concluded by Gibson (2003), migratory behaviour for marine organisms is timed and directed by a range of mechanisms involving responses to both external and internal stimuli. Moreover Harden Jones (1984) suggested that clues and cues could be used, relative to migration, for information as to where and when fish are directed. Heldt (1929; quoted by Brusle 1981) called the inward movement of grey mullets *'appel du lac'* and Myers (1949; quoted by Thomson 1955) '*osmoregulatory migration*', while Thomson (1955) stated that grey mullet are 'trophically brackish water fish' and their fry movements are '*trophic migrations*'. The same view was supported by Savchuk (1973) in the North-western part of the Black Sea, where fry migrate into coastal waters in search of feeding grounds.

Grey mullet larvae exhibit a strong association with surface waters where they are usually caught; moreover there are references for *M. cephalus* and *M. curema* off the South-eastern United States (Collins and Stender 1989) and for *Chelon labrosus* and *Liza* spp. in Ireland (Tully and O'Ceidigh 1989) that show no indication of diel vertical migration (Powles 1981, Collins and Stender 1989).

Environmental factors were the first to be taken into account in order to see whether they influence larval orientation, and the first to be investigated was the effect of salinity decrease. Even though there were authors supporting such an idea, Blaber (1987) rejected this possibility, presenting the case of juvenile *L. dumerili, Valamugil (Moolgarda) cunnesius* and bluetail mullet *V. buchanani* that do not react uniformly to salinity gradients, a finding supported by field observations of post-larvae of these species in southern Africa. Similarly, squaretail mullet *L. (Ellochelon) vaigiensis,* greenback mullet *L. dussumieri (Chelon subviridis)* and *M. cephalus* enter hypersaline systems in Australia, and in West Africa juveniles enter estuaries with inverse salinity gradients (Trape et al. 2009). Temperature seems to attract juvenile mullets, as shown in Tunis lagoon (Chauvet 1980) where fry are attracted to an area where hot water is discharged from a thermal power station. In addition, in an experiment with *M. cephalus* juveniles, Major (1978) observed that orientation was associated with temperature and not light or pressure gradients. However, this environmental parameter was also rejected by Blaber (1987), as thermal gradients in estuaries are irregular and highly variable. Temperature, nevertheless, could be a factor that increases abundance in coastal areas. As Franco et al. (2008) state, grey mullets, although widespread in Europe, are more common along warmer coasts.

Turbidity increase was also proposed as a possible 'cue' for larval orientation valid for many Natal, Australian and South-East Asian estuaries, where turbidity gradients extend into the sea and where there also exists a vertical gradient from low turbidity at the surface to high at the bottom (Blaber 1987). Fish can detect changes in the complex microstructure of the water in such perturbed areas and this has been suggested as a mechanism to facilitate orientation. Blaber (1987) reported that under experimental conditions, larvae of *L. (Chelon) macrolepis* and *M. cephalus* avoid strong illumination and congregate in places of low light intensity or high water turbidity.

With an experiment carried out with a series of juvenile *Mugil brasiliensis* (*Mugil liza*) under controlled environmental factors, Kristensen (1963) proposed that an organic compound attracts grey mullet fry to hypersaline bay water in preference to seawater, since only charcoal filtration influenced their preference. Whitfield and Blaber (1978) also supported that idea because when St. Lucia Lake in Natal (South Africa) has a reversed salinity gradient and seawater flows into the lake, mullet fry could follow a land-based organic compound. Major (1978) also presumed the existence of an 'ontogenetic biological rhythm', possibly dictated by slight changes in water temperature or photoperiod on a monthly basis, which 'prepare' these fish, in terms of biochemistry and physiology, for their future life in intertidal estuarine areas, while they are still ocean-dwellers. Other environmental factors may also have an effect. De Angelis (1960; quoted by El-Zarka et al. 1970) suggested that dissolved oxygen, alkalinity and the presence of dissolved nitrates, are important factors. The effect of low DO concentration was also supported by Koutrakis et al. (2010) as a factor inhibiting juveniles entering a coastal lagoon.

Selective tidal-stream transport has been shown to be an important mechanism for migration in some fish and Blaber (1987) assumed the same for mullet post-larvae and suggests that they use such a mechanism to aid their active movement upstream, by drifting on the flood tide and settling on the ebb. He also suggested that in many Indo-Pacific estuaries, larvae move towards the bottom when they are at the mouth of the estuary, in relation also to the change from pelagic to benthic feeding, thereby taking advantage of the stronger tidal movement of water inward that occurs in the bottom. According to Blaber (1987), this is also supported by the change from feeding on surface-dwelling zooplankton to vertically migrating zooplankton (at 10–15 mm SL), from zooplankton in the benthos to meiobenthos (at 15–20 mm) and the shift to iliophagous, ingesting sandgrains and microbenthos at 15–45 mm. Martin and Drewry (1978) also stated that flathead mullet *M. cephalus* probably spawn at sea well south of the mid-Atlantic U.S.A. states and use the western edge of the Gulf Stream to transport and nourish larva and put the juveniles within active migration range of the middle and North-eastern Atlantic coastline. Evidence supporting this hypothesis comes from the delayed peak season of *M. cephalus* along the northern U.S.A. coastline (see also Habitat and seasonal occurrence). Claridge et al. (1986) also conclude that selective tidal-stream transport was used by thin-lipped grey mullet *L. ramada*, in the Severn Estuary (U.K.) along with passive transport of the larvae.

Post-larvae migrations are probably affected by tidal movements, as shown by an experiment by Torricelli et al. (1982) to check daily variations in catches in the Arno River (Italy) in which the largest catches were made during high tide. Fry are usually observed entering estuary mouths during flood-tide, such as the South African mullet *L. richardsoni*, and they actively migrate towards the banks to avoid being swept back out to the sea (Beckley 1985). Whitfield and Blaber (1978), on the other hand, suggested that grey mullet fry enter Lake St. Lucia by swimming against the current and the same was claimed by El-Zarka and Kamel (1965) for the Mex canal (Egypt). Torricelli et al. (1982) observed two peaks during daily variations of migrating grey mullet fry in the Arno River (Italy), one in the morning and the other at sunset; fry increase slightly in the first part of the night, decrease sharply in the second part and increase near sunrise; the best moment of the day for fry recruitment is when high tide and sunset coincide.

Grant and Spain (1975) studied the influence of wind on the coastal migration of fry in Queensland. They found that the wind caused a local current bringing inshore the larvae and juveniles that had spawned offshore. Powles (1981) suggested a similar mechanism for the *M. cephalus* found off the South-eastern U.S.A. in winter, where larval recruits use a wind driven (Ekman) drift with a shoreward component, which could cause shoreward movement. Chang et al. (2000) noted that the interannual change in estuarine recruitment of *M. cephalus* in Taiwan estuaries may be influenced by both population density and environmental factors and pointed out the prevalence of coastal currents. In North Carolina (U.S.A.) estuaries Hettler et al. (1997) observed that juvenile striped mullet recruits in pulses and this was possibly related to the lunar cycle, even though this type of pulse of new recruits was not observed into South Carolina estuaries (McDonough and Wenner 2003). Climatic conditions can thus affect migration and recruitment and that is why it has been suggested that the prolonged spawning and recruitment seasons have a 'buffering' action against failure of recruitment failure as a result of adverse climatic conditions (Wallace 1975, Wallace and van der Elst 1975).

Distribution and Species Composition

Distribution of fry and juveniles in estuarine systems is usually thought to be influenced by salinity (De Angelis 1967, El-Zarka et al. 1970, Zismann and Ben-Tuvia 1975). Zydlewski and Wilkie (2013) state that mullets apparently maintain a state of physiological 'preparedness' for a wide range of salinities. Katavic (1980) reports that *M. cephalus* and *L. ramada* are more commonly found in areas under strong freshwater influence, while *L. aurata* and *Chelon labrosus,* in localities with marked maritime properties. Lasserre and Gallis (1975) observed preference for low salinity water for *L. ramada*, and the same was found by Koutrakis (2004), who however reported the same also for *L. saliens*, a species known to "flourish in more saline waters" (Bograd 1961) and to have a significant tolerance to a wide range of salinity levels, but is usually found in brackish waters (Katavic 1980, Vidy and Franc 1992). By comparing habitat availability and habitat use in the estuaries of Minorca (Spain), Cardona (2006) showed that salinity strongly affected the distribution of all grey mullet species; *M. cephalus* and *L. ramada,* that are good osmoregulators in a wide range of salinity levels, quite obviously preferred sites with salinity under 15; *L. aurata*, preferred polyhaline water, and kept clear from sites with salinity levels lower than 15 throughout the year, and the same stands for *L. saliens* in spring and summer. Rather unexpectedly, *C. labrosus* showed a strong preference for salinity levels under 15, even if it has a lower osmoregulatory competence in freshwater than *L. ramada* and despite the fact that the species can only survive in freshwater for a few months (Lasserre and Gallis 1975). Cardona (2006) explained that food availability possibly influences the habitat preferences of the species.

Salinity was also reported by El-Zarka and Kamel (1965) to influence the species that will enter a system; in the Mex canal (Egypt), where salinities are very low (near freshwater), *M. cephalus* fry were attracted in large quantities, in correlation with the discharged lake water, while the other three grey mullet species found in the area, were not attracted in noticeable quantities to the freshwater flow. The preference of *M. cephalus* for freshwater has been confirmed by many authors (Bograd 1961, Major 1978, Leber et al. 1996). Chang et al. (2004) used Sr:Ca ratios in otoliths to reconstruct the past salinity history of *M. cephalus* in Taiwanese estuaries and revealed the existence of individual differences in the habitat preference as two types of fish were found, those that migrated to freshwater and those that migrated between estuary and offshore waters, but rarely entered the freshwater habitat. With a similar method, Wang et al. (2010) found that *M. cephalus* migrated between brackish and freshwater habitats after the juvenile stage and occupied diverse environments. Moreover McDonough and Wenner (2003) in South Carolina, found that the distribution of *M. cephalus* was more abundant in mesohaline and polyhaline habitats.

Blaber and Blaber (1980) considered that the most important factors influencing the distribution of grey mullet juveniles in estuaries are shallow turbid areas and suitable food. Cyrus and Blaber (1987a,b) and Harrison and Whitfield (1995) suggested that turbidity, which is influenced by wind speed, substratum particle size and wave action, is the most important factor influencing both occurrence and distribution of species in small estuarine systems on the South-Eastern coast of Africa. Different species occur in different systems with different levels of turbidity, such as *L. dumerilii, L. macrolepis* and *Valamugil buchanani* in clear water and clear to partially turbid areas and *Valamugil (Moolgarda) cunnesius* and *M. cephalus* in intermediate and turbid waters. Turbidity was also reported as an important factor influencing species recruitment of *L. saliens* and *L. ramada* in Mediterranean salt marsh creeks (Franco et al. 2006a,b).

Other factors influencing distribution and species composition include the substratum. Franco et al. (2012) in a study in four northern Mediterranean estuaries, argue that Mugilidae typified assemblages from unvegetated habitats and in Venice lagoon (Franco et al. 2006b) most mullet species are restricted to sand habitats. Wallace and van der Elst (1975) proposed that marginal and submerged aquatic vegetation support greater densities of juvenile fish than bare sand or deeper muddy bottomed estuarine basins and this is attributed to the rich feeding regime provided by epiphytes growing on these plants and the invertebrate fauna associated with the detritus they produce. A similar conclusion was reached by Koussoroplis et al. (2010) who found that after settlement of *L. saliens* in a coastal lagoon, the benthic omnivore/detritivore diet was confirmed, with macrophyte debris and more sand and detritus. Also Franco et al. (2008) pointed out that grey mullets belong to the detritivorous species feeding guild, which is infrequent in fishes from European estuaries. Blaber (1987) also suggested that the large biomass of adult and juvenile mugilids in estuaries is attributable to the large resource of energy contained in the substratum, therefore concluding that survival of juvenile mullet recruiting to estuaries depends upon sequential changes in diet.

Migration can also be affected by predators, and mortality from fish, bird and reptile predation can be very high (Blaber 1987). The author presents a case where predation by Ambassidae (a small family of small fishes that are abundant in the inshore waters of the Indo-Pacific region) on mugilid eggs can significantly affect recruitment. Moreover, Blaber (1987) presents a list from published information of at least 11 species that include juvenile mullets in their diet, even if no mullet species forms a predominant part of the diet, and concludes that shallow waters protect fry from large fish predators during migration, which in any case have reduced numbers in most estuaries. Gibson (2003) also suggested that intertidal migration minimizes predation risk, since prey species manage to hold their position in shallow waters against the rising tide, thereby minimizing the possibility of becoming visible to migrating predators. Bell et al. (1984) also pointed out that mangrove creeks in Australia afford shelter to small fishes because they are too shallow for most larger piscivorous fish. Shallow waters on the other hand render mullet fry vulnerable to piscivorous birds, which are probably the most important predators of mugilids in South-eastern Africa (Blaber 1987). As Whitfield and Blaber (1978) describe, white pelecans *Pelecanus onocrotalus* feed on pre- and post-spawning schools of *M. cephalus* in the St. Lucia estuary and conclude that birds have a major effect on estuarine mullet populations in South-eastern Africa. Moreover Liordos and Goutner (2007) found that mugilids were the most important prey of the great cormorant *Phalacrocorax carbo* in estuarine systems in Greece.

Mullets are fundamentally schooling fish and their school size has a seasonal variability; schools are smaller during the phases of feeding and post-spawning, than in the pre-spawning accumulation and migration period (Thomson 1955). Schooling increases the chances of survival of individual mullet, especially in the presence of predators (Major 1978, Blaber 1987). Major (1978) described a successful behaviour in order to avoid predation: when a school of juvenile *M. cephalus* was attacked, it usually split into two or more groups and passed around behind the predator to reform into a single school again. Juveniles of the five Mugilidae species in the Mediterranean form mixed schools (Koutrakis 2004). Figueiredo and Silva (1983) observed an interesting behaviour of grey mullet post-larvae caught for experimental rearing of *L. ramada* and *L. aurata*; schooling was more easily achieved by the very young fry, while fry over 25 mm, when disturbed, often lost the ability to school, looking frightened without any interest in feeding. Light appears to be necessary for schooling, but also affects the ability of post-larvae to see prey (Kraul 1983) and is considered as the major abiotic factor influencing the feeding rhythm of the fry, while tides and temperature only play a secondary role (Torricelli et al. 1988).

The channels which connect the lagoons with the sea can also influence the presence of juvenile grey mullet in coastal lagoons. The limited presence of the Mugilidae juveniles inside the re-flooded Drana lagoon in North Greece was related to the prevailing tidal inlet dynamics (i.e., strong ebb and flow at the artificial inlet made by cement with vertical banks), that prevented the grey mullet juveniles from entering the lagoon (Koutrakis et al. 2009). On the other hand, there are cases, such as in South Africa (Harrison and Cooper 1991), where juvenile grey mullet manage to enter estuaries closed to the sea, that open only for short periods, even under strong current velocities, probably using a special technique, moving through the bottom of the standing waves in a series of steps, resting temporarily between successive waves.

Interaction between Species

According to Cardona (2001) who reviewed the structuring forces in aquatic communities, there are three major theories: the disturbance theory that predicts that competitive exclusion takes place only in moderately harsh environments, the top-down control theory that predicts that only the species of some trophic levels will reach carrying capacity and hence competition will affect only them, and the lottery theory that states that in some situations recruitment is independent of previous adult performance and hence competitive exclusion does not operate. Malavasi et al. (2004) reported that their research on the structure of fish assemblages and environmental parameters suggested the importance of certain physical factors, such as salinity, temperature and hydrological conditions, but also report substratum characteristics.

Even though recruitment of juvenile grey mullet species is usually balanced between species throughout the year, several grey mullet species have overlapping recruitment seasons, thus there is space and niche overlap that could create interspecific competition at the juvenile stage. Grey mullet fry are zooplanktivorous and interspecific trophic overlap is high among fry (i.e., Albertini-Berhaut 1973, 1975, Tosi and Torricelli 1988, Torricelli et al. 1988, Salvarina et al. 2010). Therefore when there is resource limitation, e.g., when zooplankton is scarce in estuaries, as in winter (Gisbert et al. 1995), this could be a limiting factor that could increase competition among species, which could in turn influence distribution, habitat preference and structuring of grey mullet assemblages.

Cardona (2006) suggested that competition could influence distribution and habitat preference of *C. labrosus* post-larvae, as it is complementary to *L. saliens*, which was more abundant, even if Gisbert et al. (1995) demonstrated that not all species are affected by coexistence with other mullet species. Fry of *L. aurata* are very negatively affected by interspecific competition with *L. ramada*, which is not affected at all, whereas the fry of *L. saliens* were only slightly affected by competition with co-occurring

M. cephalus juveniles (Gisbert et al. 1995). Salvarina et al. (2010) also found that the diet overlap between species was generally moderate to low, and only the pair *L. saliens—M. cephalus* exhibited the highest overlap in the Rihios estuarine system (Greece); the two species did not appear to compete for common resources, as there was not always a spatiotemporal overlap, enough food was available, or because they exploited different resources. Cardona (2006) also supported the hypothesis that competition between juveniles, together with salinity, are important for structuring grey mullet assemblages. In their study in three estuarine areas, Cardona et al. (2008) showed that *L. aurata* and *M. cephalus* were scarcely recorded everywhere and did not dominate the assemblage on any occasion; *Liza saliens* was the dominant species of the assemblage in areas where salinity exceeded 13; *Liza ramada* dominated the assemblage in all cases where salinity was less than 13; and the assemblage was dominated by *C. labrosus* only in the absence of common carp *Cyprinus carpio* (also an iliophagous species) and the salinity level was lower than 13. Consequently, Cardona et al. (2008) concluded that the structure of grey mullet assemblages inhabiting Mediterranean estuaries is determined by salinity and competitive interactions at the fry stage.

Interspecific competition in estuaries is however, reduced by a superabundance of food, differential recruitment, substratum preferences and the extent of penetration into freshwater (Blaber 1987). Moreover, daily variations in feeding also help to reduce competition between species occupying the same environment and depending upon the same resources (Blaber 1976) as found by Torricelli et al. (1982) in the Arno River (Italy), where the two mugilid species present simultaneously in the catches feed in different hours of the day/night (*L. saliens* during daylight and *M. cephalus* during sunset and early in the night). As Salvarina et al. (2010) pointed out, grey mullets exhibit similar patterns of feeding strategies with varying levels of specialization at an individual level and a rather generalized pattern at the population level, and this feeding strategy may permit them to coexist.

Morphology

Mugilidae are uniform in appearance, torpedo-shaped, oval in cross-section with a smoothly curving profile; some are robust and deeper bodied than others and there is some variation in inter-ocular breadth (Thomson 1966). In the young stages, even if they are similar to the adults, they also have significantly distinct morphologies; the mouth gape is steeply inclined, but later leads to less sleep and almost horizontal in some genera; the nostrils are usually above the level of the eye in young mullet, but descend below the level of the upper rim of the eye in most species; the dorsal fins are closer to each other in young mullet, the tip of the recumbent first dorsal spine almost touching the origin of the second dorsal fin and the tip of the second dorsal fin almost reaching the base of the caudal fin state (Thomson 1997).

The description of morphometric parameters and melanophore patterns in grey mullet post-larvae and juveniles is a significant source of knowledge, mainly for the characteristics that are used to differentiate species. The smaller the fish, the more difficult it is to identify the species. This is because the rapid changes in morphology and the changes in the proportion of different parts of the body, occurring during the early stages, only permit comparison among fry of the various species within narrow size ranges (Zismann 1981, Reay and Cornell 1988, Minos et al. 2002).

Of the various characters used to recognize species of grey mullet fry, some remain constant at all sizes, and some vary with the length of the fish (Zismann 1981). From the characters which remain constant at all sizes, the number and arrangement of pyloric caeca, the number of scales along the lateral line and the total number of elements in the anal fin can be used for identifying some species in their juvenile stages (Zismann 1981). Pyloric caeca which are sources of enzymes, as mentioned by Thomson (1966), are within the most important characters used for fry identification (Perlmutter et al. 1957, Cambrony 1984). The primitive number of two pyloric caeca is found throughout the Agonostominae and in *Mugil, Rhinomugil, Sicamugil, Chaenomugil* and *Myxus* (except for *Myxus capensis*), while in other genera the number of caeca varies (two–22), though it usually ranges between five and nine (Thomson 1997). The smallest size at which pyloric caeca can be counted in grey mullet post-larvae is about 8 mm, but differences in size or grouping of large and small caeca can be noted in specimens of over 12 mm, although the range in numbers is the same for all size-groups of a species (Zismann 1981, Thomson 1997).

Most of the characters of larvae and post-larvae change with size or can be seen only on larger specimens, as in the extension of the adipose tissue over the eye that in some species has not reached the adult state even at 60 mm (Thomson 1997). In juvenile *M. cephalus* the adipose eye-lid can be seen macroscopically at over 40 mm (Katavic 1980), but microscopically is noticeable at 28 mm (TL). Within the characters which change with size is the number of gillrakers on the first gill arch (Zismann 1981). Scales of mugilids also change with size. Both cycloid and ctenoid scales are found in Mugilidae (Thomson 1997). On most scales of *M. cephalus* and related species, a row or two of weak ctenii mark the posterior margin of the scale; this type of scale is distinguished as 'pavement ctenoid', while two other types of scales have been described as 'cycloid' (Thomson 1997). One type has a firm rounded posterior edge and is commonly found on the flanks of certain species of *Liza* and *Mugil*; the other type is found on juveniles of *Valamugil* and *Crenimugil* that have similar scales, which however later develop posteriorly a flexible membranous margin (Thomson 1997). Observations made by Zismann (1981) on five grey mullets showed that the scales of specimens of less than 25 mm (TL) were cycloid; ctenoid scales found on larger fish are firstly observed along the belly. Jacot (1920) gave a very detailed description of the scales of *M. cephalus* and *M. curema*, by using Mastermann's (1913) method that divides scales into areas and presents a 'circuli' formula. He also showed that *M. cephalus* scales first appear at 23 mm (TL) and in *M. curema* at 20 mm, and that most were formed on the spawning grounds. Later, Anderson (1958) showed that *M. cephalus* scales were forming at 11 mm (TL) and Ben Khemis et al. (2013) showed that *C. labrosus* scales start forming at 7.5–10 mm (TL). Ibáñez et al. (2007) in a morphometric analysis of scales of different mugilid species from the Gulf of Mexico and the Aegean Sea (northern Mediterranean) showed that scale morphology can allow discrimination by genera, species, geographic variants and local populations. Ibáñez et al. (2007) also observed that scales of the genus *Mugil* are generally relatively longer and narrower than those of other Mugilidae taxa.

Mucus canals, that is pits or grooves which are found on the scales of some Mugilidae genera, can also be used as diagnostic characters. These initially become apparent as almost circular depressions in fish of about 40 mm (TL), as in the scales of adult Agonostominae (Thomson 1997). In *Myxus* spp. and in some *Liza* species, such as *L. abu*, a short canal runs forward from the original pit. In other genera and especially in *Liza* spp., the canal may be much longer, but does not penetrate the segment of the scales posterior to the nucleus stominae (Thomson 1997). In some species a scale may have Y-shaped, T-shaped, double or even triple canals, such as in early juveniles of *Liza saliens* and *L. dumerili* where the dorsal scales on the head and on the back, anterior to the first dorsal fin, have two canals, which however become multicanaliculate, having between seven and 10 canals in the adult stage (Thomson 1997).

There are also other morphological characters of grey mullets that change during development, such as the colour of the peritoneum. Jacot (1920) studied *M. cephalus* in Florida and observed that at first appearance in December, the walls of the abdomen were orange, while the peritoneum was greyish with dark spots, or a semi-translucent blackish colour; three months later the peritoneum was black and the flesh about the viscera had assumed the more natural black colouration.

Albertini-Berhaut (1987) studied the intestine morphology at different growth stages and found that the length of *L. aurata* and *L. ramada* intestines, like those of most herbivorous fish, increase notably during the first year of life and stabilize at a length of 70–80 mm (SL). *Liza saliens* intestine, however, shows no significant intestine growth during the first year, like that of the carnivorous fish and Albertini-Berhaut (1987) explained that this perhaps happens because this species consumes more faunal elements, such as invertebrates, than the other two species. Moreover, the goblet cells are scarce in the larval stages, becoming numerous in the juvenile stage.

There are also characters which cannot be counted or measured and are often difficult to describe, even with the aid of illustrations, but sometimes help the identification of fry, such as the general appearance and colour, which mainly in live specimens, can be very helpful. Other morphological characters are also studied, such as the head, the general profile, lips, lingual teeth and the jugular region (Zismann 1981).

The colours and pigment patterns in live fish may change under stress and with the type of habitat (Zismann 1981). These colours further change on preservation; the chromatophores become more obvious, the gold or blue fading. There are also slight differences in colour patterns from the young stages to the commercial sizes. The gold patch, often quite obvious on the operculum of adult *L. aurata*, is not found on fry, although a diffused gold colour may sometimes be seen on the whole operculum and on the sides of the body, which however can also be seen on *L. saliens* specimens. Live specimens of *M. cephalus* and *L. ramada*, shortly after they are removed from the sea, are described as metallic silver in colour (Jacot 1920, Zismann 1981); while fry of *L. aurata, L. saliens* and *C. labrosus* are described as silvery grey and often metallic blue on the back (Zismann 1981). McDonough and Wenner (2003) reported that *M. cephalus* juveniles at 18–30 mm (TL) during recruitment in South Carolina estuaries had the silvery sheen of the pelagic stage and also Major (1978) described these juveniles as having a distinct silvery, countershaded colour. Also Martin and Drewry (1978) described that *M. cephalus* juveniles between 16–40 mm (TL) are silver coloured ventrally and laterally, progressively more pigmented on the dorsal surface while stripes become evident after 40 mm (SL) and Jacot (1920) described that fry after 30–35 mm become duskier.

Fry Identification

The study of grey mullet post-larvae and juveniles has shown that the keys for the identification of adult and sub-adult specimens are not applicable for fry and juveniles. Identification using visible external features is neither possible, since features typically used to identify adults are undeveloped in fry (van der Elst and Wallace 1976, Zismann 1981, Reay and Cornell 1988, Serventi et al. 1996). The characters used in keys and detailed descriptions are only valid for juveniles over than 60 mm (SL) (Thomson 1997). At this length the transformation of the third supporting element of the anal fin to a spine was observed, together with a considerable rearrangement of the mouth parts (see Development and growth).

In many areas a key is not needed since species can be distinguished by their geographic or seasonal separation. In North Carolina there are three mugilid species (*M. cephalus, M. curema, Agonostomus monticola*) two of which are similar in appearance, but can be separated by fin ray counts or measurements (Collins and Stender 1989), or the presence of ctenoid scales on *Agonostomus monticola* fry that quickly separates them from the other two species (Anderson 1957b). In Hawaii there are only two indigenous species (*M. cephalus* and *Neomyxus leuciscus*) that, apart from seasonal and habitat differences, also have morphological differences (e.g., fin counts) that allow distinction (Nishimoto et al. 2007). In the Gulf of Mexico, Ditty and Shaw (1996) proposed some diagnostic characters in order to separate larvae of the nine grey mullet species of the Gulf of Mexico, in U.S.A., based mainly on the counts of the dorsal and anal fin and their position. Vieira (1991) also give a simplified key for identification of the juvenile (< 100 mm, TL) grey mullets (*M. curema, M. gaimardianus, M. platanus*) occurring in the Lagoa dos Patos Estuary (Brazil), based on anal fin elements counts (11 for *M. platanus*) and scale characteristics (cycloid for *M. curema*, ctenoid for *M. gaimardianus*). Finally, van der Elst and Wallace (1976) conducted an amazing work on characters of dental morphology and have also drew up a key for the identification of juvenile mullets of the east coast of South Africa, based mainly on counts of the lateral line scale in combination with features of dentition.

Other criteria for the taxonomic diagnoses of mugilid fry that have been used by different authors are the electrophoretic variability of proteins and allozymes (Herzberg and Pasteur 1975, Anderson 1982, Menezes et al. 1992). It is remarkable that molecular markers (a PCR–RFLP technique, developed on the mitochondrial 16S ribosomal RNA region) were also used in recruitment studies for identification of different species of grey mullets (Trape et al. 2009), although this technique could not be applied on a large scale.

Identification, however, cannot always be done using external futures. Nine species of grey mullets have now been reported in the Mediterranean and the Black Sea. Among these species, three are not endemic. *Liza carinata* comes from the Red Sea and its distribution is restricted in the South-eastern areas around Israel (Golani et al. 2002). *Liza haematocheila* was introduced from the Japan Sea into brackish ponds in the Azov Sea in the late 1960s; later, more transfers and releases of juveniles followed and the species became rather common in lagoons of the Black Sea, replacing the stocks of the local mullets (Starushenko and Kazansky 1996), found also in freshwater systems in the northern Aegean Sea (Greece; Koutrakis and Economidis 2000, Golani et al. 2002). Saleh (2008) refers also to the presence of the bluespot mullet *Valamugil seheli* on the Mediterranean coast of Egypt, in the coastal waters of an area extending from Damyitta to the north-west of the Sinai Peninsula; its presence there, however, is rare and its fry are collected from the city of Suez near the mouth of Suez canal, the gulf of Suez and the Bitter Lakes (S. Sadek, pers. comm.). Moreover *Oedalechilus labeo* is a typically stenohaline species, found only in marine areas (Perlmutter et al. 1957).

The other five species however, which are present all around the Mediterranean, the Black Sea and the Atlantic Coast of Europe (*M. cephalus*, *L. aurata*, *L. ramada*, *L. saliens* and *C. labrosus*), have overlapping periods when fry and juveniles approach the coast and in similar sizes, thus their identification is very difficult. Perlmutter et al. (1957), following Bograd's (1955; quoted by Perlmutter et al. 1957) work for identification of grey mullets larger than 100 mm, was the first to present good descriptive information to enable the six (including *O. labeo*) Mediterranean mugilid species to be identified from as small as 20 mm (TL), including the side view of the fish and ventral view of the head and a view of the pyloric caeca. El-Zarka and Kamel (1965) also provided a key and plates with figures of the pyloric caeca, based on works of Zambriborch (1949, 1951; quoted by El-Zarka and Kamel 1965), in order to separate fry used for stocking Lake Mariut in Egypt. According to this, the easiest to separate is *M. cephalus* which has two pyloric caeca, while *C. labrosus* has six (rarely seven) caeca; *L. saliens* can also be separated by the two groups that the pyloric caeca are forming, the ventral with three or four longer than the other four or five. The most difficult species to separate by the number and form of pyloric caeca are *L. aurata* and *L. ramada*, since they both have seven to eight of the same length, even if El-Zarka and Kamel (1965) contend that the middle two caeca of *L. ramada* are somewhat smaller and thinner than the rest of the caeca, as if they were originally one.

The keys produced relied highly on the number and form of pyloric caeca, which require manipulation or sacrifice of the specimens; thus attempts have been made to find ways to enable biologists and commercial fishermen to quickly identify numerous small specimens, with minimal manipulation.

Professional fry fishermen, however, have gained enough experience and are capable of identifying them by looking at the schools without having to handle specimens. Katavic (1980) reported that *M. cephalus* fry is easily identified due to the marked silver body colour and relatively big head in relation to the body. Koutrakis et al. (1994) also reported that in autumn, young *M. cephalus* form mixed schools with *L. saliens* fry and that the two groups were easily distinguished in that the former were more vigorous with a bigger head, while the latter are longer with a characteristic line of chromatophores along the side of the body. The same was supported by Kladas et al. (1989); *M. cephalus*, which had high commercial value, can be distinguished from *L. saliens* schools (which had very low value) during late autumn, also by a white spot on the dorsal area of *M. cephalus*, while a small head with pointed snout characterises *L. saliens* fry; during spring *C. labrosus* juveniles are characterized by two white spots on the dorsal area. Moreover *L. aurata* juveniles that are found together with *L. ramada* in coastal areas can be separated because of their larger size (Villani 1987).

Pigmentation patterns constitute an alternative set of characters for fry identification. Some authors have described differences in pigmentation on the flanks of fry (Perlmutter et al. 1957, Farrugio 1975, 1977, Cambrony 1984, Serventi et al. 1996), but the differences are not distinct for fry of all stages. Perlmutter et al. (1957) also described the ventral head melanophore patterns and later other authors included information on the melanophore patterns on the ventral side of the head (e.g., Farrugio 1977, Zismann 1981). Zismann (1981) and Cambrony (1984) have proposed keys for fry of 20–60 mm (TL), based on both the arrangement of the pyloric caeca and pigmentation patterns. The five common grey mullet species that can be found around Mediterranean or Europe (*M. cephalus*, *Liza aurata*, *L. ramada*, *L. saliens* and *C. labrosus*) are shown in Fig. 12.1.

Reay and Cornell (1988) first provided a key for the British juvenile mugilid species, based on the patterns of ventral head melanophores at three different sizes (25–45 mm). Later, Serventi et al. (1996) provided a detailed description of the five Mediterranean species, with figures showing the pharyngobranchial morphology, and the dorsal and ventral view of head, showing distribution of chromatophores. Minos et al. (2002) also suggested a key for the five species, based on the shape of the lower jaw and the melanophore patterns along the edge of the lower jaw and the ventral side of the head. According to this, *M. cephalus* is characterized by lightly pigmented ventro-opercular and gular regions of the head; *L. aurata* has spots at the corners of the mouth, the gular region and the ventro-opercular region; in *L. ramada*, the pigmentation

Figure 12.1. Fry of the five common grey mullet species from the Mediterranean and the Atlantic coasts of Europe (*C. labrosus*, *L. aurata*, *L. ramada*, *L. saliens* and *M. cephalus*). Next to each species there is a photo of the ventral side of the head of each species (specimens have been preserved in buffered alcohol solution, except from *L. aurata* that was preserved in formalin) (photos by E. Koutrakis).

is darkest in the mandibular and gular regions; in *C. labrosus* there are two rows of melanophores in the ventro-opercular region and one or two rows in the gular region; finally *L. saliens* has two rows of melanophores in the ventro-opercular region and two in the gular region (Minos et al. 2002). Katselis et al. (2006) also proposed the use of analysis of morphometric variations of fry for future development of software routine for the identification of fry.

Figure 12.2 depicts a comparative description of the melanophore pattern on the ventral side of the head of the above five species of grey mullet and at length range of 20–30 mm. Drawings are based on the pigmentation patterns of each species published by Perlmutter et al. (1957), Reay and Cornell (1988) and by Minos et al. (2002) and on the observation of preserved specimens. Names of different regions are after Reay and Cornell (1988). Based only on the melanophore pattern on the ventral side, *C. labrosus* is the easiest to distinguish among other species, since it has a heavy pigmentation, both in the mandibular, the ventro-opercular and the gular regions, very pronounced in the anterior part of the head. *Mugil cephalus* can be distinguished by the inverse Y-shaped pigmentation in the gular region. *Liza* species are more difficult to discriminate. *Liza ramada* has very scarce melanophores in the gular region. On the other hand, *L. aurata* has a characteristic row of melanophores under the eye extending almost to the middle of the operculum and small size melanophores on the anterior part of the ventro operculum region. *Liza saliens*, which however can be identified by the characteristic stripe on the flanks of the body, on the ventral side of the head is mainly pigmented in the ventro-opercular regions.

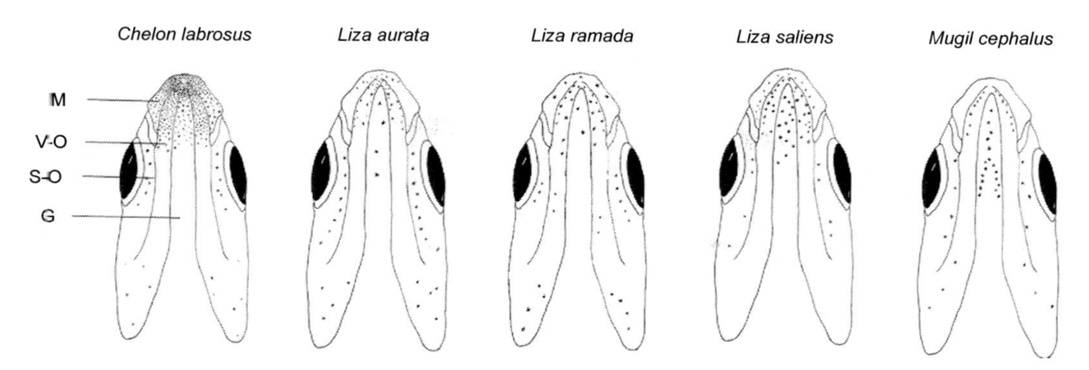

Figure 12.2. Comparative description of the melanophore pattern on the ventral side of the head in the five Mediterranean species of grey mullets (length range 20–30 mm, SL) (M = mandibular region, V-O = ventro-opercular, S-O = sub-orbital, G = gular region).

Development and Growth

The quantity of grey mullet fry ascending to coastal areas depends on the spawning success of adult mullet species in the sea and on the survival of the different development stages (embryonic, larval and juvenile) (El-Zarka and Kamel 1965, El-Zarka et al. 1970). According to Bensam (1989) the early development stages of marine fishes are egg, larva, post-larvae and juvenile: larva starts from hatching until yolk is absorbed, the mouth is formed and eyes are pigmented; post-larvae can be applied from the termination of the larval phase and includes the preflexion, flexion and post-flexion stages; among those, the post-flexion stage has the longest duration and leads to the juvenile phase. These stages are linked to the development of the caudal fin together with its supporting elements, before, during and after the posterior tip of the notochord flexes upwards (Bensam 1989).

Fish larvae, which often inhabit entirely different niches and present distinct body shape from juvenile and/or adult stages, are transitory forms possessing a dynamic and continuously changing morphology (Strauss and Bond 1990). Two critical periods are reported from Brusle (1981) during the development of the larvae of two mugilid species (*M. cephalus* and *L. macrolepis*), at age three days when the mouth is formed and feeding begins, which is a key event in the survival of all fishes (Nunn et al. 2012); and at age 11–12 days, when the larvae assume a fish-like shape, develop scales, and attain sustained swimming powers. *Mugil cephalus* is the grey mullet species whose development and growth is best studied around the world. Larvae of the species hatch with no mouth, paired fins, or branchial skeleton (Thomson 1963), range from 2.2–3.6 mm and have a yolk sac with an ovoid to oblong-ellipsoidal oil globule, while by 4 dph (days post hatching), the mouth is formed, pectoral fins are developing, gill clefts have opened, and the yolk has been absorbed; stomach, spleen, intestines, gall bladder, and swim bladder begin forming between 3.1 and 3.4 mm on the 5 dph and gill filaments begin to form at 3.4–3.8 mm at 8 dph (Martin and Drewry 1978). Growth becomes more intensive from 9 to 12 dph and by 10–15 dph the oil globule has been depleted (Kuo et al. 1973). From 5.4 to 6.6 mm the body has acquired scattered dark spots, with a silver-green colour (Martin and Drewry 1978). In *M. curema*, pigmentation also starts early; Anderson (1957a) reported that at 3.7 mm, pigmentation spots are scattered on the body while at 8.6 mm, spots are spread all over the head and sides of the body and at 14.5 mm the fish are so densely covered that they appear almost black. Powles (1981) suggested that the early development of heavy body pigmentation in *M. cephalus* and *M. curema* larvae may be an adaptation to high levels of ultraviolet radiation in the surface layer of the sea.

Chelon labrosus development was also studied, as a species with aquaculture potential, using larvae reared in mesocosms, an innovative technique for rearing fish which present good similarities to wild fish (Ben Khemis et al. 2006, 2013). The development priorities of the post-larvae at preflexion stage (4.5 mm TL, 14 dph) concerned mainly feeding efficiency, by improving detection ability (sensory system development), ingestion capacity (head growth) and assimilation performance (digestive system differentiation), together with respiration efficiency (gill development); while post-larval priorities at the post-flexion stage (8.5 mm TL, 25 dph) concerned locomotion and oxygenation performances, thus fast growth of trunk and tail, acquisition of adult axial muscle distribution and completion of gill filaments (Ben Khemis et al. 2013).

Scales start to develop at 8–11 mm, TL (24–25 dph) both in *M. cephalus* (Martin and Drewry 1978) and *C. labrosus* (Ben Khemis et al. 2013) while in *M. curema* they develop at 7.2 mm (Anderson 1957a). The largest *C. labrosus* scale was 0.40 x 0.25 mm with one–three circuli (Ben Khemis et al. 2006). Harrison and Howes (1991) studied the development of the pharyngobranchial organ of mugilids and it was at least moderately developed in 11 mm *Liza* spp. specimens. Eleven anal fin rays are present at 6 mm (Ditty and Shaw 1996). The 11 anal fin rays in *M. cephalus* fuse into a complement of two anal spines and nine anal fin rays at 19 to 23 mm TL (Martin and Drewry 1978). This stage, where mugilid post-larvae have only two anal spines is also referred as the querimana stage and at one time a genus *Querimana* was recognized for young mullets having only two anal spines (Thomson 1966). Shehadeh and Norris (1972) state that *M. cephalus* larval life lasts for 42 days, by which time a length of 17–18 mm is attained. In general, post-larvae appear in lagoons, estuaries and mangrove systems at 10–15 mm (SL), a size that corresponds to ages between four and six weeks (Thomson 1966).

According to Balon (1975), the juvenile period begins when the fins are fully differentiated and most of the temporary organs are replaced by definitive organs and Bensam (1989) defined as juvenile the stage at which all the vital meristic characters and general morphometric patterns have developed. Demir (1971) used the formation of the third anal spine to separate the post-larval stage from the juvenile (> 27 mm total length, TL) in three specimens of *Liza aurata* found off Plymouth. The development of the third anal spine from a ray, in *M. cephalus*, was already described by Jacot (1920), who observed that this ray ceases growing with the same rapidity as the true rays and becomes heavier basally until it becomes longer than the second spine, a length maintained throughout life. The diagnostic count of three anal spines and eight anal fin rays is evident in *M. cephalus* at 35 to 45 mm SL (Anderson 1958). Metamorphosis is concluded at 28.5 mm for *C. labrosus,* as was shown by Ben Khemis et al. (2013), while Zismann (1981) reported that the range of sizes at which there are two spines in the anal fin of Mugilidae may be as much as 30 mm (TL). That is why the length of 30 mm (TL) was chosen to be used for all grey mullet species as a threshold for separating post-larvae from juveniles (see also Introduction).

According to NCDENR (2005), Higgins (1927) observed that when juveniles *M. cephalus* in North Carolina (U.S.A.) arrive close to the coast by mid January, they show little growth until water temperatures reach 20ºC in mid April, and a growth rate of 20 mm per month occurred from May to October. Anderson (1958) in Georgia (U.S.A.) estimated 5 mm growth per month for juveniles of 18 to 19 mm (SL) from November until January, followed by no growth during the coldest winter months; while 10 mm growth occurred between February and March when water temperatures started rising, followed by 17 mm per month to next October. Anderson (1957a) also estimated the growth of *M. curema* at 17 mm per month during the first year, reaching 70 mm from spring to December. Kraul (1983) studied the growth of *M. cephalus* larvae in Hawaii by rearing larvae to bait size (1–2 g); individual growth averaged 4.4% weight gain per day from 40 to 130 days, while De Silva and Silva estimated a rate of 0.24 mm per day. *Mugil cephalus* juvenile growth was also described as $TL = 0.342\ (Age)^{1.039} \pm 19.4$ (McDonough and Wenner 2003).

Whitfield and Blaber (1978), by estimating the distance covered by juveniles in the St. Lucia lake system (South Africa), claimed that grey mullet species (*M. cephalus, L. macrolepis, L. dumerili, Valamugil (Moolgarda) cunnesius, V. buchanani*) entered the system at 10 mm and grew almost 10 mm per month for the first two–four months of their lives. Hickling (1970) estimated the growth of juvenile *C. labrosus* in Plymouth estuaries (U.K.) and found that, from the size of 20 mm in August, juveniles reached 40–45 mm by December; while specimens caught after one year from the first recruitment (August) grew at about 60 mm in coastal waters. He confirmed this by finding a 'winter ring' at 35–55 mm fish length, calculated from scales of thick-lipped mullet of 110–150 mm (FL). Kennedy and Fitzmaurice (1969) found similar growth for the same species in Ireland (40 mm by the first winter). Bartulović et al. (2007), in Croatia estimated the growth of *L. ramada* during the first 70 days of recruitment (+ 2.5% in terms of length and + 7.5% in terms of weight) that was low, even negative in the first 50 days and considered that this was due to the fact that recruitment was in a period of generally low temperatures (9–13°C, February), while a higher rate of growth was observed in April and later, when sea water temperature exceeded 15°C. Chervinsky (1975) compared the growth of juveniles *L. aurata* raised under experimental conditions as 0.6–1.1 g/fish/day, which was the slowest, comparing with *L. ramada* (1.07–1.70 g/fish/day) and the fastest *M. cephalus* (2.65–3.68 g/fish/day). *Mugil cephalus* grew much more in the experiments for polyculture in Israel (Pruginin et al. 1975).

Labourg et al. (1985) on the western Atlantic Coast of France estimated the weight-length relationships of juveniles in different months of their recruitment and the 'b' value ranged from 2.97 to 2.35 for *L. aurata*, 2.73 to 2.93 for *L. ramada* and 3.074 to 3.2 for *C. labrosus*. Minos et al. (1995) observed that the 'b' value of *L. ramada* is greater than that of *L. saliens*, thus the body of *L. ramada* tends to be more spherical than that of *L. saliens*, which is more slender. The condition factor was also estimated for *L. ramada* juveniles reaching 1.2 to 2.2 during the first six months; later stabilized at 1.8 during the winter season when reserves were consumed (Albertini-Berhaut 1975).

Among the factors influencing juvenile growth is certainly temperature. Rossi (1986) claims that grey mullet growth becomes apparent only when water temperature exceeds 10°C and that the yearly growth of juveniles is markedly seasonal, with a long period of no or slow growth, followed by a shorter one from April to June, when growth is highest. Albertini-Berhaut (1975) reports that autumn and winter seasons seem to slow down the growth rate of *L. ramada* in the Golfe of Marseilles, beginning in the month of November. Salinity levels may also alter the growth and food intake of juvenile mullets (Blaber 1987), as De Silva and Perera (1976) and Perera and De Silva (1978) showed with laboratory studies on *M. cephalus*, salinity dependence was highest at 30 psu; with conversion efficiencies, however, greatest at salinities lower than 10 psu, probably because the rate of digestion was slower at low salinities. Cardona (2006) suggested that growth of euryhaline species is often affected by salinity because the energy used for osmoregulation is not available for growth. As a result, there is an optimum salinity level at which many of the grey mullet species present the highest growth rate and at the same time, the lowest cost of osmoregulation; it is possible that this fact affects fish distribution in the wild (Blaber 1987). In West Africa however, in a hypersaline estuary, fry of *L. dumerili* (12 mm, SL) and *M. cephalus* (20 mm, SL) already had the osmoregulatory capacity to massively enter areas with salinities up to 78 psu (Trape et al. 2009), even if Saleh (2008) in Egypt states that only adult mullets have been found in waters ranging from zero salinity to 75, while juveniles can tolerate such wide salinity ranges only after they reach lengths of 40–70 mm. Walsh et al. (1991) investigated the combined effects of temperature and salinity on embryonic development of fertilized *M. cephalus* eggs transferred directly from the spawning tank and found that normal embryonic development occurred at water temperatures from 20 to 30°C and a salinity range of 15–36, as was also proposed by Kraul (1983), and hatching time was affected by temperature, but not by salinity. Hence their first appearance on the coast could be affected by the environmental conditions of that area in that year.

Thomson (1997) used the size of 60 mm (SL) as a threshold size from the adult stage of grey mullets, since at about this size, a degree of metamorphosis takes place, involving not only the transformation of the third supporting element of the anal fin to a spine (in the Mugilinae and a few Agonostominae genera), but also a considerable rearrangement of the mouth parts. Moreover within the range of 30 to 70 mm there are shifts in diet that also affect the growth of grey mullets (Albertini-Berhaut 1975, Koutrakis and Sinis 1994), while most grey mullets complete their first year at a length range from 90 to 130 mm (Albertini-Berhaut 1975, Koutrakis and Sinis 1994). Albertini-Berhaut (1975) in the Golfe de Marseille in France, observed a change in the length-weight relationship of *L. ramada* and *L. aurata*, at 67 mm and 72 mm (SL) respectively; Koutrakis and Sinis (1994) in northern Greece also describe a similar reduction in growth, at a length of 35–37 mm for three mugilid species (*C. labrosus, L. saliens, L. ramada*). Both studies proposed different length-weight curves for juveniles (< 67 and < 40 mm SL respectively), each significantly different from that of the adult of the same species. The aforementioned observations probably illustrate a powerful 'stress' that these fish experience, possibly because the different diets that they adopt include a different protein composition (from animal diet to mixed and finally to vegetarian diet). An ontogenic dietary shift occurring in a high number of mugilid species, when fish move from browsing on pelagic prey to grazing on benthic resources, has been shown by several authors, either by studying stomach contents (Brusle 1981, Eggold and Motta 1992, Verdiell-Cubedo et al. 2007, Sarr et al. 2013), or by the use of other techniques such as stable-isotope signatures (Koussoroplis et al. 2010, Lebreton et al. 2013). According to Lebreton et al. (2013), this shift from a pelagic to a limno-benthophagous diet is clearly related to length, and thus to the age of individuals, as shown by the sharp increases in $\delta^{13}C$ and $\delta^{15}N$ values from 20 to 30 mm (fork length, FL). Eggold and Motta (1992) however, stated that in juvenile *M. cephalus* the observed trophic shift was not related to morphological changes in the mouth and gillrakers. Some authors also support that this shift can also be detected on the scales of several young specimens of age 0+, 1+, and 2+ as a 'fry ring' (Albertini-Berhaut 1974, De Silva 1980).

It should be noted that many authors use standard length for measuring fry and juveniles, since the tail is often damaged by handling during sampling procedures. Therefore conversion equations are given from the most common commercially European grey mullet species: *M. cephalus:* TL = 1.22 SL (for larvae > 8 mm, SL; Kraul 1983); *C. labrosus*: TL = –0.565 + 1.252 SL, *L. ramada*: TL = 0.025 + 1.236 SL, *L. saliens*: TL = 0.460 + 1.249 SL (Koutrakis and Sinis 1994). In average a more simplistic estimation is considering TL = 1.2 SL.

Acknowledgements

To Maria Anagnostopoulou for editing the language of the chapter, Argyris Sapounidis for his help in preparing the figures and to Chrysanthi Koutraki for her drawings.

References

Abu El-Regal, M.A. and N.K. Ibrahim. 2014. Role of mangroves as a nursery ground for juvenile reef fishes in the southern Egyptian Red Sea. Egyptian Journal of Aquatic Research, http://dx.doi.org/10.1016/j.ejar.2014.01.001.

Albertini-Berhaut, J. 1973. Biologie des stades juvéniles de Téléostéens Mugilidae, *Mugil auratus* Risso 1810, *M. capito* Cuvier 1829 et *M. saliens*, Risso 1810. 1. Régime alimentaire. Aquaculture 2: 251–266.

Albertini-Berhaut, J. 1974. Biologie des stades juvéniles de Téléostéens Mugilidae, *Mugil auratus* Risso 1810, *Mugil capito* Cuvier 1829 et *Mugil saliens* Risso 1810. 11. Modifications du régime alimentaire en relation avec la taille. Aquaculture 4: 13–27.

Albertini-Berhaut, J. 1975. Biologie des stades juvéniles de Téléostéens Mugilidae, *Mugil auratus* Risso 1810, *Mugil capito* Cuvier 1829 et *Mugil saliens* Risso 1810. III. Croissance linéaire et pondérale de *Mugil capito* dans le Golfe de Marseille. Aquaculture 5: 179–97.

Albertini-Berhaut, J. 1987. L'intestin chez les Mugilidae (Poissons: Téléostéens) à différentes étapes de leur croissance. I. Aspects morphologiques et histologiques. J. Appl. Ichthyol. 3: 1–12.

Anderson, M. 1982. The identification of British grey mullets. J. Fish Biol. 20: 33–38.

Anderson, W.W. 1957a. Early development, spawning, growth, and occurrence of the silver mullet (*Mugil curema*) along the South Atlantic Coast of the United States. U.S. Fish Wildl. Serv. Fish Bull. 119: 397–414.

Anderson, W.W. 1957b. Larval forms of the fresh-water mullet (*Agonostomus monticola*) from the open ocean off the Bahamas and South Atlantic coast of the United States. U.S. Fish Wildl. Serv. Fish Bull. 120: 415–425.

Anderson, W.W. 1958. Larval development, growth, and spawning of striped mullet (*Mugil cephalus*) along the south Atlantic coast of the United States. Fish Bull. 58: 501–519.

Baldó, F. and P. Drake. 2002. A multivariate approach to the feeding habits of small fishes in the Guadalquivir Estuary (SW Spain). J. Fish Biol. 61 (Supplement A): 21–32.

Balon, E.K. 1975. Terminology of intervals in fish development. J. Fish Res. Board Can. 32: 1663–1670.

Bar-Ilan, M. 1975. Stocking of *Mugil capito* and *Mugil cephalus* and their commercial catch in Lake Kinneret. Aquaculture 5: 85–89.

Bartulović, V., B. Glamuzina, D. Lučić, A. Conides, N. Jasprica and J. Dulčić. 2007. Recruitment and food composition of juvenile thin-lipped grey mullet, *Liza ramada* (Risso 1826), in the Neretva River estuary (Eastern Adriatic, Croatia). Acta Adriatica 48: 25–37.

Beckley, L.E. 1983. The ichthyofauna associated with *Zostera capensis* Setchell in the Swartkops estuary, South Africa. S. Afr. J. Zool. 18: 15–24.

Beckley, L.E. 1985. Tidal exchange of ichthyoplankton in the Swartkops estuary mouth, South Africa. S. Atr. Tydskr. Dierk. 20: 15–20.

Bell, J.D., D.A. Pollard, J.J. Burchmore, B.C. Pease and M.J. Middleton. 1984. Structure of a fish community in a temperate tidal mangrove creek in Botany Bay, New South Wales. Aust. J. Mar. Freshw. Res. 35: 33–46.

Ben Khemis, I., D. Zouiten, R. Besbes and F. Kamoun. 2006. Larval rearing and weaning of thick-lipped grey mullet (*Chelon labrosus*) in mesocosm with semi-extensive technology. Aquaculture 259: 190–201.

Ben Khemis, I., E. Gisbert, C. Alcaraz, D. Zouiten, R. Besbes, A. Zouiten, A.A. Masmouti and C. Cahu. 2013. Allometric growth patterns and development in larvae and juveniles of thick-lipped grey mullet *Chelon labrosus* reared in mesocosm conditions. Aquaculture Research 44: 1872–1888.

Bensam, P. 1989. Terminology of early development stages of marine fish. Proceedings of the Summer Institute in Recent Advances on the Study of Marine Fish Eggs and Larvae, CMFRI/SI/1989/Th.VII.

Blaber, S.J.M. 1976. The food and feeding ecology of Mugilidae in the St. Lucia lake system. Biol. J. Limn. Soc. 8: 267–277.

Blaber, S.J.M. 1987. Factors affecting recruitment and survival of mugilids in estuaries and coastal waters of southern Africa. Amer. Fish Soc. Symp. 1: 507–518.

Blaber, S.J.M. 2002. 'Fish in hot water': the challenges facing fish and fisheries research in tropical estuaries. J. Fish Biol. 61 (Supplement A): 1–20.

Blaber, S.J.M. and A.K. Whitfield. 1977. The feeding ecology of juvenile mullet (Mugilidae) in south east African estuaries. Biol. J. Linn. Soc. 9: 277–284.

Blaber, S.J.M. and T.G. Blaber. 1980. Factors affecting the distribution of juvenile estuarine and inshore fish. J. Fish Biol. 17: 143–162.

Bobori, D.C., I. Rogdakis and P.S. Economidis. 1998. Some preliminary results of fish stocking with *Mugil cephalus* in several lakes of Greece. EIFAC—European Inland Fisheries Advisory Commission, Lisbon, Portugal, 23–25 June.

Bograd, L. 1961. Occurrence of *Mugil* in the rivers of Israel. Bull. Res. Counc. of Israel 9B: 169–190.

Bok, A.H. 1979. The distribution and ecology of two mullet species in some freshwater rivers in the Eastern Cape, South Africa. J. Limnol. Soc. Sth. Afr. 5: 97–102.

Brusle, J. 1981. Sexuality and biology of reproduction in grey mullets. pp. 99–154. *In*: O.H. Oren (ed.). Aquaculture of Grey Mullets. IBP 26, Cambridge University Press, Cambridge.

Cabral-Solís, E.G., M. Gallardo-Cabello, E. Espino-Barr and A.L. Ibáñez. 2010. Reproduction of *Mugil curema* (Pisces: Mugilidae) from the Cuyutlán lagoon, in the Pacific coast of México. Avances en Investigación Agropecuaria 14: 3–18.

Cambrony, M. 1984. Identification et périodicité du recrutement des juvéniles de Mugilidae dans les étangs littoraux du Languedoc-Roussillon. Vie et Milieu 34: 221–227.

Cardona, L. 2001. Non-competitive coexistence between Mediterranean grey mullet (Osteichthyes, Mugilidae): evidences from seasonal changes in food availability, niche breadth and trophic overlap. J. Fish Biol. 59: 729–744.

Cardona, L. 2006. Habitat selection by grey mullets (Osteichthyes, Mugilidae) in Mediterranean estuaries: the role of salinity. Sci. Mar. 70: 443–455.

Cardona, L., B. Hereu and X. Torras. 2008. Juvenile bottlenecks and salinity shape grey mullet assemblages in Mediterranean estuaries. Est. Coast. Shelf Sci. 77: 623–632.

Chang, C.W., W.N. Tzeng and Y.C. Lee. 2000. Recruitment and hatching dates of grey mullet (*Mugil cephalus* L.) juveniles in the Tanshui Estuary of Northwest Taiwan. Zoological Studies 39: 99–106.

Chang, C.W., Y. Iizuka and W.N. Tzeng. 2004. Migratory environmental history of the grey mullet *Mugil cephalus* as revealed by otolith Sr:Ca ratios. Mar. Ecol. Prog. Ser. 269: 277–288.

Chauvet, C. 1980. Estimation de la mortalité dans l'ichthyofaune du lac de Tunis occasionnée par les rejets de la centrale thermique. Impact sur la gestion du lac. Journées Étud. Pollutions, Cagliari, C.I.E.S.M. 749–756.

Chauvet, C. 1984. La pêcherie du lac de Tunis. Biologie des pêches et relèvement de la production par des voies autres que la réglementation. pp. 615–691. *In*: J.M. Kapetsky and G. Lasserre (eds.). Management of Coastal Lagoon Fisheries. GFCM Studies and Reviews n. 61, FAO, Rome.

Chauvet, C. 1986. Exploitation des poissons en milieu lagunaire méditerranéen. Dynamique du peuplement ichthyologique de la lagune de Tunis et des populations exploitées par les bordigues (muges, loups, daurades). Thèse, Univ. Perpignan. 549pp.

Chaves, P., H. Pichler and M. Robert. 2002. Biological, technical and socioeconomic aspects of the fishing activity in a Brazilian estuary. Journal of Fish Biology 61 (Supplement A): 52–59.

Chervinsky, J. 1975. Experimental raising of golden grey mulet (*L. aurata*) in saltwater ponds. Aquaculture 5: 91–98.

Chervinsky, J. 1982. Stocking Lake Kinneret with *L. aurata* and *C. labrosus* in the event of shortage in *M. cephalus* and *L. ramada*. Fish. Fish breeding Israel 16: 28–33.

Chessa, L.A., S. Casu, G.M. Delitala, R.A. Vacca, G. Corso, M. Pala, S. Ligios, A. Pais and S. Tola. 1988. The Calich Lagoon (NW Sardinia): general ecological observation and fry migration. Rapp. Comm. int. Mer. Médit. 31: 63.

Claridge, P.N., I.C. Potter and M.W. Hardisty. 1986. Seasonal changes in movements, abundance, size composition and diversity of the fish fauna of the Severn estuary. J. Mar. Biol. Ass. UK 66: 229–258.

Collins, M.R. and B.W. Stender. 1989. Larval striped mullet (*Mugil cephalus*) and white mullet (*M. curema*) off the southeastern United States. Bull. Mar. Sci. 45: 580–589.

Crosetti, D. 2015. Current state of grey mullet fisheries and culture. *In*: D. Crosetti and S.J.M. Blaber (eds.). Biology, Ecology and Culture of Grey Mullet (Mugilidae). CRC Press, Boca Raton, USA (this book).

Cyrus, D.P. and S.J.M. Blaber. 1987a. The influence of turbidity on juvenile marine fishes in estuaries. Part 1. Field studies at Lake St. Lucia on the southeastern coast of Africa. J. Exp. Mar. Biol. Ecol. 109: 53–70.

Cyrus, D.P. and S.J.M. Blaber. 1987b. The influence of turbidity on juvenile marine fishes in estuaries. Part 2. Laboratory studies, comparisons with field data and conclusions. J. Exp. Mar. Biol. Ecol. 109: 71–91.

De Angelis, C.M. 1967. Osservazioni sulle specie del genere *Mugil* segnalate lungo le coste del Mediterraneo. Bolletino di Pesca, Piscic. e Idrob. 22: 5–36.

De Angelis, R. 1960. Mediterranean Brackish Water Lagoons and their Exploitation. GFCM Studies & Reviews n. 12, FAO, Rome.

De Silva, S.S. 1980. Biology of the young grey mullet: a short review. Aquaculture 19: 21–37.

De Silva, S.S. and P.A.B. Perera. 1976. Studies of the young grey mullet, *Mugil cephalus* L. I. Effects of salinity on food intake, growth and food conversion. Aquaculture 7: 327–338.

De Silva, S.S. and E.I.L. Silva. 1979. Biology of young grey mullet, *Mugil cephalus* L., populations in a coastal lagoon in Sri Lanka. J. Fish Biol. 15: 9–20.

Demir, N. 1971. On the occurrence of grey mullet postlarvae off Plymouth. J. Mar. Biol. Ass. UK 51: 235–246.

Ditty, J.G. and R.F. Shaw. 1996. Spatial and temporal distribution of larval striped mullet (*Mugil cephalus*) and white mullet (*M. curema*, Family: Mugilidae) in the northern Gulf of Mexico, with notes on mountain mullet *Agonostomus monticola*. Bull. Mar. Sci. 59: 271–288.

Dodson, J.J. 1988. The nature and role of learning in the orientation and migratory behavior of fishes. Environ. Biol. Fish. 23: 161–182.

Durand, J.D. 2015. Implications of molecular phylogeny for the taxonomy of Mugilidae. *In*: D. Crosetti and S.J.M. Blaber (eds.). Biology, Ecology and Culture of Grey Mullet (Mugilidae). CRC Press, Boca Raton, USA (this book).

Durand, J.D., W.J. Chen, K.N. Shen, C. Fu and P. Borsa. 2012a. Genus-level taxonomic changes implied by the mitochondrial phylogeny of grey mullets (Teleostei: Mugilidae). Compt. R. Biol. 335: 687–697.

Durand, J.D., K.N. Shen, W.J. Chen, B.W. Jamandre, H. Blel, K. Diop, M. Nirchio, F.J. García De León, A.K. Whitfield, C.W. Chang and P. Borsa. 2012b. Systematics of the grey mullets (Teleostei: Mugiliformes: Mugilidae): Molecular phylogenetic evidence challenges two centuries of morphology-based taxonomy. Mol. Phylogenet. Evol. 64: 73–92.

Eggold, B.T. and P.J. Motta. 1992. Ontogenetic dietary shifts and morphological correlates in striped mullet, *Mugil cephalus*. Environmental Biology of Fishes 34: 139–158.

El-Zarka, S.E.D. and F. Kamel. 1965. Mullet fry transplantation and its contribution to the fisheries of inland brackish lakes in the United Arab Republic. Proc. Gen. Fish Coun. Medit. 8: 209–226.

El-Zarka, S., A.M. El-Maghraby and K. Abdel-Hamid. 1970. Studies on the distribution, growth and abundance of migrating fry and juveniles of mullet in a brackish coastal lake (Edku) in the United Arab Republic. Stud. Rev. Gen. Fish Council Mediterr. 46: 19.

Eschmeyer, W.N. and J.D. Fong (eds.). 2014. Catalog of Fishes: Species by Family/Subfamily (http://researcharchive.calacademy.org/research/ichthyology/catalog/SpeciesByFamily.asp). Electronic version accessed 20.04.2014.

Ezenwa, B.L., W.O. Alegbeleye, P.E. Anyanwu and P.U. Uzukwu. 1990. Cultivable fish seeds in Nigeria coastal waters: A research survey (second phase: 1986–1989. Nigerian Institute for Oceanography and Marine Research Lagos. Technical Paper 66: 34pp.

Fahay, M.P. 1975. An annotated list of larval and juvenile fishes captured with surface-towed meter nets in the South Atlantic Bight during four RV Dolphin cruises between May, 1967 and February, LW. NO AA Tech. Rep. NMFS SSRF. 39pp.

Farrugio, H. 1975. Les muges (Poisons téléostéens) de Tunisie. Répartition et pêche, contribution à leur étude systématique et biologique. Thesis Dissertation p. 201.

Farrugio, H. 1977. Clés commentées pour la détermination des adultes et des alevins de Mugilidae de Tunisie. Cybium (3ème Série) 2: 57–73.

Figueiredo, M.J. and J.J. Silva. 1983. Preliminary experiments on the rearing of mugilid fry from portuguese estuarine waters. Biol. Inst. Nac. Invest. Pescas, Lisboa 10: 51–63.

Franco, A., P. Franzoi, S. Malavasi, F. Riccato and P. Torricelli. 2006a. Use of shallow water habitats by fish assemblages in a Mediterranean coastal lagoon. Est. Coast. Shelf Sci. 66: 67–83.

Franco, A., P. Franzoi, S. Malavasi, F. Riccato and P. Torricelli. 2006b. Fish assemblages in different shallow water habitats of the Venice Lagoon. Hydrobiologia 555: 159–174.

Franco, A., M. Elliott, P. Franzoi and P. Torricelli. 2008. Life strategies of fishes in European estuaries: the functional guild approach. Mar. Ecol. Prog. Ser. 354: 219–228.

Franco, A., A. Pérez-Ruzafa, H. Drouineau, P. Franzoi, E.T. Koutrakis, M. Lepage, D. Verdiell-Cubedo, M. Bouchoucha, A. López-Capel, F. Riccato, A. Sapounidis, C. Marcos, J. Oliva-Paterna, M. Torralva-Forero and P. Torricelli. 2012. Assessment of fish assemblages in coastal lagoon habitats: Effect of sampling method. Est. Coast. Shelf Sci. 112: 115–125.

Gandolfi, G. 1978. La rimonta di novellame di Mugilidi (Pisces: Mugilidae) alla foce del fiume Magra (Golfo di La Spezia). Ateneo Parmense, Acta Nat. 14: 157–166.

Gandolfi, G., R. Rossi and P. Tongiorgi. 1981. Osservazioni sulla montata del pesce novello lungo le coste italiane. Quad. Lab. Tecnol. Pesca. 3: 215–232.

Gautier, D. and J. Hussenot. 2005. Les Mulets des Mers d'Europe. Synthèse des connaissances sur les bases biologiques et les techniques d'aquaculture. Editions IFREMER.

Gibson, R.N. 2003. Go with the flow: tidal migration in marine animals. pp. 153–161. *In*: M.B. Jones, A. Ingolfsson, E. Olafsson, G.V. Helgason, K. Gunnarsson and J. Svavarsson (eds.). Migrations and Dispersal of Marine Organisms. Hydrobiologia 503.

Gisbert, E., L. Cardona and F. Castelló. 1995. Competition between mullet fry. J. Fish Biol. 47: 414–420.

Golani, D., L. Orsi-Relini, E. Massuti and J.P. Quingnard. 2002. CIESM Atlas of Exotic Species in the Mediterranean. Vol. 1. Fishes. F. Briand (ed.). CIESM Publisher, Monaco. 254pp.

Grant, C.J. and A.V. Spain. 1975. Reproduction, growth and size allometry of *M. cephalus* L. (Pisces: Mugilidae) from North Queensland inshore waters. Austr. J. Zool. 23: 181–201.

Harden Jones, F.R. 1984. A view from the ocean. pp. 1–26. *In*: J.D. McCleave, G.P. Arnold, J.J. Dodson and W.H. Neill (eds.). Mechanisms of Migration in Fishes. Plenum Press, New York and London.

Harrison, I.J. and G.J. Howes. 1991. The pharyngobranchial organ of mugilid fishes; its structure, variability, ontogeny, possible function and taxonomic utility. Bull. Br. Mus. Nat. Hist. (Zool.) 57: 111–132.

Harrison, T.D. and J.A.G. Cooper. 1991. Active migration of juvenile grey mullet (Teleostei: Mugilidae) into a small lagoonal system on the Natal coast. South African Journal of Science 87: 395–396.

Harrison, T.D. and A.K. Whitfield. 1995. Fish community structure in three temporarily open/closed estuaries on the natal coast. J.L.B. Smith Institute of Ichthyology, Ichthyological Bulletin No. 64.

Heldt, H. 1929. Le lac de Tunis (partie nord); résultats des pêches au filet fin. Bull. Sm. Oceanogr. Salammbô 2: 5–74.

Herzberg, A. and R. Pasteur. 1975. The identification of grey mullet species by disc electrophoresis. Aquaculture 5: 99–106.

Hettler, W.F., D.S. Peters, D.R. Colby and E.H. Laban. 1997. Daily variability in abundance of larval fishes inside Beaufort Inlet. Fish Bull. 95: 477–493.

Hickling, C.F. 1970. A contribution to the natural history of the English grey mullets (Pisces: Mugilidae). J. Mar. Biol. Ass. UK 50: 609–633.

Higgins, E. 1927. Progress in biological inquiries, 1926. U.S. Bureau of Fisheries, Report of Commissioner of Fisheries 1029: 517–559.

Hung, C.-M. and D. Shaw. 2006. The impact of upstream catch and global warming on the grey mullet fishery in Taiwan: A non-cooperative game analysis. Mar. Res. Econ. 21: 285–300.

Ibáñez, A.L. 2015. Age and growth of Mugilidae. *In*: D. Crosetti and S.J.M. Blaber (eds.). Biology, Ecology and Culture of Grey Mullet (Mugilidae). CRC Press, Boca Raton, USA (this book).

Ibáñez, A.L., I.G. Cowx and P. O'Higgins. 2007. Geometric morphometric analysis of fish scales for identifying genera species, and local populations within Mugilidae. Can. J. Fish Aquat. Sci. 64: 1091–1100.

Ishak, M.M., S.A. Abdel-Malek and M.M. Sharik. 1982. Development of mullet fisheries (Mugilidae) in Lake Quarun, Egypt. Aquaculture 27: 251–260.

IUCN. 2007. Guide for the Sustainable Development of Mediterranean Aquaculture. Interaction between Aquaculture and the Environment. Gland, Switzerland and Malaga, Spain.

Jacot, A.P. 1920. Age, growth and scale characters of the mullets, *M. cephalus* and *M. curema*. Trans. Amer. Fish. Soc. 39: 199–229.

Katavic, I. 1980. Temporal distribution of young mugilids (Mugilidae) in the coastal waters of the Central Eastern Adriatic. Acta Adriatica 21: 137–150.

Katselis, G., G. Minos, A. Marmagas, G. Hotos and I. Ondrias. 1994. Seasonal distribution of Mugilidae fry and juveniles in Messolonghi coastal waters, western Greece. Bios (Macedonia, Greece) 2: 101–108.

Katselis, G., G. Hotos, G. Minos and K. Vidalis. 2006. Phenotypic affinities on fry of four Mediterranean grey mullet species. Turk. J. Fish. Aquat. Sci. 6: 49–55.

Kennedy, M. and P. Fitzmaurice. 1969. Age and growth of thick-lipped grey mullet *Crenimugil labrosus* (Risso) in Irish waters. J. Mar. Biol. Ass. UK 49: 683–699.

Kladas, G. and G. Rogdakis. 1988. Seasonal occurrence of some euryhaline fish fry in the southwestern coasts of Greece. pp. 26–33. 4th Panhellenic Congress Ichthyol., June 23–28, 1988, Thessaloniki, Greece (in Greek).

Kladas, G., G. Rogdakis and T. Mpaltas. 1989. Fishing wild fry. Fishing News 99: 61–67 (in Greek).

Koussoroplis, A.M., A. Bec, M.E. Perga, E. Koutrakis, C. Desvilettes and G. Bourdier. 2010. Nutritional importance of minor dietary sources for leaping grey mullet *Liza saliens* (Mugilidae) during settlement: insights from fatty acid $\delta^{13}C$ analysis. Mar. Ecol. Prog. Ser. 404: 207–217.

Koutrakis, E.T. 1994. Grey mullets and their potential introduction in internal waters. Fisheries News 161: 58–63 (in Greek).

Koutrakis, E.T. 1999. Review of the history and the systematics of the Mediterranean grey mullets (Mugilidae). Geotechnical Scientific Issues 10, 6: 365–374 (in Greek).

Koutrakis, E.T. 2004. Temporal occurrence and size distribution of grey mullet juveniles (Pisces: Mugilidae) in the estuarine systems of the Strymonikos Gulf (Greece). J. Appl. Ichthyol. 20: 76–78.

Koutrakis, E.T. and A.I. Sinis. 1994. Growth analysis of gray mullets (Pisces: Mugilidae) as related to age and site. Isr. J. Zool. 40: 37–53.

Koutrakis, E.T. and P.S. Economidis. 2000. First record in the Mediterranean (North Aegean Sea, Greece) of the Pacific *Mugil so-iuy* Basilewsky 1855 (Pisces: Mugilidae). Cybium 24: 299–302.

Koutrakis, E.T., A.I. Sinis and P.S. Economidis. 1994. Seasonal occurrence, abundance and size distribution of grey mullet fry (Pisces: Mugilidae) in the Porto-Lagos Lagoon and Lake Vistonis (Aegean Sea, Greece). The Israeli Journal of Aquaculture-Bamidgeh 46: 182–196.

Koutrakis, E.T., A.C. Tsikliras and A.I. Sinis. 2005. Temporal variability of the ichthyofauna in a Northern Aegean coastal lagoon (Greece). Influence of environmental factors. Hydrobiologia 543: 245–257.

Koutrakis, E.T., G. Sylaios, N.I. Kamidis, D. Markou and A. Sapounidis. 2009. Fish fauna recovery in a newly re-flooded Mediterranean coastal lagoon. Est. Coast. Shelf Sci. 83: 505–515.

Koutrakis, E.T., A. Sapounidis, D. Lachouvaris, D. Chariskos and A. Mirli. 2010. Fish assemblages of two adjacent coastal lagoons in river Nestos delta (NE Greece). Rapp. Comm. int. Mer. Médit. 39: 766.

Kraul, S. 1983. Results and hypothesis for the propagation of the grey mullet, *Mugil cephalus* L. Aquaculture 30: 273–284.
Kristensen, I. 1963. Hypersaline bays as an environment of young fish. Proc. Gulf Caribb. Fish Inst. 16: 139–142.
Kuo, C.M., Z.H. Shehadeh and K.K. Milisen. 1973. A preliminary report on the development, growth and survival of laboratory reared larvae of the grey mullet, *Mugil cephalus* L. J. Fish Biol. 5: 459–470.
Labourg, P.J., C. Clus and G. Lasserre. 1985. Résultats préliminaries sur la distribution des juvéniles de poissons dans un marais maritime du Bassin d'Arcachon. Oceanologica Acta 8: 331–341.
Lasserre, P. and J.L. Gallis. 1975. Osmoregulation and differential penetration of two grey mullets, *Chelon labrosus* (Risso) and *Liza ramada* (Risso) in estuarine fish ponds. Aquaculture 5: 323–344.
Leber, K.M., S.M. Arce, D.A. Sterritt and N.P. Brennan. 1996. Marine stock-enhancement potential in nursery habitats of striped mullet, *Mugil cephalus*, in Hawaii. Fish Bull. 94: 452–471.
Leber, K.M., H.L. Blankenship, S.M. Arce and N.P. Brennan. 1997. Influence of release season on size-dependent survival of cultured striped mullet, *Mugil cephalus*, in a Hawaiian estuary. Fish Bull. 95: 267–279.
Lebreton, B., P. Richard, G. Guillou and G.F. Blanchard. 2013. Trophic shift in young-of-the-year Mugilidae during salt-marsh colonization. J. Fish Biol. 82: 1297–1307.
Liao, I.C. 1981. Cultivation methods. pp. 361–390. *In*: O.H. Oren (ed.). Aquaculture of Grey Mullets. Cambridge University Press, Cambridge.
Liao, I.C., N.H. Chao and C.C. Tseng. 2015. Case study: Capture and culture of Mugilidae in Taiwan. *In*: D. Crosetti and S.J.M. Blaber (eds.). Biology, Ecology and Culture of Grey Mullets (Mugilidae). CRC Press, Boca Raton, USA (this book).
Liordos, V. and V. Goutner. 2007. Spatial patterns of winter diet of the great cormorant in coastal wetlands of Greece. Waterbirds 30: 103–111.
Loftus, W.F., J.A. Kushlan and S.A. Voorhees. 1984. Status of the mountain mullet in southern Florida. Florida Scientist 47: 256–263.
Lucas, M.C. and E. Baras. 2001. Migration of Freshwater Fishes. Blackwell Science Ltd., Oxford.
Major, P.F. 1978. Aspects of estuarine intertidal ecology of juvenile stripped mullet, *Mugil cephalus* in Hawaii. Fish Bull. 76: 299–314.
Malavasi, S., R. Fiorin, A. Franco, P. Franzoi, A. Granzotto, F. Riccato and D. Mainardi. 2004. Fish assemblages of Venice Lagoon shallow waters: an analysis based on species, families and functional guilds. J. Mar. Syst. 51: 19–31.
Martin, F.D. and G.E. Drewry. 1978. Development of fishes of the Mid-Atlantic Bight. Fish and Wildlife Service, United States Department of the Interior 6: 416.
Mastermann, A.T. 1913. Report on investigations upon the Salmon with special reference to age determination by stuffy of scales. Board of Agriculture and Fisheries of England, Fisheries Investigations, Series I I: 12.
Mathieson, S., A. Cattrijsse, M.J. Costa, P. Drake, M. Elliott, J. Gardner and J. Marchand. 2000. Fish assemblages of European tidal marshes: a comparison based on species, families and functional guilds. Mar. Ecol. Prog. Ser. 204: 225–242.
McDonough, C.J. and C.A. Wenner. 2003. Growth, recruitment, and abundance of juvenile striped mullet (*Mugil cephalus*) in South Carolina estuaries. Fish Bull. 101: 343–357.
McDowall, R.M. 1997. The evolution of diadromy in fishes (revisited) and its place in phylogenetic analysis. Rev. Fish Biol. Fish. 7: 443–462.
Menezes, M.R., M. Martins and S. Naik. 1992. Interspecific genetic divergence in grey mullets from the Goa region. Aquaculture 105: 117–129.
Mićković, B., M. Nikčević, A. Hegediš and I. Damjanović. 1994. Seasonal dynamics of fish fry populations in brackish waters of the Mrčevo Valley. Bios. 2: 143–147.
Mićković, B., M. Nikčević, A. Hegediš, S. Regner, Z. Gacic and J. Krpo-Cetkovic. 2010. Mullet fry (Mugilidae) in coastal waters of Montenegro, their spatial distribution and migration phenology. Arch. Biol. Sci., Belgrade 62: 107–114.
Minos, G., G. Katselis, P. Kaspiris and I. Ondrias. 1995. Comparison of the change in morphological pattern during the growth in length of the grey mullets, *Liza ramada* and *Liza saliens* from Western Greece. Fisheries Research 23: 143–155.
Minos, G., G. Katselis, I. Ondrias and I.J. Harrison. 2002. Use of melanophore patterns on the ventral side of the head to identify fry of grey mullet (Teleostei: Mugilidae). Israeli J. Aquat.-Bamidgeh 54: 12–26.
Minos, G., A. Imsiridou and P.S. Economidis. 2010. *Liza haematocheilus* (Pisces: Mugilidae) in the northern Aegean Sea. pp. 313–332. *In*: D. Golani and B. Appelbaum-Golani (eds.). Fish Invasions of the Mediterranean Sea: Change and Renewal. Pensoft Publishers.
Myers, G.S. 1949. Usage of anadromous, catadromous and allied terms for migratory fishes. Copeia 2: 89–96.
Nagelkerken, I., S.J.M. Blaber, S. Bouillon, P. Green, M. Haywood, L.G. Kirton, J.O. Meynecke, J. Pawlik, H.M. Penrose, A. Sasekumar and P.J. Somerfield. 2008. The habitat function of mangroves for terrestrial and marine fauna: A review. Aquatic Botany 89: 155–185.
Nash, C.E. 1978. The grey mullet (*Mugil cephalus* L.) as a marine bio-indicator. International Workshop on Monitoring Environmental Materials and Specimen Banking. October 23–28, Berlin.
Nash, C.E. and R.M. Koningsberger. 1981. Artificial propagation. pp. 265–312. *In*: O.H. Oren (ed.). Aquaculture of Grey Mullets. IBP 26, Cambridge University Press, Cambridge.
Nash, E.C. and C.M. Kuo. 1975. Hypotheses for problems impeding the mass propagation of grey mullet and other finfish. Aquaculture 5: 119–133.
NCDENR, 2005. North Carolina Fishery Management Plan: Striped Mullet. North Carolina Department of Environment and Natural Resources (NCDENR). Morehead City. U.S.A.

Nelson, J.S. 2006. Fishes of the World, 4th ed. John Wiley and Sons, New York.

Nishimoto, R.T., T.E. Shimoda and L.K. Nishiura. 2007. Mugilids in the Muliwai: a tale of two mullets. pp. 143–156. *In*: N.L. Evenhuis and J.M. Fitzsimons (eds.). Biology of Hawaiian Streams and Estuaries. Bishop Museum Bulletin in Cultural and Environmental Studies 3.

Nordlie, F.G. 2015. Adaptation to salinity and osmoregulation in Mugilidae. *In*: D. Crosetti and S.J.M. Blaber (eds.). Biology, Ecology and Culture of Grey Mullet (Mugilidae). CRC Press, Boca Raton, USA (this book).

Nunn, A.D., L.H. Tewson and I.G. Cowx. 2012. The foraging ecology of larval and juvenile fishes. Rev Fish Biol. Fisheries 22: 377–408.

Paulin, C.D. and L.J. Paoul. 2006. The Kaipara mullet fishery: nineteenth-century management issues revisited. Tuhinga 17: 1–26.

Payne, A.I. 1976. The relative abundance and feeding habits of the grey mullet species occurring in an estuary in Sierra Leone, West Africa. Mar. Biol. 35: 277–286.

Perera, P.A.B. and De Silva. 1978. Studies on the biology of young grey mullet (*Mugil cephalus*) digestion. Mar. Biol. 44: 383–387.

Perez-Ruzafa, A. and C. Marcos. 2012. Fisheries in coastal lagoons: An assumed but poorly researched aspect of the ecology and functioning of coastal lagoons. Est. Coast. Shelf Sci. 110: 15–31.

Perlmutter, A., I. Bogradand and J. Pruginin. 1957. Use of the estuarine and sea fish of the family Mugilidae (grey mullets) for ponds culture in Israel. Proc. Gen. Fish. Counc. Mediterr. 4: 289–304.

Powles, H. 1981. Distribution and movements of neustonic young of estuarine dependent (*Mugil* spp., *Pomatomus saltatrix*) and estuary independent (*Coryphaena* spp.) fishes off the southeastern United States. Rapp. P.-V. Reun. Cons. Int. Explor. Mer. 178: 207–209.

Pruginin, Y., S. Shilo and D. Mires. 1975. Grey mullet: a component in polyculture in Israel. Aquaculture 5: 291–298.

Quignard, J.P. and H. Farrugio. 1981. Age and growth of grey mullet. pp. 155–184. *In*: O.H. Oren (ed.). Aquaculture of Grey Mullets. IBP 26, Cambridge University Press, Cambridge.

Rafail, S.Z. and E.M. Hamid. 1974. The abundance of mullet fry at the sides of Mex canal. Bull. Inst. Ocean. Fish ARE 4: 97–129.

Rajyalakshmi, T. and D.M. Chandra. 1987. Recruitment in nature, and growth in brackishwater ponds of the striped mullet, *Mugil cephalus* L. in Andhra Pradesh, India. Ind. Jour. Anim. Scie. 57: 229–240.

Reay, P.J. 1992. *Mugil cephalus* L. a first British record and a further 5°N. J. Fish Biol. 40: 311–313.

Reay, P.J. and V. Cornell. 1988. Identification of grey mullet (Teleostei: Mugilidae) juveniles from British waters. J. Fish Biol. 32: 95–99.

Rossi, R. 1981. La pesca del pesce novello da semina nell'area meridionale del delta del Po. Quad. Lab. Tecnol. Pesca 3: 23–26.

Rossi, R. 1986. Occurrence, abundance and growth of fish fry in Scardovari bay, a nursery ground of the Po river delta (Italy). Archivio di Oceanografia e Limnologia 20: 259–280.

Saleh, M. 2008. Capture-based aquaculture of mullets in Egypt. pp. 109–126. *In*: A. Lovatelli and P.F. Holthus (eds.). Capture-based Aquaculture. Global Overview. FAO Fisheries Technical Paper. No. 508. Rome.

Salvarina, I., E.T. Koutrakis and I. Leonardos. 2010. Juvenile feeding habits of Mugilidae species from estuarine systems in North Aegean Sea. Rapp. Comm. Int. Mer. Médit. 39: 795.

Sarr, S.M., J.A.T. Kabre and F. Niass. 2013. Régime alimentaire du mulet jaune (*Mugil cephalus*, Linneaus 1758, Mugilidae) dans l'estuaire du fleuve Sénégal. J. Appl. Biosci. 71: 5663–5672.

Sauriau, P.G., J.P. Robin and J. Marchand. 1994. Effects of the excessive organic enrichment of the Loire Estuary on the downstream migratory patterns of the amphihaline grey mullet *Liza ramada* (Pisces: Mugilidae). pp. 349–356. *In*: K.R. Dyer and R.J. Orth (eds.). Changes in Fluxes in Estuaries (ECSA22/ERF symposium, Plymouth, September 1992). Olsen & Olsen, Fredensborg, Denmark.

Savchuk, M.Y. 1973. Feed migrations of the Mugilidae fry at the Crimea and West Caucasus coast. Gubrobiolozicheski Zhurnal 9: 28–35.

Serventi, M., I.J. Harrison, P. Torricelli and G. Gandolfi. 1996. The use of pigmentation and morphological characters to identify Italian mullet fry. J. Fish Biol. 49: 1163–1173.

Shehadeh, Z.H. and K.S. Norris. 1972. The grey mullet: induced breeding and larval rearing, 1970–1972. Report, Oceanic Institute, Waimanalo, Hawaii. pp. 202.

Starushenko, L.I. and A.B. Kazansky. 1996. Introduction of mullet haarder (*Mugil soiuy* Basilewsky) into the Black Sea and the sea of Azov. GFCM Studies and Reviews 67: 29.

Strauss, R.E. and C.E. Bond. 1990. Taxonomic methods: morphology. pp. 109–140. *In*: C.B. Schreck and P.B. Moyle (eds.). Methods for Fish Biology. American Fisheries Society, Maryland.

Strydom, N.A. 2008. Utilization of shallow subtidal bays associated with warm temperate rocky shores by the late-stage larvae of some inshore fish species, South Africa. African Zoology 43: 256–269.

Strydom, N.A. and B.D. d'Hotman. 2005. Estuary-dependence of larval fishes in a non-estuary associated South African surf zone: evidence for continuity of surf assemblages. Est. Coast. Shelf Sci. 63: 101–108.

Tang, Y.A. 1975. Collection handling and distribution of grey mullet fingerlings in Taiwan. Aquaculture 5: 81–84.

Thompson, D.W. 1947. A Glossary of Greek Fishes. Oxford University Press, Geoffrey Cumberlege. 297p.

Thomson, J.M. 1955. The movements and migrations of mullet (*Mugil cephalus* L.). Aust. J. Mar. Fresh. Res. 6: 328–347.

Thomson, J.M. 1963. Synopsis of biological data on the grey mullet *Mugil cephalus* Linnaeus 1758. CSIRO Fisheries and Oceanography. Fisheries Synopsis 1. Commonwealth Scientific and Industrial Research Organization, Melbourne, Australia. 66pp.

Thomson, J.M. 1966. The grey mullets. Oceanogr. Mar. Biol. Ann. Rev. 4: 301–355.

Thomson, J.M. 1997. The Mugilidae of the world. Mem. Queensl. Mus. 43: 457–562.

Torricelli, P., P. Tongiorgi and P. Almansi. 1982. Migration of grey mullet fry into the Arno river: seasonal appearance, daily activity, and feeding rhythms. Fish Res. 1: 219–234.

Torricelli, P., P. Tongiorgi and G. Gandolfi. 1988. Feeding habits of mullet fry in the Arno River (Tyrrhenian coast). I. Daily feeding cycle. Boll. Zool. 3: 161–169.

Tosi, P. and P. Torricelli. 1988. Feeding habits of mullet fry in the Arno River (Tyrrhenian coast). II. The diet. Boll. Zool. 3: 171–177.

Trape, S., J.D. Durand, F. Guilhaumon, L. Vigliola and J. Panfili. 2009. Recruitment patterns of young-of-the-year mugilid fishes in a West African estuary impacted by climate change. Est. Coast. Shelf Sci. 85: 357–367.

Tully, O. and P. O'Ceidigh. 1989. The ichthyoneuston of Galway Bay, Ireland. 1. The seasonal, diel and spatial distribution of larval, post-larval and juvenile fish. Marine Biology 101: 27–41.

Unsworth, R.K.F., P. Salinas De León, S.L. Garrard, J. Jompa, D.J. Smith and J.J. Bell. 2008. High connectivity of Indo-Pacific seagrass fish assemblages with mangrove and coral reef habitats. MEPS 353: 213–224.

van der Elst, R. and J.H. Wallace. 1976. Identification of the juvenile mullet of the east coast of South Africa. J. Fish Biol. 9: 371–374.

Verdiell-Cubedo, D., A. Egea-Serrano, F.J. Oliva-Paterna and M. Torralva. 2007. Biologia trofica de los juveniles del genero *Liza* (Pisces: Mugilidae) en la laguna costera del Mar Menor (SE Peninsula Iberica). Limnetica 26: 67–73.

Vidy, G. and J. Franc. 1992. Saisons de présence à la côte des alevins de muges (Mugilidae) en Tunisie. Cybium 16: 53–71.

Vieira, J.P. 1991. Juvenile mullets (Pisces: Mugilidae) in the Estuary of Lagoa dos Patos, RS, Brazil. Copeia 2: 409–418.

Villani, P. 1987. The ascent of Mugilidae fry into a coastal lagoon of the southern Adriatic sea. FAO Fish Rep. 394: 181–188.

Wallace, J.H. 1975. The estuarine fishes of the East coast of South Africa. III Reproduction. Invest. Rep. Oceanogr. Res. Inst. 41: 1–48.

Wallace, J.H. and R.P. van der Elst. 1975. The estuarine fishes of the East coast of South Africa. IV Occurrence of juveniles in estuaries. V Ecology, estuarine dependence and status. Invest. Rep. Oceanogr. Res. Inst. 42: 1–63.

Walsh, W.A., C. Swanson and C.S. Lee. 1991. Combined effects of temperature and salinity on embryonic development and hatching of striped mullet, *Mugil cephalus*. Aquaculture 97: 81–289.

Wang, C.H., C.C. Hsu, C.W. Chang, C.F. You and W.N. Tzeng. 2010. The migratory environmental history of freshwater resident flathead mullet *Mugil cephalus* L. in the Tanshui River, Northern Taiwan. Zoological studies 49: 504–514.

Whitfield, A.K. and S.J.M. Blaber. 1978. Distribution, movements and fecundity of Mugilidae at Lake St. Lucia. The Lammergeyer 26: 53–63.

Whitfield, A.K. and M. Elliot. 2002. Fishes as indicators of environmental and ecological changes within estuaries: a review of progress and some suggestions for the future. J. Fish. Biol. 61 suppl. A: 229–250.

Whitfield, A.K., J. Panfili and J.D. Durand. 2012. A global review of the cosmopolitan flathead mullet *Mugil cephalus* Linnaeus 1758 (Teleostei: Mugilidae), with emphasis on the biology, genetics, ecology and fisheries aspects of this apparent species complex. Rev. Fish Biol. Fisheries 22: 641–681.

Wijeyaratne, M.J.S. and H.H. Costa. 1987. Fishery, seasonal abundance and mortality of grey mullets (Pisces: Mugilidae) in Negombo Lagoon, Sri Lanka. J. Appl. Ichth. 3: 116–118.

Zambriborch, F.S. 1949. Time of appearance of young mullets, their species at the coast of the southwest part of the Black Sea. Reports of Odessa State University 5: 75–78.

Zambriborch, F.S. 1951. Some anatomical features of the Black Sea mullet. Zoologicheskii zhurnal 30: 143–148.

Zismann, L. 1981. Means of identification of grey mullet fry for culture. pp. 155–184. *In*: O.H. Oren (ed.). Aquaculture of Grey Mullets. IBP 26, Cambridge University Press, Cambridge.

Zismann, L. and A. Ben-Tuvia. 1975. Distribution of juvenile mugilids in the hypersaline Bardawil Lagoon January 1973–January 1974. Aquaculture 6: 143–161.

Zydlewski, J. and M.P. Wilkie. 2013. Freshwater to seawater transitions in migratory fishes. pp. 253–326. *In*: S.D. McCormick, A.P. Farrell and C.J. Brauner (eds.). Fish Physiology, Vol. 32: Euryhaline Fishes. Elsevier, New York.

CHAPTER 13

Adaptation to Salinity and Osmoregulation in Mugilidae

Frank G. Nordlie

Introduction

There are groups of teleost fishes that spend their entire lives in Fresh Waters (FW), while other groups of teleost fishes spend their entire lives in Sea Water (SW). These two groups, the stenohaline teleosts (stenohaline freshwater and stenohaline marine) are said to constitute the majority of all teleost fishes (Schultz and McCormick 2013). The remaining small fraction (< 10%) of teleost fishes is composed of those species that have the capabilities of moving between FW and SW environments, the euryhaline fishes (from the Greek words *urus* meaning wide and *halinos* meaning of salt, with first known use of the word in the late 19th century, Oxford Dictionary, 2014, on line). The family Mugilidae, the topic of this book, consists exclusively, or almost so, of euryhaline species. There are presently considered to be 71 valid species that belong to the family Mugilidae, and these are distributed among 20 genera (with two other genera questionable) (Eschmeyer 2014). This family has a worldwide distribution throughout the tropic and temperate zones (e.g., Thomson 1966, 1997, Nelson 2006). Various species of mullets inhabit waters of all salinity levels including fresh, brackish, marine, and even hypersaline waters, of lakes, rivers, estuaries, lagoons, seas, and oceans. The flathead grey mullet[1] *Mugil cephalus* has a worldwide distribution between latitude 51° N and 42° S, though it is less abundant in the tropics (Briggs 1960, Harrison 2002). Individuals of *M. cephalus* have been collected at locations spanning most of the range of salinities among surface waters. While all, or nearly all, members of the family Mugilidae have euryhaline capabilities, there is a great deal of variation among species in their degree of euryhalinity, and apparently, among populations of some widely distributed species, e.g., *M. cephalus*. However, genetic relationships among populations of this circumglobally distributed mullet are still being debated (see Chapter 1—González-Castro and Ghasemzadeh 2015, Chapter 2—Durand 2015, Chapter 15—Rossi et al. 2015).

Some species of mullets reside in FW throughout most of their lives, arriving there in early developmental stages, but return to the sea to spawn. Such species are referred to as catadromous species (a subgroup of the diadromous category). Members of other species may not spend extended periods in FW, or may only enter lower reaches of estuaries, and/or not spawn in fully marine waters. However,

Department of Biology, P.O. Box 118525, University of Florida, Gainesville, FL, USA 32611-8525.
Email: fnordlie@ufl.edu

[1] All scientific names are from Eschmeyer (2014), and common names for the various species of mullet discussed in this chapter follow listings from FishBase (Froese and Pauly 2014).

there is a significant number of mullet species that have been shown to tolerate fully FW, and a significant number of species that tolerate waters significantly more saline than that of normal SW (salinity $\approx$ 35). While it has been a general assumption that most of the mullets spawn in brackish to marine waters, there is evidence that two species of mullet (the abu mullet *Liza abu* and so-iny mullet *Liza haematocheila*) successfully reproduce in waters of salinities significantly lower than those of open seas (< 30), and in some cases of, or approaching, FW (in Ünlü et al. 2000, Luzhnyak 2007, Mohamed et al. 2009, Minos et al. 2010, Mohamed 1978 in Muhsin 2011, Sahinöz et al. 2011, Mohamed et al. 2012, Dogu et al. 2013). Hatching for most species takes place in SW, but the developing young may make their way back (or are returned by currents) to near-shore areas, some making their way into estuaries, and upstream, even into FW (see Chapter 12—Koutrakis 2015).

Osmoregulatory Problems, Mechanisms and Controls in Teleost Fishes

Euryhaline teleosts as well as stenohaline teleost fishes in marine waters maintain their body fluids at concentrations considerably lower than that of the SW in which they are residing. This is unlike members of some of the other groups of fish-like vertebrates, e.g., Chondrichthyes (sharks, skates, and rays) that maintain their body fluid concentrations at levels equal to or slightly higher than that of the SW in which they reside. Among stenohaline FW teleosts or euryhaline teleosts that are in FW, body fluid concentrations are maintained at levels considerably higher than that of their external medium, but generally somewhat lower than concentrations at which body fluids are maintained in SW by stenohaline or euryhaline teleost fishes. Examples from a group of diadromous teleosts (species that either live in FW and spawn in the sea, or vice versa) showed a mean plasma osmotic concentration of $\approx$ 308 mOsm.kg^{-1} in FW, range 228–351 mOsm.kg^{-1}, and a mean plasma osmotic concentration of $\approx$ 357 mOsm.kg^{-1} in SW, range of values, 275–484 mOsm.kg^{-1} (values were not at common temperatures, data from Nordlie 2009). Teleost fishes residing in SW are referred to as hyporegulators and teleosts in FW are referred to as hyperregulators. The phenomenon of hyperregulation is illustrated by comparing the mean plasma osmotic concentration of such fishes in FW, $\approx$ 308 mOsm.kg^{-1} (from above) with the osmotic concentration of FW ($\leq$ 15 mOsm.kg^{-1}). Hyporegulation is then illustrated by comparing the mean plasma osmotic concentration of such teleosts in normal SW, $\approx$ 357 mOsm.kg^{-1} (again, value from above), with the osmotic concentration of normal SW, $\approx$ 1,000 mOsm.kg^{-1}. It is obvious that there are considerable differences between the internal concentrations of fluids in teleost fishes and that of the waters in which they are found, but in opposite directions when in FW compared to SW.

Basic Organs and Organ Systems Involved in Osmoregulatory Functions in Teleost Fishes

A short overview of the organs and organ systems that are major sites of the iono-osmoregulatory processes in teleost fishes and the functions that maintain homeostasis in internal body fluids (e.g., regulation of levels of water and the various ions including Na^+ and Cl^-, as well as of other components) are described here. All teleost fishes must maintain osmotic homeostasis though the stenohaline species need do so only for FW or for SW, while the euryhaline species must be able to maintain osmotic homeostasis over a wide range of salinities, necessitating their being capable of both hypo- and hyperregulation, though not simultaneously. These capabilities permit euryhaline teleosts, such as members of the family Mugilidae, to survive and thrive in aquatic environments of varying salinities.

The most basic organs involved in osmoregulatory functions in fishes include the gills, alimentary tract, and kidneys, along with the circulatory system that interconnects all these organs. The fundamental problem being faced by a teleost fish living in FW with its internal osmotic concentration considerably higher than that of the medium in which it lives (e.g., *L. abu*, *L. haematocheila*, see Table 13.1) is that its body is covered by an integument that is, of necessity, somewhat permeable to O_2 and CO_2, the gases that are exchanged in respiration, and also to water and ions such as Na^+ and Cl^-. Obviously, the primary site of permeability is the gills, the major location of respiratory gas exchange in teleost fishes. Because of the semi-permeable nature of the gill epithelium and the higher concentrations of vital solutes, e.g., Na^+ and Cl^-, in the FW fish blood than in the surrounding water, there will be continuous losses of these ions to the

Table 13.1. Salinities at which live individuals of listed species of mullets have been collected.

Species	Location	Salinity	Life Stage	Authors
Agonostomus monticola	Atlantic Ocean, Florida and Georgia, USA	SW	juveniles	Anderson (1957)
Agonostomus monticola	Rio Platano, Honduras	FW	juveniles and adults	Cruz (1987)
Agonostomus monticola	Mouth of Rio Platano, Honduras	ND	juveniles	Cruz (1987)
Agonostomus monticola	Five rivers, island of Trinidad	FW	juveniles and adults	Phillip (1993)
Agonostomus monticola	Gulf of Mexico west of Long. 93.00	28.5–35.9	larvae and juveniles	Ditty and Shaw (1996)
Aldrichetta forsteri	Moore River estuary, south-western Australia	2.5–25.0	juveniles	Young et al. (1997)
Aldrichetta forsteri	Wellstead Estuary, Western Australia	14.3–122	juveniles to adults	Young and Potter (2002)
Chelon labrosus	England, Wales	FW to SW	juveniles to adults	Hickling (1970)
Chelon labrosus	Northern Sinai Peninsula, Bardawil Lagoon	38.4–74.2	ND	Zismann and Ben-Tuvia (1975)
Chelon labrosus	Island of Minorca, Mediterranean Sea	SW	juveniles to adults	Cardona (2006)
Crenimugil crenilabis	Southeastern Africa	2–35	juveniles	Blaber (1987)
Ellochelon vaigiensis	Australia	35	ND	Blaber (1987)
Joturus pichardi	Rio Platano, Honduras	FW	immatures to adults	Cruz (1987)
Liza abu	Mhejran River, Iraq	5.8–7.34	≤ 31.0 cm	Mohamed (1978) in Muhsin (2011)
Liza alata	Southeastern Africa	0.7–35	juveniles	Blaber (1987)
Liza alata	Lake St. Lucia, South Africa	FW - ~ SW	ND	Whitfield et al. (2006)
Liza aurata	Northern Sinai Peninsula, Bardawil Lagoon	38.4–74.16 most in lower range	juveniles	Zismann and Ben-Tuvia (1975)
Liza aurata	Island of Minorca, Mediterranean Sea	~ ≥ 15, to FW in fall	juveniles	Cardona (2006)
Liza bandialensis	Casamance River, Senegal	39–42	ND	Kantoussan et al. (2012)
Liza carinata	Northern Sinai Peninsula, Bardawil Lagoon	38.4–74.16	ND	Zismann and Ben-Tuvia (1975)
Liza dumerili	Southeastern Africa	0.7–72	juveniles	Blaber (1987)
Liza dumerili	Lake St. Lucia, South Africa	FW - ~ 90	ND	Whitfield et al. (2006)
Liza dumerili	Sine Saloum Estuary, Senegal	30–80	larvae	Trape et al. (2009)
Liza dumerili	Casamance River, Senegal	39–71	ND	Kantoussan et al. (2012)
Liza falcipinnis	Sine Saloum Estuary, Senegal	30–70	larvae	Trape et al. (2009)
Liza falcipinnis	Casamance River, Senegal	39–71	ND	Kantoussan et al. (2012)
Liza grandisquamis	Sine Saloum Estuary, Senegal	35–45	ND	Trape et al. (2009)

Table 13.1. contd....

Table 13.1. contd.

Species	Location	Salinity	Life Stage	Authors
Liza grandisquamis	Casamance River, Senegal	39–70	ND	Kantoussan et al. (2012)
Liza haematocheila	Aegean, Azov, Black, Mediterranean Seas	FW-45	ND	Abrosimova and Abrosimov (2002), in Minos et al. (2010)
Liza luciae	Southeastern Africa	30–36	ND	Blaber (1987)
Liza macrolepis	Southeastern Africa	0.7–72	larvae	Blaber (1987)
Liza macrolepis	Lake St. Lucia, South Africa	FW - ~ 75	ND	Whitfield et al. (2006)
Liza parmatus	Southeastern Africa	35	ND	Blaber (1987)
Liza ramada	Northern Sinai Peninsula, Bardawil Lagoon	38.4–74.16	juveniles	Zismann and Ben-Tuvia (1975)
Liza ramada	Island of Minorca, Mediterranean Sea	*polyhaline, less saline in winter	juveniles and adults	Cardona (2006)
Liza richardsonii	Southeastern Africa	0.5–35	larvae	Blaber (1987)
Liza saliens	Northern Sinai Peninsula, Bardawil Lagoon	38.4–74.16	juveniles	Zismann and Ben-Tuvia (1975)
Liza saliens	Island of Minorca, Mediterranean Sea	*polyhaline, also FW	juveniles and adults	Cardona (2006)
Liza tricuspidens	Southeastern Africa	4–35	juveniles	Blaber (1987)
Liza tricuspidens	Lake St. Lucia, South Africa	~ 2–3 - ~ 90	ND	Whitfield et al. (2006)
Moolgarda buchanani	Southeastern Africa	0.7–60	larvae	Blaber (1987)
Moolgarda buchanani	Lake St. Lucia, South Africa	FW - ~ 55	ND	Whitfield et al. (2006)
Moolgarda cunnesius	Southeastern Africa	0.7–65	larvae	Blaber (1987)
Moolgarda cunnesius	Lake St. Lucia, South Africa	FW - ~ 60	ND	Whitfield et al. (2006)
Moolgarda robustus	Southeastern Africa	0.7–39	larvae	Blaber (1987)
Moolgarda robustus	Lake St. Lucia, South Africa	FW - ~ 55	ND	Whitfield et al. (2006)
Moolgarda seheli	Southeastern Africa	5–35	larvae	Blaber (1987)
Mugil bananensis	Sine Saloum Estuary, Senegal	30–55	juveniles	Trape et al. (2009)
Mugil bananensis	Casamance River, Senegal	39–71	ND	Kantoussan et al. (2012)
Mugil cephalus	Laguna Madre, Texas	to 80	ND	Simmons (1957)
Mugil cephalus	Salton Sea, California	33.68	ND	Walker et al. (1961)
Mugil cephalus	Laguna Madre, Texas	to 75	ND	Gunter (1967)
Mugil cephalus	Baffin Bay, Texas	to 50–60	ND	Gunter (1967)
Mugil cephalus	Lake St. Lucia, South Africa	35–70	ND	Wallace (1975)
Mugil cephalus	Southeastern Africa	0–80	larvae	Blaber (1987)
Mugil cephalus	Northern Gulf of Mexico	≥ 34.0	larvae	Ditty and Shaw (1996)

Table 13.1. contd....

Table 13.1. contd.

Species	Location	Salinity	Life Stage	Authors
Mugil cephalus	Moore River estuary, South-western Australia	2.5–15.0	juveniles	Young et al. (1997)
Mugil cephalus	Salton Sea, California	45	ND	Riedel et al. (2002)
Mugil cephalus	Wellstead Estuary, Western Australia	14.3–122.0	juvenile to adult	Young and Potter (2002)
Mugil cephalus	Island of Minorca, Mediterranean Sea	*oligohaline, mesohaline, Nov. in FW and polyhaline	juvenile to adult	Cardona (2006)
Mugil cephalus	Lake St. Lucia, South Africa	FW- ~ 90	ND	Whitfield et al. (2006)
Mugil cephalus	Sine Saloum Estuary, Senegal	30–70	larvae	Trape et al. (2009)
Mugil cephalus	Casamance River, Senegal	39–71	ND	Kantoussan et al. (2012)
Mugil curema	Northern Gulf of Mexico	≥ 29.9	larvae	Ditty and Shaw (1996)
Mugil curema	Sine Saloum Estuary, Senegal	40–90	ND	Trape et al. (2009)
Mugil curema	Casamance River, Senegal	55–70	ND	Kantoussan et al. (2012)
Myxus capensis	Southeastern Africa	0–35	larvae	Blaber (1987)
Oedalechilus labeo	Northern Sinai Peninsula, Bardawil Lagoon	38.4–74.16	ND	Zismann and Ben-Tuvia (1975)

*FW < 1.0, oligohaline 1.1–5.0, mesohaline 5.1–15.0, polyhaline 15.1–30.0, euhaline 30.1–40.0 (values in PSU, Venice System, Anonymous 1958).

outside. At the same time, the internal medium, blood and tissue fluids, are more concentrated than is the surrounding water, resulting in the continuous osmotic movement of water from the external environment to the interior of the fish, diluting its body fluids. Thus, in the FW environment, the fish must have some means of replacing the ions being lost to the exterior, and at the same time eliminating excess entering water.

The problems are reversed for a teleost fish residing in SW, in that its internal concentration is significantly lower than that of it environment. Thus, there will be an ongoing problem of gaining permeating ions from the exterior, while losing water to the environment. The only water available in this situation is the surrounding seawater that contains an appreciably higher concentration of salts than does the individual fish, so drinking SW is not a simple solution to the dehydration problem. Once the fish drinks seawater, it must have a means of retaining the water while eliminating the salts thus acquired in the water being swallowed along with those entering its body through areas of permeable integument as well as in food. Some portions of the salts are excreted through gills and kidneys, others voided from the alimentary canal. Such processes, either for the acquisition of ions from the surrounding waters or expulsion of ions into a more concentrated ambient water require extensive exchange systems and the utilization of energy where ions are being moved against concentration or electrochemical gradients.

Alterations in ambient salinities may occur on a predictable seasonal basis, e.g., in diadromous fishes that move from waters of low to high salinity or vice versa. However, in other euryhaline fishes, such as those that inhabit estuaries, such changes in ambient salinity may occur in times of minutes or hours, rather than over seasons. These major ion and water exchanges and locations are illustrated in Fig. 13.1 for euryhaline teleost fishes (e.g., mullet) in FW and in SW.

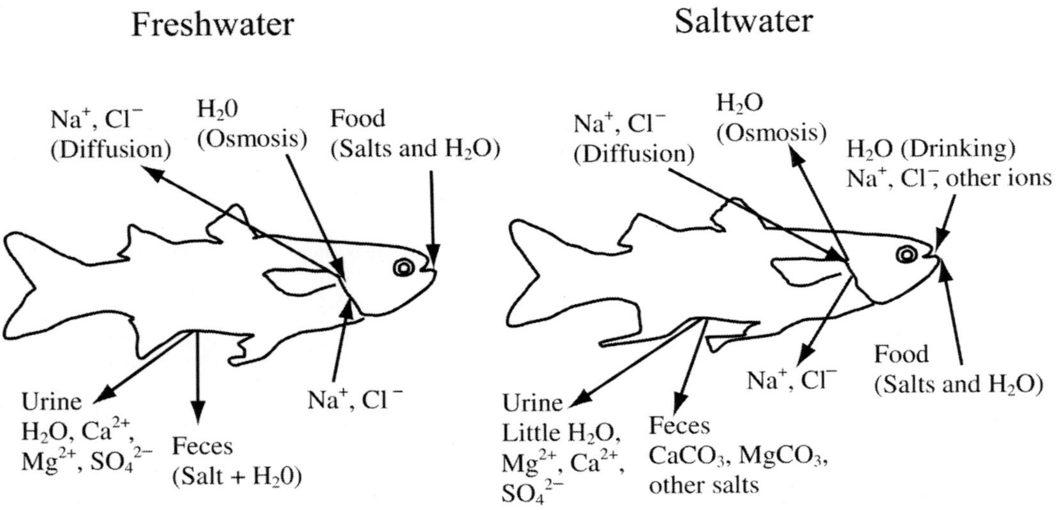

Figure 13.1. Pathways of major ion and water exchanges in euryhaline teleost fishes in freshwater and marine environments as discussed in this chapter.

Overview of Osmoregulatory Mechanisms

A short general overview of the mechanisms involved in iono-osmoregulatory functions is presented here. As very little information is available on mullets this will be a discussion of what appear to be general patterns among euryhaline teleost fishes. Much of the discussion will be based on findings from only a few species, and examples of species in which these systems have been demonstrated to function will be cited. The overview of mechanisms will encompass gills, alimentary canal, kidneys and urinary bladder. These descriptions will use material from recent literature, especially reviews as sources of information allowing the readers to examine mechanisms in greater depth and detail.

Gills

Gills of teleost fishes are major organs involved in ionic and acid-base regulation of the organism, with specialized cells referred to here as ionocytes because of the diversity of ions that are being relocated (originally referred to as chloride cells [Keys and Willmer 1932], and more recently as mitochondrion-rich cells). Activities to be discussed include Na^+ and Cl^- secretion, Na^+ uptake/acid secretion (Na^+/H^+), NH_4^+ excretion (either as NH_4^+ or as NH_3), Cl^- uptake and base secretion (Cl^-/HCO_3^-) (Evans 2011, Hwang et al. 2011, Tipsmark et al. 2011, Christensen et al. 2012, Dymowska et al. 2012, Hiroi and McCormick 2012).

The primary concern here is with models of Na^+ and Cl^- exchange activities of gills in euryhaline teleost fishes. First we will consider the situation in SW-acclimated teleosts. The transport systems of primary concern include: Na^+-K^+-ATPase (NKA) and Na^+-K^+-$2Cl^-$ co-transporter (NKCC) located in the basolateral membrane of the ionocyte; and the cystic fibrosus transmembrane conductance regulator (CFTR), an anion channel that is located in the apical membrane of the ionocyte. The elimination of Na^+ involves the paracellular space and a tight junction between the ionocyte and an adjacent epithelial cell (referred to as an accessory cell). Simplifying the activities proposes that Na^+ and Cl^- are moved into the ionocyte through activity of the NKCC, with energy provided by the NKA. The Cl^- exits the cell to the surrounding seawater through the CFTR, while the Na^+, which moves to the extracellular space in exchange for K^+, exits along a paracellular path between the ionocyte and an accessory cell, passing through the tight junction between the two cells to the surrounding seawater.

The gills are involved in the transport of ions in FW as well as in SW. Two of the current models for ion exchanges in FW-acclimated euryhaline fishes involve Na^+/H^+ exchange (NHE), or Na^+ and

Cl^- co-transport (NCC). The first of the systems is involved in Na^+ uptake, with the second involved in both Na^+ and Cl^- uptake in FW. The system being used apparently depends on the species involved and ambient salinity (Evans 2011, Hwang et al. 2011, Christensen et al. 2012, Hiroi and McCormick 2012). Another suggested pathway for Cl^- uptake is by way of Cl^-/HCO_3^- exchange, though debate continues with respect to the driving mechanism in this pathway (Hwang et al. 2011). Also, some species of fishes (e.g., the mummichog *Fundulus heteroclitus*) are not known to extract Cl^- from the ambient waters (Wood and Marshall 1994, Patrick et al. 1997, Patrick and Wood 1999, Wood 2011). Ammonia excretion can occur by way of several pathways including: diffusion of NH_3 or NH_4^+ through tight junctions in the gill epithelium; by NH_4^+, instead of H^+, on Na^+/acid exchangers (NHE); or (in SW) in exchange for K^+ on basolateral NKA (Evans et al. 2005). The primary location of these activities is again in ionocytes, though other epithelial cells have also been suggested to be involved in some exchanges. Some of these mechanisms are known to operate in only one or a few species.

Variations in concentrations of gill Na^+-K^+-ATPase activity in gills have been found in several species of euryhaline fishes including mullets relative to whether they have been acclimated to FW or to SW. An early work carried out on gills of the thicklip grey mullet *Chelon labrosus* from the Bassin d'Arcachon, France showed NKA activity to be ≈ 2X higher in the FW-acclimated individuals than in the SW group (Lasserre 1971). Fish were placed directly into SW and acclimated gradually to FW (over a period of five days). However, higher NKA intensity and activity was found in gills of the golden grey mullet *Liza aurata* that had been acclimated to artificial SW at salinities of 36 and 46 than to either Caspian Sea water (salinity of 12) or FW (Khodabandeh et al. 2009). This line of inquiry was continued by Gallis and Bourdichon (1976), who made a comparison of NKA activity in FW- and SW-acclimated individuals of *C. labrosus* and the thinlip grey mullet *Liza ramada*. Here, again the NKA activity was found to be significantly higher in the FW-acclimated group of *C. labrosus* (≈ 2X higher), than in the SW-acclimated group, while no significant difference was found in NKA activity between the FW- and SW-acclimated groups of *L. ramada*. In a subsequent study Belloc and Gallis (1980) found again that the NKA activity of gills of *C. labrosus* increased with transfer from SW to FW.

Juvenile individuals of *M. cephalus,* collected from Mississippi Sound at Ocean Springs, MS (USA), were maintained for one week in water with a salinity range of 8–14, at 20–24°C, following which groups were transferred to tanks containing FW, and water of salinities of 10, 26, and 45. NKA content of the gills, evaluated using the technique of ^{3}H-oubain binding (Hossler et al. 1979) increased with increased acclimation salinity. In a later study, again evaluating juvenile individuals of *M. cephalus* (mean SL, 28.05 ± 3.54 mm, SD), differing results were found. These fishes were collected from the Tyrrhenian Sea (Mediterranean Sea) and held at a salinity of 38 for two months (Ciccotti et al. 1994), after which SW was gradually replaced by FW. Individuals were taken for gill NKA assay at weekly intervals during the FW acclimation. It was found that the NKA activity increased in FW, so that at day 36 it was 2.6X higher in FW than it had been in SW. These results were opposite to those of Hossler et al. (1979) in the same species, though from different locations. However, the authors of the later study Ciccotti et al. (1994) asserted that results were not directly comparable because of differences in techniques used in the two studies. This question regarding levels of NKA activity in SW- vs FW-acclimated teleosts had previously been posed with respect to the waters of origin of euryhaline species, e.g., *C. labrosus*, that showed higher gill levels of NKA after FW acclimation, tolerating a wide range of ambient salinities but not surviving indefinitely in FW (Gallis and Bourdichon 1976). Species of euryhaline fishes thought to have had a marine origin (e.g., *C. labrosus*) were said to show such an increase in NKA activity in FW (Christensen et al. 2012).

Responses in NKA-positive ionocytes were compared between SW-acclimated and FW-acclimated teleosts (Hwang and Lee 2007). Results indicated that while gill ionocytes in a large number of the teleosts showed greater staining intensity for NKA after transfer from FW to SW, and the cells were found to be larger in SW than in FW, a small fraction of teleost species showed greater staining intensity and the cells were larger in size and/or in numbers in FW. Differences in patterns were also found among the species evaluated (McCormick et al. 2003, Hiroi and McCormick 2007). Lamellar ionocytes were generally found to be responsible for ion uptake in FW, while filament ionocytes were responsible for ion secretion in SW, though numerous variations of these patterns were found (McCormick et al. 2003, Hiroi and McCormick 2007, Hwang and Lee 2007, Hiroi et al. 2008). They also found that species without ionocytes on the gill lamellae were euryhaline, but not diadromous fishes, while species with ionocytes on gill lamellae in FW were (mostly) diadromous teleosts, either anadromous or catadromous species. It was also found that among teleosts there are ionocytes with different ion-transport functions in FW, and that the FW co-transporter molecule involved in ion absorption is NCC (McCormick et al. 2003, Dymowska et al. 2012, Hiroi and McCormick 2012). However, they concluded that there is only one type of ionocyte in SW teleost which is capable of salt secretion as well as of acid-base regulation and the co-transporter molecule in ion secretion in SW-acclimated fishes is NKCC.

Alimentary Canal

The alimentary tract of a 'typical' mullet is somewhat unique, so a short description is given here to help clarify some of its features using information from reviews by Thomson (1966) and Odum (1970).

The mouth opens to the pharyngeal region which functions as a filtering device, separating small food particles from larger sediment particles. The oesophagus separates the pharynx from the stomach. The stomach consists of two parts: a proximal thin-walled chamber, the cardiac stomach, and the distal pyloric stomach, a muscular 'gizzard' that functions to grind the food. According to Ishida (1935) the stomach of *M. cephalus* lacks gastric glands. The pyloric stomach is separated from the proximal end of the intestine by the pyloric valve. Pyloric ceca are located in this region of the intestine and vary in numbers among species. The intestine of mullets is a long, coiled structure. *M. cephalus* from the Mediterranean Sea with a mean body length of 25 $\pm$ 3.53 cm, showed a mean intestinal length of 82.4 $\pm$ 9.09 cm, 3.3X the body length. By comparison a group of gilthead bream *Sparus aurata* (Sparidae), shellfish eaters, with a mean body length of 20.75 $\pm$ 2.86 cm had intestinal tracts with a mean length of 18.5 $\pm$ 4.27 cm, slightly less than the mean body length (El-Bakary and El-Gammal 2010).

The area of the rectum at the terminal end of the alimentary tract shows a greater vascularization than the intestine and though sometimes difficult to distinguish using external features, can be distinguished histologically.

Teleost fishes in FW avoid drinking water though they constantly gain water with their food, as well as osmotically through permeable areas of their integuments. However, in saltwater these fishes continuously lose water by osmosis through their permeable tissues to the surrounding environment. They replace this water by drinking SW and absorbing both water and portions of the contained ions (principally Na^+, K^+, Cl^-, HCO_3^-, along with some Ca^{2+}, Mg^{2+} and SO_4^{2-}). The bulk of these absorbed ions must be eliminated from the internal fluids in order to maintain osmotic balance. The monovalent ions are lost primarily through the gills, as has already been discussed, while a portion of the divalent ions Mg^{2+} and SO_4^{2-} are excreted through the kidney. Also, most of the Ca^{2+} and HCO_3^- along with some of the Mg^{2+} and SO_4^{2-} are lost through the alimentary tract.

An overview of processes involved in absorbing water while eliminating the excess Na^+, Cl^- and other electrolytes[2] will be presented next. The processing of ingested SW is said to begin in the upper portion of the oesophagus. Seawater swallowed by eels was found to lose a large fraction of its original concentration by the time it reached the stomach (Hirano and Mayer-Gostan 1976). Two species of eels, the Japanese eel *Anguilla japonica* and the European eel *A. anguilla*, maintained in either FW or SW, showed Na^+ and Cl^- moving out of the lumen following their concentration gradients, 'desalting' the water (one-third of the total concentrations of Na^+ and Cl^- when ingesting SW). The oesophagus was found to be impermeable to water, though a small amount of water moved to the lumen in FW (Hirano and Mayer-Gostan 1976). Further work on the role of the oesophagus in desalting ingested SW involved a study of nine species of teleosts three of which, euryhaline species, were also evaluated in FW along with a FW species. While there was some variation, most showed a pattern of uptake of Na^+ and Cl^- similar to that of the eels (Kirsch and Meister 1982). As the concentration gradient of Na^+ and

[2] Primary sources of information for this section include: Shehadeh and Gordon 1969, Cataldi et al. 1988a,b, P.J. Walsh et al. 1991, Beyenbach 2000, Grosell 2006, Taylor and Grosell 2006, Grosell and Taylor 2007, Whittamore et al. 2010, Grosell 2011, and Whittamore 2012.

Cl^- between the ingested water and the oesophagus diminished, further uptake involved active transport (Nagashima and Ando 1994). Extensive capillary beds near the surface of the oesophagus pick up and transport the entering Na^+ and Cl^-.

Stomach. The water next passes through the stomach where it may be diluted by stomach fluids. Apparently there is generally little absorption of materials in the stomach (Kurita et al. 2008, Genz et al. 2011, Grosell 2011, Whittamore 2012), though there are exceptions such as the rainbow trout *Oncorhynchus mykiss* that shows a significant uptake of ions in the stomach (Bucking and Wood 2006, 2007).

Intestine. The concentration of fluid in the intestinal lumen is greatly reduced here ($\approx$ 300–400 mOsm.kg^{-1}, Grosell and Taylor 2007), only slightly higher than that of the body fluids. Absorption of the water actually begins in this region of the intestine, where the water moves against an osmotic gradient (Skadhauge 1969, Genz et al. 2011, Whittamore 2012). Water absorption in the intestine is associated with Na^+-Cl^- co-transport ('solute-linked' water transport). In this manner, a large fraction of the water can be absorbed, along with nearly all of the Na^+ and Cl^- (Genz et al. 2011, Grosell 2011, Whittamore 2012).

The principal solutes that remain in solution in the intestine include Ca^{2+}, Mg^{2+} and SO_4^{2-}, of which some of the latter two are taken up in the intestine. At this point in the alimentary tract the fluids have an alkaline pH that may be as high as nine, produced by the secretion of HCO_3^- (Shehadeh and Gordon 1969, Wilson et al. 2002, Grosell 2006, 2011). It is suggested that HCO_3^- is secreted into the intestinal lumen in exchange for Cl^-, and that HCO_3^- secretion occurs via an apical Cl^-/HCO_3^- exchanger (Kurita et al. 2008, Whittamore et al. 2010, Grosell 2011). Much of the Ca^{2+} (and also some of the Mg^{2+}) reacts with the secreted HCO_3^- in the intestinal fluid to form carbonates that precipitate. The precipitation of $CaCO_3$ removes it from solution, reducing the osmotic concentration of fluid in the intestinal lumen, allowing more water to be absorbed. The remaining fluid and solids in the intestinal lumen are voided to the exterior via the rectum (Kurita et al. 2008). Some of the Ca^{2+} in the intestinal lumen may be associated with the mucous layer that lines the intestine (Sundell and Björnsson 1988, Wilson and Grosell 2003, Wilson et al. 2009). The carbonates expelled by teleost fishes in marine waters may become a major component of the global carbon cycle (P.J. Walsh et al. 1991, Wood et al. 2008, Wilson et al. 2009, Grosell 2011).

Formation of precipitates largely of carbonates in the alimentary tracts of teleost fishes in marine waters has been recognized for a long time (e.g., Shehadeh and Gordon 1969). However, potential impacts of these carbonates in the marine environment and its chemical cycles were not immediately visualized. An early study dealing with the potential magnitude of such contributions of carbonates to a marine environment, involved an evaluation in the waters of Florida Bay (Florida, USA). A budget of carbonate contributions to the waters of Florida Bay was formulated for a single common inhabitant, the gulf toadfish *Opsanus beta* (P.J. Walsh et al. 1991). The precipitated carbonate (white pellets) being extruded by this teleost fish was identified as of calcian kutnohorite [$Ca_{0.74}(Mg, Mn)_{0.26} CO_3$]. The carbonate was being produced from HCO_3^- formed from respiratory CO_2 that was being secreted into the lumen of the alimentary tract. There it reacted with divalent cations introduced into the intestinal fluids in seawater being swallowed by the fish in hyperosmotic waters. The authors estimated that the contribution of carbonate from this single species could amount to 2.5 g m^{-2} yr^{-1} in Florida Bay. This ignored all the other teleost fishes that inhabit this water (P.J. Walsh et al. 1991).

Calculations of potential carbonate contributions to the global marine ecosystem from the resident teleost fauna was presented by Wilson et al. (2009), with a conservative estimate of 3 to 15% of the total annual new carbonate in marine waters being contributed by the resident teleost fishes. Carbonates from teleost fishes contain more magnesium than those from most other marine sources, and are thus more soluble, so are more immediately available to the chemical cycles of the marine ecosystems. The contributions of carbonates from marine fishes were suggested to rise with increases in ambient dissolved CO_2.

Kidneys

Kidney structures among fishes vary greatly in overall morphology, degree of development and function of the excretory units. Differences in kidney structures among teleost fishes led Ogawa (1961) to categorize them into five groups (I –V). Freshwater teleosts kidneys all fall into types I, II, or III. Kidneys of teleosts belonging to the family Mugilidae, along with several groups of freshwater fishes and the majority of the marine fishes, are included as having kidneys of type III (Ogawa 1961). The marine fishes identified by Ogawa as having type III kidneys included those considered to be euryhaline species by Hickman and Trump (1969). Their inclusion of teleost families in the euryhaline category was based largely on work of Gunter (1942, 1956). A brief description of the renal structure and function of a typical euryhaline teleost with a type III kidney, as described by Ogawa (1961), follows. Teleost kidneys are subdivided into the anterior head kidney and posterior trunk kidney based on features of internal structure. The overall kidney generally shows two parallel sections of the head kidney, with a single fused trunk portion (referred to as an opisthonephric kidney, Kerr [1919] quoted in Hickman and Trump 1969). A ventral view of the kidney of *M. cephalus* is found in Fig. 13.2.

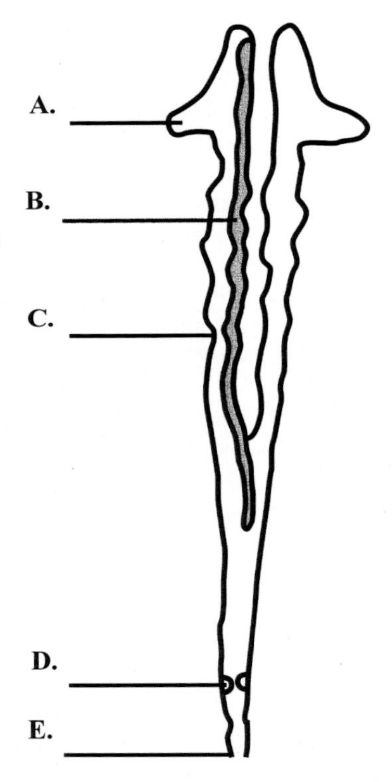

Figure 13.2. Ventral view of *M. cephalus* kidney, a type III kidney. (A) head kidney; (B) postcardinal vein; (C) kidney; (D) Corpuscles of Stannius; (E) ureter. Redrawn after Ogawa (1961).

The two archinephric ducts fuse at some point beyond the posterior end of the kidney, and form the urinary bladder. The glomeruli are made up of capillaries developed from the afferent arterioles. The glomerular capillaries rejoin to form efferent arterioles that again break up into the peritubular capillaries. The peritubular capillaries are joined by capillaries of the renal portal system that contain blood collected by the caudal and segmental veins that break down into the renal portal capillaries. The combined capillaries (peritubular plus renal portal) drain into the postcardinal veins that flow anteriorly into a common cardinal vein before reaching the heart (Hickman and Trump 1969, Satchell 1992).

Among the truly euryhaline teleosts, grey mullet included, the nephric units include a glomerulus, neck, proximal segment, distal segment and collecting tubule, with no difference between males and females (Ogawa 1961, Hickman and Trump 1969).

Kidneys of FW teleosts, which generally have well-developed glomeruli, excrete large quantities of dilute urine, a principal means of eliminating osmotically-gained water. Saltwater teleosts show less development of the excretory units, to the extent of some being aglomerular. The quantity of water being excreted in a seawater environment by teleost fishes is small, but contains large amounts of solutes including divalent ions, principally Mg^{2+} and SO_4^{2-}, but also Ca^{2+} and the trivalent PO_4^{3-} (Hickman and Trump 1969, Cataldi et al. 1991, Renfro 1999, Beyenbach 2000, 2004). The total osmotic concentration of the urine can be up to that of the blood plasma of the individual. Kidneys of fishes lack a Loop of Henle, preventing, except in unusual circumstances, production of urine that is hyperosmotic to the blood (Stanley and Fleming 1964). Mullet, among other euryhaline fishes, have kidneys that are capable of shifting from the role of a FW kidney to that of a SW fish. When in FW the kidney functions as a major organ in eliminating water being gained osmotically from the hypoosmotic environment, while conserving solutes. However, in SW or hypertonic estuarine waters, where body water is being lost by osmotic flow to the external environment, the kidney functions to conserve water while excreting primarily divalent ions, and other metabolic wastes that are not removed from the body in other locations.

It has been experimentally demonstrated that the glomerular filtration rates of euryhaline fishes decrease very significantly when individuals are moved from FW to SW, and increase when moved from SW to FW (Holmes and McBean 1963, Stanley and Fleming 1964, Beyenbach 2000, 2004). The rate of filtration of a glomerulus varies, and may cease and restart in the kidney of a euryhaline fish, based on the concentration of the environment in which the individual is located (Beyenbach 2004). This situation is referred to as 'glomerular intermittency'. A large fraction of the glomeruli are perfused and filtering in FW, though some are perfused but not filtering, and others are not perfused. However, in marine waters the fraction of the perfused and filtering glomeruli is greatly reduced. The result is a great reduction in the extent of glomerular filtration taking place in euryhaline fishes when in hypersaline waters. Thus, nephric tubular functions are the principal activities of kidneys of teleosts in marine waters. Changes in tubular functions that accompany a shift from a SW environment to that of FW involves some reabsorption of electrolytes in the proximal tubules, but the majority of monovalent ions are reabsorbed in the distal tubules (Beyenbach 2000). The resorption of monovalent ions and the maintenance of a low permeability to water were said to be the principal functional aspects of the distal segment and collecting tubule of the kidney of teleost fishes in FW (Hirano et al. 1973). Euryhaline teleost fishes were shown to eliminate, or nearly so, tubular secretion of Mg^{2+} and SO_4^{2-}, reducing tubular permeability to water (Hickman and Trump 1969).

Euryhaline species typically have significantly higher filtration rates when in a FW environment than when in SW. Generally the glomerular filtration rate and urine flow decline significantly when a FW adapted euryhaline fish is placed into SW. However, the extent of this alteration varies considerably among species. The reduction may be transient, as other slower changes occur. These include increasing water permeability of the nephric tubular epithelium (Hickman and Trump 1969). Both euryhaline and stenohaline teleosts living in the Black Sea where salinities are of 17–18, showed higher average numbers and larger sizes of glomeruli than those found in members of the same species taken from the Mediterranean Sea and the Barents Sea where the salinities are of 33–38 (Lozovik 1963 quoted in Hickman and Trump 1969). The most extreme variation in kidney structure is found in some marine fishes that have aglomerular kidneys (Marshall and Smith 1930, Lahlou et al. 1969, Baustian and Beyenbach 1999, Beyenbach 2004, McDonald and Grosell 2006, Genz et al. 2011). However, there are also some groups with aglomerular kidneys that have moved back to brackish waters and even to FW (Beyenbach 2004).

The significant problems in kidney function for euryhaline fishes when moving between FW and SW include going from a situation in which the functions are ridding the organism of large quantities of water through glomerular filtration while conserving Na^+ and Cl^- through absorption in the renal tubules, to a situation in SW in which glomerular filtration is greatly reduced and the principal ions being excreted, Mg^{2+}, SO_4^{2-} and PO_4^{3-}, along with organic acids and bases and some Na^+ and Cl^-, are being secreted in the renal tubules. The latter two ions are reabsorbed further along in the excretory system. The overall result is that there is little urine flow, but urine of a high concentration, though lower than or similar to the plasma osmotic concentration (Beyenbach 2000). Also, renal excretion of divalent ions is less than that of intestinal expulsion (Beyenbach 2004).

Urinary bladder

The urinary bladder or urinary sinus of teleost fishes is derived from the fused archinephric ducts (Hickman and Trump 1969, Marshall 1988), referred to as the mesonephric ducts (Curtis and Wood 1991). A number of studies have focused on the alterations in volume, solute concentrations, and overall osmotic concentration of urine of teleost fishes, following its passage through the urinary bladder. In some cases the passage is rapid, while in others, the urine may be retained for periods of time in the bladder before its release to the outside environment. FW individuals of *O. mykiss* show intermittent bursts of urination. The period of retention of urine in the bladder ($\approx$ 25 minutes) was thought to allow for a significant uptake of Na^+ and Cl^- (Curtis and Wood 1991).

Bladder function was evaluated in FW- and SW-acclimated individuals of European flounder *Platichthys flesus* in which FW-acclimated individuals showed reductions in Na^+ and Cl^- as well as in the overall urine osmotic concentration. Saltwater-acclimated individuals showed some uptake of Na^+ and Cl^-, but without a significant change in the urine osmotic concentration (Lahlou 1967). A more extensive study (Hirano et al. 1973) involved evaluations of urinary bladders surgically removed from individuals of a total of 17 species of teleost fishes including stenohaline FW and SW species as well as some euryhaline species. The bladders from the marine and euryhaline species had come from individuals acclimated to SW or to FW. None of the FW species could be acclimated to SW. Some of these FW species showed active uptake of Na^+ from the urine. Marine fishes in SW showed modest levels of uptake of Na^+ and Cl^-. Among a selected group of euryhaline species the uptake of Na^+ was high in both FW- and SW-acclimated individuals. Bladders of some species were concluded to serve more as urine storage areas than as osmoregulatory organs. Hirano et al. (1973) concluded that the bladders from stenohaline FW teleosts were nearly impermeable to water. Bladders of euryhaline teleosts that originated in a marine environment showed increased permeability to water in SW, but impermeability in FW with the hormone prolactin involved in the transition. However, the urinary bladders from euryhaline teleosts whose origins were in FW maintained impermeability of the urinary bladder under all environmental conditions.

Endocrine Control of Osmotic Regulation in Teleost Fishes

Fish that swim through waters of varying salinities, or live in areas where the salinity changes either in a predictable or unpredictable way need to adjust the functioning of systems that maintain homeostasis of their internal fluids. This includes maintaining control of concentrations of both water and ions that are components of blood plasma and tissue fluids. The ability to respond to changes in environmental concentrations depends on sensing of such changes and sending appropriate signals to the systems and organs involved in ionic and osmotic regulation. Signals to and controls of functional systems include those of the neuroendocrine system. Response times vary with the rate and magnitude of salinity changes (McCormick 2001).

The traditional view of hormonal influence in movements of teleost fishes between FW and SW was that cortisol was the SW-adapting hormone and prolactin was the FW-adapting hormone (Utida et al. 1972, McCormick 2001). It is now known that adaptation/acclimation to either SW or FW involves more hormones and interactions among hormones. The principal hormones involved in the regulation of osmotic regulatory functions of teleost fishes will be discussed, with respect to the general situations where they are activated.

Cortisol, a corticosteroid, is produced in the interrenal tissues in teleost fishes (Mommsen et al. 1999). Cortisol levels increased when teleost fishes were acclimated to SW (Ball et al. 1971, Assem and Hanke 1981). This hormone has been found to be involved in reversing the reduction of Na^+-K^+-ATPase activity, while stimulating other enzyme activity, thus increasing tolerance of certain species of teleost fishes to increased salinities (McCormick 1995). Cortisol concentrations also increased in a number of species of teleosts, including at least one species of mullet (*M. cephalus*), when they were acclimated to FW (McCormick 2001). This suggested that the hormone also plays a role in FW-acclimation of teleost fishes. Cortisol has also been found to increase the expression of Na^+-K^+-$2Cl^-$ cotransporter as well as of Na^+-K^+-ATPase in the gills of freshwater Atlantic salmon (Pelis and McCormick 2001). That increase in expression is based on the proliferation of ionocytes in the epithelium in Atlantic salmon and other salmonids (Madsen 1990, Pelis and McCormick 2001). The functions of cortisol were concluded to be cytogenic, involving the growth and differentiation of transport cells in the gills of both SW and FW teleosts. Cortisol appears to function jointly with prolactin in FW, but with growth hormone/insulin-like growth factor in SW (Evans et al. 2005).

The second of these hormones, prolactin (PRL), is a peptide hormone synthesized in the vertebrate pituitary (Sakamoto and McCormick 2006, Breves et al. 2014) and is involved in ion uptake and inhibition of salt secretion in euryhaline teleost fishes in FW. There is also significant evidence that prolactin interacts with cortisol in FW acclimation of teleosts (McCormick 2001). There are differences among species in actions and in the relative importance of this hormone in osmoregulatory functions (Manzon 2002). The basic function of PRL in FW is the regulation of hydromineral balances that are accomplished through reducing the uptake of water and increasing the retention of ions (particularly Na^+ and Cl^-). Such effects can be accomplished through changes in permeability of surfaces of gills, skin, kidney, intestine and urinary bladder (Hirano et al. 1986, Breves et al. 2014). PRL has a significant effect on the morphology, distribution and numbers of ionocytes in gills, and decreases Na^+-K^+-ATPase in the gill in some species, but has no effect in other species. The hormone also has a species-specific effect on glomerular size and activity. While it is known to have an effect on absorption of Na^+ and Cl^- in the intestine and water permeability of the intestine this is also species-specific (Breves et al. 2014). Prolactin is known to exert influence on ion transporters, tight junction proteins, and water channels in ionocytes. It is also a key regulator of NCC in the vertebrate kidney (Breves et al. 2014). This hormone plays a role in ion uptake in most, if not all, teleosts (Hirano et al. 1986, McCormick 1995) and is also implicated in ionocyte morphology (McCormick 1995). The hormone is involved in reductions in the sizes of ionocytes and stimulates formation of small ionocytes (reviewed by Evans et al. 2005).

The third major player involved in regulating the osmoregulation processes in teleost fishes is Growth Hormone (GH), a member of the same hormone family as prolactin. This hormone operates in concert, at least in some functions, with insulin-like growth factor-I (IGF-I, a 70-amino acid polypeptide McCormick 1995). A major effect of growth hormone appears to be stimulation of osmoregulation in SW. This action is produced, at least in part, through the combination of GH and IGFI (GH/IGF-I), in addition to cortisol, and is responsible for development and differentiation of SW-type ionocytes. The ionocytes of teleost fishes inhabiting FW are regulated by prolactin and cortisol (McCormick 2001, Sakamoto and McCormick 2006). The authors proposed that control of salinity acclimation in teleost fishes by prolactin and growth hormone is produced primarily through regulation of cell proliferation, apoptosis,[3] and differentiation, the latter including up-regulation of specific ion transporters. It has been observed that GH/IGF-I effects on salinity acclimation may be very widespread among teleost fishes (Sakamoto et al. 1993, Xu et al. 1997).

Comparative genomic analyses have shown that hormones that have Na^+ extruding and vasopressor properties are greatly diversified among teleosts. There is significant evidence that members of a number of hormone families are involved in various roles in SW-adaptation of teleost fishes (Takei 2008). Examples of recent additions to the list of hormones with influences on ionic regulation in SW fishes includes the Natriuretic Peptide (NP) family which consists of seven members; the guanylin family consisting of three members (paralogs); and the adrenomedullin (AM) family with five members and several others (Evans et al. 2005, Evans 2008).

[3] "A genetically directed process of cell self-destruction that is marked by the fragmentation of nuclear DNA, inactivated either by the presence of a stimulus or removal of a suppressing agent or stimulus, and is a normal physiological process eliminating DNA-damaged, superfluous, or unwanted cells; also called programmed cell death"–Merriam-Webster dictionary 2014, on line.

Salinity Relationships/Tolerance

The subject here involves salinity relationships of members of the family Mugilidae, including anecdotal information on ambient salinities (low and high) reported from field situations in which various species of mullets were found to survive, as well as of laboratory evaluations of salinity tolerances of embryos (fertilized eggs), larvae, and juveniles (fry and fingerlings). Some information on salinity and temperature interactions will also be included. It is unfortunate that most of the work on mullets has been carried out on a very few species.

The most studied of all mullets is probably *M. cephalus*, an important species for capture fisheries and aquaculture in many regions of the world, recently suggested, on molecular bases, to represent a species complex (Chapter 2—Durand 2015).

Salinity Tolerance of Mullets Based on Field Observations

There have been numerous reports of sightings of various species of mullets included among fishes found dead in estuarine systems. Such 'fish kill' events frequently follow periods in which an estuary is closed off from the adjacent sea by low rainfall and diminished tributary discharge and/or by winds building up sandbars that block interchange of water and of fishes. Fish kills have also been reported from such systems resulting from the input of inordinately large quantities of FW, especially from heavy rainfalls. The more useful part of field analyses in such situations is usually determines which species survive under such extremes of increased or reduced salinity (at existing temperatures). The literature contains information on a wide range of species, including information on a number of species of mullets. A sampling of these observations, especially in situations where information on body sizes was also included, is reported in Table 13.1.

Data selected from the cited sources suggest that a number of species of mullets are capable of surviving in FW, and also in waters at salinities appreciably higher than that of normal SW (salinity 36). Based on the information presented here (Table 13.1), one species, *L. abu*, has been taken exclusively from waters at very low salinities, appreciably below that of SW (Mohamed 1978 in Muhsin 2011). There are also a number of species that have been taken from waters of salinities from ≈ FW or near to it, to salinities of normal SW, but no higher. This group includes species that spend a large portion of their life cycle in FW, but also reside in SW in the life cycle. Some of these species migrate to SW and spawn there (catadromous species), with the developing offspring returning (taken by currents, or actively swimming) to nearshore and/or FW habitats. Unresolved questions have been raised as to where mountain mullet *Agonostomus monticola* and the bobo mullet *Joturus pichardi* spawn, whether at sea, in estuaries, or in FW streams. There is no direct evidence as to the spawning location, but developing juveniles of both species have been taken either at sea (*A. monticola*, Anderson 1957) or in estuarine waters (*J. pichardi*, Cruz 1987). Adults have not been taken at sea. While both of these species have sometimes been referred to as having catadromous life cycles, this has been questioned (Gilbert and Kelso 1971, Loftus et al. 1984, Cruz 1987, Phillip 1993, Ditty and Shaw 1996, Matamoros et al. 2009). It has been suggested that fertilized eggs or newly hatched larvae could be swept out from the estuary to the adjacent marine water where development continues for some period of time before the individuals return to FW (amphidromous life cycle).

The final arbitrary grouping to be considered here is that of mullet species reported from waters of salinities well below those of SW as well as from waters of salinities reaching 2X to 3X that of SW or higher in some reports (Table 13.1). The two species showing the widest ranges of ambient salinity at which individuals have been found living are yellow-eye mullet *Aldrichetta forsteri* and *M. cephalus*. Individuals of both species have been reported from waters ranging in salinity from FW (*M. cephalus*, Southeastern Africa, Blaber 1987; Lake St. Lucia, South Africa, Whitfield et al. 2006), or near to it (*A. forsteri*, salinity of 2.5, Moore River Estuary, Southwestern Australia, Young et al. 1997), to a high ambient salinity value of 122. High values for the two species were reported from the Wellstead Estuary, western Australia Young and Potter (2002).

Such observational data represent only what could be observed, not what the fishes may actually be capable of tolerating over long periods of time. Such limits may be broader including lower and/or higher salinities, or more limited at low and/or high salinity levels. Comparisons of known salinity tolerance ranges of mullets included in Table 13.1 are found in Fig. 13.3. This allows the reader to compare the patterns of tolerance among species of mullets.

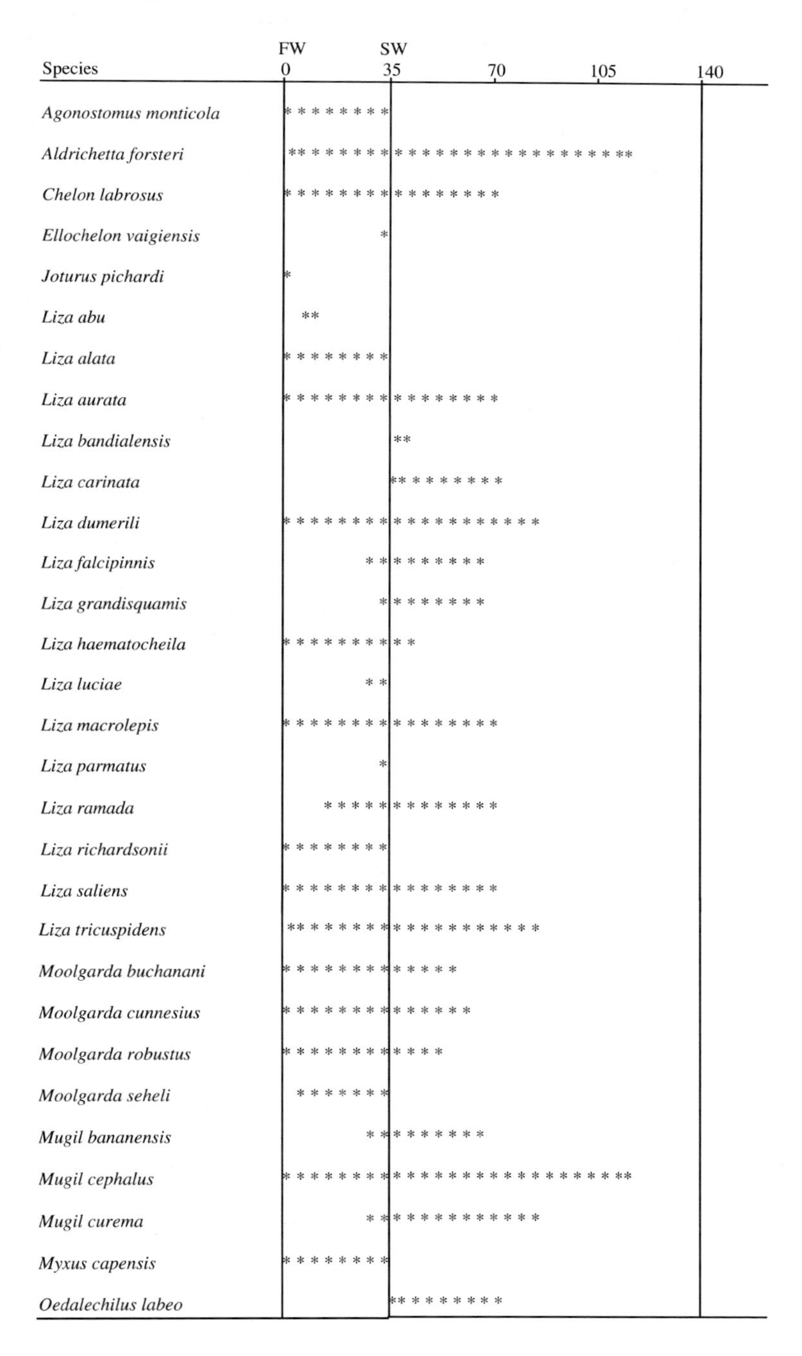

Figure 13.3. Salinity ranges over which various species of mullets have been taken alive. Values presented were approximated in the figure format from information presented in Table 13.1 where sources are also given. Several of these species are thought to occupy wider ranges of salinities than documented here.

Laboratory Determinations of Salinity Tolerance of Mullet Embryos

Various life stages of mullets are separated by use of the following set of terms: eggs referring to the fertilized and developing ova of a species; larvae referring to the newly hatched individuals up to the point where they have a full fin ray complement and have lost 'larval characteristics'; juveniles, a stage that is exited to the definitive adult stage with the attainment of adult body proportions, pigment and habits (from Kendall et al. 1984, as used by Blaxter 1988) (see Chapter 11—González-Castro and Minos 2015 and Chapter 12—Koutrakis 2015). The term fry appears to be used in a very flexible way, to refer to individuals in stages of development from recently hatched larvae to small juveniles. Fingerlings are 'small' individuals, sometimes defined as of less than a year in age. In the following accounts the terms egg or ova, larval and juvenile stages will be used.

Few, if any, species included in the family Mugilidae are known to spawn in FW, though there are reports, as mentioned above of some populations of two species that apparently spawn and whose embryos develop in waters of relatively low salinities, in some cases possibly in FW (*L. abu* [Ünlü et al. 2000, Mohamed et al. 2009, Muhsin 2011, Mohamed et al. 2012, Sahinöz et al. 2011] and *L. haematocheila* [Abrosimova and Abrosimov 2002 in Minos et al. 2010, Luzhnyak 2007, Matishov and Luzhnyak 2007 in Minos et al. 2010]). Questions have frequently been raised as to the range of salinities that can be tolerated in various life stages by eggs and larvae of mullets. This range of salinities includes survival of newly spawned ova and sperm, salinities that can allow for fertilization, and salinities that can support embryonic development and hatching. This also includes ranges of salinities that can be tolerated by the newly hatched larvae, and later by the developing juveniles as they journey in life, generally to inshore waters where salinities may range from that of FW to SW. While these questions are pertinent to understanding the biology of any species of embryonic fishes, the principal goal of gaining answers to some of these questions has been focused most heavily on species, e.g., *M. cephalus*, of interest for aquaculture. Results of laboratory studies of salinity tolerances of fertilized eggs/developing embryos are summarized in Table 13.2. Several of these studies involved laboratory fertilization of the eggs.

Salinity tolerances of embryos of two species of mullets, *L. ramada* and *M. cephalus*, were evaluated. The first of these evaluations (listed alphabetically by generic name, with studies reviewed in chronological order) involved *L. ramada* embryos from the El-Max larval collection centre, Alexandria, Egypt (on the Mediterranean Sea), with survival found to be highest (33.3%) in water of a salinity of three over a 54-days test (El-Dahhar et al. 2000).

Salinity tolerances of the *M. cephalus* embryos were evaluated in four sets of studies. The first of these studies, conducted in Hawaii, showed highest survival at a salinity of 32, using a 48-hour test period, and at a salinity of 30 using a 96-hour test period, both series of evaluations carried out at a temperature of 19.5–20.5ºC (Sylvester et al. 1975). The second set of evaluations also carried out in Hawaii showed highest survival at a salinity of 35, temperature of 24–26ºC, in a 36-hour test, and at a salinity of 25, temperature 22.0–25.5ºC, in a 40-hour test (Lee and Menu 1981). The third series, also carried out in Hawaii, involved a series of evaluations at five test temperatures with highest survival at a salinity of 30 at a temperature of 20ºC; at a salinity of 45, temperature of 24ºC; at a salinity of 30 at 28ºC; at a salinity of 45 at 30ºC; and with no significant survival at a temperature of 32ºC. Durations of these tests were based on hatching times as functions of temperature, declining from 69 hours at a temperature of 20ºC to 25 hours at a temperature of 32ºC (W.A. Walsh et al. 1991). The fourth study involved embryos from the El-Max collection centre, Alexandria, Egypt. These showed highest (but only of 8%) survival at a salinity of five in a 60-day test period (El-Dahhar et al. 2000).

Laboratory survival of *L. ramada* and *M. cephalus* embryos appeared to be most successful at salinities conditions near to those of the habitat where the individuals or their parents were captured. These were a salinity of ≈ 7 for the *L. ramada* and *M. cephalus* embryos from the El-Max, collection centre (El-Dahhar et al. 2000), and SW for the adult *M. cephalus* used in the studies carried out in Hawaii (Sylvester et al. 1975, Lee and Menu 1981, W.A. Walsh et al. 1991). Temperature or temperature x salinity interactions also appeared to influence survival in one series of evaluations (W.A. Walsh et al. 1991). Larvae, except for those of *L. ramada* and of *M. cephalus*, from the Egyptian studies (El-Dahhar et al. 2000) were hatched in the laboratory (Sylvester et al. 1975, Lee and Menu 1981, W.A. Walsh et al. 1991) with induced spawning in the latter study. Survival of *M. cephalus* embryos was shown to be high over a wide range of salinities, up to 45 or higher (Lee and Menu 1981, W.A. Walsh et al. 1991).

Table 13.2. Laboratory determinations of salinity tolerance of mullet embryos.

Species	Sal. ppt	Survival %	Temp. °C/Time	Life stage	Authors
Liza ramada[1]	3	33.3	ND/54 d	larvae	El-Dahhar et al. (2000)
	6	24.4	ND/54 d	larvae	
	9	3.3	ND/54 d	larvae	
	12	24.4	ND/54 d	larvae	
	15	13.3	ND/54 d	larvae	
Mugil cephalus[2]	24	5	19.5–20.5/48 hr	embryo	Sylvester et al. (1975)
	26	9	19.5–20.5/48 hr	embryo	
	28	38	19.5– 20.5/48 hr	embryo	
	30	79	19.5–20.5/48 hr	embryo	
	32	90	19.5–20.5/48 hr	embryo	
Mugil cephalus[3]	24	0	19.5–20.5/96 hr	larvae	Sylvester et al. (1975)
	26	98	19.5–20.5/96 hr	larvae	
	28	70	19.5–20.5/96 hr	larvae	
	30	75	19.5–20.5/96 hr	larvae	
	32	71	19.5–20.5/96 hr	larvae	
	34	42	19.5–20.5/96 hr	larvae	
	36	7	19.5–20.5/96 hr	larvae	
Mugil cephalus[4]	5	0	24–26/36 hr	embryo	Lee and Menu (1981)
	10	0	24–26/36 hr	embryo	
	15	34.9	24–26/36 hr	embryo	
	20	31.4	24–26/36 hr	embryo	
	25	62.3	24–26/36 hr	embryo	
	30	69.2	24–26/36 hr	embryo	
	35	91.2	24–26/36 hr	embryo	
	40	80.4	24–26/36 hr	embryo	
	45	75.0	24–26/36 hr	embryo	
	50	61.4	24–26/36 hr	embryo	
	55	0	24–26/36 hr	embryo	
	60	0	24–26/36 hr	embryo	
Mugil cephalus[5]	10	1.0	22.0–25.5/40 hr	embryo	Lee and Menu (1981)
	15	58.7	22.0–25.5/40 hr	embryo	
	20	64.7	22.0–25.5/40 hr	embryo	
	25	73.9	22.0–25.5/40 hr	embryo	
	30	72.9	22.0–25.5/40 hr	embryo	
	35	71.1	22.0–25.5/40 hr	embryo	
	40	69.8	22.0–25.5/40 hr	embryo	
	45	66.5	22.0–25.5/40 hr	embryo	
	50	48.8	22.0–25.5/40 hr	embryo	
Mugil cephalus[6]	15	8.1	20/69 hours	embryo	W.A. Walsh et al. (1991)
	20	41.3	20/69 hours	embryo	
	25	48.0	20/69 hours	embryo	
	30	71.6	20/69 hours	embryo	
	35	50.0	20/69 hours	embryo	
	45	21.3	20/69 hours	embryo	
Mugil cephalus[7]	15	45.2	24/43 hours	embryo	W.A. Walsh et al. (1991)
	20	82.1	24/43 hours	embryo	
	25	79.1	24/43 hours	embryo	
	30	81.4	24/43 hours	embryo	
	35	79.4	24/43 hours	embryo	
	45	84.4	24/43 hours	embryo	
Mugil cephalus[8]	15	11.9	28/30 hours	embryo	W.A. Walsh et al. (1991)
	20	63.8	28/30 hours	embryo	
	25	73.9	28/30 hours	embryo	
	30	90.4	28/30 hours	embryo	
	35	86.9	28/30 hours	embryo	
	45	89.0	28/30 hours	embryo	

Table 13.2. contd....

Table 13.2. contd.

Species	Sal. ppt	Survival %	Temp. °C/Time	Life stage	Authors
Mugil cephalus[9]	15	18.9	30/28 hours	embryo	W.A. Walsh et al. (1991)
	25	66.5	30/28 hours	embryo	
	35	74.8	30/28 hours	embryo	
	45	79.5	30/28 hours	embryo	
Mugil cephalus[10]	15	0	32/25 hours	embryo	W.A. Walsh et al. (1991)
	25	0.9	32/25 hours	embryo	
	35	1.4	32/25 hours	embryo	
	45	0	32/25 hours	embryo	
Mugil cephalus[11]	FW	1.1	ND/60 days	embryo	El-Dahhar et al. (2000)
	5	8.0	ND/60 days	embryo	
	10	0.0	ND/60 days	embryo	
	15	0.0	ND/60 days	embryo	
	20	0.0	ND/60 days	embryo	
	35	0.0	ND/60 days	embryo	

[1] and [11]—Wild-caught, from El-Max larval collection centre, Egypt.
[2–10]—All from Hawaii.
[4]—Fertilized eggs in series transferred before 2-blastomere stage, 200 in each trial.
[5]—Fertilized eggs in series transferred at gastrula stage, 400 individuals in each trial.

Laboratory Determinations of Salinity Tolerance of Juvenile Mullets

Laboratory evaluations of salinity tolerances of juvenile individuals of mullets have been carried out in several geographic areas, but primarily around the Mediterranean Sea (Table 13.3). Studies were found of salinity tolerances of juvenile mullets belonging to five species. These studies were carried out over varying ranges of salinity, temperature, and with test periods of varying duration (Table 13.3).

Juvenile *C. labrosus* were addressed in two evaluations of individuals taken from the Mediterranean Sea. The first series showed 100% survival at all salinities ranging from FW to SW, temperature of 17–19°C, and 48-hour test period (Chervinski 1977). A second evaluation carried out over a series of salinities ranging from 35–75, showed highest survival at salinities of 35 and 40, temperature of 22 ± 2°C, two-day experimental period. These *C. labrosus* were captured in the Mesolonghi-Etoliko Lagoon (Greece) (Hotos and Vlahos 1998).

Evaluations of salinity tolerances of juvenile *L. aurata* from the Mediterranean Sea showed 100% survival following acclimation to SW and testing in 10% SW, while those that were tested after direct transfer to FW all died. However, when the SW acclimated individuals were acclimated to 10% SW before being tested in FW, all survived (Chervinski 1975).

The third species group studied was of *L. ramada*, with survivorship experiments conducted in two series, both at 16°C. The first experiment in the first series involved individuals acclimated to FW transferred directly to SW, with survival of 97%; in the second experiment the sequence was FW to SW then back to FW, with survival of 85%. A third experiment involved juveniles again acclimated to FW, moved to SW, then back to FW, showing 84% survival at 9°C (Mires et al. 1974). The second series of experiments in that study involved acclimating individuals to 16°C, then transferring groups to a series of test temperatures, a 16°C control, and baths of 12, 14, and 26°C, all in FW. Survival was 100% at all temperatures (Mires et al. 1974).

Groups of *L. saliens* acclimated to SW were transferred to a series of test salinities from SW (control) to FW, temperature of 22–23°C with 100% survival at all except the FW group. However, when fish were acclimated to 5% SW for 48 hours before transfer to FW, survival was 100% (Chervinski 1977).

Evaluations of salinity tolerances of juvenile *M. cephalus* involved groups from three locations. The first of these groups was from the Naaman Estuary (Mediterranean Sea). Fish were taken from SW and transferred to a series of test salinities: SW (control); 28% SW; 3.3% SW; 2.8% SW and FW, with high survival only in the control and 28% SW groups, 16°C, 24-hour tests. A second experiment with fish

Table 13.3. Laboratory determinations of salinity tolerance of juvenile mullets.

Species	Salinity ppt	Survival (%)	Temp. °C /Time	Size (cm)	Authors
Chelon labrosus[1]	FW	100	17–19/48 hours	1.8–3.2	Chervinski (1977)
	10% SW	100	17–19/48 hours	1.8–3.2	
	30% SW	100	17–19/48 hours	1.8–3.2	
	SW	100	17–19/48 hours	1.8–3.2	
Chelon labrosus[2]	35	100	22 ± 2/2 days	2.6	Hotos and Vlahos (1998)
	40	100	22 ± 2/2 days	2.6	
	45	79.2	22 ± 2/2 days	2.6	
	50	66.7	22 ± 2/2 days	2.6	
	55	37.5	22 ± 2/2 days	2.6	
	60	37.5	22 ± 2/2 days	2.6	
	65	4.2	22 ± 2/2 days	2.6	
	70	0	22 ± 2/2 days	2.6	
	75	0	22 ± 2/2 days	2.6	
Liza aurata[3]	10% SW	100	ND/ND	2.0–4.0	Chervinski (1975)
	FW	0	ND/ND	2.0–4.0	
Liza aurata[4]	FW	100	ND/ND	2.0–4.0	Chervinski (1975)
Liza ramada[5]	FW	85	16/24 hours	1.5–1.8	Mires et al. (1974)
	SW	3	16/24 hours	1.5–1.8	
	FW	84	9/24 hours	1.5–1.8	
Liza ramada[6]	FW	100	16/24 hours	1.5–1.8	Mires et al. (1974)
	FW	100	14/24 hours	1.5–1.8	
	FW	100	12/24 hours	1.5–1.8	
	FW	100	26/24 hours	1.5–1.8	
Liza saliens[7]	FW	35	22–23/48 hours	2.5–4.3	Chervinski (1977)
	5% SW	100	22–23/48 hours	2.5–4.3	
	10% SW	100	22–23/48 hours	2.5–4.3	
	20% SW	100	22–23/48 hours	2.5–4.3	
	30% SW	100	22–23/48 hours	2.5–4.3	
	SW	100	22–23/48 hours	2.5–4.3	
Liza saliens[8]	FW	100	22–23/48 hours	2.5–4.3	Chervinski (1977)
Mugil cephalus[9]	SW	76.7	16/24 hours	1.8–2.5	Mires et al. (1974)
	28% SW	100	16/24 hours	1.8–2.5	
	3.3% SW	16.7	16/24 hours	1.8–2.5	
	2.8% SW	3	16/24 hours	1.8–2.5	
	FW	0	16/24 hours	1.8–2.5	
Mugil cephalus[10]	10% SW	93.4	16/24 hours	1.8–2.5	Mires et al. (1974)
	6.7% SW	93.4	16/24 hours	1.8–2.5	
	5.6% SW	100	16/24 hours	1.8–2.5	
	FW	13	16/24 hours	1.8–2.5	
Mugil cephalus[11]	35	100	20 ± 3/48 hours	2.6	Hotos and Vlahos (1998)
	40	100	20 ± 3/48 hours	2.6	
	45	66.7	20 ± 3/48 hours	2.6	
	50	58.3	20 ± 3/48 hours	2.6	
	55	29.2	20 ± 3/48 hours	2.6	
	60	0	20 ± 3/48 hours	2.6	
	65	0	20 ± 3/48 hours	2.6	

Table 13.3. contd....

Table 13.3. contd.

Species	Salinity ppt	Survival (%)	Temp. °C /Time	Size (cm)	Authors
Mugil cephalus[12]	FW	0	28.6–30/8 weeks	1.9–2.2	Olukolajo and Omolara (2013)
	5	30	28.6–30/8 weeks	1.9–2.2	
	10	40	28.6–30/8 weeks	1.9–2.2	
	15	40	28.6–30/8 weeks	1.9–2.2	
	20	20	28.6–30/8 weeks	1.9–2.2	
	25	20	28.6–30/8 weeks	1.9–2.2	
	30	0	28.6–30/8 weeks	1.9–2.2	
Rhinomugil corsula[13]	FW	4250 minutes	35 (37)	'fingerlings'	Kutty et al. (1980)
	7	7600 minutes	20 (37)	'fingerlings'	
	15	2380 minutes	35(37)	'fingerlings'	

[1] Mediterranean Sea. All groups other than SW transferred from SW (SW salinity of 39).
[2] Mesolonghi-Etoliko Lagoon, Ionian Sea.
[3] Mediterranean Sea. Both groups transferred from SW (salinity 39) to test salinity.
[4] Mediterranean Sea. Salinity gradually decreased from 10% SW to FW.
[5] Naaman Estuary. All groups acclimated to FW, then to SW, before transfer to test salinity.
[6] Naaman Estuary. All groups acclimated to FW at 16°C.
[7] Mediterranean Sea. All groups transferred directly from SW.
[8] Mediterranean Sea. Fish transferred to FW after 48 hours in 5% SW.
[9] Naaman Estuary. All fish caught in SW and transferred to test salinities.
[10] Naaman Estuary. All fish acclimated to 10% SW before transfer to test salinities.
[11] Mesolonghi-Etoliko Lagoon, Ionian Sea.
[12] Lagos Lagoon, Nigeria.
[13] Vaigai Reservoir, Tamil Nadu, India. Survival values are resistance times in minutes.
Temperatures are first the acclimation temperature, and second (in parentheses) test temperature.

from the Naaman Estuary involved first acclimating individuals to a salinity of 10% SW then transferring them to test salinities of 10% SW (control), 6.7% SW, 5.6% SW and FW. Survival was of 93.4% in the control, 93.4% in 6.7% SW, 100% survival at 5.6 % SW, but only 13% survival in FW, 16°C, 24-hour tests (Mires et al. 1974). Juvenile *M. cephalus* (from Mesolonghi-Etoliko lagoon) were tested at a series of salinities varying from 35 to 65, all at 20 ± °C, 48-hour tests. Highest survival was 100% at salinities of 35 and 40 (Hotos and Vlahos 1998). The final evaluation of juvenile *M. cephalus* included here involved individuals taken from the Lagos Lagoon, Nigeria. Test salinities ranged from FW to 30, with highest survival (40%) at salinities of 10 and 15, temperature 28.6–30.0°C, over an 8-week test period (Olukolajo and Omolara 2013).

Juvenile individuals of corsula, *Rhinomugil corsula*, collected from the Vaigai reservoir in India, showed longest mean resistance times to various levels of lethal temperatures at a salinity of seven, said to be the isosmotic level for the species. Acclimation temperatures ranged from 20–35°C and test temperatures ranged from 37–41°C (Kutty et al. 1980).

Studies of temperature x salinity tolerances of juvenile mullets have been primarily focused on the ability of these fishes to acclimate to ranges of salinity or of temperature x salinity conditions. Salinity or temperature x salinity ranges at which the juveniles were found to survive were broad for all of these species, generally ranging from FW to SW, and at temperatures between ≈ 20 and 30°C. However, some of these evaluations were at lower and/or higher temperatures (Mires et al. 1974, Chervinski 1977), and two studies showed high salinity tolerance into the hypersaline range (Hotos and Vlahos 1998). Juveniles of both *C. labrosus* and *M. cephalus* showed survival of > 50% of the test individuals at salinities as high as 50 (Hotos and Vlahos 1998). It also appeared that while all of these species could ultimately survive in FW for undermined periods of time, that FW survival generally required a sequential acclimation from waters of higher salinities for varying periods of time.

It is expected that the salinity x temperature combinations showing highest survival, albeit over varying periods of time, reflect the conditions in nature of areas occupied before capture by developing juveniles of these species populations. Salinities of selected habitats may reflect salinity conditions where growth is maximized and has been shown in some studies to be near to that of the isosmotic salinity level of the species (Cardona 2000, 2006).

It is difficult to generalize on the salinity (temperature x salinity) tolerance of juvenile mullets as the species involved, and even groups of a single species, come from different geographical areas and different environmental conditions. Some of the variation observed is likely also based on the differences in approach taken among studies, such as in the range of salinity and temperature over which tests were made and duration of the observation period (experimental periods varied from one day to eight weeks). While there is a solid base of information available on environmental conditions required for successful development of a few species of mullets in captive situations, much work remains to be done to include juveniles of other species of mullets.

Ionic and Osmotic Regulation in Mullets

Patterns of Plasma/Serum Sodium, Potassium, Chloride and Osmotic Concentrations

Discussions here involve concentrations of principal ions and osmotic concentrations in blood plasma/serum relative to ionic and osmotic concentrations of the plasma/serum Na^+, Cl^-, along with K^+, a tightly regulated ion of low concentration, are used as indicators of the ability of teleost fish to maintain homeostasis under varying environmental conditions, especially of salinity and temperature. Plasma/serum values of these ions along with total osmotic concentrations for five species representing three genera of mullets are found in Table 13.4. These individuals represent a range in size/age from small juveniles to mature adults.

The first of the included evaluations were for *C. labrosus* taken from a pond on the Atlantic Coast of France, following acclimation to either FW or SW, temperature 18°C (Lasserre 1971). A second set of collections made in the months of October and January was from ponds of a range of salinities at Arcachon on the Atlantic Coast of France (Lasserre and Gallis 1975). The third evaluation involved groups of fish taken from Arcachon Bay on the Atlantic Coast of France and acclimated to SW at temperatures of 4, 7, or 11°C for a period of eight days, before blood components were measured (Lasserre and Gallis 1975).

A single set of values for blood plasma Na^+, K^+ and Cl^- was found for *L. aurata* taken from the Black Sea (Natochin and Lavrova 1974). Only Na^+ values from FW- and SW-acclimated individuals were available for *L. haematocheila* (Peter the Great Bay, Sea of Japan, Serkov 2003).

Three evaluations of blood components were found for *L. ramada*, two on fish from the Mediterranean Sea, the third from the Atlantic Coast of France. Evaluations of fish from the Mediterranean Sea involved measurements of plasma osmotic concentrations of fish acclimated to FW or SW (Rabeh et al. 2010, Khalil et al. 2012). The fish from ponds on the Atlantic Coast of France entered the pond network from adjacent bay waters. These ponds differed from one another in salinities and water temperatures during the sampling periods (Lasserre and Gallis 1975).

The remaining species, *M. cephalus*, was evaluated in collections from five locations, all in the Northern Hemisphere. Several different questions were tested in this group of studies including: patterns of Na^+ and Cl^- regulation over a range of salinities from ≈ FW to hypersaline levels in fish from the Matanzas Inlet, Atlantic Coast of Florida, USA (Nordlie and Leffler 1975); temperature influence on osmotic concentrations over the range from 10 to 30°C, also in fish from the Matanzas Inlet (Nordlie 1976); osmotic regulatory abilities at a range of salinities from ≈ FW to SW in juveniles of three size/age groups of fish taken from the Gulf of Mexico, Cedar Key, Florida (Nordlie et al. 1982); overall ranges of serum Na^+, Cl^- and K^+ concentrations evaluated monthly over a period of one and one-half years in fish collected at Sabine Island, Pensacola, Florida in the Gulf of Mexico (Folmar et al. 1992); tissue osmolality as a function of acclimation time in salinities declining from SW to FW in fish from the Tyrrhenian Sea (Ciccotti et al. 1994); and Na^+, K^+, Cl^- and osmotic concentrations in fish from Korean marine waters acclimated either to FW or to SW (Lee et al. 1997).

Table 13.4. Plasma/serum sodium, potassium, chloride and osmotic concentrations of mullets.

Species	Salinity ppt/Temp. °C	Na^+*	K^+*	Cl^-*	Osm** concn	Authors
Chelon labrosus[1]	SW(33)/18	142.2	5.6		386.1	Lasserre (1971)
	FW/18	128.5	4.4		270.3	
Chelon labrosus[2]	O 18–28/13–17	172	2.5	160	372	Lasserre and Gallis (1975)
	J 17–22/4–6	212	4.1	191	437	
	J 9–12/6–9	171	2.4	148	360	
	O 12–13/11–16	179	2.8	154	369	
	J FW/7–11	100	2.0	88	233	
Chelon labrosus[3]	SW/4 (8 d)	215	3.8	190	435	Lasserre and Gallis (1975)
	SW/7 (8 d)	178	3.7	155	365	
	SW/11 (8 d)	176	2.6	166	370	
Liza aurata[4]	~ 50% SW/ND	190	1.6	184		Natochin and Lavrova (1974)
Liza haematocheila[5]	SW (32)/ND	177				Serkov (2003)
	FW/ND	138				
Liza ramada[6]	SW (35)/18–20				341	Rabeh et al. (2010)
	FW/18–20				234	
Liza ramada[7]	O 18–28/13–17	176	3.2	154	367	Lasserre and Gallis (1975)
	J 17–22/4–6	221	4.2	195	443	
	J 9–12/6–9	166	2.2	147	349	
	O 12–13/11–16	176	3.1	152	365	
	J 12–13/8–11	166	2.9	130	337	
	J FW/7–11	146	2.7	137	336	
Liza ramada (all females)[8]	FW 24 hr/ND	150	2.6			Khalil et al. (2012)
	SW 24 hr/ND	148	3.2			
	SW 60 hr/ND	162	4.0			
Mugil cephalus[9]	0.7/25	156.7		134.4		Nordlie and Leffler (1975)
	3.9/25	163.5		131.3		
	10.5/25	172.9		142.1		
	21.0/25	189.3		147.8		
	35.0/25	182.4		161.8		
	45.5/25	203.0		155.8		
	55.0/25	182.9		157.5		
Mugil cephalus[10]	~ 33/10	158.8	10.7	138.8	323	Nordlie (1976)
	~ 33/20	166.9	10.3	144.8	341	
	~ 33/30	178.0	10.5	158.2	378	
Mugil cephalus[11]	1.6/22 ± 2				314.7	Nordlie et al. (1982)
	8/22 ± 2				335.0	
	16/22 ± 2				377.9	
	24/22 ± 2				386.4	
	32/22 ± 2				472.3	
Mugil cephalus[12]	1.6/22 ± 2				321.4	Nordlie et al. (1982)
	8/22 ± 2				336.3	
	16/22 ± 2				373.7	
	24/22 ± 2				369.7	
	32/22 ± 2				432.4	
Mugil cephalus[13]	0.35/22 ± 2				314.8	Nordlie et al. (1982)
	1.6/22 ± 2				322.1	
	8/22 ± 2				296.7	
	16/22 ± 2				323.8	
	24/22 ± 2				338.2	
	32/22 ± 2				335.0	

Table 13.4. contd....

Table 13.4. contd.

Species	Salinity ppt/Temp. °C	Na[+]*	K[+]*	Cl[–]*	Osm** concn	Authors
Mugil cephalus[14]	SW/ND	184	3.1	151		Folmar et al. (1992)
	SW/ND	158	1.7	132		
Mugil cephalus[15]	SW (0 wk)/11–17				441.4	Ciccotti et al. (1994)
	SW (5 wk)/11–17				407.7	
	FW (1 wk)/11–17				387.5	
	FW (5 wk)/11–17				397.6	
Mugil cephalus[16]	SW/22.7 ± 2.3	167.1	10.5	127.6	338.1	Lee et al. (1997)
	FW/22.6 ± 2.2	155.3	8.2	131.6	314.5	

*Units in meq.L^{-1} or mmol.L^{-1}; **units in mOsm.L^{-1} or mOsm.kg^{-1}.
[1] Fish pond, Atlantic coast of France, adult fish.
[2] Bassin d'Arcachon, Atlantic coast, France; large juveniles, O, October, J, January.
[3] Arcachon Bay, Atlantic coast, France; large juveniles.
[4] Black Sea; water ~ 50% of standard SW.
[5] Peter the Great Bay, Sea of Japan; acclimated to FW for 7 d before Na^+ was measured.
[6] Tunis Gulf, Khelii entrance, Mediterranean Sea; large juveniles, in SW, to FW (24 hours).
[7] Bassin d' Arcachon, Atlantic coast, France; large juveniles, O, October, J, January.
[8] El-Serw Fish Research Station, Egypt; ≥ 2 years of age.
[9] Matanzas Inlet, Atlantic Ocean, Florida, USA; large juveniles.
[10] Matanzas Inlet, Atlantic Ocean, Florida, USA; fish of ~ 18 g.
[11] Salt marsh, Cedar Key, Florida, USA, Gulf of Mexico; fish of 2.0–2.9 cm SL.
[12] Salt marsh, Cedar Key, Florida, USA, Gulf of Mexico; fish of 3.0–3.9 cm SL.
[13] Salt marsh, Cedar Key, Florida, USA, Gulf of Mexico; fish of 4.0–6.9 cm SL.
[14] Sabine Island, Pensacola, Florida, USA, Gulf of Mexico; high and low values estimated from graphs of monthly mean values.
[15] Tyrrhenian Sea (Mediterranean Sea); all juveniles, values of tissue osmotic concentrations.
[16] Pusan, Korea; juveniles; transferred from SW to FW and maintained there for 60 days.
Values for SW fish were means from text; values for FW-acclimated fish estimated from graphs.
For reference, standard seawater (25°C), salinity of 35.17, has a concentration of Na^+ of 469.0, K^+ of 10.2, and Cl^- of 545.9, all in meq.kg^{-1} (after Millero et al. 2008).

Mean values and ranges of plasma/serum Na^+, K^+, and Cl^- were obtained from data in tables in Holmes and Donaldson (1969) for teleosts from FW and from SW, and used here as rough standards for comparison. The table did not include water temperatures. Mean values for FW teleosts were of 130.3 mmol.L^{-1}, range 115.5–140.9; 3.6 mmol.L^{-1}, range 2.1–6.2; and 117.8 mmol.L^{-1}, range 98.1–140, for Na^+, K^+, and Cl^-, respectively. Teleost fishes from SW showed mean values of 187.7 mmol.L^{-1}, range 156–253; 7.4 mmol.L^{-1}, range 1.4–26.8; and 171.2 mmol.L^{-1}, range 106–251, for Na^+, K^+, and Cl^-, respectively. Using these range values as criteria, it appears that the Na^+ value in *C. labrosus* was lower than expected in SW from the fishpond (Lasserre 1971); and values for Na^+, K^+, and Cl^- were low in *C. labrosus* from the FW pond (Lasserre and Gallis 1975). Values that were high compared to Holmes and Donaldson (1969) ranges included Na^+ in FW-acclimated *M. cephalus* from the Matanzas River, Florida (Nordlie and Leffler 1975), and both Na^+ and K^+ in *M. cephalus* in FW from Korean waters (Lee et al. 1997).

It was apparent from information on osmotic and ionic regulation in *C. labrosus*, *L. aurata*, *L. haematocheila*, *L. ramada*, and *M. cephalus*, that all of these species of mullets show characteristics of euryhalinity, though with variations in responses to ranges of salinities as well as to ambient temperatures. The comparative study of osmotic and ionic responses in *C. labrosus* and *L. ramada* demonstrated that both species are capable of coping with significant variations in ambient salinity and temperature, but in a FW environment *C. labrosus* appeared to lose control of its plasma osmotic, Na^+ and Cl^- concentrations (individuals of that species are not expected to survive indefinitely in that medium Lasserre and Gallis 1975). Individuals of both *C. labrosus* and *L. ramada* showed highest levels of Na^+, Cl^- and K^+, as well as plasma osmotic concentration in individuals taken in January from the pond with the highest salinity, but at the lowest temperature (Table 13.4). The values for all three plasma ions and osmotic concentration, obtained in a laboratory acclimation study of *C. labrosus* individuals in SW at a temperature of 4°C, were virtually identical to those obtained for the species in the January samples from the pond with a salinity range of 11–22, temperature range 4–6°C. However, the values for *C. labrosus* obtained following acclimation at 7°C in the SW group were appreciably lower than those obtained at 4°C, suggesting that at the colder temperature individuals of the species do not maintain as tight control of plasma ions and thus, osmotic concentration.

Both plasma Na^+ and Cl^- ion concentrations in *M. cephalus* showed increases with increased ambient salinities, with Na^+ increasing to a salinity of ≈ 46, while Cl^- showed increases only to a salinity of ≈ SW (Nordlie and Leffler 1975). When acclimated to a series of temperatures, 10–30°C, plasma Na^+, Cl^- and osmotic concentrations were found to increase with temperature while the K^+ concentration was held constant (Nordlie 1976). Juvenile individuals of *M. cephalus* were found to develop tolerance to FW and ≈ adult osmotic regulatory capability at a size of ≥ 40 mm SL, age of ≥ 7.5 mo (Nordlie et al. 1982). Seasonal cycles of serum components evaluated in a study of *M. cephalus* from Gulf of Mexico waters near Pensacola, Florida, revealed significant changes in mean ionic concentrations between cold and warm periods (Folmar et al. 1992). Only minor changes were found in Na^+, K^+, Cl^- and osmotic concentrations in individuals of *M. cephalus* taken from SW compared with a group from the same location acclimated to FW (Lee et al. 1997).

Calcium and Magnesium Regulation

Plasma/serum concentrations of Ca^{2+} and Mg^{2+} were found for *L. aurata* from the Black Sea (Natochin and Lavrova 1974), females of *L. ramada* from FW (El-Serw Fish Research Station, Egypt Khalil et al. 2012) and three groups of *M. cephalus* (Table 13.5). The *M. cephalus* groups included adult males and females from Hawaii (Peterson and Shehadeh 1971); a group from the Atlantic Ocean at Matanzas River, Florida USA, that were evaluated at a series of four salinities, FW–SW; and individuals from monthly collections from the Gulf of Mexico near Pensacola, Florida USA (Folmar et al. 1992), from which overall high and low mean values were estimated from graphs (Table 13.5).

Table 13.5. Plasma/serum calcium and magnesium concentrations of mullets.

Species	Salinity/Temp. °C	Ca^{2+}*	Mg^{2+}*	Size	Authors
Liza aurata[1]	~ 50% SW/ND	10.7	13.5	ND	Natochin and Lavrova (1974)
Liza ramada	FW (24 hours)/ND	27.8	1.2	30–42 cm	Khalil et al. (2012)
(all females)[2]	SW (60 hours)/ND	34.8	2.7	30–42 cm	
Mugil cephalus[3]	SW ?/ND	6.2 (F)		adult	Peterson and
	SW ?/ND	7.1 (M)		adult	Shehadeh (1971)
Mugil cephalus[4]	FW (2 weeks)/20 ± 1	5.0	2.8	10.5 cm TL,	Nordlie and Whittier
	25% SW/20 ± 1	5.2	3.3	mean	(1983)
	50% SW/20 ± 1	5.7	4.3		
	SW/20 ± 1	4.8	3.9		
Mugil cephalus[5]	SW (high mean)/ND	7.3	7.4	ND	Folmar et al. (1992)
	SW (low mean)/ND	5.0	2.3		

*Units of Ca^{2+} and Mg^{2+} are $meq.L^{-1}$.
[1] Black Sea.
[2] El-Serw Fish Research Station, Egypt.
[3] Hawaii: Salinity was not specified but was assumed to be of SW.
[4] Matanzas River, Atlantic Ocean, Florida, USA.
[5] Sabine Island, Pensacola, Florida, Gulf of Mexico. High and low values estimated from graphs of monthly means.

For reference, standard seawater (25°C), salinity of 35.17, has a concentration of Ca^{2+} of 20.6 and of Mg^{2+} of 105.6, both in meq.kg (after Millero et al. 2008).

Means and ranges of values of plasma/serum Ca^{2+} and Mg^{2+} for marine and FW teleosts (primarily stenohaline species) given in Holmes and Donaldson (1969) were used for comparisons with values cited for mullet species. Mean values for FW teleosts were of 5.8 meq.L^{-1}, range 3.6–9.8, and 2.9 meq.L^{-1}, range 2.5–3.4 for Ca^{2+} and Mg^{2}, respectively. Mean values and ranges for marine teleosts were of 8.0 meq.L^{-1}, range 4.4–16.6, and 7.7 meq.L^{-1}, range 1.6–19.4 for Ca^{2+} and Mg^{2+}, respectively. Values from fishes included in Table 13.5 that were outside of these arbitrarily determined ranges were of Ca^{2+} in FW and SW and for Mg^{2+} in FW for *L. ramada* (Khalil et al. 2012).

The values obtained by Folmar et al. (1992) were roughly equal to those of Peterson and Shehadeh (1971) and of Nordlie and Whittier (1983), suggesting that there is relatively tight regulation of Ca^{2+} and Mg^{2+} in *M. cephalus*. However, there are insufficient values available for other species of mullets to make broad generalizations at this time.

Mullets in Freshwater

It has long been known that various species of fishes including mullets, considered to be marine fishes, are found in freshwaters in many areas where they have connections to marine waters. An early suggestion of how these marine fishes were able to tolerate such low salinities came from Breder (1934) who had observed the presence of several species of marine teleosts in a lake in the Bahama Islands and found that the water had an unusually high Ca^{2+} concentration. Later he demonstrated that certain marine fishes showed extended, but not indefinite, periods of survival in New York City tap water when the Ca^{2+} concentration was enhanced. Considerable interest in this work was shown in various areas of the world including Florida, USA where numerous species of marine fishes including mullets are found in streams and rivers that receive water from springs of the Floridan Aquifer and ultimately discharge to marine waters. These spring waters contain significant concentrations of calcium gained from the aquifer that is located in limestone and dolomite deposits (Rosenau et al. 1977). Three species of mullets, *A. monticola*, *M. cephalus* and *M. curema*, have been reported from inland waters in Florida, though there are few rivers in which all three species occur together. One exception is the Apalachicola River in Northwest Florida (Ichthyology Collection, Florida Museum of Natural History, collection records, on line), where concentrations of Ca^{2+} (13 mg.L^{-1}), Na^{+} + K^{-} (6.3 mg.L^{-1}) and Cl^{-} (4.0 mg.L^{-1}) are all quite low (Black and Brown 1951). The species most commonly observed in Florida freshwaters is *M. cephalus,* which is almost ubiquitous in waters that drain to either the Atlantic Ocean or Gulf of Mexico.

Florida spring-fed rivers that flow to the Atlantic Ocean or Gulf of Mexico were surveyed with respect to their geological origins, calcium and chloride concentrations and presence of marine 'visitors'. The conclusion in that study was that the concentration of Cl^{-} (dissolved from geological strata with which the water came into contact), not of Ca^{2+}, limited the occupation of these waters by marine teleost fishes (Odum 1953). The author also concluded that *M. cephalus* appeared to have no lower Cl^{-} limit for occupation of these inland waters, being found in Ponce de Leon Spring, Holmes County, Florida where the Cl^{-} concentration was of 3 mg.L^{-1}. However, this spring has a Ca^{2+} concentration of 30 mg.L^{-1} (Black and Brown 1951).

Evans (1973) found that some stenohaline marine fishes were capable of extracting Na^{+} and Cl^{-} from FW, but survived only in those waters in which the Ca^{2+} concentration, which reduces the permeability of the fish's gill integument to diffusive loss of Na^{+} and Cl^{-}, was high enough for the fish to maintain a survivable ion balance. Subsequently, Evans et al. (2005) further emphasized that the balance between diffusive losses of ions from teleost gills, and the capabilities of cells in the gills to replace these ions from the surrounding waters are essential elements in osmoregulatory functions accompanying the movements of euryhaline teleost fishes into brackish or FW environments. Thus, Ca^{2+}, Cl^{-} and/or Na^{+} concentrations are important in the occupation of FW by euryhaline fishes such as mullets, with the necessary concentrations determined by the physiological capabilities of the species.

Acknowledgements

My sincere thanks to Kelly Barber for preparing the illustrations, to Kyle Arola and David Reddig for their help with computer problems, to the Department of Biology for work space and facilities, and to Dr. B.K. McNab for help in formatting and for the many discussions of questions in environmental physiology of vertebrate animals.

References

Anderson, W.W. 1957. Larval forms of the fresh-water mullet (*Agonostomus monticola*) from the open ocean off the Bahamas and south Atlantic coast of the United States. Fish Bull. 120. 57: 415–425.

Anonymous, 1958. Symposium on the classification of brackish waters, Venice, 8–14 April, 1958. Oikos 9: 311–312.

Assem, H. and W. Hanke. 1981. Cortisol and osmotic adjustment of the euryhaline teleost, *Sarotherodon mossambicus*. Gen. Comp. Endocrinol. 43: 370–380.

Ball, J.N., I.C. Jones, M.E. Forster, G. Hargreaves, E.F. Hawkins and K.P. Milne. 1971. Measurement of plasma cortisol levels in the eel *Anguilla anguilla* in relation to osmotic adjustments. J. Endocrinol. 50: 75–96.

Baustian, M.D. and K.W. Beyenbach. 1999. Natriuretic peptides and the acclimation of glomerular toadfish to hypoosmotic media. J. Comp. Physiol. 169: 507–514.

Belloc, F. and J.-L. Gallis. 1980. Fresh water adaptation in the euryhaline teleost, *Chelon labrosus*-III. Biochemical characterization and increase in the acid phosphatase activity in gill. Comp. Biochem. Physiol. 65A: 433–437.

Beyenbach, K.W. 2000. Renal handling of magnesium in fish: from whole animal to brush border membrane vesicles. Frontiers in Bioscience 5: 712–719.

Beyenbach, K.W. 2004. Kidneys sans glomeruli. Am. J. Physiol. Renal Physiol. 286: F811–F827.

Blaber, S.J.M. 1987. Factors affecting recruitment and survival of mugilids in estuaries and coastal waters of southern Africa. Amer. Fish Soc. Symp. 1: 507–518.

Black, A.P. and E. Brown. 1951. Chemical character of Florida's waters. Water Survey and Research, Paper No. 6, 30 November, 1951. Division of Water Survey and Research, State Board of Conservation, State of Florida, Tallahassee, FL.

Blaxter, J.H.S. 1988. Pattern and variety in development. pp. 1–58. *In*: W.S. Hoar and D.J. Randall (eds.). Fish Physiology, Vol. XI, The Physiology of Developing Fish, Part A. Academic Press Inc., San Diego, CA.

Breder, C.M. 1934. Ecology of an oceanic freshwater lake, Andros Island, Bahamas, with special reference to its fishes. Zoologica. 18: 57–88.

Breves, J.P., S.D. McCormick and R.D. Karlstrom. 2014. Prolactin and teleost ionocytes: New insights into cellular and molecular targets of prolactin in vertebrate epithelia. Gen. Comp. Endocrinol. 203: 21–28.

Briggs, J.C. 1960. Fishes of worldwide (circumtropical) distribution. Copeia 1960: 171–180.

Bucking, C. and C.M. Wood. 2006. Gastrointestinal processing of Na^+, Cl^-, and K^+ during digestion: implications for homeostatic balance in freshwater rainbow trout. Am. J. Physiol. Regul. Integr. Comp. Physiol. 291: R1764–R1772.

Bucking, C. and C.M. Wood. 2007. Gastrointestinal transport of Ca^{2+} and Mg^{2+} during the digestion of a single meal in the freshwater rainbow trout. J. Comp. Physiol. 177: 349–360.

Cardona, L. 2000. Effects of salinity on the habitat selection and growth performance of Mediterranean flathead grey mullet *Mugil cephalus* (Osteichthyes, Mugilidae). Est. Coast. Shelf Sci. 50: 727–737.

Cardona, L. 2006. Habitat selection by grey mullets (Osteichthyes, Mugilidae) in Mediterranean estuaries: the role of salinity. Scientia Marina 70: 443–455.

Cataldi, E., D. Crosetti, C. Conte, D. D'Ovidio and S. Cataudella. 1988a. Morphological changes in the oesophageal epithelium during adaptation to salinities in *Oreochromis niloticus*, *O. mossambicus* and their hybrids. J. Fish Biol. 32: 191–196.

Cataldi, E., D. Crosetti, C. Leone and S. Cataudella. 1988b. Oesophagus structure during adaptation to salinity in *Oreochromis niloticus* (Perciformes: Pisces) juveniles. Boll. Zool. 55: 59–62.

Cataldi, E., L. Garibaldi, D. Crosetti, C. Leone and S. Cataudella. 1991. Variations in renal morphology during adaptation to salinities in tilapias. Environ. Biol. Fish 31: 101–106.

Chervinski, J. 1975. *Liza aurata* (Risso) (Pisces: Mugilidae) and its adaptability to various saline conditions. (abstract). Aquaculture 5: 110.

Chervinski, J. 1977. Adaptability of *Chelon labrosus* (Risso) and *Liza saliens* (Risso) (Pisces: Mugilidae) to fresh water. Aquaculture 11: 75–79.

Christensen, A.K., J. Hiroi, E.T. Schultz and S.D. McCormick. 2012. Branchial ionocyte organization and ion-transport protein expression in juvenile alewives acclimated to freshwater or seawater. J. Exp. Biol. 215: 642–652.

Ciccotti, E., G. Marino, P. Pucci, E. Cataldi and S. Cataudella. 1994. Acclimation trial of *Mugil cephalus* juveniles to freshwater: morphological and biochemical aspects. Environ. Biol. Fishes 43: 163–170.

Cruz, G.A. 1987. Reproductive biology and feeding habits of Cuyamel, *Joturus pichardi* and Tepemechín, *Agonostomus monticola* (Pisces: Mugilidae) from Rio Plátano, Mosquitia, Honduras. Bull. Mar. Sci. 40: 63–72.

Curtis, B.J. and C.M. Wood. 1991. The function of the urinary bladder *in vivo* in the freshwater rainbow trout. J. Exp. Biol. 155: 567–583.

Ditty, J.G. and R.F. Shaw. 1996. Spatial and temporal distribution of larval striped mullet (*Mugil cephalus*) and white mullet (*M. curema*, family: Mugilidae) in the northern Gulf of Mexico, with notes on mountain mullet, *Agonostomus monticola*. Bull. Mar. Sci. 59: 271–288.

Dogu, Z., E. Sahinöz, F. Aral and R. Sevik. 2013. The growth characteristics of *Liza* (*Mugil*) *abu* (Heckel 1943) in Atatürk Dam Lake. African Journal of Agricultural Research 8: 4434–4440. DOI: 10.5897/AJAR2013.7193.

Durand, J.D. 2015. Implications of molecular phylogeny for the taxonomy of Mugilidae. *In*: D. Crosetti and S.J.M. Blaber (eds.). Biology, Ecology and Culture of Grey Mullet (Mugilidae). CRC Press, Boca Raton, USA (this book).

Dymowska, A.K., P.-P. Hwang and G.G. Goss. 2012. Structure and function of ionocytes in the freshwater fish gill. Respir. Physiol. Neurobiol. 184: 282–292.

El-Bakary, N.E.R. and H.L. El-Gammal. 2010. Comparative biological, histochemical, and ultrastructural studies on the proximal intestine of flathead grey mullet (*Mugil cephalus*) and sea bream (*Sparus aurata*). Wor. Appl. Sci. J. 8: 477–485.

El-Dahhar, A., M. Salama, Y. Moustafa and S. Zahran. 2000. Effect of salinity and salinity acclimatization on survival and growth of the wild hatched mullet larvae using different feeds in glass aquaria. Egypt. J. Aquat. Biol. & Fish. 4: 139–155.

Eschmeyer, W.N. (ed.). 2014. Catalog of Fishes: Species of Fishes by Family/Subfamily. http://researcharchive.calacademy.org/research/Ichthyology/catalog/SpeciesByFamily.asp. Electronic version accessed 20/02/2014.

Evans, D.H. 1973. Sodium uptake by the sailfin molly, *Poecilia latipinna*: kinetic analysis of a carrier system present in both freshwater-acclimated and sea-water-acclimated individuals. Comp. Biochem. Physiol. 45A: 848–850.

Evans, D.H. 2008. Teleost fish osmoregulation: what have we learned since August Krogh, Homer Smith, and Ancel Keys. Am. J. Physiol. Regul. Integr. Comp. Physiol. 295: R704–R713.

Evans, D.H. 2011. Freshwater fish gill ion transport: August Krogh to morpholinos and microprobes. Acta Physiol. 202: 349–359.

Evans, D.H., P.M. Piermarini and K.P. Choe. 2005. The multifunctional fish gill: dominant site of gas exchange, osmoregulation, acid-base regulation, and excretion of nitrogenous waste. Physiol. Rev. 85: 97–177.

Folmar, L.C., T. Moody, S. Bonomelli and J. Gibson. 1992. Annual cycle of blood chemistry parameters in striped mullet (*Mugil cephalus* L.) and pinfish (*Lagodon rhomboids* L.) from the Gulf of Mexico. J. Fish Biol. 41: 999–1011.

Froese, R. and E. Pauly (eds.). 2014. FishBase, World Wide Web electronic publication. www.fishbase.org. Electronic version accessed 20/02/2014.

Gallis, J.L. and M. Bourdichon. 1976. Changes of (Na^+-K^+) dependent ATPase activity in gills and kidneys of two mullets *Chelon labrosus* (Risso) and *Liza ramada* (Risso) during fresh water adaptation. Biochimie 58: 625–627.

Genz, J., D. McDonald and M. Grosell. 2011. Concentration of $MgSO_4$ in the intestinal lumen of *Opsanus beta* limits osmoregulation in response to acute hypersalinity stress. Am. J. Physiol. Regul. Integr. Comp. Physiol. 300: R895–R909.

Gilbert, C.R. and D.P. Kelso. 1971. Fishes of the Tortuguero area, Caribbean Costa Rica. Bull. Florida State Museum, Biol. Sci. 16: 1–54.

González-Castro, M. and G. Minos. 2015. Sexuality and reproduction of Mugilidae. *In*: D. Crosetti and S.J.M. Blaber (eds.). Biology, Ecology and Culture of Grey Mullet (Mugilidae). CRC Press, Boca Raton, USA (this book).

González-Castro, M. and J. Ghasemzadeh. 2015. Morphology and morphometry based taxonomy of Mugilidae. *In*: D. Crosetti and S.J.M. Blaber (eds.). Biology, Ecology and Culture of Grey Mullet (Mugilidae). CRC Press, Boca Raton, USA (this book).

Grosell, M. 2006. Intestinal anion exchange in marine fish osmoregulation. J. Exp. Biol. 209: 2813–2827.

Grosell, M. 2011. Intestinal anion exchange in marine teleosts is involved in osmoregulation and contributes to the oceanic inorganic carbon cycle. Acta Physiol. 202: 421–434.

Grosell, M. and J.R. Taylor. 2007. Intestinal anion exchange in teleost water balance. Comp. Biochem. Physiol. A148: 14–22.

Gunter, G. 1942. A list of the fishes of the mainland of North and Middle America recorded from both fresh water and sea water. Am. Midland Naturalist 28: 305–326.

Gunter, G. 1956. A revised list of euryhaline fishes of North and Middle America. Am. Midland Naturalist 56: 345–354.

Gunter, G. 1967. Vertebrates in hypersaline waters. pp. 230–241. *In*: B.J. Copeland (ed.). Effects of Supersaline Conditions on Aquatic Ecosystems. Contributions in Marine Science, Univ. of Texas 12.

Harrison, T.D. 2002. Preliminary assessment of the biogeography of fishes in South African estuaries. Mar. Freshwater Res. 53: 479–490.

Hickling, C.F. 1970. A contribution to the natural history of the English grey mullets (Pisces: Mugilidae). J. Mar. Biol. Assoc. UK 50: 609–633.

Hickman, C.P. and B.F. Trump. 1969. The kidney. pp. 91–239. *In*: W.S. Hoar and D.J. Randall (eds.). Fish Physiology, Vol. 1. Academic Press, New York.

Hirano, T. and N. Mayer-Gostan. 1976. Eel oesophagus as an osmoregulatory organ. Proc. Nat. Acad. Sci. USA 73: 1348–1350.

Hirano, T., D.W. Johnson, H.A. Bern and S. Utide. 1973. Studies on water and ion movements in the isolated urinary bladder of selected freshwater, marine and euryhaline teleosts. Comp. Biochem. Physiol. 45A: 529–540.

Hirano, T., T. Ogasawara, J.P. Bolton, N.L. Collie, S. Hasegawa and M. Iwata. 1986. Osmoregulatory role of prolactin in lower vertebrates. pp. 112–124. *In*: R. Kirsch and B. Lahlow (eds.). Comparative Physiology of Environmental Adaptation, Vol. I. Karger, Basel.

Hiroi, J. and S.D. McCormick. 2007. Variation in salinity tolerance, gill Na^+/K^+-ATPase, $Na^+/K^+/2Cl^-$ cotransporter and mitochondria-rich cell distribution among three salmonids *Salvelinus namaycush*, *Salvelinus fontinalis*, and *Salmo salar*. J. Exp. Biol. 210: 1015–1024.

Hiroi, J. and S.D. McCormick. 2012. New insights into gill ionocyte and ion transporter functions in euryhaline and diadromous fish. Respir. Physiol. Neurobiol. 184: 257–268.

Hiroi, J., S. Yasumasu, S.D. McCormick, P.-P. Hwang and T. Kaneko. 2008. Evidence for an apical Na-Cl cotransporter involved in ion uptake in a teleost fish. J. Exp. Biol. 211: 2584–2599.

Holmes, W.N. and R.L. McBean. 1963. Studies of the glomerular filtration rate of rainbow trout (*Salmo gairdneri*). J. Exp. Biol. 40: 335–341.

Holmes, W.N. and E.M. Donaldson. 1969. The body compartments and distribution of electrolytes. pp. 1–89. *In*: W.S. Hoar and D.J. Randall (eds.). Fish Physiology, Vol. I. Academic Press, New York.

Hossler, F.E., J.R. Ruby and T.D. McIlwain. 1979. The gill arch of the mullet, *Mugil cephalus* II. Modification in surface ultrastructure and Na, K-ATPase content during adaptation to various salinities. J. Exp. Zool. 208: 399–406.

Hotos, G.N. and N. Vlahos. 1998. Salinity tolerance of *Mugil cephalus* and *Chelon labrosus* (Pisces: Mugilidae) fry in experimental conditions. Aquaculture 167: 329–338.

Hwang, P.-P. and T.-H. Lee. 2007. New insights into fish ion regulation and mitochondrion-rich cells. Comp. Biochem. Physiol. A148: 479–497.

Hwang, P.-P., T.-H. Lee and L.-Y. Lin. 2011. Ion regulation in fish gills: recent progress in the cellular and molecular mechanisms. Am. J. Physiol. Regul. Integr. Comp. Physiol. 301: R28–R47.

Ishida, J. 1935. The stomach of *Mugil cephalus* and its digestive enzymes. Annot. Zool. Japon 15: 182–189.

Kantoussan, J., J.M. Ecoutin, M. Simier, L.T. de Morais and R. Laë. 2012. Effects of salinity on fish assemblage structure: An evaluation based on taxonomic and functional approaches in the Casamance estuary (Senegal, West Africa). Est. Coast. Shelf Sci. 113: 152–162.

Keys, A. and E.N. Willmer. 1932. "Chloride secreting cells" in the gills of fishes, with special reference to the common eel. J. Physiol. Lond. LXXVI: 368–378.

Khalil, N.A., A.M. Hashem, A.A.E. Ibrahim and M.A. Mousa. 2012. Effects of stress during handling, seawater acclimation confinement, and induced spawning on plasma ion levels and somatolactin-expressing cells in mature female *Liza ramada*. J. Exp. Zool. 317A: 410–424.

Khodabandeh, S., M.S. Moghaddam and B. Abtahi. 2009. Changes in chloride cell abundance, Na^+, K^+-ATPase immunolocalization and activity in the gills of golden grey mullet, *Liza aurata*, fry during adaptation to different salinities. Yakhteh Medical Journal 11: 49–54.

Kirsch, R. and M.F. Meister. 1982. Progressive processing of ingested water in the gut of sea-water teleosts. J. Exp. Biol. 98: 67–86.

Koutrakis, E. 2015. Biology and ecology of fry and juveniles of Mugilidae. *In*: D. Crosetti and S.J.M. Blaber (eds.). Biology, Ecology and Culture of Grey Mullet (Mugilidae). CRC Press, Boca Raton, USA (this book).

Kurita, Y., T. Nakada, A. Kato, H. Doi, A.C. Mistry, M.-H. Chang, M.F. Romero and S. Hirose. 2008. Identification of intestinal bicarbonate transporters involved in formation of carbonate precipitates to stimulate water absorption in marine teleost fish. Am. J. Physiol. Regul. Integr. Comp. Physiol. 294: R1402–R1412.

Kutty, M.N., N. Sukumaran and H.M. Kasim. 1980. Influence of temperature and salinity on survival of the freshwater mullet, *Rhinomugil corsula* (Hamilton). Aquaculture 20: 261–274.

Lahlou, B. 1967. Excrétion rénale chez au poisson euryhalin, le flet (*Platichthys flesus* L.): charactéristiques de l'urine normale en eau douce et en eau de mer et effets des changements de milieu. Comp. Biochem. Physiol. 20: 926–938.

Lahlou, B., I.W. Henderson and W.H. Sawyer. 1969. Renal adaptations by *Opsanus tau*, a euryhaline aglomerular teleost, to dilute media. Am. J. Physiol. 216: 1266–1272.

Lasserre, P. 1971. Increase of ($Na^+ + K^+$)—dependent ATPase activity in gills and kidneys of two euryhaline marine teleosts, *Crenimugil labrosus* (Risso 1826) and *Dicentrarchus labrax* (Linnaeus 1758), during adaptation to fresh water. Life Sciences 10 (part II): 113–119.

Lasserre, P. and J.-L. Gallis. 1975. Osmoregulation and differential penetration of two grey mullets, *Chelon labrosus* (Risso) and *Liza ramada* (Risso) in estuarine fish ponds. Aquaculture 5: 323–344.

Lee, C.-S. and B. Menu. 1981. Effects of salinity on egg development and hatching in grey mullet *Mugil cephalus* L. J. Fish Biol. 19: 179–188.

Lee, Y.C., Y.J. Cheng and B.K. Lee. 1997. Osmoregulation capability of juvenile grey mullets (*Mugil cephalus*) with the different salinities. J. Korean Fisheries Society 30: 216–224.

Loftus, W.F., J.A. Kushlan and S.A. Voorhees. 1984. Status of the mountain mullet in Southern Florida. Florida Scientist 47: 256–263.

Luzhnyak, V.A. 2007. New data on specific features of the ecology of reproduction of haarder *Liza haematocheilus* (Mugilidae) in the Azov-Black Sea basin. J. Ichthyol. 47: 676–679.

Madsen, S.S. 1990. The role of cortisol and growth hormone in seawater adaptation and development of hyposmoregulatory mechanisms in sea trout parr (*Salmo trutta trutta*). Gen. Comp. Endocrinol. 79: 1–11.

Manzon, L.A. 2002. The role of prolactin in fish osmoregulation: a review. Gen. Comp. Endocrinol. 125: 291–310.

Marshall, W.S. 1988. Passive solute and fluid transport in brook trout (*Salvelinus fontinalis*) urinary bladder epithelium. Can. J. Zool. 66: 912–918.

Marshall, E.K., Jr. and H.W. Smith. 1930. The glomerular development of the vertebrate kidney in relation to habitat. Biol. Bull. LIX: 135–153.

Matamoros, W.A., J. Schaefer, P. Mickle, W. Arthurs, R.J. Ikoma and R. Ragsdale. 2009. First record of *Agonostomus monticola* (Family: Mugilidae) in Mississippi freshwaters with notes of its distribution in the southern United States. Southeastern Naturalist 8: 175–178.

McCormick, S.D. 1995. Hormonal control of gill Na^+, K^+-ATPase and chloride cell function. pp. 285–315. *In*: C.M. Wood and T.T. Shuttleworth (eds.). Cellular and Molecular Approaches to Fish ion Regulation. Fish Physiology Vol. 14. Academic Press, San Diego, CA.

McCormick, S.D. 2001. Endocrine control of osmoregulation in teleost fish. Amer. Zool. 41: 781–794.

McCormick, S.D., K. Sundell, B.T. Björnsson, C.L. Brown and J. Hiroi. 2003. Influence of salinity on the localization of Na^+/K^+-ATPase, $Na^+/K^+/2Cl^-$ cotransporter (NKCC) and CFTR anion channel in chloride cells of the Hawaiian goby (*Stenogobius hawaiiensis*). J. Exp. Biol. 206: 4575–4583.

McDonald, M.D. and M. Grosell. 2006. Maintaining osmotic balance with an aglomerular kidney. Comp. Biochem. Physiol. A143: 447–458.

Millero, F.J., R. Feistel, D.G. Wright and T.J. McDougall. 2008. The composition of standard seawater and the definition of the reference composition salinity scale. Deep-Sea Research Part I 55: 50–72.

Minos, G., A. Imsiridou and P.S. Economidis. 2010. *Liza haematocheilus* (Pisces: Mugilidae) in the northern Aegean Sea. pp. 313–332. *In*: D. Golani and B. Applebaum-Golani (eds.). Fish Invasions of the Mediterranean Sea: Change and Renewal. PenSoft Publishers, Sofia-Moscow.

Mires, D., Y. Shak and S. Shilo. 1974. Further observations on the effect of salinity and temperature changes on *Mugil capito* and *Mugil cephalus* fry. Bamidgeh 26: 104–109.

Mohamed, A.-R.M., N.A. Hussain, S.S. Al-Noor, B.W. Coad and F.M. Mutlak. 2009. Status of diadromous fish species in the restored East Hammar marsh in southern Iraq. American Fisheries Society Symposium 69: 577–588.

Mohamed, A.-R.M., N.A. Hussain, S.S. Al-Noor, F.M. Mutlak, I.M. Al-Sudani and A.M. Mojer. 2012. Ecological and biological aspects of fish assemblage in the Chybayish marsh, southern Iraq. Ecohydrology & Hydrobiology 12: 65–74.

Mommsen, T.P., M.M. Vijayan and T.W. Moon. 1999. Cortisol in teleosts: dynamics, mechanisms of action, and metabolic regulation. Rev. Fish Biol. Fisheries 9: 211–268.

Muhsin, K.A. 2011. The reproductive cycle and oocyte development of Khishni female *Liza abu* (Heckel 1843) from south of Iraq. Al-Mustansiriyah J. Sci. 22: 76–92.

Nagashima, K. and M. Ando. 1994. Characterization of esophageal desalination in the seawater eel, *Anguilla japonica*. J. Comp. Physiol. B164: 47–54.

Natochin, Y.V. and E.A. Lavrova. 1974. The influence of water salinity and stage in life history on ion concentration of fish blood serum. J. Fish Biol. 6: 545–555.

Nelson, J.S. 2006. Fishes of the World. 4th ed. John Wiley and Sons, Hoboken, NJ.

Nordlie, F.G. 1976. Influence of environmental temperature on plasma ionic and osmotic concentrations in *Mugil cephalus* Lin. Comp. Biochem. Physiol. 55A: 379–381.

Nordlie, F.G. 2009. Environmental influences on regulation of blood plasma/serum components in teleost fishes: a review. Rev. Fish Biol. Fisheries 19: 481–564.

Nordlie, F.G. and C.W. Leffler. 1975. Ionic regulation and the energetics of osmoregulation in *Mugil cephalus* Lin. Comp. Biochem. Physiol. 51A: 125–131.

Nordlie, F.G. and J. Whittier. 1983. Influence of ambient salinity on plasma Ca^{2+} and Mg^{2+} levels in juvenile *Mugil cephalus* L. Comp. Biochem. Physiol. 76A: 335–338.

Nordlie, F.G., W.A. Szelistowski and W.C. Nordlie. 1982. Ontogenesis of osmotic regulation in the striped mullet, *Mugil cephalus* L. J. Fish Biol. 20: 79–86.

Odum, H.T. 1953. Factors controlling marine invasion into Florida fresh waters. Bull. Mar. Sci. Gulf and Caribbean 3: 134–156.

Odum, W.E. 1970. Utilization of the direct grazing and plant detritus food chains by the striped mullet *Mugil cephalus*. pp. 222–240. *In*: J.J. Steel (ed.). Marine Food Chains. Oliver and Boyd, Ltd., Edinburgh, Scotland.

Ogawa, M. 1961. Comparative study of the external shape of the teleostean kidney with relation to phylogeny. Sci. Rept. Tokyo Kyoiku Daigaku B10: 61–68.

Olukolajo, S.O. and L.-A.A. Omolara. 2013. Salinity tolerance of grey mullet, *Mugil cephalus* (Linnaeus) fry in the laboratory. J. Fish. Sci. 7: 292–296.

Patrick, M.L. and C.M. Wood. 1999. Ion and acid-base regulation in the freshwater mummichog (*Fundulus heteroclitus*): a departure from the standard model for freshwater teleosts. Comp. Biochem. Physiol. A122: 445–456.

Patrick, M.L., P. Pärt, W.S. Marshall and C.M. Wood. 1997. Characterization of ion and acid-base transport in the fresh water adapted mummichog (*Fundulus heteroclitus*). J. Exp. Biol. 279: 208–219.

Pelis, R.M. and S.D. McCormick. 2001. Effects of growth hormone and cortisol on Na^+- K^+-$2CL^-$ cotransporter localization and abundance in the gills of Atlantic salmon. Gen. Comp. Endocrinol. 124: 134–143.

Peterson, G.L. and Z.H. Shehadeh. 1971. Changes in blood components of the mullet, *Mugil cephalus* L., following treatment with salmon gonadotropin and methyltestosterone. Comp. Biochem. Physiol. 38B: 451–457.

Phillip, D.A.T. 1993. Reproduction and feeding of the mountain mullet, *Agonostomus monticola*, in Trinidad, West Indies. Environ. Biol. Fishes 37: 47–55.

Rabeh, I., K. Telahigue, I. Chetoui, W. Masmoudi and M. El Cafsi. 2010. The effects of low salinity on lipid composition in the gills of grey mullet *Liza ramada*. Bull. Inst. Natn. Scien. Tech. Mer de Salammbô 37: 65–73.

Renfro, J.L. 1999. Recent developments in teleost renal transport. J. Exp. Zool. 283: 653–661.

Riedel, R., L. Caskey and B.A. Costa-Pierce. 2002. Fish biology and fish ecology of the Salton Sea, California. Hydrobiologia 473: 229–244.

Rosenau, J.C., G.L. Faulkner, C.W. Hendry, Jr. and R.W. Hull. 1977. Springs of Florida. Prepared by the U.S. Geological Survey in cooperation with the Bureau of Geology, Division of Resource Management, Florida Department of Natural Resources, and Bureau of Water Resources Management, Florida Department of Environmental Regulation. Bulletin Florida Bureau of Geology No. 31, Rev.

Rossi, A.R., S. Livi and D. Crosetti. 2015. Genetics of Mugilidae. *In*: D. Crosetti and S.J.M. Blaber (eds.). Biology, Ecology and Culture of Grey Mullet (Mugilidae). CRC Press, Boca Raton, USA (this book).

Sahinöz, E., Z. Dogu, F. Aral, R. Sevik and H.H. Atar. 2011. Reproductive characteristics of mullet (*Liza abu* H., 1843) (Pisces: Mugilidae) in the Atatürk Dam Lake, Southeastern Turkey. Turk. J. Fish. Aquat. Sci. 11: 7–13.

Sakamoto, T. and S.D. McCormick. 2006. Prolactin and growth hormone in fish osmoregulation. Gen. Comp. Endocrinol. 147: 24–30.

Sakamoto, T., S.D. McCormick and T. Hirano. 1993. Osmoregulatory actions of growth hormone and its mode of action in salmonids: a review. Fish Physiol. Biochem. 11: 155–164.

Satchell, G.H. 1992. The venous system. pp. 141–183. *In*: W.S. Hoar, D.J. Randall and A.P. Farrell (eds.). Fish Physiology, Vol. XII, Part A. Academic Press, Inc. San Diego.

Schultz, E.T. and S.D. McCormick. 2013. Euryhalinity in an evolutionary context. pp. 477–533. *In*: S.D. McCormick, A.P. Farrell and C.J. Brauner (eds.). Fish Physiology, Vol. 32. Elsevier Inc., NY.

Serkov, V.M. 2003. Salinity tolerance of some teleost fishes of Peter the Great Bay, Sea of Japan. Russian Journal of Marine Biology 29: 368–371.

Shehadeh, S. and M. Gordon. 1969. The role of the intestine in salinity adaptation of the rainbow trout, *Salmo gairdneri*. Comp. Biochem. Physiol. 30: 397–418.

Simmons, E.G. 1957. An ecological survey of the upper Laguna Madre of Texas. Pub. Inst. Mar. Sci. Univ. Texas 4: 156–200.

Skadhauge, E. 1969. The mechanism of salt and water absorption in the intestine of the eel (*Anguilla anguilla*) adapted to waters of various salinities. J. Physiol. 204: 135–158.

Stanley, J.G. and W.R. Fleming. 1964. Excretion of hypertonic urine by a teleost. Science 144: 63–64.

Sundell, K. and B.T. Björnsson. 1988. Kinetics of calcium fluxes across the intestinal mucosa of the marine teleost, *Gadus morhua*, measured using an *in vitro* perfusion method. J. Exp. Biol. 140: 171–186.

Sylvester, J.R., C.E. Nash and C.R. Emberson. 1975. Salinity and oxygen tolerances of eggs and larvae of Hawaiian striped mullet, *Mugil cephalus* L. J. Fish Biol. 7: 621–629.

Takei, Y. 2008. Exploring novel hormones essential for seawater adaptation in teleost fish. Gen. Comp. Endocrinol. 157: 3–13.

Taylor, J.R. and M. Grosell. 2006. Feeding and osmoregulation: dual function of the marine teleost intestine. J. Exp. Biol. 209: 2939–2951.

Thomson, J.M. 1966. The grey mullets. Oceanogr. Mar. Biol. Ann. Rev. 4: 301–335.

Thomson, J.M. 1997. The Mugilidae of the world. Memoirs of the Queensland Museum 41: 457–562.

Tipsmark, C.K., J.P. Breves, A.P. Seale, D.T. Lerner, T. Hirano and E.G. Grau. 2011. Switching of Na^+, K^+-ATPase isoforms by salinity and prolactin in the gill of a cichlid fish. J. Endocrinol. 209: 237–244.

Trape, S., J.-D. Durand, F. Guilhaumon, L. Vigliola and J. Panfili. 2009. Recruitment patterns of young-of-the-year mugilid fishes in a West African estuary impacted by climate change. Est. Coast. Shelf Sci. 85: 357–367.

Ünlü, E., K. Balci and N. Meriç. 2000. Aspects of the biology of *Liza abu* (Mugilidae) in the Tigris River (Turkey). Cybium 24: 27–43.

Utida, S., T. Hirano, H. Oide, M. Ando, D.W. Johnson and H.A. Bern. 1972. Hormonal control of the intestine and urinary bladder in teleost osmoregulation. Gen. Comp. Endocrinol. Suppl. 3: 317–327.

Walker, B.W., R.R. Whitney and G.W. Barlow. 1961. The fishes of the Salton Sea. pp. 77–91. *In*: B.W. Walker (ed.). The Ecology of the Salton Sea, California, in Relation to the Sport Fishery. Fish Bulletin No. 113, State of California Department of Fish and Game.

Wallace, J.H. 1975. Aspects of the biology of *Mugil cephalus* in a hypersaline estuarine lake on the east coast of South Africa (abstract). Aquaculture 5: 111.

Walsh, P.J., P. Blackwelder, K.A. Gill, E. Danulat and T.P. Mommsen. 1991. Carbonate deposits in marine fish intestines: a new source of biomineralization. Limnol. Oceanogr. 36: 1227–1232.

Walsh, W.A., C. Swanson and C.-S. Lee. 1991. Combined effects of temperature and salinity on embryonic development and hatching of striped mullet, *Mugil cephalus*. Aquaculture 97: 281–289.

Whitfield, A.K., R.H. Taylor, C. Fox and D.P. Cyrus. 2006. Fishes and salinities in the St. Lucia estuarine system—a review. Rev. Fish Biol. Fisheries 16: 1–20.

Whittamore, J.M. 2012. Osmoregulation and epithelial water transport: lessons from the intestine of marine teleost fish. J. Comp. Physiol. B182: 1–39.

Whittamore, J.M., C.A. Cooper and R.W. Wilson. 2010. HCO_3^- secretion and $CaCO_3$ precipitation play major roles in intestinal water absorption in marine teleost fish *in vivo*. Am. J. Physiol. Regul. Integr. Comp. Physiol. 298: R877–R886.

Wilson, R.W. and M. Grosell. 2003. Intestinal bicarbonate secretion in marine teleost fish-source of bicarbonate, pH sensitivity, and consequences for whole animal acid-base and calcium homeostasis. Biochim. Biophys. Acta 1618: 163–174.

Wilson, R.W., J.M. Wilson and M. Grosell. 2002. Intestinal bicarbonate secretion by marine teleost fish–why and how? Biochim. Biophys. Acta 1566: 182–193.

Wilson, R.W., F.J. Millero, J.R. Taylor, P.J. Walsh, V. Christensen, S. Jennings and M. Grosell. 2009. Contribution of fish to the marine inorganic carbon cycle. Science 323: 359–362.

Wood, C.M. 2011. Rapid regulation of Na^+ and Cl^- flux rates in killifish after acute salinity challenge. J. Exp. Mar. Biol. Ecol. 409: 62–69.

Wood, C.M. and W.S. Marshall. 1994. Ion balance, acid-base regulation and chloride cell function in the common killifish, *Fundulus heteroclitus*–A euryhaline estuarine teleost. Estuaries 17: 34–52.

Wood, H.L., J.I. Spicer and S. Widdicombe. 2008. Ocean acidification may increase calcification rates, but at a cost. Proc. Biol. Sci., Royal Society 275: 1767–1773.

Xu, P., H. Miao, P. Zhang and D. Li. 1997. Osmoregulatory actions of growth hormone in juvenile tilapia (*Oreochromis niloticus*). Fish Physiol. Biochem. 17: 295–301.

Young, G.C. and I.C. Potter. 2002. Influence of exceptionally high salinities, marked variations in freshwater discharge and opening of estuary mouth on the characteristics of the ichthyofauna of a normally-closed estuary. Est. Coast. Shelf Sci. 53: 223–246.

Young, G.C., I.C. Potter, G.A. Hyndes and S. De Lestang. 1997. The ichthyofauna of an intermittently open estuary: implications of bar breaking and low salinities on faunal composition. Est. Coast. Shelf Sci. 45: 53–68.

Zismann, L. and A. Ben-Tuvia. 1975. Distribution of juvenile mugilids in hypersaline Bardawil Lagoon January 1973–January 1974. Aquaculture 6: 143–161.

CHAPTER 14

Ecological Role of Mugilidae in the Coastal Zone

Alan K. Whitfield

Introduction

Mugilids are a highly opportunistic fish family and occupy a wide variety of coastal habitats with varying temperature, salinity, turbidity, sedimentary and dissolved oxygen regimes. This diverse family can dominate fish assemblages in both Northern and Southern Hemisphere estuaries, lagoons and lakes. For example, seven species of mugilids contributed 42% towards the total fish catch in Koycegiz estuarine lagoon in Turkey (Akin et al. 2005) and 10 species have been recorded in the estuarine Lake St. Lucia in South Africa, some of which are abundant within this system (Blaber 1977). Even in those coastal ecosystems where only one or two mullet species occur, this particular family can still dominate the fish fauna, e.g., in the Berg Estuary, where the South African mullet *Liza richardsonii* is abundant when compared to other marine fish taxa entering this system (Clark et al. 2009).

Although virtually all coastal habitats are suitable for occupation by mugilids, some are particularly favoured by members of this family, e.g., in salt marsh creeks of Mont Saint-Michel Bay, France, thinlip mullet *Liza ramada* represented 87% of the total fish biomass (Laffaille et al. 2000), in salt marsh creeks in the Kariega Estuary, South Africa, mugilids comprised more than 75% of the fish biomass (Paterson and Whitfield 2003), and in salt marsh creeks of the Yangtze Estuary, China, redlip (so.iuy) mullet *Liza haematocheila* is also the dominant marine fish species (Jin et al. 2007).

Most mullet can be regarded as marine estuarine-opportunists (Potter et al. 2015) or cycle migrants (Mariani 2001), entering estuaries, lagoons and coastal catchments primarily as nursery areas, and then returning to the sea at the onset of maturity (Blaber 1987). Out of the 15 species of southern African mugilids covered by Blaber (1987), 80% of them make regular use of estuaries at some stage of their life cycle and this pattern also appears to occur globally (Fishbase 2014). However, if the connectivity of an estuary with the marine environment is compromised, then the diversity and abundance of mugilids within such a system will show a significant decline (Vivier et al. 2010).

The versatility and adaptability of grey mullet to an array of coastal environmental conditions is well illustrated by their presence in riverine, lacustrine, estuarine, marine and even hyperhaline environments (Wallace 1975a, Whitfield et al. 2012). The widespread distribution of mullet in coastal ecosystems is not only promoted by their flexible physiology, it is also a reflection of their position at the base of the aquatic food web which facilitates both large numbers and a high biomass of this family in most estuarine

South African Institute for Aquatic Biodiversity (SAIAB), Private Bag 1015, Grahamstown 6140, South Africa.
Email: A.Whitfield@saiab.ac.za

systems. Although their diets are usually dominated by microphytobenthos, particulate organic matter and meiofauna, mullet have been variously described as 'interface feeders', 'benthic microphagous omnivores', 'microbenthos and meiobenthos feeders', 'iliophagous' and 'limno-benthophagous' (Carpentier et al. 2013 and references therein). The trophic positions described for mugilids have also shown considerable variation (e.g., herbivorous, detritivorous, omnivorous and planktivorous), depending on the author's interpretations and the dominant dietary organisms recorded during particular studies (Chapter 9—Cardona 2015). Clearly this dietary flexibility, physiological adaptability and the abundance of available food resources are major factors accounting for the ecological success of this family in tropical, subtropical and warm temperate coastal zones around the world.

Breeding Biology and Ecology

Although many mugilids attain sexual maturity within estuaries (Wallace 1975b), actual breeding by virtually all these species takes place at sea (Thomson 1966) and may well be linked to the marine origins of the family, as well as the relatively stable salinity, temperature and dissolved oxygen conditions for eggs and larvae when compared to the fluctuating and unpredictable estuarine physico-chemical environment. Such a view is supported by the experimental work of Walsh et al. (1989, 1991) who found that the optimum yield for normal flathead mullet *Mugil cephalus* larvae from fertilized eggs occurred at a salinity of 36. Other studies have also shown that salinities close to that of sea water are a pre-requisite for successful mullet spawning (van der Horst and Erasmus 1981). In addition, it has been suggested that if mugilid eggs touch bottom sediments, then the survival and development of the embryo can be adversely affected (Kuo et al. 1973). Such a scenario is more likely to occur in shallow estuarine systems rather than in deeper continental shelf waters, thus providing yet another reason for marine spawning by this family.

Flushing of estuaries with river water during floods may also mitigate against estuarine spawning due to the low salinities prevailing under such circumstances. Indeed, van der Horst and Erasmus (1981) suggested that mullet spawning does not normally take place in the sea during the rainy season in high rainfall areas of the world and, where the spawning does coincide with the rainy season, these are invariably relatively dry regions. Although river flooding might pose a threat to fry recruiting into estuaries in the former areas, it is more likely that the impact of such events on coastal salinities may also be a deciding factor as to when viable spawning is possible.

Another important variable in determining when mullet spawn is water temperature. According to van der Horst and Erasmus (1981) mullets between the equator and 30° latitude mostly spawn in winter, between 30 and 40° latitude most species spawn in summer (although some breed in winter if water temperatures are above 16°C), and above 40° latitude they only breed in summer. Therefore, although the above 'general rule' seems to suggest that most mugilids spawn in relatively warm marine waters where the temperature is between 16 and 27°C, there are many species whose eggs and larvae can tolerate temperatures outside of this range (although growth rates and mortality may be adversely affected). The above pattern is well illustrated by *Mugil cephalus* which spawns during different seasons in different parts of the world, with peak breeding activity occurring at temperatures between 20–26°C (Table 14.1).

Despite the overwhelming evidence that most mullet species spawn in offshore or coastal marine waters near estuaries (Wallace 1975b, van der Horst and Erasmus 1981), there are reports of mullet spawning in polyhaline and euhaline estuarine lagoons and lakes but the reproductive success of such unusual events is unknown (Thomson 1966, Connell 1996). Although the capture of ripe running individuals in estuary mouths has been regarded as evidence of estuarine spawning (e.g., van der Horst and Erasmus 1978), it is also possible that such individuals were en route to the sea to breed or that spawning in the mouth meant that fertilized eggs carried out of the estuary on the ebb tide reduces the distance for postlarval mullet to recruit back into estuarine nursery areas at a later date.

Apart from Kurian (1975), who makes passing mention of the Corsula mullet *Rhinomugil corsula* spawning in certain rivers and lakes of India, there have been no confirmed reports of mullet breeding in freshwater areas anywhere in the world. The Abu mullet *Liza abu* may also breed in inland waters and according to Fishbase (2014) this species inhabits mainly large river systems (e.g., Tigris and Euphrates) of Iraq and Pakistan. It is of interest to note that 'freshwater' mugilids belonging to the genus

Table 14.1. Flathead mullet *Mugil cephalus* spawning seasons (months with black borderline and bold temperatures) from various parts of the world (after Whitfield et al. 2012). The numbers represent the mean monthly coastal seawater temperatures (°C) for each region (SST data source http://fr.surf-forecast.com/).

Areas	Spawning periods												References
	J	J	A	S	O	N	D	J	F	M	A	M	
Black Sea	**20**	**23**	**24**	20	16	12	8	6	5	6	9	15	Apekin and Vilenskaya (1978)
Turkey (Mediterranean)	**24**	**26**	**28**	26	24	20	18	17	16	16	18	20	Erman (1959)
Egypt (Mediterranean)	**24**	**26**	**27**	**26**	25	22	20	18	17	17	18	20	Faouzi (1938)
Morocco (Atlantic Ocean)	**21**	**23**	**23**	**23**	21	20	18	17	17	17	17	19	Ameur et al. (2003)
Caspian Sea	**20**	**24**	**24**	**21**	16	12	9	6	6	6	9	14	Avanesov (1972)
Adriatic Sea	22	**25**	**26**	**22**	**19**	16	14	12	11	11	13	18	Morovic (1963)
Greece (Agean Sea)	22	**24**	**25**	**23**	**20**	**17**	15	13	13	13	14	18	Koutrakis (2004)
Greece (Mediterranean)	22	25	**26**	**24**	**22**	19	17	15	15	15	16	19	Katselis et al. (2005)
Tunisia (Mediterranean)	21	24	**26**	**25**	**22**	19	17	15	14	15	16	18	Brusle and Brusle (1977)
USA (Atlantic Ocean)	26	28	**28**	**26**	**24**	**23**	**22**	20	19	20	21	24	Bacheler et al. (2005)
Mexico (Gulf of Mexico)	28	28	29	28	**26**	**23**	**21**	**20**	**19**	19	22	25	Ibáñez and Gutierrez-Benitez (2004)
India (Indian Ocean)	29	28	28	29	**29**	**28**	**27**	**27**	27	29	30	30	Mohanraj et al. (1994)
Mauritania (Atlantic Ocean)	23	26	27	28	27	**25**	**22**	**20**	**19**	19	19	20	Brulhet (1975)
Sri Lanka (Indian Ocean)	29	28	28	28	**28**	**28**	**28**	**27**	**28**	**29**	30	30	De Silva and Silva (1979)
USA (Atlantic Ocean)	26	28	28	27	**25**	**23**	**22**	**20**	**20**	**20**	**20**	24	McDonough et al. (2005)
USA (Atlantic Ocean)	**26**	**28**	29	27	**25**	**23**	**21**	**23**	**21**	**20**	**23**	**26**	Kilby (1955)
Australia (Pacific Ocean)	**22**	**21**	21	22	22	24	25	26	26	**26**	**25**	**24**	Kesteven (1942)
South Africa (Indian Ocean)	**22**	**22**	**21**	**21**	22	23	24	26	26	**26**	**25**	**24**	Wallace (1975b)

Agonostomus are also located in tropical American and East African islands. The adults of the mountain mullet *Agonostomus monticola* occupy freshwater rivers and streams of Central American islands, with the juveniles sometimes encountered in brackish waters (Kenny 1995). Similarly, adults of the fairy mullet *Agonostomus telfairii* inhabit mainly freshwater areas in Madagascar and the surrounding islands, with some specimens occasionally recorded in estuaries (Thomson 1984).

Most mullet species have a prolonged spawning season (e.g., Table 14.1) which suggests a lack of population synchrony (Solomon and Ramnarine 2007) but does provide an insurance against temporary unfavourable conditions for eggs and larvae at certain times or in a particular area. Nevertheless, within each spawning season there is usually a period of peak batch spawning activity which then coincides with subsequent maximum recruitment strength for that species in that particular region (Wallace and van der Elst 1975). Although the spawning and recruitment periods for the different taxa are often not the same, and can occur at any time of the year, there is often a peak in the number of mullet species recruiting into estuaries during spring and early summer (Blaber 1987). This strategy facilitates maximum growth rates for the newly recruited juveniles due to increased benthic aquatic productivity and warm littoral waters during the spring and summer months.

Marine surface water temperatures appear to be an important factor determining the period over which spawning can take place. For example, the grooved mullet *Liza dumerili* on the subtropical north-eastern South African coast spawn mainly during winter and spring (June–November) (Wallace 1975b), whereas the same species along the warm temperate south-eastern coast spawn mainly in summer between January and February (van der Horst and Erasmus 1978). Similarly, *Mugil cephalus* in subtropical north-eastern South African waters spawn mainly between May and August (winter) (Wallace 1975b), whereas in the cool-temperate south-western region spawning occurs predominantly during January and February (summer) (Brownell 1979).

Movements, Migrations and Habitats

Most mullet species exhibit a high degree of residency within estuarine and other coastal nursery systems. Major migrations to marine spawning localities occur as adults and according to the breeding cycle of each species. The spawning migration of *Mugil cephalus* has been the most documented of all the mugilids and has been described in detail for localities as far apart as Australia, South Africa and North America (Thomson 1966, Wallace 1975b). In the case of the autumn spawning migration down the 66 km long segmented Lake St. Lucia system, the timing and movement of the *M. cephalus* shoals towards the sea did not appear to be affected by the presence or absence of salinity gradients, and this led Whitfield and Blaber (1978a) to suggest that a negative response to land-based organic cues may serve to guide the mullet shoals towards the estuary mouth. Once these breeding shoals reach the Indian Ocean their movements are unknown, in contrast to information from other continents (Arnold and Thompson 1958). On the east and west coasts of Australia, and east coast North America, the spawning 'run' of mature *M. cephalus* appears to occur against the prevailing coastal currents, i.e., northward in Australia and southward in the USA (Thomson 1966). Once the spawning has been completed there is evidence of a return migration to the estuarine environment, with spent individuals of a number of mullet species being recorded in estuaries (Wallace 1975b).

Following spawning, the fertilized mugilid eggs (usually 0.6–0.9 mm diameter) and preflexion larvae tend to drift passively as part of the marine plankton, but vertical movements associated with changes in buoyancy are possible (Thomson 1966, Whitfield et al. 2012). However, once the postflexion larval stage is attained these individuals become more motile and can begin more active vertical and horizontal movements. As the postflexion larvae enter coastal waters they appear to use flood tide fronts to advance towards the surf zone and offshore estuarine plumes may well provide cues for the larvae and early juveniles in these waters that an estuary is nearby (Kingsford and Suthers 1994). Although Blaber (1987) ruled out salinity and temperature gradients as primary 'clues' to migrating mugilid larvae and early juveniles, he did suggest that turbidity gradients, both horizontal and vertical, are important for migrating mugilid early juveniles entering southern African estuaries from the marine environment. However, the available evidence (Kristensen 1963, James et al. 2008a) suggests that the early juveniles of estuary-associated marine fish species probably use olfactory cues as a primary means to find estuarine nursery areas. Once inside an estuary, it is probable that sediment composition, turbidity and salinity preferences may also play roles in determining the exact regions and habitats occupied by juvenile Mugilidae (Fig. 14.1).

Sampling of ichthyoplankton and small nekton in surf zones (Fig. 14.1a) suggests that mugilids often dominate these fish assemblages and make extensive use of this habitat as an interim nursery area for early juveniles (Whitfield 1989, Strydom and d'Hotman 2005). However, the distribution of these fishes is not uniform and depends on available microhabitats within the surf zone (Watt-Pringle and Strydom 2003, Inoue et al. 2008). The relative absence of preflexion and postflexion larvae from the surf zone

a

b

Figure 14.1. (a) Swartvlei Bay surf zone, South Africa, where ichthyonekton sampling with a plankton net revealed that postlarval mullet dominated the fish catches in this interim nursery area. (b) The nearby Swartvlei Estuary mouth through which the mugilid postlarvae migrate towards their final nursery areas (photos by A.K. Whitfield).

(Pattrick and Strydom 2008) suggests that mugilids only move into this area once the juvenile stage has been attained. The greater swimming ability and mobility of the 0+ juveniles makes it very easy for them to locate and enter nearby estuaries (Fig. 14.1b) which offer a variety of nursery habitats for these species (Fig. 14.2).

The consistent presence of dominant mugilid taxa in a range of estuaries in various parts of the world (e.g., see Cardosa et al. 2011, Harrison and Whitfield 2006) suggests that recruitment of these species

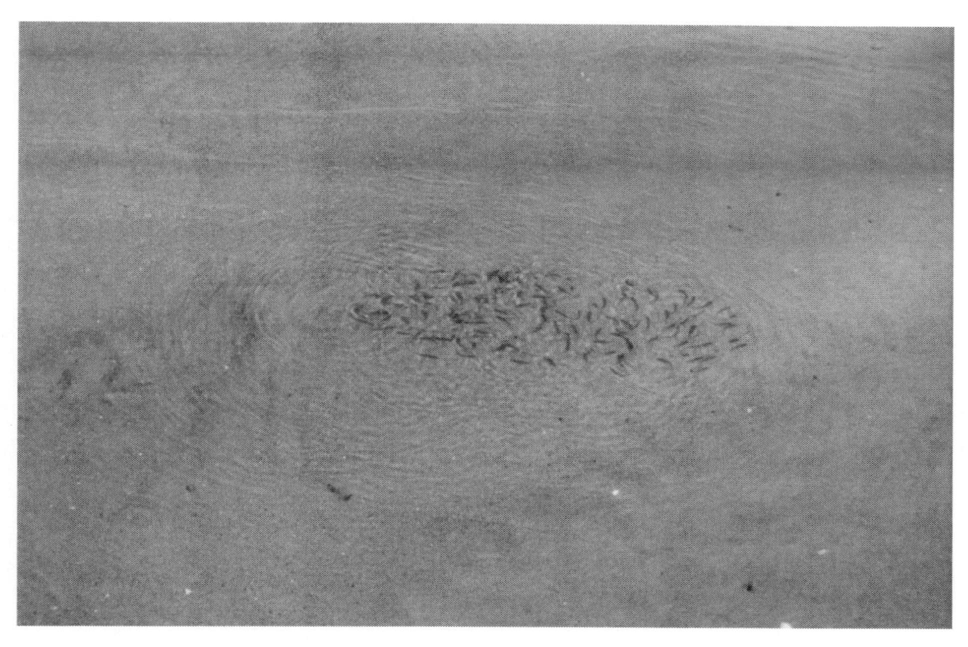

Figure 14.2. Typical sheltered shallow water nursery habitat for early mugilid juveniles (10–20 mm SL) when they first enter an estuary. (photo by A.K. Whitfield).

is always successful and there are strong indications that large numbers enter these nursery areas on an annual basis (Thomson 1966). The early juveniles are capable of actively swimming into estuaries against strong outgoing water currents (Harrison and Cooper 1991) or using flood tidal transport to assist with movement towards their upstream nursery areas (Torricelli et al. 1982, Trancart et al. 2011). Evidence from the Wilderness system in South Africa suggests that water depth is a strong driver of recruitment, with most individuals recruiting up the shallow estuarine channel during the day and when the flood tide water depth exceeds 10 cm (Hall et al. 1987). Although nocturnal recruitment into the Wilderness system is much lower than during daylight hours, a wider variety of mullet species recruited up the shallow channel and into the estuarine lakes at night.

In the absence of flood tidal transport, mullet fry are still very able at making their way upstream to their preferred nursery area. This is clearly indicated by Johnson and McClendon (1970) who recorded 31 *M. cephalus* 28–40 mm SL some 192 km up the Colorado River, implying a daily travel distance of approximately 3 km. Whitfield and Blaber (1978a) similarly recorded movements of early juvenile mullet (10–40 mm SL) belonging to five species up the St. Lucia system and estimated that average distances of 0.5–1.5 km per day were undertaken by these fry. Large juvenile and adult mullet also undertake major movements, with Whitfield and Blaber (1978a) citing a tagged *Liza dumerili* moving 28 km in 14 days (or a minimum of 2 km per day). Thomson (1955), however, described *M. cephalus* being recaptured in the same area months after tagging whereas adults have been recovered up to 720 km from the tagging site following completion of spawning activities (Thomson 1966).

Available evidence suggests that most mugilid juveniles first recruit into estuaries between 10 and 20 mm SL (Blaber 1987), but some species appear in these systems at a larger size (Wallace and van der Elst 1975). The fact that early mugilid juveniles (12–20 mm SL) recruit into West African estuaries regardless of the salinity regime suggests that olfactory cues attracting these fish to their nursery areas have an estuarine rather than a riverine origin (Trape et al. 2009). Once in the estuary, field observations and laboratory experiments using flumes indicate that juveniles of species such as *Liza ramada* make use of flood tidal transport to move up estuaries to reach catchment nursery areas, thus reducing migration energy costs (Trancart et al. 2012).

Spatio-temporal abundance and otolith Sr/Ca ratio analyses of seven sympatric mullet species in Taiwan showed that all species are spawned at sea and then recruit into estuarine waters of varying salinities. The fry of some species, such as *Mugil cephalus*, even enter freshwater areas (Fig. 14.3) whereas others do not (Chang and Iizuka 2012). In general, the juveniles of most mugilid species tend to be located in estuarine nursery areas, rather than adjacent marine or freshwater habitats (Blaber 1987). Although most mullet species can enter freshwater, estuarine and marine waters during their life cycles, the degree of association with each aquatic ecosystem varies from exclusively freshwater to almost exclusively marine (Fig. 14.4). However, overall the majority of mugilids are found mainly in estuarine waters for most of their life cycle, with a lesser occurrence in freshwaters or the sea.

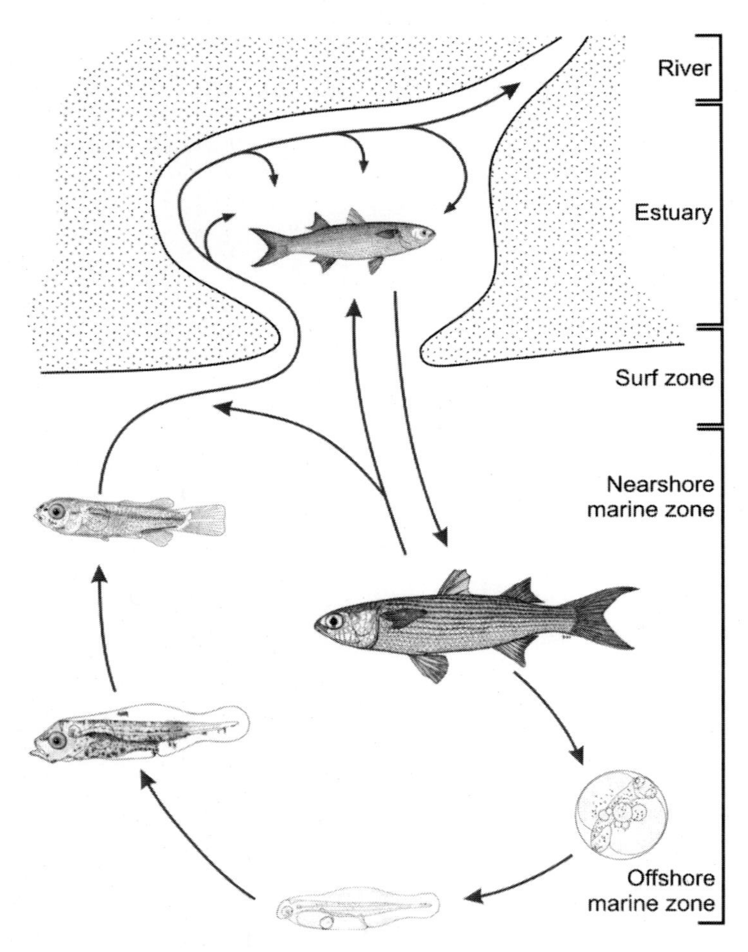

Figure 14.3. Life history of the flathead mullet *Mugil cephalus*, indicating the different aquatic environments occupied by this species during its life cycle (after Whitfield et al. 2012). This particular 'template' is valid for most mugilid species around the world.

Those estuaries that are in a temporarily closed phase represent a recruitment problem for early juveniles attempting to enter these systems. Studies in South Africa have shown that early 0+ mugilids use episodic marine wave overwash of estuarine sandbars at the mouth to recruit into such systems (Cowley et al. 2001, Kemp and Froneman 2004). Although this risky recruitment strategy can assist in providing some stability for mugilid populations in temporarily closed estuaries, interannual population variability was still evident, especially in relation to the timing and frequency of mouth opening events (James et al. 2008b).

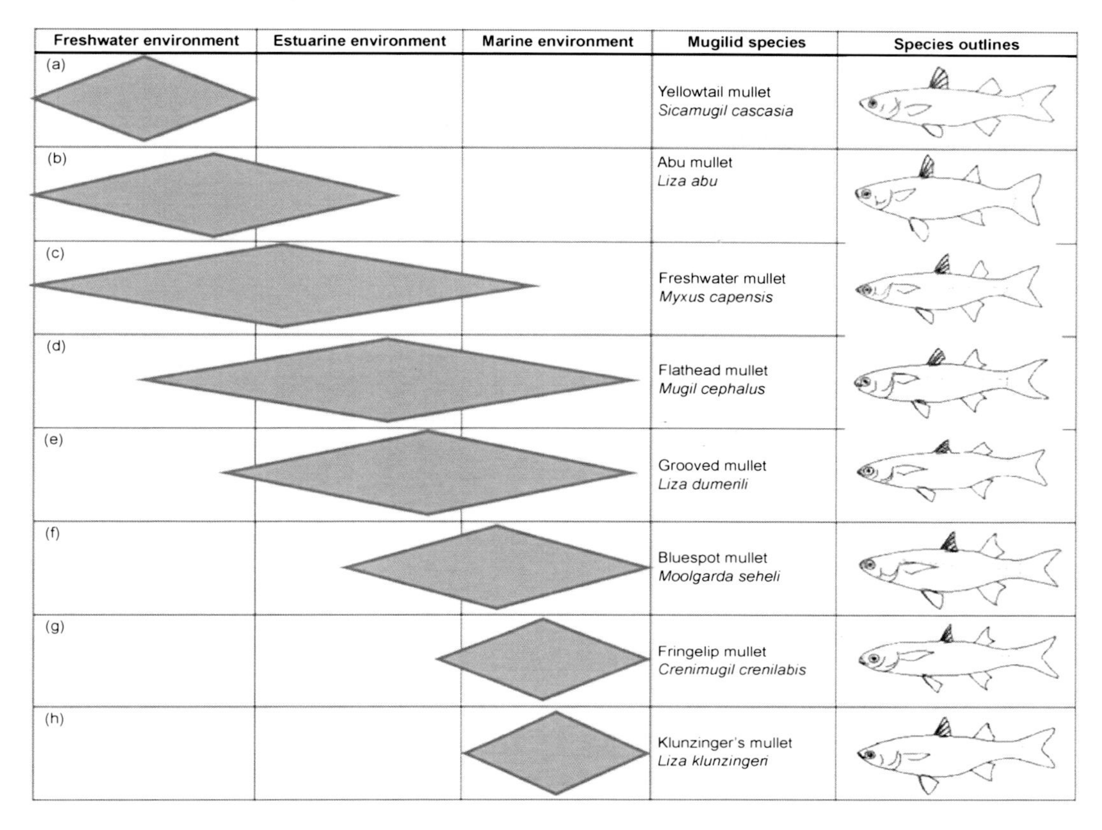

Figure 14.4. Aquatic ecosystem selection as exhibited by a range of mugilid species: (a) freshwater, yellowtail mullet *Sicamugil cascasia*; (b) mainly freshwater but extends into estuaries, Abu mullet *Liza abu*; (c) mainly freshwater and estuarine but spawns at sea, freshwater mullet *Myxus capensis*; (d) freshwater, estuarine and marine, flathead mullet *Mugil cephalus*; (e) mainly estuarine with breeding at sea, grooved mullet *Liza dumerili*; (f) mainly marine with some estuarine penetration, bluespot mullet *Moolgarda seheli*; (g) mainly marine with occasional entry into estuaries, fringelip mullet *Crenimugil crenilabis*; (h), marine, Klunzinger's mullet *Liza klunzingeri*.

Habitat preferences of grey mullet may change according to ontogeny, e.g., early juvenile *Mugil cephalus* in the Balearic Achipelago show a preference for fresh or oligohaline waters, late juveniles for oligohaline and mesohaline areas, and adults for polyhaline areas with strong avoidance of freshwater sites (Cardona 2000). The juveniles of six mugilid species were recorded in large numbers in the head waters of Eastern Cape (South Africa) estuaries, and highlights the importance of these transitional waters to juveniles of this family (Wasserman and Strydom 2011). Conversely the adults of most mullet species are usually more abundant in the lower and middle reaches of estuarine systems, especially when the spawning season approaches and they need to return to the sea.

Mugilid species tend to be more abundant in the shallow littoral rather than deeper offshore waters (Simier et al. 2004, Jin et al. 2010). In particular, juvenile *Mugil cephalus* in mangrove creeks in Zanzibar, Tanzania, are attracted to shallow waters where microphytobenthos is readily available as a food resource and the turbid water can provide a refuge from predation (Mwandya et al. 2010). In the shallow waters of Sulaibikhat Bay, Kuwait, young of the year keeled mullet *Liza carinata* dominated the fish assemblage and were closely associated with detached chlorophytes which they used as both a refuge and a temporary food source (Wright 1989). A similar situation was recorded in the surf zone of sandy beaches near Perth, Australia, where yellow-eye mullet *Aldrichetta forsteri* used detached macrophyte detritus as both a food source and shelter from predatory cormorants (Robertson and Lenanton 1984). In other littoral systems, the abundance of mugilids increased when submerged macrophyte beds disappeared and were replaced by a bare sand habitat (Whitfield 1986, Sheppard et al. 2011) which is more conducive to the feeding habits of mullet.

Although juvenile mullet tend to congregate in shallow marginal waters, subadult and adult mullet have complete freedom of movement within estuaries and occur in both littoral and deeper offshore areas, as well as any longitudinal position within a system. Acoustic studies on adult *Liza ramada* in the Mira Estuary, Portugal, showed that they move in conjunction with both the flood and ebb tide, covering a median distance of 6.25 km during a complete tidal cycle. Most time was spent in the upper reaches of the estuary and, although they moved at an average ground speed of 0.8 body lengths per second, fish were also stationary for up to at least three hours at a time (Almeida 1996).

The main seaward migrations of three mugilid species in western Greek coastal lagoons were driven by the need for marine spawning and are closely related to the full moon lunar cycle in the case of *Mugil cephalus* and new and full moon cycle for leaping mullet *Liza saliens* and golden mullet *Liza aurata* (Katselis et al. 2007). The drive to undertake spawning migrations is strong (Fig. 14.5) and, where it is prevented due to estuary mouth closure, the mature gonads are then resorbed (Wallace 1975b). There is also some evidence (empty stomachs and alimentary system) that during mullet spawning migrations feeding ceases (Thomson 1966, Whitfield and Blaber 1978a).

Figure 14.5. Mature flathead mullet *Mugil cephalus* attempting to leave the intermittently open West Kleinemonde Estuary, South Africa, as the sand bar at the mouth was in the process of being breached by high estuarine water levels caused by river flooding (photo by P.D. Cowley).

Unpredictable events also affect mugilid movements into and out of estuaries. River flooding can cause most mullets to temporarily leave estuaries affected by such occurrences and take temporary refuge in the coastal zone. For example, major freshwater inputs into the Patos estuarine lagoon, Brazil, caused the abundance (Catch Per Unit Effort—CPUE) of Lebranche mullet *Mugil liza*, white mullet *Mugil curema* and redeye mullet *Mugil gaimardianus* to plummet, only to recover once the tidal regime had been re-established (Garcia et al. 2003). Conversely, a large anoxic 'black tide' event in the coastal marine zone caused large shoals of *Liza richardsonii* to take refuge in the adjacent Berg Estuary, South Africa (Lamberth et al. 2010). Furthermore, in high latitudes, mugilid species are known to leave estuaries and bays in the autumn so as to avoid cold winter water temperatures that are characteristic of shallow waters in these regions (Thomson 1966).

Physiology and Growth Ecology

Preflexion mugilid larval growth is dependent on nutrients provided by the yolk sac, with independent feeding and growth only commencing at about three days when the mouth is formed (Kuo et al. 1973, James et al. 1983). At an age of approximately 11–12 days the larvae assume a more fish-like shape and acquire sustained swimming powers that facilitate both active feeding and extensive movement within the marine environment. Postlarvae and early juveniles first appear in surf zones and other coastal environments at about 10 mm SL when they are four–six weeks old (Thomson 1966). Entry into estuaries occurs mainly between 10 and 15 mm SL (Blaber 1987) at which stage dietary changes occur and growth increases rapidly in highly productive nursery habitats.

Salinity is an important driver influencing the abundance of different mullet species in estuaries (Chapter 13—Nordlie 2015), e.g., *Liza saliens* dominated fish assemblages in western Mediterranean estuaries where the salinity was > 13 whereas *Liza ramada* dominated assemblages where the salinity was < 13 (Cardona et al. 2008). Mugilids appear to be universally euryhaline which makes this family ideal for the colonization of estuarine and brackish waters. In addition, many species appear to be able to tolerate both oligohaline and hyperhaline conditions (Whitfield et al. 2006) but each species has optimum salinity conditions for both food consumption and assimilation, e.g., food intake by *Mugil cephalus* was highest at a salinity of 30 but conversion efficiencies were greatest when the salinity was less than 10, mainly because the rate of digestion was slower at lower salinities (De Silva and Perera 1976).

It is noteworthy that the eggs and preflexion larvae of many mugilids are not tolerant of freshwater or even oligohaline conditions (Walsh et al. 1991), yet the early juveniles of some of these species are attracted to such salinities (Bok 1979). Conversely, in a hyperhaline West African estuary, small mugilid recruits (12–20 mm SL) belonging to six species were sampled in salinities ranging from 47–78. This suggests that a euryhaline capacity within mugilids develops at an early stage of ontogenetic development and means that this family is well adapted to changing environmental conditions due to climate change (Trape et al. 2009). It is perhaps significant that in the hyperhaline (salinities > 70) Lake St. Lucia, South Africa, *Mugil cephalus* still reached sexual maturity and undertook their normal spawning migration down the system, despite a reversed salinity gradient being present at the time (Wallace 1975b).

As is the case with most fish species, there is a strong relationship between water temperature and the ability of mugilids to survive very high or low salinities. Highest survival rate of *Mugil curema* early juveniles (23–97 mm) occurs at a salinity of 8.5. Also, oxygen uptake was lowest at this salinity and highest at a salinity of 1.7 when oxygen consumption was at a maximum (Fanta-Feofiloff et al. 1986). The above fingerlings did not survive direct transfer into freshwater but this could have been an artefact of laboratory conditions since other fry such as *M. cephalus* are known to occupy freshwater catchments following an estuarine acclimation phase (Bok 1979). There are also strong indications that at least some mugilid species have an endogenous oxygen consumption cycle that may be linked to a diurnal rhythm, with *Liza dumerili* having low oxygen consumption at midday and midnight and maximum oxygen consumption just after sunrise and sunset (Marais 1978).

Mugilids are also able to survive in a wide range of temperatures, each of which is unique to the particular species but with most taxa having a preference for warmer rather than cooler waters. This is epitomized by a critical thermal maximum for squaretail mullet *Liza vaigiensis* of 44°C which enables this species to occupy hot littoral waters in the tropics and is therefore well adapted to the consequences of global warming (Eme et al. 2011). The fact that most mugilid species occur mainly in tropical and subtropical regions, and are absent from very high latitudes (e.g., much of cold temperate northern Europe, Asia and America), suggests that this family has its origin in a warm rather than cool coastal biogeographic region.

Turbidity preferences of mugilids also vary according to the species, e.g., juvenile *Liza dumerili* were attracted to clear estuarine waters (< 10 NTU) whereas bluetail mullet *Valamugil buchanani* preferred clear to partially turbid water (10–80 NTU) (Cyrus and Blaber 1987). An ability to tolerate high turbidities is a feature of most mullet species in estuaries and is a consequence of turbidities > 80 NTU arising from wind or tide generated disturbance of fine sediments and particulate matter within these systems or by river flooding. Thomson (1966) reviewed information which suggests that the photosensitivity of the retinal pigments in some mullet species (e.g., *Mugil cephalus*) is well adapted to optimizing vision in turbid water.

Despite feeding at the base of the food pyramid on material that is not as rich in fats and proteins as the prey of carnivorous fishes at higher trophic levels, mullet appear to have a relatively rapid growth rate (Chapter 10—Ibáñez 2015), often achieving at least 150 mm SL in the first year (Anderson 1958, Whitfield and Kok 1992). Growth rates, in terms of length increments, are much slower from year two onwards (Kumar et al. 2014), but body mass increments do not necessarily follow the same declining pattern. In temperate biogeographic regions growth is usually much slower in winter than summer and may even cease altogether during the colder months (Thomson 1966), with these differential growth patterns detectable in the layering patterns of deposits within mugilid otoliths (Whitfield et al. 2012). Annual growth zones in *Mugil cephalus* and freshwater mullet *Myxus capensis* in South Africa have been validated from known age specimens using otoliths (Ellender et al. 2012), with both species reaching at least 10 years of age. Similarly, Kendall et al. (2010) used validated otolith readings to determine the maximum age of flat-tail mullet *Liza argentea* (17 years) and sand mullet *Myxus elongatus* (12 years) in two South-eastern Australian estuaries.

The growth and body condition of *Liza richardsonii* was higher in South African estuaries when compared to the same species in the sea, probably as a result of the better feeding conditions in the former areas (De Decker and Bennett 1985). However, the growth of *Liza aurata* in Greek lagoons was not significantly different from that in the adjacent marine environment (Hotos and Katselis 2011). In this context it is also interesting to note that Garbin et al. (2014) recorded adult *Mugil liza* caught in the Patos Lagoon, Brazil, being smaller and younger than adult mullet from the adjacent marine environment. However, the growth rates of *M. liza* in the two environments was not recorded.

Food and Feeding Ecology

Mugilids are well adapted to benthic foraging with a protrusible mouth, sensitive lips with taste buds, a pharynx that has been modified into an efficient filtering apparatus, specialized gill rakers, a gizzard-like stomach, and a long intestine for the assimilation of a range of dietary items (Thomson 1966, Marais 1980). Not only can mullet graze on sand and mudflats, they are equally adaptable at foraging on submerged rocks and artificial hard structures, as well as water surface skimming if a microalgal rich layer is present (Chapter 9—Cardona 2015). Generally mugilids are dependent on food provided by the benthic rather than planktonic environment (Faye et al. 2011) and it has been suggested that the large biomass of juvenile and adult mugilids present in estuaries is directly attributable to the abundant epibenthic energy resources associated with the substratum (Blaber 1987).

In terms of diet, mullet larvae are initially planktonic foragers, consuming mainly microzooplankton in the marine environment and having a short gut with a simple loop due to their easily digestible prey (Thomson 1966). Once the early juveniles migrate into shallow lagoonal and estuarine waters, a dietary switch takes place (Chan and Chua 1979). Individuals of most species between 10 and 15 mm SL consume zooplankton in the water column and zooplankton migrating into the benthos. Between 10 and 20 mm SL the diet comprises mainly zooplankton in the benthos and meiobenthos, with the final transition taking place between 15 and 25 mm SL when both meiobenthos and microbenthos dominates their stomach contents (Blaber and Whitfield 1977).

There is a concomitant increase in the percentage frequency of sand in the stomachs of these early juveniles, from < 10% at a length of 10–15 mm SL to > 90% by a length of 35 mm SL. All species from South Africa have 100% frequency of occurrence of sand grains in their stomach contents by a length of 45 mm SL (Blaber and Whitfield 1977). These authors suggested that the changeover from planktonic prey to the ingestion of microbenthos and sediment can only take place in the shallow, quiet waters provided by estuaries because the high energy South African coastline would make it difficult for these small fish to consume mobile sand and associated organisms when turbulent conditions are present. In addition, the quieter, deeper offshore waters would probably support very low densities of favoured food organisms such as diatoms (Blaber and Whitfield 1977), so that environment would also be unsuitable as a nursery area for large numbers of juveniles from this family. Nevertheless, the juveniles of certain species of mullet (e.g., fringelip mullet *Crenimugil crenilabis*) do occur in the turbulent coastal zone, especially in the vicinity of rocky shores which provide a ready source of microalgae growing on the intertidal and subtidal hard surfaces.

Juvenile and adult mugilids are capable of feeding on both invertebrates and plankton but are primarily detritivorous and benthic microalgal feeders, ingesting, filtering and concentrating large quantities of organic matter using a pharyngeal filter apparatus (Odum 1968, Laffaille et al. 2002, Lin et al. 2007). Sediments and particulate matter that is rejected during the filtering process is expelled by a "characteristic coughing or spitting out of material" (Thomson 1966) and it then sinks through the water column onto the bottom. If very fine material is present it can remain in suspension, thus increasing water turbidity and identifying littoral areas where mullet shoals are actively feeding. Schools of foraging mullet are capable of transporting large quantities of surface sediment through their feeding activities and faecal production, and could therefore be termed 'ecosystem engineers'.

Swallowed material enters the muscular stomach where trituration commences, strongly aided by the inorganic sedimentary 'grinding paste' that is almost always present when food has been ingested. The alimentary canal, which differs in relative length between the species, is well adapted for the assimilation of preferred food items, some of which are difficult to digest (Marais 1980). According to Marais and Erasmus (1977a) the composition of the alimentary canal contents can differ considerably in terms of protein, ash and carbohydrate content between the mullet taxa, especially when a species deviates from its 'normal' diet. This occurred at a sampling station in the Swartkops Estuary (South Africa) where the striped mullet *Liza tricuspidens* utilized a localized concentration of the mysid *Mesopodopsis slabberi* as a temporary food source and the protein content of the alimentary canal contents increased from 14 to 41%, whilst the ash content declined from 65 to 27%.

Mullet also concentrate on major meiofauna grazers such as nematodes and harpacticoid copepods in their diet, as well as foraminiferans and ostracods (Carpentier et al. 2013). A large number of studies have shown that most mullet species obtain their food almost exclusively from the benthic trophic compartment in estuaries (Blaber 1976), a fact confirmed by the isotopic work of Pasquaud et al. (2010). Nevertheless, some mullet species (e.g., South African mullet *Liza richardsonii*) have been recorded opportunistically foraging on planktonic surf zone diatom blooms (Romer and McLachlan 1986) or mysids (Marais and Erasmus 1977a), and others (e.g., *Mugil cephalus*) on pine pollen from surface estuarine lake waters (Whitfield 1982).

Mugilids contribute significantly to the functioning of estuarine systems by using particulate organic matter (including detritus) and primary production directly, as well as accelerating microphytobenthos turnover in littoral waters (Lefeuvre et al. 1999). The dominance of this family in the Mhlanga Estuary, South Africa, was attributed by Whitfield (1980) to the large amount of benthic floc (detritus and associated micro-organisms) within this system. Unfortunately there is little definitive evidence of the relative importance of particulate organic matter and benthic microalgae to mugilids but work in the Mira Estuary, Portugal, has shown that was no significant correlation between the quantity of particulate organic matter and concentration of microalgae present in the stomach contents of *Liza ramada* (Almeida 2003). Nevertheless there appears to be general agreement that mullet populations play a significant role in organic matter fluxes in coastal waters around the world (Lebreton et al. 2011).

Carpentier et al. (2013) have highlighted the importance of mudflat biofilms and associated meiofauna to the nutrition of *Liza aurata* and *Liza ramada* in Mediterranean estuaries. They also emphasized the ability of these species to export processed sediment, biofilm and associated meiofauna from intertidal to subtidal areas. A similar conclusion was reached by Laffaille et al. (2002) and Degré et al. (2006) during studies of intertidal ecosystems in France where grazing *L. ramada* export microalgae and detrital material, together with sediment, from macrotidal salt marsh creeks and mudflats. The general paucity of mugilid shoals on open sandy coasts where turbulent wave action is a feature may also be a reflection of the general absence of benthic biofilms due to high sedimentary instability.

Inorganic sediment particles are ingested by all mugilid species but there does appear to be a size differentiation between species which was first documented by Blaber (1976) and subsequently by other authors (e.g., Marais 1980). In South African east coast estuaries there were two mullet species (diamond mullet *Liza alata*, *L. tricuspidens*) selecting mainly coarse sand (500–1000 μm) with their food, two species (largescale mullet *Liza macrolepis*, *Mugil cephalus*) selecting medium sand (250–500 μm) and four species (*Moolgarda buchanani*, *M. seheli,* longarm mullet *M. cunnesius*, robust mullet *Valamugil robustus*) selecting fine sand (125–250 μm) (Blaber 1977). However, the same species can select sediment particles of different mean sizes in different estuaries or sites, depending on sediment size composition where the fish are feeding, e.g., *Liza dumerili* in the St. Lucia system preferred sand grains of approximately 300 μm (Blaber 1976) whereas the same species in the Swartkops Estuary preferred sand grains approximately 125 μm (Marais 1980). Similarly, *Mugil cephalus* from a Sardinian lagoon were found to have a preference for fine sands (125–250 μm) (Mariani et al. 1987), whereas the same species preferred sediments of 250 μm in Lake St. Lucia (Blaber 1976) and 150–250 μm in various other KwaZulu-Natal and Pondoland estuaries (Blaber 1977).

Clearly the final size particle composition of a particular mullet species is influenced by substratum composition within an estuary or even particular sites within an estuary (Whitfield and Blaber 1978a). For example, all eight mullet species sampled by Blaber (1977) in the Mpenjati Estuary had significantly larger sand particle sizes in their diet when compared to the same species in other KwaZulu-Natal estuaries. Similarly, Marais (1980) recorded that substratum particle size composition at particular sites within the Swartkops Estuary influenced mean particle size consumed by three mullet species at these sites. He also determined that smaller *Liza dumerili* tended to ingest finer sediments than larger individuals feeding in the same area, thus indicating some buccal cavity processing separation within a species based on fish size. In contrast, Blaber (1976, 1977) found little or no difference in mean sediment particle diameter consumed by individual mullet species of different sizes (> 5 cm SL) from Lake St. Lucia and other South African estuaries.

The degree to which the teeth and gill rakers of mullet assist with the sorting of food and sediments prior to swallowing is largely unknown. However, it is perhaps significant that species with a large number of small, fine teeth (e.g., *Liza richardsonii*) ingest much finer material than those species (e.g., *Liza tricuspidens*) with fewer, large teeth (Marais 1980). Indeed, the relatively sharp tricuspid teeth of *L. tricuspidens* are ideally suited to the cropping and ingestion of macroalgae and plant detritus which are important components in the diet of this species (Blaber 1977, Marais 1980). Although gill raker differences between the mullet species may also be important, with the numbers of gill rakers increasing with growth (Smith and Smith 1986), the exact nature of their role in food selection has yet to be determined. What is known is that the numerous slender gill rakers form a perfect filter apparatus, with those on the inner three arches being close-set and meshed together by basal setiform processes to form a rigid plane surface (Smith and Smith 1986).

Apart from the range of diatom species consumed by mugilids, there is generally a low diversity of food items in the diet of *Liza ramada* from the Mira Estuary (Portugal), with the Bacillariophyceae being the dominant group (Almeida 2003). Common diatom species consumed included *Melosira*, *Ciclotella*, *Navicula*, *Nitzschia* and *Surirella*. In the St. Lucia system the large *Aptinoptychus splendens* was a dominant diatom present in the diet of *Mugil cephalus* together with *Coscinodiscus granii* (Blaber 1976) but other diatom species such *Auliscus pruinosus*, *Nitzschia punctate*, *Diploneis suborbularis*, *Cocconeis pseudomarginata* and *Biddulphia levis* were also recorded in the stomach contents of this species (Whitfield and Blaber 1978b).

The volume of ploughed sediment resulting from feeding activity by *Liza ramada* increased exponentially with fish body length (Almeida 2003). Enhanced patchiness in the biomass of microphytobenthos due to juvenile *Liza aurata* grazing may actually increase the mean microphytobenthic biomass by increasing algal growth rates due to the removal of algae by the fish (Como et al. 2013). However, food quality in the aquatic environment does not remain constant and *Liza saliens* exhibit increased daily rations when food quality decreases (Cardona 1999). Other triggers for increased food intake include preparation by mullet for breeding, e.g., increased food consumption at the beginning of the spawning migration by *Liza ramada* in the Mira Estuary, Portugal (Almeida 2003). Water temperature may also influence food intake, e.g., swimming speed during feeding and the number of foraging events by juvenile *Liza aurata* increased when the temperature was raised from 10ºC to 20ºC (Como et al. 2013).

There are strong indications that large shoals of mugilids feeding in shallow coastal areas may be ecosystem engineers in a similar manner to that of some burrowing invertebrate species (Pillay et al. 2007). If one considers that a 100 g mullet can consume 20–60 g of wet material per day (Marais 1980), then the potential transfer of sediment and organic matter from the littoral to deeper areas within estuaries and lagoons must be considerable and worthy of further investigation. Similarly, coastal and estuarine sediment recycling by burrowing macrobenthic invertebrates can have a major impact on the food and feeding activities of mugilids. This is because the surface microalgal biofilms can be partially smothered or destroyed by the activities of these invertebrates, thus reducing or eliminating potential food sources for mullet (Pillay et al. 2012).

Fat, protein and carbohydrate content of the food consumed by mugilids in the Swartkops Estuary varied according to the species, with *Liza richardsonii* being considerably more efficient at concentrating these components from the substratum than *Liza dumerili* (Marais 1980). There was also strong evidence from the above study that food selection efficiency is higher in coarse versus fine sediment areas, but that fine substrata yield richer biochemical components and energy when compared to coarse sediments. Within a species, the most nutritious material was consumed by the smallest specimens, irrespective of species (Marais and Erasmus 1977a). Despite this trend, smaller mugilids were lower in energy and fat than larger individuals. In addition, sexually mature *Liza dumerili* in this estuary showed a build-up in energy reserves just prior to the commencement of the spawning season (Marais and Erasmus 1977b).

Digestion efficiencies are likely to vary between the mullet species, as evidenced by differing numbers of pyloric caecae and Intestinal Length (IL) to fish Standard Length (SL) ratios. For example, *Mugil cephalus* had an IL:SL ratio approximately double that of three other mugilid species in the Swartkops Estuary, but *M. cephalus* had the fewest and most poorly developed pyloric caecae of the species studied (Marais 1980). Barrington (1957) proposed that pyloric caecae significantly increase the intestinal epithelium surface area, therefore aiding food digestion. It is therefore tempting to suggest that the increased intestinal length in *M. cephalus* is required by this species for complete digestion of diatom and detrital material (Odum 1970) in the absence of a large number of highly efficient pyloric caecae. In contrast, the much larger muscular stomach, well developed pyloric caecae, and shorter and wider intestine of *Liza dumerili* is more than adequate for the assimilation of diatoms which predominate in the diet of this species (Masson and Marais 1975, Marais 1980).

Mugilidae are often positioned at a higher trophic level than purely primary consumers and this position can vary on a seasonal basis if the diet changes (Vinagre et al. 2012). Some stable isotope studies (using $^{\delta15}$N) seem to indicate that mugilid growth is supported largely by primary consumers such as benthic meiofauna or small macrofauna (Lebreton et al. 2011), rather than microalgae which are also an important part of their diet. However, other stable isotope studies have suggested that benthic microalgae and macroalgal periphyton are of prime importance to this family (Lin et al. 2007) and that primary consumers such as harpacticoid copepods and meiofauna are of secondary importance in mullet diets.

Diel foraging activity by mugilids varies according to the species. For example *Liza saliens*, *L. ramada* and *L. aurata* feed mainly during daylight hours in the Arno Estuary, Italy, whereas *M. cephalus* forages mainly at dusk and during the early part of the night (Torricelli et al. 1982). In the St. Lucia estuarine lake (South Africa) most mugilid species, including *M. cephalus*, foraged mainly during the day and generally not at night (Blaber 1976). Feeding periodicity of the dominant mullet species in the Swartkops Estuary (South Africa) followed a similar diurnal trend but individuals of some species also foraged at night (Marais 1980). However, the available evidence suggests that most mullet species are mainly diurnal rather than nocturnal feeders and that this may be related to a synchronization of foraging with the photosynthetic production by benthic microalgae, especially diatoms. It is also important to note that feeding is most prevalent at high tide when most estuarine littoral areas rich in diatom 'carpets' are inundated, with food intake generally lowest at low tide, especially when such a tide coincides with the hours of darkness (Marais 1980).

The apparent lack of a definitive separation between the diet and feeding periodicity of sympatric mullet species has given rise to the notion that mugilids feed on superabundant resources and that co-existence between these species is therefore possible. This view is strongly supported by the detailed feeding study of mugilids in Lake St. Lucia (Blaber 1976) and two species of mullet from Pullamadam

Lagoon in India (Luther 1962). Not only is there a high degree of overlap in diets but the feeding zones and feeding periodicity of different mugilid species also show a high degree of overlap. This does not mean that competitive interactions between the mullet species are absent – it merely indicates that food resources do not appear to be limiting (Blaber 1976). It is also likely that major habitat and physico-chemical changes in estuarine systems (Chrystal and Scharler 2014) do not alter the base of the food pyramid to any significant extent and that there will always be an abundance of detritus and benthic microalgae for mugilids to feed upon in estuarine ecosystems. This view is supported by long-term feeding studies (e.g., Blaber 1976) which have indicated little or no seasonal variation in diet by various mullet species, and is possibly also related to the low position in the food pyramid that most dietary items are obtained.

Competition between alien common carp *Cyprinus carpio* and the juveniles of thicklip mullet *Chelon labrosus* in western Mediterranean estuaries was such that mugilid species had low catch per unit effort numbers in any system where the carp were present. This, according to Cardona et al. (2008), confirmed the juvenile competitive bottleneck hypothesis which then manifested itself in terms of low abundance by adult stages of *C. labrosus* in the invaded systems. Invasive piscivorous fishes may also negatively impact on indigenous mullet populations, e.g., introduced largemouth bass *Micropterus salmoides* in the headwaters of the Kowie Estuary (South Africa) are a major predator of juvenile *Myxus capensis* migrating to riverine nursery areas within this system (Weyl and Lewis 2006).

Potential food competition between the flathead mullet *Mugil cephalus*, Mozambique tilapia *Oreochromis mossambicus* and milkfish *Chanos chanos* in the St. Lucia system was avoided by these three iliophagous species selecting different benthic dietary items (Whitfield and Blaber 1978b). Apart from the above dietary separation there was also an absence of any overlap in major diatom species consumed by the three fish taxa. Further indications of differential feeding mechanisms among the above species was provided by the fact that 97% of *M. cephalus* had sediment in their stomach contents whereas only 38% of *C. chanos* and 35% of *O. mossambicus* had sediments in their stomachs.

Natural Predators of Grey Mullet

Predation on juvenile mullet in estuaries represents a complex balance between their longitudinal and horizontal positions within these systems. Those species such as *Mugil cephalus* and *Myxus capensis*, which frequent rivers and other freshwater areas (Bok 1979) not only avoid the piscivorous fishes present in estuaries but are also able to utilize the primary food resources found in catchments with a limited number of potential fish competitors being present. Similarly, within estuaries, juvenile mugilids can escape piscivorous fishes in these systems by occupying productive shallows (Paterson and Whitfield 2000) but in doing so they increase their vulnerability to piscivorous bird predation (Whitfield and Blaber 1979c). Indeed, the normal reaction of shoals of small mugilids to the presence of piscivorous fish on the edges of an estuary channel is to flee towards very shallow marginal waters in order to escape these predators. Furthermore, the comparative scarcity of juvenile mullet in clear intertidal pools and shallow sheltered areas of the coast when compared to adjacent turbid estuaries may well be a function of the clarity of the water in former areas which would make these fish more vulnerable to predation.

Natural predators of mugilids include piscivorous fish (Whitfield and Blaber 1978c, Blaber 1987), sharks (Thomson 1966, Blaber 1987), crocodiles (Whitfield and Blaber 1979a), birds (Whitfield and Blaber 1979b, Jackson 1984, Dias et al. 2012), otters (Clavero et al. 2006) and dolphins (Barros and Cockroft 1991, Barros et al. 2004). In the case of the Eurasian otter *Lutra lutra* in Spain and Portugal this predation can have a strong seasonal component which may be linked to changing mullet densities (Clavero et al. 2004), whereas for Nile crocodiles *Crocodylus niloticus*, bull sharks *Carcharinus leucas* and white pelicans *Pelecanus onocrotalus* in South Africa this seasonality is correlated with the annual spawning migration of *Mugil cephalus* (Whitfield and Blaber 1978a, 1979a, 1979b). Sharks have also been recorded preying on adult *M. cephalus* during spawning migrations in other parts of the world (Kesteven 1942, Darnell 1958). In addition, *Carcharinus melanopterus* has been described by Helfrich and Allen (1975) attacking a spawning aggregation of *Crenimugil crenilabis* in the Pacific Ocean.

Cech and Wohlschlag (1982) provide an interesting physiological perspective in terms of a possible ecological strategy by *Mugil cephalus* to avoid predators during the breeding season in Texas waters. These authors have shown that *M. cephalus* exhibit enhanced haemoglobin concentrations which, in association with changes in respiration and gill ventilation rates, result in more rapid and sustained swimming activity during the spawning season. This increased aerobic fitness probably reduces predator induced mortality of the adult *M. cephalus* during this vulnerable period when large schools form and are frequently attacked by piscivores.

Extensive predation on mugilid egg and early larval stages in the sea by both invertebrates and planktivorous fishes is highly likely but poorly documented (Blaber 1987). Natarajan and Patnaik (1968) did record large numbers of mugilid eggs in the diet of the bald glassy *Ambassis gymnocephalus* from Lake Chilka (India) and concluded that this predation may even have significant effects on recruitment to the local mullet fishery. In addition, predation on postlarval mugilids in surf zones by predatory isopods and other nektonic invertebrates is also a possibility since this has been observed to occur in ichthyoplankton sample bottles from such areas (Whitfield 1989).

In his review of piscivorous fish predation on mugilids in southeast African estuaries Blaber (1987) lists 15 species that have been recorded preying on juvenile and/or adult grey mullet. Most predatory fish had a percentage frequency of < 5% mugilids in their diet but those that consumed > 5% included *Carcharinus leucas*, tenpounder *Elops machnata*, leerfish *Lichia amia*, bartail flathead *Platycephalus indicus*, bluefish *Pomatomus saltatrix*, great barracuda *Sphyraena barracuda* and pickhandle barracuda *S. jello*. Apart from *P. indicus* which is more of an ambush predator, all the other piscivorous fish species are fast swimming and therefore capable of chasing down fast moving mullet.

Considering the relative abundance of mugilids in southeast African estuaries (Blaber 1977), we can conclude that this family is under-represented in the diets of piscivorous fishes from these systems. To what extent this is due to their effective schooling habit (Major 1978), or their occupation of shallow littoral areas where large fish predators are scarce (Becker et al. 2011), is unknown. What is known is that mullet form schools from a very early age, even when first entering coastal waters or estuaries (Fig. 14.2). This shoaling tendency persists as juveniles and adults, with individual foraging mullet immediately forming a school if alarmed. The shape of a mullet school may change rapidly from a compact oval mass to a long narrow band, with leadership of the shoal also changing continuously (Thomson 1966).

When attacked from below, individual mullet will often leap into the air in an apparent effort to evade the attacking piscivorous fish. Small mugilids will also tend to flee towards shallow water as an avoidance mechanism from large predatory fish. Although the jumping behaviour of certain species of mullet has been linked to predator avoidance, there is also evidence to suggest that this is not always the case (Thomson 1966, Hoese 1985). There are also strong indications that certain mullet species (e.g., *Mugil cephalus*) are 'jumpers' or 'springers' whereas others (e.g., *Liza dumerili*) are not (Smith and Smith 1986). If leaping out of the water is an effective predator avoidance mechanism for mugilids, the question then arises – why do only a few species within this family employ this attribute?

Fish eagles at St. Lucia (South Africa) prey mainly on mugilids, especially *M. cephalus*, throughout the year and this is linked to the surface swimming behaviour of these species when not feeding (Whitfield and Blaber 1978d). The high percentage frequency (20–60%) of mugilids in the diet of piscivorous birds from estuaries in South Africa (Blaber 1987) suggests that this family is heavily harvested by avian predators. Results from the Kosi Estuary, South Africa, indicate that two species of mullet (*Mugil cephalus* and *Moolgarda buchanani*) formed 76% of the energy intake of *Phalacrocorax carbo* in this system (Jackson 1984). This high incidence in the diet of cormorants may also explain why a number of trematode parasites use mullet as an intermediate host when the definitive hosts are piscivorous birds (Thomson 1966).

Although most mugilid prey fall into the 20–200 mm SL size classes, fish eagles tend to target larger specimens in the 240–540 mm SL size range (Whitfield and Blaber 1978d). Within each piscivorous bird category, including herons, kingfishers, terns and cormorants, there appears to be a separation in the size of mullet that are targeted, e.g., the grey heron *Ardea cinerea* feeds on smaller mullet when compared to the goliath heron *Ardea goliath*, with reed cormorants *Phalacrocorax africanus* preying on smaller mullet than the white-breasted cormorant *Phalacrocorax carbo* (Whitfield and Blaber 1979c).

In contrast to most other fish species, the eyes of most mugilids are angled slightly downwards from the vertical, which would suggest that their primary predatory threat comes from below if they are swimming in surface waters. Although a number of piscivorous fishes would attack surface swimming mullet shoals from below, the positioning of the eyes renders these fish very vulnerable to predation from above, especially by aerial diving piscivorous birds such as kingfishers, terns, ospreys and fish eagles (Whitfield and Blaber 1978d). Perhaps it is this 'blind spot' that contributes to making this family such a common part of the diet of piscivorous birds around the world (Blaber 1987).

Mullet may well be a 'keystone' species in many ecosystems. Indeed, it has been suggested that in the St. Lucia system predation on *Mugil cephalus* during the autumn spawning migration is important for the subsequent winter breeding success of white pelicans and the build-up of winter fat reserves for the Nile crocodile. An example of the magnitude of the energy source provided by *Mugil cephalus* migrating down the St. Lucia system can be gauged by the estimated 3.9 t of mullet consumed in eight days by Nile crocodiles and up to 200 kg per day of *M. cephalus* by white pelicans over a slightly longer period (Whitfield and Blaber 1979a, 1979b). The actual size of the migrating *M. cephalus* shoals is likely to be orders of magnitude greater than the fish that fall prey to the above predators and we can only speculate as to the total biomass of this species leaving the estuary to spawn. Once in the sea, sharks and dolphins prey on the dense mullet shoals, with some sharks becoming stranded in the surf zone as they pursue the fish into shallow waters. Bottlenose dolphins *Tursiops aduncus* entering Australian estuaries also prey on *M. cephalus* and run the risk of becoming stranded in shallow waters. The dolphins have therefore adopted a strategy of using these estuaries mainly on high and flood tides when there is a lower stranding risk (Fury and Harrison 2011).

Mugilids as Biological Indicators of Ecological Health

Mullets have been proposed as a good indicator of pollution and environmental degradation (Wang et al. 2011, Fig. 14.6). The use of mugilids as a sentinel species for the coastal zone arises from their bioaccumulation of land-based pollution, a feature greatly enhanced by their consumption of surface

Figure 14.6. A variety of dead mullet species associated with a pollution event that caused temporary anoxia in the Mdloti Estuary, South Africa. Notice the presence of invasive floating plants (*Eichhornia crassipes* and *Salvinia molesta*) which help promote dissolved oxygen depletion of the underlying waters that were responsible for this fish kill (photo by N. Forbes).

benthic sediments together with their food. Pollution uptake occurs when fish assimilate or absorb hydrophobic organic compounds or heavy metals into their tissues from the surrounding water, sediments and/or their prey. In the case of mugilids, their benthic feeding habits make them particularly vulnerable to heavy metal uptake.

Gills contain highly permeable membranes and can therefore be considered an important route for the entry of contaminants (Pereira et al. 2010). Just as pollutants can enter the body of fishes, so too can the body cleanse itself of these substances if given the opportunity to do so, e.g., juvenile *Liza ramada* are adversely affected by exposure to certain herbicides but hepatic recovery can occur if fish are transferred to clean water after exposure (Biagianti-Risbourg and Bastide 1995). However, reversal of liver lipoid degeneration by mullet in the above instance can only occur if the concentrations of the herbicide are subacute and not lethal.

Polychlorinated biphenyls (PCBs) were detected in three species of mugilids from the Mondego Estuary, Portugal, and in the case of *Liza ramada* there was an increase in the concentration of PCBs with age (Baptista et al. 2013). In addition, those species that had spent more than three years in the estuary had higher PCB concentrations than those taxa that had spent less than three years in the system. Similarly, mercury accumulation by *Liza aurata* in two southern European coastal ecosystems indicated that mercury concentrations in all mullet tissues were significantly higher in the more contaminated of the two systems (Tavares et al. 2011).

Mzimela et al. (2003) have shown that mullet in the Mhlathuze Estuary, South Africa, had a general increase in metal concentrations between 1975 and 1997 associated with a major harbour development. Similarly, analysis of gill tissue from *Liza aurata* along a gradient of pollution in the Óbidos Lagoon, Portugal, demonstrated the value of such tissue in being able to reflect water contamination (Pereira et al. 2010). Foraging habits and habitat selection also affect the uptake of pollutants, e.g., smaller *Liza ramada* were more exposed to heavy metals in the Tagus Estuary, Portugal, because of their preference for finer sediment fractions when feeding (Pedro et al. 2008).

The effects of pollutants on mugilids and other fish species can be extremely complex, e.g., Endocrine Disruptors (EDs) can alter the functioning of the endocrine system and thereby cause adverse health effects. A study by Puy-Azurmendi et al. (2013) provides strong evidence of EDs in the Urdaibai Biosphere Reserve (Bay of Biscay) and showed how these disruptors were having a negative effect on resident thicklip mullet *Chelon labrosus*. In other instances, pollution can cause deformities in fishes, e.g., higher skeletal anomalies in four mugilid species from the Adriatic region of Italy where anthropogenic impacts were highest (Boglione et al. 2006).

A number of authors have proposed that the cosmopolitan flathead mullet (*Mugil cephalus*) would be an ideal species to use for coastal biomonitoring studies (e.g., Ameur et al. 2012, Vasanthi et al. 2013, Waltham et al. 2013). Indirect support for such an approach is provided by *M. cephalus* abundance in three Tanzanian mangrove creeks, two of which included areas where mangroves had been removed for solar salt farms and recorded low mugilid densities when compared with abundances in a creek associated with an undisturbed mangrove forest (Mwandya et al. 2009). However, Shervette et al. (2007) also identified major differences in the fish assemblages of two adjacent Ecuador estuaries, one of which had intact mangrove habitat and the other had lost its mangrove wetlands. Despite the much lower fish species diversity in the estuary which had no mangroves, the white mullet *Mugil curema* was one of only two species that were collected in large numbers from both systems. This latter result indicates that *M. curema* would not have been a good indicator of habitat degradation in Ecuador estuaries.

The ability of mugilids to rapidly colonize newly inundated aquatic systems has been linked to their position at the base of the food pyramid and therefore an ability to act as primary fish colonizers. This attribute mitigates against their use as an indicator of environmental recovery where pollution is not an issue, e.g., *Liza abu* was one of the earliest and most abundant fish species to colonize restored marshes in southern Iraq, a characteristic attributed to this mugilid by virtue of it being a detritivore and therefore able to find suitable food sources almost immediately (Mohamed et al. 2012).

Stable isotope analyses provide an opportunity to assess the response of aquatic food webs to sewage effluent discharges, e.g., *Mugil cephalus* in two intermittently open estuaries in New South Wales, Australia, showed differential ^{15}N values according to the nature of wastewater inputs (Hadwen and Arthington 2007).

Mugilids are also a good biological indicator of altered sewage treatment patterns in the Ria Formosa coastal lagoon in southern Portugal, with numbers of this family declining due to improved sewage effluent treatment and a reduction in organic matter loading of the lagoon (Ribeiro et al. 2008).

Another way in which mugilids can reflect anthropogenic impacts from the catchment can be found in Cuba where mullet fishery landings over two decades decreased as the number and extent of catchment impoundments increased (Baisre and Arboleya 2006), thus depriving estuaries of riverine nutrient and organic inputs. Support for the correlation between mugilid abundance and riverine flow is also provided by the work of Gillson et al. (2009) who recorded a decline in gill net fishery catch rates of *Mugil cephalus* during a major drought in eastern Australia.

Summary and Conclusions

Most mullet enter estuaries at some stage in their life cycle, with many species occupying these systems for protracted periods during their juvenile and subadult years. Although virtually all species have at least a short marine, egg and larval phase, many then spend most of their life cycle associated with estuarine systems of various shapes and sizes. The occurrence of juvenile mullet in clear marine waters deeper than 2–3 m is a rare occurrence, with the nursery area of most species being shallow estuaries and brackish lagoons. Some species may even use riverine pools and backwaters as nursery areas but even these species return to the sea for spawning activities. Although mugilid eggs have been located in euhaline estuarine waters, it is often uncertain whether these were the product of co-ordinated and deliberate spawning activities or the result of tidal transport of eggs from a coastal spawning event. Indeed in the case of most mullet species, for which actual spawning has been recorded, the breeding seems to occur in offshore coastal waters rather than in the mouths of estuaries. Those few mugilid species that seldom enter estuaries seem to be confined to either large river systems or shallow marine waters for much of their life cycle. This may be related to a lack of euryhalinity by these particular species or the availability of suitable food resources in riverine and littoral marine areas that are not being fully utilized. However, a more probable explanation is that freshwater and marine mugilids are occupying relatively vacant detrital and benthic microalgal niches in the freshwater and marine environments respectively, and are avoiding the high densities and wide variety of mullet species found in estuaries. Another possibility is that some of these species are poorly studied and that further work will reveal that they do indeed make more extensive use of estuaries than previously recorded and therefore conform to 'coastal waters template' of most mugilids.

The extent to which the juveniles of many mugilid species are dependent on estuaries as nursery areas has been raised by many authors. Given that very few juveniles of the above species are ever found outside estuaries and their adjacent lagoonal and riverine environments, it would be logical to deduce that estuaries are critical nursery areas for these particular species. However, there are parts of the world where estuaries are scarce, but estuary-associated mugilids are still recorded – so the obligatory nature of the estuary association component in their life cycle still has to be proven. What we do know is that, in the case of some species, marine raised individuals grow slower and are in poorer condition than their estuarine raised counterparts.

Mugilids are involved in major energy fluxes both within and between coastal ecosystems such as estuaries, lagoons and the sea but this has also not been quantified. Clearly mullets appear to be particularly well adapted to a migratory life history but are also very vulnerable to over-exploitation at critical stages in that life cycle, e.g., during mass immigration of fry into estuaries when they are harvested for aquaculture or during mass emigration from estuarine systems during the breeding season when they captured primarily for the high value of their roe.

Despite the widespread abundance of mugilids and their seeming resilience to adverse physico-chemical conditions brought about by anthropogenic disturbances of their environment, it should be emphasized that all species have their limits in terms of tolerance to pollution and over-exploitation. If we do not take heed of the early warning signs on the adverse impacts we are causing to the biology and ecology of certain mullet species, then substantial ecosystem damage will be the end result.

References

Akin, S., E. Buhan, K.O. Winemillar and H. Yilmaz. 2005. Fish assemblage structure of Koycegiz Lagoon – Estuary, Turkey: Spatial and temporal distribution patterns in relation to environmental variation. Estuarine, Coastal and Shelf Science 64: 671–684.

Almeida, P.R. 1996. Estuarine movement patterns of adult thin-lipped grey mullet, *Liza ramada* (Risso) (Pisces: Mugilidae), observed by ultrasonic tracking. Journal of Experimental Marine Biology and Ecology 202: 137–150.

Almeida, P.R. 2003. Feeding ecology of *Liza ramada* (Risso 1810) (Pisces: Mugilidae) in a south-western estuary of Portugal. Estuarine, Coastal and Shelf Science 57: 313–323.

Ameur, B., A. Bayed and T. Bennazou. 2003. Rôle de la communication de la lagune de Merja Zerga (Gharb, Maroc) avec l'Océan Atlantique dans la reproduction d'une population de *Mugil cephalus* L. (Poisson Mugilidae). Bulletin de l'Institut Scientifique, Rabat, Section Sciences de la Vie 25: 77–82.

Ameur, W.B., J. de Lapuente, Y. El Megdiche, B. Barhoumi, S. Trabelsi, L. Camps, J. Serret, D. Ramos-López, J. Gonzales-Linares, M. Ridha Driss and M. Borràs. 2012. Oxidative stress, genotoxicity and histopathology biomarker responses in mullet (*Mugil cephalus*) and sea bass (*Dicentrarchus labrax*) liver from Bizerte Lagoon, Tunisia. Marine Pollution Bulletin 64: 241–251.

Anderson, W.W. 1958. Larval development, growth and spawning of striped mullet (*Mugil cephalus*) along the South Atlantic coast of the United States. Fishery Bulletin of the Fish and Wildlife Services 58: 501–519.

Apekin, V.S. and N.I. Vilenskaya. 1978. A description of the sexual cycle and state of the gonads during the spawning migration of the striped mullet, *Mugil cephalus*. Journal of Ichthyology 18: 446–456.

Arnold, E.L. and J.R. Thompson. 1958. Offshore spawning of the striped mullet, *Mugil cephalus*, in the Gulf of Mexico. Copeia 1958: 130–132.

Avanesov, E.M. 1972. Present spawning conditions of mullets (genus *Mugil*) in the Caspian Sea. Journal of Ichthyology 12: 419–425.

Bacheler, N.M., R.A. Wong and J.A. Buckel. 2005. Movements and mortality rates of striped mullet in North Carolina. North American Journal of Fisheries Management 25: 361–373.

Baisre, J.A. and Z. Arboleya. 2006. Going against the flow: effects of river damming in Cuban fisheries. Fisheries Research 81: 283–292.

Baptista, J., P. Pato, S. Tavares, A.C. Duarte and M.A. Pardal. 2013. PCB bioaccumulation in three mullet species – a comparison study. Ecotoxicology and Environmental Safety 94: 147–152.

Barrington, E.J.W. 1957. The alimentary canal and digestion. pp. 109–161. *In*: M.E. Brown (ed.). The Physiology of Fishes, Vol. 1. Academic Press, New York.

Barros, N.B. and V.G. Cockroft. 1991. Prey of humpback dolphins, *Sousa plumbea*, stranded in Eastern Cape Province, South Africa. Aquatic Mammals 17: 134–136.

Barros, N.B., T.A. Jefferson and E.C.M. Parsons. 2004. Feeding habits of Indo-Pacific humpback dolphins (*Sousa chinensis*) stranded in Hong Kong. Aquatic Mammals 30: 179–188.

Becker, A., P.D. Cowley, A.K. Whitfield, J. Järnegren and T.F. Næsje. 2011. Diel fish movements in the littoral zone of a temporarily closed South African estuary. Journal of Experimental Marine Biology and Ecology 406: 63–70.

Biagianti-Risbourg, S. and J. Bastide. 1995. Hepatic perturbations induced by a herbicide (atrazine) in juvenile grey mullet *Liza ramada* (Mugilidae: Teleostei): An ultrastructural study. Aquatic Toxicology 31: 217–229.

Blaber, S.J.M. 1976. The food and feeding ecology of Mugilidae in the St. Lucia lake system. Biological Journal of the Linnean Society 8: 267–277.

Blaber, S.J.M. 1977. The feeding ecology and relative abundance of mullet (Mugilidae) in Natal and Pondoland estuaries. Biological Journal of the Linnean Society 9: 259–275.

Blaber, S.J.M. 1987. Factors affecting recruitment and survival of mugilids in estuaries and coastal waters of southern Africa. American Fisheries Society Symposium 1: 507–518.

Blaber, S.J.M. and A.K. Whitfield. 1977. The feeding ecology of juvenile mullet (Mugilidae) in southeast African estuaries. Biological Journal of the Linnean Society 9: 277–284.

Boglione, C., C. Costa, M. Giganti, M. Cecchetti, P. Di Dato, M. Scardi and S. Cataudella. 2006. Biological monitoring of wild thicklip grey mullet (*Chelon labrosus*), golden grey mullet (*Liza aurata*) thinlip mullet (*Liza ramada*) and flathead mullet (*Mugil cephalus*) (Pisces: Mugilidae) from different Adriatic sites: Meristic counts and skeletal anomalies. Ecological Indicators 6: 712–732.

Bok, A.H. 1979. The distribution and ecology of two mullet species in some freshwater rivers in the eastern Cape, South Africa. Journal of the Limnological Society of Southern Africa 5: 97–102.

Brownell, C.L. 1979. Stages in the early development of 40 marine fish species with pelagic eggs from the Cape of Good Hope. Ichthyological Bulletin of the J.L.B. Smith Institute of Ichthyology 40: 1–84.

Brulhet, J. 1975. Observations on the biology of *Mugil cephalus ashenteensis* and the possibility of its aquaculture on the Mauritian coast. Aquaculture 5: 271–281.

Brusle, J. and S. Brusle. 1977. Les muges de Tunisie: pêche lagunaire et biologie de la reproduction de trois espèces (*Mugil capito*, *Mugil cephalus* et *Mugil chelo*) des lacs d'Ichkeul et de Tunis. Rapports Commission International pour l'Exploration Scientifique de la Mer Méditerranée 24: 101–130.

Cardona, L. 1999. Seasonal changes in the food quality, diel feeding rhythm and growth rate of juvenile leaping grey mullet *Liza saliens*. Aquatic Living Resources 12: 263–270.

Cardona, L. 2000. Effects of salinity on the habitat selection and growth performance of Mediterranean flathead grey mullet *Mugil cephalus* (Osteichthyes, Mugilidae). Estuarine, Coastal and Shelf Science 50: 727–737.

Cardona, L. 2015. Food and feeding of Mugilidae. *In*: D. Crosetti and S.J.M. Blaber (eds.). Biology, Ecology and Culture of Grey Mullet (Mugilidae). CRC Press, Boca Raton, USA (this book).

Cardona, L., B. Hereu and X. Torras. 2008. Juvenile bottlenecks and salinity shape grey mullet assemblages in Mediterranean estuaries. Estuarine, Coastal and Shelf Science 77: 623–632.

Cardosa, I., S. França, M.P. Pais, S. Henriques, L.C. da Fonseca and H.N. Cabral. 2011. Fish assemblages of small estuaries of the Portuguese coast: A functional approach. Estuarine, Coastal and Shelf Science 93: 40–46.

Carpentier, A., S. Como, C. Dupuy, C. Lefrançois and E. Feunteun. 2013. Feeding ecology of *Liza* spp. in a tidal mudflat: evidence of the importance of primary production (biofilm) and associated meiofauna. Journal of Sea Research DOI: 10.1016/j.seares.2013.10.007.

Chan, E.H. and T.E. Chua. 1979. The food and feeding habits of greenback grey mullet, *Liza subviridus* (Valenciennes), from different habitats and at various stages of growth. Journal of Fish Biology 15: 165–171.

Chang, C.-W. and Y. Iizuka. 2012. Estuarine use and movement patterns of seven sympatric Mugilidae fishes: The Tatu Creek estuary, central western Taiwan. Estuarine, Coastal and Shelf Science 106: 121–126.

Cech, J.J. and D.E. Wohlschlag. 1982. Seasonal patterns of respiration, gill ventilation, and hematological characterisitics of striped mullet, *Mugil cephalus* L. Bulletin of Marine Science 32: 130–138.

Chrystal, R.A. and U.M. Scharler. 2014. Network analysis indices reflect extreme hydrodynamic conditions in a shallow estuarine lake (Lake St. Lucia), South Africa. Ecological Indicators 38: 130–140.

Clark, B.M., K. Hutchings and S.J. Lamberth. 2009. Long-term variations in composition and abundance of fish in the Berg Estuary, South Africa. Transactions of the Royal Society of South Africa 64: 238–258.

Clavero, M., J. Prenda and M. Delibes. 2004. Influence of spatial heterogeneity on otter (*Lutra lutra*) prey consumption. Annales Zoological Fennici 41: 551–561.

Clavero, M., J. Prenda and M. Delibes. 2006. Seasonal use of coastal resources by otters: Comparing sandy and rocky stretches. Estuarine, Coastal and Shelf Science 66: 387–394.

Como, S., C. Lefrancois, E. Maggi, F. Antognarelli and C. Dupuy. 2013. Behavioral responses of juvenile gray mullet *Liza aurata* to changes in coastal temperatues and consequences for benthic food resources. Journal of Sea Research. DOI: 10.1016/j.seares.2013.10.004.

Connell, A.D. 1996. Sea fishes spawning pelagic eggs in the St. Lucia estuary. South African Journal of Zoology 31: 37–41.

Cowley, P.D., A.K. Whitfield and K.N.I. Bell. 2001. The surf zone ichthyoplankton adjacent to an intermittently open estuary, with evidence of recruitment during marine overwash events. Estuarine, Coastal and Shelf Science 52: 339–348.

Cyrus, D.P. and S.J.M. Blaber. 1987. The influence of turbidity on juvenile marine fishes in estuaries. Part 1. Field studies at Lake St. Lucia on the southeastern coast of Africa. Journal of Experimental Marine Biology and Ecology 109: 53–70.

Darnell, R.M. 1958. Food habits of fishes and larger invertebrates of Lake Pontchartrain, Louisiana, an estuarine community. Publications of the Institute of Marine Science, University of Texas 5: 353–416.

De Decker, H.P. and B.A. Bennett. 1985. A comparison of the physiological condition of the southern mullet *Liza richardsoni* (Smith), in a closed estuary and the sea. Transactions of the Royal Society of South Africa 45: 427–436.

De Silva, S.S. and P.A.B. Perera. 1976. Studies on the young grey mullet *Mugil cephalus* L. 1. Effects of salinity on food intake, growth and food conversion. Aquaculture 7: 327–338.

De Silva, S.S. and E.I.L. Silva. 1979. Biology of young grey mullet, *Mugil cephalus* L., populations of a coastal lagoon in Sri Lanka. Journal of Fish Biology 15: 9–20.

Degré, D., D. Leguerrier, E. Armynot du Chatelet, J. Rzeznik, J.-C. Auguet, C. Dupuy, E. Marquis, D. Fichet, C. Struski, E. Joyeux, P.-G. Sauriau and N. Niquil. 2006. Comparative analysis of the food webs of two intertidal mudflats during two seasons using inverse modelling: Aiguillon Cove and Brouage Mudflat, France. Estuarine, Coastal and Shelf Science 69: 107–124.

Dias, E., P. Morais, M. Leopold, J. Campos and C. Antunes. 2012. Natural born indicators: Great cormorant *Phalacrocorax carbo* (Aves: Phalacrocoridae) as monitors of river discharge influence on estuarine ichthyofauna. Journal of Sea Research 73: 101–108.

Ellender, B.R., G.C. Taylor and O.L.F. Weyl. 2012. Validation of growth zone deposition in otoliths and scales of flathead mullet *Mugil cephalus* and freshwater mullet *Myxus capensis* from fish of known age. African Journal of Marine Science 34: 455–458.

Eme, J., T.F. Dabruzzi and W.A. Bennett. 2011. Thermal responses of juvenile squaretail mullet (*Liza vaigiensis*) and juvenile crescent terapon (*Terapon jarbua*) acclimated at near-lethal temperatues, and the implications for climate change. Journal of Experimental Marine Biology and Ecology 399: 35–38.

Erman, F. 1959. Observations on the biology of the common grey mullet *Mugil cephalus* L. Proceedings of the General Fisheries Council for the Mediterranean 5: 157–169.

Fanta-Feofiloff, E., D.R. De Brito Eiras, A.T. Boscardim and M. Lacerda-Krambeck. 1986. Effect of salinity on the behavior and oxygen consumption of *Mugil curema* (Pisces: Mugilidae). Physiology and Behavior 36: 1029–1034.

Faouzi, H. 1938. Quelques aspects de la biologie des Muges en Egypte. Rapports Commission International pour l'Exploration Scientifique de la Mer Méditerranée 11: 63–68.

Faye, D., L.T. de Morais, J. Raffray, O. Sadio, O.T. Thiaw and F. Le Loc'h. 2011. Structure and seasonal variability of fish food webs in an estuarine tropical marine protected area (Senegal): Evidence from stable isotope analysis. Estuarine, Coastal and Shelf Science 92: 607–617.

FishBase. 2014. http://www.fishbase.org/.

Fury, C.A. and P.L. Harrison. 2011. Seasonal variation and tidal influences on estuarine use by bottlenose dolphins (*Tursiops aduncus*). Estuarine, Coastal and Shelf Science 93: 389–395.

Garbin, T., J.P. Castello and P.G. Kinas. 2014. Age, growth, and mortality of the mullet *Mugil liza* in Brazil's southern and southeastern coastal regions. Fisheries Research 149: 61–68.

Garcia, A.M., J.P. Vieira and K.O. Winemiller. 2003. Effects of the 1997–1998 El Niño on the dynamics of the shallow-water fish assemblage of the Patos Lagoon Estuary (Brazil). Estuarine, Coastal and Shelf Science 57: 489–500.

Gillson, J., J. Scandol and I. Suthers. 2009. Estuarine gill net fishery catch rates decline during drought in eastern Australia. Fisheries Research 99: 26–37.

Hadwen, W.L. and A.H. Arthington. 2007. Food webs of two intermittently open estuaries receiving ^{15}N-enriched sewage effluent. Estuarine, Coastal and Shelf Science 71: 347–358.

Hall, C.M., A.K. Whitfield and B.R. Allanson. 1987. Recruitment, diversity and the influence of constrictions on the distribution of fishes in the Wilderness lakes system, South Africa. South African Journal of Zoology 22: 163–169.

Harrison, T.D. and J.A.G. Cooper. 1991. Active migration of juvenile grey mullet (Teleostei: Mugilidae) into a small lagoonal system on the Natal coast. South African Journal of Science 87: 395–396.

Harrison, T.D. and A.K. Whitfield. 2006. Estuarine typology and the structuring of fish communities in South Africa. Environmental Biology of Fishes 75: 269–293.

Helfrich, P. and P.M. Allen. 1975. Observations on the spawning of mullet, *Crenimugil crenilabis* (Forsskål), at Enewetak, Marshall Islands. Micronesica 11: 219–225.

Hoese, H.D. 1985. Jumping mullet – the internal diving bell hypothesis. Environmental Biology of Fishes 13: 309–314.

Hotos, G.N. and G.N. Katselis. 2011. Age and growth of the golden grey mullet *Liza aurata* (Actinopterygii: Mugiliformes: Mugilidae), in the Messolonghi-Etoliko Lagoon and the adjacent Gulf of Patraikos, Western Greece. Acta Ichthyologica 41: 147–157.

Ibáñez, A.L. 2015. Age and growth of Mugilidae. *In*: D. Crosetti and S.J.M. Blaber (eds.). Biology, Ecology and Culture of Grey Mullet (Mugilidae). CRC Press, Boca Raton, USA (this book).

Ibáñez, A.L. and O. Gutierrez-Benitez. 2004. Climate variables and spawning migrations of the striped mullet and white mullet in the north-western area of the Gulf of Mexico. Journal of Fish Biology 65: 622–631.

Inoue, T., Y. Suda and M. Sano. 2008. Surf zone fishes in an exposed sandy beach at Sanrimatsubara, Japan: Does fish assemblage structure differ among microhabitats? Estuarine, Coastal and Shelf Science 77: 1–11.

Jackson, S. 1984. Predation by pied kingfishers and whitebreasted cormorants on fish in the Kosi estuary system. Ostrich 55: 113–132.

James, N.C., P.D. Cowley, A.K. Whitfield and H. Kaiser. 2008a. Choice chamber experiments to test the attraction of postflexion *Rhabdosargus holubi* larvae to water of estuarine and riverine origin. Estuarine, Coastal and Shelf Science 77: 143–149.

James, N.C., A.K. Whitfield and P.D. Cowley. 2008b. Long-term stability of the fish assemblages in a warm-temperate South African estuary. Estuarine, Coastal and Shelf Science 76: 723–738.

James, P.S.B.R., V.S. Rengaswamy, A. Raju, G. Mottanraj and V. Gandhi. 1983. Induced spawning and larval rearing of the grey mullet *Liza macrolepis* (Smith). Indian Journal of Fisheries 30: 185–202.

Jin, B., C. Fu, J. Zhong, B. Li, J. Chen and J. Wu. 2007. Fish utilization of a salt marsh intertidal creek in the Yangtze River estuary, China. Estuarine, Coastal and Shelf Science 73: 844–852.

Jin, B., H. Qin, W. Xu, J. Wu, J. Zhong, G. Lei, J. Chen and C. Fu. 2010. Nekton use of intertidal creek edges in low salinity salt marshes of the Yangtze River estuary along a stream-order gradient. Estuarine, Coastal and Shelf Science 88: 419–428.

Johnson, D.W. and E.L. McClendon. 1970. Differential distribution of striped mullet, *Mugil cephalus* Linnaeus. California Fish and Game 56: 138–139.

Katselis, G., C. Koutsikopoulos, Y. Rogdakis, T. Lachanas, E. Dimitriou and K. Vidalis. 2005. A model to estimate the annual production of roes (avgotaracho) of flathead mullet (*Mugil cephalus*) based on the spawning migration of species. Fisheries Research 75: 138–148.

Katselis, G., K. Koukou, E. Dimitriou and C. Koutsikopoulos. 2007. Short-term seawardfish migration in the Messolonghi-Etoliko lagoons (western Greek coast) in relation to climatic variables and the lunar cycle. Estuarine, Coastal and Shelf Science 73: 571–582.

Kemp, J.O.G. and P.W. Froneman. 2004. Recruitment of ichthyoplankton and macrozooplankton during overtopping events into a temporarily open/closed southern African estuary. Estuarine, Coastal and Shelf Science 61: 529–537.

Kendall, B.W., C.A. Gray and D. Bucher. 2010. Age validation and variation in growth, mortality and population structure of *Liza argentea* and *Myxus elongatus* (Mugilidae) in two temperate Australian estuaries. Journal of Fish Biology 75: 2788–2804.

Kenny, J.S. 1995. Views from the bridge: a memoir on the freshwater fishes of Trinidad. Maracas, Trinidad and Tobago. 98pp.

Kesteven, G.L. 1942. Studies on the biology of Australian mullet. 1. Account of the fishery and preliminary statement of the biology of *Mugil dobula* Gunther. Australian Commonwealth Scientific and Industrial Research Organisation Bulletin 157.

Kilby, J.D. 1955. The fishes of two Gulf coastal marsh areas of Florida. Tulane Studies in Zoology 2: 175–247.

Kingsford, M.J. and I.M. Suthers. 1994. Dynamic estuarine plumes and fronts: Importance to small fish and plankton in coastal waters of NSW, Australia. Continental Shelf Research 14: 655–672.

Koutrakis, E. 2004. Temporal occurrence and size distribution of grey mullet juveniles (Pisces: Mugilidae) in the estuarine systems of the Strymonikos Gulf (Greece). Journal of Applied Ichthyology 20: 76–78.

Kristensen, I. 1963. Hypersaline bays as an environment for young fish. Proceedings of the Gulf and Caribbean Fisheries Institute 16: 139–142.

Kumar, R.S., U.K. Sarkar, O. Gusain, V.K. Dubey, A. Pandey and W.S. Lakra. 2014. Age, growth, population structure and reproductive potential of a vulnerable freshwater mullet, *Rhinomugil corsula* (Hamilton 1822) from a tropical river Betwa in central India. Proceedings of the National Academy of Science, India (Section B: Biological Sciences) 84: 275–286.

Kuo, C.M., Z.H. Shedhdah and K.K. Milisen. 1973. A preliminary report on the development, growth and survival of laboratory reared larvae of the grey mullet, *Mugil cephalus* L. Journal of Fish Biology 5: 459–470.

Kurian, C.V. 1975. Mullets and mullet fisheries of India. Aquaculture 5: 114.

Laffaille, P., E. Feunteun and J.-C. Lefeuvre. 2000. Composition of fish communities in a European macrotidal salt marsh (the Mont Saint-Michel Bay, France). Estuarine, Coastal and Shelf Science 51: 429–438.

Laffaille, P., E. Feunteun, C. Lefebvre, A. Radureau, G. Sagan and J.-C. Lefeuvre. 2002. Can thin-lipped mullet directly exploit the primary and detritic production of European macrotidal salt marshes? Estuarine, Coastal and Shelf Science 54: 729–736.

Lamberth, S.J., G.M. Branch and B.M. Clark. 2010. Estuarine refugia and fish responses to a large anoxic, hydrogen sulphide, "black tide" event in the adjacent marine environment. Estuarine, Coastal and Shelf Science 86: 203–215.

Lebreton, B., P. Richard, E.P. Parlier, G. Guillou and G.F. Blanchard. 2011. Trophic ecology of mullets during their spring migration in a European saltmarsh: A stable isotope study. Estuarine, Coastal and Shelf Science 91: 502–510.

Lefeuvre, J.-C., P. Laffaille and E. Feunteun. 1999. Do fish communities function as biotic vectors of organic matter between salt marshes and marine coastal waters. Aquatic Ecology 33: 293–299.

Lin, H.-J., W.-Y. Kao and Y.-T. Wang. 2007. Analyses of stomach contents and stable isotopes reveal food sources of estuarine detritivorous fish in tropical/subtropical Taiwan. Estuarine, Coastal and Shelf Science 73: 527–537.

Luther, G. 1962. The food habits of *Liza macrolepis* (Smith) and *Mugil cephalus* L. (Mugilidae). Indian Journal of Fisheries 9: 604–626.

Major, P.F. 1978. Aspects of estuarine intertidal ecology of juvenile striped mullet, *Mugil cephalus* in Hawaii. U.S. National Marine Fisheries Service Fishery Bulletin 76: 299–314.

Marais, J.F.K. 1978. Routine oxygen consumption of *Mugil cephalus*, *Liza dumerili* and *L. richardsoni* at different temperatures and salinities. Marine Biology 50: 9–16.

Marais, J.F.K. 1980. Aspects of food intake, food selection, and alimentary canal morphology of *Mugil cephalus* (Linnaeus 1958), *Liza tricuspidens* (Smith 1935), *L. richardsoni* (Smith 1846), and *L. dumerili* (Steindachner 1869). Journal of Experimental Marine Biology and Ecology 44: 193–209.

Marais, J.F.K. and T. Erasmus. 1977a. Chemical composition of alimentary canal contents of mullet (Teleostei: Mugilidae) caught in the Swartkops Estuary near Port Elizabeth, South Africa. Aquaculture 10: 263–273.

Marais, J.F.K. and T. Erasmus. 1977b. Body composition of *Mugil cephalus*, *Liza dumerili*, *Liza richardsoni* and *Liza tricuspidens* (Teleostei: Mugilidae) caught in the Swartkops estuary. Aquaculture 10: 75–86.

Mariani, S. 2001. Can spatial distribution of ichthyofauna describe marine influence on coastal lagoons? A central Mediterranean case study. Estuarine, Coastal and Shelf Science 52: 261–267.

Mariani, A., S. Panella, G. Monaco and S. Cataudella. 1987. Size analysis of inorganic particles in the alimentary tracts of Mediterranean mullet species suitable for aquaculture. Aquaculture 62: 123–129.

Masson, H. and J.F.K. Marais. 1975. Stomach content analyses of mullet from the Swartkops Estuary. Zoologica. Africana 10: 193–207.

McDonough, C.J., W.A. Roumillat and C.A. Wenner. 2005. Sexual differentiation and gonad development in striped mullet (*Mugil cephalus* L.) from South Carolina estuaries. Fishery Bulletin–National Oceanic and Atmospheric Administration 103: 601–619.

Mohamed, A.-R.M., N.A. Hussain, S.S. Al-Noor, F.M. Mutlak, I.M. Al-Sudani and A.M. Mojer. 2012. Ecological and biological aspects of fish assemblage in the Chybayish marsh, southern Iraq. Ecohydrology and Hydrobiology 12: 65–74.

Mohanraj, G., P. Nammalwar, S. Kandaswamy and A.C. Sekar. 1994. Availability of grey mullet spawners in Adyar Estuary and Koovalum backwater around Madras, India. Journal of the Marine Biological Association of India 36: 167–180.

Morovic, D. 1963. Contribution à la connaissance du début de la première maturité sexuelle et de la période de ponte chez *Mugil cephalus* L. et *Mugil chelo* Cuv. en Adriatique (Dalmatie). Rapports Commission International pour l'Exploration Scientifique de la Mer Méditerranee 17: 779–786.

Mwandya, A.W., M. Gullström, M.C. Öhman, M.H. Andersson and Y.D. Mgaya. 2009. Fish assemblages in Tanzanian mangrove creek systems influenced by solar salt farm constructions. Estuarine, Coastal and Shelf Science 82: 193–200.

Mwandya, A.W., Y.D. Mgaya, M.C. Öhman, I. Bryceson and M. Gullström. 2010. Distribution patterns of striped mullet *Mugil cephalus* in mangrove creeks, Zanzibar, Tanzania. African Journal of Marine Science 32: 85–93.

Mzimela, H.M., V. Wepener and D.P. Cyrus. 2003. Seasonal variation of selected metals in sediment, water and tissues of the groovy mullet, *Liza dumerili* (Mugilidae) from Mhlathuze Estuary, South Africa. Marine Pollution Bulletin 46: 659–676.

Natarajan, A.V. and S. Patnaik. 1968. Occurrence of mullet eggs in gut contents of *Ambassis gymnocephalus*. Journal of the Marine Biological Association of India 9: 192–194.

Nordlie, F.G. 2015. Adaptation to salinity and osmoregulation in Mugilidae. *In*: D. Crosetti and S.J.M. Blaber (eds.). Biology, Ecology and Culture of Grey Mullet (Mugilidae). CRC Press, Boca Raton, USA (this book).

Odum, W.E. 1968. The ecological significance of fine particle selection by the striped mullet *Mugil cephalus*. Limnology and Oceanography 13: 92–97.

Odum, W.E. 1970. Utilization of the direct grazing and plant detritus food chains by the striped mullet *Mugil cephalus*. pp. 222–240. *In*: J.H. Steele (ed.). Marine Food Chains. Oliver & Boyd, Edinburgh.

Pasquaud, S., M. Pillet, V. David, B. Sautour and P. Elie. 2010. Determination of fish trophic levels in an estuarine system. Estuarine, Coastal and Shelf Science 86: 237–246.

Paterson, A.W. and A.K. Whitfield. 2000. Do shallow water habitats function as refugia for juvenile fishes? Estuarine, Coastal and Shelf Science 51: 359–364.

Paterson, A.W. and A.K. Whitfield. 2003. The fishes associated with three intertidal salt marsh creeks in a temperate southern African estuary. Wetlands Ecology and Management 11: 305–315.

Pattrick, P. and N.A. Strydom. 2008. Composition, abundance, distribution and seasonality of larval fishes in the shallow nearshore of the proposed Greater Addo Marine Reserve, Algoa Bay, South Africa. Estuarine, Coastal and Shelf Science 79: 251–262.

Pedro, S., V. Canastreiro, I. Caçador, E. Pereira and A. Duarte. 2008. Granulometric selectivity by *Liza ramada* and potential contamination resulting from heavy metal load in feeding areas. Estuarine, Coastal and Shelf Science 80: 281–288.

Pereira, P., H. de Pablo, C. Vale and M. Pacheco. 2010. Combined use of environmental data and biomarkers in fish (*Liza aurata*) inhabiting a eutrophic and metal-contaminated coastal system – Gills reflect environmental contamination. Marine Environmental Research 69: 53–62.

Pillay, D., G.M. Branch and A.T. Forbes. 2007. The influence of bioturbation by the sandprawn *Callianassa kraussi*on feeding and survival of the bivalve *Eumarcia paupercula* and the gastropod *Nassarius kraussianus*. Journal of Experimental Marine Biology and Ecology 344: 1–9.

Pillay, D., C. Williams and A.K. Whitfield. 2012. Indirect effects of bioturbation by the burrowing prawn *Callichirus kraussi* on a benthic foraging fish, *Liza richardsonii*. Marine Ecology Progress Series 453: 151–158.

Potter, I.C., J.R. Tweedley, M. Elliott and A.K. Whitfield. 2015. The ways in which fish use estuaries: A refinement and expansion of the guild approach. Fish and Fisheries 16: 230–239.

Puy-Azurmendi, E., M. Ortiz-Zarragoitia, M. Villagrasa, M. Kuster, P. Aragón, J. Atienza, R. Puchades, A. Maquieira, C. Domínguez, M. López de Alda, D. Fernandes, C. Porte, J.M. Bayona, D. Barceló and M.J. Cajaraville. 2013. Endocrine disruption in thicklip grey mullet (*Chelon labrosus*) from the Urdaibai Biosphere Reserve (Bay of Biscay, southwestern Europe). Science of the Total Environment 443: 233–244.

Ribeiro, J., C.C. Monteiro, P. Monteiro, L. Bentes, R. Coelho, J.M.S. Gonçalves, P.G. Lino and K. Erzini. 2008. Long-term changes in fish communities of the Ria Formosa coastal lagoon (southern Portugal) based on two studies made 20 years apart. Estuarine, Coastal and Shelf Science 76: 57–68.

Robertson, A.I. and R.C.J. Lenanton. 1984. Fish community structure and food chain dynamics in the surf-zone of sandy beaches: The role of detached macrophyte detritus. Journal of Experimental Marine Biology and Ecology 84: 265–283.

Romer, G.S. and A. McLachlan. 1986. Mullet grazing on surf diatom accumulations. Journal of Fish Biology 28: 93–104.

Sheppard, J.N., N.C. James, A.K. Whitfield and P.D. Cowley. 2011. What role do beds of submerged aquatic macrophytes play in structuring estuarine fish assemblages? Lessons from a warm-temperate South African estuary. Estuarine, Coastal and Shelf Science 95: 145–155.

Shervette, V.R., W.E. Aguirre, E. Blacio, R. Cevallos, M. Gonzalez, F. Pozo and F. Gelwick. 2007. Fish communities of a disturbed mangrove wetland and an adjacent tidal river in Palmar, Equador. Estuarine, Coastal and Shelf Science 72: 115–128.

Simier, M., L. Blanc, C. Aliaume, P.S. Diouf and J.J. Albaret. 2004. Spatial and temporal structure of fish assemblages in an "inverse estuary", the Sine Saloum system (Senegal). Estuarine, Coastal and Shelf Science 59: 69–86.

Smith, M.M. and J.L.B. Smith. 1986. Family No. 222: Mugilidae. pp. 714–720. *In*: M.M. Smith and P.C. Heemstra (eds.). Smiths' Sea Fishes. Macmillan South Africa, Johannesburg.

Solomon, F.N. and I.W. Ramnarine. 2007. Reproductive biology of white mullet, *Mugil curema* (Valenciennes) in the southern Caribbean. Fisheries Research 88: 133–138.

Strydom, N.A. and B.D. d'Hotman. 2005. Estuary-dependence of larval fishes in a non-estuary associated South African surf zone: Evidence for continuity of surf assemblages. Estuarine, Coastal and Shelf Science 63: 101–108.

Tavares, S., H. Oliveira, J.P. Coelho, M.E. Pereira, A.C. Duarte and M.A. Pardal. 2011. Lifespan mercury accumulation pattern in *Liza aurata*: Evidence from two southern European estuaries. Estuarine, Coastal and Shelf Science 94: 315–321.

Thomson, J.M. 1955. The movements and migrations of mullet (*Mugil cephalus* L.). Australian Journal of Marine and Freshwater Research 6: 469–485.

Thomson, J.M. 1966. The grey mullets. Oceanography and Marine Biology Annual Review 4: 301–335.

Thomson, J.M. 1984. Mugilidae. *In*: W. Fischer and G. Bianchi (eds.). FAO Species Identification Sheets for Fishery Purposes. Western Indian Ocean (Fish Area 51). FAO, Rome.

Torricelli, P., P. Tongiorgi and P. Almansi. 1982. Migration of grey mullet fry into the Arno River: Seasonal appearance, daily activity, and feeding rhythms. Fisheries Research 1: 219–234.

Trancart, T., P. Lambert, E. Rochard, F. Daverat, C. Roqueplo and J. Coustillas. 2011. Swimming activity responses to water current reversal support selective tidal stream transport hypothesis in juvenile thinlip mullet *Liza ramada*. Journal of Experimental Marine Biology and Ecology 399: 120–129.

Trancart, T., P. Lambert, E. Rochard, F. Daverat, J. Coustillas and C. Roqueplo. 2012. Alternative flood tide transport tactics in catadromous species: *Anguilla anguilla*, *Liza ramada* and *Platichthys flesus*. Estuarine, Coastal and Shelf Science 99: 191–198.

Trape, S., J.-D. Durand, F. Guilhaumon, L. Vigliola and J. Panfili. 2009. Recruitment patterns of young-of-the-year mugilid fishes in a West African estuary impacted by climate change. Estuarine, Coastal and Shelf Science 85: 357–367.

van der Horst, G. and T. Erasmus. 1978. The breeding cycle of male *Liza dumerili* (Teleostei: Mugilidae) in the mouth of the Swartkops estuary. Zoologica. Africana 13: 259–273.

van der Horst, G. and T. Erasmus. 1981. Spawning time and spawning grounds of mullet with special reference to *Liza dumerili* (Steindachner 1869). South African Journal of Science 77: 73–78.

Vasanthi, L.A., P. Revathi, J. Mini and N. Munuswamy. 2013. Integrated use of histological and ultrastructural biomarkers in *Mugil cephalus* for assessing heavy metal pollution in Ennore Estuary, Chennai. Chemosphere 91: 1156–1164.

Vinagre, C., J.P. Salgado, V. Mendonça, H. Cabral and M.J. Costa. 2012. Isotopes reveal fluctuation in trophic levels of estuarine organisms, in space and time. Journal of Sea Research 72: 49–54.

Vivier, L., D.P. Cyrus and H.L. Jerling. 2010. Fish community structure of the St. Lucia estuarine system under prolonged drought conditions and its potential for recovery after mouth breaching. Estuarine, Coastal and Shelf Science 86: 568–579.

Wallace, J.H. 1975a. The estuarine fishes of the east coast of South Africa. I. Species composition and length distribution in the estuarine and marine environments. II. Seasonal abundance and migrations. Investigational Report of the Oceanographic Research Institute 40: 1–72.

Wallace, J.H. 1975b. The estuarine fishes of the east coast of South Africa. Part 3. Reproduction. Investigational Report of the Oceanographic Research Institute 41: 1–48.

Wallace, J.H. and R.P. van der Elst. 1975. The estuarine fishes of the east coast of South Africa. Part 4. Occurrence of juveniles in estuaries. Part 5. Ecology, estuarine dependence and status. Investigational Report of the Oceanographic Research Institute 42: 1–63.

Walsh, W.A., C. Swanson, C.S. Lee, J.E. Banno and H. Eda. 1989. Oxygen consumption by eggs and larvae of the striped mullet, *Mugil cephalus*, in relation to development, salinity and temperature. Journal of Fish Biology 35: 347–358.

Walsh, W.A., C. Swanson and C.S. Lee. 1991. Combined effects of temperature and salinity on embryonic development and hatching of striped mullet, *Mugil cephalus*. Aquaculture 97: 281–289.

Waltham, N.J., P.R. Teasdale and R.M. Connolly. 2013. Use of flathead mullet (*Mugil cephalus*) in coastal biomonitoring studies: Review and recommendations for future studies. Marine Pollution Bulletin 69: 195–205.

Wang, C.-H., C.-C. Hsu, W.-N. Tzeng, C.-F. You and C.-W. Chang. 2011. Origin of the mass mortality of the flathead grey mullet (*Mugil cephalus*) in the Tanshui River, northern Taiwan, as indicated by otolith elemental signatures. Marine Pollution Bulletin 62: 1809–1813.

Wasserman, R.J. and N.A. Strydom. 2011. The importance of estuary head waters as nursery areas for young estuary- and marine-spawned fishes in temperate South Africa. Estuarine, Coastal and Shelf Science 94: 56–67.

Watt-Pringle, P. and N.A. Strydom. 2003. Habitat use by larval fishes in a temperate South African surf zone. Estuarine, Coastal and Shelf Science 58: 765–774.

Weyl, O.L.F. and H. Lewis. 2006. First record of predation by the alien invasive freshwater fish *Micropterus salmoides* L. (Centrarchidae) on migrating estuarine fishes in South Africa. African Zoology 41: 294–296.

Whitfield, A.K. 1980. A quantitative study of the trophic relationships within the fish community of the Mhlanga estuary, South Africa. Estuarine and Coastal Marine Science 10: 417–435.

Whitfield, A.K. 1982. Trophic relationships and resource utilization within the fish communities of the Mhlanga and Swartvlei estuarine systems. Ph.D. thesis, University of Natal, Pietermaritzburg.

Whitfield, A.K. 1986. Fish community structure response to major habitat changes within the littoral zone of an estuarine coastal lake. Environmental Biology of Fishes 17: 41–51.

Whitfield, A.K. 1989. Ichthyoplankton in a southern African surf zone: nursery area for the postlarvae of estuarine associated fish species? Estuarine, Coastal and Shelf Science 29: 533–547.

Whitfield, A.K. and S.J.M. Blaber. 1978a. Distribution, movements and fecundity of Mugilidae at Lake St. Lucia. Lammergeyer 26: 53–63.

Whitfield, A.K. and S.J.M. Blaber. 1978b. Resource segregation among iliophagus fish in Lake St. Lucia, Zululand. Environmental Biology of Fishes 3: 293–296.

Whitfield, A.K. and S.J.M. Blaber. 1978c. Food and feeding ecology of piscivorous fishes at Lake St. Lucia, Zululand. Journal of Fish Biology 13: 675–691.

Whitfield, A.K. and S.J.M. Blaber. 1978d. Feeding ecology of piscivorous birds at Lake St. Lucia. Part 1: Diving birds. Ostrich 49: 185–198.

Whitfield, A.K. and S.J.M. Blaber. 1979a. Predation on striped mullet (*Mugil cephalus*) by *Crocodylus niloticus* at St. Lucia, South Africa. Copeia 1979: 266–269.

Whitfield, A.K. and S.J.M. Blaber. 1979b. Feeding ecology of piscivorous birds at Lake St. Lucia. Part 3: Swimming birds. Ostrich 50: 10–20.

Whitfield, A.K. and S.J.M. Blaber. 1979c. Feeding ecology of piscivorous birds at Lake St. Lucia. Part 2: Wading birds. Ostrich 50: 1–9.

Whitfield, A.K. and H.M. Kok. 1992. Recruitment of juvenile marine fishes into permanently open and seasonally open estuarine systems on the southern coast of South Africa. Ichthyological Bulletin of the J.L.B. Smith Institute of Ichthyology 57: 1–39.

Whitfield, A.K., R.H. Taylor, C. Fox and D.P. Cyrus. 2006. Fishes and salinities in the St. Lucia system – A review. Reviews in Fish Biology and Fisheries 16: 1–20.

Whitfield, A.K., J. Panfili and J.D. Durand. 2012. A global review of the cosmopolitan flathead mullet *Mugil cephalus* Linnaeus 1758 (Teleostei: Mugilidae), with emphasis on the biology, genetics, ecology and fisheries aspects of this apparent species complex. Reviews in Fish Biology and Fisheries 22: 641–681.

Wright, J.M. 1989. Detached chlorophytes as nursery areas for fish in Sulaibikhat Bay, Kuwait. Estuarine, Coastal and Shelf Science 28: 185–193.

CHAPTER 15

Genetics of Mugilidae

Anna Rita Rossi,[1,*] *Donatella Crosetti*[2,a] and *Silvia Livi*[2,b]

Introduction

This chapter provides an overview of the research into the genetics of Mugilidae, with particular emphasis on the flathead grey mullet, *Mugil cephalus*, a worldwide species on which most research was performed over the years. Grey mullets are coastal species, found in temperate, subtropical and tropical regions of the world. Several mullet species are of interest to capture fisheries and aquaculture, for both the quality of their flesh and for high priced processed products that are considered a delicacy in many parts of the world, such as dry mullet roe (see Chapter 16—Crosetti 2015) and are locally known and appreciated in coastal zones.

The Mugilidae family, ascribed to the Order Perciformes, is composed of several genera and species, the number and validity of which have been thoroughly debated over the years (for a review see Chapter 1—González-Castro and Ghasemzadeh 2015 and Chapter 2—Durand 2015). Most taxonomic studies were based on anatomo-morphological traits, which are strongly conservative in Mugilidae. Many species which had been described on morphogical bases have been later synonymized to other species in taxonomic revisions of the family (Thomson 1997, Ghasemzadeh 1998) when other traits were considered and original morphotypes were analyzed for direct comparison of preserved samples from different regions of the world. In addition molecular markers enabled the detection of new species.

The introduction of genetic studies to fish species (Reviewed by Hauser and Seeb 2008), especially to those with a commercial value for fisheries production (Waples et al. 2008, McCusker and Benzen 2010, Nielsen et al. 2012), and the development of new molecular markers greatly contributed to defining intra and inter-specific variability in fish (Avise 2004). New technologies, based on the direct analysis of DNA, created a new interest that permitted examination of what could not be observed with the naked eye or under the microscope. A new dimension was discovered, that enabled on the one hand definition of phylogeny, and on the other, spotting of hidden genetic variability, previously undetectable through

[1] Department of Biology and Biotechnology C. Darwin, University of Rome "La Sapienza", Via Borelli 50, 00161 Roma, Italy.
Email: annarita.rossi@uniroma1.it
[2] ISPRA, Aquaculture Dept., Via Brancati 60, 00144 Roma, Italia.
[a] Email: donatella.crosetti@isprambiente.it
[b] Email: silvia.livi@isprambiente.it
* Corresponding author

morphological studies, leading to the discovery of cryptic[1] species within large species complexes, which had been ascribed to a single species using morphology based taxonomy.

The genetic analysis of mullets has been carried out since the 1970s, with over 100 publications to date.[2] In the present chapter the most important information on the genetics of Mugilidae are reviewed; it is divided in two different sections, one on phylogeography and population genetics and the other on cytogenetics.

Phylogeography and Population Genetics

Population genetics analyzes the distribution of alleles within and across populations, i.e., how genetic diversity is allocated in the populations of a species. It allows us to describe the genetic variability and the population structure of the species, to estimate gene flow within a species and/or detect hybridization and introgression between species, through the use of standard parameters of genetic variability and genetic diversity. The data for these estimates derive from molecular markers that, starting from the 1960s, have become essential tools, not only for phylogenetic and phylogeographic issues, but also in the management of natural resources, and are widely used for the identification of conservation units that deserve priority in conservation plans (Avise 2004).

In fish, the development of new markers and their use in population genetic studies assumed a pivotal role for commercially important species. In the latter, molecular marker based studies were mainly focused on the identification of fishery stocks (Antoniou and Magoulas 2014, Mariani and Bekkevold 2014) and on the determination of the impact of harvest on them (Allendorf et al. 2014, Pinsky and Palumbi 2014) for a sustainable use of resources (Ferguson and Danzmann 1998, Reiss et al. 2009, Chauhan and Rajiv 2010). The integration of genetic data in marine species management however, was slow and sporadic (Waples et al. 2008). Indeed, the comparison between captures and genetic data on 32 North-eastern Atlantic marine fish species revealed that genetic data were lacking for nine species and, most importantly, there was a mismatch between population structure and current management units for six species, including important species for capture fisheries, such as Atlantic cod *Gadus morhua*, hake *Merluccius merluccius* and herring *Clupea harengus* (Reiss et al. 2009). In cultured fish, population genetic studies aim at analyzing wild populations and at evaluating the impact of aquaculture activities on them. The Atlantic salmon *Salmo salar* is one of the most thoroughly studied cultured species at the genetic level, as it shows a strong homing behaviour in both anadromous and non-anadromous forms and a pronounced population structure (see Verspoor et al. 2007 for a review, Glover et al. 2013, Taranger et al. 2014). Genetic analyses were also useful for the identification of conservation units designation (Hansen 2010) and to explore adaptive variation (Garcia de Leaniz et al. 2007), providing basal information for sustainable management of its culture.

Applications of genetic studies to aquaculture in Mugilidae are quite limited, as mullet culture is still largely based on wild seed, induced spawning is not carried out at a commercial level and breeding programmes have never been started. Genetic analyses were applied to the identification of juveniles by molecular markers (Trape et al. 2009) and to species identification in dry mullet roe, a commercial processed product (Klossa-Kilia et al. 2002, Murgia et al. 2002). Most genetic studies are focused on phylogeographic and phylogenetic issues, as a consequence of the low level of morphological variability and thus of the presence of unsolved taxonomy obtained using morpho-anatomical characters. A summary of the genetic studies performed on the various mullet species, with indication of the sampling areas, the molecular marker used and the main topics of the paper, is reported in Tables 15.1 and 15.2.

Before discussing the most important issues reported in the literature on Mugilidae genetics, a few considerations may be necessary. The first is that different molecular markers were used over time, following the development of new techniques and more powerful genetic tools, and showed various

[1] Cryptic species: distinct evolutionary lineages with a substantial amount of genetic distinctiveness and no apparent morphological differences (Durand and Borsa 2015).

[2] A bibliographic research on Scopus and ISIWeb of Science (1 November 2014) combining the topic 'Mugilidae' and 'genetic*' yielded 87 and 119 results, respectively.

Table 15.1. Summary of genetic studies on Mugilidae (1994–2014). Grey highlight indicates surveys in which more than one mullet species were analyzed (see Table 15.2).

Species[1]	Geographic area[2]	Markers									Main topic[5]					Reference
		nDNA[3]				mtDNA[4]										
		microsat	allozyme	AFLP	other	COI	Cyt b	16S	12S	Other	Phylogeny	Phylogeogr	Mol Ident	Pop Genet	Isolation	
Agonostomus																
A. catalai	W. Indian (Comores)	-	-	-	-	Sq	Sq	Sq	-	-	+	-	-	-	-	Durand et al. 2012a,b[6]
A. monticola	E Pacific, W Atlantic (N & C America)	-	-	-	S7-1	-	Sq	-	-	-	-	+	-	-	-	McMahan et al. 2013
	E Pacific, W Atlantic (N & C America)	-	-	-	-	Sq	Sq	Sq	-	-	+	-	-	-	-	Durand et al. 2012a,b
	Caribbean Sea	15	-	-	-	-	-	-	-	-	-	-	-	-	+	Feldheim et al. 2009
Aldrichetta																
A. forsteri	S. Pacific (New Zealand, Australia)	-	-	-	-	Sq	Sq	Sq	-	-	+	-	-	-	-	Durand et al. 2012a,b
Cestraeus																
C. oxyrhyncus	S. Pacific (New Caledonia)	-	-	-	-	Sq	Sq	Sq	-	-	+	-	-	-	-	Durand et al. 2012a,b
C. goldiei	China Sea (Philippines)	-	-	-	-	Sq	Sq	Sq	-	-	+	-	-	-	-	Durand et al. 2012a,b
Chaenomugil																
C. proboscideus	E Pacific (Mexico)	-	-	-	-	Sq	Sq	Sq	-	-	+	-	-	-	-	Durand et al. 2012a,b
Chelon																
C. labrosus	E Atlantic, Mediterranean	-	-	-	-	Sq	Sq	Sq	-	-	+	-	-	-	-	Durand et al. 2012a,b
	Mediterranean	-	-	-	-	-	-	Sq	-	-	-	-	+	-	-	Erguden et al. 2010

Table 15.1. contd....

Table 15.1. contd.

Species[1]	Geographic area[2]	Markers									Main topic[5]					Reference
		nDNA[3]				mtDNA[4]										
		microsat	allozyme	AFLP	other	COI	Cyt b	16S	12S	Other	Phylogeny	Phylogeogr	Mol Ident	Pop Genet	Isolation	
Liza aurata	Mediterranean	-	-	-	-	Sq	Sq	-	Sq	-	+	-	-	-	-	Heras et al. 2009
	Mediterranean	-	-	-	-		Sq	Sq	-	-	+	-	-	-	-	Aurelle et al. 2008[7]
	Mediterranean	-	16	-	-	-	-	-	-	-	+	-	-	-	-	Blel et al. 2008
	Mediterranean	-	-	-	-	Sq		Sq	Sq		+	-	-	-	-	Papasotiropoulos et al. 2007
	Mediterranean	-	-	-	5S	-	-	-	-	-	+	-	-	-	-	Gornung et al. 2007
	Mediterranean	-	-	-	5S	-	-	R	R	R	+	-	-	-	-	Imsiridou et al. 2007
	Mediterranean	-	-	-	-	-	-	R	R	R (ND)	+	-	-	-	-	Semina et al. 2007a
	Mediterranean	-	11	-	-		-	-	-	-	+	-	-	-	-	Turan et al. 2005
	Mediterranean	-	35	-	-		-	Sq	-	-	+	-	-	-	-	Rossi et al. 2004
	Mediterranean	-	-	-	-	R	-	R	R	-	+	-	-	-	-	Papasotiropoulos et al. 2002
	Mediterranean	-	-	-	-	-	-	R	-	-	-	-	+	-	-	Klossa-Kilia et al. 2002
	Mediterranean	-	-	-	-	-	Sq	-	-	Sq (CR)	-	-	+	-	-	Murgia et al. 2002
	Mediterranean	-	-	-	-	-		-	-	-	+	-	-	-	-	Papasotiropoulos et al. 2001
Crenimugil																
C. crenilabis	Indian, S Pacific	-	-	-	-	Sq	Sq	Sq	-	-	+	-	-	-	-	Durand et al. 2012a,b
	unknown	-	-	-	-	-	Sq	Sq	-	-	+	-	-	-	-	Aurelle et al. 2008

Ellochelon																
E. vaigiensis	S Pacific	-	-	-	-	Sq	Sq	Sq	-	-	+	-	-	-	-	Durand et al. 2012a,b
	China Sea	-	-	-	-	-	-	Sq	Sq	-	+	-	-	-	-	Liu et al. 2010
Joturus																
J. pichardi	W Atlantic (Panama)	-	-	-	-	Sq	Sq	Sq	-	-	+	-	-	-	-	Durand et al. 2012a,b
Liza																
L. abu	Tigri River	-	-	-	-	Sq	Sq	Sq	-	-	+	-	-	-	-	Durand et al. 2012a,b
	Mediterranean	-	-	-	-	-	-	Sq	-	-		-	-	-	-	Erguden et al. 2010
	Mediterranean	-	-	-	-	-	-	-	-	-	+	-	-	-	-	Turan et al. 2005
L. affinis	China Sea	-	-	-	-	Sq	Sq	Sq	-	-	+	-	-	-	-	Durand et al. 2012a,b
L. alata	Indian, S Pacific	-	-	-	-	Sq	Sq	Sq	-	-	+	-	-	-	-	Durand et al. 2012a,b
L. argentea	S Pacific (Australia)	-	-	-	-	Sq	Sq	Sq	-	-	+	-	-	-	-	Durand et al. 2012a,b
L. aurata	Caspian Sea	-	-	-	-	-	-	-	-	Sq (CR)	-	-	-	+	-	Saeidi et al. 2014
	Caspian Sea	-	-	-	-	-	-	Sq	-	-	+	-	-	-	-	Nematzadeh et al. 2013a
	Caspian Sea	-	-	-	-	-	-	Sq	-	-	+	-	-	-	-	Nematzadeh et al. 2013b
	E Atlantic, Mediterranean	-	-	-	-	Sq	Sq	Sq	-	-	+	-	-	-	-	Durand et al. 2012a,b
	Mediterranean	-	-	-	-	-	-	Sq	-	-	-	-	+	-	-	Erguden et al. 2010
	Mediterranean	-	-	-	-	Sq	Sq		Sq		+	-	-	-	-	Heras et al. 2009
	Mediterranean	-	-	-	-	-	Sq	Sq	-	-	+	-	-	-	-	Aurelle et al. 2008

Table 15.1. contd....

Table 15.1. contd.

Species[1]	Geographic area[2]	Markers									Main topic[5]					Reference
		nDNA[3]				mtDNA[4]										
		microsat	allozyme	AFLP	other	COI	Cyt b	16S	12S	Other	Phylogeny	Phylogeogr	Mol Ident	Pop Genet	Isolation	
L. aurata	Mediterranean	-	16	-	-	-	-	-	-	-	+	-	-	-	-	Blel et al. 2008
	Mediterranean	-	-	-	-	-	Sq	-	-	-		-	+	-	-	Trabelsi et al. 2008
	Mediterranean	-	-	-	5S	-	-	-	-	-	+	-	-	-	-	Gornung et al. 2007
	Mediterranean	-	-	-	5S	-	-	-	-	-	+	-	-	-	-	Imsiridou et al. 2007
	Mediterranean	-	-	-	-	Sq	-	-	Sq	-	-	-	-	-	-	Papasotiropoulos et al. 2007
	Mediterranean	-	-	-	-	-	-	R	R	R (ND)	+	-	-	-	-	Semina et al. 2007a
	Sea of Azov	-	-	-	-	-	-	R	R	R (ND)	+	-	-	-	-	Semina et al. 2007b
	Mediterranean	-	11	-	-	-	-	-	-	-	+	-	-	-	-	Turan et al. 2005
	Mediterranean	-	35	-	-	-	-	Sq	-	-	+	-	-	-	-	Rossi et al. 2004
	Mediterranean	-	-	-	-	-	-	R	-	-	-	-	+	-	-	Klossa-Kilia et al. 2002
	Mediterranean	-	-	-	-	-	Sq	-	-	Sq (CR)	-	-	+	-	-	Murgia et al. 2002
	Mediterranean	-	-	-	-	R	-	-	R		+	-	-	-	-	Papasotiropoulos et al. 2002
	Mediterranean	-	22	-	-	-	-	-	-	-	+	-	-	-	-	Papasotiropoulos et al. 2001
	Mediterranean	-	-	-	-	-	-	Sq	-	-	+	-	-	-	-	Caldara et al. 1996

L. bandaliensis	E Atlantic (Africa)	-	-	-	-	Sq	Sq	Sq	-	-	+	-	-	-	-	Durand et al. 2012a,b
	E Atlantic (Africa)	-	-	-	-	-	-	-	Sq	-	-	-	+	-	-	Trape et al. 2009
L. carinata (*L. carinatus*)	China Sea	-	-	-	-	-	-	-	Sq	Sq	+	-	-	-	-	Liu et al. 2010
	Mediterranean	-	11	-	-	-	-	-	-	-	-	-	-	-	-	Turan et al. 2005
L. dumerili	Indian, E Atlantic (Africa)	-	-	-	-	Sq	Sq	Sq	-	-	+	-	-	-	-	Durand et al. 2012a,b
		-	-	-	-	-	-	-	Sq	-	-	-	+	-	-	Trape et al. 2009
L. falcipinnis	E Atlantic (Africa)	-	-	-	-	Sq	Sq	Sq	-	-	+	-	-	-	-	Durand et al. 2012a,b
L. grandisquamis	E Atlantic (Africa)	-	-	-	-	Sq	Sq	Sq	-	-	+	-	-	-	-	Durand et al. 2012a,b
	E Atlantic (Africa)	-	-	-	-	-	-	Sq	-	-	-	-	+	-	-	Trape et al. 2009
L. haematocheila (*L. haematocheilus*)	China Sea, Sea of Japan	-	-	-	-	-	-	-	-	Sq (CR)	-	+	-	+	-	Gao et al. 2014
(*L. haematocheilus*)	China Sea, Sea of Japan	-	-	5	-	-	-	-	-	-	-	+	-	-	-	Han et al. 2013
	China Sea	-	-	-	-	Sq	Sq	Sq	-	-	+	-	-	-	-	Durand et al. 2012a,b
(*M. soiuy*)	Black Sea	-	-	-	-	-	-	Sq	-	-	-	-	+	-	-	Erguden et al. 2010
	China Sea	-	-	-	-	-	-	Sq	Sq		+					Liu et al. 2010
(*M. soiuy*)	China Sea	10	-	-	-	-	-	-	-	-	-	-	-	-	+	Xu et al. 2009
(*C. haematocheilus*)	China Sea	-	14	-	-	-	-	-	-	-	-	-	-	+	-	Meng et al. 2007
(*C. haematocheilus*)	China Sea, Sea of Japan	-	-	-	-	-	-	-	-	Sq (CR)	-	+	-	+	-	Liu et al. 2007
(*L. haematocheilus*)	Sea of Azov, Sea of Japan	-	-	-	-	-	R	-	-	R (CR)	-	-	-	+	-	Salmenkova et al. 2007
(*L. haematocheilus*)	Sea of Japan	-	-	-	-	-	-	R	R	R (ND)	+	-	-	-	-	Semina et al. 2007a
	Sea of Japan	-	-	-	-	-	-	R	R	R (ND)	+	-	-	-	-	Semina et al. 2007b

Table 15.1. contd....

Table 15.1. contd.

Species[1]	Geographic area[2]	Markers									Main topic[5]					Reference
		nDNA[3]				mtDNA[4]										
		microsat	allozyme	AFLP	other	COI	Cyt b	16S	12S	Other	Phylogeny	Phylogeogr	Mol Ident	Pop Genet	Isolation	
(*M. soiuy*)	Black Sea	-	11	-	-	-	-	-	-	-	+	-	-	-	-	Turan et al. 2005
(*M. soiuy*)	Azov Sea, Sea of Japan	-	21	-	-	-	-	-	-	-	-	-	-	+	-	Omelchenko et al. 2004
L. lauvergnii (*L. affinis*)	China Sea (Taiwan)	-	19	-	-	-	-	-	-	-	+	-	-	-	-	Lee et al. 1995
L. macrolepis (*C. macrolepis*)	Indian, S. Pacific, China Sea, Oman Sea	-	-	-	-	Sq	Sq	Sq	-	-	+	-	-	-	-	Durand et al. 2012a,b
	Indian (E India)	-	-	-	-	-	-	Sq	Sq		+	-	-	-	-	Shekhar et al. 2011
	China Sea (Taiwan)	-	10	-	-	-	-	-	-	-	-	-	-	+	-	Lee et al. 1996
L. melinoptera (*C. melinopterus*)	S Pacific (Fiji)	-	-	-	-	Sq	Sq	Sq	-	-	+	-	-	-	-	Durand et al. 2012a,b
	China Sea (Taiwan)	-	19	-	-	-	-	-	-	-	+	-	-	-	-	Lee et al. 1995
L. melinoptera[8] (*P. parmatus*)	Indian (Java)	-	-	-	-	Sq	Sq	Sq	-	-	+	-	-	-	-	Durand et al. 2012a,b
L. parsia	Indian (E India)	-	-	-	-	-	-	Sq	Sq	-	+	-	-	-	-	Shekhar et al. 2011
L. ramada	Mediterranean	-	-	-	-	-	-	Sq	-	-	+	-	-	-	-	Nematzadeh et al. 2013a
(*M. capito*)	Oman Sea	-	-	-	-	-	-	Sq	-	-	+	-	-	-	-	Nematzadeh et al. 2013b
	E Atlantic, Mediterranean	-	-	-	-	Sq	Sq	Sq	-	-	+	-	-	-	-	Durand et al. 2012a,b
	Mediterranean	-	-	-	-	-	Sq	Sq	-	-	+	-	-	-	-	Aurelle et al. 2008
	Mediterranean	-	-	-	-	-	-	Sq	-	-	-	-	+	-	-	Erguden et al. 2010

	Mediterranean	-	16	-	-	-	-	-	-	-	+	-	-	-	-	Blel et al. 2008
	Mediterranean	-	-	-	-	Sq	Sq	-	Sq	-	+	-	-	-	-	Heras et al. 2009
	Mediterranean	-	-	-	5S	-	-	-	-	-	+	-	-	-	-	Gornung et al. 2007
	Mediterranean	-	-	-	5S	-	-	-	-	-	+	-	-	-	-	Imsiridou et al. 2007
	Mediterranean	-	-	-		Sq	-	Sq	Sq		+	-	-	-	-	Papasotiropoulos et al. 2007
	Mediterranean	-	-	-	-	-	-	R	R	R (ND)	+	-	-	-	-	Semina et al. 2007a
	Mediterranean	-	11	-	-	-	-	-	-	-	+	-	-	-	-	Turan et al. 2005
	Mediterranean	-	35	-	-	-	-	Sq	-	-	+	-	-	-	-	Rossi et al. 2004
	Mediterranean	-	24	15	-	-	-	-	-	-	-	-	-	+	-	Papa et al. 2003
	Mediterranean	-	-	-	-	-	-	R	-	-	-	-	+	-	-	Klossa-Kilia et al. 2002
	Mediterranean	-	-	-	-	-	Sq	-	-	Sq (CR)	-	-	+	-	-	Murgia et al. 2002
	Mediterranean	-	-	-	-	R	-	R	R	-	+	-	-	-	-	Papasotiropoulos et al. 2002
	Mediterranean	-	22	-	-	-	-	-	-	-	+	-	-	-	-	Papasotiropoulos et al. 2001
	Mediterranean	-	18	-	-	-	-	-	-	-	+	-	-	-	-	Rossi et al. 1998b
	Mediterranean	-	-	-	-	-	Sq	-	Sq	-	+	-	-	-	-	Caldara et al. 1996
L. richardsonii	E Atlantic, Indian	-	-	-	-	Sq	Sq	Sq	-	-	+	-	-	-	-	Durand et al. 2012a,b
L. saliens	Mediterranean	-	24	15	-	-	-	-	-	-	-	-	-	+	-	Papa et al. 2003
	Caspian Sea	-	-	-	-	-	-	Sq	-	-	+	-	-	-	-	Nematzadeh et al. 2013a

Table 15.1. contd....

Table 15.1. contd.

Species[1]	Geographic area[2]	Markers									Main topic[5]					Reference
		nDNA[3]				mtDNA[4]										
		microsat	allozyme	AFLP	other	COI	Cyt b	16S	12S	Other	Phylogeny	Phylogeogr	Mol Ident	Pop Genet	Isolation	
Liza saliens	Caspian Sea	-	-	-	-	-	-	Sq	-	-	+	-	-	-	-	Nematzadeh et al. 2013b
	Mediterranean	-	-	-	-	Sq	Sq	Sq	-	-	+	-	-	-	-	Durand et al. 2012a,b
	Mediterranean	-	-	-	-	-	-	Sq	-	-	-	-	+	-	-	Erguden et al. 2010
	E Atlantic (Morocco)	-	-	-	-	Sq	Sq	-	Sq	-	+	-	-	-	-	Heras et al. 2009
	Mediterranean	-	-	-	-	-	Sq	Sq	-	-	+	-	-	-	-	Aurelle et al. 2008
	Mediterranean	-	16	-	-	-	Sq	-	-	-	+	-	-	-	-	Blel et al. 2008
	Mediterranean	-	-	-	-	Sq	-	Sq	Sq	-	+	-	-	-	-	Trabelsi et al. 2008
	Mediterranean	-	-	-	5S	-	-	-	-	-	+	-	-	-	-	Gornung et al. 2007
	Mediterranean	-	11		5S	-	-	-	-	-	+	-	-	-	-	Imsiridou et al. 2007
	Mediterranean	-	-	-	-	-	-	-	-	-	+	-	-	-	-	Papasotiropoulos et al. 2007
	Mediterranean	-	-	-	-	-	-	R	R	R (ND)	+	-	-	-	-	Semina et al. 2007a
	Mediterranean	-	-	-	-	-	-	-	-	-	+	-	-	-	-	Turan et al. 2005
	Mediterranean	-	35	-	-	-	-	Sq	-	-	+	-	-	-	-	Rossi et al. 2004
	Mediterranean	-	-	-	-	-	-	R	-	-	-	-	+	-	-	Klossa-Kilia et al. 2002
	Mediterranean	-	-	-	-	R	-	R	R	-	+	-	-	-	-	Papasotiropoulos et al. 2002

	Mediterranean	-	22	-	-	-	-	-	-	-	+	-	-	-	-	Papasotiropoulos et al. 2001
	Mediterranean	-	-	-	-	-	Sq	-	Sq	-	+	-	-	-	-	Caldara et al. 1996
Liza sp.	Indian, China Sea	-	-	-	-	Sq	Sq	Sq	-	-	+	-	-	-	-	Durand et al. 2012a,b
L. subviridis	Persian Gulf	-	-	-	-	-	-	Sq	-	-	+	-	-	-	-	Nematzadeh et al. 2013a
	Indian, China Sea	-	-	-	-	Sq	Sq	Sq	-	-	+	-	-	-	-	Durand et al. 2012a,b
(*L. dussumieri*)	China Sea	-	-	-	-	-	-	Sq	Sq	-	+	-	-	-	-	Liu et al. 2010
L. tade	Indian, S Pacific	-	-	-	-	Sq	Sq	Sq	-	-	+	-	-	-	-	Durand et al. 2012a,b
	Indian (E India)	-	-	-	-	-	-	-	-	-	+	-	-	-	-	Shekhar et al. 2011
		-	-	-	-	-	-	-	-	-	+	-	-	-	-	Liu et al. 2010
L. tricuspidens	Indian (S Africa)	-	-	-	-	Sq	Sq	Sq	-	-	+	-	-	-	-	Durand et al. 2012a,b
Moolgarda																
M. buchanani (*V. buchanani*)	Indian, S Pacific	-	-	-	-	Sq	Sq	Sq	-	-	+	-	-	-	-	Durand et al. 2012a,b
M. cunnesius	Indian	-	-	-	-	Sq	Sq	Sq	-	-	+	-	-	-	-	Durand et al. 2012a,b
(*O. ophuyseni*)	China Sea	-	-	-	-	-	-	Sq	Sq	-	+	-	-	-	-	Liu et al. 2010
M. engeli	N & S Pacific	-	-	-	-	Sq	Sq	Sq	-	-	+	-	-	-	-	Durand et al. 2012a,b
M. perusii	China Sea, Indian, S Pacific	-	-	-	-	Sq	Sq	Sq	-	-	+	-	-	-	-	Durand et al. 2012a,b
M. robustus (*V. robustus*)	Indian	-	-	-	-	Sq	Sq	Sq	-	-	+	-	-	-	-	Durand et al. 2012a,b
M. seheli	China Sea, Indian, Pacific	-	-	-	-	Sq	Sq	Sq	-	-	+	-	-	-	-	Durand et al. 2012a,b

Table 15.1. contd....

Table 15.1. contd.

Species[1]	Geographic area[2]	Markers									Main topic[5]					Reference
		nDNA[3]				mtDNA[4]										
		microsat	allozyme	AFLP	other	COI	Cyt b	16S	12S	Other	Phylogeny	Phylogeogr	Mol Ident	Pop Genet	Isolation	
M. seheli (*V. formosae*)	China Sea (Taiwan)	-	19	-	-	-	-	-	-	-	+	-	-	-	-	Lee et al. 1995
Moolgarda sp. (*Valamugil* sp.)	China Sea, Indian, S Pacific	-	-	-	-	Sq	Sq	Sq	-	-	+	-	-	-	-	Durand et al. 2012a,b
Mugil																
M. bananensis	E Atlantic (Africa)	-	-	-	-	Sq	Sq	Sq	-	-	+	-	-	-	-	Durand et al. 2012a,b
	E Atlantic (Africa)	-	-	-	-	-	-	Sq	-	-	-	-	+	-	-	Trape et al. 2009
M. capurrii	E Atlantic (Africa)	-	-	-	-	Sq	Sq	Sq	-	-	+	-	-	-	-	Durand et al. 2012a,b
	E Atlantic (Africa)	-	-	-	-	-	-	Sq	-	-	-	-	+	-	-	Trape et al. 2009
M. cephalus complex	worldwide	-	-	-	-	-	-	-	-	Sq (CR)	-	+	-	-	-	Jamandre et al. 2014
	Mediterranean, Black Sea	6	-	-	Prl-1	-	Sq	-	-	-	-	-	-	+	-	Durand et al. 2013
	S Pacific (E Australia)	6	-	-	-	-	-	-	-	-	-	-	-	+	-	Huey et al. 2013
	S Pacific (E Australia)	-	-	-	SNPs	-	-	-	-	-	-	-	-	-	+	Krück et al. 2013
	Oman Sea	-	-	-	-	-	-	Sq	-	-	+	-	-	-	-	Nematzadeh et al. 2013a
	Oman Sea	-	-	-	-	-	-	Sq	-	-	+	-	-	-	-	Nematzadeh et al. 2013b
	worldwide	-	-	-	-	Sq	Sq	Sq	-	-	+	-	-	-	-	Durand et al. 2012a,b
	China Sea	-	-	-	-	Sq	-	-	-	-	-	-	-	+	-	Sun et al. 2012
	Mediterranean	-	-	-	RAPD	-	-	-	-	-	-	-	+	-	-	El-Zaeem 2011
	worldwide	-	-	-	-	-	Sq	-	-	-	-	+	-	-	-	Livi et al. 2011

	Pacific (S America)	-	-	-	5S	-	-	-	-	-	-	-	+	-	-	Rodrigues-Filho et al. 2011
	Indian (E India)	-	-	-	-	-	-	Sq	Sq	-	+	-	-	-	-	Shekhar et al. 2011
	China Sea, Sea of Japan	10	-	-	-	Sq	Sq	-	-	-	+	+	-	-	-	Shen et al. 2011
	Mediterranean	6	-	-	Prl-1	-	-	-	-	-	-	-	-	+	-	Blel et al. 2010
	Mediterranean	-	-	-	-	-	-	Sq	-	-	-	-	+	-	-	Erguden et al. 2010
	China Sea	-	-	-	-	-	-	Sq	Sq	-	+	-	-	-	-	Liu et al. 2010
	China Sea	12	-	-	-	-	-	-	-	-	-	-	-	-	+	Xu et al. 2010
	Mediterranean, W Atlantic (N & S America)	-	-	-	-	Sq	Sq	-	Sq	-	+	-	-	-	-	Heras et al. 2009
	China Sea	-	-	-	-	-	-	-	-	Sq (CR)	-	+	-	-	-	Jamandre et al. 2009
	China Sea	-	-	-	-	-	Sq	-	-	-	-	+	-	-	-	Ke et al. 2009
	China Sea	-	-	-	-	-	-	-	-	Sq (CR)	-	-	-	+	-	Liu et al. 2009a
	China Sea	-	-	5	-	-	-	-	-	-	-	-	-	+	-	Liu et al. 2009b
	E Atlantic (Africa)	-	-	-	-	-	-	Sq	-	-	-	-	+	-	-	Trape et al. 2009
	Indian, E Pacific, W Atlantic, Mediterranean	-	-	-	-	-	Sq	Sq	-	-	+	-	-	-	-	Aurelle et al. 2008
	Mediterranean	-	16	-	-	-	-	-	-	-	+	-	-	-	-	Blel et al. 2008
	W Atlantic (S America)	-	-	-	-	-	Sq	-	-	-	-	+	+	-	-	González-Castro et al. 2008
	Mediterranean	-	-	-	-	-	Sq	-	-	-	-	-	+	-	-	Trabelsi et al. 2008
	W Atlantic (S America)	-	-	-	-	-	Sq	Sq	-	-	+	-	-	-	-	Fraga et al. 2007
	Mediterranean	-	-	-	5S	-	-	-	-	-	+	-	-	-	-	Gornung et al. 2007

Table 15.1. contd....

Table 15.1. contd.

Species[1]	Geographic area[2]	Markers									Main topic[5]					Reference
		nDNA[3]				mtDNA[4]										
		microsat	allozyme	AFLP	other	COI	Cyt b	16S	12S	Other	Phylogeny	Phylogeogr	Mol Ident	Pop Genet	Isolation	
M. cephalus complex	Mediterranean	-	-	-	5S	-	-	-	-	-	+	-	-	-	-	Imsiridou et al. 2007
	Mediterranean	-	-	-	-	Sq	-	Sq	Sq	-	+	-	-	-	-	Papasotiropoulos et al. 2007
	E Pacific (Gulf of California)	-	-	-	-	Sq	-	Sq	-	-	-	-	+	-	-	Peregrino-Uriarte et al. 2007
	Mediterranean, Sea of Japan	-	-	-	-	-	-	R	R	R (ND)	+	-	-	-	-	Semina et al. 2007a
	Sea of Azov, Sea of Japan	-	-	-	-	-	-	R	R	R (ND)	+	-	-	-	-	Semina et al. 2007b
	Mediterranean	-	11	-	-	-	-	-	-	-	+	-	-	-	-	Turan et al. 2005
	Mediterranean	-	35	-	-	-	-	Sq	-	-	+	-	-	-	-	Rossi et al. 2004
	Mediterranean, S Pacific (E Australia)	11	-	-	-	-	-	-	-	-	-	-	-	-	+	Miggiano et al. 2005
	Mediterranean	-	-	-	-	-	-	R	-	-	-	-	+	-	-	Klossa-Kilia et al. 2002
	Mediterranean	-	-	-	-	-	Sq	-	-	Sq (CR)	-	-	+	-	-	Murgia et al. 2002
	Mediterranean	-	-	-	-	R	-	R	R	-	+	-	-	-	-	Papasotiropoulos et al. 2002
	China Sea	-	2	-	-	-	-	-	-	-	-	-	+	-	-	Huang et al. 2001
	Mediterranean	-	22	-	-	-	-	-	-	-	+	-	-	-	-	Papasotiropoulos et al. 2001
	W Atlantic and E Pacific (USA)	-	-	-	-	-	-	-	-	Sq (CR)	-	+	-	-	-	Rocha-Olivares et al. 2000
	worldwide	-	27	-	-	-	-	-	-	-	-	-	-	+	-	Rossi et al. 1998a

	worldwide	-	18	-	-	-	-	-	-	-	+	-	-	-	-	Rossi et al. 1998b
	Mediterranean, W Atlantic (USA)	-	-	-	-	-	Sq	-	Sq	-	+	-	-	-	-	Caldara et al. 1996
	China Sea (Taiwan)	-	19	-	-	-	-	-	-	-	+	-	-	-	-	Lee et al. 1995
	worldwide	-	-	-	-	-	-	-	-	R (tot)	-	-	-	+	-	Crosetti et al. 1994
M. curema complex	Caribbean Sea, W Atlantic (S America)	-	-	-	-	Sq	Sq	Sq	Sq	Sq (ATPs)	-	-	+	-	-	Siccha-Ramirez et al. 2014
	Atlantic, E Pacific, Caribbean Sea	-	-	-	-	Sq	Sq	Sq	-	-	+	-	-	-	-	Durand et al. 2012a,b
	W Atlantic (S America)	-	-	-	5S	-	-	-	-	-	-	-	+	-	-	Rodrigues-Filho et al. 2011
	E Atlantic (Africa)	-	-	-	-	-	-	Sq	-	-	-	-	+	-	-	Trape et al. 2009
	W Atlantic, E Pacific (N & S America)	-	-	-	-	-	Sq	Sq	-	-	+	-	-	-	-	Aurelle et al. 2008
	W Atlantic (S America)	-	20	-	-	-	Sq	Sq	-	-	+	-	-	-	-	Fraga et al. 2007
	Caribbean Sea	-	-	-	-	-	-	-	-	-	-	-	+	-	-	Nirchio et al. 2007
	E Pacific (Gulf of California)	-	-	-	-	Sq	-	Sq	-	-	-	-	+	-	-	Peregrino-Uriarte 2007
	W Atlantic (N & S America)	-	-	-	-	Sq	Sq	-	Sq	-	+	-	+	-	-	Heras et al. 2006
	W Atlantic (USA)	-	18	-	-	-	-	-	-	-	+	-	-	-	-	Rossi et al. 1998b
	W Atlantic (USA)	-	-	-	-	-	Sq	-	Sq	-	+	-	-	-	-	Caldara et al. 1996
M. hospes	W Atlantic (S America)	-	-	-	-	Sq	Sq	Sq	Sq	Sq (ATPs)	-	-	+	-	-	Siccha-Ramirez et al. 2014
	Caribbean Sea	-	-	-	-	Sq	Sq	Sq	-	-	+	-	-	-	-	Durand et al. 2012a,b
	W Atlantic (S America)	-	-	-	5S	-	-	-	-	-	-	-	+	-	-	Rodrigues-Filho et al. 2011

Table 15.1. contd....

Table 15.1. contd.

Species[1]	Geographic area[2]	Markers									Main topic[5]					Reference
		nDNA[3]				mtDNA[4]										
		microsat	allozyme	AFLP	other	COI	Cyt b	16S	12S	Other	Phylogeny	Phylogeogr	Mol Ident	Pop Genet	Isolation	
M. hospes	W Atlantic (Brazil)	-	-	-	-	-	Sq	Sq	-	-	+	-	-	-	-	Aurelle et al. 2008
	W Atlantic (S America)	-	-	-	-	-	Sq	Sq	-	-	+	-	-	-	-	Fraga et al. 2007
M. incilis	Caribbean Sea	-	-	-	-	Sq	Sq	Sq	Sq	Sq (ATPs)	-	-	+	-	-	Siccha-Ramirez et al. 2014
	Caribbean Sea	-	-	-	-	Sq	Sq	Sq	-	-	+	-	-	-	-	Durand et al. 2012a,b
	W Atlantic (S America)	-	-	-	5S	-	-	-	-	-	-	-	+	-	-	Rodrigues-Filho et al. 2011
	W Atlantic (Brazil)	-	-	-	-	-	Sq	Sq	-	-	+	-	-	-	-	Aurelle et al. 2008
	W Atlantic (S America)	-	-	-	-	-	Sq	Sq	-	-	+	-	-	-	-	Fraga et al. 2007
M. liza	Caribbean Sea and W Atlantic (S America)	-	-	-	-	Sq	Sq	Sq	Sq	Sq (ATPs)	-	+	+	-	-	Siccha-Ramirez et al. 2014
	W Atlantic, Caribbean Sea	-	-	-	-	Sq	Sq	Sq	-	-	+	-	-	-	-	Durand et al. 2012a,b
	W Atlantic (S America)	-	-	-	-	Sq	Sq	-	Sq	-	+	-	-	-	-	Heras et al. 2009[9]
(*M. platanus*)	Caribbean Sea	-	-	-	-	Sq	Sq	-	Sq	-	+	-	-	-	-	Heras et al. 2009
	W Atlantic (S America)	-	-	-	-	-	Sq	Sq	-	-	+	-	-	-	-	Aurelle et al. 2008
(*M. platanus*)	W Atlantic (S America)	-	-	-	-	-	Sq	Sq	-	-	+	-	-	-	-	Aurelle et al. 2008
(*M. platanus*)	W Atlantic (S America)	-	-	-	-	-	Sq	-	-	-	+	-	+	-		González-Castro et al. 2008
	W Atlantic (S America)	-	-	-	-	-	Sq	Sq	-	-	+	-	-	-	-	Fraga et al. 2007
(*M. platanus*)	W Atlantic (S America)	-	-	-	-	-	Sq	Sq	-	-	+	-	-	-	-	Fraga et al. 2007[9]

M. liza	W Atlantic (S America)	-	-	-	5S	-	-	-	-	-	-	-	+	-	-	Rodrigues-Filho et al. 2011
(*M. platanus*)	W Atlantic (S America)	-	-	-	5S	-	-	-	-	-	-	-	+	-	-	Rodrigues-Filho et al. 2011[9]
M. rubrioculus	Caribbean Sea, W Atlantic (S America)	-	-	-	-	Sq	Sq	Sq	Sq	Sq (ATPs)	-	-	+	-	-	Siccha-Ramirez et al. 2014
	Caribbean Sea, E Pacific	-	-	-	-	Sq	Sq	Sq	-	-	+	-	-	-	-	Durand et al. 2012a,b
	Caribbean Sea	-	20	-	-	-	-	-	-	-	-	-	+	-	-	Nirchio et al. 2007
(*M. gaimardianus*)	W Atlantic (S America)	-	-	-	-	-	Sq	Sq	-		+	-	-	-	-	Fraga et al. 2007
M. trichodon	Caribbean Sea	-	-	-	-	Sq	Sq	Sq	Sq	Sq (ATPs)	-	-	+	-		Siccha-Ramirez et al. 2014
	Caribbean Sea	-	-	-	-	Sq	Sq	Sq	-	-	+	-	-	-	-	Durand et al. 2012a,b
	W Atlantic (USA)	-	18	-	-	-	-	-	-	-	+	-	-	-	-	Rossi et al. 1998b
M. thoburni	E Pacific	-		-	-	Sq	Sq	Sq	-	-	+	-	-	-	-	Durand et al. 2012a
	E Pacific (Galapagos)	-	18	-	-	-	-	-	-	-	+	-	-	-	-	Rossi et al. 1998b
Myxus																
M. elongatus	S Pacific	-	-	-	-	Sq	Sq	Sq	-	-	+	-	-	-	-	Durand et al. 2012a,b
Neochelon																
N. falcipinnis	E Atlantic (Africa)	-	-	-	-	-	-	Sq	-	-	-	-	+	-	-	Trape et al. 2009
Neomyxus																
N. leuciscus	Pacific	-	-	-	-	Sq	Sq	Sq	-	-	+	-	-	-	-	Durand et al. 2012a,b

Table 15.1. contd....

Table 15.1. contd.

Species[1]	Geographic area[2]	Markers									Main topic[5]					Reference
		nDNA[3]				mtDNA[4]										
		microsat	allozyme	AFLP	other	COI	Cyt b	16S	12S	Other	Phylogeny	Phylogeogr	Mol Ident	Pop Genet	Isolation	
Oedalechilus																
O. labeo	Mediterranean	-	-	-	-	Sq	Sq	Sq	-	-	+	-	-	-	-	Durand et al. 2012a,b
	Mediterranean	-	-	-	-	-		Sq	-	-		-	+	-	-	Erguden et al. 2010
	Mediterranean	-	-	-	-	-	Sq	Sq	-	-	+	-	-	-	-	Aurelle et al. 2008
	Mediterranean	-	-	-	5S	-	-	-	-	-	+	-	-	-	-	Gornung et al. 2007
	Mediterranean	-	11	-	-	-	-	-	-	-	+	-	-	-	-	Turan et al. 2005
	Mediterranean	-	35	-	-	-	-	Sq	-	-	+	-	-	-	-	Rossi et al. 2004
	Mediterranean	-	-	-	-	-	Sq	-	Sq	-	+	-	-	-	-	Caldara et al. 1996
O. labiosus	China Sea	-	-	-	-	Sq	Sq	Sq	-	-	+	-	-	-	-	Durand et al. 2012a,b
Osteomugil																
Unknown[10] (*O. strongylo-cephalus*)	China Sea	-	-	-	-	-	-	Sq	Sq	-	+	-	-	-	-	Liu et al. 2010
Pseudomixus																
P. capensis (*M. capensis*)	Indian (S Africa)	-	-	-	-	Sq	Sq	Sq	-	-	+	-	-	-	-	Durand et al. 2012a,b
Rhinomugil																
R. corsula	Indian	-	-	-	-	Sq	Sq	Sq	-	-	+	-	-	-	-	Durand et al. 2012a,b

R. nasutus	S Pacific	-	-	-	-	Sq	Sq	Sq	-	-	+	-	-	-	-	Durand et al. 2012a,b
Sicamugil																
S. cascasia	Indian	-	-	-	-	Sq	Sq	Sq	-	-	+	-	-	-	-	Durand et al. 2012a,b
S. hamiltonii	Indian	-	-	-	-	Sq	Sq	Sq	-	-	+	-	-	-	-	Durand et al. 2012a,b
Trachystoma																
T. petardi	S Pacific	-	-	-	-	Sq	Sq	Sq	-	-	+	-	-	-	-	Durand et al. 2012a,b

[1] Valid name according to Eschmeyer (2015), when a different synonym was used by the authors it is reported below, in parenthesis.
[2] E = East, W = West, C = Central, S = South, N = North.
[3] nDNA = nuclear DNA, Microsat = number of microsatellite loci, Allozyme = number of allozyme loci, AFLP = number of primers combination used in Amplified Fragment Length Polymorphism, Other = other molecular markers (SNP = single nuclear polymorphism, sequences of Prl-1 = prolactin 1 first intron, S7 = ribosomal S7 first intron, 5S = 5S rDNA).
[4] mtDNA = mitochondrial DNA, COI = cytochrome oxidase subunit I, Cyt b = cytochrome b, 16 = 16S rDNA, 12S = 12S rDNA, Other = other molecular markers (ATPs = ATP 6/8 synthase, CR = control region, ND = nicotinamide dinucleotide dehydrogenase subunits 3/4L/4/5/6, total = total mtDNA). Sq = sequences, R = Restriction Fragment Length Polymorphism
[5] Phylogeny = Molecular phylogeny, Phylogeogr = Phylogeography, Mol Ident = Molecular Identification, Pop Genet = Population genetics, Isolartion = loci Isolation.
[6] The same species were analysed in the two papers (except for *M. thoburni*). The species names are reported as listed in Durand et al. 2012b (Table 1), without considering the changes of nomenclature suggested in Durand et al. 2012a.
[7] The study is based exclusively on sequences retrieved from GeneBank.
[8] Samples reported as *Paramugil parmatus* by Durand et al. 2012b was a misidentification with *L. melinoptera* as explained in Durand and Borsa 2015.
[9] The authors analysed samples identified as *M. liza* and other as *M. platanus*, that is now considered a synonymy of the former.
[10] *Osteomugil strongylocephalus* is not included in Eschmeyer (2015) catalogue (accessed 15/01/2015).

Table 15.2. Genetic studies on several mullet species (1994–2014).

Species[1]	**Synonym used by the authors**	**Reference**
M. liza[2] *(M. cephalus, M. curema, M. hospes M. incilis, M. rubrioculus)*		Siccha-Ramirez et al. 2014
L. aurata, L. ramada, L. saliens, L. subviridis, M. cephalus		Nematzadeh et al. 2013a
L. aurata, L. saliens, L. subviridis, L. ramada, M. cephalus, Mo. malabarica***	**M. capito, ** V. buchanani*	Nematzadeh et al. 2013b
53 species (20 genera)		Durand et al. 2012a
53 species (20 genera)		Durand et al. 2012b
*M. cephalus, M. curema, M. hospes, M. incilis, M. liza, M. liza**	**M. platanus*	Rodrigues-Filho et al. 2011[3]
L. macrolepis, L. parsia, L. tade, M. cephalus		Shekhar et al. 2011
L. ramada, M. cephalus		El-Zaeem 2011
C. labrosus, L. abu, L. aurata, L. haematocheila, L. ramada, L. saliens, M. cephalus, O. labeo*	**M. soiuy*	Erguden et al. 2010
*E. vaigiensis, L. carinata, L. subviridis *, L. haematocheila, L. tade, V. buchanani, M. cephalus, Mo. cunnesius***, Not found***[4]	**L. dussumieri, ** Os. ophuyseni*, ***Os. strongylocephalus*	Liu et al. 2010
*C. labrosus, L. aurata, L. ramada, L. saliens, M. cephalus, M. curema, M. liza, M. liza**	** M. platanus*	Heras et al. 2009[3]
L. bandialensis, L. dumerili, L. falcipinnis, L. grandisquamis, M. bananensis, M. capurrii, M. cephalus, M. curema		Trape et al. 2009
12 species (5 genera)		Aurelle et al. 2008[5]
C. labrosus, L. aurata, L. ramada, L. saliens, M. cephalus		Blel et al. 2008
*M. cephalus, M. liza**	**M. platanus*	González-Castro et al. 2008
M. cephalus, L. aurata, L. saliens		Trabelsi et al. 2008
M. cephalus, M. curema, M. hospes, M. incilis, M. liza, M. liza, M. rubrioculus***	**M. platanus, **M.* sp. *(M. gaimardianus)*	Fraga et al. 2007[3]
C. labrosus, L. aurata, L. ramada, L. saliens, M. cephalus, O. labeo		Gornung et al. 2007
C. labrosus, L. aurata, L. ramada, L. saliens, M. cephalus		Imsiridou et al. 2007
C. labrosus, L. aurata, L. ramada, L. saliens, M. cephalus		Papasotiropoulos et al. 2007
M. cephalus, M. curema		Peregrino-Uriarte et al. 2007
C. labrosus, L. aurata, L. haematocheila, L. ramada, L. saliens, M. cephalus*	**L. haematocheilus*	Semina et al. 2007a
L. aurata, L. haematocheila, M. cephalus*	**L. haematocheilus*	Semina et al. 2007b
C. labrosus, L. abu, L. aurata, L. carinata, L. haematocheila, L. ramada, L. saliens, M. cephalus, O. labeo*	**M. soiuy*	Turan et al. 2005
C. labrosus, L. aurata, L. ramada, L. saliens, M. cephalus, O. labeo		Rossi et al. 2004

L. ramada, L. saliens		Papa et al. 2003
C. labrosus, L. aurata, L. ramada, L. saliens, M. cephalus		Klossa-Kilia et al. 2002
C. labrosus, L. aurata, L. ramada, M. cephalus		Murgia et al. 2002
C. labrosus, L. aurata, L. ramada, L. saliens, M. cephalus		Papasotiropoulos et al. 2002
C. labrosus, L. aurata, L. ramada, L. saliens, M. cephalus		Papasotiropoulos et al. 2001
L. ramada, M.cephalus, M. curema, M. thoburni, M. trichodon***	**X. thoburni, **M. gyrans*	Rossi et al. 1998b
C. labrosus, L. aurata, L. ramada, L. saliens, M. cephalus, M. curema, O. labeo		Caldara et al.1996
*L. affinis, L. macrolepis, M. cephalus, Mo. seheli**	**V. formosae*	Lee et al. 1995

[1] valid name according to Eschmeyer (2015); the number of species/genera analysed is reported whenever it is >10. *A.* = *Agonostomus*; *C.* = *Chelon*; *E.* = *Ellochelon*; *L.* = *Liza*; *M.* = *Mugil*; *Mo.* = *Moolgarda*; *O.* = *Oedalechilus*; *Os.* = *Osteomugil*; *V.* = *Valamugil*; *X.* = *Xenomugil.*

[2] The paper deals with the phylogeography of *M. liza*, and the molecular identification of the *Mugil* species.

[3] The authors analysed samples identified as *M. liza* and other as *M. platanus*, now considered a synonym of the former.

[4] *Osteomugil strongylocephalus* is not included in Eschmeyer (2015) catalogue (accessed 15/01/2015).

[5] The study is based exclusively on sequences retrieved from GeneBank.

mutation rates, thus providing information at different levels (historical and present) of species dispersal and of demographic events (Sala-Bozano et al. 2009). However, only a few studies used multiple sets of genetic markers, i.e., nuclear and mitochondrial, on the same species (e.g., Rossi et al. 2004, Shen et al. 2011, Durand et al. 2013). The second consideration is on the thin border line which separates classical population genetic investigations from molecular taxonomy studies. Indeed, although some studies aim at identifying population structure, most of the genetic analyses of mullets are focused on solving taxonomic controversies or analyzing the phylogeographic distribution of variability that arises from population structure data. Finally, most of the genetic studies are focused on a few species, particularly on the flathead grey mullet *M. cephalus*, a cosmopolitan species living in the temperate and subtropical regions of the world, at present considered by many authors as a species complex, in which each lineage still requires to be assigned to a specific name (Durand et al. 2012a,b, see Chapter 2—Durand 2015). Genetic data on mountain mullet *Agonostomus monticola*, golden grey mullet *Liza aurata*, redlip mullet *L. haematocheila, M. cephalus* (including lebranche mullet *M. liza*), white mullet *M. curema*, that have been selected as examples of different geographic areas, are discussed hereafter.

Mountain mullet *Agonostomus monticola*

The mountain mullet *Agonostomus monticola* is a diadromous fish that typically occurs in tropical rivers, though scarce information is still available on its biology and ecology. It is distributed along the Atlantic slope in North, Central and South America from North Carolina to Venezuela, and along the Pacific slope from Baja California to Colombia and Galapagos Islands (Thomson 1997, Berra 2001, Miller et al. 2005 in McMahan et al. 2013), following the model of transisthmian geminate species pairs, separated by the Isthmus of Panama (McMahan et al. 2013).

A. monticola is one of the three species of the genus *Agonostomus,* together with the fairy mullet *Agonostomus telfairii* and the Comoro mullet *Agonostomus catalai*, both from the coast of Comoros, Madagascar, Reunion Island and Mauritius. The genus has long been neglected in systematic and biogeographical studies (McMahan et al. 2013) and only recently, on molecular evidence, Durand et al. 2012a suggested splitting the genus *Agonostomus* in two, resurrecting the genus *Dajaus* for the American species (*D. monticola*), and leaving the genus *Agonostomus* for the two Indian Ocean species, *A. telfairii* and *A. catalai* (see Chapter 2—Durand 2015 for details).

A. monticola is a controversial taxon, synonymized by Thomson (1997) to other 20 taxa, though these synonymies could partially be attributed to variations of several diagnostic morphological characters that vary as size increases (Thomson 1997). Considering the allopatric distribution of the species populations that occur in both the Pacific and the Atlantic Oceans, *A. monticola* was suggested to be a species complex (Harrison 2002, Miller et al. 2005 in McMahan et al. 2013). Indeed recent molecular studies (Durand et al. 2012a) on three mitochondrial loci (16S rRNA, cytochrome oxidase I, cytochrome b) showed that all haplotypes of *Agonostomus monticola* from four Atlantic and five Pacific sampling sites were grouped into a single clade consisting of three distinct lineages, two of which are sympatric in the Pacific off western Central America, and one from the Atlantic; the Atlantic lineage is the sister lineage of one of the Pacific lineages (Fig. 3 in Durand et al. 2012a).

These data were confirmed by a wider phylogeographic investigation on specimens collected at 40 sampling sites throughout the species range, carried out to identify cryptic diversity and biogeographical patterns (McMahan et al. 2013). Sequences analysis of the cytochrome b gene combined with the first intron of the nuclear ribosomal S7 gene provided a well resolved phylogeography in which four different lineages (three of which correspond to those reported by Durand et al. 2012a) could be identified: Caribbean, Gulf of Mexico, Pacific-A and Pacific-B (two populations, partially sympatric, in the Pacific Ocean) (Fig. 15.1). The four clades exhibit different levels of variability and do not share any haplotypes indicating the lack of gene flow among lineages. Divergence time estimates indicate early to mid-Miocene divergences for all four *A. monticola* clades, with Oligocene to Miocene divergences of internal nodes. The biogeographic structure of the *A. monticola* lineages is congruent with the main geological events that occurred in the area, such as the movement of the Chortis block and the formation of the Isthmus of Panama.

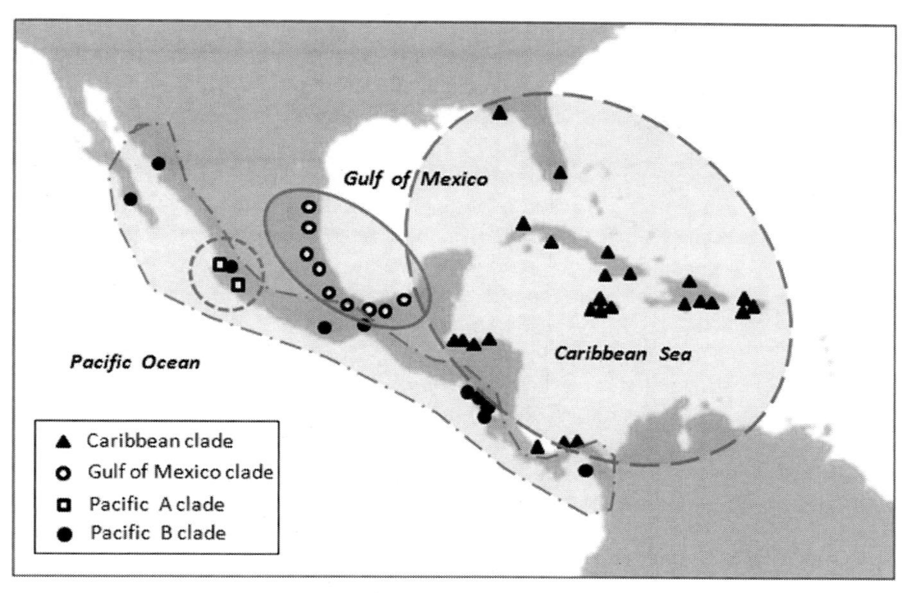

Figure 15.1. *Agonostomus monticola* lineages identified using cyt b and S7-1 sequence data (modified from McMahan et al. 2013).

Further investigations on morphological differences among the four geographical lineages of *A. monticola* are being carried out, as the high degree of genetic divergence, in combination with their geographical and genetic isolation, strongly suggest that these distinct lineages may represent distinct cryptic species within a species complex (McMahan et al. 2013).

Golden grey mullet *Liza aurata*

The golden grey mullet *Liza aurata* is distributed in the eastern Atlantic, from Norway and Scotland to Morocco and Cape Verde and in the Mediterranean and Black Seas. The species was used in many genetic analyses on Mediterranean mullets with different molecular markers (see Table 15.1), and most of these studies were aimed at establishing phylogenetic relationships among the Mediterranean Mugilidae (see Chapter 2—Durand 2015 and Chapter 7—Turan 2015). Following molecular studies, Mediterranean *Liza* spp. were found to cluster with the genus *Chelon*, in which they were synonymized by Durand et al. (2012a).

The species was introduced into the Caspian Sea in 1930–34, with three million individuals of *L. aurata*, *L. saliens* and *M. cephalus* from the Black Sea. The two *Liza* spp. adapted to local conditions and have become the target of commercial fisheries since 1942 (Saeidi et al. 2014). They migrate seasonally from the northern and central Caspian Sea to the central and southern areas to spawn (Tereshchenko 1950, Khoroshko 1981). Following a serious decline in catches in the last decade, that halved production, two population genetic studies, using microsatellites (Ghodsi et al. 2011) and CR (D-loop) mtDNA sequences (Saeidi et al. 2014), were undertaken on populations from the South Caspian Sea, to detect information on mullet stocks for sustainable management of capture fisheries. Discordant results were obtained between the two studies. The two populations (Gilan and Golestan) examined by Saeidi et al. (2014) exhibited high haplotype and nucleotide diversities and, despite the relative short distance between the two sampling sites (about 300 km), low gene flow and high genetic differentiation (Fst calculated between them 0.499). These results are in contrast with the results of Ghodsi et al. (2011), who could not discriminate among the populations he studied that were sampled over a smaller area.

Further genetic studies in different areas of the Caspian Sea, which is about 1200 km long, could clarify the genetic structure of *L. aurata* as introduced species in this region.

Redlip mullet *Liza haematocheila*

The redlip mullet *Liza haematocheila* is naturally distributed in the North West Pacific, from Japan through the Korean peninsula to the South China Sea, and occupies shallow coastal waters and rivers. The species has many synonyms which are still being used; in particular *Mugil soiuy* is cited in several papers. In the FAO fisheries and aquaculture production statistics FAO (2014), the species is reported as two different taxa: *L. haematocheilus* and *M. soiuy*, creating confusion for the reader. There is still a controversy on whether this species should be named *L. haematocheila* or *L. haematocheilus* (for details see Minos et al. 2010); the former is adopted here according to the Eschmeyer (2015) species catalogue. Durand et al. 2012a,b (and Chapter 2) suggest inserting this species in the resurrected genus *Planiliza*, which includes *Chelon* spp. and *Liza* spp. from the Indo-Pacific region.

The species was analyzed genetically in the North West Pacific and showed a pronounced genetic structure in its supposed distribution area. Genetic divergence due to geographic isolation was observed between four native populations from the South China Sea and in other sea areas of China analyzed by allozymes (Meng et al. 2007). CR sequence analysis of specimens from nine (Liu et al. 2007) and five (Gao et al. 2014) sampling sites identified three main lineages, showing differences in geographic distribution: lineage A dominates in the East China Sea, lineage B dominates in the Sea of Japan, and lineage C in the South China Sea populations; AMOVA allowed the identification of a pronounced genetic structure among the three groups. The three lineages have diverged since the Middle-Late Pleistocene, probably reflecting isolation of the three marginal seas in the North-western Pacific. This evidence allows us to infer the geographic origin of each lineage and suggests limited gene flow, far more restricted spatially than predicted by the potential dispersal capabilities of the species (Gao et al. 2014). AFLP analysis of redlip mullet collected from eight different sites (Han et al. 2013) yield results congruent with previous data on nuclear loci, but not with those obtained by mtDNA analysis: indeed AFLP identified only two population clusters, one in the north and one in the south of the distribution range. Because of the maternal inheritance of the mtDNA control region, Han et al. (2013) asserted that it is not possible to determine whether the presence of the two divergent mitochondrial lineages identified in the same sample by Liu et al. (2007) is the result of secondary contact after an extended period of isolation and/or depends on the real presence of two cryptic species. Significant genetic divergence and reduced gene flow were observed in pair-wise comparison involving all samples. Further studies on a higher number of samples, at additional localities and with different genetic markers are required to clarify the genetic structure of *L. haematocheila* in the North West Pacific.

The redlip mullet was introduced into the Azov Sea and Black Sea brackish ponds and coastal lagoons from the late 1960s to 1980 to counterbalance the decline in capture fisheries in the area (Starushenko and Kazansky 1996), and is a eurythermal and euryhaline species. *L. haematocheila* started to reproduce in the wild in the late 1980s and spread to the Sea of Marmara and the Dardanelles, reaching the Mediterranean, and the northern Aegean Sea, where it is found sporadically (Minos et al. 2010). At present, *L. haematocheila* is the most abundant mugilid species in the Black Sea, having displaced the three local mullets, *M. cephalus, L. aurata* and *L. saliens,* which are less adaptable to extreme water temperature and salinities (Starushenko and Kazansky 1996). In the new environment, the species has physiological differences compared to the fish from the Sea of Japan, its area of origin, and reveal an invader effect, with a typical population outbreak: it grows to larger size at a higher growth rate, maturity appears one year earlier and vitellogenic oocytes and eggs differ in size (Minos et al. 2010).

Several genetic analyses were performed to compare the original population in the Sea of Japan and the translocated population of the Azov Sea and the Black Sea. A comparison of two wild and four introduced populations based on allozyme analyses detected highly significant genetic differentiation between the native fish from Primorkyi (Russia, Sea of Japan) and the fish introduced into the Sea of Azov (Omelchenko et al. 2004). These results were later confirmed by cyt b and D-loop RFLP analysis: only five haplotypes were found to be common to both populations, while 13 were unique (19 in the Sea of Japan and three in the Sea of Azov) (Salmenkova et al. 2007). The Sea of Azov population showed a reduced number of polymorphic loci and of the mean number of alleles per loci, which could be explained by a severe bottleneck caused by the transplantation into the area of a limited number of specimens, from which present populations originate, and a consequent reduced effective population size, N_e (Omelchenko et al. 2004). This low genetic variability was confirmed by a low mtDNA haplotype diversity in the Sea of Azov (Salmenkova et al. 2007). No differences were detected between the two sampling sites in the Sea of Japan, whereas a genetic divergence was found among the Sea of Azov populations, probably associated with areas of different salinities (Omelchenko et al. 2004).

Flathead grey mullet *Mugil cephalus*

The flathead grey mullet *Mugil cephalus* has long been considered a cosmopolitan species with a discontinuous distribution. Its worldwide distribution in temperate, subtropical and tropical regions of the world, associated with its coastal ecology, makes it a very interesting case study from a phylogeographic point of view, giving rise to crucial questions on whether the genetic connectivity among geographically distant populations is maintained and how and when such a distribution occurred. However, the taxonomic status of *M. cephalus*, whether a valid species or a complex of species, has long been debated (Crosetti et al. 1994, Rossi et al. 1998b, Rocha-Olivares et al. 2000, 2005, Fraga et al. 2007, Heras et al. 2009, Jamandre et al. 2009). Indeed, despite the uniformity in morphological characteristics or the difficulty of interpreting morphological variations (Harrison and Howes 1991) among *M. cephalus* populations from different areas of the species distribution range, different ecological and biological features were often reported (Liu 1986, Su and Kawasaki 1995, Huang et al. 2001, Whitfield et al. 2012). Several investigations carried out on *M. cephalus* population genetics at both global and regional scales (Fig. 15.2) eventually lead to strong evidence that *M. cephalus* should be considered a complex of species (Ke et al. 2009, Shen et al. 2011). This is especially true in the North West Pacific region, where the reproductive isolation among populations could be demonstrated due to the sympatry of the different lineages. At present, the Barcode of Life Data Systems (BOLD) report 15 main clusters (Barcode Index Numbers, BINs) in the *M. cephalus* complex, obtained from 224 published records from 12 countries, whereas Durand et al.

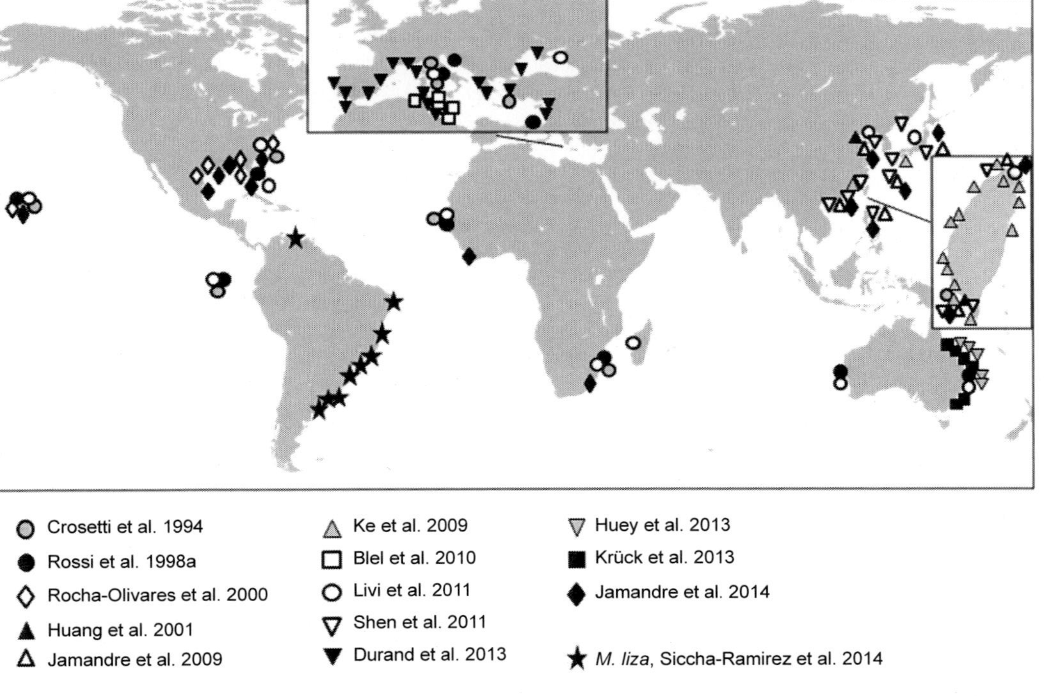

Figure 15.2. *M. cephalus* and *M. liza* sampling sites from different genetic studies. In the boxes enlargements of the Mediterranean Sea (left) and Taiwan (right), that are not scaled proportionally.

(2012b) describes 14 genetic lineages, 13 of which are currently named *M. cephalus* and one *M. liza* (and evidence of a further lineage following the addition of few *M. cephalus* specimens from the Galapagos Islands, Durand et al. 2012a). Very recently Durand and Borsa (2015) reported in details the 15 lineages identified within the *M. cephalus* species complex, each indicated with a different letter (*Mugil* sp. B-L), keeping the name *M. cephalus* for the Mediterranean Sea lineage, and the name *M. liza* for the lineage that includes samples from French Guyana, Venezuela and Uruguay.

Global Scale

Mullet samples from different locations worldwide were investigated with both nuclear and mitochondrial markers. The first studies relied on Restriction Site Length Polymorphism (RFLP) of mitochondrial DNA (mtDNA) (Crosetti et al. 1994) and allozyme loci (Rossi et al. 1998a) to investigate the genetic structure of flathead grey mullets from different geographic areas: three sampling sites in the Mediterranean (East and West coasts of Italy, Turkey), USA East Coast, Mauritania, Hawaii, Galapagos Islands, South Africa, Taiwan, East Australia and West Australia (the last two areas were only investigated in Rossi et al. 1998a). In both studies a pronounced genetic structure was found among samples from different sites. Indeed, Crosetti et al. (1994) obtained an estimated sequence divergence $P \cong 0.02–0.03$ between remote *M. cephalus* populations and a dendrogram with clusters coherent with the geographic origin of specimens; the only exception being Mediterranean locations where genotypes were shared among the three sampling sites (east and west coasts of Italy, Turkey) (Crosetti et al. 1994).

Similarly, Rossi et al. (1998a) showed Nei's (1978) unbiased genetic distance D from 0 to 0.242 between populations; several populations were characterized by fixed allelic differences as pair-wise comparisons showed up to four loci with no alleles shared between populations. In sum these first investigations showed that specimens from distant geographic locations were highly differentiated with null or very low gene flow with the markers used, and fall into the grey zone between well-differentiated conspecific populations and closely related species.

Nucleotide sequences of partial Cytochrome b (Cyt b) mitochondrial gene were used to gain insight into phylogeographic relationships and demographic history of flathead grey mullet populations, using the same samples analyzed by Crosetti et al. (1994) and Rossi et al. (1998a), with the addition of four sampling sites: the Azov Sea, Mozambique, Japan and East China (Livi et al. 2011). A pronounced genetic differentiation among sampling sites was confirmed, with most populations harbouring all private haplotypes and showing reciprocal monophyly. The high genetic differentiation among haplotypes (*p* distance ranged between 0.3 and 8.7% with an average of 5.8%) and the low genetic diversity within populations suggested a long-term reproductive isolation of flathead grey mullet populations characterized by reductions of population abundance and lineage sorting (haplotype diversity *h* ranged from 0 in North Carolina, East China, east coast of Italy and Azov Seas to 0.56 in East Australia). As observed in the North West Pacific by Jamandre et al. (2009) and Shen et al. (2011), a correlation between the distribution range of the different lineages and their nucleotide diversity might exist due to the fact that populations located at or close to the species northernmost distribution limits were more subject to demographic crashes during the Pleistocene.

The origin of *M. cephalus* is hypothesized to be in the West Pacific, with the East Australia lineage appearing the closest to the ancestral sequence. The strong genetic break between East Australia and West Australia, detected with both Cyt b (Livi et al. 2011) sequences and allozyme loci (Rossi et al. 1998a), lead to the speculation of a vicariant separation between the two lineages followed by the colonization of the East Pacific and Atlantic Ocean by dispersion. However, as these phylogeographic inferences are based exclusively on a short fragment (300bp) of a single mitochondrial marker, they have to be considered with caution and require further investigation. The differentiation among worldwide populations of flathead mullet was recently addressed by Jamandre et al. (2014) with the high variable mitochondrial Control Region (CR). Most of the sequences used by the authors were retrieved from GenBank, deposited by Rocha-Olivares et al. (2000, 2005) and Miya et al. (2001). The phylogenetic results obtained confirm the star-like topology tree described in Livi et al. (2011), and the results obtained by Durand et al. (2012b) with three concatenated mitochondrial fragments, in which phylogenetic relationships among the different lineages observed worldwide are not resolved. Indeed, CR sequences variation among lineages, mainly due to a Variable Number of Tandem Repeats (VNTR), largely exceeded intraspecific polymorphism reported in other vertebrates (Jamandre et al. 2014). Due to its high variability, the CR mtDNA region might not be suitable to highlight phylogenetic relationships due to homoplasy.

Regional Scale

Mediterranean Sea. The population genetic structure of *M. cephalus* in the Mediterranean region was investigated with both mitochondrial (Cyt b sequences, 857bp) and nuclear markers (seven microsatellite loci), in 18 geographical sites, from the Atlantic Ocean to the Black Sea (Durand et al. 2013). The Mediterranean, eastern Atlantic and Black Sea samples belong to the same *M. cephalus* lineage (thereon called Mediterranean lineage), with an average low nucleotide diversity (π 0,0013) and a haplotype diversity (*h*) ranging from 0 to 0,93. These results are in agreement with the presence of a single population cluster across the Mediterranean Sea (Crosetti et al. 1994, Rossi et al. 1998a, Blel et al. 2010) and of an exceptionally low nucleotide and haplotype diversity detected from the east coast of Italy and Azov Sea samples (Livi et al. 2011). The low mitochondrial diversity is consistent with past demographic crashes; the mismatch distribution and Bayesian Skyline Plot (BSP) model based on mtDNA sequences confirmed a recent population expansion of the Mediterranean lineage, though this is not evident at nuclear loci, perhaps due to the low number of loci tested. Within this lineage a significant differentiation was observed at mitochondrial loci between the Mediterranean and the Black Sea samples. At nuclear loci however, a stronger genetic structure was detected: the North-East Atlantic, the Mediterranean and the Black Seas were identified as the main genetically differentiated units with genetic breaks at the Almeria-Oran front and the Bosphorus. Additional barriers known to act as genetic breaks in the structure of several Mediterranean marine species (Patarnello et al. 2007) were identified by the software BARRIER; for flathead grey mullets these barriers are located in the Siculo-Tunisian Strait and south of Peloponnesus and enabled the identification of three additional clusters. The difference observed between nuclear and mitochondrial genetic differentiation patterns is either due to sex biased migration and dispersal, or to differences in effective population size, N_e, and variance in estimates of population differentiation among loci. Population subdivision estimated by TESS and BARRIER softwares indicated larger effective population size for *M. cephalus* in the eastern Mediterranean compared to more peripheral populations, suggesting the existence of a *refugium* in the eastern Mediterranean of flathead grey mullets, which then dispersed east to the Black Sea and west to the Atlantic Ocean (Durand et al. 2013).

North-West Pacific. The population genetics of flathead grey mullet were thoroughly investigated in the North-West Pacific, with both mitochondrial and nuclear molecular markers, such as the mitochondrial CR region (Jamandre et al. 2009, Liu et al. 2009a), Cyt b gene (Ke et al. 2009), Cytochrome Oxidase I gene (COI) (Shen et al. 2011, Sun et al. 2012), nuclear Amplification Fragment Length Polymorphism (AFLP) (Liu et al. 2009b) and microsatellites loci (Shen et al. 2011). All these surveys are consistent in indicating the presence of different mitochondrial lineages in this area with inter-lineage genetic distance comparable to the one observed in *M. cephalus* between far distant geographic areas at a worldwide scale; the demographic histories and the number of lineages reported vary among authors, probably due to different sampling efforts and molecular markers employed.

Some investigations carried out with different mitochondrial molecular markers support the existence of two lineages (Jamandre et al. 2009, Liu et al. 2009a, Sun et al. 2012), one dominating samples in the northern region along the eastern China Sea (lineage 1), the other mainly found in the southern China Sea (lineage 2). Both lineages were found in sympatry by Jamandre et al. (2009) in samples from Japan, an area considered by the author as a secondary contact zone. The two lineages show different levels of genetic variation associated with different demographic histories: lineage 1 shows a lower genetic variability (π = 0.005–0.008) and a mismatch distribution consistent with a recent demographic expansion, whereas lineage 2 shows a higher variability (π = 0.032–0.066) and a more stable population history. These results are in agreement with Liu et al. (2007), who found a similar trend in *L. haematocheila* (therein reported as *Chelon haematocheilus*) historical demography, i.e., one lineage characterized by population expansion in the northern region (East China Sea) and another lineage in equilibrium in the southern region

(South China Sea). High genetic divergence between marine organisms populations between East China Sea and South China Sea have also been reported for the bivalve molluscs *Tegillarca granosa* (Li et al. 2003) and *Cyclina sinensis* (Pan et al. 2005). According to Jamandre et al. (2009), Pleistocene events caused the isolation between lineages 1 and 2, showing more drastic effects in high than low latitude regions. However, the time of expansion was estimated by Liu et al. (2009a) at about 49,000–98,000 years ago for the northern region populations and much longer, about 486,000–972,000 years ago, for the southern population; the divergence of southern and northern region populations was estimated to date back about 5–11 million years, which was much earlier than the Pleistocene period. However, as already reported above, the mutation rate of the CR in *M. cephalus* is exceptionally larger than those reported for other vertebrates and its estimate is probably misleading, with an overestimation of the divergence time. When nuclear markers were applied (AFLP, Liu et al. 2009b) on the same samples investigated by Liu et al. (2009a) with CR sequences, four different lineages (instead of the two) were identified. The four lineages show a low level of gene flow ($Nm = 0.73$) and a significant association between geographic and genetic distance, which led the author to hypothesize a specific migratory model for the flathead grey mullet, enhancing genetic differentiation between populations.

A thorough investigation of flathead grey mullet populations in the North-West Pacific was carried out by Shen et al. (2011) who analyzed over 700 specimens from 12 locations using mtDNA COI sequences and 10 microsatellite loci. The phylogenetic tree obtained with nucleotide sequences and the assignment test carried out with nuclear genotypes were congruent and strongly supported the existence of three monophyletic lineages (NWP1, NWP2, NWP3) (Fig. 15.3). According to the authors only two lineages (NWP1 and NWP2) are of Pleistocene origin (about 1.607 MY), while the NWP3 lineage diverged from the NWP1 and NWP2 common ancestor during the Miocene–Pliocene epoch (about 4.2 MY). Besides, the origin of NWP3 does not seem to be related to geological events in the NW Pacific as this lineage is also observed in the Southwest Pacific (Shen et al. 2011, Durand et al. 2012b, Whitfield et al. 2012). The South China Sea is recognized as the *refugium* of the NWP2 lineage, whereas, based on estimation of divergence time between NWP1 and NWP2 and historical geological events, the origin of NWP1 is hypothesized to be due to a vicariant event such as the closure of the Tsushima Straits between Korea and Japan. This also explains the observed genetic diversity and Tajima's D and Fu's FS indices, which indicate bottleneck events in NWP1, probably due to demographic crashes that NWP1 lineage ancestors

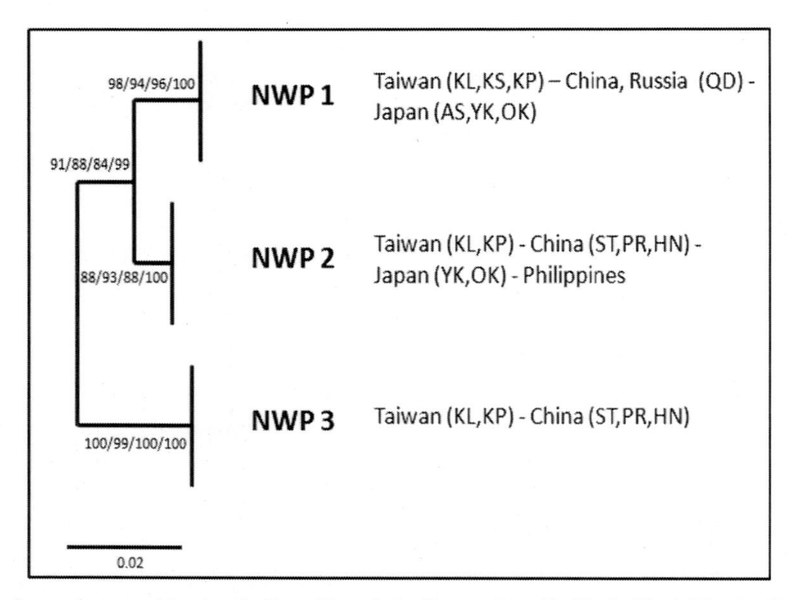

Figure 15.3. Phylogenetic tree evidencing the three *M. cephalus* lineages identified in the North West Pacific (sampling sites from north to south, see original reference). Bootstrap values obtained for NJ, ML and MP analysis and posterior probabilities of BI are reported (modified from Shen et al. 2011).

might have experienced in the Japan Sea during glacial episodes. The subdivision into three different lineages was highly supported by both markers. The COI inter-lineage divergence exceeded intra-lineage divergence by a ratio greater than 10, which is an indicator of cryptic species according to the 10 × rule (Hebert et al. 2004). The assignment tests made with nuclear genotypes with the software STRUCTURE strongly supported the existence of three groups (the mean q-values of assignment were over 0.99). The significant statistical evidences of three lineages together with the high congruence between mitochondrial and nuclear markers suggest that lineage sorting of ancestral polymorphisms has been completed providing strong evidence that the three lineages represent three different valid species. The limited gene flow estimated among sympatric individuals belonging to different mtDNA lineages and the absence of genetic introgression indicate that reproductive isolation has been reached for sympatric lineages. According to evidences of different migration patterns in the North-West Pacific (Huang et al. 2001, Ke et al. 2009), Shen et al. (2011) suggest that genetic isolation is maintained among *M. cephalus* species by spatial and temporal differences in spawning migration patterns.

East Australia. Along the eastern Australian coast different results were obtained on the population structure of *M. cephalus,* by different nuclear markers. Indeed, AFLP analysis (Huey et al. 2013) did not evidence genetic structure among five different locales across 550 km, thus suggesting that northward migrations of adults and juveniles and southward currents are sufficient to guarantee substantial gene flow along the coastline. On the contrary, Krück et al. (2013) developed SNPs able to detect a multispecies fishery stock along this coast (but over longer distances), and the presence of two clusters of *M. cephalus*, corresponding to two cryptic species, one from the southern and the other from the northern sampling area. No differentiation was observed among samples within major commercial fishing grounds south of the Great Barrier Reef, whereas strong genetic differentiation occurs north of this area. Marked differences in temperature between South and North-eastern Australian waters would be the driver for the ecological speciation of these two cryptic species.

West Atlantic. The population genetics of the flathead grey mullet was investigated along the western Atlantic coast by Rocha-Olivares et al. (2000) through the analysis of CR sequences. Specimens from five localities in the Atlantic Ocean and Gulf of Mexico coasts of the continental U.S.A. (N. Carolina, Florida, Mississippi, Louisiana, and Texas) and from one site in Hawaii were analyzed. The study revealed unusual high values of haplotype ($h = 1$ in all samples) and nucleotide diversity (π from 1,1 to 2,0%), comparable only to few other fish species (Rocha-Olivares et al. 2000). The genetic structure analysis showed that the division of fish into Pacific and Atlantic geographical regions accounted for most of the molecular variance (95%), while the remaining variation was found within, but not among putative populations within regions. Indeed all the samples from different sites along the western Atlantic coast represent a single panmictic population. Based on demographic inferences, all the samples showed a mismatch distribution significantly different from the one expected under the model of sudden expansion, though results suggest that the Gulf and Atlantic populations had a shorter period of population stability than the Hawaiian sample. The higher rate of mitochondrial evolution in *Mugil* species compared to other Mugilidae (Caldara et al. 1996) might explain the high haplotype diversity found in *M. cephalus* samples; the high variability of CR could make such a marker of limited utility in population genetics studies due to mutation saturation (Rocha-Olivares et al. 2000).

The Case of Mugil liza. The lebranche mullet *M. liza* is distributed on the East Atlantic coast, from Florida to the Bahamas, and throughout the Caribbean Sea to Argentina. Genetic analyses have been carried out by different authors, but mainly for phylogenetic purposes (Fraga et al. 2007, Heras et al. 2009, Rodrigues-Filho et al. 2011).

Within a molecular identification focused paper on Central and South American mullet species, Siccha-Ramirez et al. (2014) analyzed six different mtDNA regions in 38 specimens of *M. liza* from eight sampling sites across Venezuela, Brazil, Uruguay and Argentina. No characteristic haplotypes was observed within each population and very low genetic distance was detected among sampling sites, with an overall intraspecific genetic distance mean value of 0.2%, leading the authors to conclude that all specimens analyzed belong to a single population. Even natural barriers represented by areas at different salinities (such as the Amazon River mouth), that prevent connectivity in other marine species, do not seem to affect *M. liza* (and grey mullets in general) distribution thanks to their euryhalinity that enables them adapt to different salinities (Siccha-Ramirez et al. 2014 and references therein). These data are congruent with meristic and morphometric data taken from lebranche mullet samples collected from Venezuela to Argentina (Menezes et al. 2010), that clearly confirm the existence of a single species in the Caribbean Sea region and the Atlantic Coast of South America and that *Mugil liza* is the appropriate name. Also *M. platanus* has been recently synonymized with *M. liza* (Menezes et al. 2010).

Phylogenetic studies based on molecular markers attribute *M. liza* to the *M. cephalus* species complex (Heras et al. 2009, Durand et al. 2012a,b, Siccha-Ramirez et al. 2014), or consider it as one of its synonyms (Fraga et al. 2007). Phylogenetic considerations are beyond the scope of this chapter, but this topic is thoroughly discussed in Chapter 1—González-Castro and Ghasemzadeh 2015 and Chapter 2—Durand 2015.

Mugil cephalus*: a Species Complex?* Based on genetic evidence, *M. cephalus* should be considered a complex of species which also includes *M. liza* (Durand et al. 2012b). Species boundaries is a difficult task to address, especially when putative species are characterized by high morphological homogeneity and are allopatric as in the case of most *M. cephalus* lineages. For this reason, during the past 10 years, the scientific community started to adopt the DNA barcode to classify the biota and to uncover cryptic species (Hebert et al. 2003, Savolainen et al. 2005, Ward et al. 2005, Radulovici et al. 2010, Taberlet et al. 2012). Although the method shows some limitations as it is only based on mitochondrial sequences (more specifically partial COI gene in animals), it is certainly a practical and objective way to approach taxonomic classification which, in the case of the *M. cephalus* complex, implies so far the existence of 15 putative species. The reproductive isolation among three *M. cephalus* lineages was demonstrated in the North West Pacific where these species are sympatric. At a global scale, a strong genetic isolation among allopatric lineages has been thoroughly addressed at matrilineal mitochondrial DNA (Livi et al. 2011, Jamandre et al. 2014, Durand et al. 2012b), while at nuclear loci, hints of genetic isolations are so far based on allozyme data only (Rossi et al. 1998a). Hopefully this gap will be filled as several species specific nuclear genetic markers, such as microsatellite loci and Single Nucleotide Polymorphism (SNP), have been developed for *M. cephalus* and have already been used in regional studies (Miggiano et al. 2005, Xu et al. 2010, Molecular Ecology Resources Primer Development Consortium et al. 2010, Shen et al. 2011, Krück et al. 2013).

Genetic variability and structure within *M. cephalus* lineages is quite heterogeneous and has probably been influenced by migratory patterns and past demographic events. Indeed, in the North West Pacific, both resident and migratory lineages have been observed (Ke et al. 2009, Shen et al. 2011); besides, typical migratory patterns with the juveniles and post-spawning adults actively migrating back to the original coastal waters to feed have also been hypothesized (Chen et al. 1989, Lee 1992). Concerning past demographic events, glaciations seem to have had a stronger impact on the history of northernmost latitude populations, consistent with the lower genetic diversity shown by lineages in temperate regions.

According to the separation time estimated by Livi et al. (2011), dispersal events in flathead grey mullet should have occurred between two and four million years ago. Given the coastal and estuarine habits of *M. cephalus*, the Pleistocene epoch has probably favoured population isolation and lineage sorting, accelerating the speciation process. The hypothesis of a rapid radiation is consistent with the star-like phylogenetic tree obtained at global scale by Livi et al. (2011), Durand et al. (2012a,b) and Jamandre et al. (2014). At regional scales finer genetic and phylogeographic patterns have been addressed. For instance in the North West Pacific three distinct lineages, corresponding to valid species, coexist (Shen et al. 2011, Ke et al. 2009) and their distributional ranges appear to be shaped by the sea surface temperature and three major oceanographic current systems, namely the South China, North China Coastal and Kuroshio Currents (Shen et al. 2011). In the Mediterranean region only one putative species, genetically structured at mitochondrial and nuclear loci, is present from east in the Black Sea west to the East Atlantic Ocean. Preliminary results on East Australia samples analyzed with the nuclear SNP markers suggest the presence of two cryptic species with different temperature preferenda, which has been also suggested for the distribution of the three North West Pacific species (Krück et al. 2013, Shen et al. 2011), whereas along the USA West Atlantic Ocean coast a single panmictic population is known (Rocha-Olivares et al. 2000).

White mullet *Mugil curema*

The white mullet *Mugil curema* is a widely distributed species that inhabits both sides of the Atlantic Ocean and the eastern Pacific Ocean, along the African Coast from Senegal to Namibia, the eastern American Coasts from Cape Cod to Argentina, including the western Central Atlantic, Bermuda and the Caribbean Sea, and the western American Coasts from the Gulf of California to North Chile (Harrison 2002, González Castro et al. 2006). The white mullet can adapt to a wide range of salinities and can be found at sea (where it spawns) and also in inshore and estuarine waters, brackish coastal lagoons and freshwater lakes.

After the first evidence provided by cytogenetic analysis, i.e., different chromosome numbers found in specimens morphologically identified as *M. curema* from various localities (see below), *M. curema* was suggested not to be a single species, but a species complex: indeed, molecular identification and phylogenetic reconstruction based on mtDNA sequences provided incontrovertible evidence of the existence of different *M. curema* lineages along the American Atlantic Coasts (Fraga et al. 2007, Heras et al. 2009). Unfortunately these data were associated with the adoption of a very ambiguous nomenclature of the various lineages by the different authors, thus contributing to a chaotic picture. Fraga et al. (2007) analyzed 16S and COI sequences from samples collected in Venezuela and along the Brazilian Coast and identified two highly divergent mtDNA lineages, that they called *M. curema* type I and type II. Their type I actually corresponds to the samples of Nirchio et al. (2003) originally identified as *M. gaimardianus* and later renamed *M. rubrioculus* (Harrison et al. 2007, Nirchio et al. 2007), thus to a different species. Heras et al. (2009) analyzed 12S, COI and Cyt b sequences from samples from the USA (East Coast), Brazil and Argentina, and included in the analysis data from Fraga et al. (2007). Within what they called '*M. curema*-like species', they identified three lineages: (a) McurBra, corresponding to type I of Fraga et al. (2007) (and to the present *M. rubrioculus*), (b) McurArg, corresponding to type II of Fraga et al. (2007) and to samples from South Carolina, USA examined by Caldara et al. (1996), and (c) McurUSA that "does not correspond to any other *M. curema* sequences available". On these results they recommended an in-depth taxonomic revision of *M. curema*.

This issue was taken into account by Durand et al. (2012b) in a comprehensive molecular systematic revision of the family based on 16S, COI and Cyt b sequences. To this end, samples of white mullet were collected across the whole species range: West Pacific (Mexico, El Salvador, Panama, Ecuador, Peru), West Atlantic (USA, Mexico, Honduras, Belize, Guadalupe, Venezuela, Uruguay, Brazil) and East Atlantic (Senegal, Benin, Togo). All the samples and derived haplotypes constitute a monophyletic group that also includes *M. incilis*, being the node separating the two nominal taxa that was not resolved. At least four different *M. curema* lineages could be identified, partly on a geographic basis (Fig. 15.4). Indeed two lineages are present in sympatry in the West Atlantic: the first one includes sequences from USA to Brazil and corresponds to *M. curema* type II by Fraga et al. (2007), thus to the only lineage identified in the type locality of *M. curema*, Bahia (Brazil). This is the lineage (called West Atlantic clade in Fig. 15.4) that should keep the original name (Durand and Borsa 2015). The second one includes exclusively Caribbean sequences from Venezuela and Panama, and corresponds to the sample to which Menezes et al. (2015) attributed the new name *M. margaritae*. These two lineages had previously been cytogenetically analyzed (Nirchio et al. 2005b), differed in chromosome number and were assigned to different cytotypes by Sola et al. 2007 (see below). No specimens with intermediate chromosome numbers between the two lineages were ever detected, in spite of the high number of individuals cytogenetically analyzed, indicating the reproductive isolation of these lineages (Nirchio et al. 2005b). This hypothesis is consistent with the high genetic distance (4.7%) estimated on white mullet COI sequences of Brazil and Venezuela (Siccha-Ramirez et al. 2014), that is more than twice of the value (2%) that April et al. (2011) suggested as a threshold to identify different species. These two lineages were also confirmed by morphological features: specimens from Brazil and Venezuela showed differences in the scale counts and in the number of pectoral fin rays (Nirchio et al. 2005b) and geometric morphometric analysis on samples from Argentina and the east coast of Mexico allowed the identification of two groups, sorted according to their geographic origin (González-Castro et al. 2012). All these evidences agree on the existence of two different species of white mullet in the western Atlantic area, and thus on the validity to assign them two different species name.

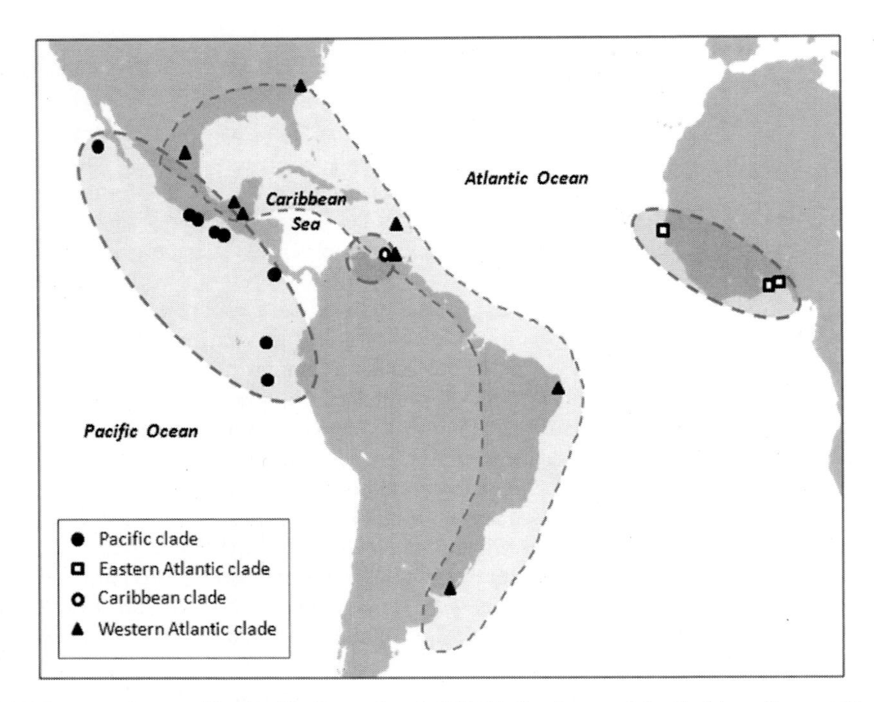

Figure 15.4. *M. curema* lineages identified by Durand et al. (2012b). Specimens of the Caribbean lineage (*M. margaritae*) were only sampled in Venezuela, though by mistake they were also reported on the East coast of Panama in Fig. 4b of the original paper by Durand et al. 2012b (Durand pers. comm.).

The other two *M. curema* lineages identified by Durand et al. (2012b) are reciprocally allopatric, one corresponding to haplotyes from the eastern Pacific Coast of North America reported as *Mugil* sp. O by Durand and Borsa (2015), and the other to sequences and haplotypes of the eastern Atlantic Coast of central Africa reported as *Mugil* sp. M by Durand and Borsa (2015). These two last lineages are still awaiting the attribution of a specific name that should be associated with the new investigations of morphological and meristic traits.

Concerning the phylogenetic relationships among the four phyletic lines, the eastern Pacific group clusters with the western Atlantic Caribbean group (*M. margaritae*), and both constitute a monophyletic group joining the eastern Atlantic African lineage in a monophyletic assemblage. This latter then joins to the other West Atlantic lineage (*M. curema*) (Durand et al. 2012b).

Cytogenetics

Cytogenetic analysis are usually aimed at identifying chromosomal markers that (a) are useful to discriminate between similar populations/species and thus in the identification of cryptic species, (b) might contribute to the reconstruction of karyotype/genome evolution within restricted or extended phyletic taxa (Bertollo et al. 2000, Artoni et al. 2009, Phillips 2013, Pansonato-Alves et al. 2014). This goal is particularly difficult when cold-blooded vertebrates are considered, as they usually lack the genome compartmentalization in GC and AT rich isochores described for warm-blooded vertebrates (Bernardi 2005) and chromosome bands (G- and R-bands), typically found for mammals, can rarely be observed. In addition fish karyotypes are generally made up of a large number of small chromosomes (up to 372 in Acipenseridae), and can be obtained only from live specimens, tissues or cells (Arai 2011); only in a few taxa are cell cultures successful and routinely applied.

Classical cytogenetic studies in fish usually reported chromosome number and morphology, and information derived from the application of standard staining techniques: C-banding (Sumner 1972) that detects the constitutive heterochromatin localization, and silver staining (Howell and Black 1980), used to visualize active Nucleolar Organizer Regions (NORs). The latter correspond to the chromosomal sites where the multigene family of ribosomal genes (rDNAs) that transcribes for 18S, 5.8S and 28S rRNA (45S rDNA) in the nucleolus, are localized, and represent the most widely investigated chromosome regions in fish. Protocols of fluorescence *in situ* hybridization (FISH) of repeated DNA sequences probes, used in human and mammal cytogenetics, were adapted and applied to fish in the last two decades, with consequent physical mapping of some genes (Gornung 2013). For all these reasons, fish karyotypes have not been studied proportionally to the number of fish species: 33,349 according to the Eschmeyer and Fong (2015) catalogue, corresponding to more than 50% of known living vertebrates. However some of the gaps are being rapidly filled. For instance, information on the evolution of sex chromosomes in different lineages of fish derives from chromosome painting data with specific probes obtained by the DNA amplification of total or partial chromosomes, selected by microdissection or sorted by flow cytometry (Diniz et al. 2008, Cioffi et al. 2011, Parise-Maltempi et al. 2013). A closer comparison between karyotypes and genomic data, now available thanks to the next generation sequences techniques, is still missing (Nirchio et al. 2014).

The analysis of chromosome number and morphology of teleost fish indicates that a higher karyotype diversification is present in freshwater species compared to marine ones (Nirchio et al. 2014), probably as a consequence of differences in population size, geographic barriers (Molina 2007) and speciation rates in the two environments (Bloom et al. 2013). In this framework, Mugilidae have all the characteristics of not being a particularly attractive taxonomic group to cytogeneticists: besides lacking R- and G-bands, mullets are marine species and are therefore expected to show limited karyotype variability. It is not surprising therefore that only a limited number of genera and species have been cytogenetically analyzed. To date, 37 scientific papers have been published on grey mullet cytogenetics, from the first studies on *Rhinomugil corsula* from the Indian Ocean (Nayyar 1966, Khuda-Bukhsh and Manna 1974) to the Mediterranean mullets (Cataudella and Capanna 1973, Cataudella et al. 1974) (Table 15.3).

However, in the last two decades, the analysis and location of repetitive sequences (telomeric sequences, satellite DNAs, and both classes of major and minor ribosomal genes, 45S rDNA and 5S rDNA, respectively) by FISH allowed the identification of chromosomal differences among mullet species and inference of evolutionary changes in the karyotype of the family.

Mugilidae Karyotypes

In their comprehensive review on cytogenetics of Mugilidae, Sola et al. (2007) reported karyological data on 16 species, belonging to five genera: one *Chelon*, five *Liza* spp., seven *Mugil* spp., one *Oedalechilus* and one *Rhinomugil*. These species represented about one fourth/one fifth of the species reported in the family, according to the different sources cited therein (Thomson 1997, Froese and Pauly 2004, respectively). Indeed according to Thomson (1997) the Mugilidae includes 14 genera, 62 valid species and 18 species *inquerenda*, whereas Froese and Pauly (2014) on the FishBase web site reported 17 genera and 75 valid species. For many of the species only the chromosome number and morphology were known; when available, data on NORs, C-banding, fluorochrome staining (chromomycin A_3, CMA_3 and Di Amino Phenyl Indol, DAPI) and FISH with different probes, were reported. Since the Sola et al. (2007) review, only a few other cytogenetic data have been added (Nirchio et al. 2007, Cipriano et al. 2008, Nirchio et al. 2009, Hett et al. 2011, Değer et al. 2013). Considering that some species have been synonymized, others changed names or were attributed to a different genus, 19 mullet species (according to the nomenclature by Eschmeyer 2015) have been cytogenetically analyzed (Table 15.3). However this number could not be deemed as final in the present taxonomic uncertainty (see Chapters 1 and 2), considering for instance that the list of karyotyped species includes taxa like *M. cephalus* and *M. curema* which are now accounted as species complexes on molecular evidence, or *M. liza,* which is considered part of the *M. cephalus* complex (Durand et al. 2012b).

Although limited in number, the data obtained after 2007 are pivotal as they provide a different perspective of the karyotypic changes within the family. Nirchio et al. (2009) reported the first data

Table 15.3. Summary of cytogenetic studies on Mugilidae.

Species[1]	Geographic area	2n[2]	Karyotype[3]	FN[4]	Cytotype	Ag-NOR	C-bands	CMA_3/DAPI	FISH[5]				References
									18S	5S	telom	sat	
Agonostomus													
A. monticola	Two sites rivers (Venezuela, Panama)	48	46a + 2st	48	B1	+	+	+	+	+	-	-	Nirchio et al. 2009
Chelon													
C. labrosus	Mediterranean Sea (Italy)	48	46a + 2st	48	B1	-	-	-	-	-	-	-	Cataudella and Capanna 1973, Cataudella et al. 1974
	Mediterranean Sea (Spain)	48	46a + 2sm	48	B1	+	+	-	-	-	-	-	Delgado et al. 1990, 1991
	Mediterranean Sea (Spain)	48	46a + 2sm	48	B1	-	-	-	-	-	-	-	Delgado et al. 1992
	Mediterranean Sea (Italy)	48	46a + 2st	48	B1	-	-	-	+	+	-	-	Gornung et al. 2001
	Mediterranean Sea (Italy)	48	46a + 2st	48	B1	-	-	-	-	-	+	-	Gornung et al. 2004
	Mediterranean Sea (Italy)	48	46a + 2st	48	B1	-	-	-	-	-	-	+	Sola et al. 2007
Liza													
L. abu	Two Rivers (Turkey)	48	46a + 2m	50	B2	+	-	-	-	-	-	-	Değer et al. 2013
L. aurata	Mediterranean Sea (Italy)	48	46a + 2st	48	B1	-	-	-	-	-	-	-	Cataudella et al. 1974
	Mediterranean Sea (Spain)	48	46a + 2st	48	B1	-	-	-	-	-	-	-	Cano et al. 1982
	Mediterranean Sea (Spain)	48	46a + 2sm	48	B1	+	+	-	-	-	-	-	Delgado et al. 1990, 1991
	Mediterranean Sea (Spain)	48	46a + 2sm	48	B1	-	-	-	-	-	-	-	Delgado et al. 1992
	Mediterranean Sea (Italy)	48	46a + 2st	48	B1	-	-	-	+	+	-	-	Gornung et al. 2001
	Mediterranean Sea (Italy)	48	46a + 2st	48	B1	-	-	-	-	-	+	-	Gornung et al. 2004
	Mediterranean Sea (Italy)	48	46a + 2st	48	B1	-	-	-	-	-	-	+	Sola et al. 2007
L. macrolepis	Indian Ocean (India)	48	48a	48	A	-	-	-	-	-	-	-	Choudhury et al. 1979
L. parmatus (*L. oligolepis*)	Indian Ocean (India)	48	48a	48	A	-	-	-	-	-	-	-	Choudhury et al. 1979
L. parsia (*Mugil parsia*)	Unknown	48	48a	48	A	-	-	-	-	-	-	-	Chatterjee and Majhi 1973
	Indian Ocean	48	48a	48		-	-	-	-	-	-	-	Khuda-Bukhsh and Manna 1974

L. ramada	Mediterranean Sea (Italy)	48	46a + 2st	48	B1	-	-	-	-	-	-	-	Cataudella and Capanna 1973, Cataudella et al. 1974
	Mediterranean Sea (Spain)	48	46a + 2sm	48	B1	+	+	-	-	-	-	-	Delgado et al. 1990, 1991
	Mediterranean Sea (Spain)	48	46a + 2sm	48	B1	-	-	-	-	-	-	-	Delgado et al. 1992
	Mediterranean Sea (Italy)	48	46a + 2st	48	B1	+	+	+	+	-	-	-	Rossi et al. 1997
	Mediterranean Sea (Italy)				B1	-	-	-	+	+	-	-	Gornung et al. 2001
	Mediterranean Sea (Italy)				B1	-	-	-	-	-	+	-	Gornung et al. 2004
	Mediterranean Sea (Italy)	48	46a + 2st	48	B1	-	-	-	-	-	-	+	Sola et al. 2007
L. saliens	Mediterranean Sea (Italy)	48	46a + 2st	48	B1	-	-	-	-	-	-	-	Cataudella et al. 1974
	Black Sea (Russia)	48	46a + 2sm	48	B1	-	-	-	-	-	-	-	Arefyev 1989
	Mediterranean Sea (Italy)	48	46a + 2st	48	B1	+	+	+	+	+	-	-	Gornung et al. 2001
	Mediterranean Sea (Italy)	48	46a + 2st	48	B1	-	-	-	-	-	+	-	Gornung et al. 2004
	Mediterranean Sea (Italy)	48	46a + 2st	48	B1	-	-	-	-	-	-	+	Sola et al. 2007
Mugil													
M. cephalus	Mediterranean Sea (Italy)	48	48a	48	A	-	-	-	-	-	-	-	Cataudella and Capanna 1973, Cataudella et al. 1974
	Indian Ocean (Sri Lanka)	48	48a	48	A	-	-	-	-	-	-	-	Natarajan and Subrahmanyam 1974
	West Atlantic Ocean (USA)	48	48a	48	A	-	-	-	-	-	-	-	Le Grande and Fitzsimons 1976
	Mediterranean Sea (Spain)	48	48a	48	A	-	-	-	-	-	-	-	Cano et al. 1982
	West Atlantic Ocean (USA)	48	48a	48	A	+	-	+	-	-	-	-	Amemiya and Gold 1986
	Black Sea (Russia)	48	48a	48	A	-	-	-	-	-	-	-	Arefyev 1989
	China Sea (Taiwan), Pacific Ocean (Hawai), West Atlantic Ocean (USA), Mediterranean Sea (Italy)	48	48a	48	A	+	+	+	-	-	-	-	Crosetti et al. 1993

Table 15.3. contd....

Table 15.3. contd.

Species[1]	Geographic area	2n[2]	Karyotype[3]	FN[4]	Cytotype	Ag-NOR	C-bands	CMA_3/DAPI	FISH[5]				References
									18S	5S	telom	sat	
M. cephalus	Indian Ocean (E Australia) China Sea (Taiwan), Pacific Ocean (Hawai), West Atlantic Ocean (USA), Mediterranean Sea (Italy)	48	48a	48	A	+	+	+	+	-	-	-	Rossi et al. 1996
	Mediterranean Sea (Italy)	48	48a	48	A	-	-	-	+	+	-	-	Gornung et al. 2001
	Mediterranean Sea (Italy)	48	48a	48	A	-	-	-	-	-	+	-	Gornung et al. 2004
	Mediterranean Sea (Italy)	48	46a + 2st	48	B1	-	-	-	-	-	-	+	Sola et al. 2007
M. curema	West Atlantic Ocean (USA)	28	4a + 4st + 20 m	48	C1	-	-	-	-	-	-	-	Le Grande and Fitzsimons 1976
	West Atlantic Ocean (Brazil)	28	4a + 4st + 20 m	48	C1	+	-	+	-	-	-	-	Cipriano et al. 2002
	West Atlantic Ocean (Brazil)	28	4a + 4st + 20 m	48	C1	+	+	-	-	-	-	-	Nirchio et al. 2005b
	West Atlantic Ocean (Brazil)	28	4a + 4st + 20 m	48	C1	-	-	-	-	-	-	-	Cipriano et al. 2008
M. incilis	Caribbean Sea (Venezuela)	48	48a	48	A	+	+	+	+	+		-	Hett et al. 2011
M. liza	W Atlantic Ocean (Brazil)	48	48a	48	A	-	-	-	-	-	-	-	Pauls and Coutinho 1990
	Caribbean Sea (Venezuela)	48	48a	48	A	-	-	-	-	-	-	-	Nirchio and Cequea 1998
	Caribbean Sea (Venezuela)	48	48a	48	A	+	-	-	-	-	-	-	Nirchio et al. 2001
	Caribbean Sea (Venezuela)	48	48a	48	A	-	-	+	+	+	+		Rossi et al. 2005
(*M. platanus*)	West Atlantic Ocean (Brazil)	48	48a	48	A	+	+	-	-	-	-	-	Jordão et al. 1992
M. margaritae (*M. curema*)	Caribbean Sea (Venezuela)	24	2 sm + 22 m	48	C2	-	-	-	-	-	-	-	Nirchio and Cequea 1998
	Caribbean Sea (Venezuela)	24	2 sm + 22 m	48	C2	+	-	-	-	-	-	-	Nirchio et al. 2001
	Caribbean Sea (Venezuela)	24	2 sm + 22 m	48	C2	-	-	-	-	-	-	-	Nirchio et al. 2003
	Caribbean Sea (Venezuela)	24	2 sm + 22 m	48	C2	+	+	-	-	-	-	-	Nirchio et al. 2005b
	Caribbean Sea (Venezuela)	24	2 sm + 22 m	48	C2	-	-	+	+	+	+	-	Rossi et al. 2005

	Caribbean Sea (Venezuela)	24	2 sm + 22 m	48	C2	+	+		+			-	Nirchio et al. 2007
M. rubrioculus	Caribbean Sea (Venezuela)	48	48a	48	A	+	+	-	+	-	-	-	Nirchio et al. 2007
	Caribbean Sea (Venezuela)	48	48a			-	-	-	-	+	-	-	Nirchio pers. commun.
(*M. gaimardianus*)	Pacific Ocean (Panama), Caribbean Sea (Venezuela)	48	48a	48	A	-	-	-	-	-	-	-	Nirchio et al. 2003
M. trichodon	Caribbean Sea (Venezuela)	48	48a	48	A	+	+	-	-	-	-	-	Nirchio et al. 2005a
	Caribbean Sea (Venezuela)	48	48a	48		-	-	-	-	+	-	-	Nirchio pers. commun.
Moolgarda													
M. speigleri (*Mugil speigleri*)	Indian Ocean (India)	48	48a	48	A	-	-	-	-	-	-	-	Rishi and Singh 1982
Oedalechilus													
O. labeo	Mediterranean Sea (Italy)	48	46a + 2st	48	B1	-	-	-	-	-	-	-	Cataudella et al. 1974
	Mediterranean Sea (Italy)	48	46a + 2st	48	B1	+	+	+	+	-	-	-	Rossi et al. 2000
	Mediterranean Sea (Italy)	48	46a + 2st	48	B1	-	-	-	-	+	-	-	Gornung et al. 2001
	Mediterranean Sea (Italy)	48	46a + 2st	48	B1	-	-	-	-	-	+	-	Gornung et al. 2004
	Mediterranean Sea (Italy)	48	46a + 2st	48	B1							+	Sola et al. 2007
Rhinomugil													
R. corsula (*Mugil corsula*)	Indian Ocean (India)	48	48a	48	A	-	-	-	-	-	-	-	Nayyar 1966
	Indian Ocean	48	48a	48	A	-	-	-	-	-	-	-	Khuda-Bukhsh and Manna 1974
	Indian Ocean	48	48a	48	A	+	+	-	-	-	-	-	Chakrabarti and Khuda-Bukhsh 2000

[1]Valid name according to Echmeyer (2015), when a different synonym was used by the authors it is reported below, in parenthesis
[2]2n = diploid number
[3]Karyotype formula as derived from original figures: a = acrocentric, st = subtelocentric, sm = submetacentric, m = metacentric chromosomes. For cytotype B1 sm chromosomes are considered as uniarmed (see text).
[4]FN = Fundamental number, indicating the number of chromosome arms
[5]Probes used in FISH: 18S = 18S rDNA, 5S = 5S rDNA, telomere = telomericsequences, sat = satellite DNA

on the karyotype of *A. monticola* allowing inferences on chromosome evolution within Mugilidae. Indeed although Thomson (1997) considered the genus *Agonostomus* as the most basal among extant mullets (although this statement is not congruent with molecular data), this karyotype shed light on the chromosomal changes that had occurred and allowed a different interpretation of previous cytogenetic data (Nirchio et al. 2009). In addition, the cytogenetic features reported for the abu mullet *Liza abu* (Değer et al. 2013) had never been reported before for any other grey mullet species and enabled the identification of a new cytotype (see next paragraph).

Identified Cytotypes

With the exception of *Mugil curema* species complex, grey mullets show a conservative diploid number of 48 chromosomes, mainly acrocentric and with a fundamental number (FN) of 48 (and FN of 50 in *Liza abu*) (Table 15.3). A karyotype showing these features has long been considered as the ancestral one for all teleosts (Ohno 1974), but more recently it has been regarded as the synapomorphic chromosome complement of Clupeomorph and Euteleostei (Brum and Galetti 1997). The analysis of more than 100 fish species, representative of 17 out of the 40 Teleost orders, indicates that a karyotype composed of 48 all acrocentric chromosomes characterizes 60% of the marine species and is particularly frequent in Perciformes (Nirchio et al. 2014).

According to chromosome morphology, Sola et al. (2007) classified mullet karyotypes in three main cytotypes. Figure 15.5 reports the main cytotypes identified by Sola et al. (2007) updated and revised with recent data, using the species nomenclature according to Eschmeyer (2015).

The first one, named cytotype A, is the most common within the family and characterizes 10 species, of the genera *Mugil* (*M. cephalus*, *M. liza*, *M. incilis*, *M. rubrioculus*, *M. trichodon*), *Rhinomugil* (*R. corsula*) and *Moolgarda* (*M. speigleri*), and the *Liza* spp. collected in the Indian Ocean (*L. macrolepis*, *L. parmatus*, *L. parsia*). In four of the species showing this cytotype (*L. macrolepis*, *L. parmatus*, *L. parsia*, *M. speigleri*; first row of Fig. 15.5), the available data date back more than 30 years and only the Giemsa karyotype, i.e., chromosome number and morphology, is known, whereas for the other species data from molecular cytogenetis are also available. In *M. cephalus,* all six populations collected worldwide, that includes different molecular lineages, showed the same cytotype (Crosetti et al. 1993, Rossi et al. 1996). Chromosome pair numbering for *R. corsula* is not reported in the text of the original paper (Chakrabarti and Khuda-Bukhsh 2000), but is based on the figures shown therein, according to Sola et al. (2007). Cytotype A includes exclusively uniarmed (acrocentric) chromosomes (Fig. 15.6a) that gradually decrease in size. Among the 24 chromosome pairs, only the first and the last can be identified by size. For its features and for being shared by the majority of mullets, cytotype A was long considered the ancestral karyotype for mugilids (Sola et al. 2007), but this hypothesis has been rejected after the description of the mountain mullet *A. monticola* karyotype (Nirchio et al. 2009), representative of the ancient genus *Agonostomus* (see below).

The second cytotype, named B by Sola et al. (2007), is very similar to the previous one: one of the chromosome pairs is not acrocentric, but is classified as submetacentric (*sm*) or subtelocentric (*st*), for the length of its short arms (Fig. 15.5 and Fig. 15.6b). Such differences in short arm lengths are very subtle and are due to the presence of heterochromatin, interspersed or associated to NORs (Rossi et al. 1997). Indeed, this heterochromatin can affect the rDNA copy number and thus produces size polymorphism of the corresponding chromosomal regions. In this cytotype, both subtelocentric and submetacentric chromosomes are therefore considered as uniarmed, although this is unusual, at least for submetacentric chromosomes. This cytotype has been renamed here as B1, to distinguish it from the one recently found in *Liza abu*, indicated as cytotype B2. Cytotype B1 is shared by four genera and six species: *Agonostomus* (*A. monticola*), *Chelon* (*C. labrosus*), three *Liza* spp. from the Mediterranean Sea (*L. aurata*, *L. ramada*, *L. saliens*) and *Oedalechilus* (*O. labeo*). For these species, besides chromosome number and morphology, data on other chromosomal markers, obtained both by standard staining techniques (silver staining, C-banding, fluorochrome staining) and FISH, are available (Gornung et al. 2001, 2004, Nirchio et al. 2009, Rossi et al. 1997, 2000). The subtelocentric or submetacentric chromosome pair is the smallest pair of the karyotype (n. 24) in all the species, except in *O. labeo,* where it is medium-sized (n. 9). According to Nirchio et al. (2009), cytotype B1 should be considered the plesiomorphic karyotype for the whole family, as it is not only shared by *A. monticola*, but also by representatives of other more derived genera

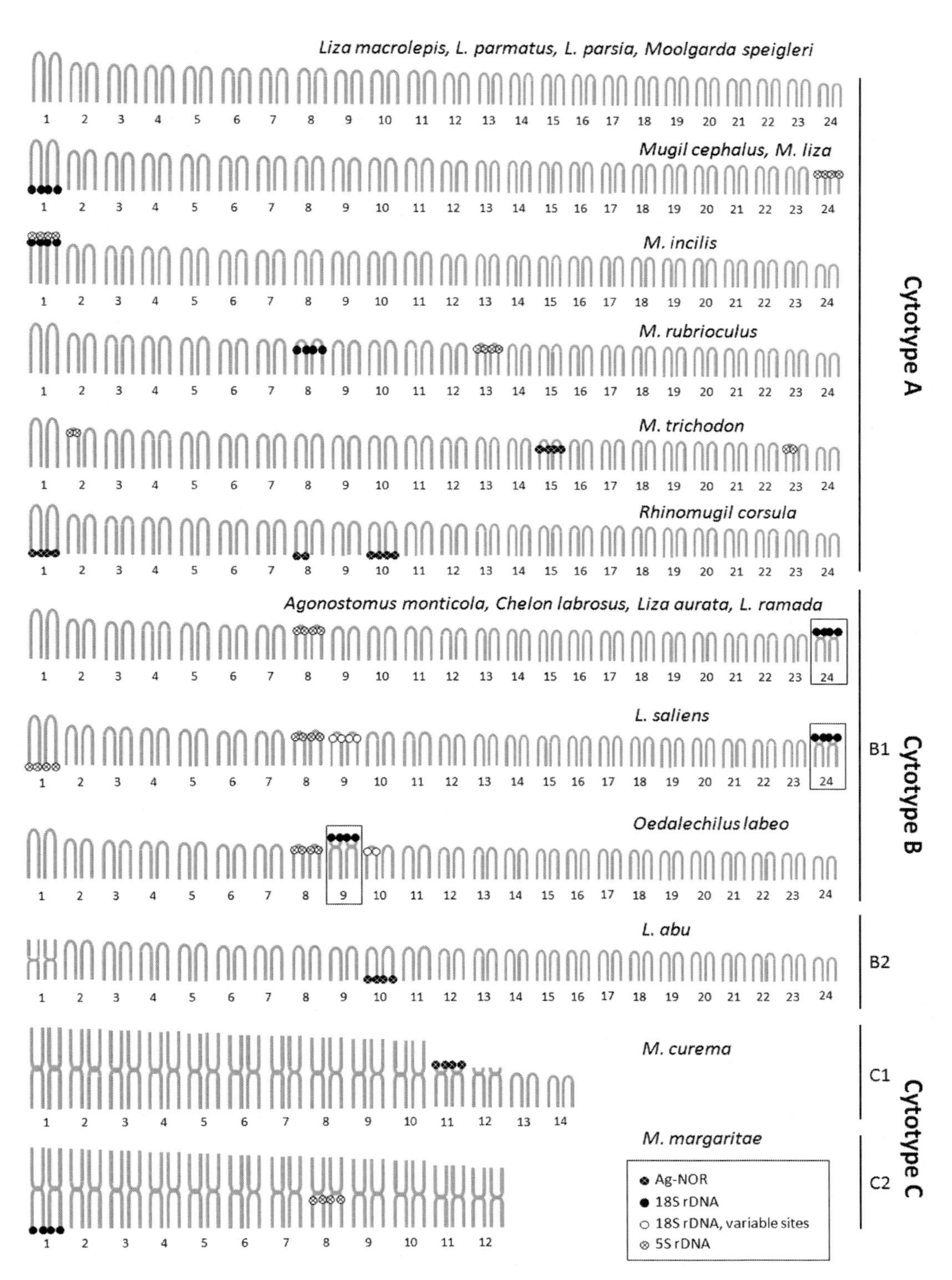

Figure 15.5. Idiograms of the karyotypes observed in Mugilidae organized according to the different cytotypes. Re-drawn and updated from Sola et al. (2007). Boxes in cytotype B1 indicate the st-sm chromosome pair.

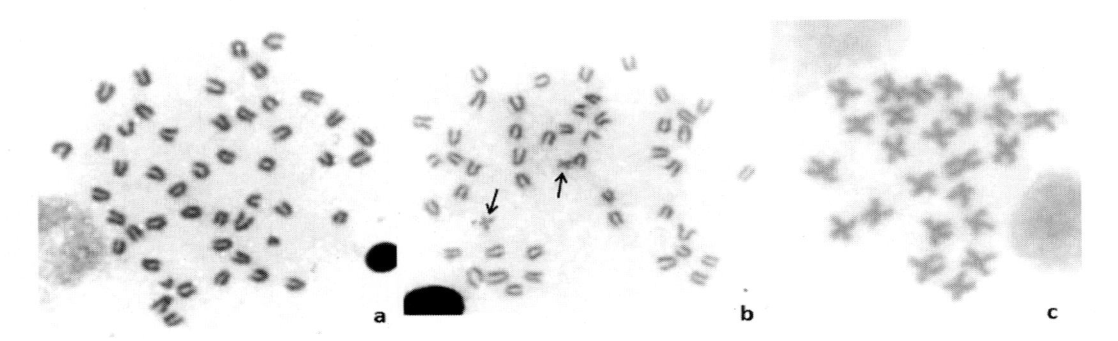

Figure 15.6. Giemsa metaphase plates showing some of the observed cytotypes in Mugilidae. (a) *M. incilis*, (b) *L. ramada*, (c) *M. margaritae* from Venezuela. Arrows in b indicate the subtelocentric chromosomes.

like *Chelon* and *Liza*. A different cytotype, in this chapter named B2, was described in Turkey for *L. abu* (Değer et al. 2013), a species known to inhabit and spawn in freshwater. Cytotype B2 is characterized by the presence of 46 acrocentric chromosomes, as cytotype B1, and two large metacentric chromosome, classified as pair number one (Değer et al. 2013). The latter is not comparable with the subtelocentric or submetacentric chromosome pair described for cytotype B1, as it differs in shape, larger size, and absence of NORs, that in *L. abu* are localized in the terminal region of a medium size chromosome pair. Data on heterochromatin distribution on this species are not available. It has a fundamental number, FN = 50, unique among mullets, though the metacentric chromosome pair could be consequent to a pericentric inversion that involves an extended chromosome region. At present it is not possible to ascertain whether this karyotype derives from cytotype B1 or from cytotype A, and further analysis on other cytogenetic markers on *L. abu* are required to infer the origin of its karyotype.

The last cytotype, named C, is only present in the white mullet *M. curema* species complex. It is characterized by having a reduced number of chromosomes compared to cytotypes A and B, and by these chromosomes being almost or exclusively bi-armed (meta/submetacentric) (Fig. 15.5). Two different sub-cytotypes were described: cytotype C1, that shows a diploid number (2n) of 28 chromosomes (20 metacentric and 8 subtelo/acrocentric) and is reported in specimens from USA and Brazil, and cytotype C2, with 2n = 24, that includes exclusively bi-armed (metacentric and submetacentric) chromosomes, typical of specimens from Venezuela (Fig. 15.6c). In spite of the different number of chromosomes, these two cytotypes show the same number of visible chromosomal arms (fundamental number, FN) = 48, shared by all mullets, except *L. abu*. This feature indicates that metacentric and submetacentric chromosomes of this cytotype derived from Robertsonian fusions of acrocentric chromosomes. LeGrande and Fitzsimons (1976), who originally described the now called cytotype C1, suggest that the *M. curema* karyotype could derive from a karyotype similar to the *M. cephalus*' (cytotype A). For cytotype C2, for which data on major and minor rDNA are available, NORs location is compatible with this hypothesis, while 5S rDNA location allowed its exclusion, unless further chromosomal re-arrangements are considered (Sola et al. 2007). The absence of interstitial telomeric sites localized at the centromeres of bi-armed chromosomes was proved by FISH in white mullets from Venezuela (Rossi et al. 2005). These data indicate that fusions are irreversible, at least in cytotype C2, as one of the prerequisite for the formation of Rb fusions is either inactivation or loss of telomeric sequences (Slijepcevic 1998), as observed in centromeric sites of Venezuelan samples. The two cytotypes were suggested to be different species (Nirchio et al. 2005b) later confirmed by Durand et al. (2012b) who combined mtDNA and cytogenetic results. Recently, Menezes et al. (2015) proposed naming species with cytotype C1 as *M. curema*, and those with the cytotype C2 as *M. margaritae*. Further cytotaxomic research could disclose the presence of different karyotypes in samples of the other two *M. curema* clades molecularly identified by Durand et al. (2012b).

Localization of Repetitive Sequences

The location of repetitive sequences, like major and minor ribosomal genes and/or telomeric sequences, as well as the location and composition of constitutive heterochromatin and satellite DNA localized therein, were investigated by different authors using FISH, but also deriving indirect information from C-banding and fluorochrome staining (Table 15.3). Differences were detected both between (Fig. 15.7) and within cytotypes, and for the same cytotype different idiograms are present.

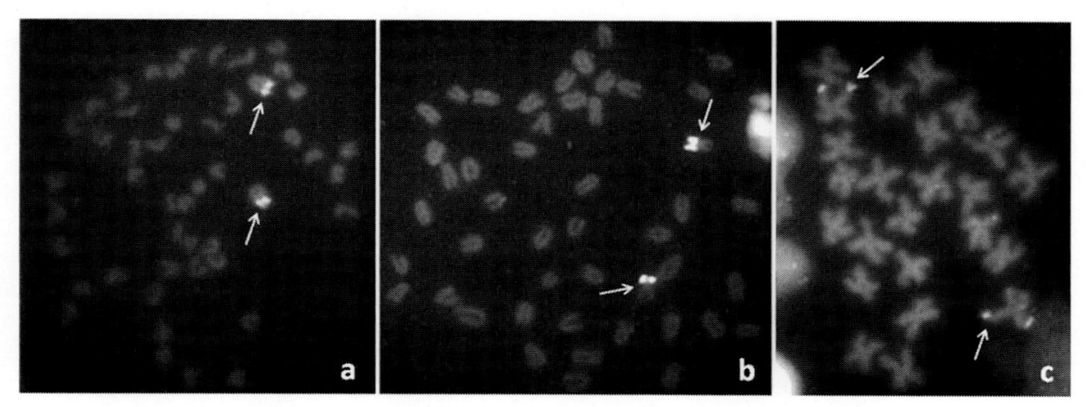

Figure 15.7. 18S rDNA FISH metaphase plates in some of the observed cytotypes in Mugilidae. (a) *M. cephalus*, (b) *L. ramada*, (c) *M. margaritae* from Venezuela. Arrows indicate 18S rDNA sites.

Data on nucleolar organizer regions derived from silver staining are available for 14 species, and for 11 of them they were verified through FISH (Table 15.3). Most of these species show a single chromosome bearing pair (a single rDNA site), according to the evidence that this is the most common condition in Teleosts, shown by 72% of the species (Gornung 2013) and to the general idea that this is the plesiomorphic condition in fishes (Amemiya and Gold 1988). NORs could be located in telomeric position on chromosomes of different sizes (five species), could be interstitial (three species) or present on the short arms of a st-sm chromosome pair (six species). Differences in NOR localization were observed within each cytotype. As an example, some of the species showing cytotype A have NORs localized on chromosome pair one in the telomeric position, like *M. cephalus* (Rossi et al. 1996) and *M. liza* (Rossi et al. 2005), while others have NORs that are interstitial on the same (*M. incilis*—Hett et al. 2011) or on a different chromosome pair (*M. trichodon*—Nirchio et al. 2005a, *M. rubrioculus*—Nirchio et al. 2007). The different localization observed for these chromosomal regions could be attributed to translocations or pericentromeric inversions that could have been facilitated by the presence of associated heterochromaic blocks. Indeed heterochromatin is composed of different types of repetitive sequences (Cioffi and Bertollo 2012) and such repetitiveness might act to promote non-homologous recombination. However other factors could be involved and could be responsible also for the presence of the additional NORs observed in *L. saliens*, *O. labeo* and *R. corsula*. In the absence of other evidence, and with the difficulty of exactly identifying homologous chromosomes, it is not possible to delineate a univocal evolutionary pathway of NORs within or among cytotypes.

Minor ribosomal gene locations, investigated using FISH with 5S rDNA probes, provided results for 12 species (Fig. 15.5), as reported in Table 15.3. Although these chromosomal regions are usually considered less variable than NORs, their localization in mullets appear to be variable, and species sharing the same cytotype and NOR position (like *L. ramada* and *L. saliens*) could show different location of 5S rDNA. In *M. incilis* these genes are co-localized with those of major ribosomal genes on chromosome pair one. According to Martins and Wasko (2004) the localization of major and minor ribosomal gene clusters in different chromosomes, as observed for the majority of vertebrates, depends on physical and evolutionary constraints, i.e., facilitation of the activity of the different RNA polymerases and allowing them to evolve independently. Thus, it is not surprising that their co-localization was observed only in one mugilid species.

Telomeric sequence localizations were investigated in the six most common Mediterranean mullet species (*C. labrosus*, *L. aurata*, *L. ramada*, *L. saliens*, *M. cephalus*, *O. labeo*) and in *M. liza* and *M. margaritae*. In all the species these sequences are present at the standard expected sites, i.e., to the ends of all chromosomes. The telomeric probes are also interspersed with the major ribosomal genes, a presence not common in teleost fishes, where Interstitial Telomeric Sequences (ITS) are only reported in 33 species, belonging to 15 different orders (Ocalewicz 2013). The meaning of the association between rDNA and ITS is still unclear, though it might be functional to the nucleolus formation (Ocalewicz 2013).

C-banding was applied to 14 species (Table 15.3) and revealed that heterochromatin is present in pericentromeric regions of all chromosomes and also interspersed with NORs, regardless of their terminal or interstitial location. In *A. monticola*, *L. ramada*, *L. saliens*, *M. cephalus* (different lineages), *M. margaritae*, *M. incilis*, *O. labeo* further information on heterochromatin composition was obtained by staining with GC-(CMA_3) and AT-(DAPI) binding fluorochromes. NOR associated heterochromatin was found to be CMA_3 positive and DAPI negative, while pericentromeric heterochromatin usually showed uniform base composition. *Liza ramada* represents an exception, with GC-rich, and thus CMA_3 positive, centromeric heterochromatin (Rossi et al. 1997). As heterochromatin includes repetitive non-coding DNA sequences, known as satellite DNA, these results suggest the existence of different types of satellite DNA at least in those species where fluorochrome staining was applied. Sola et al. (2007) reported unpublished data on the isolation and cross hybridization of two different families of satellite DNA in *L. ramada* and *M. cephalus*. The use of these satellites as probes in FISH on metaphases of the six native Mediterranean species (the two previous mentioned and *C. labrosus*, *L. aurata*, *L. saliens*, *O. labeo*) provided different patterns of positive centromeres that are congruent with the phylogenetic relationships obtained by mtDNA sequence analysis of the same species (Rossi et al. 2004), confirming the close relationships between the Mediterranean *Liza* spp. and *C. labrosus*.

Considerations and Perspectives in Mullet Cytogenetics

Chromosomal markers revealed by molecular cytogenetics delineate a variability within each cytotype and do not allow a univocal pathway of reconstruction of chromosomal re-arrangements in Mugilidae evolutionary radiation. Different translocation and independent pericentric inversions should be invoked to explain the karyotype changes that, starting from the supposed ancestral karyotype of *A. monticola*, determined the variability observed at present. By comparing these data with the phylogenies obtained by molecular markers (Durand et al. 2012b), only a partial congruence can be observed, like the already mentioned close relationships between the Mediterranean *Liza* spp. and *C. labrosus*. On the contrary *M. rubrioculus,* that shares the same cytotype with *M. cephalus* species complex and *M. trichodon*, is phylogenetically closer to *M. curema* species complex than to the other two species (Durand et al. 2012b).

A final consideration, linked to the previous one, is on the different rates of evolution in the various cytotypes and species complexes. Species of the *M. cephalus* complex collected worldwide and belonging to different molecularly-identified-lineages (not yet formally classified as species, and thus still waiting for the attribution of a specific name) share the same chromosome number and morphology as well as other karyotype features like NORs and heterochromatin localization and composition. The absence of cytogenetic markers associated with the different lineages suggests that genetic differentiation is not associated with chromosomal differentiation in this species complex. An opposite situation is observed in the *M. curema* complex, where the presence of cytogenetic differences (i.e., chromosome number, karyotype formula and NOR location) in various populations from different areas provided the first hint of the existence of cryptic species, later the object of molecular phylogeographic analysis.

Cytogenetic data need to be enhanced, to include also the analysis of other genera/species, considering the new hypotheses in Mugilidae taxonomy (Chapter 1—González-Castro and Ghasemzadeh 2015 and Chapter 2—Durand 2015) and other chromosomal markers should be applied. For instance, the isolation and chromosome mapping of transposable elements could provide pivotal information on the evolutionary diversification of genome and karyotype in mugilids. Indeed fish genome contains all different types of transposones (Volff 2005); these mobile elements might strongly influence evolutionary rates of the heterochromatic regions (Cioffi et al. 2010) and play an important role in genome evolution.

Conclusions

Genetic analysis in Mugilidae has already provided, and can further provide, fundamental information for the discrimination of taxa and for the reconstruction of phylogenetic relationships among both genera and species. Different aspects and patterns of evolution among species were unravelled by the different genetic and cytogenetic markers, according to their different rates of evolution; in addition non-homogeneous rates of karyotype and molecular changes were observed in different species and genera.

All this continuous new information gained by the application of new molecular technologies in recent years is upsetting the traditional taxonomy based on morphological and meristic characters, a situation common to other animal taxa, but here amplified due to the highly conservative morphology of the Mugilidae family. However a correlation between differences found at molecular level and the practical identification of species on the field is still missing. After the disorder of Mugilidae taxonomy by molecular geneticists, a confirmation by morphological analysis is expected and necessary, with the deposit of holotypes for the new species, and hard work directly comparing the samples from different populations/areas with the holotype. Indeed, the idea that a species has to be recognized on morphological features must be kept in mind, especially for such species as mullets which are important for commercial fisheries. If this book had been published in a few years time, the taxonomic status of most Mugilidae species would probably have been clarified.

Acknowledgements

We acknowledge J.D. Durand, V. Milana and T. Svåsand for their comments and suggestions. A.R. Rossi acknowledges M. Nirchio and L. Sola for their intense and lively collaboration on grey mullets cytogenetics in the last decade.

References

Allendorf, F.W., O. Berry and N. Ryman. 2014. Solong to genetic diversity, and thanks for all the fish. Mol. Ecol. 23: 23–25.

Amemiya, C.T. and J.R. Gold. 1986. Chromomycin A_3 stains nucleolus organizer regions of fish chromosomes. Copeia 1986: 226–231.

Amemiya, C.T. and J.R. Gold. 1988. Chromosomal NORs as taxonomic and systematic characters in North cyprinid fishes. Genetica 76: 81–90.

Antoniou, A. and A. Magoulas. 2014. Application of mitochondrial DNA in stock identification. pp. 257–295. *In*: S.X. Cadrin, L.A. Kerr and S. Mariani (eds.). Stock Identification Methods (2nd edition). Academic Press, San Diego.

Arai, R. 2011. Fish Karyotypes. A Checklist. Springer, Tokyo. 340pp.

Arefyev, V.A. 1989. The application of the method of colchicine baths to studies of karyotypes of the young of two mullet species (Mugilidae) from the Black Sea. pp. 139–149. *In*: L.A. Dushkina (ed.). Early Life History of Marine Species. Sb Nauchn Tr VNIRO, Moscow.

Artoni, R.F., M.R. Vicari, M.C. Almeida, O. Moreira-Filho and L.A.C. Bertollo. 2009. Karyotype diversity and fish conservation of southern field from South Brazil. Rev. Fish Biol. Fish 19: 393–401.

Aurelle, D., R.M. Barthelemy, J.P. Quignard, M. Trabelsi and E. Faure. 2008. Molecular phylogeny of Mugilidae (Teleostei: Perciformes). Open Mar. Biol. J. 2: 29–37.

Avise, J.C. 2004. Molecular Markers, Natural History and Evolution (2nd edition). Sinauer Associates, Sunderland, Massachusetts.

Bernardi, G. 2005. Structural and Evolutionary Genomics: Natural Selection in Genome Evolution. Elsevier. 458pp.

Berra, T.M. 2001. Freshwater Fish Distribution. Academic Press, San Diego, CA.

Bertollo, L.A.C., G.C. Born, J.A. Dergam, A.S. Fenocchio and O. Moreira-Filho. 2000. A biodiversity approach in the neotropical Erythrinidae fish *Hoplias malabaricus*. Karyotypic survey, geographic distribution of cytotypes and cytotaxonomic considerations. Chrom. Res. 8: 603–613.

Blel, H., N. Chatti, R. Besbes, S. Farjallah, A. Elouaer, H. Guerbej and K. Said. 2008. Phylogenetic relationships in grey mullets (Mugilidae) in a Tunisian lagoon. Aquacult. Res. 39: 268–275.

Blel, H., J. Panfili, B. Guinand, P. Berrebi, K. Said and J.D. Durand. 2010. Selection footprint at the first intron of the Prl gene in natural populations of the flathead mullet (*Mugil cephalus*, L. 1758). J. Exp. Mar. Biol. Ecol. 387: 60–67.

Bloom, D.D., J.T. Weir, K.R. Piller and N.R. Lovejoy. 2013. Do freshwater fishes diversify faster than marine fishes? A test using state-dependent diversification analyses and molecular phylogenetics of new world silversides (Atherinopsidae). Evolution 67: 2040–2057. doi:10.1111/evo.12074.

Brum, M.J.I. and P.M. Galetti. 1997. Teleostei ground plan karyotype. J. Comput. Biol. 2: 91–102.

Caldara, F., L. Bargelloni, L. Ostellari, E. Penzo, L. Colombo and T. Patarnello. 1996. Molecular phylogeny of grey mullets based on mitochondrial DNA sequence analysis: evidence of a differential rate of evolution at the intrafamily level. Mol. Phylogenet. Evol. 6: 416–424.

Cano, J., G. Thode and M.C. Alvarez. 1982. Karyoevolutive considerations in 29 Mediterranean teleost fishes. Vie et Milieu 32: 21–24.

Cataudella, S. and E. Capanna. 1973. Chromosome complements of three species of Mugilidae (Pisces: Perciformes). Experientia 29: 489–491.

Cataudella, S., M.V. Civitelli and E. Capanna. 1974. Chromosome complement of Mediterranean mullets (Pisces, Perciformes). Caryologia 27: 93–105.

Chakrabarti, J. and A.R. Khuda-Bukhsh. 2000. Chromosome banding studies in an Indian mullet: evidence of structural rearrangments from NOR location. Indian J. Exp. Biol. 5: 467–470.

Chauhan, T. and K. Rajiv. 2010. Molecular markers and their applications in fisheries and aquaculture. Adv. Biosc. Biotechnol. 1: 281–291.

Chatterje, K. and A. Majhi. 1973. Chromosomes of *Mugil parsia* Hamilton (Teleostei: Mugiliformes, Mugilidae). Genen en Phaenen 16: 51–54.

Chen, W.Y., W.C. Su, K.T. Shao and C.P. Lin. 1989. Morphometric studies of the grey mullet (*Mugil cephalus*) from the waters around Taiwan. J. Fish Soc. Taiwan 16: 153–163.

Choudhury, R.C., R. Prasad and C.C. Das. 1979. Chromosome of six species of marine fishes. Caryologia 32: 15–21.

Cioffi, M.B., C. Martins and L.A.C. Bertollo. 2010. Chromosome spreading of associated transposable elements and ribosomal DNA in the fish *Erythrinus erythrinus*. Implications for genome change and karyoevolution in fish. BMC Evol. Biol. 10: 271.

Cioffi, M.B., A. Sánchez, J.A. Marchal, N. Kosyakova, T. Liehr, V. Trifonov and L.A.C. Bertollo. 2011. Cross-species chromosome painting tracks the independent origin of multiple sex chromosomes in two cofamiliar Erythrinidae fishes. BMC Evol. Biol. 11: 186.

Cipriano, R.R., M.M. Cestari and A.S. Fenocchio. 2002. Levantamento citogenético de peixes marinhos do litoral do Paraná. IX Simposio de Citogenética e Genetica de Peixes. Maringa, Brasil.

Cipriano, R.R., A.S. Fenocchio, R. Ferreira Artoni, W. Molina, R. Bueno Noleto, D.L. Zanella Kantek and M.M. Cestari. 2008. Chromosomal studies of five species of the marine fishes from the Paranaguá Bay and the karyotypic diversity in the marine Teleostei of the Brazilian coast. Braz. Arch. Biol. Techn. 51: 303–314.

Crosetti, D. 2015. Current state of grey mullet fisheries and culture. *In*: D. Crosetti and S.J.M. Blaber (eds.). Biology, Ecology and Culture of Grey Mullet (Mugilidae). CRC Press, Boca Raton, USA (this book).

Crosetti, D., J.C. Avise, F. Placidi, A.R. Rossi and L. Sola. 1993. Geographic variability in the grey mullet *Mugil cephalus*: preliminary results of mtDNA and chromosome analyses. Aquaculture 111: 95–101.

Crosetti, D., W.S. Nelson and J.C. Avise. 1994. Pronounced genetic structure of mitochondrial DNA among populations of the circumglobally distributed grey mullet (*Mugil cephalus* Linnaeus). J. Fish Biol. 44: 47–58.

Değer, D., E. Ünlü and M. Gaffaroğlu. 2013. Karyotype of mullet *Liza abu* Heckel, 1846 (Pisces: Mugilidae) from the Tigris River, Turkey. J. Appl. Ichthyol. 29: 234–236.

Delgado, J.V., A. Molina, J. Lobillo, A. Alonso, D. Llanes and M.E. Camacho. 1990. Estudio citogénetico en tres especies de la familia Mugilidae, *Liza aurata, Liza ramada* y *Chelon labrosus*: morfometria chromosomica, bandas C y distribucion de los NORs. Actas III Congreso Nacional de Acuicultura: 301–305.

Delgado, J.V., G. Thode, J. Lobillo, M.E. Camacho, A. Alonso and A. Rodero. 1991. Detection of the nucleolar organizer regions in the chromosomes of the family Mugiliidae (Perciformes): technical improvements. Arch. Zootec. 40: 301–305.

Delgado, J.V., A. Molina, J. Lobillo, A. Alonso and M.E. Camacho. 1992. Morphometrical study on the chromosomes of three species of mullet (Teleostei: Mugilidae). Caryologia 45: 263–271.

Diniz, D., A. Laudicina, M.B. Cioffi and L.A.C. Bertollo. 2008. Microdissection and whole chromosome painting. Improving sex chromosome analysis in *Triportheus* (Teleostei: Characiformes). Cytogenet. Genome Res. 122: 163–168.

Durand, J.D. 2015. Implications of molecular phylogeny for the taxonomy of Mugilidae. *In*: D. Crosetti and S. Blaber (eds.). Biology, Ecology and Culture of Grey Mullet (Mugilidae). CRC Press, Boca Raton, USA (this book).

Durand, J.D. and P. Borsa. 2015. Mitochondrial phylogeny of grey mullets (Acanthopterygii: Mugilidae) suggests high proportion of cryptic species. C. R. Biologies 338: 266–277.

Durand, J.D., W.J. Chen, K.N. Shen, C. Fu and P. Borsa. 2012a. Genus-level taxonomic changes implied by the mitochondrial phylogeny of grey mullets (Teleostei: Mugilidae). C.R. Biologie 335: 687–697.

Durand, J.D., W.J. Chen, K.N. Shen, B.W. Jamandre, H. Blel, K. Diop, M. Nirchio, F.J. Garcia de León, A.K. Whitfield, C.-W. Chang and P. Borsa. 2012b. Systematics of the grey mullets (Teleostei: Mugiliformes: Mugilidae): Molecular phylogenetic evidence challenges two centuries of morphology-based taxonomy. Mol. Phylogenet. Evol. 64: 73–92.

Durand, J.D., H. Blel, K.N. Shen, E.T. Koutrakis and B. Guinand. 2013. Population genetic structure of *Mugil cephalus* in the Mediterranean and Black Seas: a single mitochondrial clade and many nuclear barriers. Mar. Ecol. Prog. Ser. 474: 243–261.

El-Zaeem, S.Y. 2011. Phenotype and genotype differentiation between flathead grey mullet [*Mugil cephalus*] and thinlip grey mullet [*Liza ramada* (Pisces: Mugilidae)]. Afr. J. Biotechnol. 10: 9485–9492.

Erguden, D., M. Gurlek, D. Yaglioglu and C. Turan. 2010. Genetic identification and taxonomic relationship of Mediterranean mugilid species based on mitochondrial 16S rDNA sequence data. J. Anim. Vet. Adv. 9: 336–341.

Eschmeyer, W.N. 2015. Catalog of Fishes: Genera, Species, References (http://research.calacademy.org/research/ichthyology/catalog/fishcatmain.asp). Electronic version accessed on 15/07/2015.

Eschmeyer, W.N. and J.D. Fong. 2015. Catalog of Fishes: Species by Family/Subfamily (http://researcharchive.calacademy.org/research/ichthyology/catalog/SpeciesByFamily.asp). Electronic version accessed on 15/01/2015.

FAO. 2014. Fishstat Aquaculture productions (1970–2012) available at http://www.fao.org/fishery/topic/166235/en, accessed on 2/11/2014.

Feldheim, K.A., P.J. Sanchez, W.A. Matamoros, J.F. Schaefer and B.R. Kreiser. 2009. Isolation and characterization of microsatellite loci for mountain mullet (*Agonostomus monticola*). Mol. Ecol. Res. 9: 1482–1484.

Ferguson, M.M. and R.G. Danzmann. 1998. Role of genetic markers in fisheries and aquaculture: Useful tools or stamp collecting? Can. J. Fish Aquat. Sci. 55: 1553–1563.

Fraga, E., H. Schneider, M. Nirchio, E. Santa-Brigida, L.F. Rodrigues-Filho and I. Sampaio. 2007. Molecular phylogenetic analyses of mullets (Mugilidae: Mugiliformes) based on two mitochondrial genes. J. Appl. Ichthyol. 23: 598–604.

Froese, R. and D. Pauly. 2014. FishBase. World Wide Web electronic publication. Available at: http//www.fishbase.org, accessed on 30/10/2014/.

Gao, T., Y. Li, C. Chen, N. Song and B. Yan. 2014. Genetic diversity and population structure in the mtDNA control region of *Liza haematocheilus* (Temminck and Schlegel 1845). J. Appl. Ichthyol. 2014: 1–7.

Garcia de Leaniz, C., I.A. Fleming, S. Einum, E. Verspoor, W.C. Jordan, S. Consuegra, N. Aubin-Horth, D. Lajus, B.H. Letcher, A.F. Youngson, J.H. Webb, L.A. Vøllestad, B. Villanueva, A. Ferguson and T.P. Quinn. 2007. A critical review of adaptive genetic variation in Atlantic salmon: implications for conservation. Biol. Rev. 82: 173–211.

Ghasemzadeh, J. 1998. Phylogeny and systematics of Indo-Pacific mullets (Teleostei: Mugilidae) with special reference to the mullets of Australia. Unpublished Ph.D. Thesis, Macquarie University, Sydney, Australia. 397pp.

Ghodsi, Z., A. Shabani and B. Shabani Pour. 2011. Study of genetic diversity of *Liza aurata* in the coasts of the Golestan Province by microsatellite method. Taxon. Biosyst. J. 3: 35–49.

Glover, K.A., C. Pertoldi, F. Besnier, V. Wennevik, M. Kent and O. Skaala. 2013. Atlantic salmon populations invaded by farmed escapees: quantifying genetic introgression with a Bayesian approach and SNPs. BMC Genet. 14: 74.

González-Castro, M., J.M. de Astarloa and M.B. Cousseau. 2006. First record of a tropical affinity mullet, *Mugil curema* (Mugilidae), in a temperate southwestern Atlantic coastal lagoon. Cybium 30: 90–91.

González-Castro, M., S. Heras, M.B. Cousseau and M.I. Roldán. 2008. Assessing species validity of *Mugil platanus* Günther 1880 in relation to *Mugil cephalus* Linnaeus 1758 (Actinopterygii). It. J. Zool. 75: 319–325.

González-Castro, M., A.L. Ibáñez, S. Heras, M.I. Roldán and M.B. Cousseau. 2012. Assessment of lineal versus landmark-based morphometry for discriminating species of Mugilidae (Actinopterygii). Zool. Studies 51: 1515–1528.

González-Castro, M. and J. Ghasemzadeh. 2015. Morphology and morphometry based taxonomy of Mugilidae. *In*: D. Crosetti and S.J.M. Blaber (eds.). Biology, Ecology and Culture of Grey Mullet (Mugilidae). CRC Press, Boca Raton, USA (this book).

Gornung, E. 2013. Twenty years of physical mapping of major ribosomal RNA genes across the Teleosts: a review of research. Cytogenet. Genome Res. 141: 90–102.

Gornung, E., C.A. Cordisco, A.R. Rossi, S. De Innocentiis, D. Crosetti and L. Sola. 2001. Chromosomal evolution in Mugilidae: karyotype characterization of *Liza saliens* and comparative localization of major and minor ribosomal genes in the six Mediterranean mullet. Mar. Biol. 139: 55–60.

Gornung, E., M.E. Mannarelli, A.R. Rossi and L. Sola. 2004. Chromosomal evolution in Mugilidae (Pisces: Mugiliformes): FISH mapping of the (TTAGGG)n telomeric repeat in the six Mediterranean mullets. Hereditas 140: 1–2.

Gornung, E., P. Colangelo and F. Annesi. 2007. 5S ribosomal RNA genes in six species of Mediterranean grey mullets: genomic organization and phylogenetic inference. Genome 50: 787–795.

Han, Z.Q., G. Han, T.X. Gao, Z.Y. Wang and B.N. Shui. 2013. Genetic population structure of *Liza haematocheilus* in north-western Pacific detected by amplified fragment length polymorphism markers. J. Mar. Biol. Assoc. U.K. 93: 373–379.

Hansen, M.M. 2010. Expression of interest: transcriptomics and the designation of conservation units. Mol. Ecol. 19: 1757–1759.

Harrison, I.J. 2002. Order Mugiliformes, Mugilidae. pp. 1071–1085. *In*: K.E. Carpenter (ed.). The Living Marine Resources of the Western Central Atlantic, Vol. 2: Bony Fishes part 1 (Acipenseridae to Grammatidae). Food and Agriculture Organization, Rome.

Harrison, I.J. and G.J. Howes. 1991. The pharyngobranchial organ of mugilid fishes; its structure, variability, ontogeny, possible function and taxonomic utility. Bull. British Mus. Natl. Hist. (Zoology Series) 57: 111–132.

Harrison, I.J., M. Nirchio, C. Oliveira, E. Ron and J. Gaviria. 2007. A new species of mullet (Teleostei: Mugilidae) from Venezuela, with a discussion on the taxonomy of *Mugil gaimardianus.* J. Fish Biol. 71 (Suppl. A): 76–97.

Hauser, L. and J.E. Seeb. 2008. Advances in molecular technology and their impact on fisheries genetics. Fish Fish 9: 473–486.

Hebert, P.D.N., A. Cywinska, S.L. Ball and J.R. de Waard. 2003. Biological identifications through DNA barcodes. Proc. R. Soc. Lond. B 270: 313–321.

Hebert, P.D.N., M.Y. Stoeckle, T.S. Zemlack and C.M. Francis. 2004. Identification of birds through DNA barcodes. PLOS Biol. 2: 1657–1663.
Heras, S., M. González-Castro and M.I. Roldan. 2006. *Mugil curema* in Argentinean waters: combined morphological and molecular approach. Aquaculture 261: 473–478.
Heras, S., M.I. Roldán and M. Gonzalez Castro. 2009. Molecular phylogeny of Mugilidae fishes revised. Rev. Fish Biol. Fish 19: 217–231.
Hett, A.K., M. Nirchio, C. Oliveira, Z.R. Siccha, A.R. Rossi and L. Sola. 2011. Karyotype characterization of *Mugil incilis* Hancock 1830 (Mugiliformes: Mugilidae), including a description of an unusual co-localization of major and minor ribosomal genes in the family. Neotrop. Ichthyol. 9: 107–112.
Howell, W.M. and D.A. Black. 1980. Controlled silver-staining of nucleolus organizer regions with a protective colloidal developer: a 1-step method. Experientia 36: 1014–1015.
Huang, C.S., C.F. Weng and S.C. Lee. 2001. Distinguishing two types of gray mullet, *Mugil cephalus* L. (Mugiliformes: Mugilidae), by using glucose-6-phosphate isomerase (GPI) allozymes with special reference to enzyme activities. J. Comp. Physiol. B171: 387–394.
Huey, J.A., T. Espinoza and J.M. Hughes. 2013. Regional panmixia in the mullet *Mugil cephalus* along the coast of eastern Queensland revealed using six highly polymorphic microsatellite loci. P. Roy. Soc. Queensland 118: 7–15.
Imsiridou, A., G. Minos, V. Katsares, N. Karaiskou and A. Tsiora. 2007. Genetic identification and phylogenetic inferences in different Mugilidae species using 5S rDNA markers. Aquaculture Res. 38: 1370–1379.
Jamandre, B.W., J.D. Durand and W.N. Tzeng. 2009. Phylogeography of the flathead mullet *Mugil cephalus* in the Northwest Pacific inferred from the mtDNA control region. J. Fish Biol. 75: 393–407.
Jamandre, B.W., J.D. Durand and W.N. Tzeng. 2014. High sequence variations in mitochondrial DNA Control Region among worldwide populations of flathead mullet *Mugil cephalus.* Int. J. Zool. 2014: ID 564105, 9 pages.
Jordão, L.C., C. Oliveira, F. Foresti and H. Godinho. 1992. Caracterizacao citogenética da tainha, *Mugil platanus* (Pisces: Mugilidae). Bol. Inst. Pesca 9: 63–66.
Ke, H.M., W.W. Lin and H.W. Kao. 2009. Genetic diversity and differentiation of grey mullet (*Mugil cephalus*) in the coastal waters of Taiwan. Zool. Sci. 26: 421–428.
Khoroshko, A.I. 1981. Population abundance and structure in the long-finned mullet (genus *Liza*, Mugilidae) during acclimation in the Caspian Sea. J. Ichthyol. 22: 62–69.
Khuda-Bukhsh, A.R. and G.K. Manna. 1974. Somatic chromosomes in seven species of teleostean fishes. Chromosome Inf. Serv. 17: 5–6.
Klossa-Kilia, E., V. Papasotiropoulos, G. Kilias and S. Alahiotis. 2002. Authentication of Messolongi (Greece) fish roe using PCR–RFLP analysis of 16s rRNA mtDNA segment. Food Control 13: 169–172.
Krück, N.C., D.I. Innes and J.R. Ovenden. 2013. New SNPs for population genetic analysis reveal possible cryptic speciation of eastern Australian sea mullet (*Mugil cephalus*). Mol. Ecol. Res. 13: 715–725.
Lee, S.C. 1992. Fish fauna and abundance of some dominant species in the estuary of Tanshui, northwestern Taiwan. J. Fish Soc. Taiwan 19: 263–271.
Lee, S.C., J.T. Chang and Y.Y. Tsu. 1995. Genetic relationships of four Taiwan mullets (Pisces: Perciformes: Mugilidae). J. Fish Biol. 46: 159–162.
Lee, S.C., H.L. Cheng and J.T. Chang. 1996. Allozyme variation in the large-scale mullet *Liza macrolepis* (Perciformes: Mugilidae) from coastal waters of western Taiwan. Zool. Stud. 35: 85–92.
LeGrande, W.H. and J.M. Fitzsimons. 1976. Karyology of the mullets *Mugil curema* and *Mugil cephalus* (Perciformes: Mugilidae) from Louisiana. Copeia 1976: 388–391.
Li, C.H., T.W. Li, L.S. Song and X.R. Su. 2003. RAPD analysis on intraspecies differentiation of *Tegillarca granosa* populations to the south and north of Fujian province. Zool. Res. 24: 362–366.
Liu, C.H. 1986. Survey of the spawning grounds of gray mullet. pp. 63–72. *In*: W.C. Su (ed.). Study on the Resource of Gray Mullet in Taiwan 1983–1985. Taiwan Fisheries Research Institute, Kaohsiung.
Liu, J.X., T.X. Gao, S.F. Wu and Y.P. Zhang. 2007. Pleistocene isolation in the Northwestern Pacific marginal seas and limited dispersal in a marine fish, *Chelon haematocheilus* (Temminck and Schlegel 1845). Mol. Ecol. 16: 275–288.
Liu, J.Y., C.L. Brown and T.B. Yang. 2009a. Population genetic structure and historical demography of grey mullet, *Mugil cephalus*, along the coast of China, inferred by analysis of the mitochondrial control region. Biochem. Syst. Ecol. 37: 556–566.
Liu, J.Y., Z.R. Lun, J.B. Zhang and T.B. Yang. 2009b. Population genetic structure of striped mullet, *Mugil cephalus*, along the coast of China, inferred by AFLP fingerprinting. Biochem. Syst. Ecol. 37: 266–274.
Liu, J.Y., C.L. Brown and T.B. Yang. 2010. Phylogenetic relationships of mullets (Mugilidae) in China Seas based on partial sequences of two mitochondrial genes. Biochem. Syst. Ecol. 38: 647–655.
Livi, S., L. Sola and D. Crosetti. 2011. Phylogeographic relationships among worldwide populations of the cosmopolitan marine species, the striped gray mullet (*Mugil cephalus*), investigated by partial cytochrome b gene sequences. Biochem. Syst. Ecol. 39: 121–131.
Mariani, S. and D. Bekkevold. 2014. The nuclear genome: neutral and adaptive markers in fisheries. pp. 297–327. *In*: S.X. Cadrin, L.A. Kerr and S. Mariani (eds.). Stock Identification Methods (2nd edition). Academic Press, San Diego.
Martins, C. and A.P. Wasko. 2004. Organization and evolution of 5S ribosomal DNA in the fish genome. pp. 335–363. *In*: C.R. Williams (ed.). Focus on Genome Research. Nova Biomedical Books, New York.

McCusker, M.R. and P. Benzen. 2010. Positive relationships between genetic diversity and abundance in fishes. Mol. Ecol. 19: 4852–4862.

McMahan, C.D., M.P. Davis, O. Domínguez-Domínguez, F.J. García-de-León, I. Doadrio and K.R. Piller. 2013. From the mountains to the sea: phylogeography and cryptic diversity within the mountain mullet, *Agonostomus monticola* (Teleostei: Mugilidae). J. Biogeogr. 40: 894–904.

Menezes, N.A., C. Oliveira and M. Nirchio. 2010. An old dilemma: the identity of the western south Atlantic lebranche mullet (Teleostei: Perciformes: Mugilidae). Zootaxa 2519: 59–68.

Menezes, N., M. Nirchio, C. de Oliveira and R. Siccha-Ramirez. 2015. Taxonomic review of the species of *Mugil* (Teleostei: Perciformes: Mugilidae) from the Atlantic South Caribbean and South America with integration of the morphological, cytogenetic and molecular data. Zootaxa 3918: 001–038.

Meng, W., T. Gao and B. Zheng. 2007. Genetic analysis of four populations of redlip mullet (*Chelon haematocheilus*) collected in China Seas. J. Ocean Univ. China 6: 72–75.

Miggiano, E., R.E. Lyons, Y. Li, L.M. Dierens, D. Crosetti and L. Sola. 2005. Isolation and characterization of microsatellite loci in the striped mullet, *Mugil cephalus*. Mol. Ecol. Notes 5: 323–326.

Miller, R.R., W.L. Minckley and S.M. Norris. 2005. Freshwater fishes of Mexico. University of Chicago Press, Chicago.

Minos, G., A. Imsiridou and P.S. Economidis. 2010. *Liza haematocheilus* (Pisces; Mugilidae) in the northern Aegean Sea. pp. 313–332. *In*: D. Golani and B. Appelbaum-Golani (eds.). Fish Invasions of the Mediterranean Sea: Change and Renewal. Pensoft Publishers, Sofia–Moscow.

Miya, M., A. Kawaguchi and M. Nishida. 2001. Mitogenomic exploration of higher teleostean phylogenies: a case study for moderate-scale evolutionary genomics with 38 newly determined complete mitochondrial DNA sequences. Mol. Biol. Evol. 18: 1993–2009.

Molecular Ecology Resources Primer Development Consortium et al. 2010. Permanent Genetic Resources added to Molecular Ecology Resources Database. Mol. Ecol. Res. 10: 1098–1105.

Molina, W.F. 2007. Chromosomal changes and stasis in marine fish groups. pp. 69–110. *In*: E. Pisano, C. Ozouf-Costaz, F. Foresti and B.G. Kapoor (eds.). Fish Cytogenetics. Science Publishers, Enfield.

Murgia, R., G. Tola, S.N. Archer, S. Vallerga and J. Hirano. 2002. Genetic identification of grey mullet species (Mugilidae) by analysis of mitochondrial DNA sequence: application to identify the origin of processed ovary products (bottarga). Mar. Biotechnol. 4: 119–126.

Natarajan, R. and K. Subrahmanyam. 1974. A karyotype study of some teleosts from Portonovo waters. Proc. Indian Acad. Sci. 79B: 173–196.

Nayyar, R.P. 1966. Karyotype studies in thirteen species of fishes. Genetica 37: 78–92.

Nei, M. 1978. Estimation of average heterozygosity and genetic distance from a small number of individuals. Genetics 89: 583–590.

Nematzadeh, M., S.R. Gillkolaei, M.K. Khalesi and F. Laloei. 2013a. Molecular phylogeny of mullets (Teleosti: Mugilidae) in Iran based on mitochondrial DNA. Biochem. Genet. 51: 334–340.

Nematzadeh, M., S. Rezvani, M.K. Khalesi, F. Laloei and A. Fahim. 2013b. A phylogeny analysis on six mullet species (Teleosti: Mugillidae) using PCR-sequencing method. Iran. J. Fish Sci. 12: 669–679.

Nielsen, E.E., A. Cariani, E. Mac Aoidh, G.E. Maes, I. Milano, R. Ogden, M. Taylor, J. Hemmer-Hansen, M. Babbucci, L. Bargelloni, D. Bekkevold, E. Diopere, L. Grenfell, S. Helyar, M.T. Limborg, J.T. Martinsohn, R. McEwing, F. Panitz, T. Patarnello, F. Tinti, J.K.J. Van Houdt, F.A.M. Volckaert, R.S. Waples, FishPopTrace Consortium and G.R. Carvalho. 2012. Gene-associated markers provide tools for tackling illegal fishing and false eco-certification. Nature Comm. 3: 851.

Nirchio, M. and H. Cequea. 1998. Karyology of *Mugil liza* and *M. curema* from Venezuela. Boletin Invest. Mar. Costeras. 27: 45–50.

Nirchio, M., D. González and J.E. Pérez. 2001. Estudio citogenético de *Mugil curema* y *M. liza* (Pisces: Mugilidae): Regiones organizadoras del nucleolo. Boletín del instituto Oceanográfico de Venezuela. Bolet. Inst. Oceanogr. Universidad de Oriente Venezuela 40: 3–7.

Nirchio, M., F. Cervigon, J.I. Rebelo Porto, J.E. Perez, J.A. Gomez and J. Villalaz. 2003. Karyotype supporting *Mugil curema* Valenciennes 1836 and *Mugil gaimardanus* Desmaret 1837 (Mugilidae: Teleostei) as two valid nominal species. Sci. Mar. 67: 113–115.

Nirchio, M., E. Ron and A.R. Rossi. 2005a. Karyological characterization of *Mugil trichodon* Poey 1876 (Pisces: Mugilidae). Sci. Mar. 69: 525–530.

Nirchio, M., R.R. Cipriano, M.M. Cestari and A.S. Fenocchio. 2005b. Cytogenetical and morphological features reveal significant differences among Venezuelan and Brazilian samples of *Mugil curema*. Neotrop. Ichthyol. 3: 107–110.

Nirchio, M., C. Oliveira, I.A. Ferreira, J.E. Pérez, J.I. Gaviria, I. Harrison, A.R. Rossi and L. Sola. 2007. Comparative cytogenetic and allozyme analysis of *Mugil rubrioculus* and *M. curema* (Teleostei: Mugilidae) from Venezuela. Interciencia. 32: 757–762.

Nirchio, M., C. Oliveira, I.A. Ferreira, C. Martins, A.R. Rossi and L. Sola. 2009. Classical and molecular cytogenetic characterization of *Agonostomus monticola*, a primitive species of Mugilidae (Mugiliformes). Genetica 135: 1–5.

Nirchio, M., A.R. Rossi, F. Foresti and C. Oliveira. 2014. Chromosome evolution in fishes: a new challenging proposal from Neotropical species. Neotrop. Ichthyol. 12:761–770.

Ocalewicz, K. 2013. Telomeres in fishes. Cytogenet. Genome Res. 141: 114–125.

Ohno, S. 1974. Protochordata, cyclostomata and pisces. pp. 1–91. *In*: B. John (ed.). Animal Cytogenetics, Vol. 4, Chordata 1. Gerbruder Borntraeger, Berlin.

Omelchenko, V.T., E.A. Salmenkova, M.A. Makhotkin, N.S. Romanov, Yu. P. Altukhov, S.I. Dudkin, V.A. Dekhta, G.A. Rubtsova and M. Yu. Kovalev. 2004. Far eastern mullet *Mugil soiuy* Basilewsky (Mugilidae; Mugiliformes): the genetic structure of populations and its change under acclimatization. Russian J. Genet. 40: 910–918.

Pan, B.P., L.S. Song, W.J. Pu and J.S. Sun. 2005. Studies on genetic diversity and differentiation between two allopatric populations of *Cyclina sinensis*. Acta. Hydrobiol. Sin. 29: 372–378.

Pansonato-Alves, J.C., E. Alves Serrano, R. Utsunomia, J.P.M. Camacho, G.J. da Costa Silva, M.R. Vicari, R. Ferreira Artoni, C. Oliveira and F. Foresti. 2014. Single origin of sex chromosomes and multiple origins of B chromosomes in fish genus Characidium. PlosOne 9: e107169.

Papa, R., F. Nonnis Marzano, V. Rossi and G. Gandolfi. 2003. Genetic diversity and adaptability of two species of Mugilidae (Teleostei: Perciformes) of the Po river delta coastal lagoons. Oceanol. Acta 26: 121–128.

Papasotiropoulos, V., E. Klossa-Kilia, G. Kilias and S. Alahiotis. 2001. Genetic divergence and phylogenetic relationships in grey mullets (Teleostei: Mugilidae) using allozyme data. Biochem. Genet. 39: 155–168.

Papasotiropoulos, V., E. Klossa-Kilia, G. Kilias and S. Alahiotis. 2002. Genetic divergence and phylogenetic relationships in grey mullets (Teleostei: Mugilidae) based on PCR–RFLP analysis of mtDNA segments. Biochem. Genet. 40: 71–86.

Papasotiropoulos, V., E. Klossa-Kilia, S. Alahiotis and G. Kilias. 2007. Molecular phylogeny of grey mullets (Teleostei: Mugilidae) in Greece. Evidence from sequence analysis of mtDNA segments. Biochem. Genet. 45: 623–636.

Parise-Maltempi, P.P., E.D. da Silva, W. Rens, F. Dearden, P.C.M. O'Brien, V. Trifonov and M.A. Ferguson-Smith. 2013. Comparative analysis of sex chromosomes in *Leporinus* species (Teleostei, Characiformes) using chromosome painting. BMC Genetics 14: 60.

Patarnello, T., F.A.M.J. Volckaert and R. Castilho. 2007. Pillars of hercules: is the Atlantic–Mediterranean transition a phylogeographical break? Mol. Ecol. 16: 4426–4444.

Pauls, E. and I.A. Coutinho. 1990. Levantamento citogenético em peixes de maior valor econômico do litoral Fluminense, RJ (23° Lat/S). Congresso Brasileiro de Zoologia 17. Universidade Estadual de Londrina. 325p.

Peregrino-Uriarte, A.B., R. Pacheco-Aguilar, A. Varela-Romero and G. Yepiz-Plascencia. 2007. Differences in the 16SrRNA and cytochrome c oxidase subunit I genes in the mullets *Mugil cephalus* and *Mugil curema*, and snooks *Centropomus viridis* and *Centropomus robalito*. Ciencias Marinas 33: 95–104.

Phillips, R.B. 2013. Evolution of sex chromosomes in Salmonid fishes. Cytogenet. Genome Res. 141: 177–185.

Pinsky, M.L. and S.R. Palumbi. 2014. Meta-analysis reveals lower genetic diversity in overfished populations. Mol. Ecol. 23: 29–39.

Radulovici, A.E., P. Archambault and F. Dufresne. 2010. DNA Barcodes for marine biodiversity: moving fast forward? Diversity 2: 450–472.

Reiss, H., G. Hoarau, M. Dickey-Collas and W.J. Wolff. 2009. Genetic population structure of marine fish: mismatch between biological and fisheries management units. Fish Fish 10: 361–395.

Rishi, K.K. and J. Singh. 1982. Karyological studies on five estuarine fishes. The nucleus 25: 178–180.

Rocha-Olivares, A., N.M. Garber and K.C. Stuck. 2000. High genetic diversity, large inter-oceanic divergence and historical demography of the striped mullet. J. Fish Biol. 57: 1134–1149.

Rocha-Olivares, A., N.M. Garber, A.F. Garber and K.C. Stuck. 2005. Structure of the mitochondrial control region and flanking region tRNA genes of *Mugil cephalus*. Hydrobiologica 15: 139–149.

Rodrigues-Filho, L.F.S., D.B. da Cunha, M. Vallinoto, H. Schneider, I. da Sampaio and E. Fraga. 2011. Polymerase chain reaction banding patterns of the 5S rDNA gene as a diagnostic tool for the discrimination of South American mullets of the genus *Mugil*. Aquacult. Res. 42: 1117–1122.

Rossi, A.R., D. Crosetti, E. Gornung and L. Sola. 1996. Cytogenetic analysis of global populations of *Mugil cephalus* (striped mullet) by different staining techniques and fluorescent *in situ* hybridization. Heredity 76: 77–82.

Rossi, A.R., E. Gornung and D. Crosetti. 1997. Cytogenetic analysis of *Liza ramada* (Pisces: Perciformes) by different staining techniques and fluorescent *in situ* hybridization. Heredity 79: 83–87.

Rossi, A.R., M. Capula, D. Crosetti, L. Sola and D.E. Campton. 1998a. Allozyme variation in global populations of striped mullet, *Mugil cephalus* (Pisces: Mugilidae). Mar. Biol. 131: 203–212.

Rossi, A.R., M. Capula, D. Crosetti, D.E. Campton and L. Sola. 1998b. Genetic divergence and phylogenetic inferences in five species of Mugilidae (Pisces: Perciformes). Mar. Biol. 131: 213–218.

Rossi, A.R., E. Gornung, D. Crosetti, S. De Innocentiis and L. Sola. 2000. Cytogenetic analysis of *Oedalechilus labeo* (Pisces: Mugilidae), with a report of NOR variability. Mar. Biol. 136: 159–162.

Rossi, A.R., A. Ungaro, S. De Innocentiis, D. Crosetti and L. Sola. 2004. Phylogenetic analysis of Mediterranean Mugilids by allozymes and 16S mt-rRNA genes investigation: are the Mediterranean species of *Liza* monophyletic? Biochem. Genet. 42: 301–315.

Rossi, A.R., E. Gornung, L. Sola and M. Nirchio. 2005. Comparative molecular cytogenetic analysis of two congeneric species, *Mugil curema* and *M. liza*, characterized by significant karyotype diversity. Genetica 125: 27–32.

Saeidi, Z., S. RezvaniGilkolaei, M. Soltani and F. Laloei. 2014. Population genetic studies of *Liza aurata* using D-Loop sequencing in the southeast and southwest coasts of the Caspian Sea. Iranian J. Fish Sci. 13: 216–227.

Sala-Bozano, M., V. Ketmaier and S. Mariani. 2009. Contrasting signals from multiple markers illuminate population connectivity in a marine fish. Mol. Ecol. 18: 4811–4826.

Salmenkova, E.A., N.V. Gordeeva, V.T. Omel'chenko, M.A. Makhotkin and S.I. Dudkin. 2007. Variation of mitochondrial DNA in far Eastern mullet pilengas, *Liza haematocheilus* Temminck and Schlegel, acclimatized in the Azov-Black Sea basin. Russian J. Genet. 43: 1062–1065.

Savolainen, V., R.S. Cowan, A.P. Vogler, G.K. Roderick and R. Lane. 2005. Towards writing the encyclopedia of life: an introduction to DNA barcoding. Philos. Trans. R. Soc. Lond., B, Biol. Sci. 360: 1805–1811.

Semina, A.V., N.E. Polyakova and V.A. Brykov. 2007a. Analysis of mitochondrial DNA: taxonomic and phylogenetic relationships in two fish taxa (Pisces: Mugilidae and Cyprinidae). Biochemistry 72(12): 1349–1355.

Semina, A.V., N.E. Polyakova, M.A. Makhotkin and V.A. Brykov. 2007b. Mitochondrial DNA divergence and phylogenetic relationships in mullets (Pisces: Mugilidae) of the Sea of Japan and the Sea of Azov revealed by PCR-RFLP analysis. Russian J. Mar. Biol. 33: 182–192.

Shekhar, M.S., M. Natarajan and K.V. Kumar. 2011. PCR-RFLP an sequence analysis of 12S and 16S rRNA mitochondrial genes of grey mullets from East coast of India. Indian J. Geo-Mar. Sci. 40: 529–534.

Shen, K.N., B.W. Jamandre, C.C. Hsu, W.N. Tzeng and J.D. Durand. 2011. Plio-Pleistocene sea level and temperature fluctuations in the northwestern Pacific promoted speciation in the globally-distributed flathead mullet *Mugil cephalus*. BMC Evol. Biol. 11: 83.

Siccha-Ramirez, R., N.A. Menezes, M. Nirchio, F. Foresti and C. Oliveira. 2014. Molecular identification of mullet species of the Atlantic South Caribbean and South America and the phylogeographic analysis of *Mugil liza*. Rev. Fish Sci. Aquacult. 22: 86–96.

Slijepcevic, P. 1998. Telomeres and mechanisms of Robertsonian fusion. Chromosoma 107: 136–140.

Sola, L., E. Gornung, M.E. Mannarelli and A.R. Rossi. 2007. Chromosomal evolution in Mugilidae, Mugilomorpha: an overview. pp. 165–194. *In*: E. Pisano, C. Ozouf-Costaz, F. Foresti and B.G. Kapoor (eds.). Fish Cytogenetics. Science Publishers, Enfield.

Starushenko, L.I. and A.B. Kazansky. 1996. Introduction of mullet harder (*Mugil so-iuy* Basilewsky) into the Black Sea and the Sea of Azov. Stud. Rev., Gen. Fish Counc. Med. 67: 29.

Su, W.C. and T. Kawasaki. 1995. Characteristics of the life history of gray mullet from Taiwan waters. Fish Sci. 61: 377–381.

Sumner, A.T. 1972. A simple technique for demonstrating centromeric heterochromatin. Exp. Cell Res. 75: 304–306.

Sun, P., Z.H. Shi, F. Yin and S.M. Peng. 2012. Genetic variation analysis of *Mugil cephalus* in China Sea based on mitochondrial COI gene sequences. Biochem. Genet. 50: 180–191.

Taberlet, P., E. Coissac, F. Pompanon, C. Brochmann and E. Willerslev. 2012. Towards next-generation biodiversity assessment using DNA metabarcoding. Mol. Ecol. 21: 2045–2050.

Taranger, G.L., Ø. Karlsen, R.J. Bannister, K.A. Glover, V. Husa, E. Karlsbakk, B.O. Kvamme, K.K. Boxaspen, P.A. Bjørn, B. Finstad, A.S. Madhun, H.C. Morton and T. Svåsand. 2014. Risk assessment of the environmental impact of Norwegian Atlantic salmon farming ICES J. Mar. Sci. doi:10.1093/icesjms/fsu132.

Tereshchenko, K.K. 1950. Materials of the Caspian Sea mullets fisheries (KASPINIRO). Ta Rybn. Kh. Va I Okeanogr. 11: 46–86.

Thomson, J.M. 1997. The Mugilidae of the world. Memoirs of the Queensland Museum 41: 457–562.

Trabelsi, M., D. Aurelle, N. Bourgia, J.P. Quignard, J.P. Casanova and E. Faure. 2008. Identification of juvenile of grey mullet species (Teleostei: Perciformes) from Kuriat Islands (Tunisia) and evidence of gene flow between Atlantic and Mediterranean *Liza aurata*. Cah. Biol. Mar. 49: 269–276.

Trape, S., H. Blel, J. Panfili and J.D. Durand. 2009. Identification of tropical Eastern Atlantic Mugilidae species by PCR-RFLP analysis of mitochondrial 16S rRNA gene fragments. Biochem. Syst. Ecol. 37: 512–518.

Turan, C. 2015. Biogeography and distribution of Mugilidae in the Mediterranean and the Black Sea, and North-East Atlantic. *In*: D. Crosetti and S. Blaber (eds.). Biology, Ecology and Culture of Grey Mullet (Mugilidae). CRC Press, Boca Raton, USA (this book).

Turan, C., M. Caliskan and H. Kucuktas. 2005. Phylogenetic relationships of nine mullet species (Mugilidae) in the Mediterranean Sea. Hydrobiologia 532: 45–51.

Verspoor, E., I. Olsen, H.B. Bentsen, K. Glover, P. McGinnity and A. Norris. 2007. Genetic effects of domestication, culture and breeding of fish and shellfish, and their impacts on wild populations. Atlantic salmon *Salmo salar*. pp. 23–31. *In*: T. Svåsand, D. Crosetti, E. García-Vázquez and E. Verspoor (eds.). Evaluation of Genetic Impact of Aquaculture Activities on Native Populations: A European Network. GENIMPACT Final Report (EU contract n. RICA-CT-2005-022802). http://genimpact.imr.no.

Volff, J.N. 2005. Genome evolution and biodiversity in teleost fish. Heredity 94: 280–294.

Waples, R., A.E. Punt and J.M. Cope. 2008. Integrating genetic data into management of marine resources: how can we do it better? Fish Fish 9: 423–449.

Ward, R.D., T.S. Zemlak, B.H. Innes, P.R. Last and P.D.N. Hebert. 2005. DNA barcoding Australia's fish species. Philos. Trans. R. Soc. Lond., B, Biol. Sci. 360: 1847–1857.

Whitfield, A.K., J. Panfili and J.D. Durand. 2012. A global review of the cosmopolitan flathead mullet *Mugil cephalus* Linnaeus, 1758 (Teleostei: Mugilidae), with emphasis on the biology, genetics, ecology and fisheries aspects of this apparent species complex. Rev. Fish Biol. Fish 22: 641–681.

Xu, G., C. Shao, X. Liao, Y. Tian and S. Chen. 2009. Isolation and characterization of polymorphic microsatellite loci from soiuy mullet (*Mugil soiuy* Basilewsky 1855). Conserv Genet. 10: 653–655.

Xu, T.J., D.Q. Sun, G. Shi and R.X. Wang. 2010. Development and characterization of polymorphic microsatellite markers in the gray mullet (*Mugil cephalus*). Genet. Mol. Res. 9: 1791–1795.

CHAPTER 16

Current State of Grey Mullet Fisheries and Culture

Donatella Crosetti

Introduction

The family Mugilidae includes various species of great ecological and economic importance for both capture fisheries and aquaculture in many regions of the world.

The schooling behaviour of many mullet species, their capacity to adapt to low salinities and hence to enter freshwaters, and their presence on the water surface, often in confined shallow waters, have made these fish an ideal target for primitive coastal populations (Cataudella and Monaco 1983). Traditional capture fisheries and culture practices were implemented by people living in coastal areas, near bays, estuaries and coastal lagoons, and who were familiar with mullet behaviour and life history. In brackish waters, mullets were abundant and represented a relatively easy prey, to be caught in rudimentary traps or more simply, to be scared and pushed to leap out of the water.

Many mullet species are cultured for human consumption, with traditions going back centuries. The extensive culture of grey mullets dates back to ancient times, and developed in coastal lagoons and flooded lands associated with river mouths and estuaries in the warm and temperate regions of the world. Primitive forms of culture carried out in confined areas originated from capture systems based on the migration of mullets, both as juveniles and as adults, towards the eutrophic feeding grounds of coastal lagoons. The capacity of mullets to use the organic matter present in coastal lagoons and their tendency to aggregate in schools and therefore be able to quickly colonize small areas with a high density certainly contributed to the origin of culture.

In the Mediterranean region, bas-reliefs in Ancient Egyptian tombs, dating from 2340 B.C., depict the capture of mullets (Fig. 20.3. in Chapter 20—Sadek 2015). Tales about the capture of mullets in ancient Roman times are narrated by Pliny, who reported peculiar fisheries techniques to capture mullets, where a male mullet collected from stocking ponds was used as bait to make females follow him to shore to be captured, or where males followed females to shore at spawning time. In stocking ponds mullets were domesticated and became acquainted with the pond owner; in Apollinaris villa in Formia, '*nomenculator mugilem citat notum*' (the *nomenculator* calls the mullet he knows by its name) (Epigrams by Martia Liber X30, in Giacopini et al. 1994).

ISPRA, Aquaculture Dept., Via Brancati 60, 00144 Roma, Italia.
Email: donatella.crosetti@isprambiente.it

Already in Roman times, capture systems, probably of Phoenician origin, were built of reeds and seasonally placed in estuaries on the waterways to the sea to prevent fish from escaping to the sea. Gaius Fannius Strabo, Columella, Pliny, and Marcus Terentius Varro left written documentation of stocking ponds ('piscinae' or *peschiere* in Italian) and capture systems, on the central East Italian Coast, where mullets were associated with more valuable species such as European sea bass and gilthead sea bream (Del Rosso 1905). Stocking ponds were built according to site characteristics, either by digging the rock or building walls of masonry, with large enclosures or tanks often equipped with pierced metal sluice gates to keep the fish in, and allow water to flow through (Giacopini et al. 1994). This very fresh fish was a luxury item, much appreciated by rich Romans, who considered it a *status symbol,* and was available on demand, independent of capture fisheries.

In his excellent book 'Pesca e peschiere antiche e moderne nell'Etruria marittima', Raffaele del Rosso (1905) reports how the knowledge that ancient Romans had of mullet biology and behaviour was applied to design the *peschiere*: "*The muggine cannot be caught for stocking with a net, if its scales are touched it soon dies. The (Roman) Emperors insisted upon having muggine at all times, and therefore stocks had somehow to be found and kept.... It was done in this way: the muggine is a great glutton of freshwater; a strong jet of such water was shot from the tank through a hole below the sea level; it attracted the muggine, who immediately followed up the fresh water track into the tanks, and could not get out again*".

Similar traditional culture strategies are still practiced in the Italian North Adriatic *valli* and the Egyptian *hoshas*, and were carried out in Hawaiian ponds until a few decades ago (Ellis 1968, Nishimoto et al. 2007). In Hawaii, traditional fishponds were located at sites that provided a protective habitat and optimal growing conditions, hence attracting great number of fingerlings (Nishimoto et al. 2007).

A history of more recent mullet culture development is thoroughly reported by Nash and Koningsberger (1981).

Despite the great enthusiasm for mullet culture in the 1970s and 1980s, well illustrated in the following sentence: "the Mugilidae have the brightest future of all the marine and brackish water finfish in the developing technology of aquaculture" by Nash and Shehadeh (1980) and also evidenced by the publication of many scientific papers and of the book 'Aquaculture of grey mullets' by Oren (1981), mullet culture never underwent the development expected in those years, and aquaculture interest and investments turned to other species. One of the main reasons was probably the difficulty in closing the mullet life cycle in captivity: although many experimental trials on the artificial propagation of mullets were carried out in different regions of the world (see below), these results have not been applied to commercial production.

This chapter seeks to provide an overview of mullet capture fisheries and aquaculture production; many topics reported herein are thoroughly described and discussed in other chapters of the present book, here cited as references. The author therefore invites the reader to refer to these other chapters for more detailed information on specific issues.

Production

The total production of Mugilidae was 698,293 tonnes in 2013, mainly (80.2%) from capture fisheries (560,150 tonnes), but with 138,143 tonnes from aquaculture (FAO 2015) (Fig. 16.1).

Production from Capture Fisheries

In terms of capture fisheries, world mullet production has steadily increased, with a similar average annual growth (AGR) of 2.7–4.1% in the four decades from 1954 to 1993, a great increase (AGR 8.6%) in the decade 1994–2003, and a slower growth (AGR 1.3%) in the last decade 2004–2013 (Fig. 16.1) (FAO 2015).

Mullet production from capture fisheries are reported for 103 countries, with over 70% of the global capture fisheries production coming from Asia, with 411,042 tonnes landed in 2013 (Fig. 16.2, Table 16.1). China is by far the largest mullet producer in Asia and in the world, with an annual production of 255,871 tonnes in 2013 (46% of the global mullet capture fisheries production and 62% of Asian mullet production), followed by Indonesia (52,254 tonnes) and India (46,489 tonnes) (Fig. 16.3, Table 16.3).

Mullets vernacular names

Grey mullets have a long tradition in coastal areas, as reflected in the variety of regional names given to the same species: examples of France and Italy are reported in Table B.1 (Gautier and Hussenot 2005) and Table B.2 (Cataudella 1974). Forty three different regional names are given in India to *M. cephalus*, 16 in the Philippines, 14 in Australia, 14 in Malaysia, 12 in Greece and nine in Croatia, just to cite a few (Froese and Pauly 2015).

In areas where grey mullets are important as commercial species, different names are often given to mullets at various growth stages, as they differ in habitat, feeding habits and consequent fishing methods (Liao 1981). Examples are reported in Table B.2 for Veneto, in Italy. In Japan, *Haku* or *Kirara* are the 23–30 mm long flathead mullets, *Oboko* the 31–180 mm long ones, *Ina* the 181–300 mm long ones, *Bora* the ones over 300 mm, *Todo* the extra size specimens and *Karasumibora* the ripe mullets in the spawning season (Liao 1981). Different size classes of the flathead mullet are recognized in the Hawaiian languages: *puo 'ama* and *pua po'ola*, fry; *kahaha* and *pahaha*, juveniles; *'ama'ama* estuary resident mullet, about 20 cm long; '*anae*, reproductive adults, over 30 cm long; and special names are given to mullets in spawning migration, *'anae-holo* and to spent specimens *'anaepali* (Nishimoto et al. 2007).

Table B.1. Regional names of mullets in France (Gautier and Hussenot 2005).

Scientific name	Common name	Regional names
M. cephalus	mulet/muge cabot	caborna (Arcachon), oile voilé (Landes) biscarlbaltza (Basuqe cosat), ramado, yol nègre (Languedoc), mujon, testu, cardou, mugo fangous (Provence) carida mujou pansard (mareseille, nice), capocchiu, mazzardu (Corsica)
Liza aurata	mulet/muge doré	saoutouss (Arcachon), limouza (Biarritz) lemasotina (Basque coast), calaga, gaouta roussa, limpousa, lou doré (languedoc), gaoouta rouso, taco jaouna, aurin (Provence), daurin, mujou de roco, mujou taco jauno (Marseille, Nice), alifranciu (Corsica)
L. saliens	mulet/muge sauteur	acucu (Corsica)
L. ramada	mulet/muge porc	buzuzabala (Basque coast), gaouta roussa, porcarella, porqué, yol négré (Languedoc), porqua, pounchudo, talugo, tusco (Provence) randao (Nice), acucu (Corsica)
C. labrosus	mulet/muge lippu	lancheou (Arcachon), batarde, canuca, chaluc, lessa, lesse, lissa negra (Laguedoc), muge labru, ueri négré (Provence), labru, muhou grosso bugo, muhou labrao (Marseille, Nice), accirita (Corsica)

Table B.2. Common names given to mullets species in the different Italian administrative regions (Cataudella 1974).

	Liguria	Toscana	Lazio	Campania	Calabria	Puglia	Abruzzo	Marche	Veneto	Sardegna	Sicilia
Mugil cephalus	mussao, muzao, massun, musai	testone, muggine, volpino, capocello, firsetta, mazzone	cefolo, mattarello, capozzo, capazzone, cefolo mazzone	cefalo verace, capuozzo	cefalu	capocefalo, capozzo, ciefl, tueppe	mujelle	mugella, baldigare, zievalo		su pisci de iscatu, lissa, muzzulu, glissà, muzzeru	mulettu, muletta, tristuni, murtareddu, cefalu
Liza aurata	taccad'oo, mussao dell'oro, musai, muzo de l'ou	lustro, muggine, firzetta	cefolo, lustro	lustro, sgarza d'oro	cefalo schiuma	vranze, ngefanu lindiu, salatuni, spririllo, cielf	vranzolo	badigia d'oro	lotregagnolo (juv), lotregan (1 year old), lotregan vecio	conchedda, conchedda de pieschera	muletto lustro, cefalu lustrinu, lustru
L. saliens	muzao, cangia, musao	muggine, firzetta, murzello	cefolo, musino	lenmmuso, appezzutiello		caval pizzute, cielf, gruta, asprune	verzelato		verzelado, mangiagiazo	birumbulo, pizzuto, musuku	mulettu pizzutu, cefalu pizzutu, tracchia
L. ramada	museu, musai	testons, acuccotto, volpino, caparello, firzetta	cefolo	cefaro, mazzone, varaco, cefalo	mazzuni	vranzulu, garzalonghe, vraute	botolo	botolo	botolo (< 1 year old) caustelo (< 2 years old), trezarino (< 3 years old), batauro (4 years old)	gevelu, lissa	cefalu i caruvana, cefaluni, varagozzu, muletto testuni
Chelon labrosus	muzao negro, ciautta	musao, testone, volpino, sciorina	cefolo, cefolo pietra, cerina	cerina, cefalo	cefaluni	cannalonga, caneluenghe, ngefanu capozza	mugella	bosiga	boseghin (< 1 year old), bosegheta (< 2 years old), bosega (> 2 years old)	lioni, lioneddu	cefalu fimmineddu

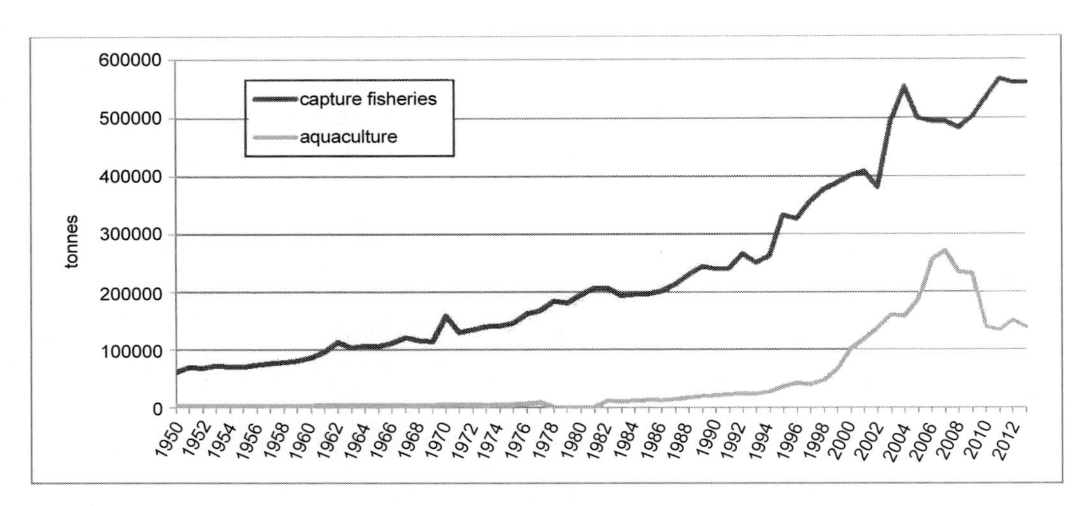

Figure 16.1. World grey mullet production from capture fisheries and aquaculture in tonnes (1950–2013) (data from FAO 2015).

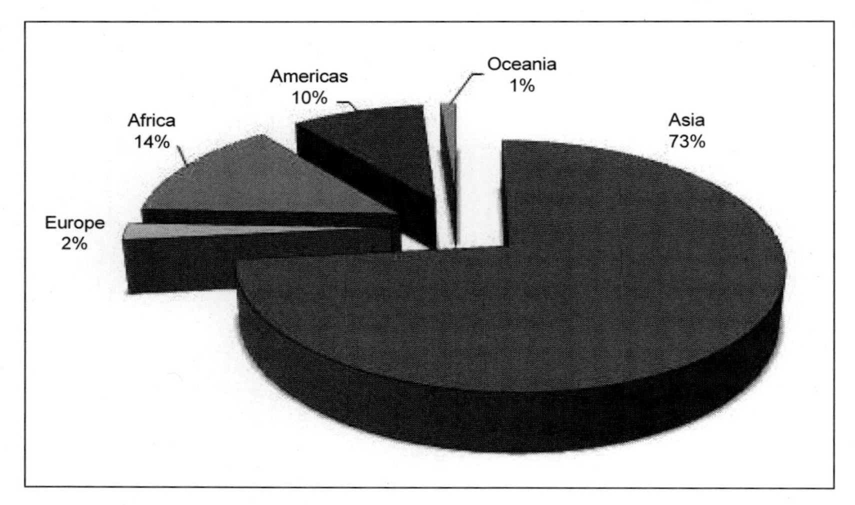

Figure 16.2. Mullet production from capture fisheries in 2013 by continent (%) (data from FAO 2015).

Africa ranks second with 76,881 tonnes, mainly due to the production in Egypt (42,428 tonnes), which ranked 4th in global capture fisheries production in 2013 (Figs. 16.2, 16.3, Table 16.3).

Mullet capture fisheries are reported for both inland and marine areas. In the Americas, Asia and Europe, most production comes from marine areas (> 99, 97 and 94%), and in Oceania only from marine areas (Table 16.1). In Africa only 47% is captured in marine waters; this figure is highly biased by Egyptian capture fisheries statistics, as only 8% of Egyptian production comes from marine areas (FAO 2015).

Among miscellaneous coastal fish, FAO (2015) 2013 statistics report production for 14[1] Mugilidae species from capture fisheries, three of which are also used in aquaculture (Tables 16.1, 16.3 and 16.4). However, as 3/4 of the countries report mullet production statistics referred to the generic term ‘mullets

[1] FAO FishStat (2015) reports one additional species in mullet production statistics, as it considers so-iny (redlip) mullet *Liza haematocheilus* and so-iuy mullet *Mugil soiuy* as two distinct species, whereas they are synonyms of one single species, red lip mullet *L. haematocheila,* according to Eschenmeyer (2014) species catalogue, the nomenclature adopted in this book. There still are many controversies on this species nomenclature and taxonomy (for details see Minos et al. 2010, Chapter 2—Durand 2015 and Chapter 15—Rossi et al. 2015).

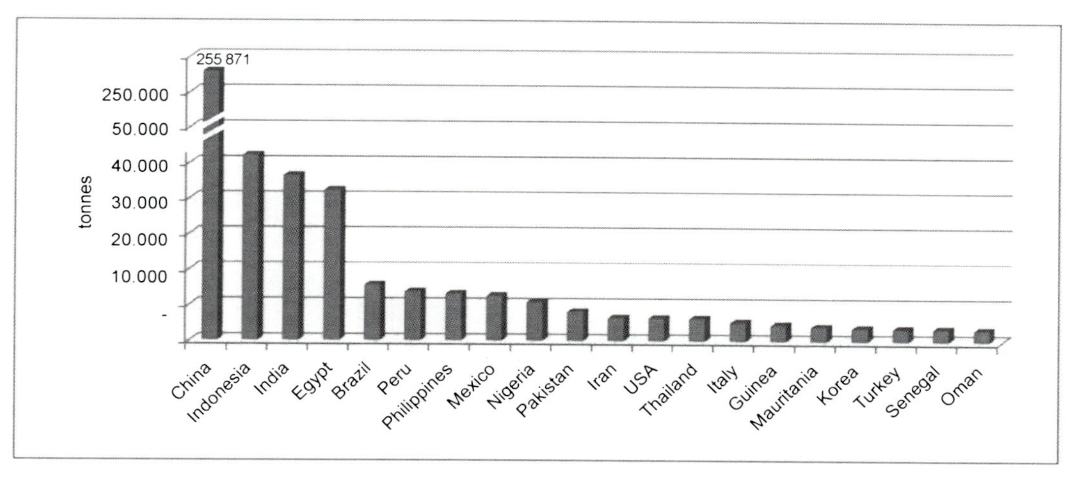

Figure 16.3. Mullet production from capture fisheries by main producing countries in 2013 in tonnes (only countries with a production > to 3,000 tonnes are reported) (data from FAO 2015).

nei[2]', without the species being identified, and 'mullet nei' represented 48% of the world capture fisheries production in 2013, it is difficult to obtain information on the production of single mullet species. More details are given for Asia, with *Liza haematocheila* (*L. haematocheilus*) contributing 35% of total Asian 2013 production, *Mugil cephalus* 29%, *Moolgarda (Valamugil) seheli* 1%, *Liza abu* < 0.2%. In Africa, the only species reported is *M. cephalus*, comprising less than 1% of the total African 2013 production. Though the generic 'mullet nei' is normally used in the Americas, *M. cephalus*, *M. curema* and *M. liza* are also reported, making up to 27, 9, 3% of the total American production in 2013 respectively; small production is reported for *M. incilis* in French Guyana, *Joturus pichardi* in Mexico, *Agonostomus monticola* in the Dominican Republic. In Europe *M. cephalus*, *Chelon labrosus*, *L. haematocheila* (*M. soiuy* + *L. haematocheilus*),[3] *Liza saliens* and *L. ramada* are reported in a few countries, adding up respectively to 9, 7, 6% and around 1% for the last two species of European production in 2013. In Oceania, only 'mullet nei' is reported in capture fisheries statistics.

Mugil cephalus (Fig. 16.4) comprised 29% of the world capture fisheries production in 2013 (FAO 2015), but almost certainly constitutes much of the production reported as 'mullet nei', probably making up to almost 50% of world production from both capture fisheries and aquaculture

a

b

Figure 16.4. *Mugil cephalus* drawn in stamps from Mauritania (a) and Albania (b). Credit to FishBase as source of the documents, and respectively to Daniel Pauly and Jongsma Henk as owners of the stamps, who have contributed these pictures to FishBase from their private stamp collection.

[2] "nei": not elsewhere included.

[3] See note 1 above.

Table 16.1. Mullet production from capture fisheries by species and continent 2013 (tonnes).

Species		Africa		Americas		Asia		Europe		Oceania	World
		Inland waters	Marine areas	Inland waters	Marine areas	Inland waters	Marine areas	Inland waters	Marine areas	Marine areas	
abu mullet	*Liza abu*					593					593
bluespot mullet§	*Moolgarda seheli*§						4,537				4,537
bobo mullet	*Joturus pichardi*				248						248
flathead grey mullet	*Mugil cephalus*	154	595		14,171		117,378	62	1,154		133,514
golden grey mullet	*Liza aurata*								170		170
klunzinger's mullet	*Liza klunzingeri*						2,544				2,544
leaping mullet	*Liza saliens*								111		111
lebranche mullet	*Mugil liza*				1,463						1 463
mountain mullet	*Agonostomus monticola*			58							58
parassi mullet	*Mugil incilis*				95						95
redlip mullet*	*Liza haematocheila*						142,318	5	739		143,062
thicklip grey mullet	*Chelon labrosus*								975		975
thinlip grey mullet	*Liza ramada*								175		175
white mullet	*Mugil curema*				4,680						4,680
mullet nei	Mugilidae	40,861	35,271		32,019	10,289	133,385	712	9,083	6,305	267,925
	Total Mugilidae	***76,881***		***52,734***		***411,044***		***13,186***		***6,305***	***560,150***

*FAO FishStat considers so-iny (redlip) mullet *L. haematocheilus* and so-iuy mullet *Mugil soiuy* as two distinct species, whereas they are synonyms of one single species: redlip mullet *Liza haematocheila*, according to the nomenclature of the Catalog of fishes by Eschmeyer 2015 used in this book.

§ the bluespot mullet *Moolgarda seheli* is reported as *Valamugil seheli* in FAO (2015).

Table 16.2. Mullet production from aquaculture by species and continent 2013 (tonnes) (FAO 2015).

Species		**Africa**		**Americas**	**Asia**			**Europe**			**World**
		Inland waters		**Marine areas**	**Inland waters**	**Marine areas**		**Inland waters**	**Marine areas**		
		BW	FW	BW	FW	BW	Marine	FW	BW	Marine	
flathead grey mullet	*Mugil cephalus*		315	8	3,711	2,077	5,364	60	700	10	12,245
lebranche mullet	*Mugil liza*			7							7
redlip mullet	*Liza haematocheila**								905	49	954
mullet nei	Mugilidae	90,151	26,000		500	8,024			262		124,937
	Total Mugilidae	***116,466***		***15***	***19,676***			***1,986***			***138,143***

* *L. haematocheilus* in FAO (2015).

Table 16.3. Grey mullet production from capture fisheries by country and species in 2013 (tonnes) (FAO 2015).

	Mullets nei	***M. cephalus*** **flathead grey mullet**	***Mugil liza*** **lebranche mullet**	***Agonostomus monticola*** **mountain mullet**	***Joturus pichardi*** **bobo mullet**	***Mugil curema*** **white mullet**	***Mugil incilis*** **parassi mullet**	***Moolgarda seheli*** **bluespot mullet**	***Liza klunzingeri*** **Klunzinger's mullet**	***Liza saliens*** **leaping mullet**	***Liza aurata*** **golden grey mullet**	***Liza ramada*** **thinlip grey mullet**	***Chelon labrosus*** **thicklip grey mullet**	***Liza haematocheila**** **redlip mullet**	***Liza abu*** **abu mullet**	***Total Country (tonnes)***
Africa																
Algeria	282	-	-	-	-	-	-	-	-	-	-	-	-	-	-	**282**
Benin	1,354	-	-	-	-	-	-	-	-	-	-	-	-	-	-	**1,354**
Congo	36	-	-	-	-	-	-	-	-	-	-	-	-	-	-	**36**
Egypt	42,428	-	-	-	-	-	-	-	-	-	-	-	-	-	-	**42,428**
Gabon	480	-	-	-	-	-	-	-	-	-	-	-	-	-	-	**480**
Gambia	1,223		-	-	-	-	-	-	-	-	-	-	-	-	-	**1,223**
Guinea	4,600	-	-	-	-	-	-	-	-	-	-	-	-	-	-	**4,600**
Guinea-Bissau	1,500	-	-	-	-	-	-	-	-	-	-	-	-	-	-	**1,500**
Kenya	220	-	-	-	-	-	-	-	-	-	-	-	-	-	-	**220**
Liberia	26	-	-	-	-	-	-	-	-	-	-	-	-	-	-	**26**
Libya		500	-	-	-	-	-	-	-	-	-	-	-	-	-	**500**
Mauritania	4,151	-	-	-	-	-	-	-	-	-	-	-	-	-	-	**4,151**
Mauritius	5	-	-	-	-	-	-	-	-	-	-	-	-	-	-	**5**
Morocco	1,026	-	-	-	-	-	-	-	-	-	-	-	-	-	-	**1,026**
Namibia	16	-	-	-	-	-	-	-	-	-	-	-	-	-	-	**16**
Nigeria	10,821	-	-	-	-	-	-	-	-	-	-	-	-	-	-	**10,821**
Senegal	3,387	154	-	-	-	-	-	-	-	-	-	-	-	-	-	**3,541**

South Africa	464	-	-	-	-	-	-	-	-	-	-	-	-	-	-	**464**
Tanzania, Un. Rep.	323	-	-	-	-	-	-	-	-	-	-	-	-	-	-	**323**
Togo	1	-	-	-	-	-	-	-	-	-	-	-	-	-	-	**1**
Tunisia	2,218	95	-	-	-	-	-	-	-	-	-	-	-	-	-	**2,313**
Zanzibar	1,571	-	-	-	-	-	-	-	-	-	-	-	-	-	-	**1,571**
Total Africa																***76,881***
Americas																
Argentina	476	-	-	-	-	-	-	-	-	-	-	-	-	-	-	**476**
Brazil	1,615	-	-	-	-	-	-	-	-	-	-	-	-	-	-	**1,615**
Chile	56	-	-	-	-	-	-	-	-	-	-	-	-	-	-	**56**
Colombia	83	-	13	-	-	-	-	-	-	-	-	-	-	-	-	**96**
Cuba	299	-	-	-	-	-	-	-	-	-	-	-	-	-	-	299
Dominican Republic	300	-	-	58	-	-	-	-	-	-	-	-	-	-	-	**358**
Ecuador	21	-	-	-	-	-	-	-	-	-	-	-	-	-	-	**21**
French Guiana	-	-	-	-	-	-	95	-	-	-	-	-	-	-	-	**95**
Mexico	1,106	6,651	-	-	248	4,580	-	-	-	-	-	-	-	-	-	**12,585**
Peru	13,781	-	-	-	-	-	-	-	-	-	-	-	-	-	-	**13,781**
Puerto Rico	4	-	-	-	-	-	-	-	-	-	-	-	-	-	-	**4**
United States of America	99	6,220	-	-	-	100	-	-	-	-	-	-	-	-	-	**6,419**
Uruguay	179	-	-	-	-	-	-	-	-	-	-	-	-	-	-	**179**
Venezuela & Bolivia Rep. of	-	1,300	1,450	-	-	-	-	-	-	-	-	-	-	-	-	**2,750**
Total Americas																***52,734***
Asia																
Azerbaijan	125	-	-	-	-	-	-	-	-	-	-	-	-	-	-	**125**
Bahrain	8	-	-	-	-	-	-	-	-	-	-	-	-	-	-	**8**
China	-	113,553	-	-	-	-	-	-	-	-	-	-	-	14.2318	-	**255,871**

Table 16.3. contd....

Table 16.3. contd.

	Mullets nei	***M. cephalus*** **flathead grey mullet**	***Mugil liza*** **lebranche mullet**	***Agonostomus monticola*** **mountain mullet**	***Joturus pichardi*** **bobo mullet**	***Mugil curema*** **white mullet**	***Mugil incilis*** **parassi mullet**	***Moolgarda seheli*** **bluespot mullet**	***Liza klunzingeri*** **Klunzinger's mullet**	***Liza saliens*** **leaping mullet**	***Liza aurata*** **golden grey mullet**	***Liza ramada*** **thinlip grey mullet**	***Chelon labrosus*** **thicklip grey mullet**	***Liza haematocheila**** **redlip mullet**	***Liza abu*** **abu mullet**	***Total Country (tonnes)***
Cyprus	3	-	-	-	-	-	-	-	-	-	-	-	-	-	-	**3**
Georgia	1	-	-	-	-	-	-	-	-	-	-	-	-	-	-	**1**
India	46,489	-	-	-	-	-	-	-	-	-	-	-	-	-	-	**46,489**
Indonesia	52,254	-	-	-	-	-	-	-	-	-	-	-	-	-	-	**52,254**
Iran (Islamic Rep. of)	2,373	-	-	-	-	-	-	4,164	1,973	-	-	-	-	-	-	**8,510**
Iraq	-	-	-	-	-	-	-	-	-	-	-	-	-	-	593	**593**
Israel	179	-	-	-	-	-	-	-	-	-	-	-	-	-	-	**179**
Kazakhstan	379	-	-	-	-	-	-	-	-	-	-	-	-	-	-	**379**
Korea, Republic of	591	3,168	-	-	-	-	-	-	-	-	-	-	-	-	-	**3,759**
Kuwait	57	-	-	-	-	-	-	-	571	-	-	-	-	-	-	**628**
Lebanon	360	-	-	-	-	-	-	-	-	-	-	-	-	-	-	**360**
Malaysia	4,914	-	-	-	-	-	-	-	-	-	-	-	-	-	-	**4,914**
Oman	3,248	-	-	-	-	-	-	-	-	-	-	-	-	-	-	**3,248**
Pakistan	8,256	-	-	-	-	-	-	-	-	-	-	-	-	-	-	**8,256**
Palestine, Occupied Tr.	8	-	-	-	-	-	-	-	-	-	-	-	-	-	-	**8**
Philippines	13,218	-	-	-	-	-	-	-	-	-	-	-	-	-	-	**13,218**
Qatar	9	-	-	-	-	-	-	-	-	-	-	-	-	-	-	**9**

Saudi Arabia	-	-	-	-	-	-	-	373	-	-	-	-	-	-	-	**373**
Singapore	16	-	-	-	-	-	-	-	-	-	-	-	-	-	-	**16**
Syrian Arab Republic	88	-	-	-	-	-	-	-	-	-	-	-	-	-	-	**88**
Taiwan Province of China	-	657	-	-	-	-	-	-	-	-	-	-	-	-	-	**657**
Thailand	6,338	-	-	-	-	-	-	-	-	-	-	-	-	-	-	**6,338**
Turkey	3,599	-	-	-	-	-	-	-	-	-	-	-	-	-	-	**3,599**
Turkmenistan	1	-	-	-	-	-	-	-	-	-	-	-	-	-	-	**1**
United Arab Emirates	750	-	-	-	-	-	-	-	-	-	-	-	-	-	-	**750**
Yemen	410	-	-	-	-	-	-	-	-	-	-	-	-	-	-	**410**
Total Asia																***411,044***
Europe																
Albania	220	-	-	-	-	-	-	-	-	-	-	-	-	-	-	**220**
Bulgaria	-	9	-	-	-	-	-	-	-	13	2	-	-	-	-	**24**
Channel Islands	4	-	-	-	-	-	-	-	-	-	-	-	-	-	-	**4**
Croatia	112	-	-	-	-	-	-	-	-	-	-	-	-	-	-	**112**
Denmark	10	-	-	-	-	-	-	-	-	-	-	-	-	-	-	**10**
France	224	167	-	-	-	-	-	-	-	98	153	175	831	-	-	**1,648**
Germany	2	-	-	-	-	-	-	-	-	-	-	-	-	-	-	**2**
Greece	980	980	-	-	-	-	-	-	-	-	-	-	-	-	-	**1,960**
Ireland	169	-	-	-	-	-	-	-	-	-	-	-	-	-	-	**169**
Italy	5,304	-	-	-	-	-	-	-	-	-	-	-	-	-	-	**5,304**
Malta	1	-	-	-	-	-	-	-	-	-	-	-	-	-	-	**1**
Montenegro	-	30	-	-	-	-	-	-	-	-	-	-	-	-	-	**30**
Netherlands	113	-	-	-	-	-	-	-	-	-	-	-	-	-	-	**113**
Portugal	1,588	-	-	-	-	-	-	-	-	-	-	-	-	-	-	**1,588**

Table 16.3. contd....

Table 16.3. contd.

	Mullets nei	*M. cephalus* flathead grey mullet	*Mugil liza* lebranche mullet	*Agonostomus monticola* mountain mullet	*Joturus pichardi* bobo mullet	*Mugil curema* white mullet	*Mugil incilis* parassi mullet	*Moolgarda seheli* bluespot mullet	*Liza klunzingeri* Klunzinger's mullet	*Liza saliens* leaping mullet	*Liza aurata* golden grey mullet	*Liza ramada* thinlip grey mullet	*Chelon labrosus* thicklip grey mullet	*Liza haematocheila** redlip mullet	*Liza abu* abu mullet	*Total Country (tonnes)*
Romania	1	-	-	-	-	-	-	-	-	-	-	-	-	-	-	**1**
Russian Federation	1,375	-	-	-	-	-	-	-	-	-	-	-	-	391	-	**1,766**
Slovenia	14	-	-	-	-	-	-	-	-	-	15	-	-	-	-	**29**
Spain	357	30	-	-	-	-	-	-	-	-	-	-	144	-	-	**531**
Ukraine	146	-	-	-	-	-	-	-	-	-	-	-	-	353	-	**499**
United Kingdom	155	-	-	-	-	-	-	-	-	-	-	-	-	-	-	**155**
Total Europe																***13,186***
Oceania																
Australia	4,746	-	-	-	-	-	-	-	-	-	-	-	-	-	-	**4,746**
Cook Islands	10	-	-	-	-	-	-	-	-	-	-	-	-	-	-	**10**
Fiji, Republic of	500	-	-	-	-	-	-	-	-	-	-	-	-	-	-	**500**
Kiribati	21	-	-	-	-	-	-	-	-	-	-	-	-	-	-	**21**
New Caledonia	99	-	-	-	-	-	-	-	-	-	-	-	-	-	-	**99**
New Zealand	929	-	-	-	-	-	-	-	-	-	-	-	-	-	-	**929**
Total Oceania																***6,305***
Total World																***560,150***

*FAO FishStat considers so-iny (redlip) mullet *L. haematocheilus* and so-iuy mullet *Mugil soiuy*, as two distinct species, whereas they are synonyms of one single species: redlip mullet *Liza haematocheila*, according to the nomenclature of the Catalog of fishes by Eschmeyer 2015 used in this book.

Table 16.4. Mullet species object of capture fisheries and aquaculture in the world in 2013 (FAO 2015).

Species		**Capture fisheries**	**Aquaculture**
abu mullet	*Liza abu*	x	
bluespot mullet	*Moolgarda seheli*§	x	
bobo mullet	*Joturus pichardi*	x	
flathead grey mullet	*Mugil cephalus*	x	x
golden grey mullet	*Liza aurata*	x	
Klunzinger's mullet	*Liza klunzingeri*	x	
leaping mullet	*Liza saliens*	x	
lebranche mullet	*Mugil liza*	x	x
mountain mullet	*Agonostomus monticola*	x	
parassi mullet	*Mugil incilis*	x	
redlip mullet* (so-iny [redlip] mullet) (so-iuy mullet)	*Liza haematocheila* *(L. haematocheilus)* *(Mugil soiuy)*	x (x) (x)	x (x)
thicklip grey mullet	*Chelon labrosus*	x	
thinlip grey mullet	*Liza ramada*	x	
white mullet	*Mugil curema*	x	
Total number of species	14	14	3

*FAO FishStat considers so-iny (redlip) mullet *L. haematocheilus* and so-iuy mullet *Mugil soiuy* as two distinct species, whereas they are synonyms of one single species: redlip mullet *Liza haematocheila,* according to the nomenclature of the Catalog of fishes by Eschmeyer 2015 used in this book.

§ the bluespot mullet *Moolgarda seheli* is reported as *Valamugil seheli* in FAO (2015).

(Whitfield et al. 2012). In Egypt, where all mullet captures are reported as 'mullet nei' in official statistics by FAO (2015), thin lipped mullet *Liza ramada* constitutes about 58% of the fisheries catch, *M. cephalus* 23% and the other three mullet species, *Moolgarda (Valamugil) seheli, L. aurata* and *L. saliens*, together constitute the rest (Saleh 2008).

On the other hand, 'mullet nei' certainly includes several other mullets, among the 71 Mugilidae species reported in Eschemeyer Catalog of fish (2015), which are the object of capture fisheries and/or aquaculture, though they do not appear in official statistics (see examples below in this chapter). Some other species with minor production were reported in previous years in the FAO statistics, but they do not appear any longer as single species; it is now likely that many mullets are included in 'mullet nei'.

Production from Aquaculture

In 2013, FAO (2015) reported aquaculture records for 15 countries, with a total of 138,143 tonnes, corresponding to 19.2% of world mullet production and a value of US$391,477 (FAO 2015) (Fig. 16.1).

Mullet culture has been reported in the FAO FishStat statistics since the first production data published in 1950, with records of *M. cephalus* and 'mullet nei' from traditional culture from South East Asia (China Hong Kong SAR, Taiwan, Japan) and the Mediterranean in Italy. The intensification of culture in the late 1990s lead to a large growth in production, with a threefold increase from 1999 to 2007, and a peak of 271,816 tonnes in 2007. Since 2010, mullet production has dropped to an annual production total of around 133,000–150,000 tonnes in the last four years (2011–2013) (FAO 2015).

Mullet culture is reported for both inland (both freshwater and brackish water) and marine areas (both brackish water and marine). Seventy four percent of mullet production is cultured in brackish waters, comprising 77% in Africa, 51% in Asia, and 94% in Europe (Table 16.2). In Africa only inland water culture is reported (FAO 2015).

Mullet culture is mainly concentrated in the Mediterranean and Black sea region and in South East Asia. Africa has the highest production among continents, comprising almost entirely the Egyptian production (Fig. 16.5). Indeed Egypt is by far the leading producer in the world, with 116,151 tonnes produced in 2013 and 84% of the world aquaculture production, followed by Indonesia (8,024 tonnes), Republic of Korea (4,810 tonnes), Taiwan (2,637 tonnes) and Israel (2,240 tonnes) (Fig. 16.6, Table 16.5,

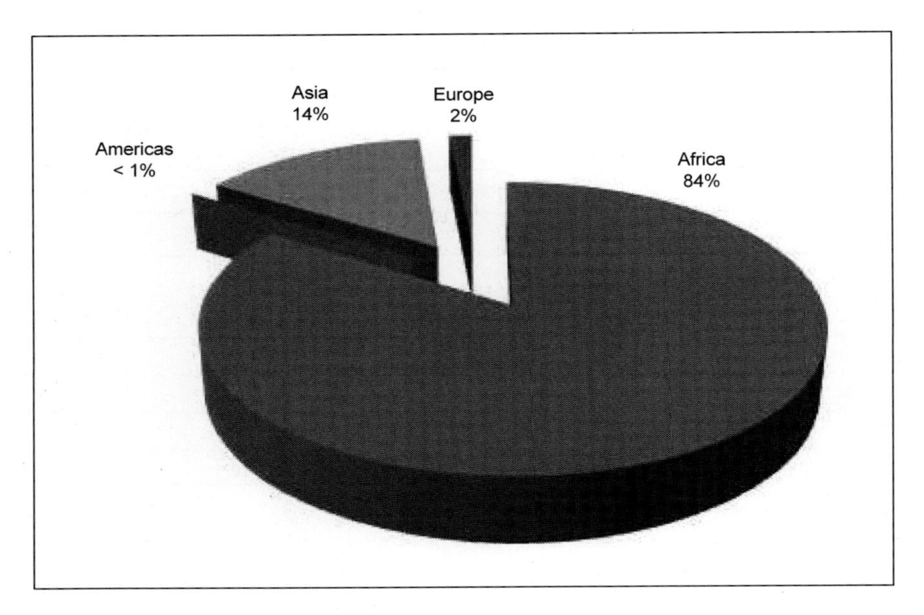

Figure 16.5. Mullet production from aquaculture in 2013 by continent (%) (data from FAO 2015).

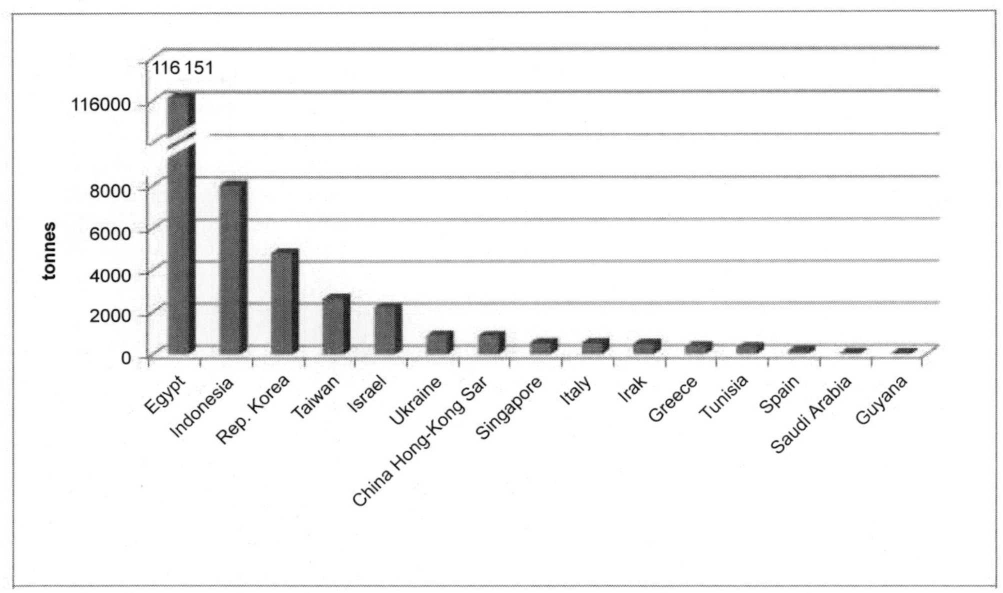

Figure 16.6. Mullet production from aquaculture in the main producing countries in 2013 in tonnes (data from FAO 2015).

Table 16.5. Mullet production (tonnes) and value (US$ 000) from aquaculture by country and species in 2013 (FAO 2015).

	Mullets nei	***Mugil cephalus* flathead grey mullet**	***Mugil liza* lebranche mullet**	***Liza haematocheila** * **redlip mullet**	**Total Country (tonnes)**	**Total Value (US$000)**
Africa						
Algeria	< 1	-	-	-	-	**2**
Egypt	116,151	-	-	-	**116,151**	**303,579**
Tunisia		315			**315**	**776**
Total Africa					***116,466***	***304,356***
Americas						
Guyana	-	8	7	-	**15**	**40**
Total Americas					***15***	***40***
Asia						
China, Hong Kong SAR	-	-	-	-	**889**	**3,198**
Indonesia	8,024	-	-	-	**8,024**	**8,491**
Iraq	500	-	-	-	**500**	**3,000**
Israel	-	2,240	-	-	**2,240**	**10,761**
Korea, Republic of	-	4,810	-	-	**4,810**	**43,446**
Saudi Arabia	-	40	-	-	**40**	**120**
Singapore	-	536	-	-	**536**	**2,274**
Taiwan	-	2,637	-	-	**2,637**	**7,042**
Total Asia					***19,676***	***78,332***
Europe					**0**	
Greece	-	390	-	-	**390**	**1,429**
Italy	150	380	-	-	**530**	**4,089**
Portugal	< 1		-	-	**< 1**	**1**
Spain	159	-	-	-	**159**	**871**
Ukraine	2	-	-	905	**907**	**1,812**
Total Europe					***1,986***	***6,390***
Total World					***138,143***	***391,477***

FAO 2015). Other producing countries are Tunisia in Africa; Guyana in the Americas; China Hong-Kong SAR, Iraq, Saudia Arabia and Singapore in Asia; Greece, Italy, Spain and Ukraine in Europe (Table 16.5). No aquaculture production is reported for Oceania. Some other producing countries, such as China or India, do not report mullet production as such in their statistical returns to FAO, and production is hidden in the general category '*Osteichthyes*' or 'other fish nei'.

Mullet production in different countries has varied greatly over time. Egypt started mullet production in 1985 with 1,500 tonnes, and had an impressive growth to 252,507 tonnes in 2007; it then decreased to the present (2103) 116,151 tonnes. Indonesia increased production in the mid 1970s, with a peak of 12,264 tonnes in 1996. Korea started mullet culture in 2010 with a production of 4,680 tonnes, Japan cultured mullets until 1964, with a maximum of 740 tonnes in 1950. The Italian production of *M. cephalus* flourished in the 1990s with up to 3,100 tonnes produced in 1996; in 2013 the production fell to 530 tonnes.

Only three mullet species are reported in the 2013 official aquaculture statistics provided by the countries (Tables 16.3 and 16.5), in addition to 'mullet nei', which represent 90% of the world aquaculture production: *M. cephalus* comprises 9%, *Liza haematocheila* < 1%, and lebranche mullet *Mugil liza* a few tonnes (FAO 2015).

However, other species are cultured in various regions of the world. In the Mediterranean region, in addition to flathead mullet *Mugil cephalus*, another four species (thicklip grey mullet *Chelon labrosus*, golden grey mullet *Liza aurata*, thinlip mullet *Liza ramada* and leaping mullet *Liza saliens*) are cultured, though they are not reported as single species in official statistics. In Egypt also the black keeled mullet *Liza carinata* is cultured, for its tolerance to high salinities (Chapter 20—Sadek 2015). The squaretail mullet *Ellochelon vaigiensis* was cultured in Thailand in the 1990s, with a maximum of 362 tonnes produced in 1996 (reported as *Liza vaigiensis* by FAO 2015), *L. ramada* culture was reported in Tunisia in the late 1980s with a few tonnes (FAO 2015). *Liza subviridis* (*Mugil dussumieri*), *Moolgarda engeli* (*Mugil engeli*), *Liza tade* (*Mugil tade*) are cultured in Indonesia (Sri Paryanti 2006). In addition to *M. cephalus, Liza macrolepis, L. tade* and *L. parsia* are grown in polyculture in brackish water earthen ponds in India (Ayyappan 2014).

Capture Fisheries

Capture fisheries still represent the main production of mullets throughout the world, with 80.2% of global mullet production. Mullets are captured with a variety of fishing gear, often very unusual, in order to counteract the peculiar behaviour of these species. A thorough review of capture methods for mullets is reported in Chapter 17 (Prossner 2015). In many cases the limit between fisheries management and capture based aquaculture is tiny, and the same fishing gear from capture fisheries is used in extensive systems where grey mullets are grown.

Fry Collection

Despite research activities and experimental trials carried out to spawn mullets in captivity since the mid-1960s (see below), and the demand for a steady supply of mullet fry for grow-out, artificial propagation has never reached a commercial scale and to date mullet culture everywhere still mainly relies on collection of fry from the wild (see Chapter 19—Liao et al. 2015, and Chapter 20—Sadek 2015). The capture of wild fry still represents a practice common to all mullet culture systems.

Mullets are among the marine migrant species that at the juvenile stage move from the open sea to coastal areas, bays, estuaries and coastal lagoons for feeding and growth (Chapter 14—Whitfield 2015). These habitats play a role as nursery areas and provide shelter from predators (Chapter 12—Koutrakis 2015). Mullet fry collection intercepts juveniles just as they reach the coast or in estuaries, outlet channels of coastal lagoons, and agriculture drainage or irrigation canals (Saleh 2008). These fry are used as seed for stocking confined portions of coastal lagoons or culture ponds.

Mullet fry school together and are therefore relatively easy to spot and to catch in huge numbers. They tend to concentrate at river mouths, swimming against the current, and often gather in groups in small indentations and in quiet spots, where they rest (Ben-Yami and Grofit 1981). Mullet catch is seasonal and each species has a different migration season, well known by fry fishers, who collect the different species at different times of the year (see Chapter 12—Koutrakis 2015). However, fry or schools of juveniles may be composed of different mullet species.

In Italy, the collection of fry by specialized fisheries (*pescenovellanti*) is a centuries-old tradition, certainly the oldest in the Mediterranean region. Its development was linked to the progress of *vallicoltura* (Bullo 1940), and was regulated by specific rules already in the *Ancient Diritto Veneto;* the first reference to mullet fry fisheries *(..pescar cefali novelli..)* dates back to March 8, 1314 (Rossi et al. 1999). This fishery enabled the concentration of young fish, dispersed throughout the whole lagoon that acted as a nursery area, into the *valli* and to stock those *valli* which were located further away from the sea, where natural spontaneous migration was often insufficient.

The collection and seeding of wild fry of mullets and of other marine migrant species, such as European sea bass and gilthead sea bream, was first used to augment natural recruitment, but assumed a key role in sustaining local stocks by the late 18th century, when slowly the North Adriatic *valli* started to be confined with permanent banks/dams/dykes. In Venice, fry collection lasted three months, with 37,000 man days, and 4,314,000 *Liza aurata*, 564,000 *L. saliens*, 5,116,000 *L. ramada* and 48,000 *Chelon labrosus* fry collected to be stocked in the *valli* in 1898 (Del Rosso 1905). An average of 21–22 million fry were estimated to be stocked annually in the Venetian *valli* at the end of the 19th century (Franzoi et al. 1999).

Fry can be collected with either passive or active gear; fixed fish traps are placed in areas where fry naturally gather and migrate, and are placed in order to guide the fry towards the trap (Chapter 17—Prossner 2015). The most commonly used fishing gear are scoop nets, seines and push seines, the latter are used to surround the fry school with a circular movement, operated by two or more fishers. Up to 20,000 fry per hour can be caught by seine during the seasonal peak of *Mugil cephalus* fry occurrence (Crosetti and Cataudella 1995). In Italy, fry are captured with a surrounding net, called '*tela da pesce novella*', 8 m long and 1.4 m high, with 2–4 mm mesh size. Two wooden poles, called '*masse*', are tied to both ends of the net and are used to keep the net open during fishing operations, while two fishers surround the fry school with the net (Rossi and Franzoi 1999, Giovanardi 2011). The seine net (*tratta or sciabica*) is 15–100 m long, with floats on the top and lead weights, with mesh increasing in size on the sides, and ropes fixed to the two poles that enable operating the seine net from a distance (Rossi and Franzoi 1999). In Israel, mullet fry are collected with mosquito-netting beach seines and hand-operated scoop-nets on the sea shore near estuaries, in estuaries, rivers and drainage outlets of fish ponds (Ben-Yami and Grofit 1981).

Great care is taken by fry fishers in the collection of fry and their transport to the grow-out site (either farm or extensive system), as stress, from both capture and handling, and eventual consequent skin bruises influence fry survival. Stress from capture and transport cause high mortalities immediately following the release of fry into the coastal lagoons, with up to 70% of the product being lost (Ravagnan 1978).

The harvested fry are transported in oxygenated containers to the site where they will be released. In some countries, anaesthetics are used during transport. No anaesthetics are used in Italy and fry fishers, *pescenovellanti*, prefer to capture fish smaller than 3.5–4 cm, in order to reduce the risk of pathologies and parasites, quite common in 5–10 cm long specimens, which are often subject to haemorrhages of the mouth region, scale loss and fin lesions subsequent to handling. If the water conditions of the destination site are different from the water of origin, fry are gradually acclimated to the temperature and salinity of the destination site. This can be carried out both during transportation or at the farm, before releasing the fry into the farm facilities.

Transport methods depend on the quantity of fry and the distance from the collecting site to destination. Small quantities of fry (< 10,000 fish) can be transported in plastic bags filled with water and topped with air or pure oxygen, depending on the length of the trip, at densities of up to 100 fry l^{-1}, though stocking densities depend on transport length, water quality and fry conditions: indeed stressed fry consume more oxygen. Fry captured early in the morning have an empty stomach and travel better (Crosetti and Cataudella 1995). Larger quantities of fry are carried in aerated tanks, both with portable battery-operated aerators or compressed oxygen cylinders, to maintain adequate oxygen concentration in the water (> 5 mg.l^{-1}). In Italy, fry are transported in trucks carrying 2 m^3 fibreglass tanks supplied with pure oxygen, with a carrying capacity of over 200,000 fry/tank.

In Italy, the collection of fry and juveniles is still currently practiced by fishers who operate in the Lagoon of Venice on a local basis, or by 'itinerant' fishers who move along the coast with trucks equipped with tanks and water oxygenation, to capture and then transport live fry to the *valli* (Rossi and Franzoi 1999). In the 1980s 200–250 fishers were involved in fry collection throughout the country, with 15–30 trucks and 50–70 boats (Franzoi et al. 1999).

The collection of fry is usually regulated by specific rules. In Italy, the collection of fry for culture or restocking purposes (*novellame per allevamento*), from both the coastal shore or the open lagoon, is considered as a specialist fishery (*pesca speciale*), and it is allowed in two periods of the year: in spring from early March to mid June and in autumn from mid September to the end of December. *Pescenovellanti* obtain a license for one or more (up to three) fisheries compartments, and are allocated annual quotas of fry catch by the (ex) General Directorate for marine fisheries and aquaculture of the Ministry of Agriculture and Forestry Policies (D.M. 7 August 1996), according to fry availability of fisheries resources and the quantities captured during the last fishery campaign. Fry collection is no longer authorized in Sicily and Sardinia. In 2014, 24 licenses for mullet fry capture were issued by the Ministry of Agriculture and Forestry Policies (MiPAAF, Ufficio Licenze, A. Maccaroni, pers. comm.). The annual captures of fry ranged from 30 to 58 million from 1988–2000, with a mean of 60% of *Liza* spp. (especially *L. ramada*), 13% of *Chelon labrosus* and 11% of *M. cephalus*, and average over this period of 25–30 million, 5 million and 5 million. year^{-1} respectively (Ciccotti and Franzoi 2001). The favoured species are the thicklip mullet, *Chelon labrosus*, and the flathead grey mullet *Mugil cephalus*, which fetch a higher market price compared to other mullets. Today most mullet fry are captured by the 'itinerant' fishers along the coast, but mullet fry are also imported from Spain, where wild fry are caught and wintered before being sold to Venetian *valli* (M. Guerrieri, pers. comm.).

In Egypt, marine fish fry for aquaculture (most of which are grey mullets: 70–80% *Liza ramada*, 12–15% *M. cephalus*) are collected by licensed fishermen under the supervision of the General Authority for Fisheries Resources Development (GAFRD) within the Fisheries law n 124/1983. Wild fry harvest is permitted at a dozen limited sites supervised and managed by the five governmental fry collection stations (see Chapter 20—Sadek 2015 for further details). Fry are then either sold directly to fish farms, or grown to fingerlings in GAFRD nursing stations. GAFRD will then supply fry or fingerlings only to licensed fish farms, according to allocated quotas (Saleh 2008). The quantity of fry supplied through this channel ranged from 40 to 109 million fry per year over the last decade (GAFRD 2013). However, these figures are greatly underestimated, especially if these data are compared with final mullet production. Indeed, parallel unreported fisheries exist and the smuggling of fry has prospered with the increasing demand for marine fish fry and restrictions from the official supply (Saleh 2008). It is very difficult to estimate the real number of fry caught and marketed illegally: considering the high mortalities (up to 96%) during transportation (Saleh 1991), often in unequipped trucks, and during acclimatization to brackish water and freshwater, the number may reach around 400–500 million fry annually (Sadek 2001, Abdel-Rahman 2015). This figure is several times greater than the recorded official trade (73 million in 2012, GAFRD 2013); around 80% of the total fry supply is therefore estimated to be purchased on the black market, with prices 2.5 to 7.5 times the official prices set by GAFRD (Sadek 2001).

In addition to Italy and Egypt, where mullet fry collection has an important role in aquaculture in *valli* and *hosha*, confined portions of coastal lagoons, in the other Mediterranean countries different situations are found. Fry collection is practiced for culture purposes in Albania, Greece, Israel, Spain and Tunisia and for stocking inland waters in Algeria and coastal lagoons in Greece and Spain, but it is not allowed in Croatia, Cyprus, France and Morocco (Massa 1999). Actual capture data are often difficult to obtain, especially in those countries where many fishers are not licensed. In Turkey the collection of wild fry for aquaculture purposes was practiced, but was forbidden by the Ministry of Food Agriculture and Livestock in 2000 (Deniz 2015). In Israel about three million *M. cephalus* fry are sown annually in freshwater lakes, about 1/3 of these fry are imported from Spain and South Africa, and the rest are collected in Israel, for a production of about 900–1,000 tonnes of market-sized fish (Massa 1999).

In Italy 205–270 fry fishers operated in 1996–1997 (Donati et al. 1999), consisting of about 1,000 workers connected with fry collection. Four hundred and sixty licensed fry fishers operated in Egypt in 2005 (Saleh 2008). The survey performed by Massa in 1999 reported about 1,050 fry fishers working in Turkey, with a capture of about 22 million fry, and about 300 fry fishers in Tunisia, with a capture of 3–5 million fry. The number of fry fishers in the whole Mediterranean has declined with the implementation of mass propagation of European sea bass and gilthead sea bream hatchery fry starting from the late 1980s, and consequent fry price decreases. Hatchery produced fry of the latter species are now used in intensive grow out systems, but wild fry are still requested whenever hatchery fry are not available or if to be stocked in extensive culture system, where wild or wild-type fry are preferred to hatchery fry as they more easily adapt to environmental conditions.

In Hawaii wild fry 25–80 mm long were captured with scoop nets along the shores to stock traditional mullet culture ponds whenever natural recruitment was too low (Ellis 1968). In Taiwan, mullet culture is based on the collection of wild *M. cephalus* fry, though no data are available on the number of mullet fry harvested annually. Several mullet species school together as fry, and fishers try to operate at sites and at times when *M. cephalus* is present in higher quantities (see Chapter 19—Liao et al. 2015). High mortality rates (up to 70%) of wild caught fry are reported (COA 2014).

The harvest of wild fry or juveniles for culture or stock enhancement purposes remains a debated issue. According to Ottolenghi et al. (2004), the supply of wild seed to the capture based aquaculture industry appears to be unsustainable in the short term and inadequate in the long term. On one side, fry capture certainly represents a reduction of recruitment to wild stocks, on the other it could be considered as an enhancement of resources, as the grow-out of wild fry and juveniles in more or less controlled culture conditions certainly increases their survival compared to the wild (Ciccotti and Franzoi 2001). Although the capture of wild grey mullet fry for culture purposes does not seem to have a negative impact on wild stocks, sustainability criteria are recommended (IUCN 2007). Cataudella et al. (1999) consider mullet fry collection in Italy as a responsible fishery, because it is not carried out for direct human consumption, it is sufficiently selective, it has a higher capacity to produce biomass in respect to recruitment, and also because mullets are not endangered species and lay a large number of eggs, hence they can be considered as r-strategist species, where few females can guarantee the successful propagation of the species. Indeed in Sicily (Italy), after the official ban of fry collection in the region in 1988, there was no evident increase of the wild resource (Cataudella et al. 1999). In Egypt, by comparing the total fisheries catch to mullet landings over the period 1980–2005, Saleh (2008) asserts that mullet fisheries do not seem to have been affected by the wild mullet fry fishery. Though often little is known about the effectiveness of stock enhancement or the survival of wild fry in the wild, the rule of thumb could be that fry survival should be higher if captured and stocked than if left in the wild. Certainly the survival of captured mullet fry in some areas could easily increase with better management and care in handling, transport and acclimatization operations, especially as far as salinities and temperature are concerned (Liao 1981, Rossi and Franzoi 1999). Practical recommendations for the handling, transportation and stocking of mullet fry to minimize stress and improve survival rate are provided by Ben-Yami (1981). The increase of mullet fry price resulted in a more rational utilization of the resources and a reduction in handling losses in Egypt (Saleh 2008).

Stock Enhancement

The mullet fry harvested in the wild or produced in a hatchery are used for stock enhancement in more or less confined water bodies or at sea.

In Hawaii, mullet stock-enhancement technology was used to help restore the declining coastal stocks and in particular to implement the recreational mullet fishery in Hilo and the commercial net fishery in Kane'ohe, Oahu island (Nishimoto et al. 2007, Chapter 18—Leber 2015). Decline in stock and in size is due to overfishing, loss or degradation of natural spawning and nursery areas, or displacement by the alien mullet *Valamugil engeli* (kanda), which was accidentally released from the South Pacific in Oahu from 1955 to 1958 together with other bait fish and has spread in many estuaries of the main Hawaiian islands (Nishimoto et al. 2007). The first batch of hatchery produced mullet were tagged and released in 1989 (K. Leber, pers. comm.) and a collaborative project of by the Division of Aquatic Resources and the Oceanic Institute followed in 1990–2000. The Hilo project verified the potential of stock enhancement as an effective tool to replenish diminishing stocks for mullets, and lead to the implementation of several management measures to improve mullet fishery management based on the results of this project (Nishimoto et al. 2007).

In the Mediterranean region, *Liza ramada* and *Mugil cephalus* fry and juveniles are stocked in freshwater lakes that do not have direct access to sea, where neither species reproduces, giving birth to new fisheries. Regular stocking programmes with both species were initiated in 1921 in the brackish water Quarun Lake and more recently in the man-made Wadi Al Raiyan Lakes in Egypt (Chapter 20—Sadek 2015) and from 1958 in Lake Kinneret in Israel, with eight% of the mullets introduced in the lake being recaptured (Sarig 1981) and about one million fry stocked at present every year (Ben Yami 1981, Sarig 1981, Ostrovsky et al. 2014). Both species were stocked since the early 20th century in the volcanic lakes of central Italy (Cataudella and Monaco 1983), with *L. ramada* growing to more than 400 g in three years, and *M. cephalus* to 800 g; new projects with *L. ramada* are being implemented for Lake Bolsena and Lake Trasimeno by the local authorities (*Province*), which are responsible for freshwater stocking programmes (M. Guerrieri, pers. comm.).

Stock enhancement is also carried out in more confined coastal lagoons (see below), though this could be considered as capture based aquaculture, according to Ottolenghi et al.'s (2004) definition.

The Culture of Grey Mullets

Mullet culture has ancient origins, and developed in extensive and semi-intensive systems in many regions of the world, often associated with coastal lagoons and flooded areas at river mouths (Crosetti and Cataudella 1995). The first 'culture activities' consisted of the confinement of juvenile mullets that migrated from the sea and were captured while going back to sea and the fidelity of mullet migrations permitted to invest in fixed traps.

Grey mullets have peculiar biological features that make them easy fish to grow. First of all, they are euryhaline species (with a few exceptions of species only living in freshwater) (see Chapter 13—Nordlie 2015 for a review), which migrate between the sea and inland waters for trophic and spawning reasons. Their euryhalinity enables them to be cultured at different salinities in marine, brackish and freshwater, often in polyculture with other mullet species or in association with other marine species (European sea bass, gilthead sea bream and eels in confined areas of Mediterranean coastal lagoons or milkfish *Chanos chanos* in south East Asia) or freshwater species, such as tilapias in Egypt, or carps and tilapias in Israel.

Second, as consumers of the lower trophic levels, mullets are able to very efficiently convert into flesh what they eat, and can therefore be extensively cultured in the most efficient and cheap way. A complex pharyngo-branchial filter associated with the gill raker system enables mullets to feed on a variety of food, from micro-organisms and decaying organic matter to larger food such as algae and insect larval stages or small molluscs. Mullets exploit the substratum and the sediments and different mullet species have evolved particular feeding mechanisms which enable them to use different grain sizes and feeding grounds (Blaber 1976, Mariani et al. 1987, see Chapter 9—Cardona 2015 for a review). When various mullets species are grown together in polyculture, each species grazes on different substratum grain sizes and they do not overlap in trophic resources. The success of mullets as cultured species also depends on their feeding habits, low in the food web. Mullet often feed on the natural food web produced in the environment where they are cultured. Manuring is often practiced to enhance natural food webs in earthen ponds where mullets are cultured, such as chicken droppings in Egypt (Saleh 2008) or pig manure and night soil in polyculture systems with Chinese carps in Taiwan and Hong Kong (Ling 1967, Tang 1970). In polyculture ponds with carps and tilapias where both manure and fish feed are provided, mullets rely totally or to a very large extent on natural food, while carps and tilapias only partially rely on the natural food web (Saleh 2008). Mullets are also supplied wheat or rice bran (Saleh 2008). Rice bran and wheat milling by-products were also used as supplemental feed in Hawaiian fish ponds (Ellis 1968). No specific feed formulated for mullets is produced, even in Egypt, which ranks first in the world in cultured mullet production.

Traditional culture in ponds and enclosures has been practiced for a long time in the Mediterranean region, South East Asia, Taiwan, Japan and Hawaii. Traditionally mullet culture can be found in different regions of the word, from the North Adriatic *valli* in Italy, the traditional Hawaiian ponds, the Indonesian *tambak*, and the Egyptian *hosha*, the so called coastal 'harbour culture' in North China, the *bheris* of Gangetic estuaries in the Indian sub-continent (Zheng 1987, Saleh 2006, Smit Ramesh Lende 2011). Today mullets are grown in extensive, semi-intensive and intensive systems, often in multispecies culture systems.

Though grey mullets represent an important fish production in some countries, and are probably among the marine species having been cultured for longer, their intensive culture did not develop as it did for other marine species with a higher market value. Indeed in many countries grey mullet cannot justify economic investments for the intensification of their culture because of their low market price. Another limiting factor to development lies in the fact that their culture is still based on fry collected from the wild, as induced spawning practices have not been transferred to commercial production level.

Producing Countries

Egypt is the first producer in the world of cultured mullets, with a production of 116,151 tonnes and a value of US$303,579 in 2013, though it ranks second in the world after China in total mullet production (from both capture fisheries and aquaculture) with 158,579 tonnes (GAFRD 2013, FAO 2015). Mullet aquaculture production represents 73% of total mullet production in the country and 11% of Egyptian aquaculture production in weight and 16% in value (FAO 2015).

The main cultured mullet species are thinlip grey mullet *Liza ramada*, flathead grey mullet *Mugil cephalus*, and black keeled mullet *L. carinata* (Chapter 20—Sadek 2015). Though no data on the aquaculture production of each single species are available, as all official statistics only report the large category of 'mullet nei', data on wild fry collection can provide some hints: in 2012, the collected wild fry were composed of 51.9% *L. ramada*, 27.7% *L. carinata* and 20.4% *M. cephalus* (GAFRD 2013).

Grey mullets are cultured in different culture systems, from the extensive systems in coastal lagoons (131–350 kg.ha^{-1}) to semi-intensive earthen ponds (1–8 tonnes.ha^{-1}) or intensive cages (3 kg.m^{-3}), usually in polyculture or in association with other species: in freshwater, with tilapias or carps, in marine water with European sea bream or sea bass (for more details see Saleh 2008 and Chapter 20—Sadek 2015). Most aquaculture activities are located in the Nile Delta region, and most fish farms can be classified as semi-intensive brackish water pond farms (Salem and Saleh 2010).

A typical brackish water culture system in Egypt, commonly practiced for many centuries in the Northern Delta Lakes region until a few decades ago, was the *hosha* system, shallow earthen ponds built by erecting low dykes made of mud and straw along the shore or around the islands of coastal lagoons, with one or more narrow openings connecting with the lagoon, where mullets and other euryhaline species were cultured. Hosha were periodically closed, water pumped out and fish harvested (Abdel-Rahman 2015). Today traditional extensive and semi-extensive aquaculture has moved towards semi-intensive systems using supplementary feeding on a routine basis, with an increase in annual production from an average of 250–400 kg to 0.7–6 tonnes ha^{-1} (Salem and Saleh 2010). Furthermore, after the conversion of around 100,000 hectares of reclaimed land to aquaculture ponds, this type of culture underwent a severe reduction in surface area during the early 1990s as a result of the competition for land and water from the expansion of land reclamation activities for agriculture (Salem and Saleh 2010). The harvest from fresh or brackish water ponds consisted of approximately 41% tilapia, 30% carp and 22% mullet in 1997 and of 55% tilapia, 30% mullet and 11% carp in 2009 (Salem and Saleh 2010). Productivity in earthen ponds is kept at the required level by adding chicken manure and/or chemical fertilizers to enhance the natural food web, sometimes integrated with supplementary food in the form of agriculture byproducts (rice and/or wheat bran) (Saleh 2008). When mullet are reared in polyculture, feeding and fertilization programmes usually target the other cultured species while mullets feed on the natural food web, detritus and feed leftovers (Saleh 2008).

Intensive pond aquaculture is now expanding to replace large areas of semi-intensive ponds, and is carried out in well designed and constructed earthen ponds, sometimes lined with polyethylene sheets, 0.3 to 0.6 hectares wide and 1.5–1.75 m deep, with annual production of 1,4–2,5 kg.m^{-2} of tilapias and mullets (Salem and Saleh 2010). Mullets in monoculture in earthen ponds and Nile cages are mainly fed tilapia compressed sinking pellet feed with low crude protein levels ranging from 18 to 25% (El-Sayed 2007). No feed specific to mullets is produced.

Indonesia is the second producer of mullets in the world, with 8,024 tonnes in 2013. Though some production was reported in 1950, culture steadily increased from the 1970s to 1999, with a peak of 13,120 tonnes in 1999, but since 2007 it decreased again to lower annual production within the range of 5,354–8,822 tonnes (FAO 2015). Mullet culture did not follow the general trend in development of Indonesian aquaculture, which had an 31% average annual growth in the last decade from 2004 to 2013, following a national policy of aquaculture development (Rimmer et al. 2013, FAO 2015): indeed today mullet production only represents 0.2% of the total annual animal production from aquaculture in the country (3,848,823 tonnes in 2013, FAO 2015). Pond-based brackish water culture of milkfish *Chanos chanos* and mullets, mostly on Java Island, is an ancient tradition in Indonesia, which has been practiced for more than 500 years (Sri Paryanti 2006). The majority of coastal aquaculture farms constructed on ex-mangrove land are extensive or 'traditional' *tambak* with little water exchange and low input of nutrients either as fertilizers or feed (Rimmer et al. 2013), and annual production of 450 kg.ha^{-1}, with milkfish as main cultured species (Sudradjat and Sugama 2010). Brackish water aquaculture is the dominant form of aquaculture by area in Indonesia, with an estimated 680,000 hectares of *tambaks*. However, although *tambaks* were initially developed to grow milkfish and mullets, and shrimps were only incidentally produced, with the boom of shrimp farming in the 1980s, shrimp became the target species for *tambak* production (Rimmer et al. 2013).

In the Republic of Korea mullet culture started in the mid 1990s, with a great development between 2000 and 2005, and has since then stabilized, with production oscillating between 4,680 tonnes in 2010 to 6,159 tonnes in 2008 (FAO 2005). Since 2012 production from aquaculture has exceeded capture fisheries landings, which showed a great decline to 3,759 tonnes in 2013, after a peak of 11,643 in 2007. The price of cultured mullet was always higher than that of the wild caught, especially out of the capture fisheries season (November–March), though mullet price has decreased in time with increased production (Park et al. 2012); however the price of cultured grey mullet witnessed a growth tendency after 2008 (Kim 2014). The flathead grey mullet *M. cephalus* is considered as an important food fish in the south-west region; it is the only mullet species officially reported to be cultured in the Republic of Korea and makes up most of the 84% of total mullet production from capture fisheries in 2013 (FAO 2015).

The culture of the flathead mullet in Taiwan was traditionally carried out in freshwater ponds in polyculture with Chinese carps at densities of 2,000–3,000 fish.ha^{-1} and in brackish water ponds in the central western part of the country. Both culture systems were based on the collection of fry in estuarine waters, and about 10 million mullet fry were required to stock these ponds (Tang 1964). Today mullets are cultured in earthen ponds, either in monoculture or in polyculture with shrimps, clams and milkfish *Chanos chanos*, and take two to three years to reach market size. The culture is mainly aimed at producing mullet egg roe, a high priced food delicacy, which is also collected from wild mullets by capture fisheries and which provides a great added value to mullet production (see below 'Products from mullets'). Despite the first pioneer work of Tang (1964) and the setting up of specific research programmes on the artificial propagation of *M. cephalus* in the 1960s-early 1970s, mullet culture still relies on wild fry (Liao 1981, Chapter 19—Liao et al. 2015).

In Israel, common carp (*Cyprinus carpio*) culture began in 1937/38 and slowly expanded through the country, first as monoculture, then as polyculture with the blue tilapia *Oreochromis aureus* and, from 1956, with the mullets *Mugil cephalus* and *Liza ramada* (Shapiro 2006). Research work on the induced spawning of *M. cephalus* in captivity was carried out in the 1960s (Abraham et al. 1966, 1968, Yashouv 1966, Yashouv and Ben-Shacahar 1967, Yashouv et al. 1969). At present mullet production derives from both aquaculture, with a few farms in northern Israel where mullets are intensively grown to market size in earthen ponds, with regular manuring and distribution of feed (Sarig 1981), and from capture fisheries in Lake Kinneret. Most farms are still stocked with wild fry due to problems in economically producing mullet fry from artificial propagation, though two hatcheries are at present producing mullet fry (G. Pagelson, pers. comm.). The Fisheries Department of the Ministry of Aquaculture stocks Lake Kinneret with hatchery produced or wild fry, to enhance fisheries and compensate for the decline in mullet capture due to overfishing (G. Pagelson, pers. comm.).

The flathead grey mullet, *'ama'ama*, is a significant species in traditional Hawaii culture (see review by Nishimoto et al. 2007). It was believed to be a supernatural fish and, because it was born of human parents, at one time it heard and understood speech, and many tales on mullets are narrated. The flathead mullet was a highly prized fish during the Hawaiian Kingdom period (1795–1893). The species was cultured in coastal fish ponds, built from small bays at a stream mouths or freshwater springs that were confined by stone walls, made of basalt rock, often cemented with chunks of corals and coralline algae, at salinities of 20–32% (Ellis 1968). Ponds varied in size from 0.4 to 212 hectares, 60–120 cm deep and walls could be up to 1 km long (Ellis 1968). The early Hawaiian people were known for their conservation practices and their aquaculture expertise (Titcomb 1972). Fry and juveniles were recruited during rising tide by opening the sluice gates, though some fry would enter through the pervious walls; further stocking was also carried out by catching wild fry along the shores. Ponds were stocked at densities of 50,000 fish per 10 hectares (25 acres) (Ellis 1968). The species was once revered by Hawaiian royalty who appreciated mullet, freshly harvested from the fish ponds (Nishimoto et al. 2007). By the late 19th century, fish ponds were slowly abandoned and in 1900 only 99 fish ponds were left, though *'ama'ama*, from both fish ponds and capture at sea, still represented 35% of the fish sold at the Honolulu fish market and was the most expensive product (Nishimoto et al. 2007). This culture practice is disappearing; the ponds have been neglected for years, and despite projects to rebuild them, most ancient ponds have been silted from the mountain sides and walls destroyed by tidal waves over time (Ellis 1968). In 2000, only two productive fish ponds survived, with an annual production of less than 500 kg of flathead grey mullets (Hawaii Division of Aquatic Resources 2004). Culturing the flathead grey mullet as a food fish has been hindered by its relatively slow growth, taking three years to reach a one pound (450 g) market size, and at present the flathead grey mullet is more significant as a recreational fishery species. In Hilo harbour there is a specialized hook-and-line recreational fishery for this species, using a wad of diatom algae as bait.

Similar culture systems, where mullets are grown in polyculture with other fish and shrimps are found in India in the traditional system of culture of *bheries* in West Bengal, where tidal water is impounded in the intertidal mudflats by man-made banks. Tidal water with fish fry and shrimp seed enters through sluice-gates during spring tides, and no manuring or feeding is carried out. Harvesting of marketable sized fish and shrimp is performed regularly during spring tides through traps placed near the sluice gates, with production levels between 500–750 $kg.ha^{-1}.year^{-1}$, with shrimp contributing 20–25% of the total production (Ayyappan 2014). In India, traditional extensive culture of mullets has been practiced in estuarine and coastal areas of Kerala, Goa and West Bengal since 1947 (Saleh 2006). The grey mullets *Mugil cephalus, Liza macrolepis, L. tade* and *L. parsia* are grown in polyculture with other fish and shrimps in brackish water earthern ponds, for an eight–nine month grow-out period with an average annual production of 2.0–2.5 $tonnes.ha^{-1}$ (Ayyappan 2014). Research work on the intensification of mullet culture began back in 1920, with some experimental trials on the culture of mullet juveniles and were further developed on various mullet species with acclimation experiments and developing polyculture practices in the 1940s (Nash and Koningsberger 1981). At present intensive farming is restricted for the non-availability of hatchery fry, though experimental induced spawning trials were carried out on *M. cephalus*, *M. macrolepis* and *M. parsia* (Sebastian and Nair 1973, 1975, Radhskrishnan et al. 1976, Chaudhuri et al. 1977, James et al. 1983, Rajhalakshmi et al. 1991, Shabanipour and Shanmugham 1992, see Table 16.6). Experimental culture of marine/brackish water fin fishes have been carried out by CMFRI (Central Marine Fisheries Research Institute) since the 1980s and significant achievements were made in the coastal pond culture of milkfish and grey mullets at Mandapam, Tuticorin, Madars, Calicut, Narkkal and Mangalore. Pen culture experiment carried out at Mandapam and Tuticorin with milkfish and grey mullets at a stocking density of 50,000.ha^{-1} yielded 400 to 800 $kg.ha^{-1}$ (Smit Ramesh Lende 2011). Experimental rearing trials with *Mugil cephalus* in monoculture or in an integrated system in marine waters were also performed (Ayyappan 2014). Mullet productions are not reported in Indian statistical return to FAO and are probably included in the general category 'finfish nei[4]' (Saleh 2008).

In Japan, feeding experiments on mullet juveniles were performed back in 1906–1907 (Nash and Koningsberger 1981). Until around 1960, earthen ponds and inlets surrounded by embankments or nets were widely utilized for extensive marine aquaculture, where mullets were also grown. Another extensive culture system is the 'kawa culture' in Aichi Prefecture, carried out in waterways where mullets were grown in polyculture with carps, crucian carps, common sea bass, perch and eels, and represented the main species, with initial stocking densities of 15,800–19,750 $mullets.ha^{-1}$, low survival rates of 5–10%, and final yields of 392–663 $kg.ha^{-1}$ (Liao 1981). However, in recent years, these culture systems have declined due to the intensification of Japanese aquaculture and interest in other species (Makino 2006), and mullets have not been reported in FAO (2015) statistics for Japan since the mid 1960s.

Although not reported in FAO (2015) aquaculture statistics as a separate species, China is a producer of mullets (reported as 'marine finfish nei') (FAO/NACA 2011). Traditional mullet culture has been practiced in China for more than four centuries, in extensive systems in brackish waters, the 'harbour culture' in northern China, often associated with marine shrimp, or in freshwater in polyculture with cyprinids

[4] not elsewhere identified.

(Zheng 1987). The artificial propagation of the red lip mullet *Liza haematocheila* has been performed since the 1970s, with 10,000 million fry produced annually (Sun 1995, Hong and Zhang 2002).

Though no mullet production is reported for Oceania (FAO 2015), a project on the sustainable development of community-based mullet culture has been recently launched in Papua New Guinea for the establishment of mullet farming systems in Loniu, Manus province, in particular for the enhancement of the wild stocks through the conservation of traditional mullet migration pathways and spawning grounds and the development of a sustainable mullet culture in cages in Loan Lagoon with best management practices (Izumi 2014).

Mediterranean Coastal Lagoons

Coastal lagoons are most probably the sites where marine aquaculture was born in different regions of the world. The possibility to more or less confine fish inside the lagoon with barriers certainly characterized coastal fish production which stands between capture fisheries and aquaculture (Cataudella et al. 2001). Capture fisheries and extensive aquaculture carried out in coastal lagoons partially overlap in statistics and it is therefore difficult to separate production from the different methods. In Egypt, for instance, mullet production is ascribed to inland waters. Extensive culture is based on the trophic resources available in the coastal ecosystems, to produce fish and shellfish with no supplementary feeding by farmers. It is sometimes difficult to draw a strict borderline between enhanced fisheries, culture-based fisheries or capture-based aquaculture (Ottolenghi et al. 2004, Lovatelli and Holthus 2008). In the case of mullets being stocked in coastal lagoons or confined parts of them (*valli* in Italy, *hosha* in Egypt), the elements to discriminate between enhanced fisheries and capture-based aquaculture lies in lagoon ownership, hydraulic control, water management and the presence of fixed trap systems. Little culture-based fisheries are carried out for mullets, as most seed comes from the wild. An example is the stock enhancement programme in Hawaii, described in Chapter 18—Leber et al. (2015).

The Mediterranean region hosts around 400 coastal lagoons, covering an area of over 624,000 hectares (Cataudella et al. 2015a,b). The fisheries management of coastal lagoons is based on the mixed culture of different species (in the Mediterranean mainly different mullet species, gilthead sea bream *Sparus aurata*, European sea bream *Dicentrarchus labrax*, eel *Anguilla anguilla*) at different sizes. Traditional extensive culture methods are based on the natural productivity of the environment: growth depends on the species and on food availability, rearing cycles are long and a low production per hectare is obtained.

Mullets are important species in Mediterranean coastal lagoons, with five species being the object of traditional capture and culture activities. The coat of arms of Orbetello, on the homonymous lagoon in central Italy, represents a "lion rampart that spears a superb mullet" (Fig. 16.7, Del Rosso 1905). Mullets often constitute a high proportion of total catches: in Greece 56% of lagoon production is composed of grey mullets and in Koycegiz lagoon (in Turkey) 78.4% (Cataudella et al. 2015). In Algeria, mullets represented 70% of the annual fish captures in Lac Mellah in 2009, with 17,337 tonnes (Seridi and Bounouni 2015). In Egypt, grey mullets represented respectively 9.8, 14.5 and 0.8% of the total captures in Lake Manzala, Burullus Lagoon, Edku Lagoon in 2009 (Abdel-Rahman 2015). In Italy mullets made up 35% of Lesina lagoon production in 2003 (Maccaroni et al. 2005) and 22 to 60% of the total fisheries production from Venice lagoon, as calculated from landings at the fish market in Chioggia from the 1945 to 2006 (Provincia di Venezia 2009), though the increased percentage of mullets in total landings is also related to the decreased eel production in the lagoon, from 20% to nearly 0 in recent years. Mullets represented 35% of landings in Orbetello Lagoon in 2014 (M. Lenzi, pers. comm.).

Fish Barriers

Capture fisheries in lagoons can be exclusively based on small-scale fisheries, as in Egypt, or include the use of fixed capture systems. These fixed traps (also called fish weirs, fish barriers, *lavorieri* in Italy, *bordigues* in France and many South Mediterranean countries, *dajlan* in Albania, *encanisadas* in Spain) have been used in various regions of the world, exploiting the migratory behaviour of mullets (Oren 1981). An accurate observation of fish behaviour by local communities must have certainly preceded

Figure 16.7. The coat of arms of Orbetello, on the homonymous lagoon (west coast of Italy) representing a "lion rampart that spears a superb mullet" (Del Rosso 1905) (photo by D. Crosetti).

the invention of fish traps, with the observation of the alternate passage of some marine fish in or out of the coastal lagoons according to seasons and the invention of the tapering inlets "always open on the way in, always closed on the way out", the *bocchini* in the Italian Tuscany tradition (Fig. 16.8b) (Del Rosso 1905).

These traps are seasonally used, exploiting the fish movements from and to the lagoon, whereas nets are used all year round, with more regular catches. The set up of fish barriers in coastal lagoons represented the passage from capture fisheries to culture: the presence of a fish barrier that confines a coastal lagoon or a portion of it enables this environment to be managed as an aquaculture system, with stocking actions to enhance production and selection in captures. The migratory fidelity of mullets has justified the investments in fixed traps, leading to the confinement of water bodies and hence to primitive forms of aquaculture.

The fish barrier was first designed in Italy and then spread to the whole Mediterranean region, in the coastal lagoons of Spain, France, Greece, Albania, Turkey, Algeria, Tunisia. In Egypt, no fish barriers are present, as by law no barrier can be placed in the channels communicating with the sea.

V-shaped traps lead the fish to the capture chamber, where they are easily harvested by fishers with seines, push seines or scoop nets (Figs. 16.9a,b,c,d). Fish barriers evolved through time and can be more or less complex (De Angelis 1960, Ardizzone et al. 1988, Ciccotti 2015). The modern fish barrier is the result of ancient traditional experience; the replacement of reeds and wooden poles by concrete and plastic or metal grids (Figs. 16.8a, 16.9d) certainly rationalized the fish barrier design and eliminated the burden and cost of regular maintenance and substitution of the decaying material every year.

Fish barriers are not only simple fishing gear that prevent fish in the lagoon from going back to sea thus enabling their capture, but are also selective fishing gear, capable of sorting species and size and enabling natural recruitment of fry and juveniles from the sea (Rossi 1995). Their location on the canals connecting

a

b

Figure 16.8. Traditional fish barrier made of reeds and wooden poles (a), with the *bocchino* (b) in Cabras Lagoon, Sardinia, Italy (photo by D. Crosetti).

the coastal lagoon to the sea enables the capture of adult mullets during their spawning migration back to sea, while the size of the grids allows juveniles to enter the lagoon.

Captures at the fish barrier often represent a large proportion of the total capture of a Mediterranean coastal lagoon. Huge captures at fish barriers are reported from the past (Del Rosso 1905): in one single dark and stormy night, 20,000 kg of fish were captured in Orbetello (Central Italy) *bondanoni*, 80,000 kg in Comacchio (North-East Italy) *lavorieri*, 40,000 kg in Mar Menor (Spain) *encanisadas*. Fish barriers account for 41% of the total fish capture in Sardinian (Italy) managed coastal lagoons (Cannas et al. 1998), and for about 30% in the Mesolonghi–Etoliko lagoon complex in Greece (Koutrakis et al. 2007), with 51.9% of these captures composed of the five Mugilidae species *L. saliens, Liza aurata, Liza ramada, Chelon labrosus* and *M. cephalus* (Katselis et al. 2003).

Massive catches of mullets are obtained at the fish barriers during the spawning migration back to sea, from August to October for the flathead great mullet. In 2014, 98% of the mullet production in Orbetello Lagoon was harvested at the three fish barriers, Nassa (92%), Ansedonia and Fibia (data from Orbetello Pesca Lagunare, M. Lenzi, pers. comm.).

Mullet Culture in Italy and Vallicoltura

In Italy mullet culture is practiced in extensive systems, and grey mullets represent an important proportion of the production of *valli*[5] and coastal lagoons, with a mean annual production of 568 tonnes in the last five years (2009–2013) (FAO 2015). Indeed semi-intensive/intensive culture systems are restricted to pre-fattening and overwintering of fry and juveniles to be later stocked in extensive systems, as even the most prized mullet species, flathead grey mullet (for the production of egg roe) or thicklip mullet (for the quality of its flesh), cannot justify economic investments for their intensive culture because of their relatively low market price.

Vallicultura represents an advanced form of coastal lagoon management, an extensive culture system based on natural cycles and dynamics. It has been traditionally practiced for centuries in the North Adriatic *valli*, and gave birth to modern Italian marine fish culture. A 100 *valli* are still used for fisheries and extensive aquaculture: 36 valli in Friuli-Venezia-Giulia, with a total surface of 1,260 ha; 50 *valli* in Veneto, with 20,373 ha; and 12 valli in Emilia Romagna, with 14,700 ha; each *valle* covering an area from a few to 1,700 ha (Rossi et al. 1999, Cataudella et al. 2001, Ciccotti 2015). Differences between *vallicoltura* and coastal lagoon management lie in history, ownership, legal and administrative aspects, technology and environmental features (Cataudella et al. 2001). The management of the *valli* for fish production is still traditional, configured as an extensive multispecies culture system with relatively low production and reduced operating costs (Ciccotti 2015). It is characterized by fry stocking, hydraulic management and most captures performed seasonally at the fish barrier, whereas coastal lagoons normally have a natural recruitment of fry and juveniles, regular artisanal fisheries activities in the water body and often lack of hydraulic management (D'Ancona 1955). Production in the *valli* is low however, and cannot compete with intensive cage culture at sea, but *vallicoltura* aims to combine environmental compatibility with economic sustainability (Pellizzato 2011) and represents one of the most interesting examples of coastal lagoon management in the world (Cataudella et al. 2001).

The management of a *valle* can be summarized in a few basic steps (Rossi 1995): natural recruitment from the sea integrated with wild fry from other areas in spring; stocking and wintering; fish captured at the fish barrier in the autumn, when migrating back to sea.

In the past the *valli* were stocked by natural recruitment of wild fry of several euryhaline species (mullets, sea breams, sea bass, eels [glass eels]) undergoing their anadromous migration into eutrophic environments represented by the *valli* or the coastal lagoons. Suitable openings at lagoon outlet channels were created and kept clean in order to facilitate recruitment. In early spring, a water flow was produced in the outlets of the lagoons to the sea to 'call' the fry and juveniles that would swim upstream into the lagoons following their rheotaxic instinct, as described by Coste in 1861 in the Comacchio lagoons.

[5] confined areas of the Lagoon of Venice or Po Delta, where '*vallicoltura*' is practiced. The name *valle* derives from the Latin '*vallum*', which means protection, hence bank/embarkment.

a

b

c

Figure 16.9. contd....

Figure 16.9. contd.

d

Figure 16.9. (a, b, c, d) Harvest of mullets in the capture chamber of a modern fish barrier in Tortoli Lagoon (Sardinia, Italy) (photo by D. Crosetti).

Whenever natural recruitment was not sufficient to cover the demand of fry and juveniles for grow out in the *valle*, wild fry collected elsewhere were used to stock the *valle*. Each species was seasonally harvested in the coastal areas and estuaries by fry fishers (see 'Fry collection' above) and transported to the *valle*. At present the *valli* are totally isolated from the lagoon by permanent soil banks, which replaced the traditional reed bundles that confined lagoon portions, maintaining the contact with the lagoon through sluice gates.

Fry and juveniles can either be released directly into a *valle* or in confined areas for a certain period to winter and to prevent predation by other species. The most expensive species, *M. cephalus* and *C. labrosus*, are released in special ponds ('*peschiere*') predator free, with deeper areas where fish can gather when temperature suddenly drops. The pre-fattening of fry and juveniles can be carried out in 1–2 ha earth ponds for three-four months with the distribution of supplementary feed, obtaining an average survival rate of 75.6% (Ardizzone et al. 1988). In those *valli* with no pre-fattening ponds, shallow portions of the *valle* can be drained and dried by lowering the water level, and 500–1,000 m^2 pens are set up, covered with nets against ichthyophagous birds and filled again with water, to obtain a predator-free confined environment where thicklip mullet fry can be stocked and fed for at least 60 days before being released into the open *valle* (Cataudella and Monaco 1983). The low value mullets, *L. saliens* and *L. aurata*, are stocked directly into the *valle* and most end up as food for predators such as the sea bass, which is a prime species in the *valle*.

The species inside the *valli* exploit the natural trophic substrata of the ecosystem, in a multispecies association. Stocking is therefore performed with fry and juveniles belonging to different fish species, trying to create a balanced ratio between omnivorous species, such as the different mullet species, and predators, such as gilthead sea bream, European sea bass or eel. Mullets represent the highest percentage in number, up to 80% (Ardizzone et al. 1988). When natural recruitment was still abundant, stocking of fry was limited to 500–1,000 mixed fry.ha^{-1} but increased parallel with the decrease of natural recruitment. An example of more recent stocking was reported by Boatto and Signora (1985) with 3,875 fish.ha^{-1}, with 80% grey mullets, 6% gilthead sea bream, 6% European sea bass and 8% yellow eels. However, the quantity of fry to be released depends on the availability of fry and on the characteristics of the environment where they are released. There are no fixed rules, and each *valle* has its peculiar ecological and environmental characteristics to be considered when stocking is performed.

More generally, the stocking of 1,000–4,000 mullet fry.ha^{-1} is performed, with a survival of 10–20%, to obtain a final production of 50 to 150 kg.ha^{-1} of mullets (Cataudella and Monaco 1983, Ravagnan 1992, Cataudella et al. 2001). Mullets have a lower survival rate compared to the other species present in the coastal lagoon waters, as they are easily preyed on by ichthyophagous birds, European sea bass and eels (Rossi 1995). Lower survival rates (2–3%) could be obtained for wild captured fry stocked in the *valle*, as up to 70% of the product could be lost immediately after release as a consequence of stress from capture and transport (Ravagnan 1978).

The stocking of the *valle* starts in January–February with *Liza saliens*, followed in May–June by gilthead sea bream, European sea bass, the other four mullets: *Mugil cephalus, L. ramada, L. aurata, Chelon labrosus*, and (when available) elvers or yellow eels of *Anguilla anguilla.*

The Venetian *valli* were annually stocked with some 40 million grey mullet fry (from different species: *cefalame—Liza* spp.*; boseghini—Chelon labrosus; meciattini—Mugil cephalus*), three million gilthead sea bream and three million European sea bass captured in the wild (Franzoi et al. 1999).

The mullet production cycle is three–five years long, according to the species. Growth depends on the biological features of each species, and on the availability of food in the water body. Most market size mullets are captured seasonally at the fish barriers in the autumn, when they tend to leave the *valle* to return to the sea in their spawning migration, or for the lowering of temperatures of the *valle* waters, though some small scale fisheries are also occasionally practiced inside the *valle* year round. Small mullets, which have not yet reached market size, are driven by water flow into wintering ponds where they are protected from low temperatures and stocked to the spring. Wintering ponds are deeper than the *valle* and oriented in order to be protected from dominant winds; they are often isolated by an ice surface layer, created by supplying a tiny freshwater flow which, being lighter than the brackish water, floats on it and freezes, which keeps temperatures above 3–4°C. In some cases, fish feed is distributed to enhance growth and increase stocking densities, creating a semi-intensive/intensive system for the length of the winter. *Liza saliens* can withstand low temperatures, and does not need to be wintered in special ponds: in northern Italy *vallicoltura* it is called '*mangiaghiaccio*' (ice eater).

The hydraulic management of the *valle*, carried out through the opening/closure of the sluice gate and the use of internal canals, is very important to ensure the correct in and outflow of marine and freshwaters and the water circulation inside the *valle*, in order to facilitate natural recruitment of fry and juveniles in the spring, to prevent dystrophic crises in the summer, to direct the fish towards the fish barrier in the autumn and to create the best environmental conditions in the wintering ponds (Ardizzone et al. 1988).

Annual productions of the Venetian *valli* range from 69 to 150 kg.ha^{-1}, with an average production around 100 kg.ha^{-1}, 50% of which are grey mullets (Ardizzone et al. 1988). Flathead grey mullet and thicklip grey mullet have the highest price and larger commercial size (0.5–1 kg), the other mullet species are sold at cheaper prices, already at a size of 200–250 g (Cataudella and Monaco 1983).

Valli production is severely affected by the massive presence of great cormorants (*Phalacrocorax carbo*) and the damage they cause in wintering ponds and during the fish migration towards the fish barrier (Volponi and Rossi 1998). Indeed in the Mediterranean region a limiting factor to the development of mullet culture in extensive systems is the presence of ichthyophagous birds, such as the great cormorant, the great crested grebe *Podiceps cristatus*, the black-necked grebe *P. nigricollis* and the red-breasted merganser *Mergus serrator*, which play an important role as top predators in the lagoon ecosystem and often create conflicts with fisheries and aquaculture. The great cormorant is a real threat to aquaculture, as its total population has grown 20-fold over the past 25 years and is estimated to comprise at least 1.7 to 1.8 million birds (Kindermann 2008). A single great cormorant can eat 400–600 g of fish day^{-1}, for a total of more than 300,000 tonnes of fish captured from European waters every year. In Sardinian lagoons (Italy), ichthyophagous birds have a density of 5 ind.ha^{-1} (Marino et al. 2009) and the predatory pressure in a Venice lagoon valle was estimated in 8.23 kg.ha^{-1} (Cherubini 1996). In Egypt up to 30,000 great cormorants were estimated to be wintering in Bardawil Lagoon and to predate more than 6% of the lagoon fish production (Abdel Rahman 2015, Khalil and Shaltout 2006). Control measures are difficult to set up, because regular hunting is not possible due to the E.U. Wild Birds Directive (79/409/EEC), except for derogation measures, restricted in time and space. This creates a serious conflict between environment conservation and aquaculture development. The conservation policy towards this predator bird leads entrepreneurs to abandon the valli for more intensive systems, which can be protected by physical barriers (normally nets) placed over fish ponds or cages (Cataudella et al. 2001). These systems are most effective in preventing predation by ichthyophagous birds, but cannot be applied to large surface areas and therefore to extensive aquaculture (Kindermann 2008).

The same problem was reported in Hawaiian fish ponds, with severe losses of mullets due to ichthyophagous birds, especially the black crowned night heron, *Nycticorax nycticorax*; to avoid predation, fry were often cultured in smaller ponds where predator bird control was possible (Ellis 1968).

Artificial Propagation

Mullet culture has always been based on natural resources for the recruitment of fry and juveniles for grow-out. However, the natural availability of fry and juveniles is subject to annual fluctuations and unpredictable variations in occurrence and abundance which are difficult to match with aquaculture planning and management, especially in more intensive systems. For instance in Italy, four years out of five the availability of *Mugil cephalus* fry was not sufficient to supply the demand for the stocking of the North Adriatic *valli* in Italy (Ravagnan 1978). Artificial propagation in hatcheries is therefore the solution to obtain a steady supply of fry and was considered by Oren (1971) as an axiom for "maximum production of mullets under controlled conditions".

Most cultured fish do not spawn spontaneously in captivity: in most cases in the captive female oocytes mature to the vitellogenesis stage then stop and without hormone treatment undergo atresia: in this case, neither ripening, ovulation or spawning occur. Prior to the natural spawning season, vitellogenesis of both wild and captive female mullet proceeds to the secondary or tertiary yolk globule stage, but it is then blocked until the final maturation leading to ovulation is triggered by the environmental stimuli of seaward migration and spawning grounds, or by the artificial stimuli of hormone treatment. In the environmental approach, the farmer tries to replicate in captivity the rearing conditions that simulate the natural conditions in which the species reproduces: some environmental parameters, such as photoperiod, temperature and salinity, are the main factors that regulate the reproductive cycle in most teleosts. In the hormonal approach, drugs are used to intervene at different levels of the hypothalamic-hypophyseal-gonadan axis (Chapter 11—González-Castro and Minos 2015). The use of hormones to induce spawning in fish has been widely practiced since Houssay (1931) first demonstrated the efficacy of hypophyseal extracts for this purpose. Various pituitary extracts were later combined with steroids and other stimulating or inhibiting drugs.

Several experimental trials with various hormonal treatments have been carried out in many regions of the world on 11 different mullet species to induce spawning in captivity and produce hatchery fry (Crosetti and Cataudella 1995, Kuo 1995, Table 16.6). The most reliable method is an acute hormonal therapy, combining different stimulating and inhibitory substances: fish pituitary homogenates, human Chorionic Gonadotropin (hCG), steroids, Lutheinizing Hormone Releasing Hormone synthetic analogues (LHRH-a), neuroleptics and dopaminic inhibiting drugs (Table 16.6). All these need to be used at very high dosages to induce mullet to spawn, if compared to other non-mullet species. Particular emphasis was given to *M. cephalus*, which has been the object of specific research programmes, mainly by a research

Table 16.6. Induced spawning trials in grey mullets.

Species	Country	Temp °C	Salinity ‰	Hormone treatment (/kg)	Females spawning/ ovulating ratio# (%)	Eggs fertilisation rate (%)	Eggs hatching rate (%)	Larval/fry survival rate (%)	Reference
Chelon labrosus	France	-	14, 30	progesterone 20 mg.d^{-1} for 6 days	-	-	-	-	Cassifour and Chambolle 1975
	Italy	23	34–39	HCG 5,000 IU	100	90–95	65–76	2.2–41.9 (20 d) 23.3–33.4 (58 d)*	Cataudella et al. 1988a,b
	Italy	14.5-17	30–35	7 carp PG + LHRH 50 µg DOM 10 + LHRH-a 100 µg	82	42–88	-	-	Crosetti 1998, Crosetti 1999, Crosetti and Cordisco 2004
	Tunisia			hCG 10,000 IU+ hCG 10,000 IU and LHRH-a 100 µg	-	-	-	-	Besbes et al. 2003
Liza aurata (Mugil auratus)	USSR	15–25	15–18	mullet PG 40 mg carp PG 70 mg	-	-	-	-	Apekin et al. 1979
Liza haemetocheila (Mugil so-iuy)	USSR			carp PG	-	-	-	-	Apekin and Kulilova 1980
	China	15–16		LHRH 125–300 µg/fish* PG (4–30)/hCG 3,500–10,000	-	-	-	-	Zheng 1987
	Russia	16.8–24.1	13	isofloxythepin§ 2 mg + GnRH-a 25 µg + GnRH-a 50 µg	100	-	-	-	Glubokov et al. 1994
	China			LHRH-a + DOM	-	-	-	-	Chen 1993a,b, Hong and Zhang 2003
Liza ramada (Mugil capito)	Israel	24	36	CPH/CPH + TSH and ACTH	stripped 25–33*	-	-	-	Abraham et al. 1967

	Italy		15–30	tuna PG 3.7–7.72 mg	stripped 62	-	0	-	Gandolfi and Orsini 1968
	Egypt			hCG 20,000 IU + hCG 40,000 IU	-	55 ± 8.4	60 ± 6.6	-	Mousa 2006, 2010
	Egypt	15 ± 2	32 ± 1.5	hCG 3,500 IU + LHRH 200–300 μg	-	52–75	-	-	Fahmy and El-Greisy 2014
Liza richardsoni	South Africa	18–24	35	*Labeo umbratus* PG or carp PG (9–13) + HCG 5,000–15,000 IU.female-1	-	> 80	> 80	0.2 (50 d)	Bok and Joenbloed 1987, Bok 1989
Mugil cephalus	Taiwan	20.4–24.8	SW	Synahorin + mullet PG	-	32	10	0 (4 d)	Tang 1964
	Taiwan	22.5–24.6	SW	Synahorin 10–60 RU + 2.5–6 mullet PG + vit. E 0–300 mg	stripping	61.7–100	51–100	-	Liao et al. 1971, 1972, Liao 1975
	Taiwan	19–20	26.6–28	LHRH and hCG	62–100	20–75	48–73	-	Cai et al. 1997
	Israel		FW-SW**	carp PG + LH (3 inj)	-	-	-	0 (4 d)	Yashouv et al. 1969
	Israel	24–28	40	carp PG 1.25–2.50 + GnRH-a 30–35 μg	-	20–98	-	-	De Monbrison et al. 1997
	Israel		SW	GnRH-a 30 μg and MET 30 mg GnRH-a 10 μg and DOM 5 mg	83–85	-	-	-	Aizen et al. 2005
	Hawaii (USA)	26	32	salmon PG 2 + Synahorin 70 RU	-	-	-	-	Shehadeh and Ellis 1970
	Hawaii (USA)	24	32	SG-G100∞ 11.9–20.9 mg	53–93 (mean 77)	60–96	yes	-	Shehadeh et al. 1973b
	Hawaii (USA)			hCG 49,000–78,000 I.U.	45–92	-	-	-	Kuo et al. 1973b
	Hawaii (USA)	26.8 ± 0.3	36.1 ± 1.0	carp PG 20 mg + LHRH-a 200 μg	94	70.4 ± 30.2	5–89	-	Lee et al. 1987
				LHRH-a 302–696 μg (2 inj)	64	0–95	0–95	-	
	Italy	23	35	hCG + carp PG	-	-	-	-	Lumare and Villani 1972

Table 16.6. contd....

Table 16.6. contd.

Species	Country	Temp °C	Salinity ‰	Hormone treatment (/kg)	Females spawning/ ovulating ratio# (%)	Eggs fertilisation rate (%)	Eggs hatching rate (%)	Larval/fry survival rate (%)	Reference
	Italy	19–24	35–45	mullet PG 7 or carp PG 3+LHRH 100/200 μg DOM 10 mg + leuprolide acetato 500/700 μg leuprolide acetato 700 μg	44	25	-	-	Crosetti 2001, Crosetti and Cordisco 2001
	URSS (Black Sea)	21–23/23–25	15–17	mullet PG 50–100 mg	-	-	-	-	Kulikova 1981
	URSS (Black Sea)	19–23	16–18	mullet PG (13–18 mg) + hCG 15,000–45,000	64	30–82	-	-	Kulikova and Gnatchenko 1987
	Egypt	25	34	carp PG 20–70 mg (or HCG 10,000 IU) + LHRH-a 200 μg	-	75–80	90	-	Meseda and Samira 2006
	Bangladesh	22.7	30	PG 5 mg, 4 mg, 3 mg	-	-	-	-	Das et al. 2008
	Iran	20–23	32	hCG	-	-	-	0.9 (50 d)	Mirhashemi Rostami et al. 2009
	United Arab Emirates	20.8 ± 1.08	37	carp PG 20 mg/tilapia PG/ hCG 1000IU + LHRH-a 200 μg	-	-	26.5–84	15.52 ± 7.32 (40 d)	Yousif et al. 2010
Mugil curema	Cuba	24–27	35–38	GS-G100 8–10 mg	-	90–94	-	0 (6 d)	Garcia and Bustamante 1981
		28–33		dry carp PG, fresh mullet PG	-	-	-	-	
Mugil liza	Brazil	20	36	HCG 10,000 IU	-	88.54	12–25	0 (29 d)	Benetti and Fagundes-Netto 1980
	Brazil	20–21.5	30.5–31.5	HCG 20,000–40,000 IU	-	-	-	0 (29 d)	Godinho et al. 1982, Godinho et al 1984
	Brazil	23	36	HCG 10,000	-	-	-	-	Monteiro-Ribas and Bonecker 2001

	Cuba	20–22	36–38	HCG (3 inj) 50,000–60,000 IU HCG 25,000 54,000 IU + mullet PG	-	50	45–98	11.1 (27 d)	Alvarez-Lajonchère et al. 1988, Alvarez-Lajonchère et al. 1991
Mugil macrolepis	India	26–30.5	29–31	dry mullet PG	40	-	-	-	Alikunhi et al. 1971, Sebastian and Nair 1975
	India	27–33.2		carp PG1,200 mg + hCG 15,000 IU carp PG 1,025 mg	-	7.2–50	-	-	James et al. 1983
Mugil parsia	India	23.3–29	0	*M. macrolepis* PG 5–25 mg (2 inj)	-	-	-	0 (16 celled morula)	Radhskrishnan et al. 1976
Mugil platanus	Brazil	19–21	30–34	hCG 20,000–40,000 IU	58	60–90	-	-	Godinho et al. 1993
				M. platanus PG 10 mg + hCG 30,000 IU	50	65–99	-	-	
				M. platanus PG 20 mg + hCG 30,000 IU	57	4–60	-	-	

DOM: Domperidone, dopamine antagonist.
MET: Metaclopramide, dopamine antagonist.
hCG: human Chorionic Gonadotropin.
LHRH-a: Lutheinizing Hormone Releasing Hormone synthetic analogue.
Synahorin: mixture of chorionic gonadotropin and mammalian pituitary extract, produced by Teikoku Zoli, Tokyo.
PG: Pituitary glands.
TSH: Thyroid-stimulating hormone.
ACTH: Adreno Cortico Tropic Hormone.
**L. haemetocheila* (*M. so-iuy*) 17–30 PG/*Carassius auratus* 4–11 PG/*Cyprinus carpio* 14 PG.
§a neuroleptic.
#: spawning ratio The number of females which ovulated (and spawned) after injection divided by the total number of injected females.
^ hand stripped.
∞SG-G100: partially purified salmon gonadotropin.

team of scientists from several agencies organized *ad hoc* in Taiwan from 1963 to 1973 (Liao et al. 1972, Liao 1985) and by the Oceanic Institute in Hawaii (USA) from the 1970s to 2000 (Kuo 1995).

The techniques implemented at the Oceanic Institute in Hawaii are summarized hereafter, and details can be found in excellent manuals (Nash and Shehadeh 1980, Tamaru et al. 1993) and reviews (Nash and Koningsberger 1981, Kuo 1995). Observations and results from artificial propagation trials described below refer to *M. cephalus*, though data from other mullet species are reported for comparison.

The first mullet eggs and larvae were obtained by Sanzo (1930), who captured ripe *M. cephalus* broodstock at sea and after stripping them, performed artificial fertilization. Tang (1964) first induced *M. cephalus* to spawn, with an injection of *M. cephalus* pituitary extracts (2 hypophysis.kg^{-1}) and Sinahorin (chorionic gonadotropin with rabbit hypophysis extract, 40 units.kg^{-1}), a fertilization rate of 32% and a hatching rate of 10%, but larvae died four days after hatching. In 1976 two hatchery produced mullets reached maturity at three years old and were induced to spawn, therefore closing the life cycle in captivity for the first time in mullets (Liao 1977, Chao and Liao 1984).

Spawning is induced in females at the tertiary yolk globule stage, at an average size of at least 600 μm in diameter, which is considered as the minimum size necessary to successively induce spawning in the Hawaiian flathead grey mullet, though in the Mediterranean region, females are considered fully mature when their oocytes reach an average diameter greater than 500 μm (De Monbrison et al. 1997). The initial oocyte diameter to induce spawning is 532–621 μm in *Mugil platanus* (Godinho et al. 1993), > 645 μm in *Mugil liza* (Benetti and Fagundes-Netto 1980, Fagundes-Netto and Benetti 1981), > 600 in *Mugil curema* (Garcia and Bustamante 1981). In Greece *Chelon labrosus* vitellogenic oocytes increased from 730 μm (February 18) to 925 μm (April 7) during the spawning season (Kokokiris et al. 2011).

The extrusion of oocytes sampled with an intraovarian catheter is the most reliable criterion to determine ovarian maturity in the field by observation alone, as identified by Shehadeh et al. (1973a), though other more subjective diagnostic criteria for female ripeness, such as 'fullness of belly' or redness of genital opening, could be used (Kuo et al. 1974).

The most reliable and cost/effective treatment obtained at the Oceanic Institute is a priming injection of commercial carp pituitary homogenate (CPH) at 20–40 mg. kg^{-1} body weight followed by a resolving injection of LHRH-a (D-Ala$_6$, Pro$_9$ NHet) at 100 mg.kg^{-1} body weight 24 hours later (Lee et al. 1987, 1988). The cost of the different hormonal treatments per number of fry produced varied six fold from the cheapest, CPH/LHRH-a, to the most expensive, 10,000 HCG/LHRH-a (Lee et al. 1988). CPH, which might be inconsistent in its potency, could be replaced by HCG, but at higher cost and with poorer results (Lee and Tamaru 1988). The high dosage required to spawn mullets is certainly a limiting factor in mullet induced spawning, both for the high cost of the treatment.kg^{-1} female body weight and for the quality of the eggs and the fry produced. In general in most experimental trials on the different mullet species, substances and dosages in various combinations have been tested in order to determine the minimal effective dosages to reduce the cost of treatment while obtaining highest ovulation and fertilization rates, highest larval survival and optimal larval quality.

In males, spermatogenesis and spermiation can be induced by androgens, and treatment of male mullets with 17s-methyltestosterone was commonly practiced at the Oceanic Institute in Hawaii to obtain mature males on demand (Shehadeh et al. 1972, Lee et al. 1986). 17-MT can be administered orally, injected or implanted into silastic capsules or cholesterol pellets, though the latter two methods are preferred (Lee et al. 1986). In Egypt, out of the reproductive season, male mullets were brought and maintained in a ripe condition in three weeks using 17-MT pellets (El-Greisi and Shaheen 2007). In Taiwan, *M. cephalus* males were only treated at the end of the spawning season (Liao 1975). In other protocols, males do not receive any hormonal treatment and during the spawning season they are selected when they are spermiating and milt flows out spontaneously with a gentle pressure on the abdomen towards the cloaca (in *M. cephalus*, Das et al. 2008; in *M. platanus*, Godinho et al. 1993; in *Chelon labrosus*, Crosetti and Cordisco 2004). In Greece, more than 80% of *Chelon labrosus* males were spermiating in mid February, at the beginning of the spawning season, but the development of an appropriate hormonal treatment is recommended by Minos et al. (2011) to enhance sperm production and quality.

Females and males are placed together in indoor tanks usually with males outnumbering females. Females spawn naturally in the tanks, and the eggs are fertilized naturally by the males (Fig. 16.10).

Figure 16.10. *Chelon labrosus* female ready to spawn in captivity, followed by a male ready to fertilise the newly spawned eggs (Photo by D. Crosetti).

Mugil cephalus has a high fecundity from 800,000 to 2.7 million eggs per female (Liao 1981), though values of up to 7,200 eggs.g^{-1} were reported (Nikolski 1954). In captivity 648 eggs.g^{-1} body weight are produced by three-year-old fish (Kuo et al. 1973) and 849 eggs.g^{-1} by older individuals (Nash et al. 1974). Relative fecundities of up to 2,024 ± 670 eggs.g^{-1} body weight of treated females were reported in *M. cephalus* from Israel by De Monbrison et al. (1997) and up to 1,649 ± 174 eggs.g^{-1} by Aizen et al. (2005). Relative fecundity in other mullets species range from 15–57 eggs.g^{-1} in *Liza cunnensis* to 581-1,243 in *Liza ramada*, 372–745 in *C. labrosus*, 200–600 in *Liza parsia*, just to cite a few (Nash and Shehadeh 1980). Individual fecundity of 0.3–3 million eggs was reported for *Liza haemetocheila* (*Mugil so-iuy*) in China (Zheng 1987).

In the wild, mullets spawn once a year during the spawning season, and ovarian maturation is group synchronous. Year-round spawning was achieved in Hawaii with the manipulation of environmental conditions in indoor tanks and using hormonal treatment, though low egg fertilization rates were observed (Tamaru et al. 1992, Nicol et al. 1993, Kuo 1995). Double spawning, about 45 days apart, was also obtained within the same spawning season in hormonally treated females held in captivity in Israel (De Mombrison et al. 1997). The influence of exogenous hormone treatments on the growth and maturation of *M. cephalus* oocytes in freshwater ponds was investigated by Zaki et al. (1998) in Egypt.

Though induced spawning of *M. cephalus* was obtained at various salinities, lower fertilization rates were observed at salinities < 30‰. A salinity of 100% sea water is recommended to induce spawning in the flathead grey mullet, especially for the activation of mullet sperm mobility, which is much lower at salinities < 20 ppt (Lee et al. 1992). Salinity and temperature are two most important factors in egg incubation and larval rearing. As in all fish, time to hatching is inversely proportional to the incubation temperature and at higher temperature, development time is also reduced. Various combinations of temperature and salinity were tested to determine the best combination for the hatching of *M. cephalus* embryos: an over 80% hatching rate was obtained at salinities of 35–37‰ and temperatures of 26–27°C (Walsh et al. 1991).

Mugil cephalus larvae are quite small (2.65 ± 0.23 mm at hatching, Kuo and Nash 1975) and need to be fed live food. In Hawaii, mullet larvae are stocked at densities of 20–25 larvae.l^{-1} and start to be fed small sized rotifers (S-type) on the second day post hatching (p.h.), at densities of 10–20 rotifers.mm^{-1}, while the egg yolk is being re-absorbed; on day 12, *Artemia* nauplii are introduced at an initial density of

0.1.ml^{-1}, which is gradually increased to a maximum of two nauplii.ml^{-1}. Artificial feed is introduced at 15 days p.h. and distributed *at libitum*. The addition of phytoplankton (*Nannochloris oculata*), which serves as food for both rotifers and the larvae, and the boosting of *Brachionus* spp. and *Artemia* nauplii with HUFA result in better growth and higher survival rates (Tamaru et al. 1995, Ye et al. 2003, Yousif et al. 2010).

Size at hatching differs greatly among mullet species and affects the type and size of their first food: just to cite some extreme examples, it ranges from 1.43 mm in *Mugil macrolepis* (*Liza macrolepis*, James et al. 1983) to 4 mm in thicklip mullet *Chelon labrosus* (Boglione et al. 1992). The large size of the latter larvae enables them to feed directly on brine shrimp nauplii, avoiding expensive parallel culture of phytoplankton and rotifers (Cataudella et al. 1988b).

The larval rearing and weaning of thicklip grey mullet was also carried out in mesocosm with semi-extensive technology in Italy and fry were transferred at 20 days to earthern ponds to reach 3.0–3.5 cm long juveniles at 58 days old (Cataudella et al. 1988b). 25,000 metamorphozed and weaned *Chelon labrosus* juveniles were produced in Tunisia in similar culture systems (Ben Khemis et al. 2006, Besbes et al. 2010, Ben Khemis et al. 2013). Green waters were used to culture *Mugil macrolepis* larvae up to 1 cm long (Sebastian and Nair 1975).

In China, *Liza haematocheila* (*Mugil so-iuy*) larvae were fed fertilized eggs, trochophore larvae and *Artemia* nauplii, then transferred into manured earthern ponds with supplementary feed of soybean milk, followed by a paste of soybean or peanut cake, and artificial feed later on, to be transferred at 30–40 days old to grow out ponds (Zheng 1987). In the northern Sea of Azov region, larvae of the same species started to actively feed at 3 d p.h., at a 3.2 cm long, and were given ciliates (*Euplotes*) as starter feed, followed by rotifers and at 8 d p.h. adult copepods and *Artemia nauplii* (Saifulina 1991).

Larval survival rates of 0.2 and 5% at 30 days p.h. were obtained in Hawaii by Kuo et al. (1973b), with two well defined critical periods, associated with high larval mortalities on the 2nd–3rd and 8th–11th days p.h., and each of the two periods were preceded by larvae sinking to the bottom of the culture tank; larval survival was highest at 22°C. Optimal initial stocking densities was 10–20 larvae.l^{-1}, with survival of 19,7–52.2% at 60 d p.h., as higher densities led to higher mortalities (Eda et al. 1990).

The quality of hatchery produced fry should be thoroughly investigated, especially when these fry are to be grown out in extensive systems. *Chelon labrosus* hatchery fry produced by induced spawning showed a high frequency of malformed individuals (85.1%) and of individuals with at least one severe morphological anomaly (21%), respectively four and five times as much as in thicklip mullet fry captured in the wild, and were therefore poorly adapted to survive after release (Boglione et al. 2006). More recent rearing trials on the same species performed in 'green' or 'clear' waters showed similar survival rates at 57 days p.h. (respectively 22 ± 2% and 18 ± 2%) in both culture conditions, but faster growth (+ 30% for linear growth and + 38% for weight growth) in 'green' waters (Besbes et al. 2010).

A note on the species *Mugil cephalus* should be made before ending this section, as differences in biological characteristics can be found in reports on artificial propagation trials carried out on *M. cephalus* specimens from different regions of the world; the observations and results reported from Hawaiian flathead grey mullet may not fully apply to other populations from distant regions of the world. Indeed, several differences were evidenced between the Mediterranean and the Hawaiian populations of *M. cephalus*: age and size at maturity, diameter of postvitellogenetic oocytes and ovulated eggs (De Monbrison et al. 1997), and behaviour (pers. observation). Great differences in growth rates were found in an experimental trial carried at the Guam Aquaculture Development and Training Centre in 1992, performed by growing in the same rearing conditions three different flathead grey mullet strains: hatchery fry produced in Hawaii and two wild caught fry from Taiwan, the so called 'migratory' strain and the so-called 'local' strain. AGR ranged from 201.9 g +/– 44.7 g body weight.year^{-1} in the Hawaiian strain to 702.8 +/– 129.5 g.year^{-1} in the Taiwan 'migratory' strain. These 'astonishing' differences in growth are consistent with the population groupings, though no meristic or morphometric differences were observed among the three strains (Tamaru et al. 1995). All these results confirm the hypothesis that *M. cephalus* may not be a single species, but a complex of species, as recently evidenced by molecular genetic markers (Chapter 2—Durand 2015).

Despite many experimental trials carried out to induce mullets to spawn in captivity since the 1970s, the artificial propagation of mullets has never reached a commercial scale, apart from some local hatcheries which produce mullet fry to locally grow-out, and hatchery produced mullet fry are not likely to become available in sufficient quantity in the near future to cover the requirements of mullet culture in many producing countries. The main causes lie in the technical difficulties of obtaining good quality eggs and larvae, mainly due to the high doses of hormones required to induce grey mullets to spawn, in the high mortalities at larval stages, in the high cost of producing hatchery fry compared to the price of the market size mullets, in the much higher price of the hatchery produced fry compared to the wild fry and in the abundance of wild fry. The use of hatchery produced fry is not economically viable for farmers at present, as the cost of hatchery produced mullet fry are similar or higher than the cost of other marine species, such as European sea bream or gilthead sea bream in the Mediterranean, which have a much higher commercial price for market-size product (Saleh 2008). Indeed, although a reliable technique for induced spawning of *M. cephalus* has been developed, further refinement of the technique of induced spawning is still necessary in order to obtain good quality mullet seed in reliable quantities to stock grow-out systems.

Products from Mullets

Mullets are commercial fish used as food in all tropical and subtropical regions of the world, though prices depend on the species, size, fishing site and area of the world. In addition to being sold for its flesh, other products from mullets such as ovaries, testes and gizzards, are appreciated as traditional delicacies in different parts of the world.

Egg Roe[6]

Among grey mullets, the flathead grey mullet *M. cephalus* is a highly commercially important species especially for its salted and dried egg roe, which represents a traditional luxury item in many regions of the world, sold at 'extravagantly high price' (Liao 1981): top quality mullet dried egg roes from Orbetello Lagoon in Italy are sold at 230€.kg^{-1} retail price (2015) (Fig. 16.11a). In the Mediterranean mullet egg roe is claimed to have aphrodisiac qualities (Katselis et al. 2005).

In the Mediterranean region, the production of dried mullet egg roe dates back to the Phoenicians, and its use was spread throughout the region in the Middle Ages by the Arabs. The present names *boutargue, poutargue, bottarga* derive from the Arab term *bot-ah-rik* for dried fish eggs (Monfort 2002). In S.E. Asia, mullet roe (*karasumi*) has been considered a delicacy in Japan since ancient times.

The traditional use of the flathead mullet to produce dried egg roe in far apart regions of the world can also be explained by the very high gonadosomatic index (GSI) of the species: 16–17% in ripe females (Katselis et al. 2005, Macdonough et al. 2005). Ripe females at the secondary or tertiary yolk stage are captured during their spawning migrations along the coastal shore (as in Australia or Mauritania, Figs. 16.12, 16.13) or at coastal lagoons fish barriers while swimming back to sea, or are cultured to maturation size (as in Taiwan).

In the Paleopotamos lagoon (Messolonghi-Etokito lagoon complex, Greece) during the spawning migration, out of 2,808 specimens of flathead mullet collected at the lagoon fish barrier (mean weight of 809.1 g), 34.4% were suitable[7] females, 8.9% unsuitable females, 44.8% were identified as males and 11.8% as immature (Katselis et al. 2005). In Orbetello Lagoon, ripe *M. cephalus* females are captured at the fish barriers in August and September, and represented 16% of the total annual mullet capture in 2014, also composed of 22% of mullets > 250 g and 61% 180–230 g mullets, mainly *Liza ramada* (data from Orbetello Pesca Lagunare, M. Lenzi, pers. comm.).

Ripe females are also cultured in ponds in Taiwan: it usually takes more than three years for females to grow and develop to these stages (Liao 1981, Chapter 19—Liao et al. 2015). In Tuscany, there is a project of stocking earthen ponds with *M. cephalus* fry to produce mullet roe and smoked fillets, as in Orbetello Lagoon (M. Guerrieri, pers. comm.).

[6] Roe: generic term for female fish eggs (hard roe) or the milt of male fish (soft roe) (Monfort 2002).

[7] According to the descriptive scale of the gonad stages of teleost fishes (Kesteven 1960), the suitable ovaries are from those that were classified at the gravid and spawning stage; unsuitable ovaries are from ovaries that were classified at the developing late stage, as partially spent and spent. Immature gonads include those classified at the virgin, mature virgin and early development stages.

a

b

Figure 16.11. Vacuum packed *bottarga* (a) and smoked mullet fillets (b) from Orbetello Lagoon (west coast of Italy) (photo by Orbetello Pesca Lagunare).

In Greece, mullet roe is still processed in the traditional way as follows: the whole ovaries are dissected from the fish, washed with water, salted with natural sea salt, dried under the sun and submerged in melted bees wax (Rogdakis 1994, Katselis et al. 2005). A comparison between proximate composition and quality changes during refrigerated storage of fresh and dried flathead grey mullet egg roe was performed by Çelik et al. (2012). In Taiwan, roe processing follows a precise standard procedure, and the final product (*wuyuzi*) is vacuum packed (Chapter 19—Liao et al. 2015). The optimum weight of fresh mullet roe is over 300 g, from three-year old females, though 250 g roe from two-year old females are also used (Chapter 19—Liao et al. 2015).

To cope with demand and insufficient local production, frozen raw mullet roe is imported from Mauritania, Mexico and Taiwan by Italy and France to be processed into *bottarga* in local plants; in Europe, Italy, France and Spain are the main producers (Monfort 2002).

Information about annual production of mullet roe is often scarce, as part of the production is sold directly as a traditional local product without going through wholesale traders and is therefore often not officially recorded.

Figure 16.12. Flathead grey mullets caught during their spawning migration along the Mauritania coast (photo by D. Crosetti).

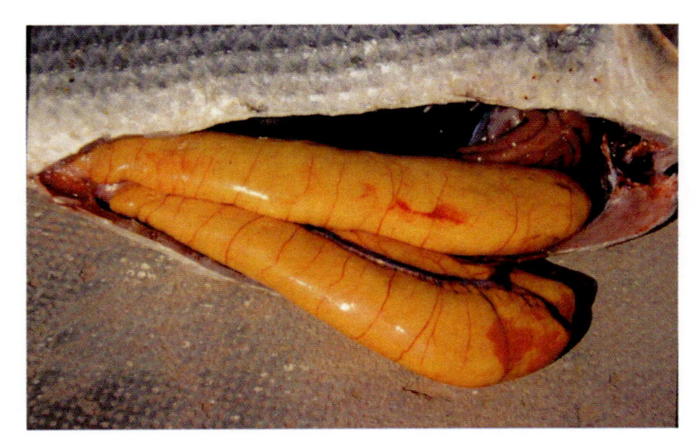

Figure 16.13. Ripe ovaries in a freshly caught female (photo by D. Crosetti).

In the Messolonghi–Etoliko lagoon complex (Greece) the weight of egg roes (*avgotaracho*) ranged from 52.9 to 326.6 g with a mean value of about 165 g. The GSI (gonadosomatic index) showed that the egg roe constituted roughly 16% of the total body weight of a female (Katselis et al. 2005). A model developed by the same authors estimated the production of *avgotaracho* in the Messolonghi–Etoliko Lagoon complex at about 7% of the landings of flathead mullet caught during their spawning migration period. A simulation showed that the error estimation of *avgotaracho* provided from the quantity of the available landings of the flathead grey mullet was about 50% of the annual *avgotaracho* production.

In Egyptian lagoons, some 1,500–2,000 tonnes of mullet egg roe (locally called *batarekh*) are estimated to be produced annually (S. Sadek, pers. comm.); the egg roe from Bardawil Lagoon (60 tonnes per year) has the highest quality (Abdel Rahman 2015). *Batarekh* is also produced from cultured *Liza ramada* in large quantities in winter and sold on the local market (Abdel Rahman 2015). In Koycecíz Lagoon in Turkey some three tonnes of mullet dry roe (*haviar*) are produced annually (Deniz 2015). Orbetello Lagoon (Italy) has an annual production of 500–1,000 kg, with a tradition dating back to the 15th century (M. Lenzi, pers. comm.). In the Tsoukalio-Rodia lagoon complex in Greece, mullet egg roe annual production fluctuated between 173 and 2,692 kg from 1975 to 2000 (Reizopoulou 2015), and in the Messolonghi-Etoliko lagoon system it averaged 600–700 kg, with a production of 920 kg and a value of 43,117€ in 2013 (E. Koutrakis, pers. comm.). In Taiwan, annual mullet roe production ranged from 17 to 90 tonnes from 2002 to 2012 (Fisheries Agency 2003–2013).

Mullet roe is sold as the whole roe sac, either bees-waxed or vacuum packed, but also as grated dry *bottarga* packed in glass jars or vacuum packed, or vacuum packed *bottarga* slices (Fig. 16.10a). Mullet roe is a stable natural source of n-3 PUFA (poly-unsaturated fatty acids), but PUFA stability decreases from raw roe to whole bottarga down to grated bottarga (Rosa et al. 2009). The quality, and hence the price, of mullet roe depends on many factors, such as size, shape, colour, salinity, oil content and origin. In Italy, whole dried mullet roe certified from Orbetello Lagoon (Tuscany) are sold locally at a retail price of 230€ kg^{-1} (February 2015), but mullet dry roe of unknown origin are much cheaper (8O€ kg^{-1}).

In Greece, a retail commercial network uses 'the good reputation of the product' to attain higher sale values than the wholesale trade; indeed local fishermen assert that processing plants often use the ovaries of other mullet species captured locally, and/or of flathead grey mullet imported frozen from other regions of the world (Katselis et al. 2005). This was confirmed by an investigation carried out in Sardinia (Italy) with mitochondrial DNA sequences (partial cytochrome b and D-loop) on mullet roe: of the five commercial mullet roe being sold in Sardinia which were analyzed, five formed a distinct separate branch within the *M. cephalus* group, indicating a non Sardinian origin, and two were attributed to the thicklip mullet *Chelon labrosus* (Murgia et al. 2002).

The dried, salted and bees-waxed ripe mullet ovaries from the Messolonghi Lagoon (Greece), *Avgotaracho Messolonghiou*, are a European and Greek protected designation of origin (PDO[8]), one of the few seafood products with a PDO. A PCR-RFL method was developed for the authentication of fish roe in the Missolonghi-Etoliko lagoon complex with 16s rRNA gene segment, in order to discriminate *M. cephalus* ovaries from the roe of the other four mullet species present in the lagoon complex (Klossa-Kilia et al. 2002), and to avoid unfair competition and assure consumers of accurate labelling of this PDO product. Lord Byron, who died in Messolonghi, is reported to have spread Messolonghi *avgotaracho* consumption throughout Europe.

In Australia, the Centre for Food Technology developed a project for the production of sun dried mullet roe (*karasumi*) in 1994 and several Australian companies are now producing *karasumi*, mainly to export to Japan. However, due to the strict specifications applied to *karasumi* by the extremely traditional Japanese market, only 35 to 40% is exported, the remainder is relegated to the less demanding (and less profitable) domestic market. Indeed mullet roe is graded into three main categories: premium, first grade and second grade: premium roe is exported fresh or frozen, first grade roe is used in the production of *karasumi* and second grade roe (approximately 15%) is sold at a loss or discarded (Hankock et al. 2000).

In Korea, the roe of flathead mullet (or pollock) is used to produce *myeongran-jeot*, a typical condiment made of salted fermented seafood seasoned with chili pepper sauce.

In Taiwan, male testes (soft roe) are also appreciated: they are generally eaten raw and are considered a tonic food (Chapter 19—Liao et al. 2015).

Other Products from Mullets

In Orbetello Lagoon, ripe females, from which ovaries are removed to be salted and dried, are filleted, hot smoked, and packed into a nice high priced product (36€.kg^{-1}) (Fig. 16.11b), an excellent system to provide added value to a by-product of egg roe processing. Mullets fillets are sun dried in Mauritania (Fig. 16.14).

[8] The term of *protected designation of origin* (PDO) is one of three European Union schemes of geographical indications and traditional specialities, used to promote and protect names of quality agricultural products and foodstuffs. PDO covers agricultural products and foodstuffs which are produced, processed and prepared in a given geographical area using recognized know-how.

Figure 16.14. Mullet fillets drying in the sun in Mauritania (photo by D. Crosetti).

Traditional methods to preserve mullet gave rise to typical plates in Sardinia (Italy): '*sa merca*' is a steamed mullet which is preserved with salt and lagoon herbs (sea purslane, *Halimione portulacoides*), '*su mugheddu*' is a gutted mullet salted in brine and later smoked.

In Egypt, salted mullets (*feseekh*) are sold in special shops that process the fish and only sell salted fish, and represent a traditional dish at Easter time (Chapter 20—Sadek 2015).

Mullet gizzards (muscular stomach) are eaten raw or processed in Taiwan and South East Asia, in Mexico by local fishermen (A. Ibañez, pers. comm.) and deep fried by Italian fishermen from South Latium. They are also used as bait by recreational fishers in Australia (Chapter 17—Prossner 2015).

Raw mullet fingerlings are considered a delicacy in Hawaii and are often poached from fish ponds (Ellis 1968).

Future Prospects for Mullet Culture

Mullets are important fish in both capture fisheries and aquaculture in several regions of the world. Their global production from capture fisheries is steadily increasing, whereas their production from aquaculture, after the increase in the late 1990s-early 2000, decreased to the present production, which is only about half of the production peak in 2006. However, despite the fact that Mugilidae, and in particular *M. cephalus*, are attractive candidates for culture because of their euryhalinity, their feeding at the lowest trophic levels and their flesh quality, their culture has never really evolved much from traditional forms of extensive culture and has not followed the development that other marine fish underwent, with the intensification of culture and the set up of artificial propagation techniques at a commercial level that enabled a programmed and stable supply of fry and juveniles to be outgrown in different culture systems. Unfortunately, many countries which have grown mullets in the past have abandoned their culture and turned to other more profit-making species.

The reasons are manifold. One of them lies in the fact that to date, mullet culture is still based on the capture of fry in the wild, as hatchery production of fry is still not carried out at a commercial level, despite several artificial propagation trials carried out in many regions of the world and protocols of hormone treatments defined in experimental trials, especially for the flathead grey mullet *M. cephalus*. The need for a steady supply of fry, independent from environmental conditions and availability of wild fry, has been highlighted in all the regions where mullets are cultured. On the other hand, the hatchery production of mullet fry has not been implemented at commercial level because of the high production cost per unit fry, mainly due to the high cost of hormone treatment and high larval mortalities, and the availability of abundant wild fry (and specialized fisheries capturing them) at a much lower cost. The use of large volumes and mesocosms with elevated hydro-dynamism in larval culture could provide interesting opportunities as far as seed quality is concerned, in order to produce wild-like mullet fry and juveniles to be grown in natural water bodies.

Despite the differences among species, mullets show a slow growth compared to other marine, brackish or freshwater species that are cultured around the world, and can take up to two-three years to reach market size, especially if reared in earthen ponds or natural water bodies where little or no supplementary food is provided. However, at present the supplementary feed given to mullets in more intensive systems is commercial feed produced for other species (tilapias, carps, etc.) associated with mullets in multispecies culture systems, but no feed has ever being formulated and commercially produced for the specific dietary requirements of mullets. Different feeds, specific to each mullet species and to the different growth stages, should be developed for the benefit of increased production. Mullet culture could also be aimed at the local production of high priced delicacies, such as egg roe, testes and gizzards, or other processed products, such as smoked fillets for instance, which would certainly provide an added value to mullet production.

Another factor certainly affecting the demand for (and the price of) mullets on the market is their reputation as scavengers: in some regions, mullets are considered low grade fish; in coastal areas they can be found schooling in harbours or at sewage water outlets. Being surface filterers and bottom grazers, they readily smell of hydrocarbons around harbour areas, or taste of organic mud. In some countries, the low price of mullets on the market is a major limiting factor in the development of more intensive practices in mullet culture. On the contrary, in other areas, some mullet species are very much appreciated for the quality of their flesh: in the Venice central fish market, thicklip mullet from the *valli* may be more expensive than European sea bass.

Finally, with the decline in the availability of fish meal and fish oil, and the increased share of these food commodities used in aquaculture to feed carnivorous fish, mullets surely appear as optimal candidates for culture because of their feeding at a low trophic level. Furthermore, the production of marine species which are substratum feeders could make a comeback in the future for the lack of marine proteins. The production potential for mullets lies in the use of low-level technology; improvement in grow-out techniques that can result in improved productivity per unit area under culture, especially when grow out activities are implemented in underutilized bodies of water. Extensive and semi-intensive systems, where the natural food web is integrated with supplementary feed specific for mullets, still seem to be the best culture systems for mullets, eventually associated with phases of intensive rearing, for pre-fattening or wintering in cold areas. Indeed the ecology of mullets and their role in the community enables successful imitations in captivity according to ecological aquaculture principles within an ecosystem approach.

Acknowledgements

My acknowledgements to Eleonora Ciccotti and Stefano Cataudella for their critical revision of the MS and their suggestions; to Massimo Guerrieri, Ana Ibañez, Manos Koutrakis, Ken Leber, Mauro Lenzi, Tonino Maccaroni, George Minos, Glen Pagelson and Sherif Sadek, for the useful, and often unpublished, information they sent me; to Wayne Wentz for a first revision of the MS: to ISPRA librarians and Armand Gribling from the FAO Library for their support in gathering the bibliographic references used in this chapter and in finding impossible-to-find publications. I would like to give credit to FishBase as source of stamps and respectively to Daniel Pauly and Jongsma Henk as owners of the stamps, who have contributed these pictures to FishBase from their private stamp collection. A particular thank to Steve Blaber, for his editing and encouragement.

References

Abdel-Rahman, S. 2015. Egypt country report. *In*: S. Cataudella, D. Crosetti and E. Massa (eds.). Mediterranean Coastal Lagoons: Sustainable Management and Interactions among Aquaculture, Capture Fisheries and Environment. GFCM Studies and Reviews n. 95. FAO, Rome.

Abraham, M., N. Blanc and A. Yashouv. 1966. Oogenesis in five species of grey mullets (Teleostei: Mugilidae) from natural and land locked habitats. Israel J. of Zool. 15: 155–172.

Abraham, M., A. Yashouv and N. Blanc. 1967. Induction expérimentale de la ponte chez *Mugil capito* confiné en eau douce. C.R. Acad. Sc. Paris 265: 818–821.

Abraham, M., N. Blanc and A. Yashouv. 1968. Persistent yolk nuclei in the oocytes of *Mugil cephalus* when confined to freshwater environment. Ann. Embriol. Morphol. 1: 169–178.

Aizen, J., I. Meiri, I. Tzchori, B. Levavi-Sivan and H. Rosenfeld. 2005. Enhancing spawning in the grey mullet (*Mugil cephalus*) by removal of dopaminergic inhibition. General and Comparative Endocrinology 142: 212–221.

Alikunhi, K.H., M.J. Sebastian, K.K. Sukumarau, V.A. Nair and T.J. Vincent. 1971. Induced spawning of the grey mullet (*Mugil macrolepis*) and observations on rearing hatchings to fingerling stage. Workshop on induced breeding of carps. Inst. Fish Educ., Working Paper 33: 2p.

Alvarez-Lajonchère, L., J.A. Arritola, O.L. Averhoff and S.D. Bellido. 1988. Positive results of induced spawning and larval rearing experiments with *Mugil liza* Valenciennes, a grey mullet from Cuban waters. Aquaculture 73: 349–355.

Alvarez-Lajonchère, L., O.G. Hernádez Molejón and L. Pérez Sánchez. 1991. Production of juveniles of the mullet *Mugil liza* Valenciennes 1836, by controlled reproduction in Cuba. Ciencias Marina 17: 47–56.

Apekin, V.S. and N.I. Kulilova. 1980. The technique of the commercial receiving mullet eggs with the purpose of artificial breeding, Moscow, 16pp. (in Russian), cited *in*: Glubokow, A.I., Kouril, J. Mikodina E.V. and Barth. T. 1994. Effect of synthetic GnRH analogues and dopamine antagonists on the maturation of Pacific mullet, *Mugil so-iuy* Bas. Aquaculture and Fisheries management 25: 419–425.

Apekin, V.S., L.G. Gnatchenko and G.A. Val'ter. 1979. Induction of maturation of long-finned mullet (*Mugil auratus* Risso) with pituitaries of mullet and carp. pp. 33–39. *In*: N.E. Sal'nikov (ed.). Problems in Mariculture. Pishchevaya Promyshlennost, Moskow, USSR.

Ardizzone, G.D., S. Cataudella and R. Rossi. 1988. Management of coastal lagoon fisheries and aquaculture in Italy. FAO Fisheries Technical Paper 293, FAO, Rome. 103p.

Ayyappan, S. 2014. National Aquaculture Sector Overview. India. National Aquaculture Sector Overview Fact Sheets. *In*: FAO Fisheries and Aquaculture Department [online]. Rome. Updated 4 April 2014 [Cited 27 March 2015]. http://www.fao.org/fishery/countrysector/naso_india/en.

Ben-Khemis, I., D. Zouiten, R. Besbess and F. Kamoun. 2006. Larval rearing and weaning of thick lipped grey mullet (*Chelon labrosus*) in mesocosm with semi-extensive technology. Aquaculture 259: 190.

Ben Khemis, I., E. Gisbert, C. Alcaraz, D. Zouiten, R. Besbes, A. Zouiten, A.S. Masmoudi and C. Cahu. 2013. Allometric growth patterns and development in larvae and juveniles of thick-lipped mullet *Chelon labrosus* reared in mesocosms conditions. Aquaculture Research 44: 1873–1888.

Benetti, D.D. and E.B. Fagundes-Netto. 1980. Considerações sobre desova e alevinagem de tainha (*Mugil liza* Valenciennes 1836) em laboratorio. Ist. Pesq. Mar, Publicação, Rio de Janeiro, Brazil 135: 1–26.

Ben-Yami, M. 1981. Handling, transportation and stocking of fry. pp. 335–359. *In*: O.H. Oren (ed.). Aquaculture of Grey Mullets, IBP 26, Cambridge, Cambridge University Press, International Biological Programme, 26.

Ben-Yami, M. and E. Grofit. 1981. Methods of capture of grey mullets. pp. 313–333. *In*: O.H. Oren (ed.). Aquaculture of Grey Mullets. IBP 26, Cambridge, Cambridge University Press, International Biological Programme, 26.

Besbes, R., C. Fauvel, H. Guerbej, A. Benseddik Besbes, A. El Ouaer, M.M. Kraiem and A. El Abed. 2003. Contribution à l'étude de la maturation et de la ponte en captivité du mulet lippu *Chelon labrosus* (Cuvier 1829). Résultats préliminaires de pontes par stimulation hormonale. Actes des 5èmes Journées Tunisiennes des Sciences de la Mer, 21–24 December 2002, Aïn Draham, Tunisia. Bulletin de l'INSTOM 7: 40–43.

Besbes, R., A. Besbes Benseddik, I. Ben Khemis, D. Zouiten, S. Zaafrane, K. Matouk, A. El Abed and R. M'rabet. 2010. Développement et croissance comparés des larves du mulet lippu *Chelon labrosus* (Mugilidae) élevées en conditions intensives: eau verte et eau claire. Cybium 34: 145–150.

Blaber, S.J.M. 1976. The food and feeding ecology of Mugilidae in the St. Lucia Lake system. Biological Journal of the Linnean Society, London 8: 267–277.

Boatto, V. and W. Signora. 1985. Le valli da pesca nella Laguna di Venezia. Università degli Studi di Padova, Istituto di economia e politica agraria, Padua. 260p.

Boglione, C., B. Bertolini, M. Russiello and S. Cataudella. 1992. Embryonic and larval development of the thick lipped mullet (*Chelon labrosus*) under controlled reproduction conditions. Aquaculture 101: 349–359.

Boglione, C., C. Costa, M. Giganti, M. Cecchetti, P. Di Dato, M. Scardi and S. Cataudella. 2006. Biological monitoring of wild thicklip grey mullet (*Chelon labrosus*), golden grey mullet (*Liza aurata*), thinlip mullet (*Liza ramada*) and flathead mullet *(Mugil cephalus*) (Pisces: Mugilidae) from different Adriatic sites: meristic counts and skeletal anomalies. Ecological Indicators 6: 712–732.

Bok, A.H. 1989. Rearing of artificially spawned southern mullet. S. Afr. J. Wildl. Res. 19: 31–34.

Bok, A.H. and H. Jomgbloed. 1987. A preliminary study on induced spawning of southern mullet. S. Afr. J. Wildl. Res. 17: 82–85.

Bullo, G. 1940. Le valli da pesca e la vallicoltura. Officine grafiche C. Ferrari, Venezia. 186p.

Cai, L., J. Ye, Z. Zheng, X. Lin and P. Wen. 1997. Studies on artificial propagation of grey mullet (*Mugil cephalus*). Journal of Oceanography in Taiwan Strait/Taiwan Haixia 16: 223–228.

Cannas, A., S. Cataudella and R. Rossi. 1998. Gli stagni della Sardegna. Quaderni C.I.R.S.P.E. 96pp.

Cardona, L. 2015. Food and feeding of Mugilidae. *In*: D. Crosetti and S.J.M. Blaber (eds.). Biology, Ecology and Culture of Grey Mullet (Mugilidae). CRC Press, Boca Raton, USA (this book).
Cassifour, P. and P. Chambolle. 1975. Induction de la ponte par injection de progestérone chez *Crenimugil labrosus* (Risso), poisson téléostéen, en milieu saumâtre. J. Physiol., Paris 70: 565–570.
Cataudella, S. 1974. Contributi alla biologia dei Mugilidi (Pisces: Perciformes): morfoecologia e citotassonomia. Tesi di Laurea, Università di Roma Roma, Italia. 160p.
Cataudella, S. and G. Monaco. 1983. Allevamento dei Mugilidi. *In*: Acquacoltura, ed CLESAV, 117–148.
Cataudella, S., D. Crosetti, F. Massa and M. Rampacci. 1988a. The propagation of juvenile mullets (*Chelon labrosus*) in earthen ponds. Aquaculture 68: 321–323.
Cataudella, S., F. Massa, M. Rampacci and D. Crosetti. 1988b. Artificial reproduction and larval rearing of the thick lipped mullet (*Chelon labrosus*). J. Appl. Ichthyol. 4: 130–139.
Cataudella, S., P. Franzoi, A. Mazzola and R. Rossi. 1999. Pesca del novellame da allevamento: valutazione di un'attività e sue prospettive. *In*: La pesca del novellame. Laguna suppl. 6: 129–135.
Cataudella, S., L. Tancioni and A. Cannas. 2001. L'acquacoltura estensiva. pp. 283–306. *In*: S. Cataudella and P. Bronzi (eds.). Acquacoltura Responsabile. Unimar-Uniprom, Roma.
Cataudella, S., D. Crosetti and F. Massa (eds.). 2015a. Mediterranean coastal lagoons: sustainable management and interactions among aquaculture, capture fisheries and environment. GFCM Studies and Reviews n. 95, FAO, Rome.
Cataudella, S., D. Crosetti, E. Ciccotti and F. Massa. 2015b. Sustainable management in Mediterranean coastal lagoons: interactions among aquaculture, capture fisheries and environment. *In*: S. Cataudella, D. Crosetti and F. Massa (eds.). Mediterranean Coastal Lagoons: Sustainable Management and Interactions among Aquaculture, Capture Fisheries and Environment. GFCM Studies and Reviews n. 95, FAO, Rome.
Çelik, U., C. Altınelataman, T. Dinçer and D. Acarlı. 2012. Comparison of fresh and dried flathead grey mullet (*Mugil cephalus*, Linnaeus 1758) caviar by means of proximate composition and quality changes during refrigerated storage at 4 ± 2°C. Turkish Journal of Fisheries and Aquatic Sciences 12: 1–5.
Chao, N.-H. and I.-C. Liao. 1984. Status and problems of propagation of marine finfish in Taiwan. pp. 33–49. *In*: I.C. Liao and R. Hirano (eds.). Proceedings of ROC Japan Symposium on Mariculture. TML Conference Proceedings 1.
Chaudhuri, H., R.M. Bhowmick, G.V. Kowtal, M.M. Bagchi, R.K. Jana and S.D. Guptha. 1977. Experiments in artificial propagation and larval development of *Mugil cephalus* in India. J. Inland Fish Soc. India 9: 30–41.
Chen, H.B. 1993a. A survey of the artificial propagation of Mugilidae species and eels in Mainland China. China Fish (Taiwan), 481: 61–64 (in Chinese)., cited *in*: Hong, W.-S. and Zhang. Q.-Y. 2002. Artificial propagation and breeding of marine fish in China. Chinese Journal of Oceanology and Limnology 20: 41–51.
Chen, H.B. 1993b. Status of artificial propagation of mullets (*Mugil* spp.) in mainland China. *In*: I.C. Liao, J.H. Chen, W.u. M.C and J.J. Guo (eds.). Proc. Symp. on Aquaculture 21–23 Dec. 1992. Bejing, Keelling Taiwan, Taiwan Fisheries Research Institute 1993: 185–189.
Cherubini, G. 1996. Composizione della dieta ed entità del prelievo del Cormorano in Laguna di Venezia. pp. 40–53. *In*: Atti del Convegno Interregionale "Il Cormorano nelle lagune venete", San Donà di Piave, 23 aprile 1996, Provincia di Venezia, Assessorato alla Caccia, Pesca, Vigilanza e Protezione civile.
Ciccotti, E. 2015. Italy country report. *In*: S. Cataudella, D. Crosetti and F. Massa (eds.). Mediterranean Coastal Lagoons: Sustainable Management and Interactions among Aquaculture, Capture Fisheries and Environment. GFCM Studies and Reviews n. 95. FAO, Rome.
Ciccotti, E. and P. Franzoi. 2001. La pesca del novellame. pp. 309–318. *In*: S. Cataudella and P. Bronzi (eds.). Acquacoltura responsabile. Unimar-Uniprom, Roma.
COA. 2014. The topic of grey mullet. Council of Agriculture, Executive Yuan. Taipei, Taiwan. from http://kmweb.coa.gov.tw/subject/mp.asp?mp=315. Electronic version accessed 27/12/2014 (In Taiwanese). cited in: Liao, I. C., N. H. Chao and C.C. Tseng. 2015. Case study: capture and culture of Mugilidae in Taiwan. *In*: D. Crosetti and S.J.M. Blaber (eds.). Biology, Ecology and Culture of Grey Mullets (Mugilidae). CRC Press, Boca Raton, USA.
Coste, M. 1861. Voyage d'exploration sur le littoral de la France et de l'Italie. Paris, Imprimerie Impériale.
Crosetti, D. 1998. Ruolo ecologico e produttivo di Mugilidi in acquacoltura, come componente di abbattimento delle cariche organiche di reflui nei sistemi vallivi integrati. Progetto di ricerca MiPAAF, Direzione Generale della Pesca e dell'Acquacoltura, 3° Piano Triennale Pesca Marittima ed Acquacoltura, L.41/82, 108p. Final report.
Crosetti, D. 1999. Induced breeding of the thick lipped mullet, *Chelon labrosus* (Mugilidae). Congresso internazionale "Verso il 2000: cosa cambia in acquacoltura?". Verona, 11-12 febbraio 1999, 33.
Crosetti, D. 2001. Messa a punto di una nuova tecnica di riproduzione controllata nei Mugilidi. Progetto di ricerca MiPA, Direzione Generale della Pesca e dell'Acquacoltura, 4° Piano Triennale Pesca Marittima ed Acquacoltura, L.41/82, 110p. Final report.
Crosetti, D. and S. Cataudella. 1995. The mullets. pp. 253–268. *In*: C.E. Nash and A.J. Novotny (eds.). Production of Aquatic Animals, -Fish-. Elsevier, Amsterdam, The Netherlands.
Crosetti, D. and C.A. Cordisco. 2001. Prove di riproduzione controllata della volpina, *Mugil cephalus*. Conferenza Internazionale dell' Acquacoltura "Dove va l'acquacoltura del 2001 nei paesi del Sud Europa?", Verona, 26-27 april 2001, 56.
Crosetti, D. and C.A. Cordisco. 2004. Induced spawning of the thick-lipped mullet (*Chelon labrosus*). Marine Life 14: 37–43.
D'Ancona, U. 1955. Raffronti economici fra stagnicoltura e vallicoltura. Boll. Pesca (nuova ser. 10) 31: 4p.

Das, N., G. Hossain, S. Bhattacharjee and P. Barua. 2008. Comparative study for broodstock management of grey mullet (*Mugil cephalus*) in cages and earthen ponds with hormone treatment. Aquaculture Asia 13: 30–33.

De Angelis, R. 1960. Mediterranean brackish water lagoons and their exploitation. Studies and Reviews. GFCM/ Etud.Rev. CGPM, n. 12. FAO, Rome.

De Monbrison, D., I. Tzchori, M.C. Holland, Y. Zohar, Z. Yaron and A. Elizur. 1997. Acceleration of gonadal development and spawning induction in the Mediterranean grey mullet, *Mugil cephalus*: Preliminary studies. Israeli Journal of Aquaculture/Bamidgeh 49: 214–221.

Del Rosso, R. 1905. Pesca e peschiere antiche e moderne nell'Etruria marittima. Paggi, Firenze, Italia. 764p.

Deniz, H. 2015. Turkey country report. *In*: S. Cataudella, D. Crosetti and F. Massa (eds.). Mediterranean Coastal Lagoons: Sustainable Management and Interactions among Aquaculture, Capture Fisheries and Environment. GFCM Studies and Reviews n. 95. FAO, Rome.

Donati, F., M. Vasciaveo and M. Zoppelletto. 1999. Valutazione dell'impatto socio-economico della pesca del novellame nel contesto produttivo delle valli da pesca. *In*: La pesca del novellame, Laguna suppl. 6: 79–93.

Durand, J.D. 2015. Implications of molecular phylogeny for the taxonomy of Mugilidae. *In*: D. Crosetti and S.J.M. Blaber (eds.). Biology, Ecology and Culture of Grey Mullets (Mugilidae). CRC Press, Boca Raton, USA (this book).

Eda, H., R. Murashinge, Y. Ooseki, A. Hagiwara, B. Eastham, P. Bass, C.S. Tamaru and C.S. Lee. 1990. Factors affecting intensive larval rearing of striped mullet, *Mugil cephalus*. Aquaculture 91: 281–294.

El-Greisi, Z.A. and A.A. Shaheen. 2007. Comparative differences between the effects of LHRH-a, 17 A-Methyltestosterone pellets and HCG on reproductive performance of *Mugil cephalus*. J. Applied Science Research 3: 890–895.

Ellis, J.N. 1968. Notes on cultivation of mullets in Hawaiian fish ponds. South Pacific Commission, Third Technical Meeting on Fisheries, Koror, Palau, 3–14 June 1968.

El-Sayed, A.-F.M. 2007. Analysis of feeds and fertilizers for sustainable aquaculture development in Egypt. pp. 401–422. *In:* M.R. Hasan, T. Hecht, S.S. De Silva and A.G.J. Tacon (eds.). Study and Analysis of Feeds and Fertilizers for Sustainable Aquaculture Development. FAO Fisheries Technical Paper. No. 497. FAO, Rome.

Eschmeyer, W.N. 2015. Catalog of fishes: genera, species, references (http://research.calacademy.org/research/ichthyology/catalog/fishcatmain.asp). Electronic version accessed on 15/01/2015.

Fagundes-Netto, E.B. and D.D. Benetti. 1981. Contribução ao conhecimento da reprodução da tainha (*Mugil liza* Valenciennes 1836). Ist. Pesq. Mar, Publicação, Rio de Janeiro, Brazil. 140. 1–23, presented at the II Congreso Brasileiro de Eng. De Pesca, July 1981, Recife, Brazil.

Fahmy, A.F. and Z.A.-B. El-Greisy. 2014. Induced spawning of *Liza ramada* using three different protocols of hormones with respect to their different effects on egg quality. Afr. J. Biotechn. 13: 4028–4039.

FAO. 2015. FishStat Aquaculture production 1950–2013 and Capture fisheries production 1950–2013. Available at http://www.fao.org/fishery/topic/16140/en, accessed 27 March 2015.

FAO/Network of Aquaculture Centres in Asia-Pacific (NACA). 2011. Regional Review on Status and Trends in Aquaculture Development in Asia-Pacific—2010 FAO Fisheries and Aquaculture Circular No. 1061/5. Rome, FAO. 89pp.

Fisheries Agency. 2003–2012. Fisheries statistical yearbook Taiwan, Kinmen and Matsu area. Fisheries Agency, Council of Agriculture, Executive Yuan. Taipei, Taiwan, cited in: Liao, I.C., N.H. Chao and C.C. Tseng. 2015. Capture and culture of Mugilidae in Taiwan. *In*: D. Crosetti and S.J.M. Blaber (eds.). Biology, Ecology and Culture of Grey Mullets (Mugilidae). CRC Press, Boca Raton, USA (this book).

Franzoi, P., R. Trisolini and R. Rossi. 1999. La pesca del novellame del pesce bianco da semina in Italia. *In*: La pesca del novellame, Laguna suppl. 6: 38–58.

Froese, R. and D. Pauly. 2015. Fish base. Available at http://www fishbase. Accessed 15 March 2015.

GAFRD. 2013. Statistics of fish production of year 2012. GAFRD, Ministry of Agriculture and Land Reclamation. 94pp.

Gandolfi, G. and P. Orsini. 1968. Fecondazione artificiale in *Mugil capito* trattati con estratti ipofisari ganadotropi di tonno (*Thunnus thynnus*). Rend. Sc. Ist. Lombardo B102: 15–22.

Garcia, A. and G. Bustamante. 1981. Resultados preliminares del desove inducido de lisa (*Mugil curema* Valenciennes) en Cuba. Ist. Ocean., Informe cientifico tecnico 158: 17–26.

Gautier, D. and J. Hussenot. 2005. Les mulets des mers d'Europe. Synthèses des connaissances sur des bases biologiques et de l'aquaculture, IFREMER.

Giacopini, L., B.B. Marchesini and L. Rustico. 1994. L'Itticoltura nell'Antichità. ENEL. IGER, Roma. 275p.

Giovanardi, O., T. Fortibuoni and T Raicevich. 2011. Fishing techniques and traditions in Italian administrative regions. pp. 206–208. *In*: S. Cataudella and M. Spagnolo (eds.). The State of Italian Marine Fisheries and Aquaculture. Ministero delle Politiche Agricole, Alimentari e Forestali (MiPAAF), Roma, Italia.

Glubokov, A.I., J. Kouril, E.V. Mikodina and T. Barth. 1994. Effect of synthetic GnRH analogues and dopamine antagonists on the maturation of Pacific mullet, *Mugil so-iuy* Bas. Aquaculture and Fisheries management 25: 419–425.

Godinho, H.E., E.R. Dias and O. Jacobsen. 1982. The effect of hormones in the induced spawning of the *Mugil liza* Val. from the lagunar region of Cananeia (25° 21' S). Atlantica Rio Grande 5: 48–49.

Godinho, H.M., E.R.A. Dias, O. Jacobsen and N. Yamanaka. 1984. Reprodução induzida de tainha *Mugil liza* Val. 1836 da região de Cananéia, SP, Brasil (25° 21' S). Simposio Brasileiro de Aquicoltura, 3., 1984, São Paulo. Anais. São Paulo: 1984a. p. 661–671.

Godinho, H.M., E.T. Kavamoto, E.F. Andrade-Talmelle, P.C. da Silva Serralheiro, P. De Paiva and E. de M Ferraz. 1993. Induced spawning of the mullet *Mugil platanus* Gunther 1880, in Cananeia, Sao Paulo, Brazil, B. Ist. Pesca 20: 59–66.

González-Castro, M. and G. Minos. 2015. Sexuality and reproduction of Mugilidae. *In*: Crosetti, D. and S.J.M. Blaber (eds.). Biology, Ecology and Culture of Grey Mullet (Mugilidae). CRC Press, Boca Raton, USA (this book).

Guerrieri, M., personal communication

Hancock, J., J. McDonald, L. Bond and C. Gore. 2000. Development of a smoked *karasumi* and a *karasumi* sauce. Seafood Services for Australia. Project Number 97/416, Final report.

Hawaii Division of Aquatic Resources. 2004. cited in: Nishimoto, R.T., T.E. Shimoda and L.K. Nishiura. 2007. Mugilids in the Muliwai: a tale of two mullets. pp. 143–156. *In*: N.L. Evenhuis and J.M. Fitzsimons (eds.). Biology of Hawaiian Streams and Estuaries. Bishop Museum Bulletin in Cultural and Environmental Studies. Volume 3.

Hong, W. and Q. Zhang. 2002. Artificial propagation and breeding of marine fish in China. Chinese J. Oceanogr. Limnol. 20: 41–51.

Hong, W. and Q. Zhang. 2003. Review of captive bred species and fry production of marine fish in China. Aquaculture 227: 305–118.

Houssay, B.A. 1931. Action sexuelle de l'hypophyse sur les poissons et les reptiles. C.R. Soc. Biol. Paris 106: 377–8.

Ibáñez, A., personal communication.

IUCN. 2007. Guide for the Sustainable Development of Mediterranean Aquaculture: n 1. Interaction between Aquaculture and the Environment. IUCN, Gland, Switzerland and Malaga, Spain. 107 pages.

Izumi, M. 2014. Community-based mullet farming in Papua New Guinea (TCP/PNG/3401). FAO Aquaculture Newsletter 52: 16.

James, P.S.B.R., V.S. Rengaswamy, A. Raju, G. Mohanraj and V. Gandhi. 1983. Induced spawning and larval rearing of the grey mullet, *Liza macrolepis* (Smith). Indian J. Fish 30: 185–202.

Katselis, G., C. Koutsikopoulos, E. Dimitriou and Y. Rogdakis. 2003. Spatial and temporal trends in the composition of the fish barriers fisheries production of the Mesolonghi–Etoliko lagoon (western Greek coast). Sci. Marina 67: 501–511.

Katselis, G., C. Koutsikopoulos, I. Rogdakis, E. Dimitriou, A. Lachanas and K. Vidalis. 2005. A model to estimate the annual production of roes (*avgotaracho*) of flathead mullet (*Mugil cephalus*) based on the spawning migration of species. Fish Res. 75: 138–148.

Kesteven, G.L. (ed.). 1960. Manual of field methods in fisheries biology. FAO Manuals Fish Sci. no. 1, 152pp.

Khalil, M.T. and K.H. Shaltout. 2006. Lake Bardawil and Zaranik Protected Area. State Ministry of Environment, Arab Republic of Egypt. Publication of Biodiversity Unit, No. 15, 559p.

Kim, D.H. 2014. Market interactions for farmed fish species on the Korean market. Ocean and Polar Res. 36: 71–76.

Kindermann, H. 2008. Draft report on the adoption of a European Cormorant Management Plan to minimise the increasing impact of cormorants on fish stocks, fishing and aquaculture (2008/2177(INI)), Committee on Fisheries, 11p, [online] Published 2008. [cited 3-4-2012]. Accessible from http://www.europarl.europa.eu/sides/getDoc.do?pubRef=-//EP//NONSGML+COMPARL+PE-409.389+01+DOC+PDF+V0//EN&language=EN.

Klossa-Kilia, E., V. Papasotiropoulos, G. Kilias and S. Alahiotis. 2002. Authentication of Messolongi (Greece) fish roe using PCR–RFLP analysis of 16s rRNA mtDNA segment. Food Control 13: 169–172.

Kokokiris, L., G. Minos, C. Simeonidis, M. Alexandrou, G. Tosounidis and T. Karidas. 2011. Ovarian development and sperm quality assessment of the thick lipped grey mullet (*Chelon labrosus*, Risso 1827) in captivity. 4th International Symposium on Hydrobiology and Fisheries, 9–11 June Volos, Greece 4p.

Koutrakis, E. 2015. Biology and ecology of fry and juveniles of Mugilidae. *In*: D. Crosetti and S.J.M. Blaber (eds.). Biology, Ecology and Culture of Grey Mullets (Mugilidae). CRC Press, Boca Raton, USA (this book).

Koutrakis, E., personal communication.

Koutrakis, E.T., A. Conides, A.C. Parpoura, E.H. van Ham, G. Katselis and C. Koutsikopoulos. 2007. Lagoon fisheries resources in Hellas. pp. 223–233. *In*: C. Papaconstantinou, A. Zenetos, V. Vasilopoulou and G. Tserpes (eds.). State of Hellenic Fisheries, HCMR.

Kulikova, N.I. 1981. Response of Black Sea striped mullet, *Mugil cephalus,* to pituitary injections in different periods of the spawning migration. pp. 35–52. *In*: N.E. Sal'nikov (ed.). Ecolo-physiological foundations of aquaculture in the Black Sea. AzcherNIRO, Kerch, USSR. (In Russian), cited in Kulikova N.I. and L.G. Gnatchenko. 1987. Response of prespawning female Black Sea striped mullet, *Mugil cephalus,* to chorionic gondatropin. J. Ichthyol. 27: 44–53.

Kulikova, N.I. and L.G. Gnatchenko. 1987. Response of prespawning female Black Sea striped mullet, *Mugil cephalus,* to chorionic gondatropin. J. Ichthyol. 27: 44–53.

Kuo, C.M. 1995. Manipulation of ovarian development and spawning in grey mullet, *Mugil cephalus* L. Israeli Journal of Aquaculture/Bamidgeh 47: 43–58.

Kuo, C.M., Z.H. Shehadeh and C.E. Nash. 1973a. Induced spawning of captive grey mullet (*Mugil cephalus* L.) females by injection of human chorionic gonadotropin (HCG). Aquaculture 1: 429–432.

Kuo, C.M., Z.H. Shehadeh and K.K. Milisen. 1973b. A preliminary report on the development, growth and survival of laboratory reared larvae of the grey mullet (*Mugil cephalus* L.). J. Fish Biol. 5: 459–470.

Kuo, C.M., C.E. Nash and Z.K. Shehadeh. 1974. A procedural guide to induce spawning in grey mullet (*Mugil cephalus* L.) females by injection of human chorionic gonadotropin. Aquaculture 3: 1–14.

Kuo, C.M. and C.E. Nash. 1975. Recent progress on the control of ovarian development and induced spawning of the grey mullet (*Mugil cephalus*). Aquaculture 5: 19–29.

Leber, K.M., personal communication.

Leber, K.M., C.-S. Lee, N.P. Brennan, S.M. Arce, C. Tamaru, L. Blankenship and R.T. Nishimoto. 2015. Stock enhancement of Mugilidae in Hawaii (USA). *In*: D. Crosetti and S.J.M. Blaber (eds.). Biology, Ecology and Culture of Grey Mullet (Mugilidae). CRC Press, Boca Raton, USA (this book).
Lee, C.S. and C.S. Tamaru. 1988. Advances and future prospects of controlled maturation and spawning of grey mullet (*Mugil cephalus*) in captivity. Aquaculture 74: 63–74.
Lee, C.S., C.S. Tamaru and C.D. Kelley. 1986. Technique for making chronic-release LHRH-a and 17αMethyltestosterone pellets for intramuscular implantation in fishes. Aquaculture 59: 161–168.
Lee, C.S., C.S. Tamaru, G.T. Miyamoto and C.D. Kelley. 1987. Induced spawning of grey mullet (*Mugil cephalus*) by LHRH-a. Aquaculture 62: 327–336.
Lee, C.S., C.S. Tamaru and C.D. Kelley. 1988. The cost and effectiveness of CPH, HCG and LHRH-a in the induced spawning of grey mullet, *Mugil cephalus*. Aquaculture 73: 341–347.
Lee, C.-S., C.S. Tamaru, C.D. Kelley, A. Moriwake and G.T. Miyamoto. 1992. The effect of salinity on the induction of spawning and fertilization in the striped mullet, *Mugil cephalus*. Aquaculture 102: 289–296.
Lenzi, M., personal communication.
Liao, I.C. 1975. Experiments on the induced breeding of the grey mullet in Taiwan from 1963–1973. Aquaculture 6: 31–58.
Liao, I.C. 1977. On completing a generation cycle of the grey mullet, *Mugil cephalus*, in captivity. J. Fish. Soc. Taiwan 5: 1–10.
Liao, I.C. 1981. Cultivation methods. pp. 361–390. *In*: O.H. Oren (ed.). Aquaculture of Grey Mullets. IBP 26 Cambridge University Press, Cambridge, UK.
Liao, I.C., Y.J. Lu, T.L. Huang and M.C. Lin. 1971. Experiments on induced breeding of the grey mullet, *Mugil cephalus* Linnaeus. Aquaculture 1: 15–34.
Liao, I.C., C.S. Cheng, L.C. Tseng, M.Y. Lim, L.S. Hsieh and H.P. Chen. 1972. Preliminary report on the mass propagation of grey mullet, *Mugil cephalus* Linnaeus. Fishery Ser. Chin.-Am. jt Comm. Rur. Reconstr. 12: 1–14.
Liao, I.C., N.H. Chao and C.C. Tseng. 2015. Capture and culture of Mugilidae in Taiwan. *In*: D. Crosetti and S.J.M. Blaber (eds.). Biology, Ecology and Culture of Grey Mullet (Mugilidae). CRC Press, Boca Raton, USA (this book).
Ling, S.W. 1967. Feeds and feeding of warm water fishes in ponds in Asia and the Far East. FAO Fish Rep. 44: 291–309.
Lovatelli, A. and P.F. Holthus (eds.). 2008. Capture-based aquaculture. Global overview. FAO Fisheries Technical Paper. n. 508. Rome, FAO. 298p.
Lumare, F. and P. Villani. 1972. Contributo alla fecondazione artificiale di *Mugil cephalus* (L.). Boll. Pesca Piscic. Idrobiol. 27: 255–261.
Maccaroni, A., M. Rampacci, R. D'Ambra, P. De Angelis, A. Fusari, M.F. Gravina, R. D'Adamo, A. Eboli and T. Russo. 2005. Valorizzazione delle produzioni ittiche lagunari attraverso pratiche di etichettatura. Progetto VI Piano triennale n. 6D18 legge 41/82 Direzione Generale Pesca e Acquacoltura. Final Report.
Maccaroni, A., personal communication
Makino, M. 2006. National Aquaculture Sector Overview. Japan. National Aquaculture Sector Overview Fact Sheets. *In*: FAO Fisheries and Aquaculture Department [online]. Rome. Updated 1 February 2006 [Cited 23 March 2015]. http://www.fao.org/fishery/countrysector/naso_japan/en.
Mariani, A., S. Panella, G. Monaco and S. Cataudella. 1987. Size analysis of inorganic particles in the alimentary tracts of Mediterranean mullet species suitable for aquaculture. Aquaculture 62: 123–129.
Marino, G., C. Boglione, S. Livi and S. Cataudella. 2009. National report of extensive and semi-intensive production practices in Italy. EU Funded project n. 044483, SEA CASE project deliverable n.20, 88p.
Massa, F. 1999. La pesca di novellame di pesce bianco nei paesi del bacino del Mediterraneo. *In*: La pesca del novellame, Laguna suppl. 6: 72–78.
McDonough, C.J., W.A. Roumillat and C. Wenner. 2005. Sexual differentiation and gonad development in striped mullet (*Mugil cephalus*) from South Carolina estuaries. Fish Bull. 103: 601–619.
Meseda, M.E.-G. and S.A. Samira. 2006. Spawning induction in the Mediterranean grey mullet *Mugil cephalus* and larval developmental stages. African J. Biotechnology 5: 1836–1845.
Minos, G., A. Imsiridou and P.S. Economidis. 2010. *Liza haematocheilus* (Pisces: Mugilidae) in Northern Aegean Sea. pp. 313–332. *In*: D. Golani and B. Appelbaum-Golani (eds.). Fish Invasions in the Mediterranean Sea: Change and Renewal. Pensoft Publishers.
Monfort, M.C. 2002. Fish roe in Europe: supply and demand conditions. FAO/GLOBEFISH Research Programme, Vol. 72. Rome, FAO. 47p.
Monteiro-Ribas, W. and A. Bonecker. 2001. Artificial fertilization and development in laboratory of *Mugil liza* (Valenciennes 1836) (Osteichthyes, Mugilidae). Bulletin of Marine Science 68: 427–433.
Mousa, M.A. 2006. Involvement of corticotrophin-releasing factor and adrenocorticotropic hormone in the ovarian maturation, seawater acclimation and induced spawning of *Liza ramada*. General and Comparative Endocrinology 146: 167–179.
Mousa, M.A. 2010. Induced spawning and embryonic development of *Liza ramada* reared in freshwater ponds. Animal Reproduction Science 119: 115–122.
Murgia, R., G. Tola, S.N. Archer, S. Vallerga and J. Hirano. 2002. Genetic identification of grey mullet species (Mugilidae) by analysis of mitochondrial DNA sequence: application to identify the origin of processed ovary products (*bottarga*). Mar. Biotechnol. 4: 119–126.

Nash, C.E. and Z.H. Shehadeh. 1980. Review of breeding and propagation techniques for grey mullet, *Mugil cephalus* L. ICLARM Studies and Reviews 3, International Center for Living Aquatic Resources Management, Manila, Philippines. 87pp.

Nash, C.E. and R.M. Koningsberger. 1981. Artificial propagation. pp. 265–312. *In*: O.H. Oren (ed.). Aquaculture of Grey Mullets. IBP 26 Cambridge University Press, Cambridge, UK.

Nash, C.E., C.M. Kuo and S.C. Mc Connell. 1974. Operational procedures for rearing larvae of the grey mullet (*Mugil cephalus* L.). Aquaculture 3: 15–24.

Nicol, V., C.D. Kelley, G.T. Miyamoto, A. Moriwake, G. Karimoto and D. Klotzback. 1993. Offseason maturation, spawning, and larval rearing of the striped mullet, *Mugil cephalus.* International Conference, World Aquaculture '93, p. 421. Oostende, Belgium: European Aquaculture Society, Special Publication No. 19 0993.

Nikovski, G. 1954. Special Ichthyology. Sovietskay naula. Moscow. 2nd edition translated from Russian, Jerusalem, 1961, 538p., cited in: Nash, C.E. and Shehadeh, Z.H. 1980. Review of breeding and propagation techniques for grey mullet, *Mugil cephalus* L. ICLARM Studies and Reviews, 3, International Center for Living Aquatic Resources Management, Manila, Philippines. 87pp.

Nishimoto, R.T., T.E. Shimoda and L.K. Nishiura. 2007. Mugilids in the Muliwai: a tale of two mullets. pp. 143–156. *In*: N.L. Evenhuis and J.M. Fitzsimons (eds.). Biology of Hawaiian Streams and Estuaries. Bishop Museum Bulletin in Cultural and Environmental Studies, Vol. 3.

Nordlie, F.G. 2015. Adaptation to salinity and osmoregulation in Mugilidae. *In*: D. Crosetti and S.J.M. Blaber (eds.). Biology, Ecology and Culture of Grey Mullet (Mugilidae). CRC Press, Boca Raton, USA (this book).

Oren, O.H. 1971. International biological programme: coordinated studies on grey mullets. Mar. Biol. 10: 30–33.

Oren, O.H. (ed.). Aquaculture of Grey Mullets. IBP 26 Cambridge University Press, Cambridge, UK.

Ostrovsky, I., T. Zohary, J. Shapiro, G. Snovsky and D. Markel. 2014. Fisheries management. pp. 635–653. *In*: T. Zohary, A. Sukenik, T. Berman and A. Nishri (eds.). Lake Kinneret, Ecology and Management. Springer Science.

Ottolenghi, F., C. Silvestri, P. Giordano, A. Lovatelli and M.B. New. 2004. Capture-based Aquaculture. The Fattening of Eels, Groupers, Tunas and Yellowtails. Rome, FAO.

Pagelson, G., personal communication.

Park, S.K., K. Davidson and M. Pan. 2012. Economic relationships between aquaculture and capture fisheries in the Republic of Korea. Aquaculture Economics & Management 16: 102–116.

Pellizzato, M. 2011. Classic and modern valliculture. pp. 237–238. *In*: S. Cataudella and M. Spagnolo (eds.). The State of Italian Marine Fisheries and Aquaculture. Ministero delle Politiche Agricole, Alimentari e Forestali (MiPAAF), Roma, Italy.

Prossner, A. 2015. Capture methods and commercial fisheries for Mugilidae. *In*: D. Crosetti and S.J.M. Blaber (eds.). Biology, Ecology and Culture of Grey Mullet (Mugilidae). CRC Press, Boca Raton, USA (this book).

Provincia di Venezia. 2009. Piano per la gestione delle risorse alieutiche delle lagune di Venezia e Caorle. A cura di: Torricelli P., Boatto V., Franzoi P., Pellizzato M., Silvestri S, AGRI.TE.CO sc., Ed. Arti Grafiche Zotelli, Dosson di Casier (TV). 203pp.

Radhakrishnan, S., K.V. Ramakrishna, G.R.M. Rao and K. Raman. 1976. Breeding of mullets by hormone stimulation. Matsya 2: 28–31.

Rajhalakshmi, T., S.M. Pillail and P. Ravichandran. 1991. Experiments on induced breeding and larval rearing of grey mullets and sea bream at Chilka Lake. J. Inland Fish Soc. India 23: 16–26.

Ravagnan, G. 1978. Elementi di vallicoltura moderna, Edizione Edagricole, Bologna. 283p.

Ravagnan, G. 1992. Vallicoltura integrata. Edizioni Edagricole, Bologna. 502pp.

Reizopoulou, S. 2015. Greece country report. *In*: S. Cataudella, D. Crosetti and F. Massa (eds.). Mediterranean coastal lagoons: sustainable management and interactions among aquaculture, capture fisheries and environment. GFCM Studies and Reviews n. 98, in press.

Rimmer, M.A., K. Sugama, D. Rakhmawati, R. Rofiq and R.H. Habgood. 2013. A review and SWOT analysis of aquaculture development in Indonesia. Reviews in Aquaculture 5: 255–279.

Rogdakis, Y. 1994. *Avgotaracho Messolonghiou*: a seafood product with designation of origin and protection. Fishing News, 153: 90–97 (in Greek). cited in: Katselis G., C. Koutsikopoulos, I. Rogdakis, E. Dimitriou, A. Lachanas and K. Vidalis. 2005. A model to estimate the annual production of roes (*avgotaracho*) of flathead mullet (*Mugil cephalus*) based on the spawning migration of species. Fish Res. 75: 138–148.

Rosa, A., P. Scano, M.P. Melis, M. Deiana, A. Atzeri and M.A. Dessì. 2009. Oxidative stability of lipid components of mullet (*Mugil cephalus*) roe and its product "*bottarga*". Food Chemistry 115: 891–896.

Rossi, A.R., S. Livi and D. Crosetti. 2015. Genetics of Mugilidae. *In*: D. Crosetti and S.J.M. Blaber (eds.). Biology, Ecology and Culture of Grey Mullets (Mugilidae). CRC Press, Boca Raton, USA (this book).

Rossi, R. 1995. Pesca ed acquacoltura nelle lagune costiere italiane modelli di gestione trasferibili nella realta' mediterranea. http://web.unife.it/utenti/remigio.rossi/didattica/ecologia/ecobase/91Ceedoclagunemedit.htm.

Rossi, R. and P. Franzoi. 1999. La tecnica di pesca del pesce novello da semina. *In*: La pesca del novellame, Laguna suppl. 6: 31–37.

Rossi, R., P. Franzoi and S. Cataudella. 1999. Pesca del pesce novello per la vallicoltura: un'esperienza nord-adriatica per la salvaguardia delle zone umide. *In*: La pesca del novellame, Laguna suppl. 6: 6–20.

Sadek, S., personal communication.

Sadek, S. 2001. Marine aquaculture in Egypt. Megapesca Lda, Nov. 2001, 33p.

Sadek, S. 2015. Culture of Mugilidae in Egypt. *In*: D. Crosetti and S.J.M. Blaber (eds.). Biology, Ecology and Culture of Grey Mullet (Mugilidae). CRC Press, Boca Raton, USA (this book).

Saifulina, E. 1991. Larval rearing of the mullet *Mugil so-iuy.* Larvi 91. Fish and Crustacean Larviculture Symposium. EAS Sp. Publ. 15, Gent., Belgium 317–318.

Saleh, M.A. 1991. Rearing of Mugilidae. A study on low salinity tolerance by thinlip grey mullet (*Liza ramada* Risso, 1826). In the proceeding of the FAO Workshop on Diversification of Aquaculture Production, Valetta, 1–6 July 1991.

Saleh, M. 2008. Capture-based aquaculture of mullets in Egypt. pp. 109–126. *In*: A. Lovatelli and P.F. Holthus (eds.). Capture-based Aquaculture. Global overview. FAO Fisheries Technical Paper. n. 508. Rome, FAO.

Saleh, M.A. 2006. Cultured Aquatic Species Information Programme. *Mugil cephalus*. Cultured Aquatic Species Information Programme. *In*: FAO Fisheries and Aquaculture Department [online]. Rome. Updated 7 April 2006 [Cited 26 March 2015]. http://www.fao.org/fishery/culturedspecies/Mugil_cephalus/en.

Salem, A.M. and M.A. Saleh. 2010. National Aquaculture Sector Overview. Egypt. National Aquaculture Sector Overview Fact Sheets. *In*: FAO Fisheries and Aquaculture Department [online]. Rome. Updated 16 November 2010 [Cited 25 March 2015]. http://www.fao.org/fishery/countrysector/naso_egypt/en.

Sanzo, L. 1930. Uova e larve di *Mugil cephalus*. Cuv. ottenute per fecondazione artificiale. Mem. R. Comm. Talossogr. Ital. 179: 1–3.

Sarig, S. 1981. The Mugilidae in polyculture in fresh and brackish water fish ponds. pp. 391–410. *In*: O.H. Oren (ed.). Aquaculture of Grey Mullets. IBP 26 Cambridge University Press, Cambridge, UK.

Sebastian, M.J. and V.A. Nair. 1973. The induced spawning of the grey mullets. Seafood Exp. J. 5: 1–4.

Sebastian, M.J. and V.A. Nair. 1975. The induced spawning of the grey mullet, *Mugil macrolepis* (Aguas) Smith, and the large-scale rearing of its larvae. Aquaculture 5: 41–52.

Seridi, F. and A. Bounouni. 2015. Algeria country report. *In*: S. Cataudella, D. Crosetti and F. Massa (eds.). Mediterranean coastal lagoons: sustainable management and interactions among aquaculture, capture fisheries and environment. GFCM Studies and Reviews n. 95. FAO, Rome.

Shabanipour, N. and K. Shanmugham. 1992. The effect of carp pituitary on gonadal maturation and ovulation in the marine teleost, *Mugil cephalus.* Indian Zool. 16: 129–133.

Shapiro, J. 2006. National Aquaculture Sector Overview. Israel. National Aquaculture Sector Overview Fact Sheets. *In*: FAO Fisheries and Aquaculture Department [online]. Rome. Updated 6 July 2006. [Cited 20 March 2015]. http://www.fao.org/fishery/countrysector/naso_israel/en.

Shehadeh, Z.H. and J.N. Ellis. 1970. Induced spawning of the striped mullet (*Mugil cephalus L.*). J. Fish Biol. 2: 355–360.

Shehadeh, Z.H., W.D. Madden and T.P. Dohl. 1972. The effect of exogenous hormone treatment on spermiation and vitellogenesis in the grey mullet, *Mugil cephalus.* L. J. Fish Biol. 5: 479–487.

Shehadeh, Z.H., C.M. Kuo and K.K. Milisen. 1973a. Validation of method for monitoring ovarian development in the grey mullet (*Mugil cephalus*). J. Fish Biol. 5: 489–496.

Shehadeh, Z.H., C.M. Kuo and K.K. Milisen. 1973b. Induced spawning of grey mullet (*Mugil cephalus* L.) with fractionated salmon pituitary extract. J. Fish Biol. 5: 471–478.

Smit Ramesh Lende. 2011. Present status of marine fish seed production in world with special reference to India Reg. No.: J4-00950-2011, http://aquafind.com/articles/Status_of_marine_fish_seed_production.php.

Sri Paryanti, T. 2006. National Aquaculture Sector Overview. Indonesia. National Aquaculture Sector Overview Fact Sheets. *In*: FAO Fisheries and Aquaculture Department [online]. Rome. Updated 9 February 2006 [Cited 27 March 2015]. http://www.fao.org/fishery/countrysector/naso_indonesia/en.

Sudradjat, A. and K. Sugama. 2010. Aquaculture of milkfish (*bandeng*) in Indonesia: grow-out culture. pp. 17–30. *In*: I.C. Liao and E.M. Leaño (eds.). Milkfish aquaculture in Asia. Asian Fisheries Society. World Aquaculture Society, The Fisheries Society of Taiwn and National Taiwn Ocean University, Quezon City, Philippines, Baton Rouge, USA, Keelung, Taiwan, cited in: Rimmer, M. A., K. Sugama, D. Rakhmawati, R. Rofiq and R.H. Habgood. 2013. A review and SWOT analysis of aquaculture development in Indonesia. Reviews in Aquaculture 5: 255–279.

Sun, C.B. 1995. A study on the industrialized breeding technique for mullet, *Liza haematocheila* (Temmick et Schlegel). Mo.d Fish Inf. 10: 14–16 (in Chinese), cited in: Hong, W. and Zhang Q. 2003. Review of captive bred species and fry production of marine fish in China. Aquaculture 227: 305–118.

Tamaru, C.S., H. Ako and C.-S. Lee. 1992. Fatty acid and amino acid profiles of spawned eggs of striped mullet, *Mugil cephalus*. Aquaculture 105: 83–94.

Tamaru, C.S., W. FitzGerald and V.S. Sato. 1993. Hatchery manual for the artificial propagation of striped mullet (*Mugil cephalus* L). Department of Commerce, Guam Aquaculture Development and Training Center and Oceanic Institute of Hawaii.

Tamaru, C.S., F. Cholik, J.C.-M. Kuo and W. FitzGerald. 1995. Status of the culture for striped mullet (*Mugil cephalus*) milkfish (*Chanos chanos*) and grouper (*Epinephelus* sp.). Reviews in Fisheries Science 33: 249–273.

Tang, I.H. 1970. Evaluation of balance between fishes and available fish food in multispecies fish culture in Taiwan. Trans. Am. Fish Soc. 99: 708–718.

Tang, Y.A. 1964. Induced spawning of striped mullet by hormone injection. Jap. J. Ichthyol. 12: 23–28.

Titcomb, M. 1972. Native Use of Fish in Hawaii. The University Press of Hawaii.

Volponi S. and R. Rossi. 1998. Predazione degli uccelli ittiofagi in acquacoltura estensiva: valutazione dell'impatto e sperimentazione di mezzi di dissuasione incruenta. Biologia Marina Mediterranea 5: 1375–1384.

Walsh, W.A., C. Swanson and C.-S. Lee. 1991. Combined effects of temperature and salinity on embryonic development and hatching of striped mullet, *Mugil cephalus*. Aquaculture 97: 281–289.

Whitfield, A.K. 2015. Ecological role of Mugilidae in the coastal zone. *In*: D. Crosetti and S.J.M. Blaber (eds.). Biology, Ecology and Culture of Grey Mullet (Mugilidae). CRC Press, Boca Raton, USA (this book).

Whitfield, A.K., J. Panfili and J.D. Durand. 2012. A global review of the cosmopolitan flathead mullet *Mugil cephalus* Linnaeus 1758 (Teleostei: Mugilidae), with emphasis on the biology, genetics, ecology and fisheries aspects of this apparent species complex. Rev. Fish Biol. Fisheries 22: 641–681.

Yashouv, A. 1966. Breeding and growth of grey mullet (*Mugil cephalus*). Bamidgeh 18: 3–13.

Yashouv, A. and A. Ben-Shacahar. 1967. Breeding and growth of Mugilidae. 2. Feeding experiments under laboratory conditions with *Mugil cephalus* and *M. capito* Cuv. Bamidgeh 19: 50–66.

Yashouv, A., E. Berner-Samsonov and M. Karamosta. 1969. Preliminary report on induced spawning of *M. cephalus* (L.) reared in captivity in freshwater ponds. Bamidgeh 21: 19–24.

Ye, J., L. Cai, X. Lin and P. Wen. 2003. Role of enriched rotifer playing on larval rearing of grey mullet, *Mugil cephalus*. Journal of Oceanography in Taiwan Strait/Taiwan Haixia 22: 53–58.

Yousif, O.M., A.A. Fatah, K. Krishna, D.V. Minh and B.V. Hung. 2010. Induced spawning and larviculture of grey mullet, *Mugil cephalus* (Linnaeus 1758) in the Emirate of Abu Dhabi. Aquaculture Asia Magazine 15: 41–43.

Zaki, M.I., M. Mousa, S. Kamel and El-Banhawy. 1998. Effects of exogenous hormone injection on growth and maturation of *Mugil cephalus* oocytes in captivity. pp. 149–161. *In*: J. Tanacredi, J. Loret and S. Earle (eds.). Ocean Pulse. Plenum Publishing Corp., 233 Spring St. New York NY 10013-1578 USA.

Zheng, C.W. 1987. Cultivation and propagation of mullet (*Mugil so-iuy*) in China. NAGA 10: 18.

CHAPTER 17

Capture Methods and Commercial Fisheries for Mugilidae

Andrew Prosser

Introduction

The grey mullets (Mugilidae) are numerous in species and found throughout the world. Because of their circumglobal distribution, high abundances, schooling behaviour and palatability, grey mullet species form a popular target of commercial, subsistence and recreational fishers. Grey mullets are harvested using a variety of methods and are caught commercially around the world. Although in excess of 70 grey mullet species exist (Eschmeyer 2014) and fishers have targeted them since ancient times (Koutrakis 1999), the general principles behind catching these fish have generally remained the same amid the evolution of gear and technology.

Grey mullets inhabit coastal, estuarine and freshwater environments in tropical, subtropical and temperate areas around the world, proximal to areas of high human populations (Blaber 1997, Whitfield et al. 2012). Adult grey mullets range in sizes from Broussonnett's mullet *Mugil broussonnettii* which grow to a maximum of 20 cm, to the bluetail mullet *Moolgarda buchanani*, which can grow to 100 cm (Froese and Pauly 2014). The flathead grey mullet *Mugil cephalus* is the most predominant grey mullet species in the world and comprise more than half of the world's total grey mullet harvest (FAO 2015). Generally, grey mullets school in high abundances in shallow, near-shore waters, and exhibit predictable spawning migrations following environmental cues (Whitfield et al. 2012). These life-history features make them susceptible to being caught in large quantities in similar locations and times each year. Grey mullets are agile, have good vision and react quickly to avoid predation (Ben-Yami and Grofit 1981). Because many grey mullet species leap (Fig. 17.1) fishers have invented unusual and ingenious methods of capture and gear adaptations to successfully capture their target.

This chapter describes the methods employed, both common and unique, and harvest principles underlying grey mullet fisheries around the world. Previous detailed descriptions of the methods specific to grey mullet capture have been written by Ben-Yami and Grofit (1981). A description of global commercial capture fisheries that target grey mullets, with a detailed example of commercial fishing in Australia is also included. Adaptations to methods that increase the catch or are successful for specific species may exist, but in general the methods described incorporate the predominant ways that grey mullets are caught around the world.

Level 1A Ecosciences Precinct, Dutton Park, PO Box 267, Brisbane, Queensland, Australia 4001.
Email: andrew.prosser@daf.qld.gov.au

Figure 17.1. A leaping flathead grey mullet *Mugil cephalus* attempting to escape a tunnel net in Moreton Bay, Australia (photo by A. Prosser).

Capture Methods

Fishermen are notoriously ingenious as their next meal or livelihood may depend on the success of their fishing activities. Capture methods exploit physical features or behaviours that make fish susceptible to capture. The methods employed for fishing of grey mullets around the world have been separated into active and passive methods, as per Gabriel et al. (2005). Active methods require the fisher to lead fishing gear into the path of the fish, driving the fish into the gear to be caught. Conversely, passive methods rely on the fish entering the gear or taking a baited hook voluntarily. It should be noted that the same type of gear could be fished passively or actively depending upon how the fisher operates the gear (Gabriel et al. 2005). For example, a gillnet may be used to catch grey mullets both actively when deployed around a school of fish, and passively as a drifting gillnet. Additionally, active and passive methods may be combined to maximize the efficiency of operation. Although similar methods of fishing occur between different geographic areas, specific adaptations between methods or particular to a specific species also exist.

Fishers target grey mullets for various reasons. Grey mullets are caught by commercial fishers as a source of income, by subsistence fishers as a source of food for their families and communities, and by recreational fishers for food, bait or fun (FAO 2005). Scientists capture planktonic, juvenile and adult grey mullets using a variety of gears and methods for reasons including investigating species assemblages (Blaber and Blaber 1980), habitat assessment (Trape et al. 2009), and collection of broodstock and fry for aquaculture purposes (Ben-Yami and Grofit 1981). Although different types of fishers may rely on the same or similar methods of catching, commercial gear and methods will typically be more efficient, involve more capital investment and used on a larger scale than recreational gear and methods, given that they are driven by catching fish for profit (FAO 2005). For example, an individual recreational fisher may use a small seine net to catch grey mullets for bait whereas a commercial fisher may use a large seine net with multiple fishers to catch a large school of fish from a vessel. Where the method is specifically employed by a sector, this is noted. Most methods discussed are used by commercial and subsistence fishers as these methods account for the greatest harvest of grey mullets around the world.

Active Methods

Seining, Encircling and Surrounding

Seine netting is a capture method that uses large nets to encircle or surround schools of fish and is a popular method of harvesting grey mullets around the world (Gabriel et al. 2005). Seine nets (also called beach haul or beach seine nets) are an efficient, low technology and inexpensive way of catching aggregated fish from the shore or a boat (Gabriel et al. 2005, Nedelec and Prado 1999). Schools of fish spotted close to the shore can be caught by seine netting which makes the method suitable for many grey mullet species as they aggregate in shallow, near-shore waters when feeding or migrating for spawning. Countries where seine nets are used to harvest grey mullets include Australia (Broadhurst et al. 2007), Hawaii (Leber and Arce 1996), USA (Bacheler et al. 2005), South Africa (Hutchings and Lamberth 2003), Egypt (Ben-Yami and Grofit 1981), Italy (Ardizzone et al. 1988) and Senegal (Trape et al. 2009); but being a simple and inexpensive method of catching fish, this method exists worldwide.

Seine nets used to capture grey mullet species typically consist of two long rectangular panels of monofilament mesh (known as wings) affixed between a floatline and a leadline, with a reinforced bunt positioned in the middle or end of the net (Fig. 17.2). Nets may also incorporate a section of mesh that forms a bag, called a codend, to concentrate fish in the bunt [for examples see Gabriel et al. (2005), Gray and Kennelly (2003), and Flood et al. (2012)]. The net is attached to long hauling lines of up to 2000 m that allow the net to be hauled from the shore. Nets are deployed by small row or powered boats perpendicular to the shoreline in front of the path of the fish, in a semicircle that encloses the school between the net and the beach. Using seine nets is dependent upon weather conditions; large waves and strong winds prevent boats being launched and the gear being deployed. While in the water, the net is hauled by hand with fishers ensuring the leadline remains on the sea floor to stop fish from escaping underneath. Fish that escape by leaping over the floatline can be caught by adding a backing net behind the codend, or attaching a horizontal net known as a verandah net to the seine net (Ben-Yami and Grofit 1981). The net is hauled from the water until the bunt is on the shore by hand or with the aid of vehicles. Seine nets may vary in size: large seine nets, such as those used in Hawaii, are 700 m long with an 18 m drop (Leber and Arce 1996); and those used in Australia can be 1,000 m long with 2,000 m of hauling rope (Gray and Kennelly 2003). Smaller seine nets called bait nets are about 20 m long and have a drop less than two m and are used by recreational fishers in Australia. Subsistence fishers also use seine nets as they are relatively inexpensive way of catching fish (Ben-Yami and Grofit 1981, Hutchings and Lamberth 2003). The mesh size depends upon the species targeted, but for the common flathead grey mullet mesh sizes used are between 5 and 7 cm. Scientists use small seine nets to catch planktonic grey mullets. Trape et al. (2009) used 10 m long, two m drop beach seines with 5 mm mesh to catch six planktonic grey mullet species including the flathead grey mullet, grooved mullet *Liza dumerili*, sicklefin mullet *Neochelon falcipinnis*, largescaled mullet *Parachelon grandisquamis*, banana mullet *Mugil bananensis* and white mullet *Mugil curema* adjacent to the shoreline of a West African estuary.

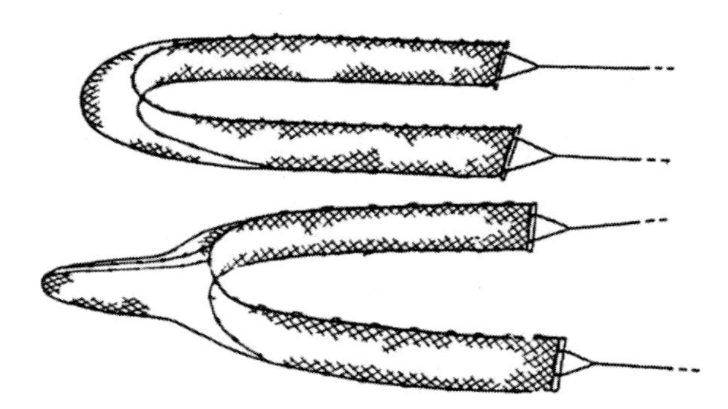

Figure 17.2. Beach seine nets without (top) and with codend (bottom). Source: Nedelec and Prado (1999).

Historically, seine nets were made with natural fibres like cotton, but have been replaced with synthetic materials as they became more widely available. Synthetic materials like polyethylene and polyamide (nylon) are stronger and more durable than natural fibres (Haverford Pty Ltd. 2014). Synthetic netting material may either be composed of single strands (monofilament) or multiple strands (multifilament) of material (Davies 1992). Combinations of different materials and mesh sizes may be used in the same net to maximize the strength, minimize the visibility and increase the efficiency of the net (Gray and Kennelly 2003). The length of the seine net used depends upon the size of the schools and species fished and the number of people available to operate them (Gabriel et al. 2005). Seine nets cannot be deployed over rock or reef substrata as the net may get snagged on uneven features, impacting the success of the operation.

Purse seining is a specialized and efficient vessel-based method for catching schooling fish. Using a vessel allows purse seining to operate in water deeper than using a beach haul seine. As with seine nets, the net is designed with a floatline and leadline, but has a separate purse line that is connected to the leadline on a series of rings (Gabriel et al. 2005). The design of the net allows it to surround the fish around the school and encloses underneath the school when the purse line is retrieved, preventing fish escaping by swimming under the net (Gabriel et al. 2005). Given the mobility of schooling fish in deeper water, purse seine nets need to be deployed around the schools quickly, and this is done by a second boat known as a skiff. To account for large quantities of fish caught, the purse seine net may have a strengthened bunt where the catch is aggregated as it is hauled on board the primary vessel. Like most gear, purse seines can be operated by a range of sizes of vessels in a range of operations, from small-scale purse seines deployed from canoes and row boats, to vessels in excess of 50 m. Historically, a Taiwanese purse-seine fishery for flathead grey mullet operated in Taiwan Strait targeting aggregated pre-spawning fish for their valuable roe (Hung and Shaw 2006). Purse seining for flathead grey mullet has also occurred in the USA (Leard et al. 1995) and for other grey mullet species in Africa (Panfili et al. 2006).

Gilling, Meshing and Entangling

Gillnetting or meshing is a method of capturing fish in nets so they become entangled in the mesh. Gillnetting is a very common method of catching many species of grey mullet, as it is cost-effective and may be undertaken by an individual fisher or small groups of fishers. Most gillnetting for grey mullets is an active method of catching fish as fishers actively set the gear in front of or around schools of fish and manipulate the net to force fish into it, but gillnets can also be fished passively as drifting gillnets. The meshing or gilling of individual species is dependent on the relationship between the fish size and the mesh size (Gabriel et al. 2005). Gillnets are usually composed of a single panel of net up to 1000 m long, made of low visibility knotted monofilament nylon mesh affixed between a floatline and a leadline, with a drop greater than the depth of water that the net will be used in. Gillnets are highly selective (Gray et al. 2004, Halliday et al. 2001), and in order to successfully gill grey mullets, mesh sizes between 7 and 10 cm are used, depending on the size of the target species.

Before the net is deployed, the movements of dense schools of fish are noted by the fisher. The direction the fish are moving depends on wind and current factors and determines where the fisher will deploy the net to prevent the school escaping. The net is deployed by boat, firstly into a tight semi-circle at one end and then the remainder of the net is deployed quickly to encircle the school of fish (Nedelec and Prado 1999). The circle of net is closed (Fig. 17.3), and then the remaining net is shot inside the circle to maximize the net area to gill the school of fish. To increase meshing success, various methods may be used by fishers to scare or drive the fish into the net, including hitting the water with sticks, making noise or using torches at night (Ben-Yami and Grofit 1981, Gabriel et al. 2005). Gillnets can only be used effectively over sand, mud or seagrass areas, as rocks or reef habitats will snag the leadline and allow fish to escape underneath. Speed is essential when deploying gillnets around schools of grey mullets to prevent the fish escaping. Powered vessels that can operate in shallow water at speeds of 20 knots or more are ideal for deploying gillnets to catch grey mullets.

Trammel nets (Fig. 17.4). are composed of multiple panels of different sized mesh which both entangle and gill encountered species, allowing a greater range of size-classes to be captured (Gabriel et al. 2005). Gillnets and trammel nets can also be towed by a single boat or between multiple boats to capture fish that

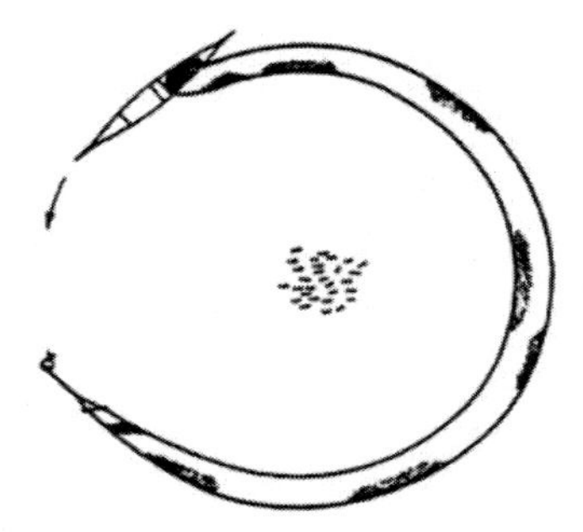
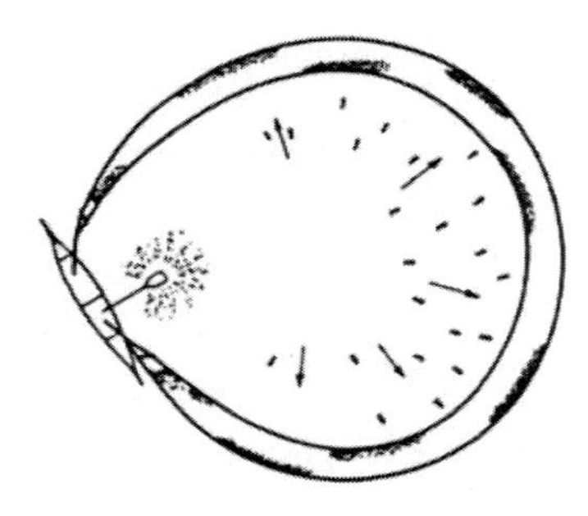

Figure 17.3. Encircling gillnet used to surround schools of grey mullet in shallow water. Arrows show direction of fish movement when scared by fisherman. Source: Nedelec and Prado (1999).

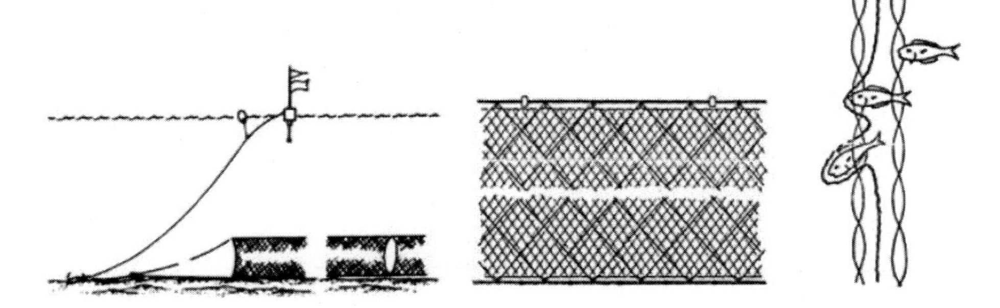

Figure 17.4. Trammel net deployment (left), design (centre) and entangling action of the differently sized mesh panels (right). Source: Nedelec and Prado (1999).

are sparsely distributed. Towing gillnets and trammel nets allows the fishers to cover more area and catch fish even if they are not actively schooling (Gabriel et al. 2005). Gillnets may also be used passively; for example, drifting gillnets are used in South Africa to catch fish in deep water (Hutchings and Lamberth 2003) and in Taiwan to catch fish more cost-effectively than using purse-seiners (Lan et al. 2014).

Dipping, Lifting and Falling

Dip nets and scoop nets are small, hand-held framed nets used to catch grey mullet when they are within reach of the operator or have been trapped by other fishing gear. They are used worldwide including in Asia, the Pacific and Mediterranean (Ben-Yami and Grofit 1981, Katselis et al. 2003). Lift nets are large framed nets that are lowered under the path of the fish and left for some time until fish enter the area and are then lifted out of the water (Nedelec and Prado 1999). They are simple, easily operated fishing gears that rely on fish being aggregated in a small area, either by bathymetry, food or light sources. Lift nets may be permanent or temporary structures (Gabriel et al. 2005).

A popular method of catching fish used by recreational, subsistence and commercial fishers is falling gear known as a cast net. Cast nets consist of a conical shape of fine mesh with lead weights attached to the base of the net, and a line that encloses the bottom of the net. The fine mesh allows a wide range of lengths of fish to be caught. To deploy, the cast net is thrown over the area where a school of fish are seen or thought to be, and allowed to sink covering the fish from above (Fig. 17.5). The fisher encloses the base of the net using the line attached to the leadline. The fine mesh does not entangle the fish and is ideal for catching fish that are intended to be kept alive. In Laguna, Brazil, commercial fishers use large cast nets (30 to 50 m perimeter, 5 to 7 cm mesh) to catch grey mullets aggregated by common bottlenose

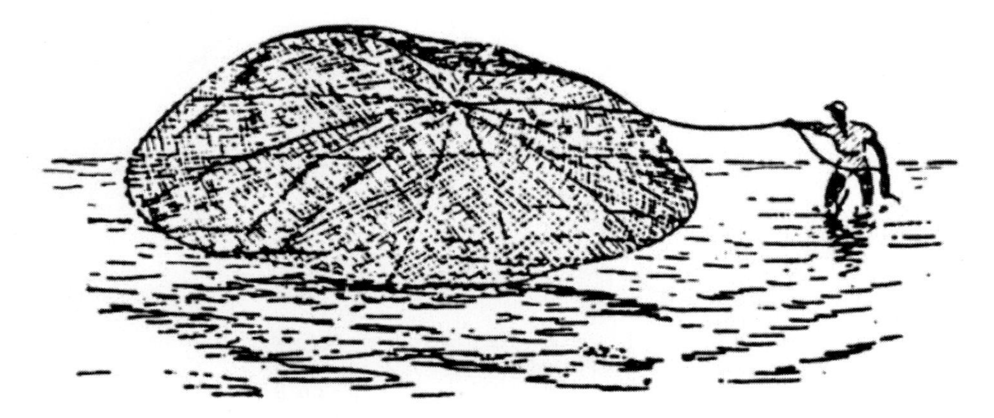

Figure 17.5. Cast net used by fisherman to fall over visible schools of grey mullet. Source: Nedelec and Prado (1999).

dolphins *Tursiops truncatus* (Peterson et al. 2008). As with gillnets, cast nets were traditionally made from natural plant fibres, but recently the change to more durable, synthetic nylon means that less time is spent repairing nets (Peterson et al. 2008).

Trawling

Trawling for grey mullets is less common compared to other netting methods, but does occur. Fish can be harvested by either beam trawling or pelagic trawling. In Germany, beam trawlers have been used to catch the thicklip grey mullet *Chelon labrosus* (Mohr 1981). Pelagic trawling requires high capital investment (Gabriel et al. 2005), which probably explains why it is not common for targeting grey mullets when other methods are effective and less expensive. In addition to targeted trawl fisheries, grey mullets are also captured as bycatch by fishers trawling for other species. For example, an estimated 15,000 t of flathead grey mullet and narrowhead grey mullet *Mugil capurii* are caught as bycatch by pelagic trawlers in Mauritania targeting other pelagic fish annually (Panfili et al. 2006).

Jagging, Gigging, Spearing and Shooting

Numerous methods of catching grey mullets involve impaling the fish using sharp hooks, spears or arrows. Schooling grey mullets can be caught by rod and line by a method known as jagging. Jagging is a process whereby a large hook, usually a treble (i.e., a hook with three points and a single shaft) is tied to a line, cast over a school of mullet and retrieved as quickly as possible. Upon retrieval, the hook impales the fish and the fish is fought until capture.

Gigging uses a spear with multiple tips to pierce the fish through the flesh. Fish are spotted from boats, chased and speared by the fisher. It is a common way of fishing recreationally in the United States of America and is similar to traditional spearing methods of the Ancient Greeks (Koutrakis 1999), Indigenous Australians (Schnierer 2011) and others still practiced around the world (Bacheler et al. 2005).

Spearfishing, where a fisher enters the water with a spear gun to hunt fish, is a selective method of catching grey mullets. Compared to other methods such as netting, spearfishing is a relatively inefficient way of fishing as low numbers of fish are captured (Gabriel et al. 2005).

Shooting grey mullets with crossbows has been reported in the United States of America where bow fishing is a popular recreational activity (Florida Sportsman 2011). Fish are spotted on the surface usually from a boat and shot with a bow and arrow connected to the bow with line. If the shot is successful, the arrow is retrieved with the fish attached.

Fishing in the Air

Some unusual but effective methods of catching grey mullet species exploits their leaping behaviour. Ben-Yami and Grofit (1981) describe a number of methods fishers have developed to catch leaping grey mullets. The methods described include driving boats through dense schools in shallow water leading them to jump into the boat, catching jumping fish with a hand-held dip nets and also and adding a 'verandah' to the top of gill (Fig. 17.6), seine or barrier nets (Ben-Yami and Grofit 1981). Several types of verandah nets are described and illustrated in Gabriel et al. (2005). The Ancient Greeks took advantage of the leaping behaviour by using boats as barriers to block the path of schools of fish, forcing them to leap into the boat as they attempted to pass the barrier (Koutrakis 1999).

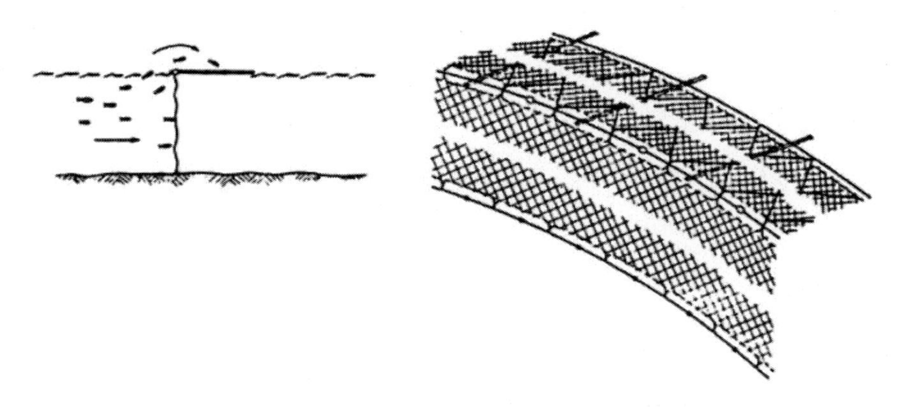

Figure 17.6. Verandah net attached above a gillnet to capture leaping fish. Source: Nedelec and Prado (1999).

Using Animals to Assist

There is evidence that people have used marine animals to help fish for grey mullets, including the Ancient Greeks (Koutrakis 1999), Indigenous Australians (Neil 2002) and artisanal fishermen in Brazil (Peterson et al. 2008). The Ancient Greeks used female grey mullet individuals tied to a line held by the fisher to attract and aggregate male fish for capture, probably by nets or spears (Koutrakis 1999). Indigenous Australians reportedly attracted common bottlenose dolphins by hitting the water with sticks and used the dolphins to drive grey mullets (probably flathead grey mullet) close to the shore so that the fish could be caught with traditional wooden spears and nets made of natural fibres (Neil 2002). The dolphins would be rewarded with a 'fish off a spear' or a fish directly handed to them by the fisher. In Brazil, a similar undertaking also occurs with common bottlenose dolphins, but aggregated fish are caught using large cast nets (Peterson et al. 2008).

Catching Larvae, Post-larvae and Juveniles

Larval, post-larval and juvenile grey mullets are a desirable capture for scientists investigating species distribution and abundance information, and especially for the collection of fry for aquaculture. Indeed most grey mullet culture in the world is still based on the use of wild caught fry and juveniles (see Chapters 12—Koutrakis 2015, 16—Crosetti 2015, 20—Sadek 2015).

Due to the size of the targets, specialized gear is required to catch these fish. Neuston nets, bongo nets, ichthyoplankton nets and fry nets are examples of gear that can catch larvae, post-larvae and juveniles. For example, the SEAMAP project sampled three species of grey mullet in the plankton off the Pacific Coast of Mexico (Ditty and Shaw 1996). The sampling programme used 61 cm bongo nets and one by 2 m neuston nets both with mesh smaller than 1 mm towed through the water column. Larvae of flathead grey mullet, white mullet and mountain mullet *Agonostomus monticola* were caught (Ditty and Shaw 1996).

To catch juvenile flathead grey mullet, fantail mullet *Liza georgii* and goldspot mullet *Liza argentea* in shallow water in Moreton Bay, Australia, Blaber and Blaber (1980) used 3 m long, 1 and 1/2 m high fry nets with 2 mm mesh.

Set plankton nets can also be used passively to catch mullet fry migrating with the tidal movements. Scoop nets are then used to collect the trapped fry (Torricelli et al. 1981), thus combining passive and active gears to catch grey mullets.

Passive Methods

Trapping and Barriers

Fish traps can either be permanent or temporary structures, and many examples designed to catch grey mullets exist around the world. Trapping relies on the fisher knowing the behaviour of the target fish and setting the trap in an appropriate place. Using fish traps takes advantage of grey mullet migrations for spawning or feeding, as well as their propensity to aggregate in shallow water. Traps can take on a number of forms, from simple traps made of thorned sticks used in Oceania, to more complicated chamber traps of the Mediterranean region (Gabriel et al. 2005). Ben-Yami and Grofit (1981) describe a fish trap used in Italy known as a '*saltarello*', which is a net that is staked in a circle, trapping fish that enter and cannot escape the design of the staked net. The *saltarello*'s effectiveness may be improved by attaching verandah nets strengthened with lengths of cane on top to capture leaping fish (Ben-Yami and Grofit 1981).

Barriers that concentrate and enclose fish are an ancient form of fishing gear and exist in a number of forms and regions worldwide (Gabriel et al. 2005). Barriers may consist of piled up stones, mud, grass, wood, concrete and steel and are designed to trap fish moving to and from areas with tides or in spawning migration. Fish concentrated by barriers are collected by hand, spear, nets or other methods.

In the Mediterranean, complex permanent barrier traps, derived from the original Italian design of *lavorieri*, are placed on narrow openings of estuarine lagoons and are used to capture and harvest migrating grey mullets (Figs. 16.8, 16.9 in Chapter 16—Crosetti 2015) (Cataudella et al. 2014, Kapetsky and Lasserre 1984, Katselis et al. 2003). At present these barriers are made of concrete and metal or plastic gates and force fish into slit-like non-return devices made of steel (Gabriel et al. 2005). The fish barriers capture flathead grey mullet, thicklip grey mullet, golden grey mullet *Liza aurata*, thinlip grey mullet *Liza ramada*, and leaping mullet *Liza saliens* exiting the lagoon areas on their spawning migrations, and have been so effective at targeting and catching the pre-spawning fish that this method is blamed for reducing the stock in certain areas (Katselis et al. 2003).

Tunnel nets are a specialized form of temporary fish trap used to catch grey mullets in Australia. Tunnel nets are a type of barrier net set in shallow water that catches grey mullets as they move with the tide (Figs. 17.7, 17.8 and Flood et al. 2012). Wings of a tunnel net are staked in shallow water near sandbanks, creek mouths, mangroves or seagrass beds, with the tunnel deployed in an area that retains

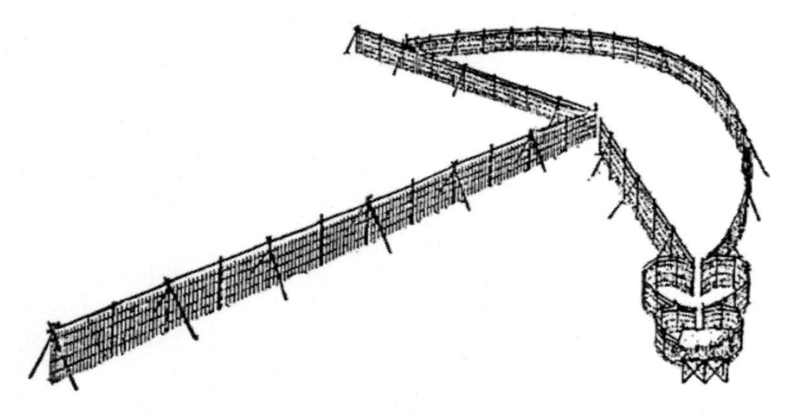

Figure 17.7. An example of a fence barrier net similar to the tunnel nets used in Australia. Source: Nedelec and Prado (1999).

Figure 17.8. A commercial tunnel netting operation in Moreton Bay, Australia, showing fishermen enclosing the wings of the net and flathead grey mullet *Mugil cephalus* trapped in the tunnel (photo by A. Prosser).

water at low tide. Tunnel nets have an advantage over gillnets in that they can be deployed in areas with uneven benthic structure. The wings of the tunnel net may be 100s of metres long, forming the mouth of the net. The wings of the net are positioned to surround the fish when the fish are in areas covered in water at high tide. Fish encounter the wings of the net as they attempt to move to deeper water as the tide recedes, and are corralled into the water remaining where the tunnel is positioned. The mesh size of the net is small enough so that fish are not gilled, reducing the harm to bycatch. A tunnel net operation from eastern Australia is described in detail below in 'Commercial Fisheries in Australia'.

A specialized floating fish trap used in the Aegean Sea catches grey mullets surrounding fish culture cages (Akyol and Ertosluk 2010). Thicklip and golden grey mullets are attracted to the cages by excess fish food or fish waste (Arechavala-Lopez et al. 2010) and those caught in the trap are cleared using trammel nets (Akyol and Ertosluk 2010).

Fyke and stake nets are two more examples of barrier nets used to catch grey mullets (Katselis et al. 2003, Ramamurthy and Muthu 1969). Like tunnel nets, fyke and stake nets take advantage of tidal movements to catch fish, but capture fish in the mesh rather than in a specialized tunnel.

To catch juvenile grey mullet, recreational fishers use small, baited traps. Traps are made by many different materials and take many different forms. For example, a simple fish trap used to catch juvenile mullet is a clear plastic bottle with bread as bait to attract fish.

Bait Fishing

Some grey mullet species can be caught using rod and line with baited hooks, and line fishing is a popular method used by recreational fishers to target these species (Bacheler et al. 2005, Broadhurst et al. 2011, Hutchings and Lamberth 2003, Leber and Arce 1996). Typically, bait is attached to small hooks, as many grey mullet species are small-mouthed. Common baits used to catch grey mullets include worms, bread, cheese, algae, whole fish and artificial lures. The Ancient Greeks knew that grey mullet species, including the flathead grey mullet, thinlip grey mullet, golden grey mullet and thicklip grey mullet, did not feed on other fish, had small mouths and would tentatively take baits (Koutrakis 1999). They used mixtures of flour, cheese and lime or bread, milk, cheese and thyme as baits to catch these 'noble' fish (Koutrakis 1999). An example of a grey mullet species that takes baited hooks is the sand mullet *Myxus elongatus* as they are aggressive feeders on small worms (Broadhurst et al. 2011, Grant 2004). Fly fishers also target grey mullet species by using flies that imitate small crustaceans or even small pieces of bread. In Japan, kites are used to deploy baited hooks on long lines to target grey mullets. The kite allows fishers to cover a wide area of a large lake or coastal estuary without the need for using a boat (Gabriel et al. 2005).

Commercial Fisheries

Commercial Fisheries around the World

Grey mullets are widely targeted by commercial fisheries around the world. In 2013, 698,293 t of grey mullet were harvested by 103 countries (FAO 2015). Of the 698,293 t, capture fisheries contributed 560,150 t to the total harvest, with the remaining 138,143 t produced by aquaculture (FAO 2015). Capture fisheries harvest wild stocks as well as wild stocks which are enhanced by the addition of fry and juveniles captured elsewhere (Chapter 16—Crosetti 2015). This section will refer specifically to commercial capture fisheries.

Grey mullet capture fisheries are economically important to communities and provide a source of food and employment. The importance, scale and operation of fisheries for grey mullets vary greatly around the world (Panfili et al. 2006). The majority of the annual global catch in 2013 was from developing parts of the world including parts of Asia, Africa, Oceania and the Americas (FAO 2015). In developing countries, grey mullets are generally caught by small-scale operations using low cost technology and are consumed locally (Gomez et al. 1995, Hutchings and Lamberth 2003, Panfili et al. 2006, Trape et al. 2009). In developed countries like Australia and the USA, larger scale commercial operations may operate and the harvested product is both consumed domestically and exported (Leard et al. 1995, Virgona et al. 1998).

Species Harvested

Grey mullet fisheries are commonly multi-specific fisheries because different species of grey mullets inhabit the same types of habitats. The dominant grey mullet species, the flathead grey mullet, contributed at least 135,000 t (or 24%) of the 560,150 t wild-capture grey mullet catch in 2013 (FAO 2015). As grey mullet are often reported by grouping rather than species, flathead grey mullet probably make up the majority of the 216,063 t reported as 'mullet nei[1]' (FAO 2015), and probably accounts for 50% of the total harvest of Mugilidae worldwide (Whitfield et al. 2012). The reason this species is so readily harvested is due to their abundance within tropical, subtropical and temperate regions of the world (Whitfield et al. 2012). Other important grey mullet species that are harvested by commercial fisheries include redlip mullet *Liza haematocheila* (Indian Ocean), white mullet *Mugil curema* (Atlantic and eastern Pacific), thicklip grey mullet *Chelon labrosus* (Atlantic and Mediterranean), bluespot mullet *Moolgarda seheli* (Indo-Pacific) and squaretail mullet *Ellochelon vaigiensis* (Indo-Pacific), but any species of grey mullets may be important to commercial local fisheries.

Countries that Harvest Grey Mullets

Between 1980 and 2013 grey mullets were harvested by more than 100 countries worldwide (FAO 2015). In 2013, the highest harvest of grey mullets occurred in Asian countries (FAO 2015). Between 1995 and 2013, China recorded the highest total capture, followed by Indonesia, Egypt, India, Brazil and Peru (FAO 2015).

---

[1] nei: not elsewhere included.

Commercial Fisheries in Australia

History

Fishing for grey mullets is one of the oldest forms of commercial fishing known in Australia, and occurs in coastal areas around the country. Indigenous Australians have harvested grey mullets for generations, using the fish to barter for goods and services (Schnierer 2011). Examples of grey mullet rock drawings by Aboriginal Australians also exist (Fig. 17.9) indicating the importance of mullet to the people. Records of commercial grey mullet harvest date back to the 19th century (Parke 2013). Since the mid-1940s, quantities in excess of 4,000 t have been taken annually, peaking in the 1990s at 8,000 t per year on the east coast of Australia alone (Bell et al. 2005). Grey mullet fisheries are economically and socially important to coastal communities, and the harvest comprises the highest amount of finfish by weight in Australia. Fishing for grey mullet pre-dates other commercial fishing methods including prawn trawling. It was a common occupation in coastal towns, and sustained fishing families for generations.

Figure 17.9. Grey mullet rock drawings from Kakadu National Park, an example of x-ray art by Aboriginal Australians [(c) Nature Connect Pty Ltd.—photographer Steve Parish].

Species Harvested

Twenty species of grey mullets inhabit Australian waters (Harrison and Senou 1999, Chapter 5—Ghasemzadeh 2015). The dominant species targeted by commercial fisheries is the flathead grey mullet *M. cephalus*, known locally as 'sea mullet'. This is due to the flathead grey mullet's abundance, schooling behaviour and its tendency to aggregate within coastal waters on predictable spawning runs in the region. Of the grey mullet species found in Australia, about half of them are commercially important. Apart from flathead grey mullet, species harvested include the bluespot mullet, fantail mullet, goldspot mullet, leaping mullet, squaretail mullet and yellow-eye mullet *Aldrichetta forsteri*. Other species may be harvested with catches of flathead grey mullet, with other grey mullets, estuarine and coastal species or targeted individually. Squaretail, bluespot and goldspot mullet are commercially targeted due to their flesh being preferred over flathead grey mullet, commanding higher prices (Grant 2004). An estuarine fishery for the yellow-eye mullet occurs in South Australia (MSC 2014).

Fishing Seasons and Locations

Commercial fishing for grey mullets in Australia occurs year round within estuaries and coastal embayments. During spawning months (March–August), large schools of migrating fish are caught along ocean beaches, and at the mouths of estuaries as the fish prepare to migrate to the ocean. Although they inhabit freshwater areas, fisheries management regulations prevent commercial fishing within many freshwater areas. The majority of the annual catch of flathead grey mullet harvest (95%) comes from waters along the eastern coast, and less than 5% of the catch is reported from the west coast (Rowling et al. 2012). In 2010, a total of 5,604 t of flathead grey mullet was harvested (Rowling et al. 2012).

Grey Mullet Products

Grey mullets provide a diverse source of products including fresh fish fillets for human consumption, processed roe products and bait.

Grey Mullets as Food. In Australia, flathead grey mullet form a staple for local seafood consumers up and down the coast in eastern Australia. 'Sea mullet' is a readily available product that provides an inexpensive source of protein and is consumed in a number of ways including smoked, crumbed, barbequed and fried. Fresh fillets are sold in local seafood outlets and supermarkets. The roe of pre-spawning flathead grey mullet is a highly valued export product (Bell et al. 2005, Virgona et al. 1998). Whole fish are sent to processors to process fish for local consumption and roe products for export to Europe and Asia. The roe is dried and smoked into products known as *bottarga* or *boutargue* in Europe or *karasumi* in Asia (Bledsoe et al. 2003, Rosa et al. 2009). In peak season, processers within Australia can process several tonnes of fish per day, employing up to 50 staff filleting, processing roe and packing grey mullet products for a number of months.

Grey Mullets as Bait. Grey mullet fillets of most species found in Australia are widely used as bait by recreational and commercial fishers. Fillets are used as bait for line fishing because the firm, oily flesh stays on the hook for extended periods and attracts a wide variety of targeted fish. The muscular stomach, called the 'onion' due to it being similar in shape to an onion, forms a popular bait used by recreational fishers and is marketed as 'mullet gut'. Whole fish or fish skeletons are used as bait in crab pots or to attract Onuphid worms for capture.

Capture Methods

There are three main ways that mullet are caught commercially in Australia; seining, gillnetting and tunnel netting. Seine nets are used on ocean beaches and in estuaries and are called haul or beach haul nets. Gillnets and tunnel nets are used in estuaries and embayments. The method and gear used depends on the habitat of the species that is targeted and also fisheries management regulations in place for specific areas.

Haul Netting

When schools of fish are seen close to ocean beaches and estuaries, haul nets are used to harvest fish. Schools of fish in depths less than 5 m are harvested by this method, as previously described in 'Active Methods'. Captured fish are placed in fish trays on the back of vehicles, or sorted on the beach beforehand if sorting occurs. They are then transferred to large fibreglass or plastic fish bins that hold up to 1 tonne of fish and ice and are transported to seafood processors. Catches in excess of 20 t are common over multiple hauls and days. Because of the manpower required for large catches, often competing teams of fishermen will work together with several nets and boats in combination known as a 'combine' to maximize the catch, then share the catch accordingly.

Gillnetting

The most common method of catching mullet in estuaries and coastal embayments is by gillnetting. Nets are made of clear monofilament nylon in a variety of thicknesses and mesh sizes depending upon the size class and species of fish being targeted. For example, fishers targeting flathead grey mullet usually use gillnets with mesh sizes between 7 and 9 cm. However, if fishers are targeting the larger-bodied squaretail mullet, they will deploy a net with 10 cm mesh or greater. Although the majority of nets are deployed with single sized mesh, different sized mesh may be used to increase the effectiveness of the net to catch a wider range of size classes.

Gillnetting has been described previously in 'Active Methods'. Once the fish have been captured in the net, the nets are hauled by hand or using an electronic winch and the hauled net is placed in the vessel or on a net reel. Harvested grey mullets are put on ice and transported to market for sale.

Tunnel Netting

A tunnel net (previously described in 'Passive Methods') is a temporary fish trap that uses tidal changes to catch fish. A typical tunnel net used in Australia consists of three sections; anterior wings, posterior wings and the tunnel, staked in position using 2 m wooden stakes and anchors. The anterior wings set around the area to be fished can be up to 800 m long and made of 5 cm monofilament mesh with a floatline and leadline. Posterior wings made of 3 cm multifilament polyethylene also consisting of a floatline and leadline are set in the shape of a funnel approximately 30 m in front of the tunnel. The tunnel can be between 10 and 20 m long depending upon the bathymetry of the area fished, and is composed of 2 and a ½ cm multifilament polyethylene mesh that has been dipped in tar resin to allow the base of the tunnel to sink. The walls of the tunnel need to be higher than the level of water at low tide to prevent mullet escaping by jumping (see Figs. 17.1, 17.8). The posterior wings aggregate the fish and the funnel-like design retains them in this area before they enter the tunnel. Stainless steel grids 3 m wide and 1 m high with 30 cm grid squares are used as excluding devices and are placed at the front of the tunnel to exclude turtles and large elasmobranchs, but allow fish to move into the tunnel.

As the tide recedes, the wings are collected in tub-like dories and the shortened wings are re-staked periodically such that the fish cannot swim around the net. This is done until all fish are inside the posterior wings, and can be herded into the tunnel by enclosing the posterior wings and forcing them through the grid into the tunnel. The wings are then picked up and all fish are forced into the tunnel. To harvest the captured fish, the base of the tunnel is gathered enveloping the catch in a concentrated area so they can be scooped out with a brail, sorted, and placed immediately on ice. Unwanted bycatch is released alive. Commercial tunnel net fishers in Queensland, Australia have developed an industry code of practice to minimize bycatch and maximize the quality of the landed product (MBSIA 2012).

Evolution of Fisheries

Although Australian grey mullet fisheries have evolved over time as technology has improved, the methods of fishing remain largely unchanged. The changes within fishing operations have been predominantly with gear and technology. Natural fibres (e.g., cotton) which were used historically were replaced in the 1960s and 1970s as synthetic fibres became cheaper, stronger, more durable and more readily available (Haverford Pty Ltd. 2014). Polyvinyl alcohol netting (called *kuralon*) originally replaced natural fibres but degraded with exposure to sunlight and salt water. Monofilament and multifilament polyamides (nylon) and polyethylene are the main netting materials used today. Technological advances include changing from rowed vessels to outboard and jet powered vessels, but traditional rowed boats are still used in some haul netting operations. Other technologies that have been adopted include the use of Global Positioning Systems (GPS) and depth sounders, which improve fish-finding capabilities in estuaries. Streamlining logistics including transportation of fish to market and the availability of ice have also improved the quality of the product at market.

The advent of mobile phones has allowed fishers to conduct reconnaissance of fishing grounds using multiple crew members as spotters. This has resulted in more flexible and effective fishing operations, as multiple spotters can search a wide area for fish without the entire fishing operation committing to a single location.

Regulation of commercial fishing operations has been implemented by government agencies over the last 50 years to sustainably manage marine resources. Fisheries are managed by input controls including licences and gear restrictions, and output controls including species restrictions and minimum legal sizes. Between 1995 and 2010, the number of commercial licences reporting grey mullet harvest reduced from 400 to 283 in Queensland (Rowling et al. 2012, Williams 1997). The decline can be attributed a number of factors including natural attrition, a reduced economic incentive to fish, the reduction of access to fishing areas and licence buyback schemes aimed to reduce fishing effort. Routine biological monitoring of flathead grey mullet stocks occurs throughout Australia to assess the sustainability of the resource.

Fisheries in the Future

Grey mullet fisheries in Australia face a number of challenges. The ability to recruit younger fishers as older fishers retire has reduced and it is difficult for existing operations to employ enough crew members to successfully undertake their fishing operations. The competition for resource allocation between commercial and recreational sectors continues to be an issue. The visibility of the operations to the general public, especially haul netting, often results in conflict between fishers and the general public. However, with continuing and cooperative measures between commercial fishers and fishery managers, and better education of the public about sustainability of fisheries resources, the grey mullet resource in Australia is likely to remain a key resource into the future.

Acknowledgements

The author would like to thank John Page, Dave Thomson, David Mainwaring, Geoff Orr, Robin Passmore, Dr. Manos Koutrakis and others consulted during the composition of the chapter. Dr. Adrian Gutteridge, Jason McGilvray and Dr. Lenore Litherland provided valuable feedback and suggestions when reviewing the chapter. Thanks to Steve Parish and Nature Connect Pty Ltd. for granting permission to publish the grey mullet rock art image used in Fig. 17.9.

References

Akyol, O. and O. Ertosluk. 2010. Fishing near sea-cage farms along the coast of the Turkish Aegean Sea. J. Appl. Ichthyol. 26: 11–15.

Ardizzone, G.D., S. Cataudella and R. Rossi. 1988. Management of coastal lagoon fisheries and aquaculture in Italy. FAO Fish. Techn. Pap. n. 293. FAO, Rome.

Arechavala-Lopez, P., I. Uglem, P. Sanchez-Jerez, D. Fernandez-Jover, J.T. Bayle-Sempere and R. Nilsen. 2010. Movements of grey mullet *Liza aurata* and *Chelon labrosus* associated with coastal fish farms in the western Mediterranean Sea. Aquacult. Env. Interac. 1: 127–136.

Bacheler, N.M., R.A. Wong and J.A. Buckel. 2005. Movements and mortality rates of striped mullet in North Carolina. N. Am. J. Fish Manage. 25: 361–373.

Bell, P.A., M.F. O'Neill, G.M. Leigh, A.J. Courtney and S.L. Peel. 2005. Stock assessment of the Queensland—New South Wales sea mullet fishery (*Mugil cephalus*). Queensland Department of Primary Industries and Fisheries, Queensland, Australia.

Ben-Yami, M. and E. Grofit. 1981. Methods of capture of grey mullets. pp. 313–334. *In*: O.H. Oren (ed.). Aquaculture of Grey Mullets. Cambridge University Press, United Kingdom.

Blaber, S.J.M. 1997. Fish and Fisheries of Tropical Estuaries. Chapman and Hall, UK.

Blaber, S.J.M. and T.G. Blaber. 1980. Factors affecting the distribution of juvenile estuarine and inshore fish J. Fish Biol. 17: 143–162.

Bledsoe, G.E., C.D. Bledsoe and B. Rasco. 2003. Caviars and fish roe products. Crit. Rev. Food Sci. Nutr. 43: 317–356.

Broadhurst, M.K., M.E.L. Wooden and R.B. Millar. 2007. Isolating selection mechanisms in beach seines. Fish Res. 88: 56–69.

Broadhurst, M.K., P.A. Butcher and B.R. Cullis. 2011. Post-release mortality of angled sand mullet (*Myxus elongatus*: Mugilidae). Fish Res. 107: 272–275.

Cataudella, S., D. Crosetti and F. Massa. 2015. Mediterranean coastal lagoons: sustainable management and interactions among aquaculture, capture fisheries and the environment. GFCM Studies and Reviews, n. 95. FAO, Rome.

Crosetti, D. 2015. Current state of grey mullet fisheries and culture. *In*: D. Crosetti and S.J.M. Blaber (eds.). Biology, Ecology and Culture of Grey Mullet (Mugilidae). CRC Press, Boca Raton, USA (this book).

Davies, P. 1992. Professional Fisherman's Guide to Making Prawn Trawl Nets. Queensland Fishing Industry Training Council, Brisbane, Australia.

Ditty, J.G. and R.F. Shaw. 1996. Spatial and temporal distribution of larval striped mullet (*Mugil cephalus*) and white mullet (*M. curema*, family: Mugilidae) in the northern Gulf of Mexico, with notes on mountain mullet, *Agonostomus monticola*. Bull. Mar. Sci. 59: 271–288.

Eschmeyer, W.N. (ed.). 2014. Catalog of Fishes: Genera, Species, References. (http://research.calacademy.org/research/ichthyology/catalog/fishcatmain.asp). Electronic version accessed 27/11/2014.

FAO. 2005. Fisheries and Aquaculture Topics: Types of Fisheries. FAO Fisheries and Aquaculture Department online. Available from http://www.fao.org/fishery/topic/12306/en.

FAO. 2015. Fisheries and aquaculture software FishstatJ—software for fishery statistical time series. Food and Agriculture Organisation of the United Nations, Rome, Italy. Electronic version accessed 24-3-2015.

Flood, M., I. Stobutzki, J. Andrews, G. Begg, W. Fletcher, C. Gardner, J. Kemp, A. Moore, A. O'Brien, R. Quinn, J. Roach, K. Rowling, K. Sainsbury, T. Saunders, T. Ward and M. Winning. 2012. Status of key Australian fish stocks reports 2012. Fisheries Research and Development Corporation, Canberra, Australia.

Florida Sportsman. 2011. Bowfishing made easier. Available from http://www.floridasportsman.com/2011/01/01/features_bowfishing_made_easy/.

Froese, R. and D. Pauly (eds.). 2014. FishBase. World Wide Web electronic publication.www.fishbase.org, Electronic version accessed 27-11-2014.

Gabriel, O., K. Lance, E. Dahm and T. Wendt. 2005. Von Brandt's Fish Catching Methods of the World, Fourth Edition. Blackwell, Oxford, UK.

Ghasemzadeh, J. 2015. Biogeography and distribution of Mugilidae in Australia and Oceania. *In*: D. Crosetti and S.J.M. Blaber (eds.). Biology, Ecology and Culture of Grey Mullet (Mugilidae). CRC Press, Boca Raton, USA (this book).

Gomez, E., F. Paredes and A. Chipollini. 1995. Fishery biology aspects of flathead grey mullet *Mugil cephalus* L. on the Peruvian coast. IMARPE, CALLAO, Peru.

Grant, E.M. 2004. Grant's Guide to Fishes. E.M. Grant Pty Ltd., Redcliffe, Australia.

Gray, C.A. and S.J. Kennelly. 2003. Catch characteristics of the commercial beach-seine fisheries in two Australian barrier estuaries. Fish Res. 63: 405–422.

Gray, C.A., D.D. Johnson, D.J. Young and M.K. Broadhurst. 2004. Discards from the commercial gillnet fishery for dusky flathead, *Platycephalus fuscus*, in New South Wales, Australia: spatial variability and initial effects of change in minimum legal length of target species. Fish Manage. Ecol. 11: 323–333.

Halliday, I., J. Ley, A. Tobin, R. Garrett, N. Gribble and D. Mayer. 2001. The effects of net fishing: addressing biodiversity and bycatch issues in Queensland inshore waters (FRDC Project no. 136 97/206). Queensland Department of Primary Industries, Queensland, Australia.

Harrison, I.J. and H. Senou. 1999. Order Mugiliformes, Mugilidae, Mullets. pp. 2069–2790. *In*: K.E. Carpenter and V.H. Niem (eds.). FAO Species Identification Guide for Fishery Purposes. The Living Marine Resources of the Western Central Pacific. Volume 4. Bony Fishes part 2 (Mugilidae to Carangidae). Food and Agriculture Organisation of the United Nations, Rome, Italy.

Haverford Pty Ltd. 2014. Fishing net materials. Available from http://www.haverford.com.au/.

Hung, C.M. and D. Shaw. 2006. The impact of upstream catch and global warming on the grey mullet fishery in Taiwan: a non-cooperative game analysis. Mar. Resour. Econ. 21: 285–300.

Hutchings, K. and S.J. Lamberth. 2003. Likely impacts of an eastward expansion of the inshore gill-net fishery in the Western Cape, South Africa: implications for management. Mar. Freshwat. Res. 54: 39–56.

Kapetsky, J.M. and G. Lasserre (eds.). 1984. Management of coastal lagoon fisheries. Aménagement des lagunes côtières. Studies and Reviews. GFCM/Etud.Rev.CGPM, n. 61, vol. 1 and 2, 776p. FAO, Rome, Italy.

Katselis, G., C. Koutsikopoulos, E. Dimitriou and Y. Rogdakis. 2003. Spatial patterns and temporal trends in the fishery landings of the Messolonghi-Etoliko lagoon system (western Greek coast). Sci. Mar. 67: 501–511.

Koutrakis, E. 1999. Review of the history and systematics of the Mediterranean grey mullets (Mugilidae). Geotechnical Scientific Issues 10: 365–374.

Koutrakis, E. 2015. Biology and ecology of fry and juveniles of Mugilidae. *In*: D. Crosetti and S.J.M. Blaber (eds.). Biology, Ecology and Culture of Grey Mullet (Mugilidae). CRC Press, Boca Raton, USA (this book).

Lan, K.-W., M.-A. Lee, C.I. Zhang, P.-Y. Wang, L.-J. Wu and K.-T. Lee. 2014. Effects of climate variability and climate change on the fishing conditions for grey mullet (*Mugil cephalus* L.) in the Taiwan Strait. Climatic Change DOI 10.1007/s10584-014-1208-y.

Leard, R., B. Mahmoudi, H. Blanchet, H. Lazauski, K. Spiller, M. Buchanan, C. Dyer and W. Keithly. 1995. The striped mullet fishery of the Gulf of Mexico, United States: a regional management plan. National Oceanic and Atmospheric Administration, Mississippi, United States of America.

Leber, K.M. and S.M. Arce. 1996. Stock enhancement in a commercial mullet, *Mugil cephalus* L., fishery in Hawaii. Fish Manage. Ecol. 3: 261–278.

MBSIA. 2012. Moreton Bay tunnel net fishery code of best practice. Moreton Bay Seafood Industry Association, Brisbane, Australia.

Mohr, H. 1981. The mullet an interesting, but uncertain fishery object. Inf. Fischwirtsch. 28: 194–196.

MSC. 2014. Certified fisheries—Lakes and Coorong, South Australia. Marine Stewardship Council online. Available from http://www.msc.org/track-a-fishery/fisheries-in-the-program/in-assessment/southern-ocean/lakes-and-coorong-south-australia.

Nedelec, C. and J. Prado. 1999. Definition and classification of fishing gear categories. FAO Fisheries Technical Paper 222. Food and Agriculture Organisation of the United Nations, Rome, Italy.

Neil, D.T. 2002. Cooperative fishing interactions between Aboriginal Australians and dolphins in eastern Australia. Anthrozoos 15: 3–18.

Panfili, J., C. Aliaume, P. Berrebi, C. Casellas, C.W. Chang, P.S. Diouf, J.-D. Durand, D. Flores Hernandez, F. Garcia de Leon, P. Laleye, B. Morales-Nin, J. Tomas, W.N. Tzeng, V. Vassilopoulou, C.W. Wang and A.K. Whitfield. 2006. State of the art of *Mugil* research. European Commission 6th Framework Programme, INCO-CT-2006-026180, MUGIL Deliverable 1.

Parke, J. 2013. Against the Tide: Queensland's Moreton Bay Fishing Industry since 1824. 5Word Productions, Lytton, Queensland.

Peterson, D., N. Hanazaki and P.C. Simões-Lopes. 2008. Natural resource appropriation in cooperative artisanal fishing between fishermen and dolphins (*Tursiops truncatus*) in Laguna, Brazil. Ocean. Coast. Manage. 51: 469–475.

Ramamurthy, S. and M. Muthu. 1969. Prawn Fishing Methods. Central Marine Fisheries Research Institute, Mandapam, India.

Rosa, A., P. Scano, M.P. Melis, M. Deiana, A. Atzeri and M.A. Dessì. 2009. Oxidative stability of lipid components of mullet (*Mugil cephalus*) roe and its product "bottarga". Food Chem. 115: 891–896.

Rowling, K., A. Roelofs and K. Smith. 2012. Sea Mullet *Mugil cephalus*. pp. 280–284. *In*: M. Flood, I. Stobutzki, J. Andrews, G. Begg, W. Fletcher, C. Gardner, J. Kemp, A. Moore, A. O'Brien, R. Quinn, J. Roach, K. Rowling, K. Sainsbury, T. Saunders, T. Ward and M. Winning (eds.). Status of Key Australian Fish Stocks Reports 2012. Fisheries Research and Development Corporation, Canberra, Australia.

Sadek, S. 2015. Culture of Mugilidae in Egypt. *In*: D. Crosetti and S.J.M. Blaber (eds.). Biology, Ecology and Culture of Grey Mullet (Mugilidae). CRC Press, Boca Raton, USA (this book).

Schnierer, S. 2011. Aboriginal Fisheries in New South Wales: Determining Catch, Cultural Significance of Species and Traditional Fishing Knowledge Needs. Fisheries Research and Development Corporation, Canberra, Australia.

Torricelli, P., P. Tongiorgi and P. Almansi. 1981. Migration of grey mullet fry into the Arno river: Seasonal appearance, daily activity, and feeding rhythms. Fish Res. 1: 219–234.

Trape, S., J.-D. Durand, F. Guilhaumon, L. Vigliola and J. Panfili. 2009. Recruitment patterns of young-of-the-year mugilid fishes in a West African estuary impacted by climate change. Estuar. Coast. Shelf Sci. 85: 357–367.

Virgona, J., K. Deguara, D. Sullings, I. Halliday and K. Kelly. 1998. Assessment of the Stocks of Sea Mullet in New South Wales and Queensland Waters. FRDC Project No. 90/024. NSW Fisheries, Cronulla, Australia.

Whitfield, A.K., J. Panfili and J.D. Durand. 2012. A global review of the cosmopolitan flathead mullet *Mugil cephalus* Linnaeus 1758 (Teleostei: Mugilidae), with emphasis on the biology, genetics, ecology and fisheries aspects of this apparent species complex. Rev. Fish Biol. Fish 22: 641–681.

Williams, L.E. 1997. Queensland's Fisheries Resources: Current Condition and Recent Trends 1988–1995. Information Series QI97007, Queensland, Australia.

CHAPTER 18

Stock Enhancement of Mugilidae in Hawaii (USA)

Kenneth M. Leber,[1,*] *Cheng-Sheng Lee,*[2] *Nathan P. Brennan,*[1] *Steve M. Arce,*[2] *Clyde S. Tamaru,*[3] *H. Lee Blankenship*[4] and *Robert T. Nishimoto*[5]

Introduction

Aquaculture-based marine fisheries enhancements have a long history, dating back to the late 19th century when releasing cultured fry into the marine environment was the principal fishery management tool. Stocking fish eggs and larvae was regarded as the way to save what was generally perceived as a declining resource, the causes of which were not well understood. By the early decades of the 20th century, billions of unmarked, newly-hatched fry had been released into the coastal environments (Radonski and Martin 1986). In the United States, Atlantic cod (*Gadus morhua*), haddock (*Melanogrammus aeglejinus*), pollack (*Pollachius virens*), winter flounder (*Pseudopleuronectes americanus*) and Atlantic mackerel (*Scomber scombrus*) were stocked (Richards and Edwards 1986). No attempt was made to evaluate stocking strategies and success was measured by numbers released rather than numbers surviving. By the early 1930s, after a half century of releases had produced no evidence of an enhancement impact (except for some salmonid stocking programs), stocking programs were largely curtailed in the US and harvest management was established as the principal means to manage marine fisheries. In the 1980s, some states in the US began new stock enhancement programs, following advances in marine fish culture and fish tagging technologies. Most of these new programs were established primarily for research on the efficacy of marine stock enhancement, with a goal of developing more effective stock enhancement strategies.

Efforts to enhance marine fisheries are limited to relatively few marine species. Except for stocking of salmon in the US Pacific Northwest, Japan and China have the largest hatchery-based marine fisheries enhancement programs. Norway began releasing tagged, hatchery-raised Atlantic cod (*Gadus morhua*) in 1983 (Svasand et al. 1990). These efforts were followed in the 1990s and beyond by numerous additional stocking programs around the world, many of which are now chronicled in the proceedings of the International Symposium on Stock Enhancement and Sea Ranching (ISSESR) (see www.SeaRanching.org).

[1] Directorate of Fisheries and Aquaculture, Mote Marine Laboratory, Sarasota, FL 34236 USA.
[2] The Oceanic Institute of Hawaii Pacific University, Waimanalo, HI 96795 USA.
[3] College of Tropical Agriculture and Human Resources, University of Hawaii, Honolulu, HI 96822 USA.
[4] Director of Biological Services, Northwest Marine Technology, Inc., Shaw Island, WA 98501 USA.
[5] Hawaii Division of Aquatic Resources, Hilo, HI 96720 USA (Retired).
* Corresponding author: KLeber@mote.org

The science underlying enhancements is relatively recent. There were no published accounts of the fate of stocked fishes until empirical studies of anadromous salmonids appeared in the mid-1970s (Hager and Noble 1976, Bilton et al. 1982), followed by the first published studies of stocked marine invertebrates in 1983 (Appeldoorn and Ballentine 1983) and marine fishes in 1989 and 1990 (Tsukamoto et al. 1989, Svasand et al. 1990). Two universal problems restricted the early development of marine stock enhancement science: 1) lack of a marking method for assessing whether hatchery releases are successful and 2) inability to culture marine fishes through the juvenile (fingerling and larger) life stage. Breakthroughs in marine finfish aquaculture technology and new benign tagging methods have led to resurgence in marine stock enhancement efforts worldwide. Emphasis is now placed on a responsible approach to stocking, emphasizing planning, fisheries management, modeling, genetics, health, pilot experiments to increase survival of released fish, evaluating contributions to wild populations and use of adaptive management (Blankenship and Leber 1995, Walters and Martell 2004, Lorenzen et al. 2010, Sass and Allen 2014). The technology has progressed to the stage where marine stock enhancement is now considered a *bona fide* fisheries management tool (Sass and Allen 2014).

The flathead grey mullet *Mugil cephalus* has a unique role in the modern development of marine fisheries enhancements. *M. cephalus* was the test species chosen for one of the first systematic series of empirical studies to evaluate effectiveness of aquaculture-based marine fisheries enhancements. Beginning in 1988, the Oceanic Institute (OI), located on Oahu, Hawaii (USA) conducted several years of experimental pilot releases with grey mullet and collaborated with the Hawaii Division of Aquatic Resources (DAR) to transfer mullet stock-enhancement technology to the state for implementation in a recreational mullet fishery in Hilo, Hawaii (Leber 1994, Nishimoto et al. 2007).

Grey mullet was also used in the Hawaii studies in the early 1990s to demonstrate the effectiveness of using pilot-release experiments to optimize release strategies; such pilot releases are a fundamental aspect of a 'Responsible Approach' to marine enhancements (Blankenship and Leber 1995, Lorenzen et al. 2010), which was partly inspired by successful results achieved in OI's mullet stock enhancement experiments and by pioneering studies in Japan (Tsukamoto et al. 1989), Norway (Svasand et al. 1990), China (Wang et al. 2006) and the US (Hager and Nobel 1976, Bilton et al. 1982). The Responsible Approach concepts have helped advance this branch of fisheries science (Sass and Allen 2014).

Responsible Approach to Marine Stock Enhancement

The modern generation of marine fisheries enhancement scientists is cultivating an integrative, quantitative and careful approach for developing and managing effective hatchery-based fisheries enhancements. The concepts were originally envisioned by an International Working Group on Stock Enhancement, formed in Torremolinos (Spain) in 1993, and published in 1995 as a platform paper by two members of the Working Group (who with several colleagues later formed the Science Consortium for Ocean Replenishment, SCORE, www.StockEnhancement.org, to help foster and refine the Responsible Approach). The International Working Group and the origin and expansion of these ideas are discussed in Leber (2013).

These concepts are presented in two publications—'A responsible approach to marine stock enhancement' (Blankenship and Leber 1995) and 'Responsible approach to marine stock enhancement: an update' (Lorenzen et al. 2010). The principles for developing, evaluating, and managing marine stock enhancement programs set out in Blankenship and Leber (1995) and Lorenzen et al. (2010) have gained widespread acceptance (Sass and Allen 2014) as a 'responsible approach' to stocking, with basic recommendations for how to make stocking work effectively (and see Cowx 1994, which emphasizes decision-making frameworks for stocking). The 'responsible approach' has been widely cited and provided a key conceptual framework for several subsequent publications (Munro and Bell 1997, Hilborn 1999, Bell et al. 2005, 2006, 2008, Taylor et al. 2005, Zohar et al. 2008). More importantly, it has been used to help guide hatchery development and reform processes in Australia, China, Denmark, Japan, New Caledonia, the Philippines and the USA (Lorenzen et al. 2010). At the same time, there has been a rapid increase in peer-reviewed literature on effects and effectiveness of stocking.

The 10 principles in the original 'responsible approach' (Blankenship and Leber 1995)

1) prioritize and select target species for enhancement by applying criteria for species selection; once selected, assess reasons for decline of the wild population
2) develop a management plan that identifies how stock enhancement fits with the regional plan for managing stocks
3) define quantitative measures of success
4) use genetic resource management to avoid deleterious genetic effects on wild stocks
5) implement a disease and health management plan
6) consider ecological, biological and life-history patterns in forming enhancement objectives and tactics; seek to understand behavioral, biological and ecological requirements of released and wild fish
7) identify released hatchery fish and assess stocking effects on fishery and on wild stock abundance
8) use an empirical process for defining optimal release strategies
9) identify economic objectives and policy guidelines, and educate stakeholders about the need for a responsible approach and the time frame required to develop a successful enhancement program
10) use adaptive management to refine production and stocking plans and to control the effectiveness of stocking.

The updated 'responsible approach' (Lorenzen et al. 2010)

Fisheries science and management in general, and many aspects of fisheries enhancement, have developed rapidly since the 'responsible approach' was first formulated. These developments made it necessary to revise the 'responsible approach' to take into account, in particular, the paradigm shift towards analyzing and managing enhancements from a fisheries management perspective (Lorenzen 2005). The developments also provided the tools for implementing the shift.

Most enhancements remain weak in at least four particular areas (Lorenzen et al. 2010):

1) Fishery stock assessments and modeling are integral to exploring the potential contribution of stocking to fisheries management goals; yet both are found lacking in most stock enhancement efforts in coastal systems
2) Establishing a governance framework for enhancements is largely ignored in stocking programs, thus, diminishing opportunities for integrating enhancement into fishery management
3) Involvement of stakeholders in planning and execution of stocking programs is key from the start, but they are rarely made an integral part of program development
4) Adaptive management of stocking is not well integrated into enhancement plans, yet is critical to achieving goals, improving efficiencies, and understanding and controlling the effects of stocking on fisheries and on wild stocks.

Lorenzen et al. (2010) expanded on these points and emphasized the importance of their inclusion in the 'responsible approach' (see updated list below). The updated approach is staged in order to ensure that key elements are implemented in the appropriate phases of development or reform processes. In particular, it is important to conduct broad-based and rigorous appraisal of enhancement contributions to fisheries management goals prior to more detailed research and technology development and operational implementation. This basic requirement applies to both development of new and/or reform of existing enhancements.

Stage I: Initial appraisal and goal setting

1) Understand the role of enhancement within the fishery system [NEW[1]]
2) Engage stakeholders and develop a rigorous and accountable decision-making process [NEW[1]]

[1] New points added by Lorenzen et al. 2010; not in the original 1995 version.

3) Quantitatively assess contributions of enhancement to fisheries management goals
4) Prioritize and select target species and stocks for enhancement
5) Assess economic and social benefits and costs of enhancement

Stage II: Research and technology development including pilot studies

6) Define enhancement system designs suitable for the fishery and management objectives [NEW[1]]
7) Develop appropriate aquaculture systems and rearing practices [NEW[1]]
8) Use genetic resource management to avoid deleterious genetic effects
9) Use disease and health management
10) Ensure that released hatchery fish can be identified
11) Use an empirical process for defining optimal release strategies

Stage III: Operational implementation and adaptive management

12) Devise effective governance arrangements [NEW[1]]
13) Define a stock management plan with clear goals, measures of success and decision rules
14) Assess and manage ecological impacts
15) Use adaptive management

Knowledge gained through research on the kinds of issues presented here is now being used and expanded upon by scientists in this field worldwide to evaluate marine fisheries enhancements in fundamentally different habitats and conditions and at different spatial and temporal scales. Since 1990, science and knowledge in this field have expanded exponentially. Collectively, this work has begun to demonstrate how and under what conditions marine fisheries enhancements can complement current approaches to sustaining, restoring, conserving and enhancing marine and estuarine fisheries and fish (and invertebrate) populations.

Aquaculture-Based Fisheries Enhancement Terminology

Confusion about the terms used in this field reflect one of the signs of a new science—lack of consensus on terminology. Stock enhancement has often been used as a generic term referring to all forms of hatchery-based fisheries enhancement. Bell et al. (2008) and Lorenzen et al. (2010) classified the intent of stocking cultured organisms in aquatic ecosystems into various basic objectives. Together, they considered five basic types, listed here from the most production-oriented to the most conservation-oriented:

1. *Sea ranching/Lake ranching*. Recurring release of cultured juveniles into marine, estuarine and lacustrine environments for harvest at a larger size in 'put, grow, and take' operations. The intent here is to maximize production for commercial or recreational fisheries.
2. *Stock enhancement*. Recurring release of cultured juveniles into wild population(s) to augment the natural supply of juveniles and optimize harvests by overcoming recruitment limitation in the face of intensive exploitation and/or habitat degradation.
3. *Re-stocking*. Time-limited release of cultured juveniles into wild population(s) to restore severely depleted spawning biomass to a level where it can once again provide regular, substantial yields (Bell et al. 2005).
4. *Supplementation*. Moderate releases of cultured fish into very small and declining populations, with the aim of reducing extinction risk and conserving genetic diversity (Hedrick et al. 2000, Hilderbrand 2002).
5. *Re-introduction*. Temporary releases with the aim of re-establishing a locally extinct population (Reisenbichler et al. 2003).

Capture-Based Enhancement of Mullet Fisheries

In many countries, mullet fry and fingerlings are captured from the sea and stocked in inland lakes and reservoirs as a form of fisheries enhancement (Lovatelli and Holthus 2008). Wild caught post-larvae and fingerling *M. cephalus* and other mullets have long been used to create fishpond, lagoon and lake fisheries, dating back to ancient Roman civilization in the Mediterranean region (Basurco and Lovatelli 2003). They have been stocked into inland water lakes of the El Fayyum area of Egypt since the 1920s, and into the Black Sea and Caspian Sea regions of Russia since 1930 (FAO 2006). Modern examples include capture based mullet fisheries in some countries in the Mediterranean region, Asia, and in Hawaii, USA (Ellis 1968, FAO 2006, Nishimoto et al. 2007, Saleh 2008, Snovsky and Ostrovsky 2014).

Stocking wild-caught mullet in inland lakes has been known in Egypt for more than eight decades. The importance of wild seed collection increased with recent aquaculture developments. In 2005, 69.4 million mullet fry were collected for both aquaculture and lake ranching and 156,400 tonnes of mullet were produced in lakes, semi-intensive ponds and coastal net pens (20% of Egypt's annual aquaculture production). As aquaculture of mullet has become more profitable, pressure on wild-caught fry has increased. The high cost of hatchery produced mullet seed has limited expansion of hatcheries in Egypt. The effect of capture-based fisheries on wild stocks of mullet is not well studied and this has become a subject of debate between aquaculture farmers and capture fisheries communities (Saleh 2008).

In Israel, two species of mullets, *Liza ramada* and *Mugil cephalus*, are stocked each year in Lake Kinneret. Neither of these reproduces in the lake. Commercial catch rates show that 27–28% of introduced fish of each species are landed. *M. cephalus* has greater impact per introduced fish than *L. ramada* (Snovsky and Ostrovsky 2014). Regular stocking programs were initiated in the 1950s. These stocking programs and fisheries on Lake Kinneret are controlled by the Israel Water Authority and by the Israel Fisheries Department of the Ministry of Agriculture. Grey mullets have the highest value of any commercial fish caught in Lake Kinneret. After collapse of the tilapia fishery in 2008, grey mullets comprised the primary income-generating commercial catch. Currently, the number of stocked fry is limited to one million fingerlings of these two species combined per year (Ostrovsky et al. 2014).

In the 1960–80s, *Mugil cephalus* and *Liza ramada* fry were used to stock the volcanic freshwater lakes of Central Italy which did not have any direct connection with the sea and were the base of a specific fishery, using the '*cefalare*' (nets especially designed for mullets - *cefali* in Italian) when adult mullets would gather in large schools (Cataudella and Monaco 1983).

There are smaller capture-based fisheries in some areas of Greece and Italy, where extensive culture systems are used to farm grey mullet in more or less confined brackish coastal lagoons, relying on wild fry that are collected and grown naturally each year. Wild caught grey mullet are also stocked in enclosed areas in Korea, Hong Kong, Taiwan and Singapore (Lovatelli and Holthus 2008).

Ancient Hawaiian people built and operated fish ponds along the shores of all the principal Hawaiian Islands. These ponds were stocked with a variety of marine species. Today, some of these ancient Hawaiian fishponds, once destroyed by natural causes, have been rebuilt and many are stocked with grey mullet (Nishimoto et al. 2007).

Looking to the future: can wild fry support expansion of capture-based mullet fisheries? Clearly, concerns about overfishing wild mullet stocks are already starting to limit current capture-based mullet fisheries. For example, in response to fisheries declines, some countries in the Mediterranean are already considering restrictions on collections of wild mullet juveniles stocked to support lake fisheries (Vasilakopoulos et al. 2014, A. Tandler, pers. comm.); what is needed is commercial-scale hatchery production of *M. cephalus* fry for grow-out and for supporting lake ranching and marine fisheries enhancements. In 2007, FAO held an "international workshop on technical guidelines for the responsible use of wild fish and fishery resources for capture-based aquaculture production" in Viet Nam and produced technical guidelines on capture-based aquaculture (Lovatelli and Holthus 2008).

Aquaculture-Based Enhancement of Mullet Fisheries

While capture-based fisheries for mullets have supported subsistence fisheries in Asia, Hawaii and the Mediterranean region for centuries, reliance on wild-caught fry cannot continue at the current pace, much less expand to meet demand (Saleh 2008). Thus, several countries are considering adding culture-based fisheries enhancements to their mullet fisheries management strategy. Following the first induced spawn in captivity (Tang et al. 1964) in Taiwan, and early larval-rearing successes in Taiwan (Liao et al. 1971, Liao 1975), attempts to close the life cycle with levels of fry production large enough to enable commercial scale aquaculture of *M. cephalus* made great progress in the 1980s and '90s (Nash and Shehadeh 1980, Lee et al. 1988, 1992, Tamaru et al. 1991, 1993, 1994, Liu and Kelley 1994, Lee and Ostrowski 2001). These developments in striped mullet aquaculture technology enabled our studies of the effectiveness of aquaculture-based mullet stock enhancement in Hawaii.

Case Study: Research in Hawaii to Develop Flathead Grey Mullet Stock Enhancement Technologies

Despite well over a century of stocking marine organisms into the sea to enhance fishery stocks, prior to the 1990s, very little effort had been allocated to assessing the effectiveness of stocking programs (Leber 2013). However, in the late 1980s, systematic studies began to develop and assess marine stock enhancement in Norway, Japan, China and the US. One of these efforts was a program launched in the US in Hawaii, which initially used flathead mullet *M. cephalus* as a test species. This work was one of the pioneering efforts worldwide to evaluate the potential of marine fisheries enhancement. The Hawaii project has been the only stock-enhancement assessment project conducted with *M. cephalus* that has employed experimental pilot releases and adaptive management to improve the outcome of stocking. Thus, the Hawaii work is presented here as a case study of fundamental aspects of conducting effective culture-based marine fisheries enhancement with *M. cephalus*.

Beginning in 1988, the Oceanic Institute (OI) began to receive federal funding from NOAA-Fisheries (US Department of Commerce) to examine the feasibility of replenishing declining marine fish populations in Hawaii using releases of cultured fish. OI's Stock Enhancement research was focused on developing effective stock enhancement strategies and transferring that technology to the state of Hawaii. The NMFS project, titled Stock Enhancement of Marine Fish in the State of Hawaii (SEMFISH), funded the primary research to develop and test enhancement strategies. In 1990, the Hawaii Division of Aquatic Resources (DAR) funded a collaborative project with OI to enable training and transfer of OI's marine stock enhancement technology to DAR. DAR aimed to restore to former abundance species whose numbers had become depleted, at least in part, by loss or degradation of natural spawning and nursery habitats. The OI and DAR projects were eventually curtailed in the early 2000s, owing to funding constraints.

Selection of Flathead Grey Mullet M. cephalus *as the Top Candidate for Stock Enhancement Research in Hawaii*

Initially, the Hawaii researchers convened a series of public workshops to identify species that were potential candidates for stock enhancement research in Hawaii (Leber 1994). A formal, semi-quantitative decision-making process was used to develop criteria and rank species. Based on the results of two workshops to prioritize species *M. cephalus* was selected, along with Pacific threadfin *Polydactylus sexfilis*, for a multi-year study to assess the potential to create culture-based fisheries in the sea.

Production of M. cephalus *Fry for Stock Enhancement Research on Oahu, Hawaii*

Successful stock enhancement is dependent on appropriate numbers and sizes of healthy, fingerlings being available for release at the appropriate time of year. This requires careful planning of production goals several months in advance of releases, and monitoring abundances and growth rates of cultured fish throughout the production process. In addition, specific measures must be taken in the hatchery to prevent disease and parasite outbreaks, and to ensure that genetic integrity of wild stocks is not degraded by hatchery releases (Tringali and Leber 1999, Tringali et al. 2007, Lorenzen et al. 2012).

Production of grey mullet for the Hawaii stock enhancement research followed protocols developed by OI for mullet broodstock acquisition, maturation and spawning, larval rearing, and nursery (Liu and Kelley 1994). Production of fingerlings for stock enhancement releases also required close attention to production parameters (growth rate, size distribution and population size), disease management, and genetic protocols for minimizing negative interactions between hatchery and wild fish (Blankenship and Leber 1995, Lorenzen et al. 2010). A detailed description of the rearing process is found in Tamaru et al. (1993) and Liu and Kelley's (1994) mullet culture manual.

Planning fish production for stock enhancement releases should begin with clear objectives about the intent of stocking and the numbers and sizes of fingerlings needed to be released into specific nursery habitats at specific times of the year (see below). In addition, recommended genetic protocols for hatchery releases require that sufficient mature broodstock from each habitat in which releases are to take place are available for spawning (Blankenship and Leber 1995, Lorenzen et al. 2010, Tringali et al. 2007). Because fish production levels need to satisfy requirements for hatchery releases, release numbers should be set with a 'window' (i.e., a range in which release numbers can fall and still meet release goals to allow for potential losses of fish from disease or parasitic outbreaks, or unexpected malfunctions in the culture system). In Hawaii, the intent of stocking was to develop and evaluate effective stock enhancement strategies. Rearing ~ 30,000 fingerlings per year for pilot release experiments was the target for aquaculture production.

Identifying Released M. cephalus *to Enable Evaluation of Stocking Impact*

Selecting a high-information content tag to identify hatchery reared fish to quantify success or failure of stocking is one of the most critical components of any enhancement efforts (Blankenship and Leber 1995, Lorenzen et al. 2010, Leber and Blankenship 2012, Leber 2013). Without some form of assessment, one has no idea of the success of a particular approach. Natural fluctuations in wild fish abundance can mask successes and failures and further necessitate a proper monitoring and evaluation system coupled with adaptive management (Walters and Hilborn 1978, Leber 2013).

Tagging technology provides the basis for quantitatively assessing survival, growth and dispersal of released fish, and their contribution to wild populations. Recaptures of tagged, hatchery-raised fish through regular sampling enables fishery managers to evaluate and refine release strategies, giving them control over the impact of hatchery releases on the fishery.

Selecting a Tagging System

Several methods are available for tagging fish including the Coded Wire Tag (CWT), Visible Implant Elastomer (VIE), Visible Implant Alpha (VIA), Passive Integrated Transponder (PIT) tags, acoustic tags and genetic fingerprinting. The CWT is considered the most suitable tagging method for stock enhancement programs, as it enables high capacity tagging of large numbers of small fish, high information control and reasonable cost considerations (Leber and Blankenship 2012).

Coded-wire tagging is done by implanting a single micro tag (1 mm long x 0.25 mm diameter) into fish tissue (usually nose or cheek tissue) beneath the skin using a technique that has no negative effect on fish health or behavior or human health. The CWT is a stainless steel tiny wire, marked with rows of numbers denoting codes of batches of fish and individuals. The tagging is done using an injector in a professional way. The tagging of large numbers of fish for stock enhancement programs requires an automatic injector and qualified operator. In addition, detectors that sense small changes in the magnetic field caused by the CWT while passing next to it are used to detect the presence of a CWT in tagged fish. Reading the codes on the tag is performed with a typical dissecting microscope.

CWTs appear to have negligible effects on tagged animals, and they are relatively cost-efficient for large-scale tagging programs (Nielsen 1992). Tags can remain in the animals indefinitely, enabling scientists to identify hatchery-raised fish at any stage of their life cycle. One of the greatest advantages of CWTs is the high information content provided by an almost unlimited number of possible codes.

CWTs have a numerical code etched onto the surface of each tag. For stocking programs, the code is often used to identify batches of fish, such as a release lot, but it is also possible to use sequentially coded CWTs to identify individual fish. These tags are automatically magnetized before insertion into cartilaginous, connective or muscular tissue. CWTs can be injected into the snout of grey mullet using head molds specifically designed by OI and Northwest Marine Technology, Inc. biologists for various sizes of juvenile grey mullet (45–130 mm Total Length [TL]). The head molds enable rapid tagging (~ 800 to 1,000 fish per hour by an experienced tagger) and are critical for correct placement of the tag. CWTs are typically implanted in the snout region of grey mullet and thus accurate placement from head molds prevents the tag from being injected into sinus cavities and eventually ejected from the fish. Because of the small size of the tags, minimal tissue damage occurs during tag insertion, and insertion wounds heal rapidly. Thus, CWTs can be used effectively with small juvenile fish (as small as 45 to 60 mm TL; Leber et al. 1996). Testing CWTs on 27 different genera of fish has shown tissue interaction to be minimal (Bergman et al. 1968, Fletcher et al. 1987).

For grey mullet, CWT retention rates averaged at least 97% over a period of several years (Leber 1995, unpubl. data). Initially, the Hawaii researchers had very poor tag retention in the snout region of grey mullet. The problem was solved by sectioning tagged fish and identifying tag placement using scanning electronic microscopy. This revealed that the tags were being injected into sinus cavities. Acceptable CWT retention rates were achieved with grey mullet after redesigning head molds to specifically target cartilaginous tissue in the head region.

CWTs are detected electronically by their magnetic field using a tag-detecting wand (NMT, Inc.). Tagged fish are returned to the laboratory where tags are dissected from fish using a binary search. Codes are read using a standard binocular microscope. CWTs injected into hatchery-released fingerlings have been recovered years later from adult mullet captured in Hawaii's mullet fishery (Leber and Arce 1996).

Evaluation of Release-Strategy Effects on Success of Mullet Stocking

Pilot releases should always be conducted prior to launching full-scale enhancement programs (Blankenship and Leber 1995, Lorenzen et al. 2010, Leber 2013). Experiments to optimize release strategies, by understanding the interactive effects of stocking variables (size at release, release habitat, release season, release magnitude) on survival of cultured fish released into the wild, are a critical step in identifying enhancement capabilities and limitations and in determining effective release strategies (Leber 1995, 2013, Leber et al. 1995, 1996, 1997, 2005). Pilot releases also provide the empirical data needed to plan enhancement objectives, test assumptions about survival and cost-effectiveness, and improve models predictions of enhancement potential.

To design effective pilot releases, critical variables that could affect survival of hatchery-released fish in the wild should be identified and then tested using an appropriate experimental design. Stocking variables that typically impact survival of stocked fish include release habitat, size-at-release, release season, release magnitude; these variables also have interactive effects (Fig. 18.1; Leber 1995, Leber et al. 1995, 1996, 1997). Acclimation and acclimatization prior to release should also be tested to determine impact on post-release survival of stocked fish (Brennan et al. 2006).

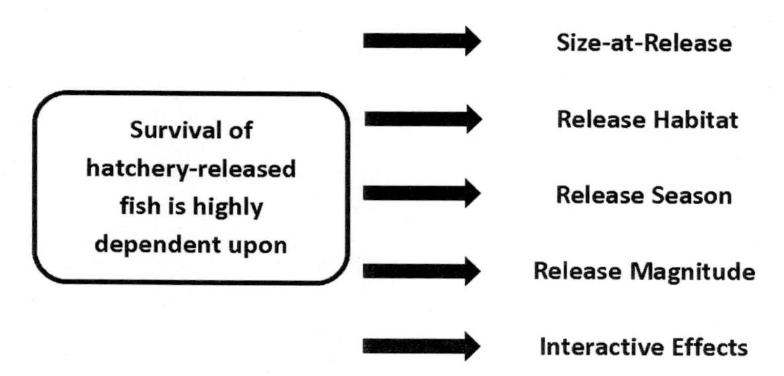

Figure 18.1. Key release variables that can strongly affect survival of released fingerlings. Evaluating the effects of these was the primary focus of the stock enhancement research in Hawaii.

Following pilot releases, sampling should be conducted to monitor survival of released hatchery fish and the effects of the chosen release strategies. Early indicators of stock enhancement effect include recovery rates, density of cultured and wild fish captured in samples and proportion of hatchery fish in collections (release contribution). Sequential pilot releases can be used to maximize enhancement benefits in a full-scale enhancement program.

Evaluation of Size-at-Release Effects on Success of Mullet Stocking

Several Size-At-Release (SAR) intervals should be evaluated during initial pilot experiments to identify optimal SAR (SAR resulting in the highest survival to production cost ratio; Leber et al. 2005) prior to large-scale hatchery releases. In the pilot experiments with grey mullet in Hawaii, a range of SAR groups were tagged and released into prime nursery habitats (Leber 1995, Leber et al. 1995, 1996, 1997). A range of fish sizes were produced by rearing eggs from several spawns, each spawn being about six weeks apart.

During summer 1990, 85,848 juvenile mullet were graded into five size groups (ranging from 45 to 130 mm total length), identified with binary-coded wire tags, and released into two estuaries (2 x 5 factorial design) on Oahu, Hawaii as part of an OI experiment to evaluate size-at-release and release habitat impacts on recruitment and survival of hatchery-released mullet (Leber 1995). Forty two thousand eight hundred and twenty two of the tagged fish were released into Kaneohe Bay on the east (windward) coast of Oahu and 43,026 were released simultaneously into Maunalua Bay on Oahu's drier south shore. To replicate experimental treatment groups, releases were blocked in time across five release lots.

To evaluate effects of size-at-release on and survival rates of released mullet, both bay systems were sampled monthly with cast nets over a 10 month period after releases. Researchers captured 733 tagged grey mullet, 277 from Kaneohe Bay and 456 from Maunalua Bay. Within six weeks after releases, recapture frequencies were clearly skewed in favor of fish that were larger at the time of release. Fish smaller than 70 mm when released were rare or absent in collections within 15 weeks after their release into Maunalua Bay, and within 25 weeks in Kaneohe Bay. This study confirmed results of a smaller-scale 1989 pilot study in Maunalua Bay and showed that fish size-at-release can have a critical impact on survival of cultured mullet in the wild. Pilot studies to identify minimum size-at-release should be conducted at each site targeted for marine hatchery releases.

Interactive Effects of Size-at-Release and Release Season

In a series of pilot releases over three years, Leber et al. (1995, 1996, 1997) showed the effectiveness of the adaptive management process at work, increasing the effectiveness of stocking by over 400% by modifying release strategies based on results from the pilot releases. Such 'active' adaptive management needs to be put into practice in existing stocking programs (Leber 2013).

Size-at-release (SAR) markedly impacted survival of stocked mullet (Leber 1995). Release season is another important variable to evaluate in pilot release experiments prior to conducting large-scale hatchery releases. Releases should be conducted during the natural recruitment time for the target species and those results compared with releases in other seasons. The Hawaii studies revealed that release season can clearly impact survival of grey mullet released into the wild by affecting size-at-release effects (Leber et al. 1997).

Hatchery-raised grey mullet were released into Kaneohe Bay, Hawaii during the spring and summer of 1991 as part of a pilot experiment to evaluate the impact of release season on recapture rates of released fish (Leber et al. 1997). Ninety thousand eight hundred and seventeen cultured grey mullet fingerlings were tagged and released into two replicate nursery habitats (Kahaluu stream and Kaneohe stream). During each season, three replicate lots of five size intervals (ranging from 45 to 130-mm total length) were released at both nursery habitats (3-way factorial design). Released fish were identified with binary-coded wire tags. Close attention was paid to releasing roughly identical numbers of fish among release lots for each season-SAR-site combination.

Survival, movement, and growth of released fish were monitored monthly over 45 weeks with a sampling program established at six nursery habitats in Kaneohe Bay. Results showed that survival and growth of released mullet were directly affected by the interactive effects of release season and size-at-release. Recapture frequencies, based on the number of individuals released within treatment groups each season, revealed an obvious and direct relationship between size-at-release and recapture rate (Fig. 18.2); when fish were released in the summer, recapture frequencies were directly proportional to size-at-release within a month after release. In contrast, size-at-release had little effect on recapture frequencies for fish released 10 weeks earlier, in the spring (Fig. 18.2). Spring was the only season tested in which 45–60 mm grey mullet have contributed significantly to abundances in the wild. However, larger fingerlings apparently had better survival rates in the wild when held in the hatchery until summer.

These data highlight how futile it would be to conduct summer releases in Kaneohe Bay of individuals that were smaller than 60 mm. As hypothesized by Leber et al. (1997), in habitats where survival of released fish is strongly impacted by fish size-at-release, survival of grey mullet will be greater when releases are timed so that fish size-at-release coincides with modes in population size structures of wild stocks. A corollary to this is the fewer wild fish in a particular size interval, the lower survival will be of released fish in that size interval.

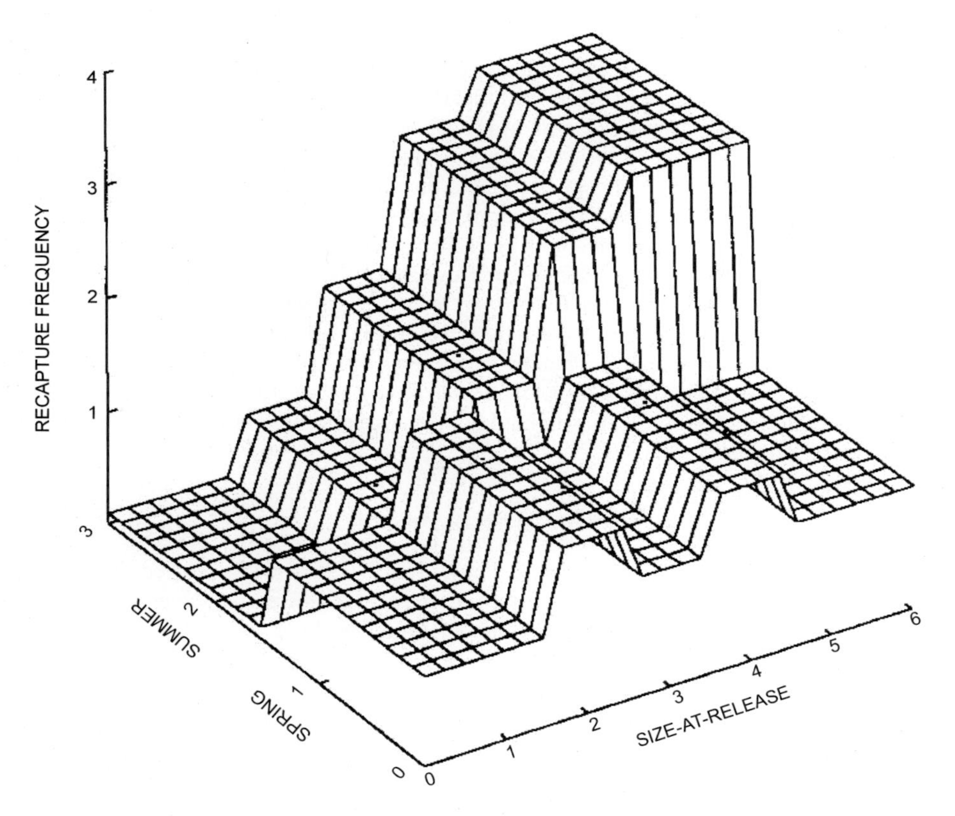

Figure 18.2. Percent of tagged, hatchery-released grey mullet recaptured in cast net samples following spring and summer releases, given for each of five size classes released (1 = 40 to 60 mm total length; 2 = 60 to 70 mm; 3 = 70 to 85 mm; 4 = 85 to 110 mm; and 5 = 110 to 130 mm total length). Data are percent recaptured of total fish released per treatment group.

Release Microhabitat Effects on Success

Mullet stocking programs should question the choice of release microhabitats carefully. In the study below (Leber 1995, Leber and Arce 1996), the Hawaii team realized the apparent loss of 30,000 *M. cephalus* fingerlings by stocking them in the 'wrong' habitat.

Pilot experiments with grey mullet reveal the importance of evaluating the effect of release habitat on survival of hatchery fish in the wild (Leber 1995, Leber and Arce 1996). In 1990, habitat preferences for juvenile grey mullet in Hawaii were not well understood. The literature describes *Mugil cephalus* as euryhaline, catadromous fish, well adapted to low salinities in rivers and estuaries (Blaber 1987). It seemed reasonable in 1990 to expect that releases anywhere in Kaneohe Bay would result in comparable survival, but this was not the case.

In 1990, only around 12,000 grey mullet were released in Kaneohe Bay at Kahaluu stream, but over 31,000 grey mullet were also released that year along the shoreline near the Hawaii Institute of Marine Biology pier in south Kaneohe Bay (HIMB) (Leber 1995). In 1991, 45,000 fish were released near the inlets at Kahaluu stream and an equal number at Kaneohe stream in the southern portion of the bay (Leber et al. 1997; Fig. 18.3).

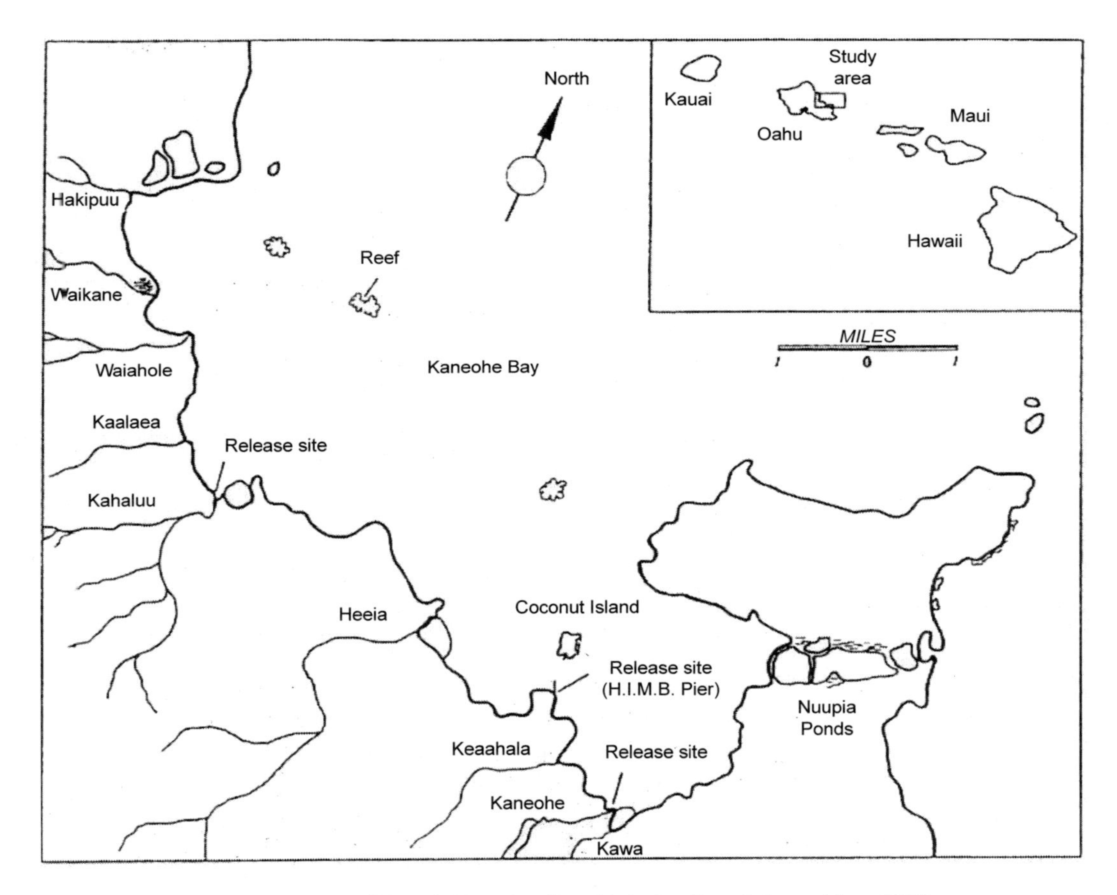

Figure 18.3. Map of Kaneohe Bay, showing release sites (from Leber and Arce 1996).

Fish released in the vicinity of streams showed similar performance in 1990 and 1991. But the 1990 release of 31,000 fish near HIMB resulted in few fish recaptured after week 16. None of the 31,000 fish released in 1990 near HIMB have been retrieved from the commercial mullet fishery in Kaneohe Bay, yet at least 20 individuals from the 1990 release at Kahaluu were recovered from that fishery during contact interviews with fishermen (Leber and Arce 1996).

Clearly, to optimize stocking success, each habitat targeted as a release site should be evaluated with pilot tag-release-recapture trials prior to large-scale hatchery releases. Initially, hatchery fish should be released into at least two to three different sites to compare differential effects of these habitats on fish survival.

Release site assessments should be conducted prior to selecting release sites, to locate primary nursery habitats containing ample food and other necessary resources, where juvenile wild fish of the target species occur naturally. Information on preferred nursery habitats for a target species can be obtained from the literature, from scientists who may have access to unpublished data, and from fishermen and others knowledgeable about the target species' ecology. Net sampling at potential nursery habitats may also be necessary before pilot releases begin, to obtain data on the target species' distribution and abundance in the wild, to clarify preferred nursery habitats. In addition, other release-site considerations include whether the site is accessible to release equipment (a truck or trailer with live fish hauling tank, hoses, etc.), and whether or not the site is frequented by fishermen, as (illegal) fishing pressure on released juveniles can threaten enhancement efforts.

Leber and Arce (1996) hypothesized that refuge from predators afforded by mangroves and other shoreline vegetation in the north end of Kaneohe Bay accounted for better survival of mullet released at Kahaluu inlet than of those released near HIMB. Mangroves are extensive along the northern shoreline of the bay from Kahaluu stream to Waiahole stream, whereas much of the shoreline in the southern portion of the bay near HIMB lacks mangroves. Also, the shoreline near HIMB is largely fronted by seawalls and mudflat-coral rubble habitat. Leber et al. (1996, 1997) showed relatively good survival following releases at Kahaluu stream, regardless of any subsequent movement along the shoreline towards adjacent streams.

Other factors besides ecological characteristics at a particular release site can impact fish survival and should be considered in determining optimal release habitat. For example, whether or not fishing regulations are enforced at a site can significantly impact survival of released fingerlings.

Test of Concept: **M. cephalus** *Stocking Impact on Juvenile Recruitment*

In the Hawaii pilot experiments with flathead grey mullet, hatchery-release variables were steadily refined to maximize grey mullet enhancement potential. Based on results from two years of pilot hatchery releases in Kaneohe Bay, a pilot experiment was designed to incorporate improved release strategies in a test of the marine stock enhancement concept (Leber et al. 1996). This study employed release strategies that had been steadily refined through the adaptive management process (in this case, with information learned through the pilot releases) to evaluate the real potential to use hatchery releases to significantly increase juvenile grey mullet recruitment in Kaneohe Bay, the largest estuary in Hawaii.

Essentially, this experiment evaluated the first assumption of the marine stock-enhancement concept: that cultured fishes released into coastal waters actually survive, grow and contribute substantially to recruitment. The criteria for success were: (1) cultured fish released in this study comprise at least 20% of the juvenile grey mullet in net samples four months after release; (2) cultured fish persist in net samples throughout the study; and (3) growth of cultured fish is comparable with measured rates in wild juveniles. If these criteria were met, it was reasonable to assume that cultured fish had substantially affected juvenile recruitment at the study site.

Eighty thousand five hundred and seven cultured grey mullet were tagged with coded wire tags and released during spring and summer into the Kahaluu stream, the principal mullet nursery in Kaneohe Bay (Fig. 18.4). For each release season-SAR combination, the experiment was replicated with three release lots at each of two release locations at Kahaluu stream (stream mouth and upper stream lagoon). SAR determinations were based on the 1991 study that revealed a strong effect between release season effects and SAR effects on survival (Leber et al. 1997). The seasonal timing of releases (spring and summer) was based on results from the previous pilot releases (Leber 1995, Leber et al. 1997). In the test-of-concept study, all five SAR intervals were released in spring. Only the three largest size groups were released in the summer (no fish smaller than 70 mm TL).

Recapture rate was six-fold greater than recapture rates had been after initial releases in Kaneohe Bay. The ~ 600% increase was a direct result of modifying release habitat and size-at-release (SAR) protocol based on recapture rates in pilot releases (releases in this experiment were confined to the vicinity of freshwater streams, and a minimum size of 70 mm total length was used during summer releases).

After 11 months, cultured fish comprised 50% of the grey mullet in collections at the release site, 20% in a nursery habitat 1 km to the north, and 10% in a nursery 3 km north of the release site. The location

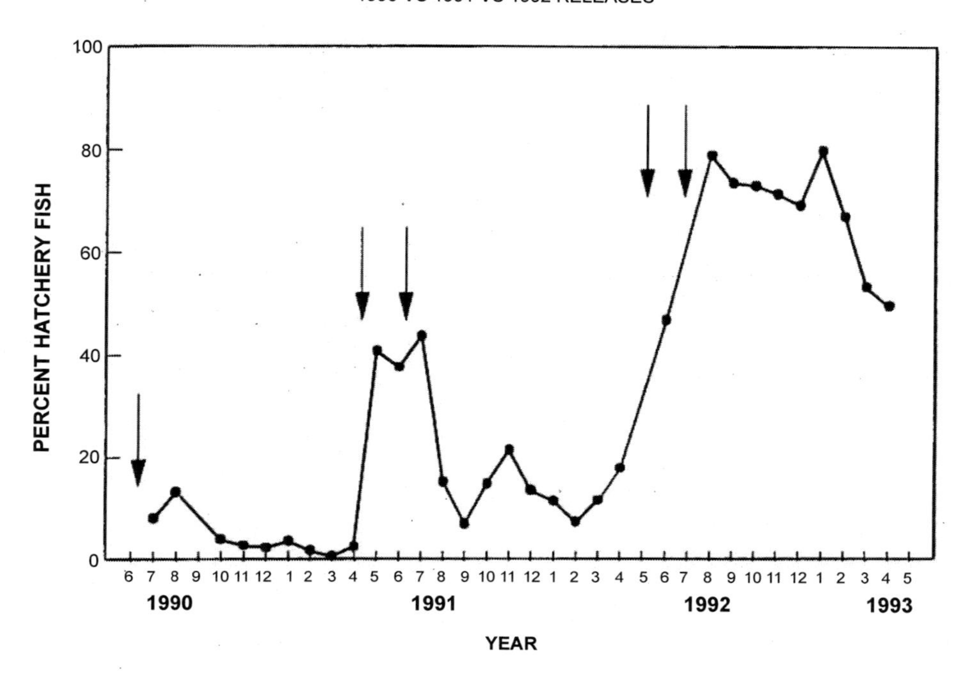

Figure 18.4. Contribution of juvenile cultured grey mullet to wild stock abundance in cast-net samples over the three year period of pilot release experiments in Kaneohe Bay. Arrows identify stocking events, showing a cumulative effect of adaptive management of release strategies—an increase of ~ 600% in the contribution of stocked mullet to overall mullet abundance in mullet nursery habitats in Kaneohe Bay (from Leber et al. 1996).

of releases at Kahaluu (stream mouth vs. upstream lagoon) significantly affected post-release dispersal patterns of cultured fish, but not growth or relative survival. SAR effects on recapture rates corroborated earlier results showing that the smallest (45 to 60 mm) fish released could survive spring releases better than summer releases. There was also a trend towards better survival of larger individuals when they were released in the summer.

Stocking effect on mullet abundances in nursery habitats was remarkable after adjusting release strategy to incorporate findings from pilot releases in Kaneohe Bay. There was a substantially greater impact on juvenile abundances in Kaneohe Bay following the 1992 releases than after pilot releases in 1990 (Leber 1995) and in 1991 (Leber et al. 1997). Proportions of cultured fish at Kahaluu 10 months after releases increased from around 3% following 1990 releases, to around 10% after 1991 releases, to around 50% in this study (Fig. 18.4).

Clearly, pilot experiments are crucial for managing enhancement impact. What is clear here is that the results realized from the 1990 releases, prior to any adaptive management, pale in comparison to what was achieved following the releases in 1992. The results from the 1992 releases were made possible by steady refinement in release strategies, based on posing hypotheses with each pilot release, monitoring the results, then making appropriate changes in release strategies based on the results accumulated from successive pilot releases.

Cost-Effectiveness of Size-at-Release of Hatchery Fish Recovered in the Mullet Fishery

How should one decide what is the optimal size hatchery fish to release? For mullet, hatchery costs to rear *M. cephalus* to various fingerling sizes in Hawaii were evaluated and compared with relative yields in the fishery of the various sizes of *M. cephalus* stocked in Kaneohe Bay. Those results were used by Leber et al. (2005) to select optimal size-at-release.

To determine unit cost to produce the various size-at-release groups, a bioeconomic model, originally developed to evaluate shrimp aquaculture production (Leung and Rowland 1989), was adapted to grey mullet production. The model specified costs associated with using existing facilities, established culture methods, and following hatchery guidelines needed to prevent deleterious genetic effects in the hatchery, as recommended by Shaklee et al. (1993), Kapuscinski and Jacobson (1987) and Busack and Currens (1995). The model determined the operating costs to produce and rear around 90,000 grey mullet to the median size within each of the five SAR intervals used in pilot release experiments in Kaneohe Bay.

Fishery contribution rates and production costs were determined for cultured fish released in 1990–1992 pilot studies that were subsequently landed in the commercial mullet fishery in Kaneohe Bay (Leber and Arce 1996). Recovery in the fishery of fish that were smaller than 60 mm when released was very poor relative to recovery from larger SAR intervals, particularly when releases were conducted in summer (Fig. 18.5).

To identify the most cost-effective (optimal) size of mullet to release, Leber et al. (2005) developed a simple mathematical model to determine the optimal SAR. The production-related cost of an enhancement effect (dollars spent in the hatchery to achieve a hatchery fish contribution to the fishery) was least for fish that were 85–110 mm TL when stocked. Although the cheapest fish to rear among the size intervals that were produced were those in the 45–60 mm interval, these results revealed that releasing larger mullet can result in greater cost-efficiency when the increase in yield (because of the increase in survival afforded by releasing larger fish) more than offsets the increase in production costs of rearing larger mullet. In Kaneohe Bay, stocked mullet afforded a greater fishery contribution per dollar spent on production when intermediate-size, not small, grey mullet fingerlings were stocked. These results do not suggest that intermediate size fingerlings should always be stocked by stocking programs; the point was that one should identify the most cost-effective size to stock, which may vary among systems based on local environmental and ecological conditions at release sites. The optimal SAR may be small fish in some systems (e.g., Tringali et al. 2008) and larger fish in others.

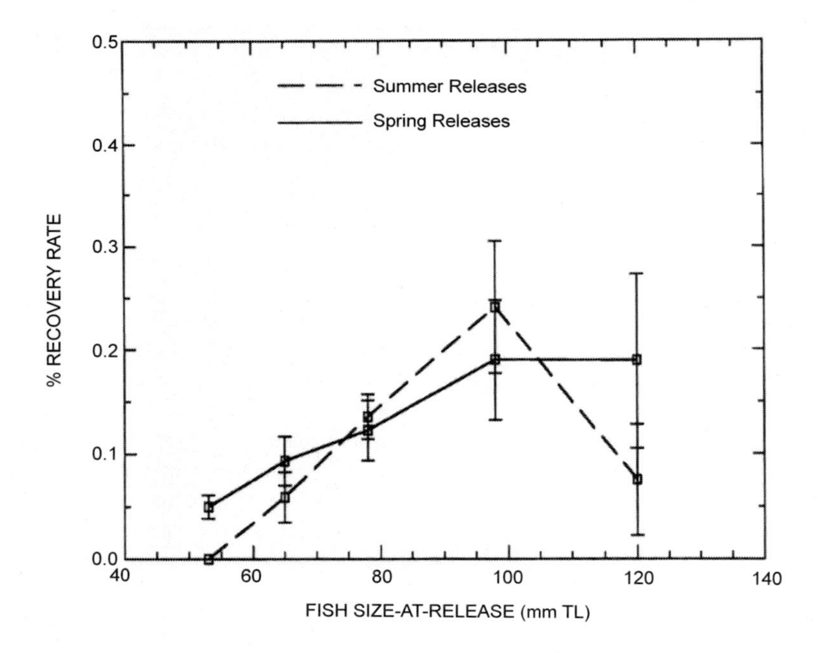

Figure 18.5. Relationship between mean percent recovery rate ([number recaptured/number released] x 100) and fish size at release (SAR) for 214 cultured grey mullet recovered from the fishery in Kaneohe Bay, Hawaii (Leber and Arce 1996, Leber et al. 2005).

Production Cost per Recruit. Using the production and cost data from this study, fish production costs in the hatchery were distributed across fishery recruitment levels for hatchery fish, assuming a release of 91,286 individuals in the optimal SAR interval, 85–110 mm TL (Fig. 18.6). This models hatchery production cost per fish landed for fishery recovery values ranging from 2 to 100% (for 91,286 fish produced and subsequently caught in a fishery). With this model, total hatchery production costs averaged over the number of landed hatchery fish decreased logarithmically from around US$30 (in 1993 dollars) per hatchery fish landed if only 2% of the released grey mullet are caught in the fishery, to US$12 if 5% are caught, US$6 if 10% are caught, US$3 if 20% are caught, US$1.20 if 50% are caught, and US$0.60 if 100% are caught.

Thus, pilot-release studies that reveal ways to maximize survival of stocked fishes without necessarily increasing rearing costs can improve cost efficiency in stocking programs. A primary concern for cost efficiency of enhancement should be "how do stocking variables affect optimal size at release"; for example, how does timing of releases (seasonal, tidal, time of day), release habitat (and microhabitat), stocking magnitude, and acclimation prior to release affect post-release survival and optimum size-at-release? These factors can all be examined in ongoing stocking programs by adopting an adaptive management approach.

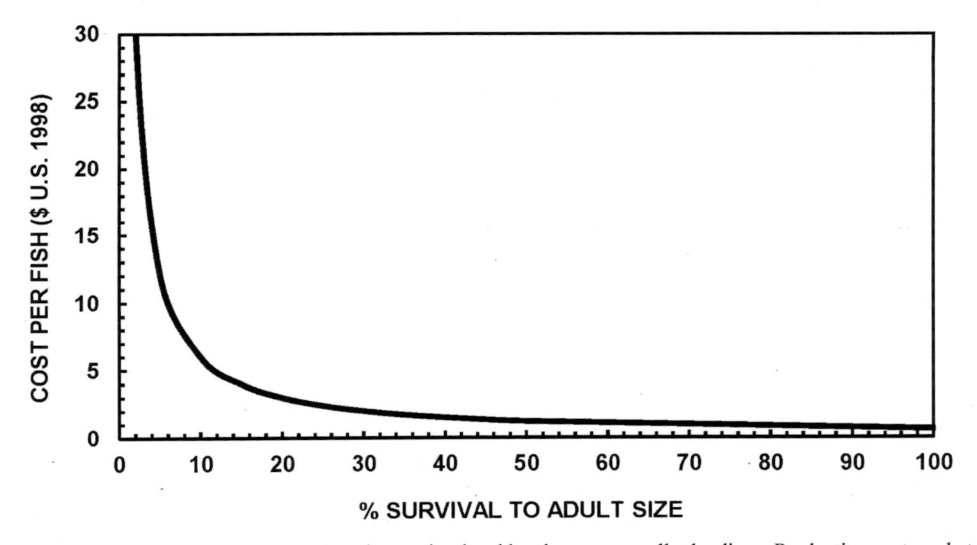

Figure 18.6. Unit production costs apportioned over simulated hatchery grey mullet landings. Production cost per hatchery recruit in the fishery is for 85–110-mm TL fingerlings stocked into Kaneohe Bay (Leber et al. 2005).

Does Stocking Cultured M. cephalus *Enhance or Displace Wild* M. cephalus*?*

Assessment of release impact should go farther than evaluation of survival and contribution rates of hatchery fish. Evaluation of hatchery fish interactions with wild stocks is also critical. Such an evaluation can address the second corollary of the marine stock enhancement concept: that hatchery fish enhance rather than displace wild stocks.

The first empirical stocking study to evaluate effect of stocking magnitude on hatchery-wild fish interactions was conducted by the Hawaii project, which documented that released mullet could indeed increase abundances of juvenile recruits in a principal nursery habitat in Hawaii without displacing wild individuals (Leber et al. 1995). In the summer of 1993, 6,000 wild mullet fingerlings were captured, tagged, and released at two of the most productive nursery habitats in Kaneohe Bay (Kaneohe stream and Kahaluu tributary). The release of wild fish established a pre-treatment condition to gather baseline data on the dispersal patterns of wild fish. A month after the wild releases, 30,000 hatchery fish were released into the Kahaluu tributary to establish the primary treatment condition, leaving Kaneohe stream as the control site.

Hawaiian researchers evaluated the hatchery impact by comparing dispersal patterns for wild fish released at the treatment site (Kahaluu tributary) with dispersal patterns of wild fish released at the control site (Kaneohe stream). Results show that recapture rates for wild fish were nearly identical between treatment and control sites, indicating that wild fish were not displaced from their natural nursery sites by hatchery releases (Fig. 18.7).

Initial dispersal patterns of the cultured fish released into Kaneohe Bay showed greater movement of cultured fish out of the release habitat than was expected, based on results from previous smaller scale releases at that site.

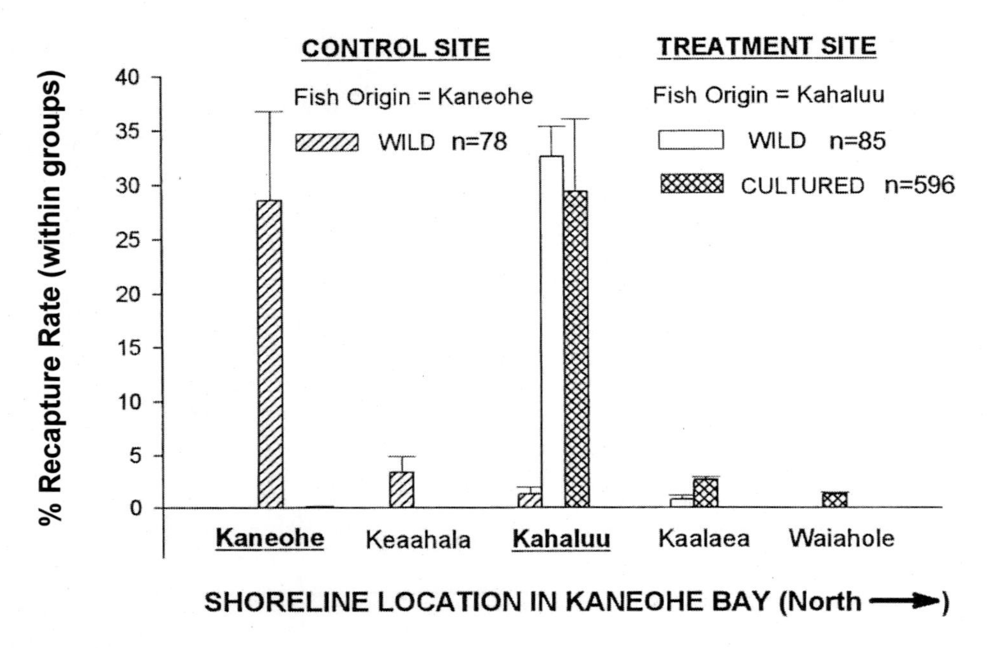

Figure 18.7. Post-stocking dispersal patterns of wild and cultured mullet from release sites, with (treatment site) and without (control site) hatchery releases of cultured mullet. Mean percent recaptured (within treatment groups) is shown with standard errors for wild fish tagged and released at the control site, wild fish released at the treatment site, and cultured fish released at the treatment site (from Leber et al. 1995).

Putting Culture-Based Mullet Enhancement into Practice in Hilo Hawaii

In Hawaii, mullet was prized as a food fish for royalty, and in modern times, it is also targeted by the recreational fishery, particularly in Hilo Harbor on Hawaii Island. Today, pole-and-line fishing for mullet is becoming a dying art (Nishimoto et al. 2007). Mullet fishing, once easily recognized by the numerous, small wooden platforms, called stilt chairs, dotting the tidal flats (Hosaka 1944) in Kaneohe Bay and Ala Wai Canal on Oahu Island, is gone. These platforms are now prohibited because of environmental regulations. Rather, small skiffs now replace the platforms for mullet fishermen. Hilo Harbor, especially the Waiākea Public Fishing Area (PFA), is one of the last strongholds of this type of mullet fishing (Fig. 18.8). Fishers use a system of a delicately balanced bobber and tandem hooks baited with algae, primarily the chain diatom, *Melosira tropicalis* (Julius et al. 2002). Fishing for grey mullet *M. cephalus* in Hilo is the only fishery in the world where diatoms are used as bait (Nishimoto et al. 2007).

In 1990, the Hawaii Division of Aquatic Resources (DAR) and Oceanic Institute (OI) partnered to develop a collaborative project to help restore the declining coastal mullet stocks using hatchery-based fisheries enhancement. Mullet fingerlings were cultured at OI (and later at DAR's hatchery facility in Hilo

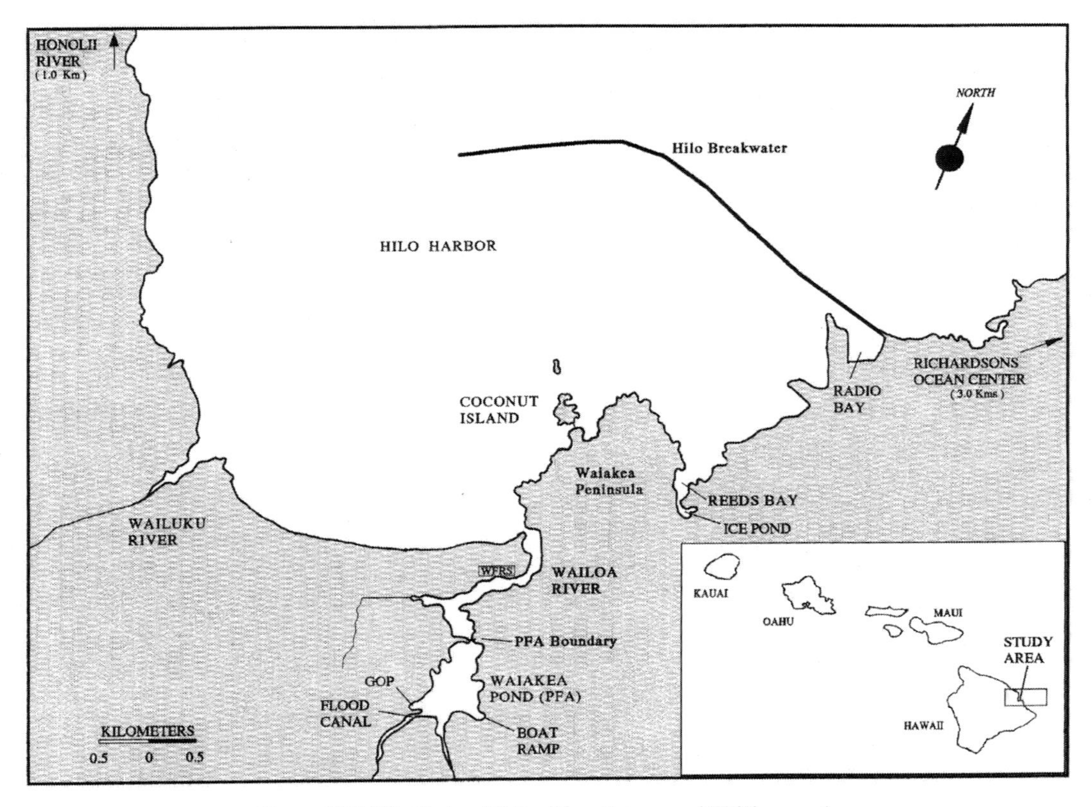

Figure 18.8. Hilo Harbor, Wailoa River Estuary and Waiäkea pond.

Harbor) and shipped to the State Fisheries Research Station in Hilo for growout. Fingerlings of various sizes were batch tagged with internal Coded Wire Tags (CWT). Tagged fishes were kept for several days to allow recuperation from tagging stress.

A total of 268,228 CWT mullet fingerlings were released at various locations in Hilo Bay from August 1990 to September 2000, except for 1996 when none were released. Hatchery release impact was assessed by creel sampling the recreational fisheries (starting in 1991) and by conducting bimonthly cast-net sampling (starting in 1990) at fixed stations in Waiäkea Pond, Wailoa River, and Reeds Bay, all located within Hilo Harbor (Fig. 18.8).

The results were significant: (1) The prototype marine stock enhancement experiment demonstrated that even small-scale releases can have a significant impact on wild stock abundance in the mullet fishery in Hilo; (2) The number of mullet entering the fishery was significant and was achieved annually; and (3) The Wailoa River Estuary, especially the boat launching ramp, was found to be an excellent release site (Nishimoto et al. 2007).

The number of CWT identified hatchery-released mullet in the fisher's creel ranged from a low of 3.9% in 2003 to as high as 61.1% during 1999 (Nishimoto et al. 2007). The overall average increase in the recreational mullet fishery after nine years of releasing hatchery-raised mullet was 21.7%.

The Hilo mullet project verified the potential of stock enhancement as an effective tool to replenish diminishing stocks. Based on the results of this project, several management measures were implemented to further DAR's mission of replenishing and conserving native fish stocks (Nishimoto et al. 2007).

Conclusions

The *M. cephalus* fisheries enhancement studies in Hawaii showed that recovery rates identified during the juvenile phase of the life cycle were a reasonably good indicator of the effects of release strategies on post-release survival patterns of hatchery fish caught in the fishery. Consistent with the results from studies of juveniles, the studies of hatchery mullet caught in local fisheries showed: (1) a direct relationship between size-at-release and recapture rate after summer releases; (2) higher recovery of individuals > 70 mm when released in the spring, rather than summer, and zero recovery of fish < 60 mm if released in summer; (3) that release habitat had an important effect (especially when fish were released away from the vicinity of their freshwater nursery habitats)—for example, shoreline releases in Kaneohe Bay near HIMB pier resulted in very poor (zero) recovery of any of the five size ranges of hatchery fish stocked, whereas releases within documented *M. cephalus* nursery habitats always resulted in recaptures when fish size was timed to coincide with size modes of wild mullet.

Such information from pilot experiments, about how release strategies affect survival and recruitment of cultured fish to nursery habitats and eventually to the fishery, is clearly needed to plan effective stock enhancement programs.

Marine fisheries enhancement appears to have high potential as one of the tools in the Hawaii fishery-management toolbox, if used responsibly and with a focus on managing the stocking program to achieve the stated goals of stocking and with ample attention to all of the factors that need to be considered in managing enhancement programs for success (Blankenship and Leber 1995, Lorenzen et al. 2010, Leber 2013, Sass and Allen 2014).

Acknowledgements

Funding was provided by the NOAA-Fisheries (United States Department of Commerce). The authors extend grateful thanks to the Oceanic Institute Stock Enhancement Program, especially Marcus Boland, Robert N. Cantrell, Glenn Karimoto, Bong Kim, Anthony Morano, Dave Sterritt, Ryan Takushi, and James West. We thank all the researchers in OI's Finfish Program in the 1990s for their contributions in developing mullet production technology, and using that technology to produce the fish for our pilot release experiments. Thanks to Laurie Peterson and Maala K. Allen for editing assistance.

Daniel Thompson and Ray Buckley, from Washington Department of Fish and Wildlife (WDFW), provided their expert assistance during the development of OI's stock enhancement technology and in transferring tagging techniques to OI. Discussions with Richard Lincoln (WDFW, ret.) became the genesis of much of the 'Responsible Approach to Marine Enhancement' during its development and demonstration in these studies.

References

Appeldoorn, R.S. and D.L. Ballentine. 1983. Field release of cultured conchs in Puerto Rico: implications for stock restoration. Proc. Gulf. Carib. Fish Inst. 35: 89–98.

Basurco, B. and A. Lovatelli. 2003. The aquaculture situation in the Mediterranean Sea—predictions for the future. Ocean Docs. http://www.oceandocs.org/handle/1834/543. Electronic version accessed 5/11/2014.

Bell, J.D., P.C. Rothlisberg, J.L. Munro, N.R. Loneragan, W.J. Nash, R.D. Ward and N.L. Andrew. 2005. Restocking and stock enhancement of marine invertebrate fisheries. Adv. Mar. Biol. 49: 1–370.

Bell, J.D., D.M. Bartley, K. Lorenzen and N.R. Loneragan. 2006. Restocking and stock enhancement of coastal fisheries: Potential, problems and progress. Fish. Res. 80: 1–8.

Bell, J.D., K.M. Leber, H.L. Blankenship, N.R. Loneragan and R. Masuda. 2008. A new era for restocking, stock enhancement and sea ranching of coastal fisheries resources. Rev. Fish. Sci. 16: 1–9.

Bergman, P.K., K.B. Jefferts, H.F. Fiscus and R.L. Hager. 1968. A preliminary evaluation of an implanted coded wire fish tag. Washington Department of Fisheries, Fish. Res. Pap. 3: 63–84.

Bilton, H.T., D.F. Alderdice and J.T. Schnute. 1982. Influence of time and size at release of juvenile coho salmon (*Oncorhynchus kisutch*) on returns at maturity. Can. J. Fish. Aquat. Sci. 39: 426–447.

Blaber, S.J.M. 1987. Factors affecting recruitment and survival of mugilids in estuaries and coastal waters in Southeastern Africa. Am. Fish. Soc. Symp. 1: 507–518.

Blankenship, H.L. and K.M. Leber. 1995. A responsible approach to marine stock enhancement. Am. Fish. Soc. Symp. 15: 167–175.

Brennan, N.P., M.C. Darcy and K.M. Leber. 2006. Predator-free enclosures improve post-release survival of stocked common snook. J. Exp. Mar. Biol. Ecol. 335: 302–311.

Busack, C.A. and K.P. Currens. 1995. Genetic risks and hazards in hatchery operations: fundamental concepts and issues. Am. Fish. Soc. Symp. 15: 71–80.

Cataudella, S. and G. Monaco. 1983. Allevamento dei Mugilidi. Acquacultura. Milano, Ed. Clesav, 117-148.

Cowx, I.G. 1994. Stocking strategies. Fisheries Manag. Ecol. 1: 15–31.

Ellis, N. James. 1968. Notes on cultivation of mullet in Hawaiian fish ponds. Third Technical Meeting on Fisheries, South Pacific Commission, Koror, Palau Trust Territory of the Pacific Islands. June 3–4, 1968. 7p.

FAO 2006. Cultured Aquatic Species Information Programme. *Mugil cephalus*. Cultured Aquatic Species Information Programme. Text by Saleh, M.A. In: FAO Fisheries and Aquaculture Department [online]. Rome. Updated 7 April 2006. [Cited 18 August 2015]. http://www.fao.org/fishery/culturedspecies/Mugil_cephalus/en.

Fletcher, H.D., F. Haw and P.K. Bergman. 1987. Retention of coded wire tags implanted into cheek musculature of largemouth bass. N. Am. J. Fish. Manag. 7: 436–439.

Hager, R.C. and R.E. Noble. 1976. Relation of size at release of hatchery-reared coho salmon to age, size and sex composition of returning adults. Prog. Fish Cult. 38: 144–147.

Hedrick, P.W., D. Hedgecock, S. Hamelberg and S.J. Croci. 2000. The impact of supplementation in winter-run Chinook salmon on effective population size. J. Hered. 91: 112–116.

Hilborn, R. Confessions of a reformed hatchery basher. 1999. Fisheries 24: 30–31.

Hilderbrand, R.H. 2002. Simulating supplementation strategies for restoring and maintaining stream resident cutthroat trout populations. N. Am. J. Fish. Manag. 22: 879–887.

Hosaka, E.Y. 1944. Sport Fishing in Hawaii. Bond's Honolulu Publishers. 198pp.

Julius, M.L., R. Nishimoto, B. Kaya, L. Nishiura and T. Shimoda. 2002. Using diatoms to catch fish. Abstract. World Aquaculture Society Conference. Madrid, Spain.

Kapuscinski, A.R. and L.D. Jacobson. 1987. Genetic guidelines for fisheries management. Sea-Grant Research Report 17 (Minnesota Sea-Grant, St. Paul). 66pp.

Leber, K.M. 1994. Prioritizing Marine Fishes for Stock Enhancement in Hawaii. The Oceanic Institute, Honolulu. 46pp.

Leber, K.M. 1995. Significance of fish size-at-release on enhancement of striped mullet fisheries in Hawaii. J. World Aquacult. Soc. 26(2): 143–153.

Leber, K.M. 2013. Marine fisheries enhancement: coming of age in the new millennium. pp. 1139–1157. *In*: P. Christou, R. Savin, B.A. Costa-Pierce, I. Misztal and C.B.A. Whitelaw (eds.). Sustainable Food Production. DOI 10.1007/978-1-4614-5797-8, Springer Science+Business Media New York.

Leber, K.M. and S.M. Arce. 1996. Stock enhancement effect in a commercial mullet *Mugil cephalus* fishery in Hawaii. Fisheries Manag. Ecol. 3: 261–278.

Leber, K.M. and H.L. Blankenship. 2012. How advances in tagging technology improved progress in a new science: marine stock enhancement. Am. Fish. Soc. Symp. 76: 3–14.

Leber, K.M., N.P. Brennan and S.M. Arce. 1995. Marine enhancement with striped mullet: are hatchery releases replenishing or displacing wild stocks? Am. Fish. Soc. Symp. 15: 376–387.

Leber, K.M., S.M. Arce, D.A. Sterritt and N.P. Brennan. 1996. Marine stock-enhancement potential in nursery habitats of striped mullet, *Mugil cephalus*, in Hawaii. Fish. B.-NOAA 94: 452–471.

Leber, K.M., H.L. Blankenship, S.M. Arce and N.P. Brennan. 1997. Influence of release season on size-dependent survival of cultured striped mullet, *Mugil cephalus*, in a Hawaiian estuary. Fish. B.-NOAA 95: 267–279.

Leber, K.M., R.N. Cantrell and P.-S. Leung. 2005. Optimizing cost-effectiveness of size at release in stock enhancement programs. N. Am. J. Fish. Manag. 25: 1596–1608.

Lee, C.S. and A.T. Ostrowski. 2001. Current status of marine finfish larviculture in the United States. Aquaculture 200: 89–109.

Lee, C.S., C.S. Tamaru and C.D. Kelley. 1988. The cost and effectiveness of CPU HCG and LHRH-a on the induced spawning of grey mullet, *Mugil cephalus.* Aquaculture 73: 341–347.

Lee, C.S., C.S. Tamaru, C.D. Kelley, A. Morinaka and G.T. Miyamoto. 1992. The effect of salinity on the induction of spawning and fertilization in the striped mullet, *Mugil cephalus.* Aquaculture 102: 289–296.

Leung, P.S. and L.W. Rowland. 1989. Financial analysis of shrimp production: an electronic spreadsheet model. Comput. Electron. Agr. 3: 287–304.

Liao, I.C. 1975. Experiments on induced breeding of the grey mullet in Taiwan from 1963 to 1973. Aquaculture 6: 31–58.

Liao, I.C., D.L. Lee, M.Y. Lim and M.C. Lo. 1971. Preliminary report on induced breeding of pond reared mullet *(Mugil cephalus* L.). Fish Ser. Chin-Am. Jt. Comm. Rur. Reconst. 11: 30–35.

Liu, K.K.M. and C.D. Kelley. 1994. The Oceanic Institute Hatchery Manual Series—Striped mullet (*Mugil cephalus*). Waimanalo, HI, USA, Honolulu. 88pp.

Lorenzen, K. 2005. Population dynamics and potential of fisheries stock enhancement: Practical theory for assessment and policy analysis. Philos. T. R. Soc. Lon. B 260: 171–189.

Lorenzen, K., K.M. Leber and H.L. Blankenship. 2010. Responsible approach to marine stock enhancement: an update. 2010. Rev. Fish. Sci. 18: 189–210.

Lorenzen, K., M.C.M. Beveridge and M. Mangel. 2012. Cultured fish: integrative biology and management of domestication and interactions with wild fish. Biol. Rev. 87: 639–660.

Lovatelli, A. and P.F. Holthus. 2008. Capture-based aquaculture: Global overview. FAO Fisheries Technical Paper n. 508. FAO, Rome.

Munro, J.L. and J.D. Bell. 1997. Enhancement of marine fisheries resources. Rev. Fish. Sci. 5: 185–222.

Nash, C.E. and Z.H. Shehadeh (eds.). 1980. Review of Breeding and Propagation Techniques for Grey Mullet, *Mugil cephalus* L. International Center for Living Aquatic Resources Management, Manila. 87pp.

Nielsen, L.A. 1992. Methods of marking fish and shellfish. Am. Fish. Soc. Spec. Pub. 23, Bethesda, MD.

Nishimoto, R.T., T.E. Shimoda and L.K. Nishiura. 2007. Mugilids in the Muliwai: a tale of two mullets. pp. 143–156. *In*: N.L. Evenhuis and J.M. Fitzsimons (eds.). Biology of Hawaiian Streams and Estuaries. Bishop Museum Bulletin in Cultural and Environmental Studies, Volume 3. Bishop Museum Press, Honolulu.

Ostrovsky, I., T. Zohary, J. Shapiro, G. Snovsky and D. Markel. 2014. Fisheries management. pp. 635–653. *In*: T. Zohary, A. Sukenik, T. Berman and A. Nishri (eds.). Lake Kinneret, Ecology and Management. Springer Science, Heidelberg.

Radonski, G.C. and R.G. Martin. 1986. Fish culture is a tool, not a panacea. pp. 7–15. *In*: R.H. Stroud (ed.). Fish Culture in Fisheries Management. American Fisheries Society, Bethesda, MD.

Reisenbichler, R.R., F.M. Utter and C.C. Krueger. 2003. Genetic concepts and uncertainties in restoring fish populations and species. pp. 149–183. *In*: R.C. Wissmar and P.A. Bisson (eds.). Strategies for Restoring River Ecosystems: Sources of Variability and Uncertainty in Natural and Managed Systems. American Fisheries Society, Bethesda, MD.

Richards, W.J. and R.E. Edwards. 1986. Stocking to restore or enhance marine fisheries. pp. 75–80. *In*: R.H. Stroud (ed.). Fish Culture in Fisheries Management. American Fisheries Society, Bethesda, Maryland.

Saleh, M. 2008. Capture-based aquaculture of mullets in Egypt. pp. 109–126. *In*: A. Lovatelli and P.F. Holthus (eds.). Capture-based Aquaculture. Global Overview. FAO Fisheries Technical Paper n. 508. Rome, FAO.

Sass, G.G. and M.S. Allen (eds.). 2014. Foundations of Fisheries Science. American Fisheries Society, Bethesda, Maryland, USA. 801pp.

Shaklee, J.B., C.A. Busack and C.W.J. Hopley. 1993. Conservation genetics programs for Pacific salmon at the Washington Department of Fisheries: Living with and learning from the past, looking to the future. pp. 110–141. *In*: K.L. Main and E. Reynolds (eds.). Selective Breeding of Fishes in Asia and the United States. The Oceanic Institute, Honolulu, Hawaii.

Snovsky, G. and I. Ostrovsky. 2014. The grey mullets in Lake Kinneret: the success of introduction and possible effect on water quality. Water and Irrigation 533: 32–35 (in Hebrew).

Svasand, T., K.E. Jorstad and T.S. Kristiansen. 1990. Enhancement studies of coastal cod in western Norway. Part 1. Recruitment of wild and reared cod to a local spawning stock. J. Cons. int. Explor. Mer. 47: 5–12.

Tamaru, C.S., C.S. Lee and H. Ako. 1991. Improving the larval rearing of striped mullet (*Mugil cephalus*) by manipulating quality and quantity of the rotifer, *Brachionus plicatilis*. pp. 89–103. *In*: W. Fulks and K. Main (eds.). Rotifer and Microalgae Culture Systems. Proc. of a U.S.—Asia Workshop. The Oceanic Institute, Honolulu.

Tamaru, C.S., W.J.F. Gerald, Jr. and V. Sato. 1993. Hatchery manual for the artificial propagation of striped mullet (*Mugil cephalus*). Department of Commerce, Suite 601, G.1.T.C Bldg. 590 South Marine Drive, Tamuning, Guam- 96911. 167pp.

Tamaru, C.S., C.S. Lee, C.D. Kelley, G. Miyamoto and A. Moriwake. 1994. Oocyte growth in striped mullet, *Mugil cephalus* (L.) maturing at different salinites. J. World Aquacult. Soc. 25: 109–115.

Tang, Y.A., I.H. Tung, C.S. Cheng, Y.W. Hou, Y.W. Hwang and Y.Y. Ting. 1964. Induced spawning of striped mullet by hormone injection. Jap. J. Ichthyol. 12: 23–28.

Taylor, M.D., P.J. Palmer, D.S. Fielder and I.M. Suthers. 2005. Responsible estuarine finfish stock enhancement: an Australian perspective. J. Fish. Biol. 67: 299–331.

Tringali, M.D. and K.M. Leber. 1999. Genetic considerations during the experimental and expanded phases of snook stock enhancement. Bull. Natl. Res. Inst. Aquacult. (Japan) Suppl. 1: 109–119.

Tringali, M.D., T.M. Bert, F. Cross, J.W. Dodrill, L.M. Gregg, W.G. Halstead, R.A. Krause, K.M. Leber, K. Mesner, W. Porak, D. Roberts, R. Stout and D. Yeager. 2007. Genetic Policy for the release of finfishes in Florida. Florida Fish and Wildlife Conservation Commission, Florida Fish and Wildlife Research Institute Publication No. IHR-2007-001.

Tringali, M.D., K.M. Leber, W.G. Halstead, R. McMichael, J. O'Hop, B. Winner, R. Cody, C. Young, C. Neidig, H. Wolfe, A. Forstchen and L. Barbieri. 2008. Marine stock enhancement in Florida: a multi-disciplinary, stakeholder-supported, accountability-based approach. Rev. Fish. Sci. 16: 51–57.

Tsukamoto, K., H. Kuwada, J. Hirokawa, M. Oya, S. Sekiya, H. Fujimoto and K. Imaizumi. 1989. Size-dependent mortality of red sea bream, *Pagrus major*, juveniles released with fluorescent otolith-tags in News Bay, Japan. J. Fish. Biol. 35(Suppl.A): 59–69.

Vasilakopoulos, P., C.D. Maravelias and G. Tserpes. 2014. The alarming decline of Mediterranean fish stocks. Curr. Biol. 24: 1643–1648.

Wang, Q., Z. Zhuang, J. Deng and Y. Ye. 2006. Stock enhancement and translocation of the shrimp *Penaeus chinensis* in China. Fish. Res. 80: 67–79.

Walters, C.J. and R. Hilborn. 1978. Ecological optimization and adaptive management. Ann. Rev. Ecol. Syst. 9: 157–188.

Walters, C.J. and S.J.D. Martell. 2004. Fisheries Ecology and Management. Princeton University Press, Princeton, NJ.

Zohar, Y., A.H. Hines, O. Zmora, E.G. Johnson, R.N. Lipcius, R.D. Seitz, D.B. Eggleston, A.R. Place, E.J. Schott, J.D. Stubblefield and J.S. Chung. 2008. The Chesapeake Bay blue crab (*Callinectes sapidus*): a multidisciplinary approach to responsible stock replenishment. Rev. Fish. Sci. 16: 24–34.

CHAPTER 19

Capture and Culture of Mugilidae in Taiwan

I Chiu Liao,[1] *Nai Hsien Chao*[2] and *Chien Chang Tseng*[3,*]

Introduction

The world fisheries production was about 158,000,000 t in 2012 (FAO 2014, Fisheries Agency 2013). Capture fisheries and aquaculture contributed about 91,300,000 t (58%) and 66,600,000 t (42%) respectively to the total fishery production. Capture production has been relatively stable in the past decade, but the contribution of aquaculture has been rising continuously and total fishery production is estimated to reach 179,000,000 t by 2021, an increase of 15% over the average production of 2009 to 2011. This increase is mainly due to aquaculture, and Asia as a whole plays a key role in the aquaculture industry, though it needs a well-rounded plan to keep the balance between capture fisheries and aquaculture in order to increase the supply of fishery products and maintain sustainable development.

Flathead grey mullet *Mugil cephalus* is an important commercial species of both capture fisheries and aquaculture in Taiwan. Other mullet species, such as large scale mullet *Chelon macrolepis*, are sold on the Taiwanese market, but do not have detailed records of production. In general, all mullet statistics in Taiwan are taken from *Mugil cephalus*. In 2012, the yield of flathead grey mullets was 2,992 t, and accounted for 0.24% of the total Taiwanese fisheries production of 1,256,082 t, valued at US$ 17,210,000, about 0.49% of the total Taiwanese fisheries production value of US$3,539,140,000 (Fisheries Agency 2013). Although the share is not significant in economic terms, the fact that the contribution of production value is twice that of production quantity indicates that its economic benefits should not be ignored. Moreover, the added value of flathead grey mullet processed products is more than their fishery value. Hence the importance of the management and development of the flathead grey mullet industry.

In the waters around Taiwan, six genera and 12 species of Mugilidae (Mugiliformes, Actinopterygii) have been recorded: *Crenimugil crenilabis, Chelon affinis, Chelon alatus, Chelon haematocheilus, Chelon macrolepis, Chelon subviridis, Ellochelon vaigiensis, Mugil cephalus, Oedalechilus labiosus, Moolgarda cunnesius, Moolgarda perusii, Moolgarda seheli* (Shao 2014). Among these species, the flathead grey

[1] National Taiwan Ocean University, 2 Pei-Ning Rd., Chung-Cheng District, Keelung 20224, Taiwan.
Email: icliao@mail.ntou.edu.tw
[2] National Cheng Kung University, No.1, University Rd., Tainan City 70101, Taiwan.
Email: chaoliao@gmail.com
[3] Department of Aquaculture, National Penghu University of Science and Technology, 300 Liu-Ho Rd., Magong, Penghu 88046, Taiwan.
* Corresponding author: jjtzeng@npu.edu.tw

mullet is the most prevalent and economically important. It is eurythermal and euryhaline, and is widely distributed in coastal waters. During the oviposition period, a female mullet can produce millions of eggs, which are isolated and pelagic. At the juvenile stage, flathead grey mullet grows in brackish waters, estuaries and mangroves, and when mature moves offshore. At different stages of its development the flathead grey mullet undergoes changes in its feeding habits, mainly feeding on plankton, benthic animals, organic detritus, and algae (Masuda and Kobayashi 1994, Shao 2014). Due to their high economic value and delicious taste, mullet products such as roe, testes, and stomach (gizzard), are highly commercial. Prior to spawning, mullet takes in large quantities of food, causing the flesh to taste better, and the roe and testes to be fatter.

Fishers in Taiwan have made a living catching flathead grey mullet since the early 17th century. In 1896, a mullet-roe producer from Nagasaki, Japan, brought mullet-roe processing technology to Taiwan and established factories in Lukang and Kaohsiung. With the increasing export of mullet roe from Taiwan, tax revenue was generated. Flathead grey mullet has been an important commercial species in the Taiwanese Strait for hundreds of years and has contributed substantial tax income to the Government (Nakamura 1997, Wu 2008). In the 1950s, fishing vessels became motorized with improved fishing gear and fishing methods. Fishery production of the flathead grey mullet has increased gradually. Since 1977, fishing vessels have been permitted to use interphone communication. When fishers of one vessel found a stock of fish, they could notify fishers of other vessels to join in. Consequently, fishery production reached a peak. During 1977–1986, the average yearly production yielded almost two million fish (corresponding to approximately 5,000 tonnes) and significantly increased fishers' income. After the 1990s, yearly production greatly decreased, probably due to pollution, overfishing and global warming (Wu and Wu 2013). Because of high market demand and economic value, Taiwan has started to import flathead grey mullet from other countries, such as the United States, Brazil, Australia, and Republic of Mauritius, and has invested heavily in the aquaculture industry for flathead grey mullet (COA 2014).

Capture Fisheries

Fishing Gear and Production from Capture Fisheries

The catching of wild flathead grey mullets starts in the winter solstice. When seawater temperature drops in November, flathead grey mullet migrate to the western coastal waters of Taiwan and spawn in the southernmost coastal waters. After spawning, they return to northern Taiwan. Late November to early January is the flathead grey mullet fishing season in Taiwan and the best fishing period lasts about 10 days around the winter solstice when the fish are in the best condition with full ovaries and tender flesh, and also have the best market price (Ochiai and Tanaka 1998, Shao and Chen 2003). Experienced fishers know well how to watch the southward moving cold water current, which brings migrating mullet to the fishing grounds.

The main fishing gears used to catch flathead grey mullet are purse seines and drift nets. In the purse seine fishery two large fishing vessels work as a team and move quickly. The boats are tied to each end of the net, and tow the net in a circle around the target fish. When the net is completely released, the two vessels approach each other. The net is then retrieved mechanically and the fish are harvested. The drift net is used by smaller vessels and *sampans*. Although these boats are smaller with fewer fishers, they are very efficient if they can find the exact migration route of the flathead grey mullet. There are now fewer vessels using purse seines, and drift nets have become the principal fishing gear for the flathead grey mullet fishery.

The annual fishery production and monetary return of flathead grey mullets were highly unstable from 2002 to 2012, with the highest production (1,328 t) in 2012 and the lowest (111 t) in 2007, corresponding to the highest monetary return (US$10.9 million) in 2012 and the lowest (US$0.8 million) in 2007 (Fig. 19.1, Fisheries Agency 2003–2013). This variation in fishery production and monetary return has undoubtedly affected the development of this industry.

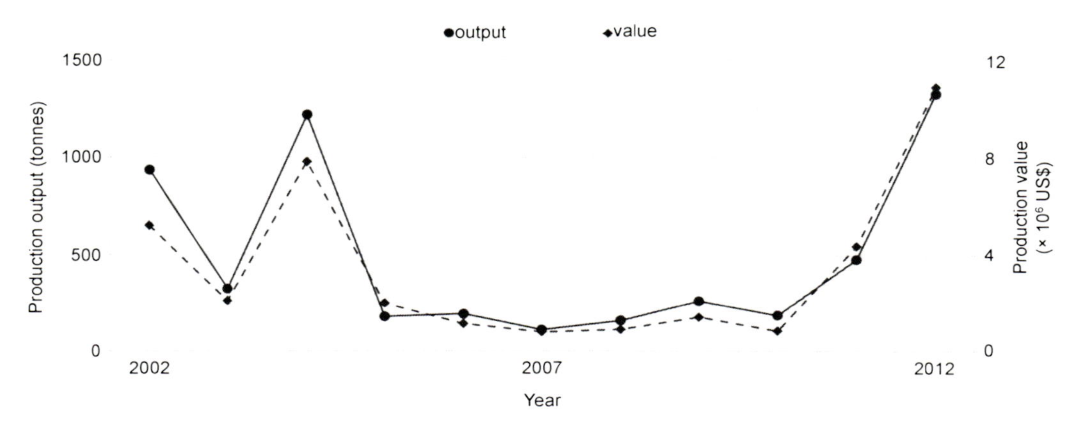

Figure 19.1. Annual production and value of captured flathead grey mullet from 2002 to 2012 in Taiwan (Fisheries Agency 2003–2013).

Processed Products from Mullets

Ovaries, testes, and stomach are the three delicacy products of the flathead grey mullet fishery (Fig. 19.2). Fresh wild flathead grey mullets are sold in the harbour landing sites, or processed. Mullet testes are usually eaten raw, and are mostly sold in restaurants. When the production increases, testes are also sold in traditional markets. Because the production season is in winter, testes are suitable to cook with sesame oil and considered a tonic ingredient. The processed ovary is the most valuable part from flathead grey mullet. Mullet stomachs are edible either raw or processed. Although the processed stomach is highly priced, its economic value is not comparable to that of ovaries. Without these three delicacies the flathead grey mullet is just an ordinary edible fish without high commercial value.

With the improvement of freezing and processing techniques and studies of the texture and flavour, the flathead grey mullet industry is facing a transition to increasing added-value on the product. Therefore, the balance between exploitation and conservation of the flathead grey mullet resource becomes a key issue for the sustainability of this industry.

Figure 19.2. Fresh flathead grey mullet sold in the market and its three valuable parts (roe, testes, and stomach) (photo by C.C. Tseng).

It takes several steps to process flathead grey mullet roe. The normal process is described in Table 19.1. In step 7, when the roe becomes hard and dried, it is ready for final processing (Fig. 19.3). Mullet roe can generally be preserved for one month in a cooler. To extend the storage to over one year, the packed mullet roe should be wrapped with towels and placed in a cardboard box, then stored in a freezer. The whole processing procedure takes 10 to 14 days. If the ovary is from cultured flathead grey mullet, it has to be frozen for one week after Step 3 because it is rich in oil. The quality of mullet roe is determined by shape, size, colour, smell, thickness uniformity, stain, salinity, oil content and viscosity, which are the key factors that set the market price. Therefore, to upgrade the industry, investment in research and development in the traditional flathead grey mullet industry is needed to produce added economic value to this product.

Table 19.1. The production procedures of flathead grey mullet roe.

Step	Processing method	Details
1	Sorting females from males	Select the females from a wild catch. When pressing on a fish belly a flow of white fluid is released from the gonopore, the fish then is male. However if the fluid is golden, the fish is female.
2	Removing ovaries	The fisherman uses a specialized knife to procure the ovary from the female flathead grey mullet. A ball is connected to the point of the knife (Fig. 19.4) to avoid damaging the ovaries. During this step, the ovaries are sorted according to their size, shape, and the quantity of oil.
3	Tying and cleaning	Secure the opening of the ovary with a wire to prevent the eggs from leaking out of the ovary. Scales or spoons are used to remove the blood from the ovarian blood vessels.
4	Salt curing	Cover the entire ovary with salt at a weight ratio of 4:1.
5	Desalting	Rinse the mullet roe with clear water and dry the roe by laying the roe on a bed of salt for two to three hours.
6	Compression moulding	Place the ovary on a wooden board and cover it with a cloth and another wooden board, then add a weight on top of the upper board to shape the roe.
7	Drying under the sun	Expose under the sun. Repeat cleaning, drying, and shaping processes. The roe, when hard and dried, is ready for the final process.
8	Product packaging	Preserve and sell the mullet roe after vacuum packaging.

Figure 19.3. Flathead grey mullet roe to dry under the sun during process (photo by C.C. Tseng).

Figure 19.4. Specialized knife (with a ball indicated by an arrow) to procure the ovary (photo by C.C. Tseng).

Furthermore, for industry transition and upgrading, it is essential to keep improving fishing techniques, processing techniques, and marketing strategy. In Taiwan, there are studies on the use of all parts of flathead grey mullets, such as extracting collagen from scales, processing fish heads for special Chinese cooking, stewing bones for soup-stock, processing flesh to fish floss, and even the use of the inedible fins as painting material to show the cultural features of fishing villages. In terms of community planning of fishing villages, renaissance can help to get rid of the stiff image of smelly and humid environments. Cleaning up the environment, re-organizing the traffic, and planning tour packages, will be beneficial to industry transition. Introducing the concept of a tourism and leisure industry and industrial collaborations are encouraged and will assist in upgrading the industry.

Flathead Grey Mullet Aquaculture Industry

History of Flathead Grey Mullet Aquaculture

The aquaculture of the flathead grey mullet started in Taiwan in the 1970s. As a stable supply of fry and juveniles is the key factor for the industry, the initial development of induced spawning techniques was a response to insufficient wild fry and juveniles. A research team of scientists from several agencies was organized in 1963 and succeeded in the artificial propagation and mass production of flathead grey mullet fry (Liao et al. 1972a), leading the Taiwanese aquaculture industry to a new horizon (Liao 1977, 1981). Despite these successful experimental trials, the production of hatchery fry has languished since then due to the long culture schedule, the high risk of failure, insufficient number of farmers, insufficient supply of wild specimens to be used as broodstocks, and the low price of hatchery produced fry. Only a few farmers were willing to invest in mullet induced spawning, and research activities related to grey mullet artificial propagation slowed down and closed. The techniques of the production of hatchery fry have been ignored for long time, but fortunately this unfavourable atmosphere has now been reversed. The price of mullet roe has recently risen, but the production of captured flathead grey mullet remains unstable. With the improvement of frozen and processing technology and better marketing strategy, interest in flathead grey mullet culture is reviving. Therefore, the use of hatchery fry for farming, restocking and restoration needs should be re-evaluated.

Recently, the aquaculture of flathead grey mullet has become profitable again because of promotion by government and related organizations. It takes more than two years to grow flathead grey mullet from fry to an adult fish. Timing of roe production from cultivated flathead grey mullet is critical. If fishes are harvested early in the season the roe will be too small, and if the harvest is later the roe will be too oily to process. The optimum wet weight of fresh mullet roe is over 300 g, when the fish are about three years old. Generally speaking, roe from two year old fish will be smaller and weigh 250 g (Fig. 19.5). After drying, the weight is reduced to 70% of wet weight. Extending grow out from two to three years has the risk linked to the increasing likelihood of typhoons which may cause floods and loss of fish. The low risk of two year old flathead grey mullet thus has the advantage of greater market demands.

Fish farmers have gradually shifted from extensive culture to intensive culture in order to increase production. Most aquaculture farms are located in Hsinchu (northern Taiwan), Yilan (North-eastern Taiwan), and the coastal area of the middle and southern parts of western Taiwan. According to the Fisheries Agency (2012–2014), the surface area of flathead grey mullet aquaculture has gradually increased, from 565 hectares in 2011 to 755 hectares in 2013. The annual yield in production and

Figure 19.5. Roe of two-year old cultured flathead grey mullet (photo by C.C. Tseng).

monetary return varied greatly (Fisheries Agency 2003–2013). Annual production ranged from 2,894 t in 2002 to 1,264 t in 2008, with 1,664 t in 2012. On the other hand, the annual monetary return ranged from US$9.9 million in 2004, to US$4.3 million in 2008, with about US$6.3 million in 2012 (Fig. 19.6). In order to maintain stability in the aquaculture industry, it is therefore necessary to plan an efficient production management.

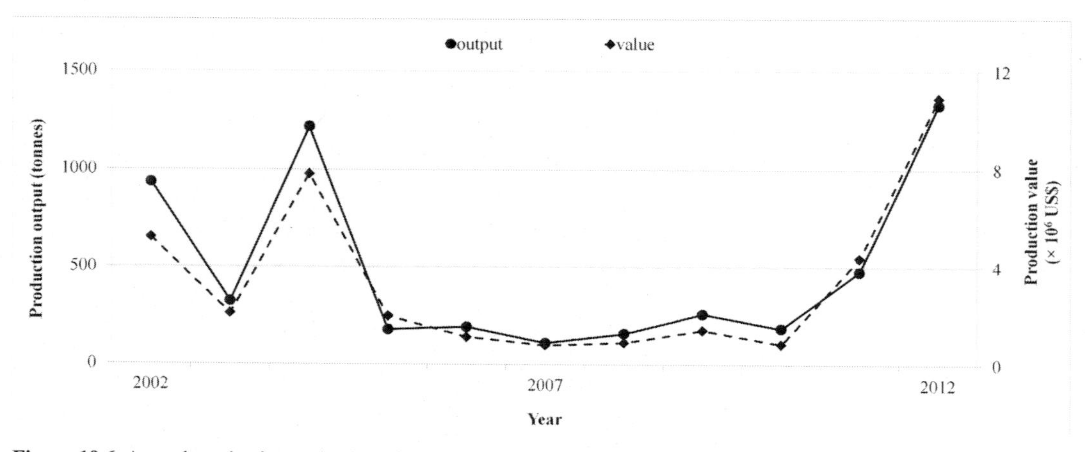

Figure 19.6. Annual production and value of cultured flathead grey mullet from 2002 to 2012 in Taiwan (Fisheries Agency 2003–2013).

Culture Methods

The species cultured for roe in Taiwan is the flathead grey mullet, *Mugil cephalus* (Fig. 19.7). Since the capture fishery production of flathead grey mullet is limited by season and by declining catches, aquaculture is the only alternative to make up the production shortfall and meet the demand for mullet roe. Due to fluctuations in demand and price, there are farmers who invested heavily and others who quit the market rapidly, resulting in a variation of aquaculture scale. For creating a healthy industry, stable production and proper marketing strategy are required.

Figure 19.7. Cultured flathead grey mullet (photo by C.C. Tseng).

The Wild Seed Resource

In order to industrialize flathead grey mullet culture, the source and quantity of fry must be stable. In Taiwan, supply of fry still relies on the capture of wild fry from the sea, and there is no commercial production of hatchery fry. Wild fry are captured mainly from coastal waters and estuaries in Fulong and Keelung, and the estuaries of Tamsui, Hsinchu, Miaoli, Chiayi, and Tainan and then transferred to fish farms. No data are available on the number of mullet fry annually harvested in Taiwan. The west coast is the principal culture area of Taiwan, and flathead grey mullet is mostly cultured on fish ponds. In general, fry captured around the winter solstice have the best quality, especially those from the northern coast and estuaries.

During the spawning season, fishermen gather at the freshwater area of the estuary with netting gear to capture the newly spawned fry transported from the spawning ground to the estuary. The survival of the fry depends on several conditions: (1) sufficient number of spawners of flathead grey mullet to migrate to the spawning ground; (2) extent of hindrance on the migration route due to oceanic currents, seawater temperature, and predators; (3) environmental conditions (such as salinity, natural feed, etc.) of the habitat where the wild fry is caught. There are several basic requirements to guarantee the appearance of fry in the traditional catch area.

The most essential is not to overfish the flathead grey mullet populations and to allow a certain amount of spawning fish to migrate to, and spawn, in their traditional spawning ground. Therefore, the government should carry out an accurate study on the resource stocks and the annual yield. International fishery cooperation should be promoted to reach the goal of coexistence and maximum sustainability of the natural resources. Next, under the threat of global warming, seawater temperatures will be affected. The migration route, predators and the change in the environment of flathead grey mullet will be an important subject to study. The influence of global climate change on the flathead grey mullet industry cannot be ignored. Although there is no clear evidence to prove that this change has affected wild flathead grey mullet fry, this problem should be monitored. The estuary where the wild flathead grey mullet fry settle is an area where people live and human activities could affect fry survival. Rivers bring freshwater and nutrients from the land to the sea and enrich the estuary with feed organisms for flathead grey mullet fry. River pollution and human construction in the estuaries may change the food webs and habitats, or even poison the fry directly. Prevention of adverse changes to the coastal estuarine environment has become a key factor for the continuous existence of wild flathead grey mullet fry.

According to information from fishers, the fry of other mullet species are mixed with flathead grey mullet fry in the wild. Few studies (COA 2014) have reported the composition of these other mullet species. Okiyama (1988) showed that the morphology of the larvae and juveniles are similar among species of the Mugilidae, therefore increasing the potential for bycatch fry of other species. In addition, fishers have indicated that with the quantity of the catch, it is too costly to practice one-by-one selection. Instead, fishermen use the information of the collecting site and time of appearance of the mullet fry in previous years to anticipate when and where to capture the fry, possibly with the maximum quantity of flathead grey mullets.

Wild flathead grey mullet fry may be available as early as the end of October, but due to the mixture with fry of other species at this time, the efficiency for species selection is low. Generally speaking, the wild fry captured during the 10 days before the winter solstice are the most desirable. Recent studies using molecular biology (Durand et al. 2012) recognized three cryptic species of mullets within the *M. cephalus* species complex, NWP1, NWP2, and NWP3, in the North-western Pacific (Shen et al. 2011, Chapter 2—Durand 2015). Based on this study and further communication with the author, it is understood that the adult of flathead grey mullet captured before and after winter solstice belong to NWP1. This cryptic species migrates southward with the China Coastal Current, and spawns in the coastal water of Pintung and then returns to the north. NWP2 is distributed in the waters surrounding the Philippines and Hainan Island. It migrates northward with the Kuroshio Current passing Taiwan and heading to Tokyo Bay, Japan. The NWP3 migrates mainly within the South China Sea. These three cryptic species overlap in their distribution in the area west of Taiwan because of their seasonal distribution patterns and the seasonal change of water temperature, and consequently the timing of appearance and spawning of the flathead grey mullet. Shen et al. (2011) further suggested that the wild fry captured in Taiwan are likely to be NWP2 and NWP3. Future research should be focused on accurate prediction of place and time to catch the wild fry, to reduce the proportions of fry of other species, and to improve the cost-efficiency of aquaculture.

Artificial Propagation

The shortage of wild flathead grey mullet fry is an unsolved problem and the high mortality rate (70%) of the captured wild fry is a waste of natural resources (COA 2014). Fry production by induced spawning can help to solve these problems. A team of scientists from the Taiwan Fisheries Research Institute, Taiwan Fisheries Bureau, and the Institute of Biology of National Taiwan University was founded in 1963 to carry out a research project on the artificial propagation of the flathead grey mullet. A series of propagation trials, including selection, transportation, culture, hormonal treatment of broodstocks and fry culture had been carried out in 1966, but experiments were discontinued in 1967 (Liao et al. 1969). In 1968, the project was transferred to Tungkang Marine Laboratory (now known as Tungkang Biotechnology Research Center, Fisheries Research Institute). These studies focused on handling techniques (Liao et al. 1969), the applicability of freshwater fish farms for culture of spawners (Liao et al. 1971), the differences between artificial insemination and natural spawning (Liao et al. 1972b), the adaptation to environmental conditions of hatched fertilized eggs and larvae at early stages and artificial propagation (Liao et al. 1972a, Liao et al. 1973), the preservation techniques for flathead grey mullet sperm (Hwang et al. 1972, Chao et al. 1973, Chao et al. 1975), hormones and their effects on gonads (Pien and Liao 1975), and analysis and discussion of key techniques and unsolved problems in the artificial propagation of flathead grey mullet seeds. The research team, based on their years of experience, reviewed the successful mass production technology (Liao et al. 1972a). At present, no induced spawning of flathead mullets is performed in Taiwan.

Liao (1975, 1976) reported using large mullets captured in waters off the south-west coast of Taiwan in winter as the source of broodstock in mullet aquaculture. The large mullets are in their oviposition period, and their maturity can be accelerated with hormone treatment to produce sperm and eggs. In addition, wild or artificially propagated fry and juveniles are cultured for two to three years to produce brood fish for farms. The dosages and frequencies of hormone and vitamin treatments to accelerate maturity and ovulation were reported in the above-mentioned two papers.

The diameter of a mature egg is between 0.93 and 0.95 mm (Liao 1975). Hatching time of fertilized eggs depends on the water temperature and salinity. When the salinity is between 30.1‰ and 33.7‰ and the temperature between 23 and 24.5°C, hatching requires 34 to 38 hours. With a salinity of 30.1‰–33.7‰, and a temperature of 22.5–23.7°C hatching requires 49–54 hours. A water temperature between 20 and 24°C is highly recommended for hatching. The newly hatched larva is approximately 2.5 to 3.5 mm long, and has an ocular system, but no pigment. The mouth is not formed and the anus is not opened. The newly hatched larva cannot swim, and their nutrition is provided by the yolk and oil globule. The mouth opens on the third day after hatching and allows the larvae to start feeding on fertilized oyster eggs or trochophores. The yolk sac is absorbed completely by the fifth day, whereas the oil globule is not absorbed completely until the eighth day. The absorption of the yolk sac and oil globule signals that larvae have begun to rely on external nutrition. Rotifers and copepods are fed to the mullet five to six days after hatching, and *Artemia* nauplii 20 days after hatching. The whole body is covered with scales on the 28th day after hatching. At this stage, artificial diets can be introduced. After 45 days, their standard length reaches 30 mm, at which time they are transferred to grow-out ponds, where the water salinity is lowered to provide more favourable conditions for the mullet fry.

Like other marine fishes that spawn isolated pelagic eggs, the flathead grey mullet larva is small-sized and possesses undifferentiated organs after incubation. Larvae draw nutrients from the yolk sac and oil globule, but consume nutrients rapidly and tolerate hunger poorly. During their early life (including the incubation period, initial feeding period, and exogenous feeding period), larval fish often die from starvation if they cannot obtain sufficient nutrients. This is a crucial factor for the early mortality of fish and substantially reduces the effectiveness of fry production. Therefore, for mullet fry production, farmers should draw information and experience from other marine fishes being artificially propagated. Particularly, in Taiwan, only wild mullet fry are used as seed for grow out, as the production of hatchery fry has been discontinued. Recently, only few studies on the flathead grey mullet, compared to those on grouper, snapper, cobia, etc., have been conducted, and aquaculture is flourishing and new marine fry continue to appear (Yu 2012). Therefore, there is a need for mullet fry production in aquaculture to develop parallel to the production of fry of other species.

Possession of spawners is the prerequisite to start artificial seed propagation. Sufficient numbers of spawners and a specific sex ratio are required. A database about the quality of spawners based on their external aspect (such as size and colour) and mortality should be established. Suitable environmental conditions for spawners such as breeding density, temperature, salinity, type of feed, and feeding method, should be understood in advance. The management of spawners during the spawning period and the collection method of fertilized eggs are other key factors. Spawning time needs to be recorded, and fertilized eggs should be handled with care to avoid damage caused by egg harvest. The management of fertilized eggs is the next step for artificial seed propagation. The correct water temperature is crucial to hatching and the later culture stages. There are many critical times for marine fish at the early developmental stage, such as the time from fertilized egg to hatching, the time when the hatched larvae start to take endogenous nutrition from yolk sac and oil globule, and the time when the feeding organ forms and larvae start to take exogenous nutrition (Tzeng and Ho 2004). There are several criteria for judging the quality of the fertilized eggs, such as the diameter of fertilized eggs, the colour of eggs, the ratio of floating eggs to sinking eggs, the hatching rate of fertilized eggs, and the survival time for hatched larvae without feeding (Liu 2010). When there is a sign indicating bad quality of fertilized eggs, farmers should make a decision whether or not to continue breeding.

The culture of feed organisms is also a key factor for artificial seed propagation. Recently, the seed propagation of marine fishes has been greatly improved. Koiso (2010) provided information on rotifers, including the types, morphology, cultivation requirements, preferred environment, nutritional value, and techniques for mass production of rotifers. Rotifers have contributed substantially to the propagation of sea fish seeds. For the past 50 years, rotifers have been a necessary feed organism. Improved rotifer cultivation techniques can foster the development as well as increase the survival rate and quality of sea fish seeds. High-quality rotifers play a crucial role in artificial mullet propagation.

Huang (2009) and Tsai (2010) respectively studied the morphological development of crimson snapper *Lutjanus erythropterus* and cobia *Rachycentron canadum* and examined salinity, temperature, density, and other environmental conditions, improvement of selection of feed, mass production technique, nutrition, and feed organism cultivation, and the feeding method related to schedule, frequency, and quantity of feeding to enhance the utilization percentage of feed and prevention of worsening water quality. The concepts and techniques found for these fish could be applied to the establishment of a suitable environment for mullet aquaculture. With a friendly environment and suitable culture conditions, the operational pressure on fish farms can be reduced. It can also lower the need to use drugs. Finally greater profit will result from the application of the concept of healthy fry in artificial propagation. With quantity no longer being the main concern for the production and culture of hatchery fry, quality will be the key to upgrade the industry.

Grow out

In Taiwan, flathead grey mullet are cultured mainly in earthen ponds, which should be periodically cleaned and managed (Fig. 19.8). There are two types of culture, monoculture with the release of 20,000–100,000 fry (2–3 cm) per hectare, and polyculture with the release of 1,200 juveniles (about 15 cm) per hectare, both mainly in brackish water.

From a strategic point of view, aquaculture risk can be reduced by implementing mixed culture with other economically important species, such as black tiger prawn *Penaeus monodon*, white shrimp *Litopenaeus vannamei*, clam *Meretrix lusoria*, and milkfish *Chanos chanos*. Flathead grey mullet is poor

Figure 19.8. A flathead grey mullet farm (photo by C.C. Tseng).

in forage ability, but capable of ingesting food residue from other species. This strategy shortens the time to harvest shrimps and enhances the economic performance of the industry. Using natural feed and efficient spacing facilitate maintaining the environment of fish farms. However, farmers must control the density of the fish population at different developmental stages, the percentages of each species, and the salinity for polyculture.

The site and water supply are important factors in culture. The next concern of fish farmers is the cleaning of fish ponds for culture. The bottom of the pond should be sterilized with calcium oxide or tea seed meal and exposed to sunshine in order to reduce the number of bacteria in the pond. The fish pond is left to grow algae and plankton as natural foods for the fry and the salinity of the water is adjusted.

After capture, the wild flathead grey mullet fry are transported and released to fish farms that are fully prepared with feed organisms for the first year of extensive culture. It is close to the winter solstice with low temperatures and strong monsoon. After the onset of winter, fry can feed on the natural food web, including algae from algal blooms, with no extra feed. During this culture period, it is most important not to disturb the fry to avoid stress and subsequent damage to the fish. After one year of culture, farmers are able to distinguish flathead grey mullet fry from fry of non-flathead grey mullet species by their external morphology. Juveniles are now strong enough to survive the size (50–60 g) selection process. After selection, the flathead grey mullet of different sizes are cultured in separate ponds equipped with auto feeding systems to supply artificial feed. The culture of flathead grey mullet juveniles in new fish ponds improves survival rates due to the change to a new culture environment and avoids deteriorating the bottom of the initial fish pond which might become unfavourable for aquaculture. After transfer to the ponds, the fry take 23–35 months to reach market size (Fig. 19.9).

To improve the quality of roe by reducing its oil content, feeding should be regulated or stopped before harvesting. Normally, the two-year old flathead grey mullet is ready for harvesting, but the size and quality of their ovaries is inferior to those of three-year old fish. Based on market demand and risk however, the two-year old mullet roe still forms part of the sale production. When fish farmers release a large quantity of two-year old roe to the market in one year, the quantity of three-year old mullet decreases substantially

Figure 19.9. Harvest of flathead grey mullets from an earthen pond (photo by C.C. Tseng).

in the following year, resulting in a shortage of high-quality mullet roe on the market. The price fluctuates considerably according to quality, demand, and quantity.

The production for the year must be adjusted to satisfy market demand for the following two to three years. However, most mullet farmers are self-employed. An absence of an agreement among farmers prevents a production and marketing strategy from being implemented. Consequently, the yearly output varies considerably, causing the farmers to be indecisive. In addition, as the quantity of captured wild mullet is unpredictable, it is important to develop good planning in order to maintain a sustainable roe production industry.

Marketing Strategy and Promotion of the Aquaculture Industry

Flathead grey mullet is one of the most popular edible fish in many countries, and mullet roe is considered a great delicacy by gourmands. Annual mullet roe production ranged from 17 to 90 tonnes from 2002 to 2012, with a production of 70 tonnes and a value of US$3,43 million in 2012 (Fig. 19.10, Fisheries Agency 2003–2013). Although wild flathead grey mullet can satisfy part of the market demand, cultured mullet play an important role in Taiwan's market. The limitations caused by the unstable supply of wild flathead grey mullet fry, the length of time and natural risk in culture, and the general belief that the quality of wild fish (both as adult for roe, and as fry for culture) is much better than that of the cultured one, are the main challenges faced by farmers. The government of Taiwan has integrated the resources of a central institution,

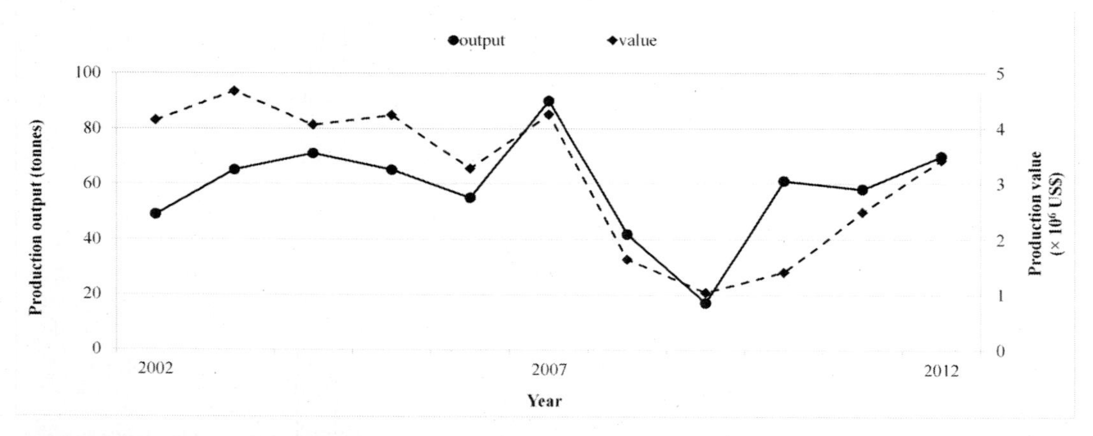

Figure 19.10. Annual production and value of flathead grey mullet roe from 2002 to 2012 in Taiwan (Fisheries Agency 2003–2013).

local associations, and individual farmers in order to break the bottleneck faced by this industry. Recently and gradually, there are promotion activities and marketing strategies not only for the integration of the industry, but also for fish farmers to bring forward the development of flathead grey mullet aquaculture industry. For example, the Fishery Agency of the Council of Agriculture, through the Taiwan Aquaculture Development Foundation (TADF), held a national contest of Taiwan TOP 10 Mullet Roe in 2009. This event is a promotional activity to encourage the aquaculture industry to improve mullet rearing technologies and mullet roe quality and to establish brand image and status. The contest provided a platform for farmers to compete and communicate with each other. The number of participants has been growing parallel to the quality of processed mullet roe and has had a positive effect on industrial promotion and marketing.

Local fishery associations create a consistent production system based on local product quantity and quality. For example, Kezailiao Harbour, managed by the fishery association of Ziguan, Kaohsiung, is a distribution centre for inshore fish farms and rich in fishery products. During the flathead grey mullet season, this harbour becomes an important distribution centre for captured grey mullet. The local fishery association has developed a number of products related to flathead grey mullet, such as processed foods, and also some health foods made from collagen extracted from flathead grey mullet. Hence this local association has developed and upgraded their production to a higher level.

Finally, individual farmers have made important contributions to the success of the promotion and marketing of the flathead grey mullet culture industry. They produce and sell their products themselves or through wholesalers to the market and families. This is still the mainstream of local business in villages before the product reaches a larger economic scale. From the quality point of view, internet marketing will be a major tool for individual farmers to promote their products. If the production of cultured flathead grey mullet can remain stable and quality can be improved, these individual farmers can help to create a reputation for future *Made-in-Taiwan* flathead grey mullet products.

Conclusions and Future Prospects of the Grey Mullet Industry

The natural resource of marine species, especially edible fishes, is decreasing hence it is vital to develop an aquaculture industry to meet the increasing market demand. It is the consumer's preference for certain species of fish that determines a farmer's choice of carnivorous or omnivorous fishes for their fish farms, but in order to grow 1 kg of fish in a fish farm, more than 1 kg of other edible species may be required as feed. An aquaculture industry that is eco-friendly has to be designed to have less effect on the environment, but the metabolic waste from intensive feeding may cause serious eutrophication.

Aquaculture must consider the environmental effects of the selection of cultured species, location of fish farms, rearing densities, feed and drugs. Marine fish farms are always located near the coast and the ponds and their inlets and outlets as well as oyster sheds are human built structures and add to the damage to coastal wetlands, particularly mangroves in the tropics. Aquaculture structures may interfere with the migration routes of spawners to the coastal area and therefore reduce the resource of flathead grey mullet fry. Using the concept of the food chain, modern fish farmers should try and reduce feed residues and recycle the metabolic waste in their operation. Aquatic plants are good for carbon fixation and can be used to slow down the deterioration of the environment, but government help is needed to encourage environmental research and development in aquaculture to reduce damage to the environment. It is a win-win situation to achieve the balance between economic benefit from aquaculture and environmental preservation. This approach can only profit the flathead grey mullet culture industry as well as all aquaculture industries. The long-term goal of the aquaculture industry must be to maximize the sustainability and economic efficiency of natural and artificial resources.

Mullets are seafood favoured in many countries, especially in areas along their migration route, where they have become a valuable source of fishery income. The following three conclusions can be drawn: (1) For centuries, the mullet has been a valuable target fish of the traditional fishery industry in Taiwan. The production of wild mullet is affected by fishing gear and methods as well as by climate change. Currently, the annual production and commercial value of wild mullet fluctuates widely. The industry relies on international cooperation for its development and sustainability. (2) Aquaculture can cope with the rising market demand, but a first step is to integrate artificial seed propagation in order to recover the loss of wild seeds before culture. Furthermore, improvements in brood fish rearing, methods of hormone treatments, methods for obtaining eggs, production of feeds, and culture techniques to supply more high-quality seeds for mullet grow out are required. (3) Sustainable mullet fry sources are currently the foundation of the mullet aquaculture industry, so it is essential to maintain the spawning of wild mullet, to preserve habitats and to keep migration routes unobstructed. A continuation of a balanced production between capture fisheries and aquaculture will benefit the mullet industry in Taiwan.

Acknowledgements

The authors wish to thank many colleagues who gave us tremendous help and advice.

References

COA. 2014. The Topic of Grey Mullet. Council of Agriculture, Executive Yuan. Taipei, Taiwan. From http://kmweb.coa.gov.tw/subject/mp.asp?mp=315. Electronic version accessed 27/12/2014.

Chao, N.H., H.P. Chen, L.C. Tseng, Y.M. Su and I C. Liao. 1973. Experiment on cryogenic preservation of grey mullet sperm I. Study on the biological characteristics and effect of protectants. J. Fish Soc. Taiwan 2: 31–41.

Chao, N.H., H.P. Chen and I C. Liao. 1975. Study on cryogenic preservation of grey mullet sperm. Aquaculture 5: 389–406.

Durand, J.D. 2015. Implications of molecular phylogeny for the taxonomy of Mugilidae. *In*: D. Crosetti and S.J.M. Blaber (eds.). Biology, Ecology and Culture of Grey Mullet (Mugilidae). CRC Press, Boca Raton, USA (this book).

Durand, J.D., K.N. Shen, W.J. Chen, B.W. Jamandre, H. Blel, K. Diop, M. Nirchio, F.J. Garcia de León, A.K. Whitfield, C.W. Chang and P. Borsa. 2012. Systematics of the grey mullets (Teleostei: Mugiliformes: Mugilidae): molecular phylogenetic evidence challenges two centuries of morphology-based taxonomy. Mol. Phylogenet. Evol. 64: 73–92.

FAO. 2014. The State of World Fisheries and Aquaculture. FAO, Rome.

Fisheries Agency. 2003–2013. Fisheries statistical yearbook Taiwan, Kinmen and Matsu area. Fisheries Agency, Council of Agriculture, Executive Yuan. Taipei, Taiwan.

Fisheries Agency. 2012. 2011 Statistical table for amount of stocking of aquaculture. Fisheries Agency, Council of Agriculture, Executive Yuan. Taipei, Taiwan.

Fisheries Agency. 2013. Annual report 2012. Fisheries Agency, Council of Agriculture, Executive Yuan. Kaohsiung, Taiwan.

Fisheries Agency. 2013. 2012 Statistical table for amount of stocking of aquaculture. Fisheries Agency, Council of Agriculture, Executive Yuan. Taipei, Taiwan.

Fisheries Agency. 2014. 2013 Statistical table for amount of stocking of aquaculture. Fisheries Agency, Council of Agriculture, Executive Yuan. Taipei, Taiwan.

Huang, C.C. 2009. The early development and seed production of artificial rearing *Lutjanus erythropterus*. Unpublished M.S. Thesis, National Penghu University, Penghu, Taiwan.

Hwang, S.W., H.P. Chen, D.L. Lee and I C. Liao. 1972. Preliminary results in the cryogenic preservation of grey mullet (*Mugil cephalus*) sperm. J. Fish Soc. Taiwan 1: 1–7.

Koiso, M. 2010. Rotifer Studies. From http://ncse.fra.affrc.go.jp/15kouza/index.html. Electronic version accessed 27/12/2014.

Liao, I C. 1975. Experiments on induced breeding of the grey mullet in Taiwan from 1963 to 1973. Aquaculture 6: 31–58.

Liao, I C. 1976. A brief review of the experiments done on the induced breeding of grey mullet in Taiwan. J. Agric. Assoc. China 96: 61–72.

Liao, I C. 1977. On completing a generation cycle of the grey mullet, *Mugil cephalus*, in captivity. J. Fish Soc. Taiwan 5: 1–10.

Liao, I C. 1981. Aquaculture of grey mullets cultivation methods. pp. 361–389. *In*: O.H. Oren (ed.). Aquaculture of Grey Mullets. Cambridge University Press, Great Britain.

Liao, I C., C.P. Hung, M.C. Lin, Y.W. Hou, T.L. Huang and I H. Tung. 1969. Artificial propagation of grey mullet, *Mugil cephalus* Linnaeus. JCRR Fish Ser. 8: 10–20.

Liao, I C., D.L. Lee, M.Y. Lim and M.C. Lo. 1971. Preliminary report on induced breeding of pond-reared mullet, *Mugil cephalus* Linnaeus. JCRR Fish Ser. 11: 30–35.

Liao, I C., C.S. Cheng, LC. Tseng, M.Y. Lim, L.S. Hsieh and H.P. Chen. 1972a. Preliminary report on the mass propagation of grey mullet, *Mugil cephalus* Linnaeus. JCRR Fish Ser. 12: 1–14.

Liao, I C., L.C. Tseng and C.S. Cheng. 1972b. Preliminary report on induced natural fertilization of grey mullet, *Mugil cephalus* Linnaeus. Aquaculture 2: 17–21.

Liao, I C., Y.J. Lu, T.L. Huang and M.C. Lin. 1973. Experiments on induced breeding of the grey mullet, *Mugil cephalus* Linnaeus. pp. 213–243. *In*: T.V.R. Pillay (ed.). Coastal Aquaculture in the Indo-Pacific Region. Fishing News (books), Ltd., Surrey, England.

Liu, S.H. 2010. Study on feeding ecology of three species of marine fish larvae. Unpublished M.S. Thesis, National Penghu University, Penghu, Taiwan.

Masuda, H. and Y. Kobayashi. 1994. Grand Atlas of Fish Life Modes. Tokai University Press, Tokyo.

Nakamura, T. 1997. Taiwan Historical Research of Dutch Ruled Period. Daw Shiang Inc., Taipei, Taiwan.

Okiyama, M. 1988. An Atlas of the Early Stage Fishes in Japan. Tokai University Press, Tokyo.

Ochiai, A. and M. Tanaka. 1998. Mugilidae. pp. 639–650. *In*: A. Ochiai and M. Tanaka (eds.). Ichthyology, 2nd edn. Kouseisha koseikaku, Tokyo.

Pien, P.C. and I C. Liao. 1975. Preliminary report of histological studies on the grey mullet gonad related to hormone treatment. Aquaculture 5: 31–39.

Shao, K.T. 2014. Taiwan Fish Database. WWW Web electronic publication. http://fishdb.sinica.edu.tw. Electronic version accessed 28/12/2014.

Shao, K.T. and C.Y. Chen. 2003. An Atlas of Fishes in Taiwan. Yuan-Liou Inc., Taipei, Taiwan.

Shen, K.N., B.W. Jamandre, C.C. Hsu, W.N. Tzeng and J.D. Durand. 2011. Plio-Pleistocene sea level and temperature fluctuations in the northwestern Pacific promoted speciation in the globally-distributed flathead mullet *Mugil cephalus*. BMC Evol. Biol. 11: 83.

Tsai, K.Y. 2010. Study on feeding ecology of cobia *Rachycentron canadum* larvae and juveniles. Unpublished M.S. Thesis, National Penghu University, Penghu, Taiwan.

Tzeng, J.J. and C.H. Ho. 2004. Transition from endogenous to exogenous nutritional sources in larvae malabar grouper, *Epinephelus malabaricus*. J. of Penghu Inst. of Technol. 8: 97–110.

Yu, N.H. 2012. Fishes of Aquaculture in Taiwan. Fish Breeding Association Taiwan. Kaohsiung, Taiwan.

Wu, C.C. and C.L. Wu. 2013. FRI e-paper no. 83. from http://www.tfrin.gov.tw/friweb/frienews/enews0083/p2.html. Electronic version accessed 28/12/14.

Wu, T.M. 2008. Village franchise system in Dutch Taiwan. Taiwan Historical Res. 15: 1–29.

CHAPTER 20

Culture of Mugilidae in Egypt

Sherif Sadek

Introduction

Egyptian aquaculture produced over 1,018,000 tonnes of finfish in 2012, about 74% of the country's total freshwater and marine fish production (1,371,975 tonnes), with a total market value of about US$1.354.65 million (1 US$ = 5.99 Egyptian pounds), providing a cheap source of protein for the country's 82.3 million people in 2012 (GAFRD 2013a). The remaining 26% (354,200 tonnes) comes from capture fisheries in the Nile, coastal lagoons and inland lakes. In the last 10 years (2003–2012) aquaculture increased 2.3 fold, from 444,800 t in 2003 to 1,018,000 t in 2012, whereas capture fisheries decreased 20%, from 431,200 t in 2003 to 354,200 t in 2012 (Fig. 20.1).

In the 1970s tilapias and mullets were the main species reared in extensive earthen ponds. Today more species are cultured: among finfish, Nile tilapia (*Oreochromis niloticus*), grey mullets (*Mugil cephalus, Liza ramada, Liza carinata, Liza saliens* and *Liza aurata*), carps (*Cyprinus carpio, Ctenopharyngodon idella, Hypophthalmichthys molitrix, Hypophthalmichthys nobilis, Mylopharyngodon piceus*), African catfish (*Clarias gariepinus*), bayad (*Bagrus bayad*), gilthead seabream (*Sparus aurata*), European seabass (*Dicentrarchus labrax*), meagre (*Argyrosomus regius*) and sole (*Solea vulgaris*); among crustaceans, freshwater prawn (*Macrobrachium rosenbergii*), green tiger shrimp (*Penaeus semisulcatus*), kuruma prawn (*P. japonicus*) and Indian white shrimp (*Fenneropenaeus indicus*). In 2012 tilapias shared 75.6% of the total aquaculture production, followed by mullets (12.8%), carps (6.6%) and other species (5.0%). The four different Egyptian aquaculture production system types are earthen ponds, cages, paddy fields and intensive tanks, representing 71.7, 24.5, 3.4 and 0.4% of the total production respectively (GAFRD 2013a, Fig. 20.2).

Grey mullets have been important in Egyptian fisheries since ancient time (Fig. 20.3). At present most mullet production comes from aquaculture. The Egyptian mullet aquaculture industry has grown dramatically over the past 10 years to around 130,000 t in 2012, with a mean annual production of 145,883 t over the decade 2003–2012. The rapid growth of mullet aquaculture in Egypt has been driven by a variety of factors, including pre-existing aquaculture practices, availability of brackish water, saline soil unsuitable for agriculture and wild fry sources near the coastal lagoons and the Suez Canal (Fig. 20.4). The mean Egyptian mullet capture fisheries production during the last decade (2003–2012) is estimated at around 30,000 t per year (73% from coastal lakes, 18% from the Mediterranean and the Red Seas, 5% from inland lakes, < 4% from Suez Canal/Bitter and Timsah Lakes and < 1% from the Nile, respectively) (Table 20.1).

Creator and General Manager, Aquaculture Consultant Office, 9 Road 256 11435 Maadi, Cairo, Egypt.
Email: Sadek_egypt35@hotmail.com; aco_egypt35@yahoo.com

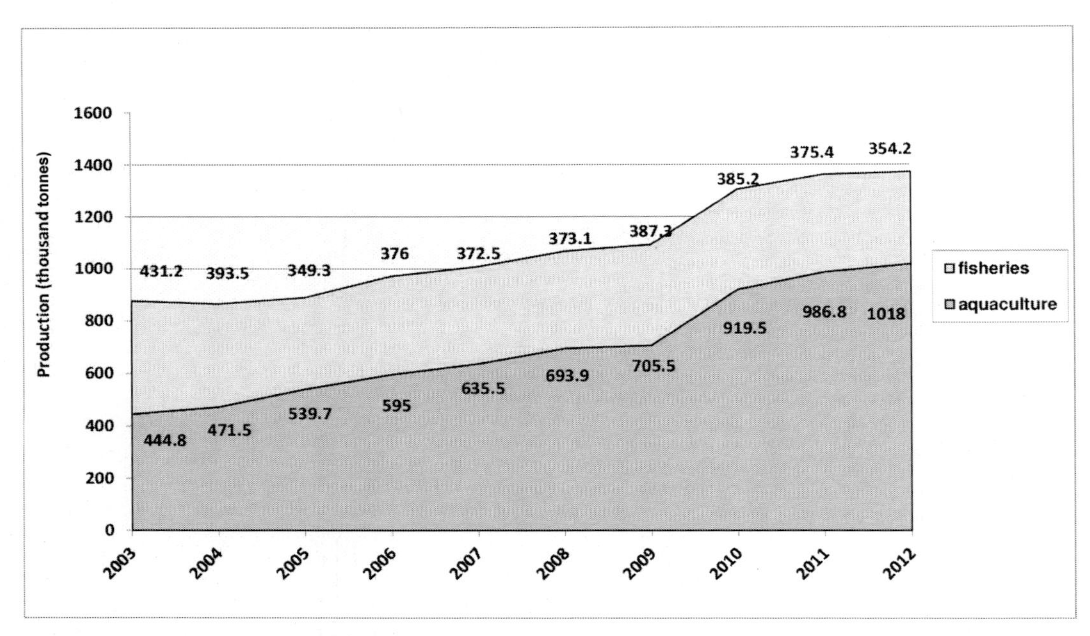

Figure 20.1. Egyptian capture fisheries and aquaculture production (x 1,000 tonnes) (2003–2012) (GAFRD 2013a).

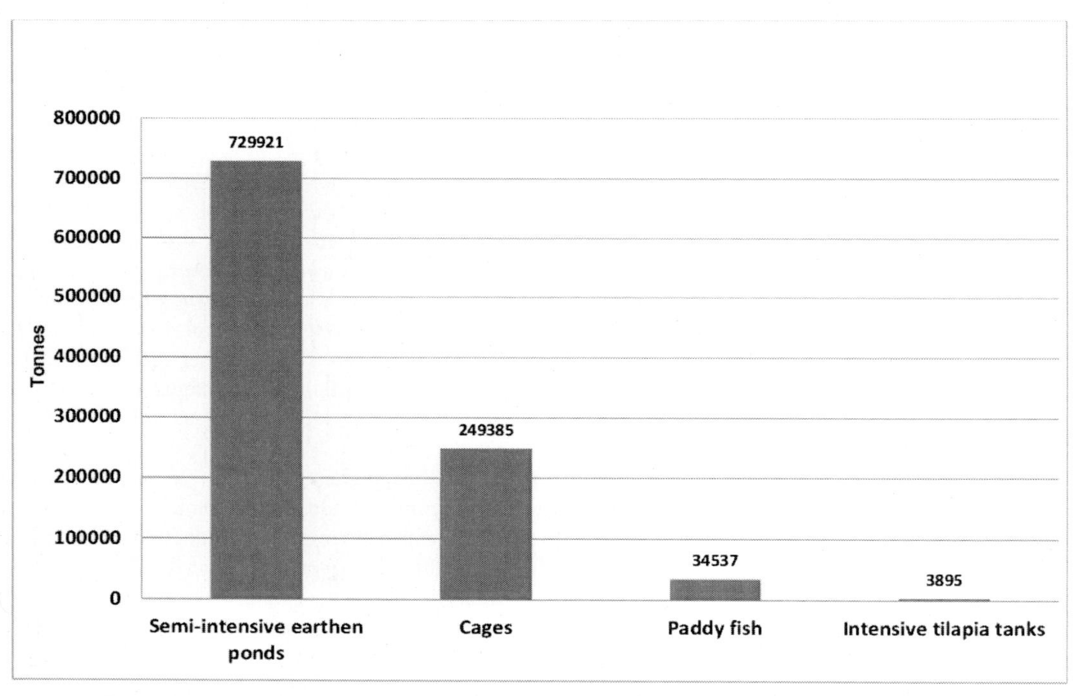

Figure 20.2. The Egyptian aquaculture production (tonnes) per system types in 2012 (GAFRD 2013a).

Today Egypt is the world leader in mullet culture. During the last decade (2003–2012), the mean annual production of farmed mullet was approximately 182,540 t. Different mullet Egyptian farming production systems are generally distinguished in the country according to their degree of intensification from extensive inland lake fisheries to semi-intensive in earthen ponds and intensive cages. In the last

a

b

Figure 20.3. Grey mullets in scenes of fishing from the Tomb of Kagemni, Saqqara, Egypt (2340 B.C.) (photos by S. Sadek).

decade mullet aquaculture production increased in 2007 to reach 252,507 t, then decreased to 129,651 t in 2012, from both earthen ponds (72.7%) and cage culture in the Nile (27.3%) (Fig. 20.5, GAFRD 2013a).

Presently, six different native species of grey mullets are cultured in Egypt; flathead grey mullet *Mugil cephalus*, thicklip grey mullet *Chelon labrosus*, golden grey mullet *Liza aurata*, black keeled mullet *Liza carinata*, thinlip mullet *Liza ramada* and leaping mullet *Liza saliens*.

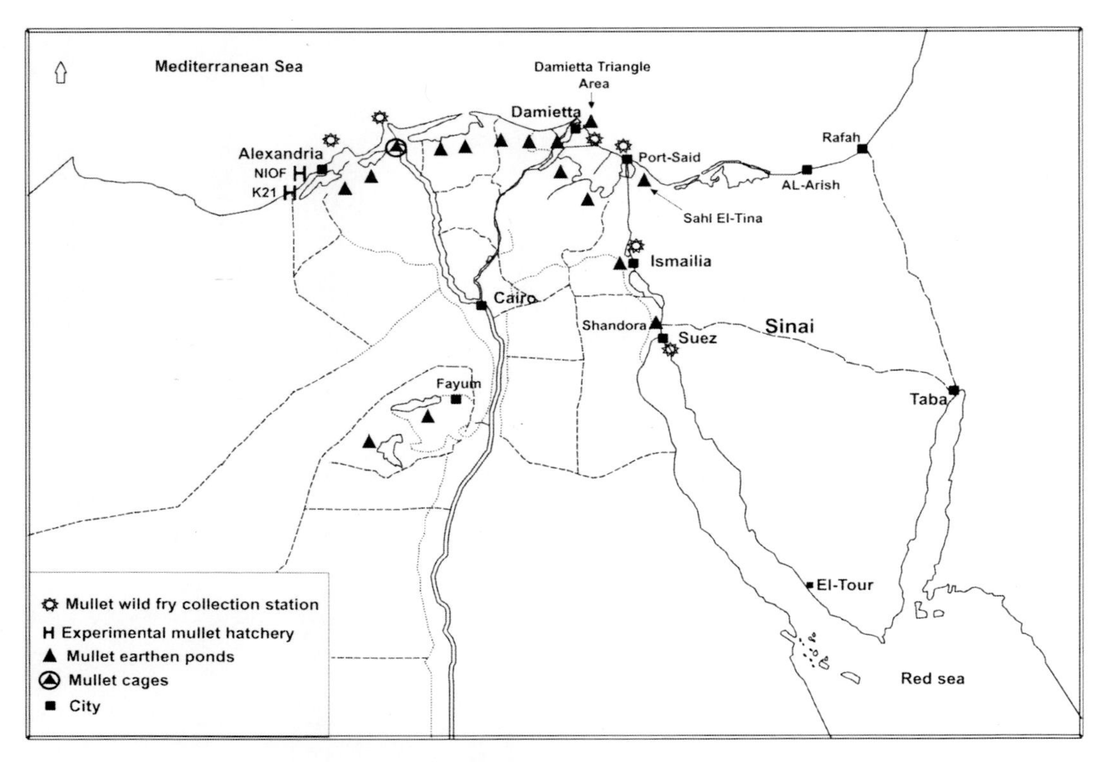

Figure 20.4. Mullet culture activities in Egypt.

Table 20.1. Egyptian capture fisheries production (tonnes) of grey mullet (2003–2012) (GAFRD 2013a).

	2003	2004	2005	2006	2007	2008	2009	2010	2011	2012
Mediterranean and Red Seas	7,898	6,857	5,324	4,256	5,668	5,819	5,332	5,052	4,534	3,855
Coastal lakes	26,676	26,548	20,961	29,026	19,601	16,215	13,649	23,507	16,100	26,549
Inland lakes	26,676	26,548	20,961	29,026	19,601	16,215	13,649	23,507	16,100	26,549
Suez Canal, Bitter and Timsah Lakes	573	429	997	846	1,251	1,255	1,890	1,972	2,116	2,087
Nile	1,868	1,702	2,184	1,613	989	1,075	991	846	538	547
Mediterranean and Red Seas	181	182	127	306	182	226	49	188	254	487
Total	**39,199**	**37,722**	**31,598**	**38,053**	**29,698**	**26,598**	**23,920**	**33,575**	**25,553**	**35,537**

Origin of Fry

Collection of Wild Fry

Egyptian mullet aquaculture has relied on the collection of wild seed since the practice began in the early 1920s to restock saline inland lakes (Wimpenny and Faouzi 1935, Faouzi 1936). Mullet seed reach the estuaries and shallow coastal waters as 12–20 mm fry. The shallow coastal lagoons, the Suez Canal, the Nile effluents and discharge canals leading to the Mediterranean have been the main source of the seasonal mullet and other euryhaline fish fry catches (El-Zarka and Kamel 1965).

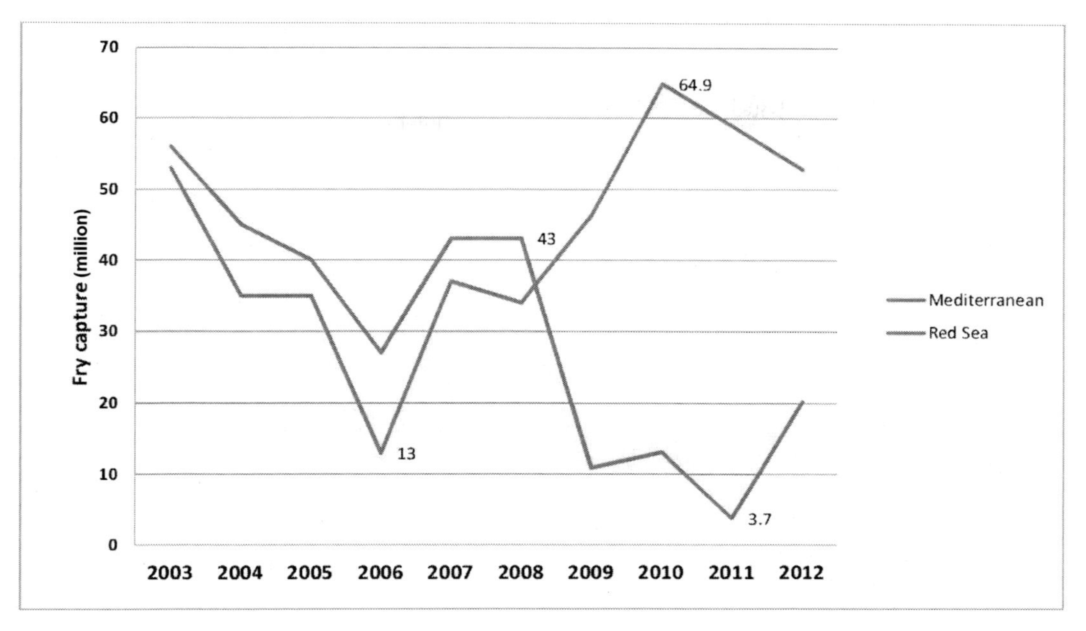

Figure 20.5. Egyptian grey mullet aquaculture production (tonnes) per sector (2003–2012) (GAFRD 2013a).

Each year, the GAFRD issues permits for fry collection at some dozen designated sites along the Mediterranean Coast, the coastal lagoons and the Gulf of Suez. Fry are collected from estuaries using seine nets and sold directly to licensed fish farms, or transported to state nurseries where they are grown to later be sold as fingerlings. The authority sets the prices and establishes a quota for purchases.

The collection of wild fry as seed for aquaculture has long been a source of contention among environmentalists and fishers. In the 1980s GAFRD established 10 fry collection stations. Today the number of these stations is only five: three are allocated on the Mediterranean Sea, one by the Suez Canal near Ismailai and one near Suez city in the Red Sea. The mean annual mullet fry collected during the last decade (2003–2012) was estimated to 73.2 million mullet fry from the Mediterranean (59%) and the Red Sea (41%) (Fig. 20.6, GAFRD 2013a).

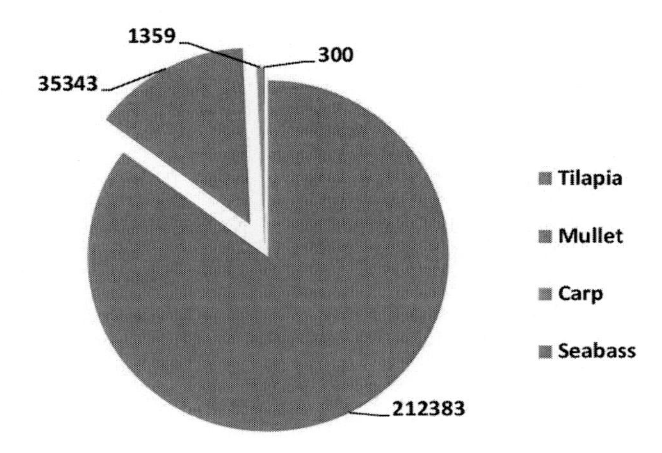

Figure 20.6. Wild mullet fry capture in Egypt (2003–2012) in million individuals (official data by GAFRD 2013a).

In 2012 around 73 million mullet fry (51.9% *L. ramada,* 27.7% *L. carinata* and 20.4% *M. cephalus*) were collected from the five wild fry collection centres through the GAFRD. *Liza carinata* is the most common species from the wild fry collection stations by the Suez Canal and near Suez city in the Red Sea, due to its tolerance to high salinity. In the three other collection stations on the Mediterranean Coast (El-Max near Alexandria, El-Gerbi near Damietta and E-Tafriaa near Port-Said) other mullet fry species (*M. cephalus* and *L. ramada*) are collected in higher percentages. Damietta fry collection station ranks first with 42.4 million, followed with Suez, Port-Said, Ismailai and western region (Alexandria), with 15.7, 7.3, 4.5 and 3.1 million respectively (Table 20.2).

Table 20.2. Wild mullet fry collected from the official wild fry stations during 2012 in millions (GAFRD 2013a).

Wild fry collection station	Number (in millions)
Damietta	42.4
Suez	15.7
Port-Said	7.3
Ismailia	4.5
Western region	3.1
Total	**73.0**

The low quota and perceived high price of government mullet seed has led many aquaculture farms to seek unlicensed suppliers. Moreover, restrictions that prohibit aquaculture on reclaimed agricultural land mean that nearly half of all fish farms, the non authorized ones, must entirely depend on illegal sources for their required seed.

Reliance on the collection of wild seed is a result of both economic considerations and technological barriers to hatchery production. Mullet farming still depends on naturally collected fry as the induced spawning of mullets is only carried out at an experimental level in Egypt. Wild seed will continue to form the basis of mullet culture "until captive breeding is developed to commercial levels". As there is really no scientific evidence that wild fry collection is harmful to the environment, and as wild mullet fry are abundant and easy to collect, there is really little to encourage artificial propagation in a hatchery (Sadek and Mires 2000).

The proiority should be to stamp out illegal fry collection and train fish farmers on how to reduce the high fry mortality during capture and transport, and later due to inadequate pond preparation (Saleh 2008, Sadek and Mires 2000). Mortalities of up to 96% of the transported fry may occur during the first seven days if the fry are directly transferred from seawater to the nursing ponds with a salinity < 2.6 ppt, though mortality is reduced to 6% through gradual acclimation (with a salinity < 2.6) (Saleh 1991).

Artificial Propagation

In 2011–2012, Egyptian commercial hatcheries produced around 411 million finfish fry and crustacean post-larvae, 97.3% of freshwater species, mainly tilapias and carps, 2.7% of marine species, mainly gilthead seabream, European seabass, sole and marine shrimp (GAFRD 2013a). Grey mullets fry are only produced at an experimental level.

Grey mullets in captivity do not spawn spontaneously and require hormone treatment to do so. The first trials were carried out on induced spawning in mullets in Egypt in the early 1990s by USAID which funded an experimental marine hatchery programme in the Km21 Marine Hatchery (K21MH) on the Alexandria-Marssa-Matrouh coastal road to produce about three million fry of flathead grey mullet annually (Abd El-Hakim et al. 1995). Despite the successful production of one million fry during the summer of 1994, production was not feasible on a commercial scale due to the high cost of hatchery mullet fry, 15 times the cost of wild-caught fry. Km21 Marine Hatchery fry production eventually shifted to other marine finfish species (such as gilthead seabream and European seabass) with lower production costs and higher market value. The same authors have described the economic impact assessment of the K21MH, and offered a projection for an economically feasible large mullet hatchery (three million juveniles/year) using an indoor building facility and a small size hatchery with a green house system with a yearly production of 300 thousand juveniles per year.

El-Gharabawy and Assem (2006) carried out induced spawning of *M. cephalus* in two steps. The first treatment consisted of a priming dose of carp pituitary extract with 20 to 70 mg, or 10,000 IU of Human Chorionic Gonadotropin (HCG) per fish, while the second dose given 24 hours later was the resolving dose of 200 $\mu g.kg^{-1}$ luteinizing and releasing hormone (LHRH-a). After the resolving the dose females were maintained together with males and spawning occurred after 17 to 24 hours on average. In all spawning trials running ripe males rated (+2) were used in 3:1 male to female sex ratio to optimize fertilization rates. The fecundity varied between 1.00 to 2.70 million eggs/spawn which represents 1,395.3 ± 190 $eggs.g^{-1}$ body weight.

The ripe unfertilized eggs of *M. cephalus* are rounded, colourless and transparent with one oil globule. The surface of the fertilized egg shell is smooth and the yolk is unsegmented. After 20 minutes from fertilization the egg membrane swells up and separates from the perivitelline space. The embryo begins to be evident after 13 ± 1 hours post fertilization. The egg diameter is 870 ± 30 μm and oil globule was 350 μm. After about 26 ± 4 hours post fertilization, hatching starts at 25 ± 1°C and water salinity of 34 ppt. Fertilization success varies between 75 to 80% and the hatching rate is about 90%. The newly hatched *M. cephalus* larvae are about 1.97 ± 0.23 mm long and begin feeding three–five days after hatching. The time of eye completion is nearly that of mouth opening and the body axis becomes straight, which helps the larvae to identify their food as internal-external feeding starts.

In 2007 Italy signed a cooperation agreement with the government of Egypt for a project entitled 'Marine Aquaculture Development in Egypt' (MADE). This was financed by the Italian Egyptian Debt for Development Swap Programme/Phase n.2 (2007/2012) aimed at developing and consolidating the aquaculture sector. In 2012 the MADE project carried out some experimental trials of induced spawning of the flathead grey mullet *Mugil cephalus* in captivity and larval rearing in K21MH. The trials were carried out for a period of 91 days from 15/10/2012 to 15/1/2013 to define the constraints and solutions for the coming new seasons. Induced spawning by hormone treatment was successful and occurred between November 2012 and January 2013. A total of 1.7 million eggs were produced from different hormonal treatments, with a fertilization rate ranging from 10 to 90%. The mean fertilized egg diameter was 0.8 ± 0.95 mm. Hatching took place after 36 hours at 23°C and the mean total length of the newly-hatched larvae was 2.0 ± 0.179 mm.

Mugil cephalus larval rearing was successful in indoor tanks using a flow-through system from days 1 to 55 after hatching. Transformation from larval to fry stage occurred between day 45 and day 52 after hatching. The maximum size of larvae and the minimum size of fry which appeared during the transitional period were 0.9 and 10.1 mm TL, respectively. By day 55, all the larvae had changed into fry with a mean TL of 29.3 ± 6.429 mm. The fry started to change into young adults with three anal spines by day 88 at a TL of 62 mm, with a survival rate of 5% (GAFRD 2013b).

Culture

Lake ranching in brackish water lakes

In Egypt mullet fry and fingerlings are stocked in inland lakes and reservoirs as a form of fisheries enhancement (culture-based fisheries). In early 1921, *Mugil cephalus* and *Liza ramada* wild fry were introduced to Lake Qarun in the province of El-Fayoum (Wimpenny and Faouzi 1935) and have formed a successful fishery (Faouzi 1936). Lake Qarun is a saline remnant of the historical freshwater Lake Moeris, with a salinity ranging from 10 ppt near the agriculture drainage canal mouth to 40 ppt in the north side of the lake. Wadi Al Raiyan Lakes are two man-made lakes in a depression connected to the agricultural drainage system of El Fayoum Province near Cairo, with a salinity of 3 to 5 ppt for the first and 16 to 20 ppt for the second lake, also received mullet fry to enhance the fisheries.

Extensive Culture in Earthen Ponds

Starting from the 1930s, a famous regime for an extensive aquaculture system called 'hosha system' has commonly been practiced on coastal lagoon shores. The farmer dug an earthen ponds and let the water from the lagoon flow in, with no control on species or size of the fish coming in with the water, and added agriculture byproducts as food (wheat and rice bran), or sometimes organic fertilizers for two–three months, then pumped the water out of the pond and made a complete harvest. In such extensive culture, natural food produced through pond fertilization is considered an important element for fish growth during early growth stages. In fattening stages, supplementary feeding with agriculture byproducts (wheat brain, rice brain, etc.) takes place. *Mugil cephalus* production ranges from 192 to 350 kg.ha^{-1} in fertilized ponds and 131 kg ha^{-1} without fertilization (El-Zarka and Fahmy 1968). In the last three decades the annual production per ha, in polyculture with Nile tilapia *Oreochromis niloticus*, various species of carps and various species of mullets and/or gilthead seabream, European sea bass and mullets, has fluctuated from 500 kg to 1 t.ha^{-1}. The extensive system is more popular, where farmers stock ponds at low densities, and fish derive most of their nutrition from the natural food web present in the culture ponds.

Semi-Intensive Culture in Earthen Ponds

Modern Egyptian aquaculture activities started in the late 1970s when the government established two big pilot projects in Kafer-El-Sheikh governorate to produce Nile tilapia, carps and mullets. The most common culture system is the semi-intensive culture in brackish water earthen ponds, in monoculture (tilapia or meagre) or polyculture (Nile tilapia associated to 10–20% mullets) (Sadek 2013), with a total production of around 730,000 t in 2012, corresponding to 72% of the total Egyptian aquaculture production. Governmental farms produced 9,509 t (73.3% tilapias, 10.2% carps, 9.4% mullets and 7.1% African catfish and European seabass) and private farms 720,412 t (73.3% tilapias, 13.0% mullets, 8.1% carps and 5.6% African catfish, gilthead seabream, European seabass and meagre, Table 20.3).

Most farms are located in the northern and eastern parts of the Nile Delta, with large ponds 2–8 ha wide. Stocking densities, energy input, level of management as well as size and type of infrastructure greatly vary among different farms. This type of aquaculture covers a land surface area of 159,191 hectares, with annual productions in polyculture in low salinity brackish waters (3–12 ppt) ranging from 1.6 t.ha^{-1} in governmental farms, with < 10% of mullets, to 4.7–8.2.ha^{-1} in private farms, with 7–15% mullets. A few private farms grow mullets in monoculture at higher salinities (10–45 ppt), with annual production s of 1–1.5 t.ha^{-1} with fry captured in the wild and wheat or rice bran as food (GAFRD 2013a, Sadek 2013).

The flathead grey mullet *Mugil cephalus* shows the highest growth rate among mullet species. Because of early collection of its fry in Egypt (September–December), it reaches market size with an average of 300–400 g in one single growing season. It is usually farmed in polyculture with tilapias. Thanks to its tolerance to a wide range of salinities, the species can be farmed in different environments from sea water (or above) to freshwater.

Table 20.3. Finfish aquaculture production from Egyptian governmental and private fish farms in 2012 (percentage) (GAFRD 2013a).

Species	Government %	Private %
Tilapias	73.3	73.3
Carps	10.2	8.1
Mullets	9.4	13.0
African catfish	4.8	0.4
European seabass	2.2	1.8
gilthead seabream	-	2.0
Meagre	-	1.2
others	0.1	0.2

Egyptian mullet farmers acclimatize wild mullet fry and stock in earthen nurseries using 2,000–4,500 m^2 earthen ponds at high densities (up to 125 m^{-2}), where they depend mainly on natural food. From 2.5 to 5.0 $t.ha^{-1}$ of animal manure are added to the soil before adding water; then chicken manure and chemical fertilizers (usually phosphate and nitrate) are added in suitable amounts on a weekly basis to keep Secchi disc readings of 20–30 cm. Rice or wheat bran is sometimes used as an additional source of food. Fry are kept in the nursery ponds for four–six months (from August–November till April) until they are about 10 g in weight. The fingerlings are then harvested, either by draining the nursery ponds into capture ponds or by netting. Over-wintered mullet fingerlings are sold for on-growing in various culture systems, but especially in semi-intensive ponds. When fry supply exceeds demand, they are retained and grown-on to market size in the nurseries.

Prior to stocking, fish ponds are prepared by drying, ploughing and manuring with 2.5–5.0 $t.ha^1$ of cow dung. Ponds are then filled to a depth of 25–30 cm and kept at that level for seven–10 days to build up the natural food web. The water level is then raised to 1.5–1.75 m and fingerlings are stocked. Productivity is kept at the required level by adding chicken manure and/or chemical fertilizers. Extruded feed is supplied to semi-intensive polyculture ponds to cover only the feeding requirements of both tilapias and carps grown in the same ponds; mullets profit from the remains of artificial feed for tilapias and carps.

Due to the optimum temperatures (20–26ºC), the growing season is normally about seven–eight months long. If mullet are monocultured, manuring may be sufficient to reach the required food level. In many cases, mullets feed directly on chicken manure and good levels of production have been recorded. Growth is checked by sampling, and if growth rates are not as expected, rice bran, wheat bran and or compressed pellet feed with a low crude protein level (18%) is added daily in amounts of 0.5–1% of biomass to supplement the pond natural food web. In the North Nile Delta the mullets are reared in polyculture near brackish-water at low salinity (3 to 7 ppt), they are usually stocked with tilapia, common carp *Cyprinus carpio* and silver carp *Hypophthalmichthys molitrix*. In this case, feeding and manuring programmes usually target these species, whereas mullets feed on the natural food web, detritus and feed leftovers. In the North NileDelta, near the North-western end of Manzala Lake, where the salinity is over 10 ppt, mullet are cultured with gilthead seabream and European seabass during a two year growing phase with an average production of 1.3 $t.ha^{-1}$. The mullets make up about 20 to 30% of the pond production (Sadek 2013).

Mullet Cage Culture in the Nile

In 2012 Nile cage numbers reached 37,371 with a total water volume of 18.4 million m^3 and an annual production of 249,000 t (85.2% tilapia, 14.2% mullet and 0.6% silver carp respectively), and an average production of 13.6 $kg.m^{-3}$. The production of mullet (mainly *L. ramada*) in cage aquaculture began in 2006 with a yearly production 19,180 t and reaching 35,343 t in 2012. In Egypt, water resources, both fresh and low salinity brackish-water are the major constraints to further development of aquaculture, with use for drinking water and land crop production having priority over aquaculture activities (Sadek et al. 2006). The Egyptian authorities removed all the Nile cages in the Nile freshwater located behind the two dams in the two Nile branches (Edfina and Faraskour) waters were polluted in varying degrees by inorganic nitrogen organic substances, phosphorus, and heavy metals and there was also conflict between fisheries and other activities. Today most of the Nile fish cages are located in two governorates, Kafr-El-Sheik and Beheira, near the Rashid branch at the end of the Nile mouth, with brackish-water (salinity ranging from 3 to over 20 ppt).

As a Mediterranean country, Egypt is characterized by cold winters and winter water temperatures can go down to 10ºC or less; too low to grow tilapias. For this reason most of the Nile cages near the mouth of Rashid are now producing mullets. The mullet Nile cage producers are using the same cages previously used for producing Nile tilapia and/or silver carp with two sizes of small nursery cages (5 x 5 x 4 m depth and 10 x 10 x 6 m depth). The producers usually stock 20 to 50 thousand mullet fry (0.1 g) in the nursery cages (5 x 5 m and net depth 3 m) for the first year and distribute them in the second year in larger cages (10 x 10 m x net depth 5 m) with a stocking density of 15–20 4 g-fingerlings $.m^{-3}$. They can produce 1.5 t of mullet/cage with an average weight of 4–5 $fish.kg^{-1}$. In the first year the farmers feed mullets with wheat bran and stale bread. During the last three months they are fed compressed sinking pellets

(25% CP) and remnants of raw dried pasta The mullet Nile cage farmers use 0.5 t of wheat bran, 1 t raw pasta and 1 t compressed fish pellets feed 25% CP to produce 1.5 t of mullet per cage (2.5–3 kg.m^{-3}).

The effect of stocking density on growth performance, feed utilization, production and economic feasibility was studied in an experimental trial where thinlip grey mullet *Liza ramada* fingerlings (81.1 g) were reared in floating net cages located in the brackish waters of the Rosetta Branch of the River Nile (Essa et al. 2012). These cages (700 m^3 water volume) were used to test three stocking densities (9.2, 11.6 and 13.9 kg.m^{-3}) for 293 days. All fish were fed a commercial diet with 22.5% crude protein and 4,357 kcal gross energy.kg^{-1} diets at a rate of 2% of total biomass.d^{-1} (six days.week^{-1}) twice a day. In the experimental cages water parameters averages temperature 24.2°C, salinity 15.9 mg.l^{-1} and total dissolved solids from 17.1 mg.l^{-1}. The highest growth was observed at the lowest density, 11.4 fish.m^{-3}, with individual total weight gain of 242.8 g, average daily gain of 0.83 g fish.d^{-1} and growth rate of 0.47%.d^{-1}, while the cages stocked at 17.1 fish.m^{-3} recorded the lowest values (150.9 g, 0.52 g fish.d^{-1} and 0.36%.d^{-1}, respectively). The best mean biomass at harvest were obtained at the highest stocking density group (4.0 kg.m^{-3}) and followed by lowest stocking density group (3.7 kg.m^{-3}cage) while medium stocking density group result was only 3.5 kg.m^{-3}. Meanwhile, the highest total weight gain (2.7 kg.m^{-3}) was found at appropriate low stocking density and followed by high stocking density (2.6 kg.m^{-3}) and medium stocking density (2.4 kg.m^{-3}).

Diet and Feeding

Egyptian aquaculture is one of the fastest growing animal production sectors in Egypt. At the turn of 2014 there were 50 fish feed mills in operation producing 600,000 t per annum, with half of this production as compressed sinking feed pellets and the rest extruded; by 2016 this figure will increase to over 60 licenced private mills with an estimated annual production of 770,000 t (Osman and Sadek 2015). Farm-made feeds are rarely used by Egyptian fish farmers, and mullet monoculture producers in earthen ponds and Nile cages use mainly tilapia compressed sinking pellet feed with low crude protein levels ranging from 18 to 25% (El-Sayed 2007).

The effects of different feeding regimes of the flathead grey mullet *Mugil cephalus* stocked at a rate of 1 fish.m^{-2} for 105 days, with different pond manuring treatments on water quality and phytoplankton fluctuations in earthen fishponds were studied by Abdel-Tawwab et al. (2005), who showed that flathead grey mullet could use both supplemental feed and/or natural food, and could completely consume artificial feed. Twenty six percent dietary crude protein level corresponds to maximum growth and feed utilization in 0.2 g flathead grey mullet fry (El-Dahhar 2000b). The performance of flathead grey mullet maintained at diets containing 175 to 265 kcal Metabolizable Energy (ME)/100 g was higher than 250 kcal ME/100 g compared to other treatments when dietary protein level was stable at 26% (El-Dahhar 2000a). When flathead grey mullet are fed at three dietary protein levels (18, 22 and 26%) and three dietary ME levels (200, 225 and 250 kcal.100 g^{-1} diet), weight gain, survival and feed efficiency improved as dietary energy increased up to 250 kcal.100 g^{-1} diet (El-Dahhar 2000a). Flathead grey mullets naturally do not consume food at random, but select preferred food from the natural food web, though they can efficiently use supplementary feed, as observed in an experimental trial on 35.1 ± 1 g flathead grey mullet cultured in ponds fertilized with different organic and inorganic fertilizers (El-Marakby et al. 2006). Significantly higher growth was observed in fishponds with feed supplementation compared to manuring alone.

Harvesting, Processing, Trade and Market

Egyptian fish farmers can partially and/or completely harvest their fish from their earthen ponds and cages. Partial harvesting is undertaken according to the market demand using gillnets of suitable mesh size, and also to catch only mullets in polyculture ponds with other marine fish (gilthead seabream and European seabass). For the complete harvest the mullet are collected in the catch ponds by a large purse seine and by scoop nets, transferred into plastic boxes, washed in running water, and then sorted according to species and sizes. Sorted fish are weighed and packed in plastic boxes (20 to 25 kg fish/box) with crushed ice. In Egypt, mullet is usually marketed daily whole and fresh, under three marketing grades for *M. cephalus* and *Liza ramada*: grade I: 2–4 fish.kg^{-1}, grade II: > 1 fish.kg^{-1} and Grade III: 5–8 fish.kg^{-1} are marketed ungutted for grilled cooking. The other species such as *L. carinata, L. saliens* and *L. aurata* are marketed under two grades: grade I: 12–16 fish .kg^{-1} and grade II: > 1 fish.kg^{-1}. In general, large size *M. cephalus* and *Liza ramada* (> 1 kg) are considered of lower quality and do not usually gain a good price compared to 2–4 fish.kg^{-1}. The smaller size of *L. carinata* (12–16 fish.kg^{-1}) has a higher market value than the larger *M. cephalus* and *Liza ramada* (1–4 fish.kg^{-1}) due to the high content of fat in *Liza carinata*, which is appreciated for grilling. Frozen mullet is considered of much lower value. Salted mullet either from the capture fishery or culture is consumed year round, with consumption reaching a traditional peak during Easter. Salted mullets under the name *feseekh* are sold in special shops that carry out the salting process and only sell salted fish. The egg roe of the flathead grey mullet is washed, salted, pressed, dried, and compressed according to a centuries-old tradition, to make a speciality food called *boutargue.*

At the retail level, farmed mullets fish are considered to be lower in quality by most consumers although they are usually unable to differentiate between farmed and wild caught fish of the same species. The market is controlled by a limited number of large wholesalers, based mainly in Cairo, Kafr-El-Sheik and Alexandria governorates, who determine the market price according to supply and demand. Farmers are free to sell their products either through wholesalers or directly to retailers. Farmers have agreements with wholesalers who purchase their harvest directly from the farm site. In the last 10 years (2002–2012), mullet prices fluctuated from US\$2.54 kg^{-1} in 2003 to US\$4.11 kg^{-1} in 2012 for grade 1 (3–4 fish.kg^{-1}) and from US\$1.98 to 2.57.kg^{-1} for grade 2 (5–8 fish.kg^{-1}) (Fig. 20.7).

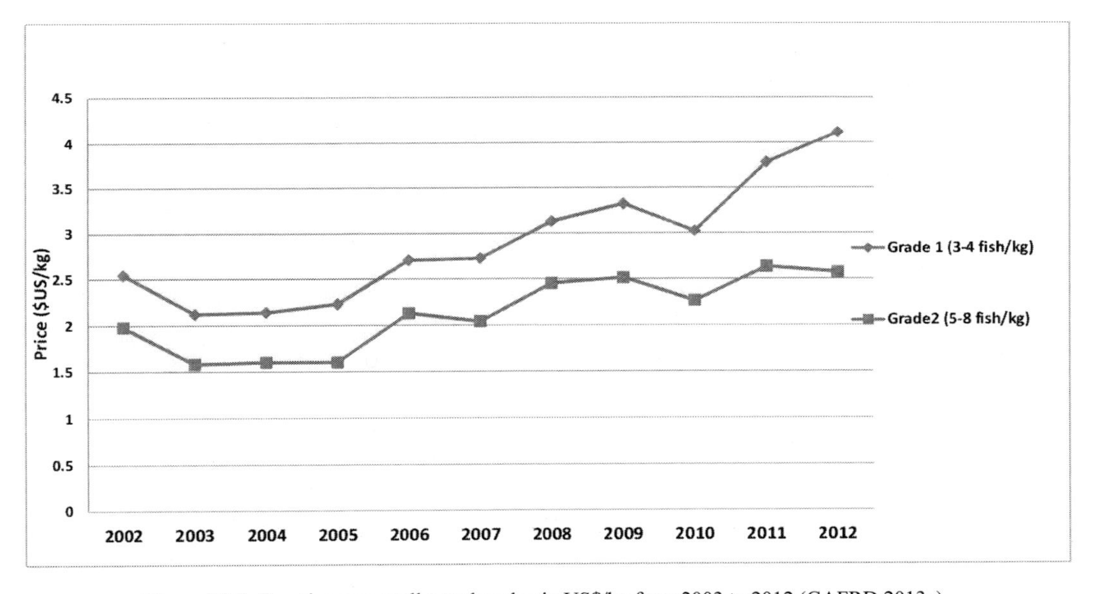

Figure 20.7. Egyptian grey mullet trade value in US\$/kg from 2003 to 2012 (GAFRD 2013a).

Production costs vary considerably, depending on the aquaculture culture system, geographical area and the level of technology applied (earthen ponds *vs* cages). Production costs also depend on the use of wild fry or nursed fingerlings, or on the culture system, whether in monoculture or polyculture. In Egypt in 2014, the free market price of just-captured wild fry was US\$110 per thousand fry for *M. cephalus*, US\$15 for *L. ramada* and US\$10 for *L. carinata*. The economic evaluation of mullet production in the Nile cages at different densities reported that all density treatments were economically profitable and the revenue could cover production costs; with the appropriate density (8,000 fish.cage^{-1}, 11.4 fish.m^{-3}) the best growth (0.83 g fish.d^{-1}), survival rate (99.0%), production (2,565 kg.cage^{-1}, 3.7 kg.m^{-3} over 293 days), average net income (US\$7.5 m^{-3} or US\$5,212 cage^{-1}) and an economic surplus of 61.0% of the total revenue obtained (Essa et al. 2012).

Bakeer (2006) has studied the growth performance, yields and economic benefits of *M. cephalus* reared in earthen ponds using fingerlings (10 ± 0.20 g) at a density of 1 to 3 fish.m^{-3}. He calculated that the density of 2 fish.m^{-3} brings the highest net return (34.6%) and the lowest production cost (US$1.3 kg^{-1}) (Table 20.4).

Table 20.4. Growth performance, yields and economic benefits of *M. cephalus* cultured in earthen ponds (Bakeer 2006).

Density (fish/m^3)	Final mean weigh (g)	Daily gain (g/day)	Net return (percent)	Production cost (US$/kg)
1	210	0.89	27.7	1.5
2	206	0.87	34.6	1.3
3	151	0.63	16.0	1.8

Conclusions and Future Trends

Mullets are economically important species for both commercial fisheries and aquaculture in Egypt. The demand for mullet in Egypt is high and the supply is moderate, so the market could accept a considerably larger production of mullets, especially from brackish-water (over 5 ppt) and high salinity ponds (over 30 ppt). Grey mullet culture could expand immediately, but is limited because it still depends mainly on wild fry, which affects the sustainability of the capture fisheries. It is therefore a priority for Egyptian mullet culture to transfer to commercial scale the mullet fry production techniques presently carried out at an experimental level.

Mullets could be the first candidates, rather than tilapias, for developing Egyptian aquaculture and for producing a reasonably priced product; in this sense aquaculture nutritionists are trying to reduce both mullet feed costs by enhancing the natural food web in the ponds through manuring and the requirements for manufactured feed by improved conversion efficiency, and to minimize environment impacts. Total nutrient input required to produce one tonne of herbivorous/omnivorous mullets is much lower than for the carnivorous cultured fish (Atlantic salmon, gilthead seabream, European seabass, meagre, etc.). It is necessary to formulate a specific mullet commercial feed using the known energy and protein demands of mullet. In conclusion, it is noteworthy that for intensifying mullet culture in ponds or cages that mullet are able to utilize both natural and/or supplemental food. Further research is however needed on the quality and quantity of supplementary feed required to successfully culture mullet in intensive systems.

References

Abd El-Hakim, N.F., O. Salama, A. El-Hindy, A. Abd El-Fattah and T. Younis. 1995. Economic impact assessment of the Maryut finfish hatchery. Project funded by USAID. Ministry of Agriculture and Land Reclamation, Agricultural Research Center (ARC), National Agricultural Research Project (NARP), New Initiatives (NIC). 175p.

Abdel-Tawwab, M., A. Eid, A. Abdelghany and H. El-Marakby. 2005. The assessment of water quality and primary productivity in earthen fishponds stocked with stripped mullet (*Mugil cephalus*) and subjected to different feeding regimes. Turkish Journal of Fisheries and Aquatic Sciences 5: 1–10.

Bakeer, M.N. 2006. Performance of grey mullet (*Mugil cephalus*) reared in monoculture in desert areas. Journal of Arabian Aquaculture Society 1: 44–56.

El-Dahhar, A.A. 2000a. Effect of dietary energy and protein levels on survival, growth and feed utilization of striped mullet *Mugil cephalus* larvae. J. Aric. Sci. Mansoura Univ. 25: 4987–5000.

El-Dahhar, A.A. 2000b. Protein and energy requirements of striped mullet *Mugil cephalus* larvae. J. Aric. Sci. Mansoura Univ. 25: 4923–4937.

El-Gharabawy, M.M. and S.S. Assem. 2006. Spawning induction in the Mediterranean grey mullet *Mugil cephalus* and larval development stages. African Journal of Biotechnology 5: 1836–1845.

El-Marakby, H.I., A.M. Eid, A.E. Abdelghany and M. Abdel-Tawwab. 2006. The impact of striped mullet, *Mugil cephalus* on natural food and phytoplankton selectivity at different feeding regimes in earthen fishponds. Journal of Fisheries and Aquatic Science 1: 87–96.

El-Sayed, A.-F.M. 2007. Analysis of feeds and fertilizers for sustainable aquaculture development in Egypt. pp. 401–422. *In*: M.R. Hasan, T. Hecht, S.S. De Silva and A.G.J. Tacon (eds.). Study and Analysis of Feeds and Fertilizers for Sustainable Aquaculture Development. FAO Fisheries Technical Paper n. 497. FAO, Rome.

El-Zarka, S. and F. Kamel. 1965. Mullet fry transplantation and its contribution to the fishery of the inland brackish lakes in Egypt, U.A.R. Proc. Con. Fish Counc. Medit. 8: 209–226.

El-Zarka, S. and F.K. Fahmy. 1968. Experiments in the culture of grey mullets *Mugil cephalus* in brackish water ponds in the U.A.R. FAO Fish Rep. 5: 255–266.

Essa, M.A., E.A. Omar, T.M. Srour, A.M. Helal and M.A. Elokaby. 2012. Effect of stocking density on growth performance, feed utilization, production and economic feasibility of thinlip grey mullet (*Liza ramada*) fingerlings, reared in floating net cages at the end of River Nile of Egypt. J. Applied Agricultural Research 17: 105–121.

Faouzi, H. 1936. Successful stocking of Lake Qarum with mullets (*Mugil cephalus* and *Mugil capita* Cuv. and Val.) from the Mediterranean. Int. Rev. Gesamten Hydrobiol. Hydrogr. 35: 434–439.

GAFRD. 2013a. Statistics of fish production of year 2012. GAFRD, Ministry of Agriculture and Land Reclamation. 94pp.

GAFRD. 2013b. Grey mullet (*Mugil cephalus*) reproduction and larvae rearing in K21 marine hatchery, Abou-Talaat, Alexandria, season 2012, Alexandria, Egypt. Government of the Arabic Republic of Egypt and Government of Italian Republic II Debt for Development Swap Agreement IEDS 2. 40pp.

Osman, M.F. and S. Sadek. 2015. Challenging the Egyptian aquaculture industry on sustainability. Aquaculture Europe, in press. 11pp.

Sadek, S. 2013. Aquaculture site selection and carrying capacity estimates for inland and coastal aquaculture in the Arab Republic of Egypt. pp. 183–196. *In*: L.G. Ross, T.C. Telfer, L. Falconer, D. Soto and J. Aguilar-Manjarrez (eds.). Site Selection and Carrying Capacities for Inland and Coastal Aquaculture. FAO/Institute of Aquaculture, University of Stirling, Expert Workshop, 6–8 December 2010. Stirling, the United Kingdom of Great Britain and Northern Ireland. FAO Fisheries and Aquaculture Proceedings n. 21. Rome, FAO.

Sadek, S. and M. Mires. 2000. Capture of wild finfish fry in Mediterranean coastal areas and possible impact on aquaculture development and marine genetic resources. The Israeli Journal of Aquaculture—Bamidgeh 52: 77–88.

Sadek, S., M.F. Osman and A. Mezayen. 2006. Aquaculture in Egypt: a fragile colossus. AQUA 2006. International Conference and Exhibition, Firenze (Florence), Italy, May 9–13, 2006.

Saleh, M.A. 1991. Rearing of Mugilidae. A study on low salinity tolerance by thinlip grey mullet (*Liza ramada* Risso 1826). Proc. of the FAO Workshop on Diversification of Aquaculture Production, Valetta, 1–6 July 1991.

Saleh, M. 2008. Capture-based aquaculture of mullets in Egypt. pp. 109–112. *In*: A. Lovatelli and P.F. Holthus (eds.). Capture-based Aquaculture. Global overview. FAO Fisheries Technical Paper. n. 508. FAO, Rome.

Wimpenny, R.S. and H. Faouzi. 1935. The breeding of the grey mullet, *L. ramada*, in Lake Quarun, Egypt. Nature Lond. 135: 1041.

CHAPTER 21

Grey Mullet as Possible Indicator of Coastal Environmental Changes: the MUGIL Project

Jacques Panfili,[1,a,*] *Catherine Aliaume,*[1,b]
Aikaterini Anastasopoulou,[2] *Patrick Berrebi,*[3] *Claude Casellas,*[4]
Chih-Wei Chang,[5] *Papa Samba Diouf,*[6]
Jean-Dominique Durand,[1,c] *Domingo Flores Hernandez,*[7,e]
Francisco J. García de León,[8] *Philippe Lalèyè,*[9]
Beatriz Morales-Nin,[10] *Julia Ramos Miranda,*[7,f]
Jaime Rendon von Osten,[7,g] *Kang-Ning Shen,*[11,h] *Javier Tomas,*[1,d]
Wann-Nian Tzeng,[12] *Vassiliki Vassilopoulou,*[13] *Chia-Hui Wang*[11,i]
and *Alan K. Whitfield*[14]

Introduction—General Overview of the MUGIL Project

The position of estuaries, deltas and lagoons in coastal areas at the interface of marine and riverine influences, results in highly variable environmental and ecological conditions that shift over both space and time. In addition, the effects of global change also have a tremendous impact on these ecosystems. The health and conservation of these environments is one of the biggest challenges facing humanity and, in order to achieve integrated management, scientists, ecologists and managers need to select relevant indicators which could be used as tracers for the state of coastal areas. These indicators are generally chosen from living species or physico-chemical parameters or a combination of both. Among the fish species living in estuaries, very few occupy these ecosystems in more than one oceanic region, but there is one particular species among the mullets, the flathead mullet *Mugil cephalus* which is found worldwide in almost all tropical, subtropical and warm temperate coastal zones. This fish is able to live in widely different habitats, but the mechanisms which drive the life cycle of *M. cephalus* are poorly known or have been studied separately in each area.

The aim of the MUGIL Project (Main Uses of the Grey mullet as an Indicator of Littoral environmental changes), financed by the European Commission between 2006 and 2009 (INCO-CT-2006-026180), was to

Authors' affiliations given at the end of the chapter.

build a collaborative network to coordinate research across the world using *Mugil cephalus* as an indicator of the state of coastal environments and to standardize methodologies for further studies. Since the flathead mullet is distributed worldwide, from tropical to temperate seas, and is of great commercial importance to fisheries in certain developing countries, it represented a good candidate for an indicator species. The project covered four global areas (Europe, Africa, Asia, and America) and involved collaborators from southern Europe (Spain, France, and Greece) and sub-tropical and tropical countries (Mexico, Senegal, Benin, South Africa, and Taiwan). The project was structured around two seminars and six workshops dealing with specific topics (life history traits, migration, genetics, biomarkers, databases). Each of these meetings involved mullet specialists and produced a report with a synthesis of the presentations, discussions and recommendations, available at http://www.mugil.univ-montp2.fr/.

Life History Trait Studies

Only a few studies have been carried out on the life history traits of *Mugil cephalus* on a global scale (Whitfield 1990). Studies on growth and/or reproduction around the world have rarely used the same methods or protocols, making international comparisons difficult. In addition, growth data on *M. cephalus* in the primary literature is scarce and even in studies focusing on age validation, growth rate data are not given (Smith and Deguara 2003). A few studies involving growth assessment of *M. cephalus* have highlighted the very variable growth rates and appear dependent on the environment (Ibáñez-Aguirre et al. 1999), but no global comparisons have been undertaken.

The reproductive behaviour of *M. cephalus* is seasonal, usually occurring when water temperature is appropriate for reproduction, but even this appears to be very variable from one area to another (Whitfield et al. 2012). More integrated studies on the reproductive cycle of *M. cephalus* throughout the world could highlight the relationship between reproductive season, environmental physico-chemical parameters and phylogenetic structure. Similarly, studies need to be undertaken on the possible effects of the environment on reproductive traits such as fecundity, size at first maturation, oocyte size, etc.

As no standardized methods for age and reproduction of *M. cephalus* are available in the literature, the MUGIL Project focused on their standardization. For age estimations, it recommended that otoliths and not scales be used. Comparisons were undertaken on otolith samples collected from different parts of the world (Mauritania, Senegal, Spain, Benin, South Africa, Greece and Taiwan). The shapes of the otoliths from the different areas were generally the same and measurements along the otolith axis were similar, thus reinforcing the idea that the species is cosmopolitan. For age estimation, the MUGIL consortium recommended the use of growth rings on transverse otolith sections of 300 μm thickness, preferably observed under transmitted light in order to reveal thin bands (Panfili et al. 2007). Nevertheless, a validation stage (i.e., verification of the timing of band deposition) should be obligatory in each study. The MUGIL consortium also defined a common maturation scale for gonad examination, which could be used when working on reproduction (Panfili et al. 2007).

Migration Studies

Mugil cephalus is diadromous and often migrates between continental and marine environments during its life cycle. Juveniles and sub-adults grow in freshwater and/or estuarine habitats, but adults undertake off-shore migrations for spawning, usually in the form of large schools (Bacheler et al. 2005). Among the tools available to study fish migration, otolith microchemistry has proved successful in resolving habitat occupation between freshwater and seawater environments by diadromous fish (e.g., Jessop et al. 2002, Tomás et al. 2005, Arai 2007). Otolith microchemistry has also been successfully applied to the identification of saline habitats occupied by flathead mullet, but these studies were scarce and confined to Taiwan (Chang et al. 2004a, Chang et al. 2004b). In the absence of detailed studies on saline habitat use by *M. cephalus* in other parts of the world, otolith microchemistry constitutes a very efficient approach to determine the migration habits of this species.

The MUGIL Project defined the best methods for otolith collection, storage and preparation for microchemical analysis to study the transhaline migrations of *M. cephalus*. The guidelines were presented in a specific deliverable (Tomás et al. 2008), which enables all future research on the subject to follow the same standard protocols for the use of otolith microchemistry, thereby facilitating global comparisons. The MUGIL consortium recommended chemical analysis of transverse sections of otoliths using Laser Ablation—Inductively Coupled Plasma Mass Spectrometry along the axis of the otolith with visible and interpretable growth rings. It is acknowledged that strontium and especially the strontium:calcium ratios are the best candidate elements to track transhaline fish migrations. The possibility exists however that other elements may bring additional information to identify dispersal within freshwater environments by using strontium isotopes, or to identify fidelity to certain coastal areas using Mn, Ni, Zn and Ba. The MUGIL consortium strongly recommended that validation experiments be conducted under controlled rearing conditions before applying these microchemistry approaches.

Genetics Studies

The taxonomic status of *Mugil cephalus* remains uncertain and no definitive conclusions have been reached regarding the possible existence of one circumtropical species or, alternatively, a complex of more regional cryptic species (Briggs 1960, Crosetti et al. 1994). As stated earlier in this book (see Chapter 1—González-Castro and Ghasemzadeh 2015 and Chapter 2—Durand 2015), one of the reasons for this may be the lack of basic information on the phylogeny of Mugilidae. Understanding the origin and evolution of grey mullets requires general knowledge of the evolutionary background within the family. Furthermore, studies that investigate the evolutionary history of *M. cephalus* have usually failed to reach conclusions about species boundaries due to loose sampling across the geographical range. During the MUGIL Project, specific genetic methodologies were discussed from four different perspectives: phylogeny, phylogeography, population genetics and adaptation. The compilation of specific guidelines for genetic studies allows scientists to follow the same protocol and therefore facilitate comparison of results (Durand et al. 2008).

Allozymes appear inappropriate for phylogenetic studies that involve numerous species from different parts of the world, due to their low polymorphism and sample handling requirements (i.e., the need for live tissues preserved at –20ºC). In contrast, DNA sequencing should be the most appropriate method and numerous genes are available to investigate the molecular phylogeny of Mugilidae. 16S RNA with 12S RNA have a low mutation rate that should be useful for the inference of deep phylogenetic relationships (Rossi et al. 2004, Heras et al. 2006, Papasotiropoulos et al. 2007) whereas cytochrome *b* and cytochrome oxydase I, with higher mutation rates, are more appropriate to estimate phylogenetic relationships of closely related and more divergent species (Heras et al. 2006, Blel et al. 2010).

Among different approaches, EPIC and semi-multiplex PCR analysis appear the relevant methods for species identification of numerous samples within a specific area. These methods are less expensive than PCR-RFLP and more reliable than D-loop sequences of mitochondrial DNA. If species identification has to be routinely performed for a large series of samples however, it would be more appropriate to set up a semi-multiplex PCR analysis.

The phylogeographic approach is particularly suited for studying the origin, dispersal and historical demography of *M. cephalus*. Mitochondrial DNA D-loop sequences are likely to be ideal for regional phylogeographic analyses and could be complemented by the analysis of slower evolving genes to resolve deeper phylogenetic divergences. D-loop sequence analyses may pose some multiple alignment challenges at large geographic scales where large sequence divergences have been documented. Lastly, for studying adaptation and selection, different strategies could be possible in order to find polymorphism under different selective pressures (e.g., genome scan approach, AFLP or Rad-Seq, link among polymorphisms of candidate genes and individual fitness, etc.).

Biomarker Studies

The MUGIL Project provided an opportunity to identify interest in developing biomarker studies on *Mugil cephalus* from different countries (Casellas et al. 2009). The objectives were to define a standard protocol for the use of biomarkers. *M. cephalus* possesses several characteristics required in a sentinel or indicator species, such as wide salinity and temperature tolerances, which enable them to occupy most coastal waters.

The use of biomarkers could offer an integrated evaluation of the effects of pollutants on the aquatic biota and also provide an early warning of potential changes at the ecosystem level.

The MUGIL consortium recommended that field sampling samples for biomarker analyses should include at least 15 males and 15 females per site. Different equipment would be required for processing the fish and storing the various sample tissues for later laboratory analysis, depending on the biomarkers to be used (Table 21.1). Experimental procedures for biomarkers studies in *M. cephalus* have been described by Casellas et al. (2009). Biomarkers tested since the turn of the century were listed (Corsi et al. 2003, Ferreira et al. 2004, Barucca et al. 2006, Ferreira et al. 2006, Neves et al. 2007) and evaluated

Table 21.1. Equipment required for processing the fish and tissues and storing the various samples for later laboratory in the case of biomarker analysis for mullets. AchE (acetyl cholinesterase).

Sample collection	Type of tube	Minimum number of tubes	Biomarker use
Blood	Heparin/Li tubes for blood sampling	1 heparin tube + 2 minitubes	Vitellogenin
Bile	Eppendorf 1.5 ml	2	*In vitro*
Liver	Microtube w/beats	2	EROD + biochemical dosages
Brain	Eppendorf 1.5 ml	2	AchE
Scales	Envelope	1	Age/DNA
Otoliths	Eppendorf 0.5 ml	1	
Muscle	Microtube w/beats	1	AchE

(Casellas et al. 2009). It was also emphasized that future biomarker analyses in different parts of the world should only proceed once an inter-calibration exercise had been carried out to reference the results.

Mugil cephalus Databases

An important activity during the MUGIL Project comprised exchanges between the participants and the FishBase consortium (www.fishbase.org) in order to (a) update the information on the *Mugil cephalus* webpage on FishBase, (b) build a bibliographic reference and *pdf* database on this species and (c) collect original data from different parts of the world to facilitate the building of an individual database (Aliaume et al. 2007, Aliaume et al. 2008). The information available online in FishBase for *M. cephalus* was reviewed in 2008 and some webpages were edited and updated.

A database with bibliographic references on *M. cephalus* (both peer-reviewed articles and grey literature), together with *pdf* copies of these articles, was built and is available to the MUGIL consortium through its intranet website (www.mugil.univ-montp2.fr). The list included 288 references, classified by world areas and thematic fields of research and applications.

The structure of a database that includes specific individual biological and ecological data was developed during the MUGIL Project. This database was built using Microsoft Access software with several tables (Fig. 21.1). The MUGIL consortium compiled individual data from each geographical area and more than 2800 individual *M. cephalus* data sets were gathered, mainly providing information on length, weight and location. Recorded salinities ranged from 0 to 89, and water temperatures from 19 to 34ºC. A large size range was represented in the dataset, with a peak in composition between 200 and 300 mm fork length. Males predominated in the database (more than 60%), followed by females (20%) and immature fish (15%). The database has not been updated since 2009.

Conclusions

The MUGIL Project reinforced the idea that the flathead mullet could be used as a bioindicator of coastal environmental states, and the project also gave guidance for different methods and protocols that could be used when studying life history traits, genetics, migrations and specific biomarkers for *M. cephalus*.

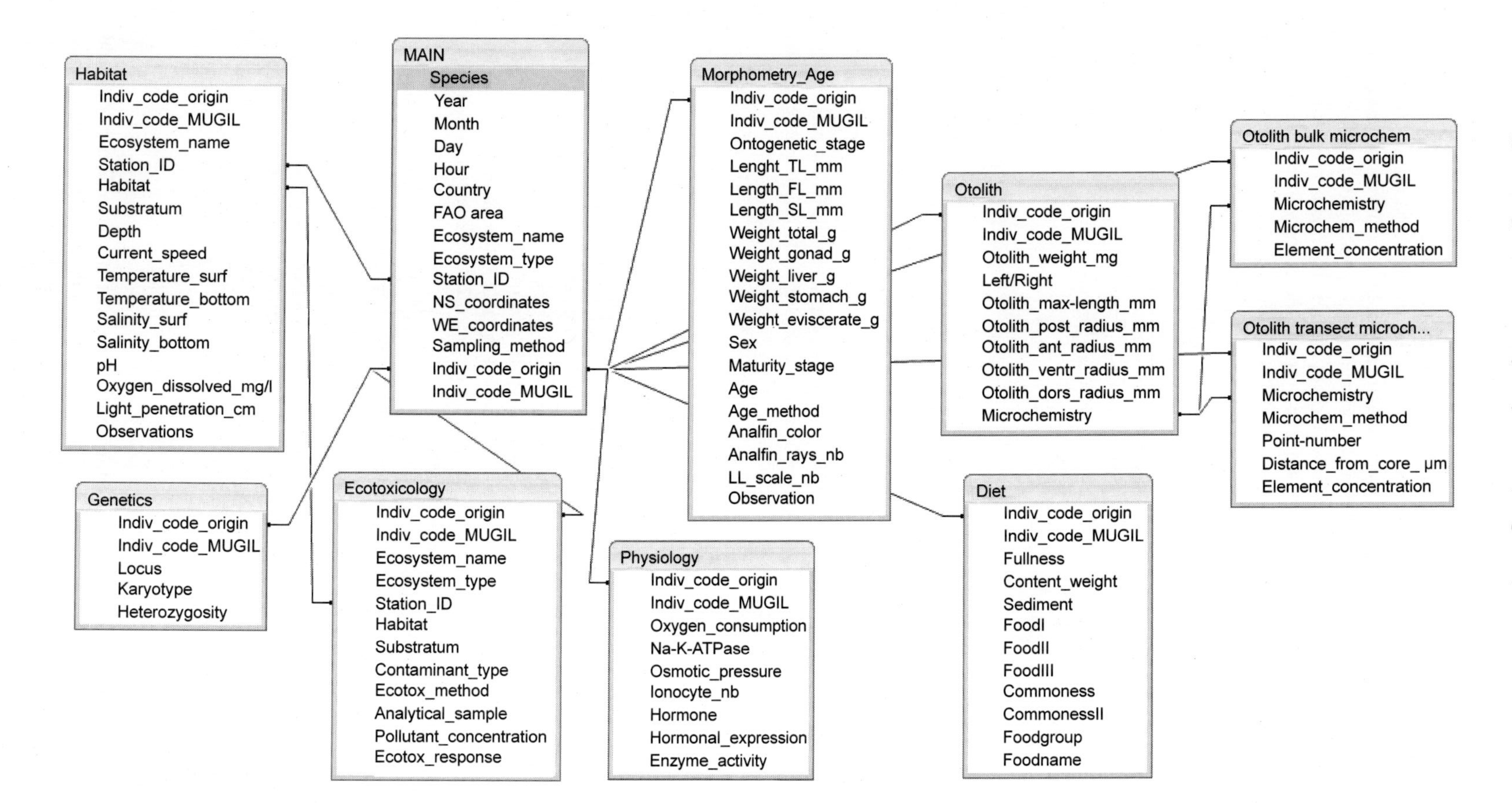

Figure 21.1. Diagrammatic representation of the database for individual *Mugil cephalus* developed during the MUGIL Project. The table 'Main' includes general information on each sample and other tables are specific to other thematic fields (Habitat, Genetics, Ecotoxicology, Physiology, Morphology_Age, Diet, and three tables are dedicated to otolith information). The Individual_MUGIL code is composed of 10 digits: two letters for the country of capture (according to the internet country code), two figures for the year (e.g., 05 for 2005), two figures for the month, and four figures for the fish specimen.

Since the end of the project, new technological advances and case studies in genetics have demonstrated that *M. cephalus* is a cryptic species complex (Chapter 2—Durand 2015, Shen et al. 2011, Krueck et al. 2013), and migrations have been studied in Mexican waters (Ibáñez et al. 2012).

The project produced several outcomes and among them peer-reviewed publications which provide additional detailed information on the flathead mullet (Hsu et al. 2007, Hsu et al. 2009, Hsu and Tzeng 2009, Jamandre et al. 2009, Trape et al. 2009a, Trape et al. 2009b, Blel et al. 2010, Jamandre et al. 2010, Shen et al. 2010, Wang et al. 2010, Shen et al. 2011, Wang et al. 2011, Chang and Lizuka 2012, Durand et al. 2012a,b, Ibáñez et al. 2012, Whitfield et al. 2012, Durand et al. 2013).

Acknowledgements

The MUGIL Project was financially supported by the European Commission (INCO-CT-2006-026180-MUGIL) from November 2006 to March 2009.

References

Aliaume, C., J. Panfili, C.W. Chang, J.-D. Durand, D. Flores Hernández, P. Lalèyè, B. Morales-Nin, W.N. Tzeng and A.K. Whitfield. 2007. MUGIL database. INCO-CT-2006-026180—MUGIL, Deliverable 2, European Commission: 27p. + Annex, available at http://www.mugil.univ-montp22.fr/.

Aliaume, C., J.-D. Durand, D. Flores Hernández, P. Lalèyè, J. Ramos Miranda, K.N. Shen, W.N. Tzeng, V. Vassilopoulou, A.K. Whitfield and J. Panfili. 2008. MUGIL database [2]. INCO-CT-2006-026180—MUGIL, Deliverable 6, European Commission: 28p + Annex, available at http://www.mugil.univ-montp22.fr/.

Arai, T. 2007. Studies on life history and migration in fish by otolith analyses. Nipp. Suis. Gakk. 73: 652–655.

Bacheler, N.M., R.A. Wong and J.A. Buckel. 2005. Movements and mortality rates of striped mullet in north Carolina. North Am. J. Fish Manage. 25: 361–373.

Barucca, M., A. Canapa, E. Olmo and F. Regoli. 2006. Analysis of vitellogenin gene induction as a valuable biomarker of estrogenic exposure in various Mediterranean fish species. Environ. Res. 101: 68–73.

Blel, H., J. Panfili, B. Guinand, P. Berrebi, K. Said and J.-D. Durand. 2010. Selection footprint at the first intron of the Prl gene in natural populations of the flathead mullet (*Mugil cephalus*, L. 1758). J. Exp. Mar. Biol. Ecol. 387: 60–67.

Briggs, J.C. 1960. Fishes of worldwide (circumtropical) distribution. Copeia 1960: 171–180.

Casellas, C., C. Aliaume, A. Anastasopoulou, J.-D. Durand, F. García de León, T. Katsiki, J. Redon Von Osten, K.N. Shen, C. Tsangaris, W.N. Tzeng, V. Vassilopoulou, A.K. Whitfield and J. Panfili. 2009. MUGIL biomarkers. INCO-CT-2006-026180—MUGIL, Deliverable 7, European Commission: 7p + Annex, available at http://www.mugil.univ-montp2.fr/.

Chang, C.W. and Y. Lizuka. 2012. Estuarine use and movement patterns of seven sympatric Mugilidae fishes in the Tatu Creek estuary, central western Taiwan. Est. Coast. Shelf Sci. 106: 121–126.

Chang, C.W., Y. Iizuka and W.N. Tzeng. 2004a. Migratory environmental history of the grey mullet *Mugil cephalus* as revealed by otolith Sr:Ca ratios. Mar. Ecol. Prog. Ser. 269: 277–288.

Chang, C.W., S.H. Lin, Y. Iizuka and W.N. Tzeng. 2004b. Relationship between Sr:Ca ratios in otoliths of grey mullet *Mugil cephalus* and ambient salinity: validation, mechanisms, and applications. Zool. Stud. 43: 74–85.

Corsi, I., M. Mariottini, C. Sensini, L. Lancini and S. Focardi. 2003. Cytochrome P450, acetylcholinesterase and gonadal histology for evaluating contaminant exposure levels in fishes from a highly eutrophic brackish ecosystem: the Orbetello Lagoon, Italy. Mar. Pol. Bull. 46: 203–212.

Crosetti, D., W.S. Nelson and J. Avise. 1994. Pronounced genetic structure of mitochondrial DNA among populations of the circumglobally distributed grey mullet (*Mugil cephalus* Linnaeus). J. Fish Biol. 44: 47–58.

Durand, J.D. 2015. Implications of molecular phylogeny for the taxonomy of Mugilidae. *In*: D. Crosetti and S. Blaber (eds.). Biology, Ecology and Culture of Grey Mullet (Mugilidae). CRC Press, Boca Raton, USA (this book).

Durand, J.-D., P. Berrebi, B.W. Jamandre, J. Panfili, A. Rocha-Olivares, K.N. Shen, C.S. Tsigenopoulos, W.N. Tzeng and F. García de León. 2008. Mugil genetics studies. INCO-CT-2006-026180—MUGIL, Deliverable 5, European Commission: 11p. + Annex, available at http://www.mugil.univ-montp12.fr/.

Durand, J.-D., W.-J. Chen, K.-N. Shen, C. Fu and P. Borsa. 2012a. Genus-level taxonomic changes implied by the mitochondrial phylogeny of grey mullets (Teleostei: Mugilidae). Compt. R. Biol. 335: 687–697.

Durand, J.D., K.N. Shen, W.J. Chen, B.W. Jamandre, H. Blel, K. Diop, M. Nirchio, F.J. Garcia de Leon, A.K. Whitfield, C.W. Chang and P. Borsa. 2012b. Systematics of the grey mullets (Teleostei: Mugiliformes: Mugilidae): molecular phylogenetic evidence challenges two centuries of morphology-based taxonomy. Mol. Phyl. Evol. 64: 73–92.

Durand, J.D., H. Blel, K.N. Shen, E.T. Koutrakis and B. Guinand. 2013. Population genetic structure of *Mugil cephalus* in the Mediterranean and Black Seas: a single mitochondrial clade and many nuclear barriers. Mar. Ecol. Prog. Ser. 474: 243–261.

Ferreira, M., P. Antunes, O. Gil, C. Vale and M.A. Reis Henriques. 2004. Organochlorine contaminants in flounder (*Platichthys flesus*) and mullet (*Mugil cephalus*) from Douro estuary, and their use as sentinel species for environmental monitoring. Aquat. Toxicol. 69: 347–357.

Ferreira, M., P. Moradas Ferreira and M.A. Reis Henriques. 2006. The effect of long-term depuration on phase I and phase II biotransformation in mullets (*Mugil cephalus*) chronically exposed to pollutants in River Douro estuary, Portugal. Mar. Environ. Res. 61: 326–338.

González-Castro, M. and J. Ghasemzadeh. 2015. Morphology and morphometry based taxonomy of Mugilidae. *In*: D. Crosetti and S.J.M. Blaber (eds.). Biology, Ecology and Culture of Grey Mullet (Mugilidae). CRC Press, Boca Raton, USA (this book).

Heras, S., M.G. Castro and M.I. Roldan. 2006. *Mugil curema* in Argentinean waters: combined morphological and molecular approach. Aquaculture 261: 473–478.

Hsu, C.-C. and W.-N. Tzeng. 2009. Validation of annular deposition in scales and otoliths of flathead mullet *Mugil cephalus*. Zool. Stud. 48: 640–648.

Hsu, C.-C., Y.-S. Han and W.-N. Tzeng. 2007. Evidence of flathead mullet *Mugil cephalus* L. spawning in waters northeast of Taiwan. Zool. Stud. 46: 717–725.

Hsu, C.-C., C.-W. Chang, Y. Iizuka and W.-N. Tzeng. 2009. A growth check deposited at estuarine arrival in otoliths of juvenile flathead mullet (*Mugil cephalus* L.). Zool. Stud. 48: 315–324.

Ibáñez-Aguirre, A.L., M. Gallardo-Cabello and X.C. Carrara. 1999. Growth analysis of striped mullet, *Mugil cephalus*, and white mullet, *M. curema* (Pisces: Mugilidae), in the Gulf of Mexico. Fish Bull. 97: 861–872.

Ibáñez, A.L., C.W. Chang, C.C. Hsu, C.H. Wang, Y. Iizuka and W.N. Tzeng. 2012. Diversity of migratory environmental history of the mullets *Mugil cephalus* and *M. curema* in Mexican coastal waters as indicated by otolith Sr:Ca ratios. Cienc. Mar. 38: 73–87.

Jamandre, B.W., J.D. Durand and W.N. Tzeng. 2009. Phylogeography of the flathead mullet *Mugil cephalus* in the north-west Pacific as inferred from the mtDNA control region. J. Fish Biol. 75: 393–407.

Jamandre, B.W., J.D. Durand, K.N. Shen and W.N. Tzeng. 2010. Differences in evolutionary patterns and variability between mtDNA cytochrome b and control region in two types of *Mugil cephalus* species complex in northwest Pacific. J. Fish Soc. Taiwan 37: 163–172.

Jessop, B.M., J.C. Shiao, Y. Iizuka and W.N. Tzeng. 2002. Migratory behaviour and habitat use by American eels *Anguilla rostrata* as revealed by otolith microchemistry. Mar. Ecol. Prog. Ser. 233: 217–229.

Krueck, N.C., D.I. Innes and J.R. Ovenden. 2013. New SNPs for population genetic analysis reveal possible cryptic speciation of eastern Australian sea mullet (*Mugil cephalus*). Mol. Ecol. Res. 13: 715–725.

Neves, R.L.S., T.F. Oliveira and R.L. Ziolli. 2007. Polycyclic aromatic hydrocarbons (PAHs) in fish bile (*Mugil liza*) as biomarkers for environmental monitoring in oil contaminated areas. Mar. Pol. Bull. 54: 1813–1838.

Panfili, J., C. Aliaume, D. Flores Hernández, C.C. Hsu, A. Jacquart, P. Lalèyè, B. Morales-Nin, J. Ramos Miranda, J. Tomás, W.N. Tzeng, V. Vassilopoulou and A.K. Whitfield. 2007. Mugil life history trait studies. INCO-CT-2006-026180—MUGIL, Deliverable 3, European Commission: 19p. + Annex, available at http://www.mugil.univ-montp12.fr/.

Papasotiropoulos, V., E. Klossa-Kilia, S. Alahiotis and G. Kilias. 2007. Molecular phylogeny of grey mullets (Teleostei: Mugilidae) in Greece: evidence from sequence analysis of mtDNA segments. Biochem. Genet. 45: 623–636.

Rossi, A.R., A. Ungaro, S. De Innocentiis, D. Crosetti and L. Sola. 2004. Phylogenetic analysis of Mediterranean Mugilids by allozymes and 16S mt-rRNA genes investigation: are the Mediterranean species of *Liza* monophyletic? Biochem. Genet. 42: 301–315.

Shen, K.N., C.Y. Chen, W.N. Tzeng, J.D. Chen, W. Knibb and J.D. Durand. 2010. Development and characterization of 13 GT/CA microsatellite loci in cosmopolitan flathead mullet *Mugil cephalus*. In Permanent Genetic Resources added to Molecular Ecology Resources Database 1 April 2010–31 May 2010. Mol. Ecol. Res. 10: 1098–1105.

Shen, K.-N., B.W. Jamandre, C.-C. Hsu, W.-N. Tzeng and J.-D. Durand. 2011. Plio-Pleistocene sea level and temperature fluctuations in the northwestern Pacific promoted speciation in the globally-distributed flathead mullet *Mugil cephalus*. Bmc Evol. Biol. 11: 83.

Smith, K.A. and K. Deguara. 2003. Formation and annual periodicity of opaque zones in sagittal otoliths of *Mugil cephalus* (Pisces: Mugilidae). Mar. Freshwat. Res. 54: 57–67.

Tomás, J., S. Augagneur and E. Rochard. 2005. Discrimination of the natal origin of young-of-the-year Allis shad (*Alosa alosa*) in the Garonne-Dordogne basin (south-west France) using otolith chemistry. Ecol. Freshwat. Fish 14: 185–190.

Tomás, J., A. Anastasopoulou, P. Berrebi, P.D. Cowley, A. Darnaude, P.S. Diouf, J.-D. Durand, D. Flores Hernández, B. Morales-Nin, W.N. Tzeng, C.-H. Wang, A.K. Whitfield and J. Panfili. 2008. Mugil migration studies. INCO-CT-2006-026180—MUGIL, Deliverable 4, European Commission: 15p. + Annex, available at http://www.mugil.univ-montp12.fr/.

Trape, S., H. Blel, J. Panfili and J.D. Durand. 2009a. Identification of tropical Eastern Atlantic Mugilidae species by PCR-RFLP analysis of mitochondrial 16S rRNA gene fragments. Biochem. Syst. Ecol. 37: 512–518.

Trape, S., J.-D. Durand, F. Guilhaumon, L. Vigliola and J. Panfili. 2009b. Recruitment patterns of young-of-the-year mugilid fishes in a West African estuary impacted by climate change. Est. Coast. Shelf Sci. 85: 357–367.

Wang, C.-H., C.-C. Hsu, C.-W. Chang, C.-F. You and W.-N. Tzeng. 2010. The migratory environmental history of freshwater resident flathead mullet *Mugil cephalus* L. in the Tanshui river, Northern Taiwan Zool. Stud. 49: 504–514.

Wang, C.-H., C.-C. Hsu, W.-N. Tzeng, C.-F. You and C.-W. Chang. 2011. Origin of the mass mortality of the flathead grey mullet (*Mugil cephalus*) in the Tanshui River, northern Taiwan, as indicated by otolith elemental signatures. Mar. Pol. Bull. 62: 1809–1813.

Whitfield, A.K. 1990. Life history styles of fishes in South African estuaries. Env. Biol. Fish 28: 295–308.

Whitfield, A.K., J. Panfili and J.-D. Durand. 2012. A global review of the cosmopolitan flathead mullet *Mugil cephalus* Linnaeus 1758 (Teleostei: Mugilidae), with emphasis on the biology, genetics, ecology and fisheries aspects of this apparent species complex. Rev. Fish Biol. Fish 22: 641–681.

[1] IRD, UMR 5119 ECOSYM, Université Montpellier 2, CC 093, Place Eugène Bataillon, 34095 Montpellier Cedex 5, France.
[a] Email: jacques.panfili@ird.fr
[b] Email: catherine.aliaume@univ-montp2.fr
[c] Email: jean-dominique.durand@ird.fr
[d] Email: javier.tomas@mac.com

[2] Hellenic Centre for Marine Research, Agios Kosmas, Helliniko, 16777, Greece.
Email: kanast@hcmr.gr

[3] Institut des Sciences de l'Evolution, Université Montpellier 2, CNRS-IRD, CC 065, Place Eugène Bataillon, 34095 Montpellier cedex 05, France.
Email: patrick.berrebi@univ-montp2.fr

[4] Département Sciences de l'Environnement et Santé Publique, Faculté de Pharmacie, Av. Charles Flahault, BP 14493, 34093 Montpellier Cedex 05, France.
Email: casellas@univ-montp2.fr

[5] National Museum of Marine Biology and Aquarium, Pingtung 944, Taiwan ROC/Graduate Institute of Marine Biology, National Dong Hwa University, Pingtung 944, Taiwan ROC.
Email: changcw@nmmba.gov.tw

[6] WWF-WAMER, Sacré Cœur III, No 9639, Dakar, Senegal.
Email: psdiouf@gmail.com

[7] Instituto EPOMEX, Universidad Autónoma de Campeche, Av. Héroe de Nacozari Núm. 480, C.P. 24029. Campeche, Cam., México.
[e] Email: doflores@uacam.mx
[f] Email: ramosmiran@gmail.com
[g] Email: jarendon@uacam.mx

[8] Instituto Politécnico Nacional 195, Playa Palo de Santa Rita Sur, La Paz, B.C.S., C.P. 23096, México.
Email: fgarciadl@cibnor.mx

[9] Faculté des Sciences Agronomiques de l'Université d'Abomey-Calavi, 01 BP 526, Cotonou, Bénin.
Email: laleyephilippe@gmail.com

[10] Institut Mediterrani d'Estudis Avançats (IMEDEA-CSIC/UIB), Miquel Marques 21, 07190 Esporles, Illes Balears, España.
Email: beatriz@imedea.uib-csic.es

[11] Department of Environmental Biology and Fisheries Science, National Taiwan Ocean University, Keelung 20224, Taiwan.
[h] Email: knshen@mail.ntou.edu.tw
[i] Email: chwang99@mail.ntou.edu.tw

[12] Department of Environmental Biology and Fisheries Science, National Taiwan Ocean University, Keelung, Taiwan, ROC/ Institute of Fisheries Science, National Taiwan University, Taipei, Taiwan, ROC.
Email: wnt@ntu.edu.tw

[13] Hellenic Centre for Marine Research, Agios Kosmas, Helliniko, 16777, Greece.
Email: celia@hcmr.gr

[14] South African Institute for Aquatic Biodiversity (SAIAB), Private Bag 1015, Grahamstown 6140, South Africa.
Email: a.whitfield@saiab.ac.za

* Corresponding author

General Index

J

K

L

M

N

O

P

R

S

T

Taxonomic Index*

A

B

C

D

E

F

G

* The nomenclature used in the Eschmeyer 'Catalog of fishes' (2015) was adopted for this book for conformity, but the same species may have been reported in the literature under different synonyms. Indeed, Mugilidae taxonomy and nomenclature has been revised several times, and a critical revision is ongoing at present.